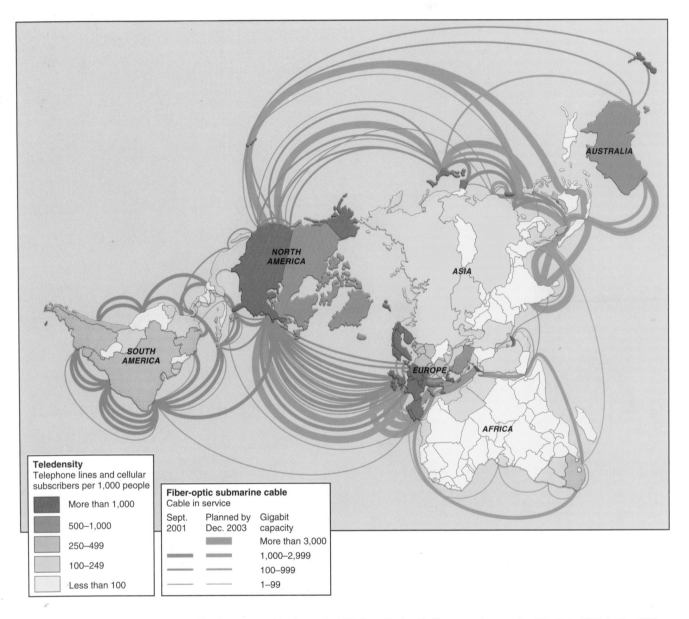

Teledensity
Telephone lines and cellular subscribers per 1,000 people

- More than 1,000
- 500–1,000
- 250–499
- 100–249
- Less than 100

Fiber-optic submarine cable
Cable in service

Sept. 2001	Planned by Dec. 2003	Gigabit capacity
		More than 3,000
		1,000–2,999
		100–999
		1–99

Global connections, in place and planned. The first ocean wiring began in 1850 from England to France and across the Atlantic in 1858. In the 1900s competition from radio and satellites made cable almost redundant until in the mid-1990s, when fiber optics provided greater speed and capacity for telephone calls. Satellites continue to be used for video (TV news).

Source: National Geographic Society, Picture ID 759480 Reprinted by permission.

Second Edition

CONTEMPORARY

World Regional

GEOGRAPHY

Global Connections, Local Voices

Michael Bradshaw
College of St. Mark and John

George W. White
Frostburg State University

Joseph P. Dymond
The George Washington University

Elizabeth Chacko
The George Washington University

 **Higher Education**

Boston Burr Ridge, IL Dubuque, IA Madison, WI New York
San Francisco St. Louis Bangkok Bogotá Caracas Kuala Lumpur
Lisbon London Madrid Mexico City Milan Montreal New Delhi
Santiago Seoul Singapore Sydney Taipei Toronto

Higher Education

CONTEMPORARY WORLD REGIONAL GEOGRAPHY: GLOBAL CONNECTIONS, LOCAL VOICES
SECOND EDITION

Published by McGraw-Hill, a business unit of The McGraw-Hill Companies, Inc., 1221 Avenue of the
Americas, New York, NY 10020. Copyright © 2007 by The McGraw-Hill Companies, Inc. All rights reserved.

Some ancillaries, including electronic and print components, may not be available to customers outside the
United States.

This book is printed on acid-free paper.

2 3 4 5 6 7 8 9 0 VNH/VNH 0 9 8 7 6

ISBN-13   978–0–07–282683–8
ISBN-10   0–07–282683–5

Publisher: *Margaret J. Kemp*
Senior Developmental Editor: *Joan M. Weber*
Marketing Manager: *Todd L. Turner*
Lead Project Manager: *Joyce M. Berendes*
Senior Production Supervisor: *Laura Fuller*
Lead Media Project Manager: *Judi David*
Media Producer: *Daniel M. Wallace*
Designer: *Rick D. Noel*
Cover/Interior Designer: *Jamie E. O'Neal*
(USE) Cover Images: © *Getty Images*
    Front cover: *Cambodia, Prah Dah, Family #200015287-001*
    Back cover: *Three monks riding on roller coaster #10197329*
Lead Photo Research Coordinator: *Carrie K. Burger*
Photo Research: *Chris Hammond/PhotoFind, LLC*
Cartography: *Electronic Publishing Services Inc., NYC*
Supplement Producer: *Melissa M. Leick*
Compositor: *Electronic Publishing Services Inc., NYC*
Typeface: *10/12 Times Roman*
Printer: *Von Hoffmann Corporation*

Library of Congress Cataloging-in-Publication Data

Contemporary world regional geography : global connections, local voices / Michael Bradshaw . . . [et al.]. —
2nd ed.
    p.   cm.
  Rev. ed. of: Contemporary world regional geography / Michael Bradshaw, George W. White, Joseph P.
Dymond. ©2004.
  Includes index.
  ISBN 978–0–07–282683–8 — ISBN 0–07–282683–5 (hard copy : alk. paper)
  1. Geography—Textbooks. I. Bradshaw, Michael J. (Michael John), 1935–. II. Bradshaw, Michael J.
(Michael John), 1935–. Contemporary world regional geography.

G116.C665   2007
910—dc22   2005022876
                                            CIP

www.mhhe.com

to Valerie,
Emily,
Maureen and Madison,
and Mary and P. Thomas Chacko

Brief Contents

Contents

Chapter 10

Latin America 435

Chapter 11

North America 499

Chapter 12

A World of Geography 557

About the Authors

Michael Bradshaw

Michael Bradshaw and his wife live in Canterbury, England, and have two sons and three grandchildren. Michael taught for 25 years at the College of St. Mark and St. John, Plymouth, as Geography Department chair and dean of the humanities course. He has written texts for British high schools and colleges since the 1960s. In 1985 he was awarded a Ph.D. from Leicester University for his study on the impacts of federal grant aid in Appalachia. His book, *The Appalachian Regional Commission: Twenty-Five Years of Government Policy,* was published in 1992. Since 1991 he has written for U.S. students and has been responsible for two physical geography texts and the successful world regional geography text, *The New Global Order.* Michael believes that we should all be better equipped to live in the modern, increasingly global world. Understanding geographic differences should make us more able to assess crucial issues and value other people who bring varied resources and who face pressures that we find difficult to imagine.

In developing the next-phase text for world regional geography courses, Michael has extended the experience and expertise of the writing team by adding new coauthors. *Contemporary World Regional Geography: Global Connections, Local Voices* is the outcome of this new collaboration. He is lead author for the first two and the last chapters, as well as for the regional chapters on East Asia and South Asia (both with Elizabeth Chacko) and Africa South of the Sahara.

George W. White

George W. White grew up in Oakland, California. He pursued graduate work in Eugene, Oregon, completing a Ph.D. at the University of Oregon. He then moved to Frostburg, Maryland, where he met his wife. George is currently an associate professor and chair of the Department of Geography at Frostburg State University. Political geography and Europe are two of his primary interests. He wrote a book titled *Nationalism and Territory: Constructing Group Identity in Southeastern Europe* (2000) and another titled *Nation, State, and Territory, Vol. 1: Origins, Evolutions, and Developments* (2004).

After meeting Michael Bradshaw, George was impressed by Michael's long and distinguished career of teaching, research, and publication. He accepted the opportunity to join Michael in his plans to write a new world regional geography text, taking lead authorship for the chapters on Europe, Russia and Neighboring Countries, and Northern Africa and Southwestern Asia, as well as contributing to other areas of the text.

George became a geographer because he believes that the field of geography is alive and dynamic, attuned to our ever-changing world and its great diversity. The world regional approach represents the breadth of the field of geography, and world regional geography texts are the epitome of the geographer's art. George White chose to collaborate with Michael Bradshaw on this project because the text combines local practices with global processes and explains interactions between the two as they shape each other.

Joe Dymond

Joe Dymond earned a Master of Science degree from the Pennsylvania State University in 1994, and a Master of Natural Sciences degree from Louisiana State University in 1999. He taught world regional geography courses for the Louisiana State University Department of Geography and Anthropology from 1995 through 2000. During Joe's six years at LSU, he instructed thousands of students and was recognized in the spring of 1997, fall of 1999, and fall of 2000 for superior instruction to freshman students by the Louisiana State University Freshman Honor Society, Alpha Lambda Delta. Joe currently lives in suburban Washington, D.C., with his wife and daughter, and is an assistant adjunct professor in the Department of Geography at The George Washington University. In the spring of 2004 he was nominated for The George Washington University's Bender Teaching Award. In this edition Joe is the lead author for the chapters on Southeast Asia and South Pacific, Latin America, and North America, and he contributed to other chapters.

Joe is interested in providing students with the geographic tools that will help them to better understand the human and environmental patterns present in their world. His greatest concern for geography students is that they obtain a comprehensive and fair perspective when learning about the people and places comprising the regions of the world. The style of this text, including the Point–Counterpoint sections, attempts to tell the regional geographic story from many perspectives to students. This structure permits students to better analyze geographic characteristics, connections, and relationships around the world and to think critically about important global issues. *Contemporary World Regional Geography: Global Connections, Local Voices* **teaches** rather than **lectures**. It provides students with the opportunity to think on their own by pointing out geographic relationships and guiding them through geographic data sets so they may establish their own informed opinions.

Elizabeth Chacko

Elizabeth Chacko is a native of Calcutta, India. She received her undergraduate degree in geography from the University of Calcutta. Moving to the United States for further study, she got a master's degree in geography from Miami University, Ohio. She also obtained a graduate degree in public health and a Ph.D. in geography, both from UCLA. Elizabeth has taught geography at various institutions including Loreto College, Calcutta; the University of California at Los Angeles; and The George Washington University, where she is Associate Professor of Geography and International Affairs. She teaches courses on South Asia, globalization, medical and population geography, and development. Elizabeth's research interests include women's health and the role of culture in health and health care. She is currently engaged in research on the Ethiopian immigrant community in the United States. Elizabeth is on the editorial board of the *Journal of Cultural Geography.* In this edition, Elizabeth worked with Michael on the East Asia and South Asia chapters and contributed to other chapters.

Preface

A New Edition and a New Coauthor

Following the award-winning first edition of *Contemporary World Regional Geography,* our second edition takes the text forward in major ways. It retains what reviewers liked and makes changes in response to trends within our subject. It also introduces important new developments, ensuring that the text continues to lead the way where others follow! This text remains the only one to fully develop a comparative approach to world regional geography.

Michael Bradshaw continues to lead the author team; George White, Frostburg State University, and Joe Dymond, The George Washington University, continue to make crucial contributions from their teaching experience in U.S. universities. Each has responsibility for three of the regional chapters. We are pleased to introduce Elizabeth Chacko, The George Washington University, as a new member of our team. She and Michael enrich the text with perspectives of world geography from outside the United States.

All the authors share a concern for and involvement in the wider global issues that affect world geography. They also share a passion for teaching that spurs them to write in a way that is readable, attracts students to a geographical understanding of local and world events, and causes them to engage further with geographical studies.

Features of the Second Edition

- **Improved maps.** Because maps are essential tools in understanding geography, all of our physical features maps have been redrawn, and every map has been evaluated for size, labeling, and color consistency.
- **Point–Counterpoint boxes** provide critical assessment of controversial issues that engage reader interest and support analytical skills. All now include a debate.
- **Photos** were carefully selected to depict landscapes and cultures of the world regions. In addition to numerous text photos, an additional 400 photos can be found on the "Digital Content Manager" CD-ROM and DVD.
- **Geography at Work boxes** are new to this edition and demonstrate the contributions to society and personal career developments arising from geographical study and research. In Chapter 1 two geographers carry out investigations based on computer mapping and geographic information systems. In Chapter 2 there is a study of ecological research in China. In Chapter 3 a geographer is involved in tourism development in the Bergen area of Norway. In Chapter 4 a geographer, who is a retired cat-

aloguer in the U.S. Library of Congress, became an authority on Russian wines and is writing a book about Ukrainian wines with a local winemaker. In Chapter 5 and Chapter 9 examples are taken from the "Interactive World Issues of Place and Planet" CD-ROM bundled with the text, focusing on the arid frontier in China and land reform in South Africa. The Chapter 6 study shows how geographers reacted to the Indian Ocean tsunami at the end of 2004. In Chapter 7 geographic information systems are used in the study of disease, first for the Bangladesh government and then from a world center of research in Seoul, South Korea. In Chapter 8 a geographer headed a U.S. Movement Control Battalion responsible for transportation in the recent Iraq war. Chapter 10 focuses on a study of rivers in eastern Mexico. In Chapter 11 geography is important in the White House and in U.S. government decision making. The Chapter 12 contribution follows a former President of the Association of American Geographers to a conference in Iran.

- **Personal View boxes** provide a glimpse into the lives of persons who are native to various world regions.
- **Chapter organization.** The internal arrangement of each chapter has changed. Each chapter opener features a regional map enhanced by photo insets linked to specific locations. Each chapter is introduced with a list of themes and photographs or a table. This is closely followed by a new, full-size map of major physical features, country boundaries, and capital cities that students can easily reference while reading the chapter.

Chapter Revisions

Chapter 1 begins with a consideration of the significance of geography to an understanding of variety, change, and closer links in the world. Defining *geography,* it links it to a spatial view, principles of location, maps, and geographic information systems. The nature of regional geography is discussed, with sections on regions and globalization, the definition of world regions, and a short account of the ways in which historic changes led to the present world regions.

Chapter 2 introduces integrative themes and concepts of human and physical geography that are basic to world regional geography. Issues of people and land, political freedom, economic inequality, cultural freedom and discrimination, and natural environmental issues are followed in this edition by a new discussion of human development and human rights. Each section introduces ideas, diagrams, and maps that are featured in the following chapters.

Chapters 3–11 are each devoted to a world region. Our focus remains comparative, based on a consistent approach in which an outline of the distinctive features of each region is followed by a discussion of the cultural history that affects modern geography and the natural environmental conditions in the region. A new feature is a further expanded general section that considers each region as a whole in the global context. The final section of each of these chapters covers the most distinctive features of subregions and select countries, including political geography, ethnic composition, and economic structure.

Chapter 12 is unique. It provides both a summary and a forward view, summarizing the features of globalization and localization that make the study of world regional geography such an integral and applicable tool in forming a more complete global comprehension. It shows how the global–local view became increasingly significant for world regional geography through developments in transportation, communications, economics, and politics—both in general and in relation to one part of the world, southern China. This final chapter takes us to a geography-centered view of changes that affect every part of our world, now and into the future.

"Global Connections, Local Voices"

We adopted this subtitle because globalization is an essentially geographical concept, and one where geographers provide a significant analysis that refers to each part of the world. Few places or people eliminate local elements as a result of increasing global connections. Political, demographic, economic, cultural, and environmental trends show some similarities from region to region, but also many—and often increasing—differences. Local voices tell us about the responses to global connections and the significance of local identities in a changing world. More than ever, each part of the world has a character and identity that make world regional geography not only interesting but vital for understanding how people live in other parts of the world. Such understanding informs our own daily decisions.

World Region Tour

The student reading this text embarks on a world tour, beginning in *Europe* (Chapter 3), where many modern global processes and innovations began. European technologies often built on previous Arab, Asian, and African achievements. Although the European roles were seminal, they were not always globally beneficial.

Moving eastward to *Russia and Neighboring Countries* (Chapter 4), we study the application of European-origin Communist principles adopted by governments for most of the 1900s and the changes following the breakup of the Soviet Union in 1991. The region continues to struggle with the shift from the self-sufficient central command system toward global economic and political systems; the "Russian Empire" remains a major political and economic legacy.

The student's world regional tour then enters *East Asia* (Chapter 5), where internal contrasts result from modern impacts on the ancient cultures of Japan and China. These two countries have taken leading global roles. Also part of the Pacific Rim, *Southeast Asia and the South Pacific* (Chapter 6) experience internal links through increasing intraregional trade and political issues that unite the contrasting human geographies of Southeast Asia, Australia, New Zealand, and the Pacific islands.

South Asia (Chapter 7) is a region of distinctive cultural background and colonial experiences. At the end of the British Raj in 1947, the newly independent countries attempted self-sufficiency, avoiding close relations with other regions. However, Cold War politics set up rivalries. Since the 1990s changed attitudes led the countries of this region to become more globally involved.

Next the tour moves farther westward into *Northern Africa and Southwestern Asia* (Chapter 8), the central part of the Islamic world. Here oil and water play central roles in political, economic, cultural, and environmental concerns. Cultural differences and natural resource competition between Israel and surrounding countries contribute to significant conflicts within the region.

The world regions tour then reaches *Africa South of the Sahara* (Chapter 9). Scientists believe *Homo sapiens* originated there. Medieval regional empires were supplanted by Arab merchants and European colonists, who came to establish global trade networks for slaves, minerals, and plantation crops. Since independence, mostly in the 1950s and 1960s, countries struggle to move forward.

Crossing the South Atlantic Ocean to *Latin America* (Chapter 10) brings us to a continent where early 1800s independence was followed by economic and political turmoil that fashioned today's geographies. Finally, the study of the geography of *North America* (Chapter 11) enables some of the world regional comparisons and contrasts to be appreciated from the world's most affluent society.

A Text for Students

Students are encouraged to think what it means to be part of a global community and to develop their geographical understanding of world events. The authors have the aim of writing a text that is well organized, well written, readable, and appropriate to where students begin and where they might end after the course.

The features arising from this aim include

- A readable text. Reviewers comment on the accessible style and clarity of writing, clear definition of terms, and up-to-date information and illustrations.

- A consistent structure to each chapter that becomes familiar and encourages comparisons.

- Straightforward maps and diagrams with styles that are repeated so students can compare regions.

- Pedagogical tools such as the Test Your Understanding boxes and the Point–Counterpoint debates.

- A fold-out world map at the back of the book that provides a useful reference tool.

Acknowledgments

The professional team involved in producing this text has vested many hours in making it the best world regional geography text possible. Even an expanded author team with a wide collective expertise relies on the massive feedback arranged by the publisher—through individual reviews, focus groups, and special critiques. Over 30 reviews of the first edition were followed by the art review and special focus group reports, as well as queries and comments forwarded from the McGraw-Hill sales force. The authors greatly appreciate the help of all involved—committed teachers who also have the same aims as the authors and editors.

In addition to the reviewers listed here, the authors wish to express special thanks to McGraw-Hill for editorial support through Marge Kemp, Donna Nemmers, and Joan Weber; the marketing expertise of Todd Turner; and the production team led by Joyce Berendes and Mary Powers, Rick Noel, and Carrie Burger. We have used many new photos supplied by Michael Camille, Alasdair Drysdale, Parvinder Sethi, Bill Westermeyer, and David Zurick.

We all depend on our families and friends for support and patience in this enterprise and express our huge debt to those to whom we dedicated this edition.

Geography at Work Contributors

Chapter 1, Mapmakers and GIS Analysts: Elio Spinello, *RPM Consulting.* Elio Spinello is a Californian who has been involved with the writing and management of the AtlasGIS software package. He shows how the combination of maps and data helps those bringing support to areas susceptible to river blindness in Mozambique, Africa.

Chapter 2, China's Landscapes and Global Change: Erle Ellis, *University of Maryland—Baltimore County.* Professor Ellis works with a team in China to understand local and global environmental changes in the context of sustainable agriculture. Focusing on the limiting nature of nitrogen in ecosystems, he found that historic high yields were linked to local nitrogen sources and is extending this understanding globally.

Chapter 3, Business Development: Anne Hunderi. Anne Hunderi, who lives in Norway, works primarily on developing the tourism industry and helping small businesses obtain start-up loans. For example, Anne has helped catering businesses and tailors to establish themselves.

Chapter 4, Wine Industry: Robert W. Hutton, *a retired subject cataloger from the Library of Congress.* Mr. Hutton researches the Russian wine industry. Robert finds that each wine is unique to the area where it is grown, and geography's concern with both the physical and human worlds takes into account all the factors explaining where and how wine is made.

Chapter 5, Frontier Forces. Geographers are at work to understand the issues relating to living on the frontier between arid and humid parts of China. This study is based on the CD accompanying this text, "Interactive World Issues of Place and Planet."

Chapter 6, The Indian Ocean Tsunami of December 2004. Instead of reviewing the work of geographers, this account demonstrates how geographers get involved at the time of a major natural disaster.

Chapter 7, Battling Infectious Diseases: Mohammed Ali, *International Vaccine Institute, Seoul, South Korea.* Mohammed Ali, a geographer from Bangladesh, used geographic information systems to identify environmental risk factors affecting cholera. He then moved to the International Vaccine Institute in Seoul, South Korea, to investigate the factors that make the application of specific vaccines successful.

Chapter 8, Operation Iraqi Freedom: Mark Corson, *Northwest Missouri State University.* Mark Corson is a lieutenant colonel (LTC) in the U.S. Army Reserve and a member of the 450th Movement Control Battalion (MCB), which helped to coordinate military transportation in Operation Iraqi Freedom (OIF).

Chapter 9, Social Justice in South Africa: Brent McCusker, *University of West Virginia.* Professor McCusker studies land reform in South Africa as a means of assessing the new involvements of local African communities. This is another study available on the CD accompanying this text, "Interactive World Issues of Place and Planet."

Chapter 10, Geographical Research on Rivers in Eastern Mexico: Dr. Paul Hudson, *University of Texas at Austin.* Dr. Hudson is a physical geographer who carried his studies of rivers and the formation of floodplains from the Gulf Coast into eastern Mexico. He discovered strong links between floodplain formation, present agriculture, and the historic settlement of the Huasteca people.

Chapter 11, Using Geography to Aid Public Policy: The White House: Dr. Bryan Hannegan, *Associate Director for Energy and Transportation in the U.S. Council on Environmental Quality.* Dr. Hannegan uses his skills acquired as a geographer in studies such as the routing of oil and gas pipelines in ecologically sensitive areas. GIS systems are also used in such areas as drug use incidence and census data analysis.

Chapter 12, AAG President in Iran: Reconciling Differences: Alexander Murphy, *University of Oregon.* Professor Murphy visited the Second International Congress of Geographers in the Islamic World in 2004 during his year as president of the Association of American Geographers. He visited universities and discussed world issues with informed local people.

Second Edition Reviewers

J. Anthony Abbott
Central Washington University

Toni Alexander
Kansas State University

Brad Baltensperger
Michigan Technological University

Royal Berglee
Morehead State University

Michelle Calvarese
California State University–Fresno

Michael A. Camille
The University of Louisiana at Monroe

Brian Ceh
Indiana State University

Sean Chenoweth
The University of Louisiana at Monroe

Deborah Corcoran
Southwest Missouri State University

Carlos E. Cordova
Oklahoma State University

James Craine
California State University–Northridge

Stephen Cunha
Humboldt State University

Bruce E. Davis
Eastern Kentucky University

Hank Dillon
George Mason University

Daniel P. Donaldson
University of Central Oklahoma

Alasdair Drysdale
University of New Hampshire

Dennis Ehrhardt
The University of Louisiana at Lafayette

Kenneth W. Engelbrecht
Metropolitan State College of Denver

Jim Engstrom
Georgia Perimeter College

John Florin
University of North Carolina–Chapel Hill

Sherry Goddicksen
California State University–Fullerton

Steve Graves
California State University–Northridge

Raymond Greene
Western Illinois University

John Grimes
Louisiana State University

Reuel Hanks
Oklahoma State University

Ellen R. Hansen
Emporia State University

Douglas Heffington
Middle Tennessee State University

Billy D. Higgins
University of Arkansas–Fort Smith

Walter Jung
University of Central Oklahoma

David Lee
Florida Atlantic University

Elizabeth J. Leppman
St. Cloud State University

Elena Lioubimtseva
Grand Valley State University

Max Lu
Kansas State University

Brent McCusker
West Virginia University

Petrina R. Medley
*Oklahoma City Community College and
    Oklahoma State University*

Bill Preston
*California Polytechnic Institute–San Luis
    Obispo*

Patricia L. Price
Florida International University

Joel Quam
College of DuPage

Ben Richason III
St. Cloud State University

Paul F. Robbins
University of Arizona

Paul A. Rollinson
Southwest Missouri State University

Deborah Salazar
Texas Tech University

Rick Sambrook
Eastern Kentucky University

Tom C. Schafer
Fort Hays State University

Daniel A. Selwa
Coastal Carolina University

J. Duncan Shaeffer
Arizona State University

Sinclair A. Sheers
George Mason University

Nancy L. G. Shirley
Southern Connecticut State University

Michelle L. Shuey
Northern Illinois University

Susan D. Siemens
Ozarks Technical Community College

Kevin F. Sims
Cedarville University

Dean Sinclair
Northwestern State University

Bonnie R. Sines
University of Northern Iowa

Andrew Sluyter
Louisiana State University–Baton Rouge

Jacob R. Sowers
Kansas State University

Selima Sultana
Auburn University

Ray Sumner
Long Beach City College

Christopher J. Sutton
Western Illinois University

Theresa Tarlos
Orange Coast College

Samuel Thompson
Western Illinois University

Jeffrey S. Torguson
St. Cloud State University

Charles L. Wax
Mississippi State University

Thomas Whitmore
University of North Carolina–Chapel Hill

Mark Wiljanen
Eastern Kentucky University

Keith Yearman
College of DuPage

First Edition Reviewers

Edward Addo
St. Cloud State University

Brad Baltensperger
Michigan Technological University

John Bowen
University of Wisconsin–Oshkosh

John Boyer
Virginia Tech

Robert F. Brinson, Jr.
Santa Fe Community College

Michelle Calvarese
California State University–Fresno

Michael A. Camille
The University of Louisiana at Monroe

Gabe Cherem
Eastern Michigan University

Jonathan C. Comer
Oklahoma State University

Carlos E. Cordova
Oklahoma State University

Mark W. Corson
Northwest Missouri State University

Marcelo Cruz
University of Wisconsin–Green Bay

Jose A. da Cruz
Ozarks Technical Community College

Robert S. Dilley
Lakehead University

Francis H. Dillon, III
George Mason University

Harold M. Elliott
Weber State University

James D. Engstrom
Georgia Perimeter College

Allen Finchum
Oklahoma State University

Jay D. Gatrell
Indiana State University

Sherry Goddicksen
California State University–Fullerton

Thomas O. Graff
University of Arkansas

John R. Grimes
Louisiana State University

Mark Guizlo
Lakeland Community College

Gregory Haddock
Northwest Missouri State University

Reuel R. Hanks
Oklahoma State University

Douglas Heffington
Middle Tennessee State University

David C. Johnson
University of Louisiana, Lafayette

Richard L. Kroll
Kean University

James Leonard
Marshall University

Elizabeth J. Leppman
St. Cloud State University

Max Lu
Kansas State University

Donald Lyons
University of North Texas

Taylor E. Mack
Mississippi State University

Susan R. Martin
Michigan Technological University

Calvin O. Masilela
Indiana University of Pennsylvania

Brent McCusker
West Virginia University

Monica Nyamwange
William Paterson University

Bimal K. Paul
Kansas State University

William Preston
California Polytechnic State University

Joel Quam
College of DuPage

Paul Robbins
Ohio State University

Paul A. Rollinson
Southwest Missouri State University

Deborah A. Salazar
Texas Tech University

Steven M. Schnell
Kutztown University of Pennsylvania

Burl Self
Southwest Missouri State University

J. Duncan Shaeffer
Arizona State University

Jonathan Taylor
California State University–Fullerton

Jeffrey S. Torguson
St. Cloud State University

Alice L. Tym
University of Tennessee–Chattanooga

Richard Ulack
University of Kentucky

Kelly Ann Victor
Eastern Michigan University

Charles L. Wax
Mississippi State University

David Welk
Reedley College–Clovis Center

Naim Zeibak
Indiana State University

Guided Tour

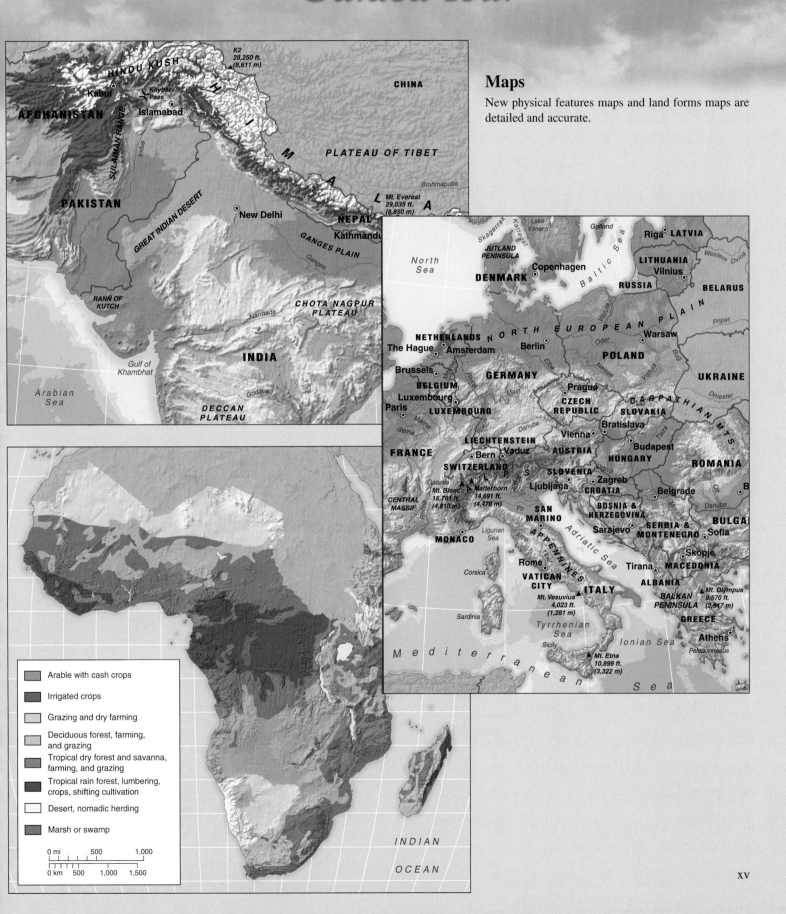

Maps

New physical features maps and land forms maps are detailed and accurate.

Legend:
- Arable with cash crops
- Irrigated crops
- Grazing and dry farming
- Deciduous forest, farming, and grazing
- Tropical dry forest and savanna, farming, and grazing
- Tropical rain forest, lumbering, crops, shifting cultivation
- Desert, nomadic herding
- Marsh or swamp

0 mi   500   1,000
0 km   500   1,000   1,500

Fold-out world maps can be easily referenced from any chapter in the text.

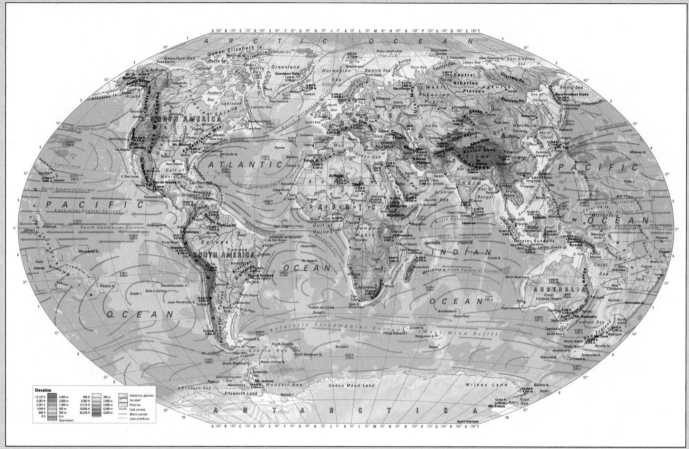

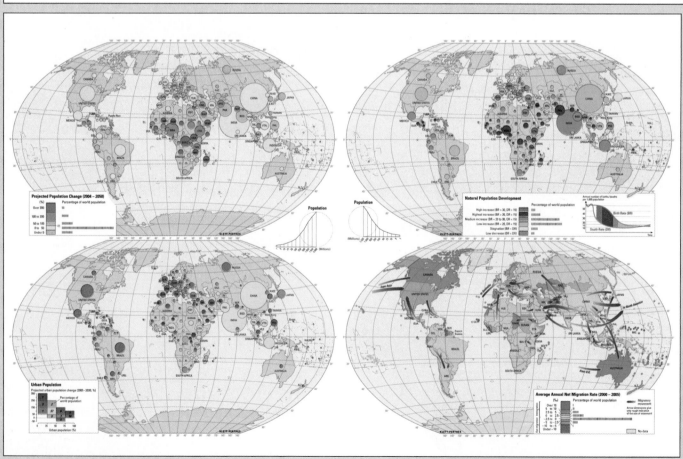

New **Geography at Work boxes** provide valuable insights regarding the ways geographical understandings and skills are used in a wide range of jobs.

Geography at Work

CHINA'S LANDSCAPES AND GLOBAL CHANGE

Over the past 50 years, population growth combined with economic development and industrial technologies such as fossil fuels and synthetic nitrogen fertilizers have been driving unprecedented ecological changes across China's ancient agricultural landscapes. Current research by Erle Ellis and his team of collaborators in China is demonstrating that the environmental impacts of these changes extend far beyond the region's 2 million square kilometers and 800 million people, impacting both global climate and stratospheric ozone. By a multiscale approach integrating regional data on land cover, terrain, and climate with higher-resolution local measurements from historical aerial photographs, high-resolution satellite imagery, household surveys, interviews with village elders, and field measurements of soils and vegetation, Ellis and his collaborators are now measuring these changes with unique precision at five rural field sites selected across environmentally distinct regions of China (Box Figure 1).

In the late 1980s, while finishing his dissertation on the physiology of crop yield at Cornell University, Ellis became convinced that sustainable agricultural development required a more holistic understanding of long-term agricultural productivity. He set out to investigate the ecology of the world's most productive and long-sustained agricultural systems. China stood out as one of the few places where agricultural systems that had sustained high yields continuously for centuries might still be available for study.

In the spring of 1990, Ellis joined a sustainable agriculture delegation to China and was immediately inspired to begin research there. In the absence of postdoctoral opportunities for foreign researchers, he accepted an English teaching position at Nanjing Agricultural University, and spent the year there learning Mandarin, familiarizing himself with China, and searching for the right project and collaborators. Returning to the United States a year later, he applied for and received a (U.S.) National Science Foundation Environmental Biology postdoctoral fellowship to investigate the nitrogen cycles of traditional versus industrial villages in China.

In 1993 Ellis began field research, believing that a traditional agricultural village might still exist in the Tai Lake Region—the area around Tai Lake bounded by Shanghai, Hangzhou, and Changzhou, long one of the most productive agricultural regions of the world and known in China as the "land of fish and rice," the Chinese equivalent of the "land of milk and honey." He discovered that the region's traditional agriculture had been transformed since the 1970s into one of the most input-intensive agricultural systems in the world and adapted his study to a comparison of the historic (circa 1930) versus contemporary (1995) state of the same village in Wujin County.

Given that nitrogen limits the productivity of most ecosystems, Ellis had hypothesized that the secret to sustaining high long-term productivity was to overcome nitrogen limitation by planting nitrogen-fixing legume "green manure" crops and by efficient nitrogen

Box Figure 1. Carrying out the humification study in the field

To read more about this study, refer to Ellis, E.C., "Long-term Ecological Cha in the Densely Populated Rural Landscapes of China," in R.S. DeFries et al., *Ecosystems and Land Use Change*, American Geophysical Union, Washin D.C. (2004).

recycling. Historical records confirmed that the region's tradit farmers sustained rice yields for more than 800 years at levels ove percent higher than possible without fertilizers. They accomplis this by means of external sources of nitrogen fertility, including s ments harvested from canals and purchased oilcakes (residues pressing oil from imported soybeans and rapeseed). Nitrogen fix and recycling proved to be only minor sources of fertility.

His initial hypothesis disproved, Ellis made another disco Because of the ready availability of synthetic nitrogen, sedi was now left to fill in village canals and had increased soil nitr concentration over time, producing a dramatic net increas nitrogen storage across the region's rural landscapes. Potent this "sink effect" had global implications, helping to explain "missing sink" for atmospheric carbon dioxide that continue complicate our understanding of anthropogenic climate cha (Carbon storage is tightly linked to nitrogen storage in soils.)

Ellis's research was pushed in its current direction by observation that fine-scale local changes in lands management could have major regional and global conseque that were observable only by detailed site-based measurem In 2004, in the final year of a five-year NSF project investigatin global impacts of long-term ecological changes across Ch densely populated rural landscapes, his work yielded surpri results, including long-term increases in tree cover that pa major increases in population density, with simultaneous incre in impervious surface areas that rival those of urban areas.

Geography at Work

AAG PRESIDENT IN IRAN: RECONCILING DIFFERENCES

Geography provides fundamental insights into the nuances of globalization, whether economic, political, or cultural. Globalization is creating tensions between peoples as some cultural practices spread and challenge local cultures and voices. Here too, geography not only concerns itself with this phenomenon but also can build bridges and help ease cultural tensions and misunderstandings. For example, Alexander B. Murphy, from the Department of Geography at the University of Oregon, traveled to Iran in 2004 when he was president of the Association of American Geographers (Box Figure 1). While in Iran, he addressed the "Second International Congress of Geographers of the Islamic World"; he talked to students and faculty at two of Tehran's major universities; and he visited different parts of the country with Iranian geographers. His visit was covered in the *Tehran Times*, and he was a guest on an Iranian television interview show.

The governments of Iran and the United States may be at odds with one another, but Professor Murphy had the chance to see why the Iranians have a reputation for being so hospitable. Everywhere he went, people were extraordinarily nice, helpful, and friendly. The trip also reminded him of how distinctive Iran is compared to its neighbors. Murphy had traveled in other parts of the Muslim Middle East—especially Egypt, Jordan, and Palestine. There are some continuities with these places, but Iran is clearly different—in language, in culture, in social norms, and much more. To visit Iran is to understand the fallacy of treating the "Islamic world" as if it were a monolith.

The presentations Murphy gave at the two universities led to interesting academic discussions with faculty and students. Very few Americans go to Iran these days, and this opportunity allowed Murphy to give Americans a human face in Iran. At the same time, he gained insight into the diversity of opinion in Iran on political and social issues. Iranians express views that span the political spectrum. This is not surprising given the rapid transformations we are all confronting in the contemporary world. What particularly impressed Murphy, however, was how well informed Iranians seem

Box Figure 1 Professor Murphy, back row, third from the left, with colleagues at the Iranian geographers' conference.

to be—gaining information not just from government news sources but also from the Internet and from friends and relatives in other parts of the world. Many Iranians are strongly critical of stances taken by the U.S. government on particular issues, but Murphy was never held personally responsible for those stances.

The Western press sometimes paints a picture of Iran as an isolated place with little personal freedom and much cultural and political radicalism. Iranians are often confronted with an image of the United States as a country where Donald Rumsfeld represents the political norm and Eminem represents the cultural norm. Murphy's trip revealed how shallow these stereotypes are. Murphy saw how much Iranian society has opened up in recent years, and his discussions with people and his media appearances highlighted a side of America that Iranians rarely see.

Point–Counterpoint boxes provide critical assessment of controversial issues that engage reader interest and support analytical skills.

POINT COUNTER POINT

FACETS OF GLOBALIZATION

Mentions of globalization raise anger in some places; mobs of people demonstrate with violence against it at meetings of international leaders. Antiglobalization protesters often shout louder than those who tell us that whatever the growing pains, globalization is good for us in terms of better incomes, better life quality, and better understanding of other people. Some of these views are posed here for you to debate. We have quoted from some writers who have gone into print since 1990. The alternative views are not necessarily shared by the authors of this text, but seek to summarize points of view in opposition to the quotations. After discussing these alternatives, add others or assess the views people might have in different parts of the world.

Globalization is good for you.	Globalization is bad for you.
Barber overstated his case to make a point in 1992. Fifteen years later changes have taken place, but not all fit his description of "bleak, neither democratic, . . . largescale bloodshed, global uniformity, . . . a planet precipitantly falling apart." In fact, many throughout the world live under better conditions with a greater sense of local identity than they did 15 years ago. Although jihad and mcworld are with us, responses to both prevent the direst outcomes.	Just beyond the horizon of current events lie two possible futures—both bleak, neither democratic. The first is a redistribution of large swaths of humankind by war and bloodshed, in which culture is pitted against culture, people against people, tribe against tribe—a *jihad* in the name of a hundred narrowly conceived faiths. . . . The second is being borne in on us by the onrush of economic and ecological forces that demand integration and uniformity and that mesmerize the world with fast music, fast computers, and fast food—with MTV, Macintosh, and McDonald's, pressing nations into one commercially homogeneous global network: one *mcworld* tied together by technology, ecology, communications, and commerce. The planet is falling precipitantly apart and coming reluctantly together at the very same moment. Benjamin R. Barber, *The Atlantic Monthly*, 1992
Fourteen years after being published, Huntington's thesis remains a prescription for a divided world and continuing conflict on a global basis. Certainly, some of the conflicts during that time occurred where different cultures or civilizations clash, but others occurred elsewhere, and many parts of the world remain peaceful despite internal country differences or outlook or terrorist threats.	World politics is entering a new phase. . . . It is my hypothesis that the fundamental source of conflict in this new world will not be primarily ideological or primarily economic. The great divisions among humankind and the dominating source of conflict will be cultural. [Countries] will remain the most powerful actors in world affairs, but the principal conflicts of global politics will occur between them and groups of different civilizations. The clash of civilizations will dominate global politics. The fault lines between civilizations will be the battle lines of the future. Samuel P. Huntington, *Foreign Affairs*, 1993
It is clear that international peace and security cannot be guaranteed: the more global our activities, the less prospect there is for overall control by countries—the present location of law-giving and security.	Many thought the new world order proclaimed by George H. Bush [in 1990] was the promise of 1945 fulfilled, a world in which international institutions, led by the United Nations, guaranteed international peace and security and the active support of the world's major powers. That world is a chimera. Even as a liberal internationalist ideal, it is infeasible at best and dangerous at worst. It requires a centralized rule-making authority, a hierarchy of institutions, and universal membership. Equally to the point, efforts to create such an order have failed. The United Nations cannot function effectively independent of the major powers that compose it, nor will those nations cede their power and sovereignty to an international institution. Anne-Marie Slaughter, *Foreign Affairs*, 1997

12

(Continued)

Globalization is good for you.	Globalization is bad for you.
The mall at Singapore's airport has a food court with 15 food outlets, all but one of which offer menus that cater to local tastes; the lone standout, McDonald's, is also the only one crowded with customers. In New York City, experts in feng shui, an ancient Chinese craft aimed at harmonizing the placement of man-made structures in nature, are sought after by real estate developers in order to attract a growing influx of Asian buyers. Globalization refers to processes that evolve as people go about their daily tasks; these processes are not hindered or prevented by territorial or jurisdictional barriers, they can readily spread in many directions across national boundaries, and are capable of reaching any community anywhere in the world. James N. Roseneau, *Current History*, 1997	Globalization is unlikely to result in a greater richness of experience for people, but will move toward global products that are the same everywhere, reducing choices and understanding of other peoples.
Many plants from other parts of the world have been planted and grown successfully without danger to local ecosystems. At times this has brought great benefits, as in the global adoption of crops that were once limited to specific cultural hearths. More globalization should lead to greater, not less, control of Bright's "small percentage of cases" of potentially harmful plant and animal invasions, as the United States rules against entrants bringing plants. Insects are more difficult to control, but seem to migrate to new areas whatever the system.	World trade has become the primary driver of one of the most dangerous and least visible forms of environmental decline: thousands of foreign, invasive species are hitchhiking through the global trading network aboard ships, planes, and railroad cars, while hundreds of others are traveling as commodities. This "biological pollution" is degrading ecosystems, threatening public health, and costing billions of dollars annually. Bioinvasion occurs when a species finds its way into an ecosystem where it did not evolve. In a small percentage of cases, the exotic finds everything it needs—and nothing capable of controlling it. In South and Central America, the growth of specialty export crops—upscale vegetables and fruits—has spurred the spread of whiteflies, which are capable of transmitting at least 60 plant viruses. In southern India, a tropical American shrub is strangling rice paddies and ruining fish habitat throughout the Cauvery River basin. Christopher Bright, *Foreign Policy*, 1999
The last few years have shown that the United States, despite its huge world presence, cannot run roughshod over other countries. Many countries in Europe disassociated themselves from the 2003 Iraq war, and there are many other voices in Europe, East and South Asia, and in the Arab world. Globalization requires a greater working together by countries.	In the past year (1999), many global issues have been framed in terms of the notion of sovereignty, that is, the right of political entities to follow their own course—which may be benign or may be ugly—and to do so free from external interference. In the real world that means interference by highly concentrated power, with its major center in the United States. This concentrated global power is called by various terms, depending on which aspect of sovereignty and freedom one has in mind. So sometimes it's called the Washington consensus, or the Wall Street/Treasury complex, or NATO, or the international economic bureaucracy (the World Trade Organization, World Bank, and IMF), or G-7 (the rich, Western industrial countries), or G-3 or, more accurately, usually G-1. Noam Chomsky, *"Rogue States: The Rule of Force in World Affairs,"* 2000
The Seattle Ministerial Conference of the World Trade Organization (WTO) demonstrated with disturbing force the huge confusions that haunt the public mind about . . . the process now known as globalization. The notion that globalization is an international conspiracy on the part of industrial country governments and large firms to marginalize the poorest nations, to exploit low wages and social costs wherever they may be found, to diminish cultures in the interests of an Anglo-Saxon model of lifestyle and language, and even to undermine human rights and cut away democratic processes that stand in the way of ever more open markets is, of course, utter nonsense. The outpouring of misconceived, ill-understood propaganda against a system that has brought vast gains to most nations over the past few decades is extraordinarily dangerous. Peter D. Sutherland, *Harvard International Review*, 2000	It is difficult to believe that there are benefits from globalization. We see city centers around the world being taken over by Starbucks, Kentucky Fried, and McDonalds. We are told that the poor are getting poorer and lagging further behind the materially wealthy. We should all go back to living within the resources of our own countries and not bother about anyone elsewhere.

13

Media Supplements

The new **Sights & Sounds CD-ROM** by David Zurick, Eastern Kentucky University, offers a unique opportunity to see and hear the music and cultural perspectives of 10 regions:

- North America: Appalachia
- Central America: Oaxaca, Mexico
- South America: Ecuador
- Europe: British Isles
- Africa South of the Sahara: Tanzania

- South Asia: Nepal
- Middle East
- Insular Southeast Asia: Bali, Indonesia
- East Asia: China
- South Pacific: Samoa

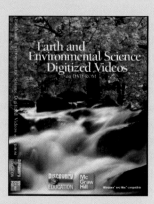

The **Digital Content Manager CD-ROM and DVD** provides unequaled instructional support. All maps, illustrations, and photos are available in jpg format, as well as PPT. Chapter-specific PPT outlines give an overview of the structure of each chapter. Interactive base maps for all world regions are useful tools to help students learn about cities and countries within the world regions.

The exciting **Earth and Environmental Science DVD by Discovery Channel Education** offers 50 short (3–5 minute) videos on topics ranging from conservation to volcanoes. Search by topic and download into your PowerPoint lecture. Available to colleges and universities. Begin your class with a quick peek at science in action. See your McGraw-Hill sales representative for a detailed listing.

The **Instructor's Testing and Resource CD-ROM** contains a wealth of cross-platform (Windows and Macintosh) resources for the instructor. Supplements featured on this CD-ROM include a computerized test bank utilizing EZ Test software to quickly create customized exams. This flexible, user-friendly program allows instructors to search for questions by topic, format, or difficulty level, as well as edit existing questions or add new ones. Multiple versions of the test can be created; and any test can be exported for use with course management systems such as WebCT, BlackBoard, or PageOut. Word files of the test bank are included for instructors who prefer to work outside the test-generator software. Other assets on the Instructor's Testing and Resource CD-ROM are grouped within easy-to-use folders.

McGraw-Hill's **ARIS—Assessment, Review, and Instruction System** (http://www.mhhe.com/bradshaw2e)—for *Contemporary World Regional Geography* is a complete online tutorial, electronic homework, and course management system, designed for greater ease of use than any other system available. On adoption of *Contemporary World Regional Geography,* instructors can create and share course materials and assignments with colleagues with a few clicks of the mouse. All PowerPoint lectures, assignments, quizzes, tutorials, and interactives are directly tied to text-specific materials in *Contemporary World Regional Geography,* but instructors can also edit questions, import their own content, and create announcements and due dates for assignments. ARIS has automatic grading and reporting of easy-to-assign homework, quizzing, and testing. All student activity within McGraw-Hill's ARIS is automatically recorded and available to the instructor through a fully integrated grade book that can be downloaded to Excel.

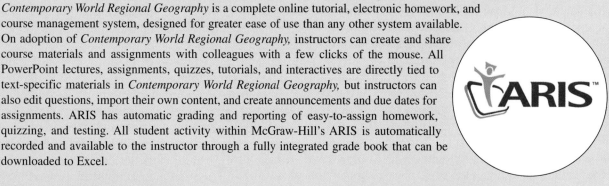

The Power of Place: Geography for the 21st Century is a revised and updated version of the World Regional Geography telecourse funded by Annenberg/CPB Projects. McGraw-Hill offers a revised Faculty Guide and Student Study Guide that are correlated to *Contemporary World Regional Geography* by Bradshaw et al.

Free **Interactive World Issues CD-ROM** set is a two-CD set for student exploration of controversial issues in water rights (Columbia River); migration (Mexico); urban spread (Chicago); population (China); and sustainability (South Africa).

Print Supplements

A new partnership with **Klett-Perthes** offers McGraw-Hill customers an exclusive 30% discount on all orders from the Klett-Perthes 2005 catalog. Contact your McGraw-Hill sales representative for more information.

A set of over 100 overhead **transparencies** includes key illustrations and figures from the text. The images are printed with better visibility and contrast than ever before, and labels are large and bold for clear projection.

Three **Geographic Perspectives** titles provide concise information on the physical environment and culture of the country. By examining the region through many of the sub-disciplines of geography—including historical, political, economic, urban, and medical—these guides serve as a framework to better understand current events. This series includes countries covered in the daily headlines:

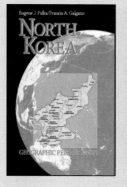

- *North Korea* by Eugene J. Palka and Francis A. Galgano.
- *Afghanistan* by Eugene J. Palka.
- *Iraq* by Jon C. Malinowski.

Students of geography and other disciplines, as well as the general reader, will find these unique guides invaluable to their understanding of current world countries and events.

Military Geography: From Peace to War by Eugene J. Palka and Francis A. Galgano involves the application of geographic information, tools, and technologies to military problems—across the spectrum of military operations from peacetime to wartime. History is replete with examples of the influence of terrain, weather, climate and culture on combat operations during war. Military problems, however, are immutably linked to geography regardless of the context in which they occur. This book retains the wartime focus of "traditional" military geography, yet broadens the scope of the subfield to incorporate a wide range of Stability and Support Operations (SASO), as well as peacetime endeavors. Notwithstanding its purpose, the conduct of any military enterprise is conditioned by the character of the area of operations—the military operating environment. **Military Geography: From Peace to War** focuses on the synergy between geography and military operations wherever they occur.

The **Annual Editions** series is designed to provide students with convenient, inexpensive access to current, carefully selected articles from the public press. They are updated regularly through continuous monitoring of over 300 periodicals. Each volume presents over 40 articles written for a general audience by experts and authorities in their fields. Organizational features include an annotated listing of selected World Wide Web sites, an annotated table of contents, a topic guide, a general introduction, and brief overviews for each section. Each title offers an instructor's resource guide containing test questions and a helpful user's guide called "Using Annual Editions in the Classroom."

The **Taking Sides** volumes present current issues in a debate-style format designed to stimulate student interest and develop critical thinking skills. Each issue is thoughtfully framed with an issue summary, an issue introduction, and a postscript. The pro and con essays—selected for their liveliness and substance—represent the arguments of leading scholars and commentators in their fields. Taking Sides readers feature annotated listings of selected World Wide Web sites. An instructor's resource guide with testing materials is available with each volume. To help instructors incorporate this effective approach in the classroom, an excellent resource called "Using Taking Sides in the Classroom" is also offered.

The **Student Atlas** series combines full-color maps and data sets to introduce students to the importance of the connections between geography and other areas of study, such as world politics, environmental issues, and economic development. In particular, the *Student Atlas: World Geography,* **4/e,** by John Allen combines over 100 full-color maps and data sets to give students a clear picture of the recent agricultural, industrial, demographic, environmental, economic, and political changes in every world region. This concise, affordable resource provides the most recent geographic data for geography students.

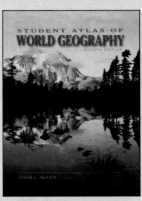

Global Studies is a unique series designed to provide comprehensive background information as well as vital current information regarding events that are shaping the cultures of the regions and countries of the world today. Each Global Studies volume features country reports in essay format and includes detailed maps and statistics. These essays examine the social, political, and economic significance of each country. In addition, relevant and carefully selected articles from worldwide newspapers and magazines are included to further foster international understanding.

Chapter 1

Globalization and World Regions

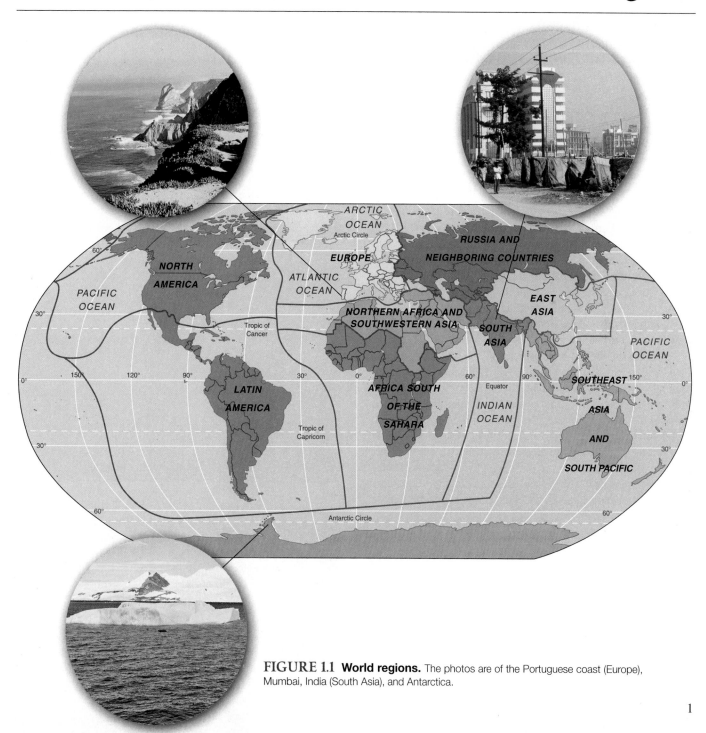

FIGURE 1.1 World regions. The photos are of the Portuguese coast (Europe), Mumbai, India (South Asia), and Antarctica.

(a)

(b)

Planet Earth: A World of Variety, Change, and Closer Links

Living on planet Earth makes geographic studies important to all of us. First, Earth's surface has great variety (Figure 1.1) in its human and physical features, giving rise to many different places. World regional geography provides insights to how places differ from each other, how people live in different parts of the world, and how they connect with each other. Equally important is the greater understanding we develop of other peoples and the increased awareness we gain of the diversity and sensitivity of our natural environments.

Second, the media, telecommunications, the Internet, and the global travel industry (Figure 1.2) enable increasing connections among peoples and places. We receive instant news of current events as they unfold across Earth's surface. Millions watched TV coverage of the Olympic Games of 2004, the U.S. presidential election of 2004, and the Indian Ocean tsunami at the end of 2004. Rapidly increasing access to the Internet exposes people from different parts of the world to one another's ideas and actions. Each year, greater numbers of business and tourist travelers make global connections by reaching more

(c)

(d)

FIGURE 1.2 **Places of tourist interest.** (a) Ecology and natural features. The rocky coast west of Lisbon, Portugal, from which the Portuguese navigators set out in the 1400s.

(b) Exciting cities. Rio de Janeiro from the Christ statue platform, looking over the harbor and Sugar Loaf with Copacabana and other famous beaches beyond the hill to the right.

(c) Exploring. The Antarctic: Zodiac boats from a cruise liner investigate a large iceberg.

(d) Emerging China. Shanghai, China. Traffic inside the rebuilt old city with its tourist shops; new hotels rise just outside the old city.

Photos: © Michael Bradshaw

2

parts of the world. As travelers increasingly interact with Earth's varied human and natural environments, the positive benefits from an enhanced awareness of cultural difference and the exchange of ideas may be challenged by a resultant corruption of host cultures or alteration of the environments of travel destinations. Poverty, oppression, natural disasters, disease, and conflict, for example, are often hidden or overlooked, but are an integral part of world regional geography (Figure 1.3).

Third, the modern world is constantly changing. Rapid shifts in our political, economic, and social experiences and expectations can bring confusion and conflict. The same changes that may end oppression or cure a deadly disease can create a newly disenfranchised population or degrade the environment. We are often unable to control, or even forecast, the outcomes. We cannot prevent change! Global connections can enhance our awareness of the fine balance between the potential for success and the production of tragedy. The contemporary study of world regional geography helps us make sense of our world. The authors chose the subtitle of this text to reflect what we observe: "global connections, local voices."

Geography in Today's World

What Is Geography?

Geographic Subject Matter

Geography is a discipline that studies spatial patterns in the human and physical world. Geographers examine where and how the human and natural features of Earth's surface are distributed, how they relate to each other, and how they change over time. Geographers examine patterns that exist on Earth's surface, such as the distribution of language or vegetation cover. They attempt to explain such patterns, how they are changing, and what they

might become. Many jobs require a geographic understanding. For example, urban planners need to be aware that cities contain people with varied preferences, traditions, fears, and desires; they are constantly moving around, interacting with one another, and having impacts on their urban environment. Disciplines such as history, sociology, economics, politics, and environmental science increasingly view differences among places as a crucial feature of understanding the human condition. They give significance to the research carried out by geographers.

Geographic studies are sometimes viewed in two parts. **Physical geography** includes natural environmental processes across Earth's surface that result in the distribution of climate varieties, plant ecologies, soil types, mountain building, or river action, among other patterns. In addition, physical geography increasingly examines the impacts of human actions on Earth's natural environments. **Human geography** is the study of the distribution of people and their activities (economies, cultures, politics, and urban changes). World regional geography integrates both aspects.

Consider what it would be like if you grew up in a different country. How might that affect the language you speak, your family's religious preference, the food you eat, the music you listen to, or the schools you attend? How might the weather and other environmental conditions affect you? What might be different about you? Might your views of world issues and possible solutions differ from what they are now? What might be the same? These are the sorts of questions geographers ask about people and the places where they live. Geographers find that understanding other peoples and places becomes increasingly important as cooperation, competition, and conflict among them become more intense.

Four geographers with similar interests, but living far apart (England, D.C., and Maryland), created this text. We enjoy almost daily e-mail contact with each other and universal access to the latest data and reports housed in various Internet websites. We could

FIGURE 1.3 **Places of difficulty.** (a) Slum area in New Delhi, India. (b) Hurricane Katrina hits New Orleans. Flooding on Canal Street on August 30, 2005, the day after the hurricane struck and water broke through the canal levees. Water levels rose to 4 meters (12 feet) in places, as well as huge destruction by winds and waves, causing the evacuation of the city, pollution of the waters, and many deaths. This was the greatest natural disaster to hit a city in the United States for a century. **Photos:** (a) © Parvinder Sethi; (b) © Chris Graythen/Getty Images.

(a)

(b)

not have done that 20 years ago. We believe that now, more than ever, geographic literacy is essential because people are connected and interacting on unprecedented and still increasing levels.

Study Methods Used by Geographers

As their horizons have widened, geographers now use a range of data and methods of analysis to achieve their goals. Modern geographers select from the following approaches in pursuit of their research goals:

- Attempts to establish scientific "laws" fitting their observations and interpretations about places after testing theories in several situations.

- Studies of individual and group perceptions of decisions—for example, in setting up a business or buying a home in a particular place.

- Efforts to understand the meanings and identities people give to places and how people derive their identities from these places.

- Investigations and analyses of the forces involved in changes resulting from technological and sociological innovations, including power relations such as social classes that bind people to their areas and societies.

- Links between the significance of individual decisions and the roles of social institutions.

- Assessments of the complexities of the modern world through the varied "stories" told by individuals and groups, believing that all knowledge is relative to each person's experience.

- Assessments of the complexities of the modern world through a combination of approaches, assuming some common basis that unites the scientifically observed mechanisms, the actual world based on extensions of precise observations, and how people experience events.

In this text we use varied data sources. The data tables are based on collections published and updated annually by the Population Reference Bureau, World Bank, and United Nations. The sections on the natural environment report scientific studies of such phenomena as weather, surface landforms, and soils. The statements made about population, economic events, politics, and cultural traditions derive partly from general overviews and partly from individual stories of people living in specific countries. After reading this section, you may have ideas concerning the information you would need to produce a geographic account of your own area that views it as distinctive or makes it possible to compare it with other places.

Geography as the Study of Places and Flows among Them

Geographers study places on Earth's surface as the environments or spaces where people live and through which they make life meaningful. Geography thus provides a place- or space-related **spatial view** of the human experience.

When we go to visit a **place,** it might be an individual building (convenience store), small town (Freeport, Maine), large city (New York), rural area (western Iowa), another state, or another country. Places may be perceived as points on a map or as large areas. Places are often identified by their position, or **location**, on the globe. Places have different relationships to each other in terms of location, direction, distance, and scale of size. Increasingly, the greater interconnectedness of people and places puts an emphasis on the movements of people, goods, and ideas among places and the consequent interactions. Geographers draw maps to represent the features of places on Earth's surface and flows among them.

Latitude and Longitude

Absolute location is the precise position of places on Earth's surface. The most universally accepted means of determining absolute location is by calculating latitude and longitude. For example, many cars are now equipped with global positioning systems (GPS) that define their exact latitude–longitude locations as they drive from place to place.

Latitude and longitude form the framework of the international reference system that pinpoints absolute location (Figure 1.4). **Latitude** describes how far north or south of the equator a place is, measured in degrees. The north pole is at 90°N and the south pole at 90°S. Although the equator is an imaginary line, its position has a direct physical relationship between Earth and the sun as the line along which the most direct radiation from the sun reaches Earth. The equator encircles the globe midway between the north and south poles and is the 0° (zero degree) line of latitude. The almost spherical Earth's circumference is around 40,000 km (25,000 mi.) at the equator. A circle that joins places of the same latitude at Earth's surface is called a **parallel of latitude.** The ground distance from one degree of latitude to the next is approximately 110 km (69 mi.) on Earth's surface. For a long time, latitude was found by measuring the angle of the sun above the horizon at noon.

Longitude measures position east or west of an imaginary line drawn from the north pole to the south pole—a half circle—that passes through the former Royal Observatory at Greenwich, London, United Kingdom. Lines joining places of the same longitude are called **meridians of longitude.** The position of the prime meridian passing through Greenwich (0°) was chosen by an international conference in 1884, when London was the world's most powerful decision-making city. Methods of determining longitude, especially when charting the position of a ship, were more complex and took longer to evolve than latitude measurements. Longitude lines are not parallel like their latitude counterparts, and thus they do not provide equal measurements of ground distance. Longitude lines are farther apart at the equator and closer together at the poles. Instead of measuring sun angle at noon, people had to create tables of planetary positions and needed accurate clocks (chronometers first made in the late 1700s). In the late 1900s, radio beacons and satellites provided standard reference points by emitting radio pulses that could be timed and interpreted rapidly in computerized navigation systems to give accurate position fixes in global positioning system devices.

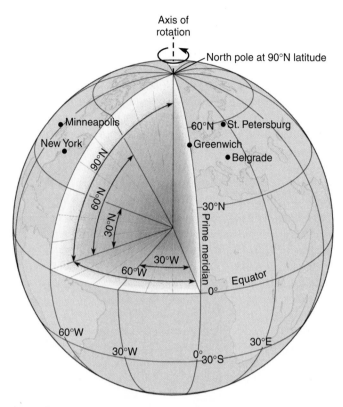

FIGURE 1.4 Location: latitude and longitude. The coordinate system is used for locating places on Earth's surface through the network of parallels of latitude and meridians of longitude. Give the latitude and longitude of each place marked on the globe. The degrees of latitude and longitude result from angles focused at Earth's center. Define the differences between 60°N and 30°N, and between the 0° meridian and 60°W in terms of angles at Earth's center. **Source:** From *Human and Cultural Geography* by Shelley & Clarke. Copyright © 1994 McGraw-Hill Company.

Distance and Direction

Direction and distance help to define the **relative location** of one place with reference to another. Natural obstacles such as mountains and oceans, political factors such as country boundaries, and cultural factors such as Latino and Anglo differences slow relative location. Increasing speeds of transportation and communications bring places relatively closer to each other. Geographers give **directions** by the cardinal points: north, east, south, west.

Physical **distance** between places is usually measured in kilometers or miles. However, travel time or travel cost (such as the cost of gasoline) may be of the greatest significance. The relative measures of time-distance and cost-distance are often substituted for measured distance in geographic studies. The increasing cost and time of distance between places gives rise to the idea of the **friction of distance.** Interaction between people is likely to be less across a distance where costs are higher or journey time is longer. For example, the friction of distance between New York and Chicago was reduced in the 1800s, when time for the journey was cut from weeks to days by building the Erie Canal and railroads.

Today air travel between these places takes a couple of hours. The increasing availability of rapid transportation facilities and the "global information highway" (the Internet) bring people into easier contact with each other, making them relatively—but not physically—closer.

Scale of Size

Places range in size from small to large; the differences are related to each other by a **scale.** There are, for example, within-country, country, world regional, and global scales. As connections among places become more intensive, places at different scales relate more closely to each other; the within-country areas connect to each other, to those in other countries, and to worldwide networks.

Maps and Geographic Information Systems

Geographers use maps to present information about location, distance, direction, and other characteristics of places. **Maps** are relatively small representations on paper of much larger areas of Earth's surface. On maps, scale also implies a mathematical relationship between actual distance on the ground and its representation on a map. Map scales vary with the size of the area to be mapped and the purpose of the map. Small-scale maps usually show areas at fractions of 1:250,000 or smaller (such as 1/1 million or a ratio of 1:1 million). They provide much less detail about larger areas (Figure 1.5). The world maps used in this text (see, for example, Figure 1.8) are small-scale maps, in which the scale along the equator is approximately 1:120 million. Large-scale maps usually have map-to-ground ratios ranging from 1:10,000 to 1:250,000. For the same size of map, they cover smaller areas, as in town maps, so they include more details. Not everything can be drawn to scale on maps (the features would be too small to be seen), so roads, rivers, buildings, and other features are replaced by symbols.

Geographic information systems (GIS) combine maps and aerial and satellite images with data relevant to the area (Figure 1.6, and see Geography at Work: Mapmakers and GIS Analysts, page 8). This has been a huge area of development since 1970, and most maps now relate to satellite images. GIS systems aid geographers in **spatial analysis** studies that examine links among places.

Geography of Regions

Our world of closer connections among peoples and places and increasingly rapid changes focuses attention on regional geography. A **region** is an area of Earth's surface with physical and human characteristics that distinguish it from other regions. Regions vary in geographic scale from major divisions of the world (world regions) to single countries and parts of countries; metropolitan regions, for example, are areas that focus on large cities and their suburbs. **Regional geography** evaluates differences and similarities within and between defined areas, or regions, of Earth's surface.

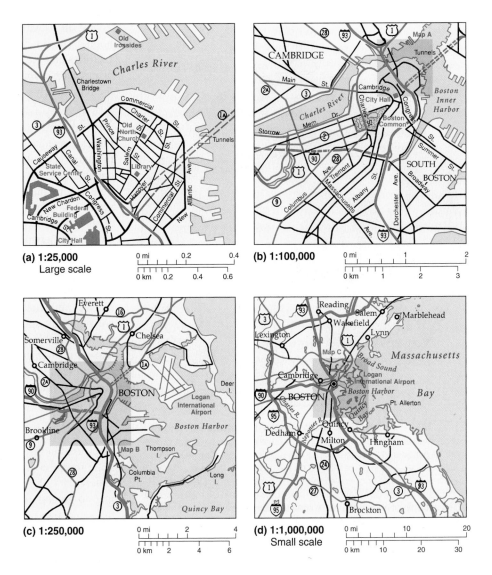

(a) 1:25,000
Large scale

(b) 1:100,000

(c) 1:250,000

(d) 1:1,000,000
Small scale

FIGURE 1.5 The effect of map scale on area and detail. Central Boston and the greater Boston area. The larger the scale, the smaller the area shown on maps of the same size and the greater the number of features that may be included. Different scales are shown by a representative fraction and a graphic scale line. **Source:** From *Human Geography*, 8th edition by Fellman, Getis and Getis. Copyright © 2005 McGraw-Hill Company.

Dynamic Concepts of Regions

Regional geography involves three basic sets of ideas.

First, regions are defined and created by geographers and others for specific purposes. Definitions of a region by different people may vary or change. Thus no region persists forever or for all purposes because the world is continually changing and so are people's perceptions of places. For example, the country of Sudan is included with Egypt in the Nile River Valley subregion in Chapter 8 of this text, and all world regional geography texts follow this pattern. However, U.S. military and foreign affairs handbooks have a different basis and place Sudan with Ethiopia in eastern Africa.

Second, regions so defined have a number of basic characteristics, including an areal extent and boundaries. In this sense, regions are "spaces of places." Regional boundaries may be based on physical features (such as coasts, mountain ranges, or rivers), political boundaries (such as administrative regions), economic characteristics (such as marketing regions), or cultural practices (such as language or religion). However, some regional boundaries are less precise because there is a broad transition to the next region. For example, Sudan's frontiers enclose a transition from the arid north to the humid south. Regions such as the "developing world" are less formally defined and may have several sets of overlapping boundaries proposed by different scholars. Regional scale varies from local-within-country to country, world regional, and global.

The patterns of spatial arrangements within regions include physical features, people, buildings, land use, and transportation links. The relationships between human and natural environmental features are basic to an understanding of regional geography. In the early 1900s, such studies focused mainly on the impacts of climate, mountains and lowlands, and soil types on human affairs. More recent studies assess the impacts of human activities on the physical environment.

Third, regions are essentially dynamic entities, marked by internal and external flow patterns of people, goods, and ideas.

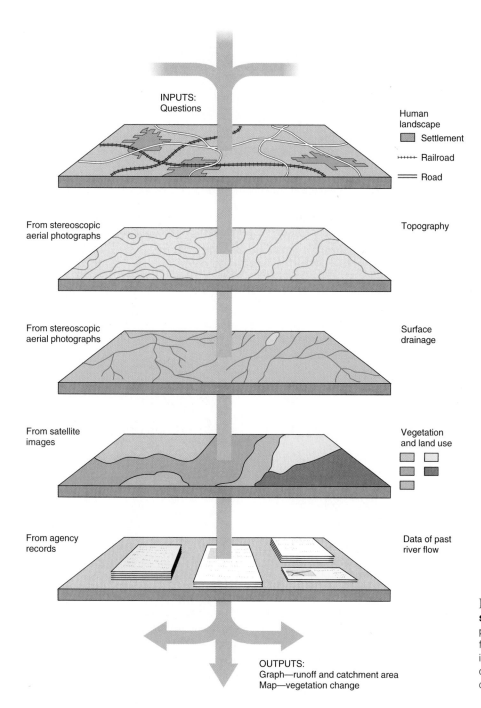

INPUTS:
Questions

Human landscape
⬛ Settlement
╫╫╫ Railroad
══ Road

From stereoscopic aerial photographs

Topography

From stereoscopic aerial photographs

Surface drainage

From satellite images

Vegetation and land use

From agency records

Data of past river flow

OUTPUTS:
Graph—runoff and catchment area
Map—vegetation change

FIGURE 1.6 Geographic information system. The layers of information in this example could be used to monitor a river system that feeds a reservoir. Outputs from the system might include graphs of seasonal stream flow divided by drainage basin area and dominant vegetation cover type, and maps of changing land use.

In this sense, regions are "spaces of flows." Flows within and among regions include

- Population migrations.
- Goods and services, imported and exported.
- Media information, Internet exchanges, or publication circulation.
- Movements of money.
- Technological innovations and transfers in manufacturing, information processing, or new transportation modes.

- Ideological messages through political and religious beliefs.

Interactions and flows between places affect the lives of people in regions. Consider a typical metropolitan region. People living there practice different religions, come from different places, speak different languages, are employed in various jobs, and go to different schools. Yet they use the same transit system and airport, the same hospitals and other emergency services, the same TV stations and newspapers. They visit family in

MAPMAKERS AND GIS ANALYSTS

Elio Spinello and Steve Lackow are Californians who have been involved with the distribution and management of AtlasGIS, a software package that combines mapping with a geographic information systems approach. In the 1990s, the huge GIS software company ESRI took over AtlasGIS, but has recently placed marketing and distribution back with Elio and Steve's company, RPM Consulting.

As an example of their work, Elio completed a project that was an epidemiological assessment of river blindness (onchocerciasis) in Mozambique. He was commissioned by Aircare International, which wanted to know the most important areas for delivering medical care. The disease does not kill, but is chronic and widespread and affects the lives of many people. The African blackfly inhabits areas with fast-flowing streams, and the female carries the microscopic worm (microfilariae) that causes spread of the disease from an infected person to an uninfected one. Fibrous nodules are produced in the infected persons with microfilariae that attack skin pigmentation, causing skin atrophy and eventually blindness. Surgery is sometimes needed to remove the nodules, but new medicines make it possible to treat many more people and prevent blindness without bad side effects.

Elio's study began by mapping the factors that encourage a concentration of African blackflies: the density of population, the concentration of rivers, and the presence of steep slopes that produce faster river flow. He came up with a composite index of these factors that gave greater weight to the population and waterway density than the slope steepness (Box Figure 1(a) and (b)). This was used to assess and map the risk of exposure to onchocerciasis in various parts of Mozambique.

Mapping data and the use of aerial photography and satellite images forms the basis of increasing amounts of geographic work. Making the basic maps with true distances, as in the United States Geological Survey, is mainly a matter of using aerial photography with checking features on the ground. Satellite images are used for monitoring land use, geology and soils, vegetation cover, and urban expansion.

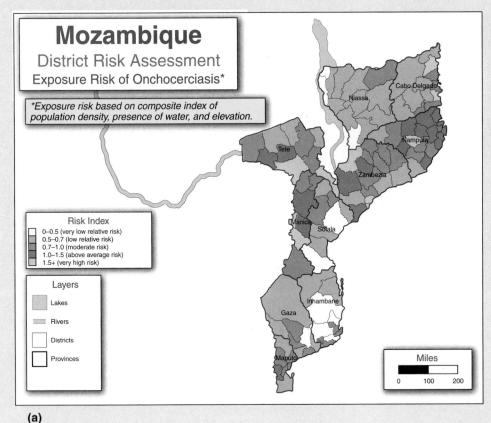

(a)

(b)

BOX FIGURE 1 (a) Map of risk of river blindness in Mozambique, the result of a map-based survey. (b) Elio Spinello. **Source:** Elio Spinello, RPM Consulting

other regions. The products they make are sold in other regions. There are flows into the metropolitan region, within the region, and out from it.

The dynamic elements of such flows within and among regions inside a country affect the regions' prominence, while flows among countries influence the countries' roles in the wider world. **Nodes** are places where flows begin, intersect, or end. The directions and extent of flows from a node may define the boundary of a region. For example, a store or restaurant can be considered a node, and the extent of the node's region is the area from which people come to shop at the store or eat at the restaurant. The capital city of a country is its political node.

Making and Remaking Regions

Interactions and flows between places produce five dynamic aspects of regional evolution.

People Create Regions

People living in a region determine its characteristics. Human actions at crucial phases of history set a region apart from other regions. People living inside or outside a region, often helped by the media or government propaganda, generate their own images of a region's role or identity. Inhabitants may define relations with other regions as friendly or "other."

Although physical environments often give character to a region through climate, natural resources, or topographic features, human actions are more important in defining regions. For example, people live in different ways in similar natural environments. Dry land inhabitants include the wandering herding tribes of the Sahara, farmers in Pakistan using the irrigation canal systems built by British engineers in the late 1800s, and those living today in the urban sprawl of Los Angeles. Looked at from another perspective, similar city landscapes in the United States occur across different types of natural environments. The regional studies in this text reflect this by considering cultural history before environmental conditions.

Regions Shape People's Activities

A region is an environment for human activities. It affects events both inside and outside its boundaries. People living in Los Angeles are constrained by how the city is built and its neighborhoods developed. They benefit from intensive links to other parts of the world. People living in very cold environments, or under harsh dictatorships, have more limited options.

So a two-way process is at work. People are the main forces in creating distinctive regions but are affected by the regional characteristics that others before them established. Over long periods, some regional characteristics are perpetuated, causing local cultural traits and social habits to emerge.

People Remake Regions

Regions—and their boundaries—may change over time, instigated by shifting flows of people, information, capital, technology, or political ideas. For example, in the 1880s, European colonial powers divided Africa into the country units that, by and large, still exist. These boundaries had nothing to do with traditional African land units (see Chapter 9); but once established, they proved difficult to change. Indeed, they provide new shapes to how Africans see themselves and behave.

Individuals and small groups often influence the course of changes in regions and countries. For example, the former Soviet Union arose out of a 1917 revolution that brought Vladimir Lenin and later Joseph Stalin to power. In the early 1930s, Stalin's practices industrialized the Soviet Union and made independent farmers work in cooperative state farms. The centralized, Moscow-based government dictated the products to

be made and the crops to be grown. Huge shifts of economic activities and massive population migrations followed, altering regions within the Soviet Union. In the late 1980s breakup of the Soviet bloc and the development of the new Russia in the 1990s, Mikhail Gorbachev and then Boris Yeltsin had major roles.

Regions Interact with Other Regions

No region is an isolated entity. For example, people living in the coastal towns of China make goods that are sold in the United States. Many migrated from inland rural areas to work in factories built since the 1980s, often financed by money from multinational corporations based in Taiwan, Hong Kong, Singapore, or Japan. Such workers still depend on local farms for their food and are subject to a cultural heritage handed down in their community and to Chinese government controls determined in Beijing. Their lives depend on the flows of people, information, capital, technology, and ideology among regions inside China.

Events such as media news reports, human rights legislation, or action on gender issues spread their influence from major global centers to local regions and sometimes rebound in the other direction. The world's remotest regions, even the upper Amazon River basin of Brazil, are affected by external demands such as the search for mineral resources, influxes of settlers looking for land, and visits by enterprising tourists. Changes in one region affect other regions: when world markets for peanuts and cotton collapsed and drought struck northern Nigeria, many people abandoned their farms and moved into urban areas in other parts of the country.

Some places affect surrounding regions by their role in funneling trade through narrow ocean-route throughways. For example, the materially wealthier countries defend the passage of their ships through **global choke points** such as the Strait of Gibraltar, the Suez (Egypt) and Panama Canals, the Straits of Hormuz (entrance to the Persian Gulf), Dardanelles, and Bosporus (entrance to the Black Sea in Turkey), and the Malacca Strait (near Singapore) to maintain access to their markets and raw material sources.

Interactions among specific pairs or groups of countries may result in close alliances in trade and defense. Conflicts occur within or among countries or regional groups. For example, different cultural groups within a country each define their territorial claims, and may fight against their country's government for what they believe are their rights. The Zapatistas in southern Mexico and the Muslims of Chechnya in southern Russia are examples.

Regions Are Used by Those in Power

Regional character may reflect the actions of powerful governments or nongovernmental institutions such as multinational corporations. Central planning under Communist governments in the former Soviet Union industrialized rural areas on a major scale, even north of the Arctic Circle. Groups of people were moved to other parts of the country for state security reasons. In the European Union, the outlying countries, such as Portugal and the Republic of Ireland, have their economies stimulated by EU investments in new roads, port facilities, and airports that help to attract new jobs in factories, often built for American and Asian multinational corporations.

In the mid-1900s, despite the end of the political colonialism that mainly European countries imposed on poorer parts of the world, larger and more affluent countries still placed political and economic pressures on poorer ones. The dominance of the United States, Europe, and Japan led them to trade cheaper raw materials for more expensive manufactured goods and services, resulting in new patterns of dependence. Other large but materially poor countries, such as India, China, Brazil, and Mexico, also exert pressures on their neighbors. At times intrusions have overcome sovereign country wishes. For example, in 1991 the United States went to war when Iraq annexed Kuwait and threatened other Persian Gulf countries and U.S. access to the world's largest oil fields. In 2001, action led by the United States against the Taliban and al-Qaeda terrorist network in Afghanistan disrupted their bases in that country. The United States obtained support for its actions by assembling a group of other countries appalled at the terrorist atrocities of September 11, 2001, or willing to take political advantage of the situation. It proved more difficult to obtain such support in 2003 when the United States and a few allies invaded Iraq to end Saddam Hussein's dictatorship.

Test Your Understanding
1A

Summary

We live in a world of variety, connectivity, and change, where geographical understanding is increasingly vital. Geographers study spatial patterns in the human and physical world and attempt to explain the patterns they identify, comparing places and linkages among them. Geographers use a variety of approaches to understand how and why people live in the places they do. Location of places is defined by latitude and longitude (absolute location) and by direction and distance (relative location).

Regional geography studies the diversity of geographic distributions and human activities. People create regions, which then affect the lives of their inhabitants; people re-create regions in times of change; regions interact with each other and are used by those in power.

Questions to Think About

1A.1   Is your definition of *geography* different than that given here?

1A.2   Why is regional geography so relevant to understanding our world today?

1A.3   What features make regional geography a study of dynamic human situations?

1A.4   What human and physical characteristics help us to define world regions? What might we consider in differentiating between one world region and another?

1A.5   How do we use direction, distance, and location to help us better understand the great variety of physical and human environmental conditions present on Earth's surface?

Key Terms

geography	direction
physical geography	distance
human geography	friction of distance
spatial view	scale
place	map
location	geographic information system (GIS)
absolute location	
latitude	spatial analysis
parallel of latitude	region
longitude	regional geography
meridian of longitude	node
relative location	global choke point

Regions and Globalization

Our world of closer connections among peoples and places and increasingly rapid changes focuses attention on regional geography in the context of global trends. Older studies of the "spaces of places" are now combining with studies of the "spaces of flows."

Globalization and Localization

Two geographic trends help us understand what makes regions different as a result of the increasing flows among them at the start of the 2000s: globalization and localization.

Globalization, in its simplest form, is the increasing level of interconnection among people and places throughout the world. Economic globalization involves the integration and exchange of capital, technology, and information across country borders. But it also affects social, cultural, and political affairs, and ecological issues arise from migrations of people, plants, and animals—including global warming, ocean pollution, global epidemics, and species extinction. It could be argued that globalization began with the European discoveries in the late 1400s or even earlier; but the speed and intensity of globalization, especially in terms of world trade and the flow of financial investments, increased markedly from the 1990s. Few people used the term "globalization" before 1990, but it is now mentioned daily in the media. It is an essentially geographic phenomenon.

Localization stems from long-established local identities that existed before globalization forces intruded. Such identities respond to globalization. As global exchanges and flows of information, ideas, people, money, and technology highlight increasing worldwide economic exchanges, political solutions, cultural attitudes, and environmental concerns, people at local levels face questions about maintaining their distinctive identities. The "local" scale is any place less than global in size, including countries, regions within countries, and wider regions that include groups of countries.

The rapid and widespread acceptance of the "globalization" theme by politicians, members of the media, special interest groups, and academia, among many others, created substantial confusion over precise meanings. In broad terms, there are three views about the phenomenon of globalization (a range of them are collected in Point–Counterpoint: Facets of Globalization on page 12). First, some people assume globalization is already in place, overriding country and smaller community boundaries and interests. They consider globalization as either a great opportunity for a more cohesive world or a danger to cultures, economies, politics, and environments. Second, others think the term is overused and unjustified when country governments remain the dominant political entities, and when cultural awareness and identities remain strong. Third, another group is sure that something significant is happening, but because it is at an early stage, this group awaits a fuller understanding of what is going on and where it will lead us. Thus we need to examine the concepts more fully.

Facets of Globalization

Increasing global connections highlight intensified flows of ideas, goods, and people worldwide. The variety of such flows produces increasingly complex societies, in which trends of global significance may override local experiences.

- *The spread of ideas, technologies, crime, and diseases.* Most materially poor countries wish to modernize to catch up with the materially wealthy countries. Technology transfer through improved communications and transportation is basic to such progress. The forces that ease the flow of modernizing technology, however, also facilitate terrorism, together with drug, armament, and slave trafficking and the spread of HIV/AIDS and other diseases.

- *Flows of goods and services.* Trade in raw materials, food products, manufactured goods, and services increased rapidly in the amount and speed of transactions in the late 1900s. Space–time compression in transportation and communications enhanced the central control of economic activities in both goods and services. Supermarket foods, clothing, electronic products, and vehicles come from an increasing number of countries, often marketed by multinational corporations of U.S., Asian, or European origin.

- *Long-term migration for work, political asylum, and family consolidation; short-term flows of people for long-distance business links and tourism.* Millions of people displaced by warfare and oppression gather in refugee camps, or become asylum seekers in foreign countries. Executives of multinational companies "commute" worldwide. Wealthy people vacation at exotic places. Materially wealthier countries attract those willing to work for low wages, while their better-paying jobs attract people with good education. Such workers from materially poorer countries often send part of their income to their relatives at home. However, this migration also produces a **brain and skills drain** because of the loss of potential leaders and skilled workers.

- *Shifts in dominant ideologies, especially religious or political beliefs.* The increasing openness of the Chinese economy from the late 1970s and the breakup of the Soviet Union in 1991 discredited Communism. Military dictators in other countries also fell from power. Their centrally controlled governments were inward-looking and resistant to external influences. Through the 1990s, democratic ideas spread and more countries had open elections. More were increasingly involved in international affairs. Furthermore, religious and other cultural forces such as language and tradition became as significant as political dependence, particularly in many materially poorer countries and those in economic transition from centrally controlled governments and economies.

- *The spread of images and messages through the media of TV, film, the Internet, and print.* In the entertainment industry, worldwide screening of U.S. movies and TV programs may disseminate images of the Western "good life." Increasingly, however, challenges to Western dominance come from the film industries of India and other countries such as Egypt and South Korea. The Internet and access to the World Wide Web enable international and local firms, political groups, and leisure interests to have exposure outside the control of individual countries. Print media (newspapers, magazines, and books) are increasingly controlled by international consortia.

- *Uncontrollable negatives.* The global spread of ideas, communication, faster transportation, and the resultant lower levels of control by countries leads to an expansion of arms, drugs, and people trafficking that have huge financial rewards, encourage and fund terrorist activities, and take advantage of complex and changeable routes to market.

Facets of Localization

In many parts of the world, local voices and identities have become more confident and significant in reacting to globalization forces in the last 20 years. Globalization changes relationships between geographic scales of places as actions move to subcountry or supracountry levels, making geographic interactions more complex. Geographic fragmentation is more frequent than global amalgamations. Many facets of localization, including material polarization between globalized activity and

FACETS OF GLOBALIZATION

Mentions of globalization raise anger in some places; mobs of people demonstrate with violence against it at meetings of international leaders. Antiglobalization protesters often shout louder than those who tell us that whatever the growing pains, globalization is good for us in terms of better incomes, better life quality, and better understanding of other people. Some of these views are posed here for you to debate. We have quoted from some writers who have gone into print since 1990. The alternative views are not necessarily shared by the authors of this text, but seek to summarize points of view in opposition to the quotations. After discussing these alternatives, add others or assess the views people might have in different parts of the world.

Globalization is good for you.

Barber overstated his case to make a point in 1992. Fifteen years later changes have taken place, but not all fit his description of "bleak, neither democratic, . . . largescale bloodshed, global uniformity, . . . a planet precipitantly falling apart." In fact, many throughout the world live under better conditions with a greater sense of local identity than they did 15 years ago. Although jihad and mcworld are with us, responses to both prevent the direst outcomes.

Fourteen years after being published, Huntington's thesis remains a prescription for a divided world and continuing conflict on a global basis. Certainly, some of the conflicts during that time occurred where different cultures or civilizations clash, but others occurred elsewhere, and many parts of the world remain peaceful despite internal country differences or outlook or terrorist threats.

It is clear that international peace and security cannot be guaranteed: the more global our activities, the less prospect there is for overall control by countries—the present location of law-giving and security.

Globalization is bad for you.

Just beyond the horizon of current events lie two possible futures—both bleak, neither democratic. The first is a redistribution of large swaths of humankind by war and bloodshed, in which culture is pitted against culture, people against people, tribe against tribe—a *jihad* in the name of a hundred narrowly conceived faiths. . . . The second is being borne in on us by the onrush of economic and ecological forces that demand integration and uniformity and that mesmerize the world with fast music, fast computers, and fast food—with MTV, Macintosh, and McDonald's, pressing nations into one commercially homogeneous global network: one *mcworld* tied together by technology, ecology, communications, and commerce. The planet is falling precipitantly apart and coming reluctantly together at the very same moment.

Benjamin R. Barber, *The Atlantic Monthly,* 1992

World politics is entering a new phase. . . . It is my hypothesis that the fundamental source of conflict in this new world will not be primarily ideological or primarily economic. The great divisions among humankind and the dominating source of conflict will be cultural. [Countries] will remain the most powerful actors in world affairs, but the principal conflicts of global politics will occur between them and groups of different civilizations. The clash of civilizations will dominate global politics. The fault lines between civilizations will be the battle lines of the future.

Samuel P. Huntington, *Foreign Affairs,* 1993

Many thought the new world order proclaimed by George H. Bush [in 1990] was the promise of 1945 fulfilled, a world in which international institutions, led by the United Nations, guaranteed international peace and security and the active support of the world's major powers. That world is a chimera. Even as a liberal internationalist ideal, it is infeasible at best and dangerous at worst. It requires a centralized rule-making authority, a hierarchy of institutions, and universal membership. Equally to the point, efforts to create such an order have failed. The United Nations cannot function effectively independent of the major powers that compose it, nor will those nations cede their power and sovereignty to an international institution.

Anne-Marie Slaughter, *Foreign Affairs,* 1997

(Continued)

| **Globalization is good for you.** | **Globalization is bad for you.** |

Globalization is good for you.

The mall at Singapore's airport has a food court with 15 food outlets, all but one of which offer menus that cater to local tastes; the lone standout, McDonald's, is also the only one crowded with customers. In New York City, experts in feng shui, an ancient Chinese craft aimed at harmonizing the placement of man-made structures in nature, are sought after by real estate developers in order to attract a growing influx of Asian buyers. Globalization refers to processes that evolve as people go about their daily tasks; these processes are not hindered or prevented by territorial or jurisdictional barriers, they can readily spread in many directions across national boundaries, and are capable of reaching any community anywhere in the world.

James N. Roseneau, *Current History,* 1997

Many plants from other parts of the world have been planted and grown successfully without danger to local ecosystems. At times this has brought great benefits, as in the global adoption of crops that were once limited to specific cultural hearths. More globalization should lead to greater, not less, control of Bright's "small percentage of cases" of potentially harmful plant and animal invasions, as the United States rules against entrants bringing plants. Insects are more difficult to control, but seem to migrate to new areas whatever the system.

The last few years have shown that the United States, despite its huge world presence, cannot run roughshod over other countries. Many countries in Europe disassociated themselves from the 2003 Iraq war, and there are many other voices in Europe, East and South Asia, and in the Arab world. Globalization requires a greater working together by countries.

The Seattle Ministerial Conference of the World Trade Organization (WTO) demonstrated with disturbing force the huge confusions that haunt the public mind about . . . the process now known as globalization. The notion that globalization is an international conspiracy on the part of industrial country governments and large firms to marginalize the poorest nations, to exploit low wages and social costs wherever they may be found, to diminish cultures in the interests of an Anglo-Saxon model of lifestyle and language, and even to undermine human rights and cut away democratic processes that stand in the way of ever more open markets is, of course, utter nonsense. The outpouring of misconceived, ill-understood propaganda against a system that has brought vast gains to most nations over the past few decades is extraordinarily dangerous.

Peter D. Sutherland, *Harvard International Review,* 2000

Globalization is bad for you.

Globalization is unlikely to result in a greater richness of experience for people, but will move toward global products that are the same everywhere, reducing choices and understanding of other peoples.

World trade has become the primary driver of one of the most dangerous and least visible forms of environmental decline: thousands of foreign, invasive species are hitchhiking through the global trading network aboard ships, planes, and railroad cars, while hundreds of others are traveling as commodities. This "biological pollution" is degrading ecosystems, threatening public health, and costing billions of dollars annually.

Bioinvasion occurs when a species finds its way into an ecosystem where it did not evolve. In a small percentage of cases, the exotic finds everything it needs—and nothing capable of controlling it. In South and Central America, the growth of specialty export crops—upscale vegetables and fruits—has spurred the spread of whiteflies, which are capable of transmitting at least 60 plant viruses. In southern India, a tropical American shrub is strangling rice paddies and ruining fish habitat throughout the Cauvery River basin.

Christopher Bright, *Foreign Policy,* 1999

In the past year (1999), many global issues have been framed in terms of the notion of sovereignty, that is, the right of political entities to follow their own course—which may be benign or may be ugly—and to do so free from external interference. In the real world that means interference by highly concentrated power, with its major center in the United States. This concentrated global power is called by various terms, depending on which aspect of sovereignty and freedom one has in mind. So sometimes it's called the Washington consensus, or the Wall Street/Treasury complex, or NATO, or the international economic bureaucracy (the World Trade Organization, World Bank, and IMF), or G-7 (the rich, Western industrial countries), or G-3 or, more accurately, usually G-1.

Noam Chomsky, *"Rogue States: The Rule of Force in World Affairs,"* 2000

It is difficult to believe that there are benefits from globalization. We see city centers around the world being taken over by Starbucks, Kentucky Fried Chicken, and McDonalds. We are told that the poor are getting poorer and lagging further behind the materially wealthy. We should all go back to living within the resources of our own countries and not bother about anyone elsewhere.

the "black holes" of rural marginalization and exclusion, maintain and enhance differences among places. Even within materially wealthy countries, contrasts remain between richer and poorer people and in particular between male and female opportunities. For much of the 1900s, democracy, welfare programs, and decolonization, together with pressure from workers and consumers, made the world more equal; but this began to change in the 1990s as economic activities moved from the country (place) basis to realms of wider interaction (flows). Such moves to inequality work against any trends toward "one world." Features of localization include

- **Political nationalism** that seeks to preserve the uniqueness of countries, groups of countries, and smaller regions within countries. Within the countries of the European Union, national identities slow political integration.

- **Separatist groups** within countries and across borders, such as the Basque people on the Spanish–French border, work for independence in smaller country units rather than incorporation in larger-scale global linkages.

- **Customs and practices** preserve local identities despite globalizing forces. Western pop music, for example, has a worldwide circulation, but variants in Africa, Asia, and Latin America strongly reflect local attitudes and traditions. Indeed, fear of globalization's effects has caused a resurgence and newfound pride in local customs and practices.

- **Religious differences** among Christian, Muslim, Jewish, Buddhist, and Hindu countries continue to be significant in marking local communities and making allies of groups of countries. For example, new immigrants in foreign cities commonly elect to live close to others of similar backgrounds. This may build new pressure groups as each group establishes its places of worship and ethnic food outlets. As immigrant groups take over housing in nearby streets, this often leads to the fragmentation of living areas in cities.

- **Resistance to the visible economic penetration** of countries around the world affects global media (such as CNN and the Murdoch group) and retail corporation policies (such as McDonalds, Starbucks, Toyota, and Nike). They are criticized as contributing to poverty or loss of local identity. This spawns local opposition groups.

Localization commonly involves political, economic, social, cultural, and environmental factors. The global and local scales of human activity are closely linked; it is increasingly difficult to establish total isolation from global influences. In the 1990s, the Taliban regime in Afghanistan attempted to establish a culture in isolation from the rest of the world. Its leaders took political control of most of Afghanistan and imposed a strict form of Islam, forcing out other religious groups and Western influences. Using the isolation to support al-Qaeda terrorists, however, put the Taliban into conflict with wider global interests and led to its downfall in 2001. The tensions between Iraq and the rest of the world over reports of its weapons of mass destruction led to the invasion of Iraq in 2003, but was followed by increasing involvement of local and foreign groups of fighters resisting the "victorious" forces.

Some commentators are optimistic about global processes improving the lot of poor countries. Others are pessimistic and see a growing global disorder. Our world regional studies will take this debate as a context, and we will assess the outcomes in Chapter 12.

Measuring Globalization

A globalization index provides a measure of the degree to which a country is involved in global forces. One approach ranks countries by four groups of data sets:

- **Political engagement indicators,** including membership in international organizations, personnel and financial contributions to United Nations Security Council missions, ratification of multinational international treaties, and government transfer (payments and receipts).

- **Technology measures,** such as the number of Internet users, Internet hosts, and secure servers for encrypted transactions.

- **Personal contact indicators,** such as international travel and tourism, international telephone traffic, cross-border nongovernmental remittances, and personal transfers (including from workers abroad).

- **Economic integration measures,** such as trade flows, portfolio capital flows, foreign direct investment, and investment income.

Figure 1.7 shows the 2004 ranks based on 2002 values for the top 20 countries out of the 62 ranked. The values reflect specific conditions for that year, including post-9/11 drops in travel and trade. Of the top 20, 14 are in Europe, accompanied by Canada, the United States, Australia, New Zealand, Singapore, and Malaysia. The United States' rank was raised by its dominance of technology, but was held back by a low level of trade proportional to overall output, few remittances from abroad, and poor implementation of international treaties. The high ranks of Malaysia and Singapore highlight the absence of any countries from Africa, East and South Asia, Latin America, or Southwest Asia in the top 20. The top 20 countries have major involvement in the economic, personal contact, and technology groups as well as the political, while the lowest ranking 10 countries have little outside the political involvement. This study also detected strong links between globalization rank, longer life expectancies, and better lives for women.

Major World Regions

In this text, the world is divided into nine major world regions (Figure 1.8) that are the subjects of Chapters 3 through 11. Each **world region** contains a group of countries linked by cultural, political, economic, and natural criteria. Global and local linkages help to define the major regions.

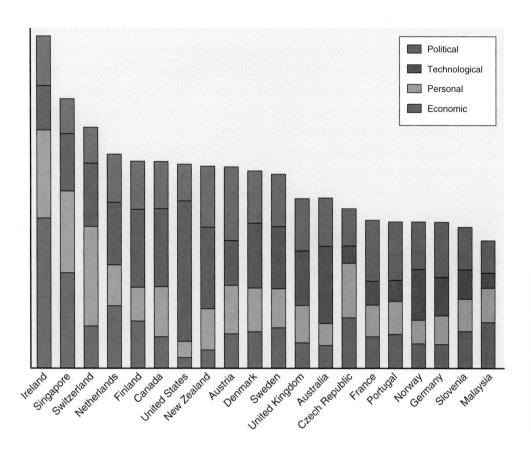

Political
Technological
Personal
Economic

Ireland
Singapore
Switzerland
Netherlands
Finland
Canada
United States
New Zealand
Austria
Denmark
Sweden
United Kingdom
Australia
Czech Republic
France
Portugal
Norway
Germany
Slovenia
Malaysia

FIGURE 1.7 The Global Top 20, 2004. *Foreign Policy* magazine works with A. T. Kearney to produce an annual index of globalization based on political engagement, technology, level of personal contact, and economic integration. Few of the Top 20 lie outside Europe and North America. **Source:** Randolph Kluver, Wayne Ru in *Foreign Policy,* March/April 2004, p. 54. Reprinted by permission of Foreign Policy via Copyright Clearance Center.

The main part of this text is a journey through the nine major world regions. Our trip begins in *Europe,* then moves east through *Russia and Neighboring Countries* into *East Asia* and *Southeast Asia and South Pacific.* Returning westward, *South Asia* leads to *Northern Africa and Southwestern Asia,* limited southward by the Sahara Desert. *Africa South of the Sahara* is the next destination before a hop across the Atlantic Ocean to *Latin America.* We finish in *North America,* taking on board the many influences brought to it from around the world and the many ways in which the United States and Canada impact other world regions.

Europe

Europe (Chapter 3) is the source of many Western trends in political, cultural, and economic areas. The region is defined as those countries that are members of the European Union or are likely to be in the next few years. It is marked by moderate midlatitude environments and advanced economies. Its cultures continue to be affected by the past dominance of Roman Catholic, Protestant, or Orthodox Christian groups, increasingly challenged by secular ideologies and the religious cultures of immigrants. Europe is marked by the variety of languages spoken, while increased immigration in the late 1900s made it home to a growing number of people from diverse world regions.

Russia and Neighboring Countries

Russia and Neighboring Countries (Chapter 4) includes all the countries that emerged from the breakup of the Soviet Union apart from the three small Baltic countries of Estonia, Latvia, and Lithuania (now part of Europe, Chapter 3). The region extends from easternmost Europe across northern Asia, a huge area that was brought together by the expansion of the Russian Empire in the last 400 years. It extended European cultures into central and northern Asia, incorporating a wide range of people, cultures, and languages. Through much of the 1900s, the centralizing Communist government of the former Soviet Union subdued the long-term clashes between those of Orthodox Christian and Muslim faith and between the Russian and other peoples. These conflicts emerged again after the Soviet Union breakup as part of the difficult transition from Communist to free-market economies.

East Asia

East Asia (Chapter 5) includes Japan, the Koreas, Mongolia, and a resurgent China. They are some of the world's most successful countries in recent economic growth, challenging the dominance of the United States and Europe. In economic

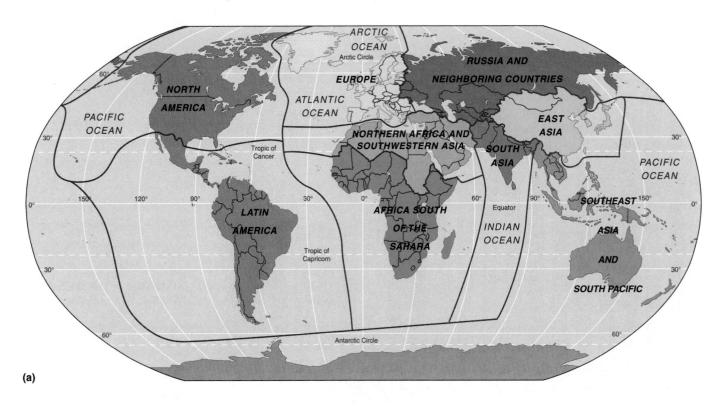

(a)

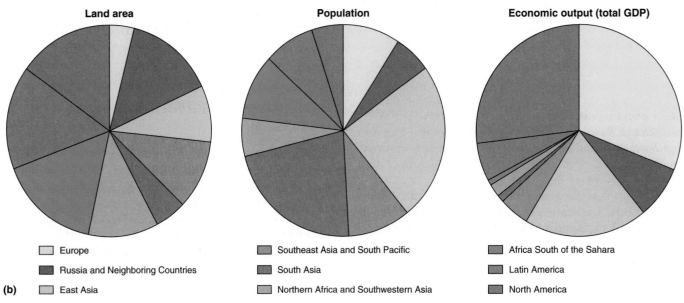

Land area

Population

Economic output (total GDP)

☐ Europe

■ Russia and Neighboring Countries

☐ East Asia

■ Southeast Asia and South Pacific

■ South Asia

■ Northern Africa and Southwestern Asia

■ Africa South of the Sahara

■ Latin America

■ North America

(b)

FIGURE 1.8 **Major world regions based mainly on cultural characteristics.** (a) These regions form the subjects of Chapters 3 through 11. World maps in Chapter 2 are divided on this basis so that comparisons may be made. (b) Comparisons of area, populations, and economic output (gross domestic product, GDP). Pie charts show the variations among world regions. Which major regions have more of the world's land area than its population? Which major regions have more of the world's economic output than its population?

terms, China now has the second highest total output by value in the world after the United States, and Japan is third. The historic and modern influences of the Chinese kingdoms and related cultures were less affected by European colonization than other world regions and bring special character to this region and to surrounding countries, where many families and business personnel live.

Southeast Asia and South Pacific

Southeast Asia and South Pacific (Chapter 6) includes the independent kingdom of Thailand and the former European colonies of Myanmar (or Burma), Malaysia, Australia, and New Zealand (all ex-British colonies); Indonesia (Dutch); the Philippines (Spanish, taken over by the United States);

Vietnam, Laos, and Cambodia (French Indochina); East Timor (Portuguese, taken over by Indonesia); and thousands of islands distributed across the South Pacific Ocean (mainly former British, French, German, and United States colonies, possessions, or trusteeships). Independence brought varied experiences. Having lost many of their trade links with Europe, Australia and New Zealand increasingly look to Asia for trading opportunities, although their essentially European outlooks at times bring them into conflict with Southeast Asian attitudes. We also include Antarctica in this world region, inhabited solely by scientific colonies—the only part of Earth's land surface that is not divided into countries.

South Asia

Over 1 billion people live in *South Asia*. Large numbers of poor people exist alongside growing technocratic and wealthy elites. South Asia (Chapter 7) experiences clashes among major religions including Islam, Hinduism, and Buddhism. Today Hinduism (the dominant religion in India) counts 80 percent of India's population as its adherents; Islam is the national religion in Bangladesh and Pakistan; Buddhism is dominant in some smaller countries. The British Empire occupation was followed since independence in 1947 by only partly successful policies of self-sufficiency. Such policies brought slow economic growth before the increases of foreign trade and investment in the 1990s.

Northern Africa and Southwestern Asia

Northern Africa and Southwestern Asia (Chapter 8) is characterized by a position at the junction of Europe, Africa, and Asia. It includes the birthplaces of the world's three monotheistic (believing in one God) religions: Judaism, Christianity, and Islam. The Islamic religion is dominant today, partly paralleled in its extent by the Arabic language—although the two Muslim countries with the largest populations in this region, Iran and Turkey, have their own languages. The presence of Jewish Israel in the predominantly Muslim region raises cultural, economic, and political tensions that result in almost continuous hostilities. The location of the world's largest oil reserves and the water shortages in a largely arid natural environment of much of this region pose internal problems of uneven resource availability.

Africa South of the Sahara

Africa South of the Sahara (Chapter 9) was the cradle of the human race, and its current population has many ethnic groups. The region contains great mineral riches but has underdeveloped economic potential. Colonial settlement by Europeans occurred much later and on a smaller scale than in the Americas or India. They introduced commercial farming and mining. Decolonization began in the 1950s, but was often followed by poor dictatorship governments. The people of this region are among the world's most deprived today.

Latin America

Latin America (Chapter 10) has many political and economic issues that relate to the region's history of colonization and its position as the nearest neighbor of the United States. From the 1500s, the indigenous peoples were reduced in numbers by the Spanish and Portuguese colonization of Latin America, but they still form major portions of the population in the Andes and Central American highlands. The Latin-based Romance languages and Roman Catholic religious culture brought by Spanish and Portuguese settlers influence most of this region. There are enclaves of other European languages, including French, Dutch, and English, particularly in the Caribbean.

North America

North America, comprising the United States of America and Canada (Chapter 11), is the world's materially wealthiest region, containing the only current world superpower (the United States). The two countries are dominated by cultures first brought by European settlers beginning in the 1500s. French and Spanish are spoken locally, but English is the most widely spoken language in North America. Almost wiped out by disease and warfare as a result of the European influx, the indigenous peoples formed smaller and smaller proportions of the population, but some lands and facilities were restored to them in the later 1900s.

Globalization and the Origins of World Regions

A quick look at the way today's diverse world geography developed historically identifies distinct phases of regional creation and re-creation, during which increased flows of people, information, wealth, technology, and ideas spread out of early centers of innovation to incorporate more and more of our world. For thousands of years, areas of growing material wealth and greater population density occupied small parts of the world, connected fitfully to each other. Later, countries could be linked in major world regions, and today many globalizing trends are effective.

Prehistory

It is now widely accepted that modern humans spread out of Africa to other parts of the world: a strong possibility, based on DNA evidence, is shown in Figure 1.9. This peopling of the world can be seen as the first phase of globalization. Until around 5000 B.C., most humans lived in small family-based groups, gaining a livelihood by hunting and gathering. The separate groups had little interaction with each other, although kin alliances linked groups across broad territories. People lived in caves or temporary dwellings that could be occupied or rebuilt as they moved to find new resources. The areas needed to support groups varied in size depending on the nature of the local environment and its resources, but population densities were low.

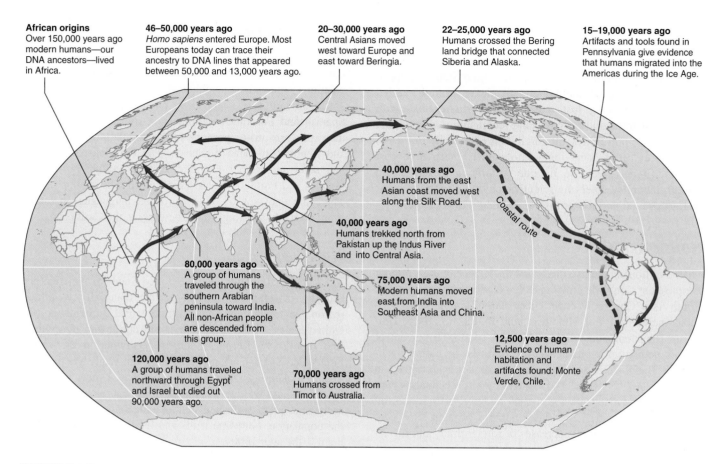

African origins Over 150,000 years ago modern humans—our DNA ancestors—lived in Africa.

46–50,000 years ago *Homo sapiens* entered Europe. Most Europeans today can trace their ancestry to DNA lines that appeared between 50,000 and 13,000 years ago.

20–30,000 years ago Central Asians moved west toward Europe and east toward Beringia.

22–25,000 years ago Humans crossed the Bering land bridge that connected Siberia and Alaska.

15–19,000 years ago Artifacts and tools found in Pennsylvania give evidence that humans migrated into the Americas during the Ice Age.

40,000 years ago Humans from the east Asian coast moved west along the Silk Road.

40,000 years ago Humans trekked north from Pakistan up the Indus River and into Central Asia.

80,000 years ago A group of humans traveled through the southern Arabian peninsula toward India. All non-African people are descended from this group.

75,000 years ago Modern humans moved east from India into Southeast Asia and China.

Coastal route

120,000 years ago A group of humans traveled northward through Egypt and Israel but died out 90,000 years ago.

70,000 years ago Humans crossed from Timor to Australia.

12,500 years ago Evidence of human habitation and artifacts found: Monte Verde, Chile.

FIGURE 1.9 Peopling of Earth by *Homo sapiens.* A study based on mitochondrial DNA evidence provides one version of this process. Mitochondrial DNA passes through the female line. **Source:** Map from *Out of Eden* by Stephen Oppenheimer, inside cover, created by Stephen Oppenheimer. © 2005 Stephen Oppenheimer. Reprinted by permission of Blake Friedmann Literary Agency Ltd.

Settled Farming

The first settled farming began in a few areas of southwestern Asia as long ago as 9000 B.C., followed by parts of China, other Asian areas, and the Americas. Settled farming came much later to western Europe and Africa.

Distinctive regions were created as geographic variations grew out of the human choices among local plants and animals for domestication. These regions shaped the lives of people living there. The new skills and ideas generated in these places spread to surrounding areas as people, techniques, religions, and languages diffused outward. Each product had daily and seasonal work patterns and linked community organization. Wheat and barley were the most important crops in southwestern Asia, rice and millet in China, a wide range of corn, squashes, beans, potatoes, tomatoes, and peppers in the Americas, and sorghums and yams in Africa. Animals also varied. Domestic herding groups (nomads) traveling seasonally to pastures for the animals generated another set of cultures, whose ways often conflicted with the settled farmers and their villages. Some of the settlements grew into towns, such as Tell es-Sultan (Jericho, c. 8000 B.C.) in the Jordan River valley and Çatal Hüyük in central modern Turkey. The early centers of innovation in agriculture, writing, and other technologies are often referred to as **culture hearths.**

City–States and Empires

From 2500 to 1000 B.C., massive local surpluses of farm products formed the basis of the first civilizations, causing their peoples to remake regions into something new and distinctive. Sun and water combined to concentrate productive irrigation farming along the lower Tigris–Euphrates Rivers (Mesopotamia), lower Nile River (Egypt), the Indus River valley (modern Pakistan), the Huang He (or Yellow River) valley in China, coastal Peru, Middle America, and western Africa (Figure 1.10). Political systems, varying from city–states to extensive empires, developed through the control of irrigation and the use of the surplus in trade.

The centers that experienced a huge accumulation of wealth often expressed it in major buildings. "High-tech" developments included the first writing (to inventory trade transactions), the organization of water distribution and land ownership, the foundations of complex mathematics, innovations in metalworking (this was the Bronze Age), pottery, construction materials, and the wheel. Artistic representations became more sophisticated, systems of religious beliefs developed with links to mythical prehistories, and laws were created. Recorded history began.

Trading networks linked the materially wealthy growth regions. Local exchanges of food, raw materials, and craft

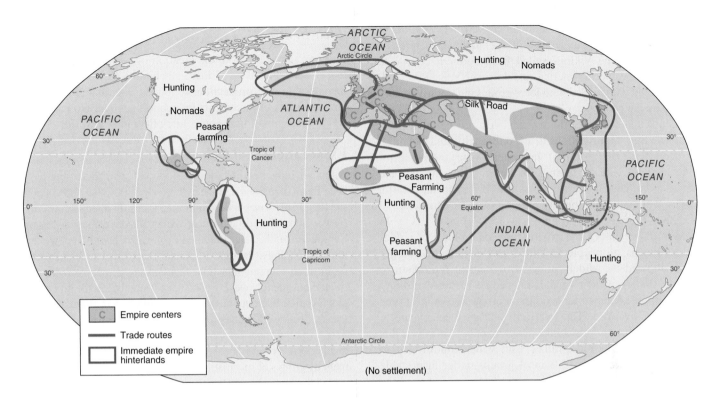

FIGURE 1.10 Feudal empires. Up to around A.D. 1450, disruptions in Europe included invasions from the Eurasian steppes and Viking homelands. New empires developed in the Americas and western Africa in addition to the longer-established empires in South and East Asia. The Silk Road connected Europe to Asia and provided a major transportation route for high-value goods. Outside the empire regions, people depended on local resources for hunting, simple cultivation, and nomadic lifestyles that had limited geographic interactions.

products complemented longer routes dealing in luxury goods such as precious metals, jewels, and spices. In most areas of the world, however, people still lived by subsistence on local products and remained separate from the areas of major wealth creation and growing trade interactions.

Trading Empires and "Classical" Civilizations

From 1000 B.C. to A.D. 600, centers of wealth accumulation extended to new and larger areas. Increased levels of trade and economic activity spread from western Asia along the Mediterranean into Europe; Chinese dynasties expanded their empire westward and southward. As mobility increased, flows of people and goods led to greater levels of interaction among regions. By A.D. 200, trading routes for high-value goods extended between Rome and China along the Silk Road route through Central Asia. In the Americas, the Maya developed their own writing system and built cities based around huge public buildings by A.D. 250.

This period saw the rise of the "classical" civilizations. At the outset, these civilizations were often led by a charismatic leader or linked to a pantheon of gods. The main centers were in Persia (Zoroaster), Greece and Rome (each with their pantheon of gods), India (Hinduism developing from more ancient

traditions; Buddhism), China (Confucius; Lao-Zi and Daoism), southwestern Asia (Judaism, Jesus Christ, Muhammad), central Europe (Celtic druids), northern Europe (Wotan and Norse gods), Egypt (Isis and other gods), and the Americas (nature gods). Large-scale movements of people resulted in more widespread and intensive warfare and other changes in regional character. For example, invasions of the Indian subcontinent and of the rich centers of China and the Mediterranean arose out of the emerging knowledge of the locations of accumulating material wealth.

Disruptions, Migrations, and Feudalism

Increased disruptions of the established empires occurred from A.D. 600 to around 1450, often termed the "Dark Ages." The mobility of larger armed groups across greater spaces resulted in some older empires giving way to new. Other empires extended their spheres of influence. Regional changes and interactions became increasingly influenced by political power.

In the Mediterranean, the Roman Empire in the west succumbed to attacks; but its eastern Byzantine wing, centered in Constantinople (modern Istanbul), lasted longer. By 650, soon after Muhammad's visions generated the new religion of Islam, Muslim people spread out of Arabia across northern Africa and into southernmost Europe, central Asia, and India.

From 800, hordes of horse-mounted warriors swept out of the Asian grasslands into central Europe, India, and China. At the same time, the Vikings sailed from northern Europe to western Europe, North America, and down the Volga River valley, where they interacted with local groups in the early history of the Russian people.

China again became the world's most prosperous empire and made huge contributions to art, philosophy, and technology (such as the first printing). In western Africa, new empires controlled the Sahara crossings. In the Americas, the Inca and Aztecs dominated large regions—the former in the central Andes Mountains and the latter taking over the Maya area in Middle America after around A.D. 1300 (Figure 1.11). In many of these places, the growing personal power of nobles and shortages of people to fulfill production and defense needs allowed the widely effective system of **feudalism** to replace slavery. Under feudalism, people received military protection in exchange for working a proportion of their time for a local overlord. The overlord then gave allegiance to higher orders of knights, kings, and emperors. However, this system resulted in local rather than wider control, led to changing combinations of control groups rather than stable political boundaries, fostered few new ideas, and slowed the expansion of trade.

The Modern, Globalizing World

From around A.D. 1450, the creation of new types of regions that shaped the lives of their inhabitants flowed from huge increases in the production of goods and material wealth. Greater levels of interregion interaction, and the growth in the significance of political power mark the modern world.

Explorations and Colonies

Although Indian and Chinese sailors and merchants may have been first to explore the globe in the early 1400s, they did not pursue their discoveries and withdrew to focus on internal concerns. New systems of trade and wealth expansion developed in western Europe by 1450. They combined exploration based on new maritime technology, the zeal to spread the Christian faith, a new profit motive among merchants, and groups of people willing to resettle outside Europe. European expansion began the process of globalization that continues today. Spaniards and Portuguese led the way, followed by the French, Dutch, and British.

An initial focus on sponsorship by European rulers wishing to add new territories and sources of income gave way to investment by groups of speculating merchants. At first, monarchs supported the Spanish and Portuguese explorers, who claimed lands in the Americas and traded around Africa into the Indian Ocean,

FIGURE 1.11 Early civilization achievements. Among these were huge temple buildings and palaces and advanced decoration and written scripts. The Mayan Kukulkan Pyramid at Chichén-Itzá, built in the A.D. 900s, and details of sculpting on the same complex. **Photos:** © Vol. 146/Corbis.

reaching China. In northwestern Europe, merchant-funded companies sent out settlers, invested in mines and plantations, and increased trade with eastern North America and India. From the 1600s, the slave trade shipped millions of Africans to plantations and mines in the Americas. Later the governments of the emerging countries in western Europe claimed colonies in the lands opened up by merchant-funded groups.

Industrialization and Colonization

From the mid-1700s, new manufacturing technologies in cloth making, leather tanning, and metal refining and fashioning provided marketing and investment opportunities in Europe. These often brought better returns than gathering produce from the overseas mines and plantations. New technologies developed, based on the raw materials from those mines and plantations. The increasing efficiency and technological breadth of manufacturing processes geographically concentrated increased production in Europe and later in the United States, making larger ships, steam engines, commercial chemicals, and then airplanes, automobiles, and a multitude of consumer goods. Producing such goods had major impacts on world regional geography. Colonies supplied European and U.S. manufacturers with raw materials, and the colonial elites bought the manufactured goods. However, the areas brought within the European political orbit as colonies soon wished to control their own destinies. Most of North and South

America achieved independence from European countries by the early 1800s. European countries colonized much of the rest of the world (Figure 1.12), but by the 1960s most of these colonies established their independence.

Globalization, Countries, and Protectionism

In the first phase of European expansion, from around 1450 to the early 1800s, trade and piracy combined on the unpoliced oceans. Much of the wealth from distant lands was used by kings in Europe to gain more power by funding armies to fight for their interests in civil and continentwide wars.

Particularly in Britain, merchants invested their new wealth from overseas ventures in the factory-based concentrations of manufacturing, giving that country advantages in increasing its political standing through widespread world trade, military power, and colonial dominance. During the late 1800s, the globalization of trade and Western cultures spread through developments in communication (the telegraph) and transportation (railroads, larger ships), financial investments in newly opened lands of the southern continents, and huge migrations of people to settle the "new lands."

Toward the late 1800s, Western countries reduced their international linkages as they defined their own and their colonies' frontiers more strictly, developed nationalist sentiments among their people, imposed tariffs on traded goods,

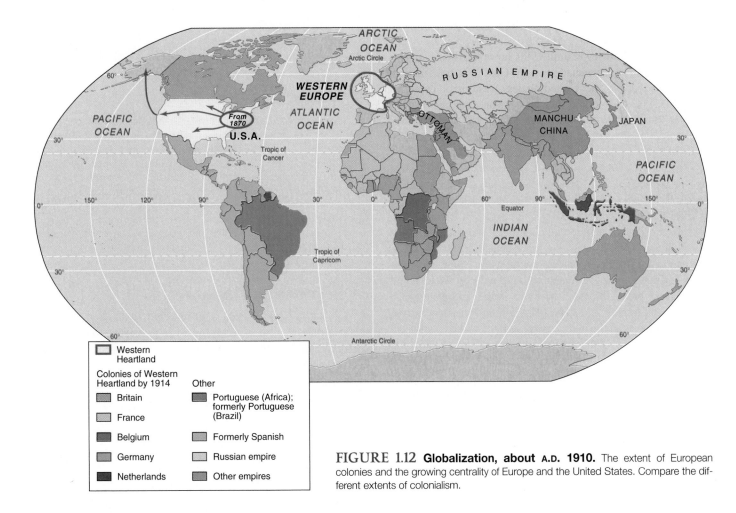

FIGURE 1.12 Globalization, about A.D. 1910. The extent of European colonies and the growing centrality of Europe and the United States. Compare the different extents of colonialism.

began to control the growth and power of large corporations, and restricted levels of immigration. Globalization trends were interrupted, or at least slowed.

Such intense international competition over trade and colonies built political hatred, and militarism led to World War I (1914–1918). After that war, new countries were carved out of old empires in Europe and western Asia. Although international bodies, such as the League of Nations, were established, actions by individual countries weakened the League's authority. The major economic crisis in the 1930s (the Great Depression) resulted from defensive and isolationist attitudes that increased restrictions on trade, migration, and movement of money.

In the 1930s and 1940s, extreme nationalism in Italy, Germany, and Japan, the closing of the Soviet Union to foreign contacts, movements for independence by colonial territories, and attempts at self-sufficiency by other poorer countries all diverted attention from global connections. Italian, German, and Japanese expansionism led to World War II, disrupting world trade and causing countries outside Europe and Asia to retreat further inside their borders and build up their own industries.

After World War II and the Cold War

Following World War II, new international institutions came into being, including the United Nations (UN), the International Monetary Fund (IMF), the World Bank, and the General Agreement on Trade and Tariffs (GATT) with its successor the World Trade Organization (WTO). Their purposes included mediating in political issues, assisting poorer countries, and liberalizing trade. From 1950 until 1990, however, their globalizing functions were overshadowed by Cold War polarization. During this period, the United States, western or capitalist Europe, Japan, and some other East Asian countries experienced economic growth. They became the world's most materially wealthy areas and controlled most trade. However, there was almost no economic growth in the rest of the world.

After the 1978 opening of China to trade and external investment and the 1991 breakup of the Soviet Union and its eastern European satellite countries, the international movement of ideas, capital, goods, and people increased significantly. Globalization trends affected all countries. Some of the materially poorer countries began to grow economically. And yet, international government made slow progress. Some commentators say that the weak attempts to bring order to Rwanda and Somalia, to break impasses between Israeli and Palestinian interests, and to remove Saddam Hussein's government in Iraq in 2003 demonstrate the ineffectiveness of the United Nations in global governance. Others argue that the United States and some European countries took matters into their own hands without giving the United Nations a chance to act. This resulted in concerns that the most powerful countries would intervene in the affairs of smaller countries. Global political control remains limited and largely unwanted by most countries.

Globalization moves forward slowly in contentious, poorly understood, and often contradictory trends; local regions may resist these trends or try to ride the globalization wave. The study of world regions emphasizes the interaction of global forces and local conditions over time. Sao Vicente in the Cape Verde Islands is an example. Long an uninhabited, arid island off the west Africa coast, the Portuguese valued its large protected harbor and colonized it as a base for their voyages to Southern Africa and South America from the late 1400s. They brought Africans to work as slaves and sold them in the New World. In the age of steamships the island became a fuel depot and then a focus of submarine telephone cables, for which an English workforce moved onto the island. During the period of sanctions against apartheid-ruled South Africa, the island offered an airfield where airliners could refuel without overflying other African countries. After 1975, the Cape Verde islands became independent from Portugal and now have a distinct currency. Today Sao Vicente is a port of call for cruise ships and has an international tourist industry. All of these events left distinctive marks on the people, main town, prosperity, and culture.

The world regional chapters in this text focus on the cultural background, natural environment, and recent events so that you can assess their significance for contemporary geography. As global connections increase, local voices are heard more clearly articulating views that are alternative to the mainstream of global-focused trends and are often linked to traditions and long-established ways of personal and community life.

Test Your Understanding
1B

Summary

Globalization is the increasing level of interconnection among peoples on Earth. Its impact on places is varied at different levels of geographic scale—global, world regional, country, and local. Localization reflects a set of responses to global trends at smaller geographical scales based on the intentions of local people to preserve their way of life. The effects of globalization and localization differ relative to political, economic, and cultural activities and with respect to the environment. Cultural and physical characteristics define the basis of nine major world regions. The current map of major world regions evolved through the history of human occupation of the Earth. Changes continue.

Questions to Think About

1B.1 What are the main elements of globalization and localization? Illustrate your answer by referring to places known to you.

1B.2 Summarize the differences among the major world regions defined in this text.

1B.3 Which world region interests you most?

Key Terms

globalization	world region
localization	cultural hearth
brain and skills drain	feudalism

Chapter 2

Concepts in World Regional Geography

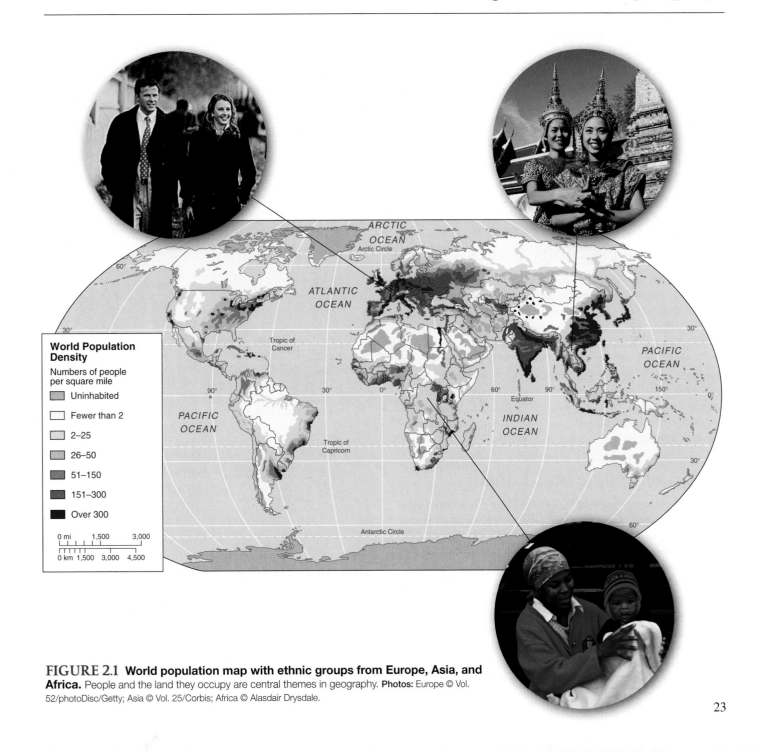

World Population Density

Numbers of people per square mile

- Uninhabited
- Fewer than 2
- 2–25
- 26–50
- 51–150
- 151–300
- Over 300

0 mi   1,500   3,000

0 km 1,500 3,000 4,500

FIGURE 2.1 World population map with ethnic groups from Europe, Asia, and Africa. People and the land they occupy are central themes in geography. **Photos:** Europe © Vol. 52/photoDisc/Getty; Asia © Vol. 25/Corbis; Africa © Alasdair Drysdale.

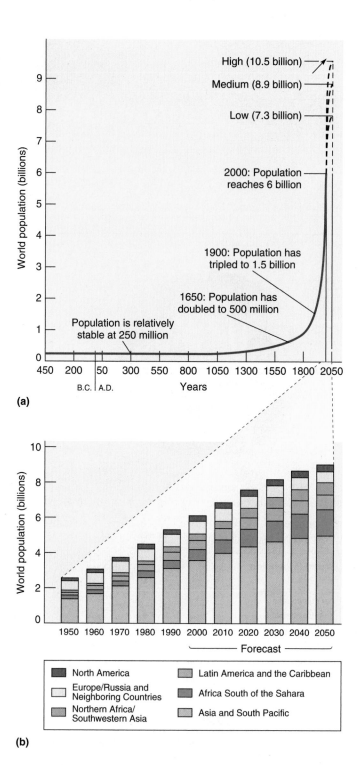

FIGURE 2.2 World population growth. (a) For most of the human occupation of Earth, population growth was slow compared to the last 300 years. The population took 1,300 years to double from 250 to 500 million, then doubled again in 200 years. In the 1900s, world population quadrupled. The high, medium, and low projections are from the United Nations, 1998. (b) When the population increase is broken down by region, all major regions outside North America and Europe show big increases. **Sources:** (a) United Nations; (b) data from *The Economist,* 1999.

Regional Geography Basics

In this chapter we study the concepts that are basic to regional geography in relation to specific issues that we meet daily. How is population changing? How do countries relate to globalizing trends? What controls the distribution of material wealth? How do people and cultural conditions vary from place to place (Figure 2.1)? What makes it possible to improve levels of human development and sustain them for the future? Within the general headings of human and physical geography, geographers identify population changes (Figure 2.2), political organization, economic activities, cultural conditions, and the natural environment as basic to their studies (Figure 2.3).

Issues of People and Land

People are central to regional geography. They create, live in the context of, and re-create geographic regions. Population dynamics such as growth, settlement patterns, and migration, combined with natural resource distribution and utilization, are major causes of differences among regions. In late 1999, the world population passed 6 billion, rising at nearly 80 million people per year; by mid-2005 it reached 6.5 billion. Although the rate of population increase is slowing after 50 years of rapid increases, the world's total population will continue to expand and place pressures on resources. Geographers examine where people live and relate that to the distribution of fresh water and good soil conditions in the context of how they use available natural resources.

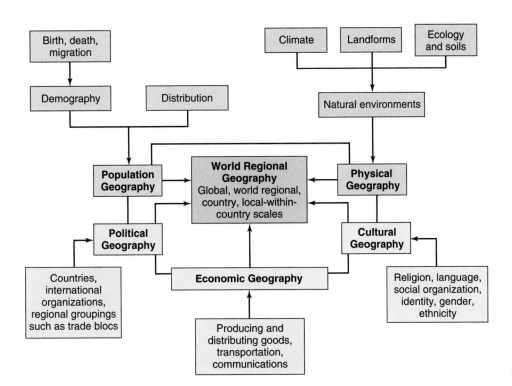

FIGURE 2.3 Regional geography. Aspects of human experience that contribute to differences among regions.

Population Distribution

Population densities—the numbers of people per given area (such as square kilometer or square mile)—vary greatly around the world (Figure 2.4). Density is a common measure of population distribution. Data for other measures are not so easily available. Such measures include **physiological density**—the population numbers per unit of cultivable land—and the **population–resource ratio,** an even more sophisticated measure based on the quantity and quality of each area's material resources and the size and technical competence of its population. Such measures emphasize the need to consider population distribution in relation to natural and human resources.

Only 29 percent of Earth's surface is land, and much of that is uninhabitable (desert, ice cap, mountain). The scarce habitable areas vary from low to high densities of people. The highest densities are in fertile lowlands and especially along coasts, where urban areas expand over good farmland. It is estimated that 97 percent of world population growth from 2000 to 2050 will be in the larger, more densely populated, and often poorer, countries. For example, 35 percent of the total is expected to be in China and India—already the world's most populous countries; Pakistan, Indonesia, Nigeria, Brazil, Bangladesh, Mexico, the Philippines, and the United States are likely to contribute another 25 percent. Reasons for the variable and changing distributions of people are complex and will be explored for each major region.

One of the most basic population and landscape differences in many countries is that between **urban** and **rural areas.** At present, half the world's people are concentrated in urban areas with very high densities of people, buildings, transportation lines, and human activities. This proportion rises to over 70 percent in the materially wealthier countries. And the largest cities get larger. Estimates for 2015 of total metropolitan area populations over 20 million raise metropolitan Tokyo to 29 million, followed by Mumbai (Bombay, India, 27 million), Lagos (Nigeria, 24 million), and Shanghai (China, 23 million), while Jakarta (Indonesia), Karachi (Pakistan), and São Paulo (Brazil) may each have 21 million. Cities combine rapid economic and social changes, bringing together peoples of varied incomes and ethnic backgrounds.

Rural areas lie outside the urban areas and contain most of the populations of materially poorer countries – around 70 percent in India, for example. Most rural areas are dominated by low-wage jobs in farming, mining, and basic manufacturing and do not have good access to educational, health, and other services. Many rural areas are dominated by traditional regimes that resist change. However, distinctions between urban and rural are becoming less clear in many countries as new industries and housing are built outside the established urban limits, as in eastern and southern China.

Population Ups and Downs

Demography is the study of population structure and change. Most major world regions have increasing populations; but Europe shows little change, while Russia and Neighboring Countries has declining totals. Increase or decline in the population of a place focuses attention on aspects that should be

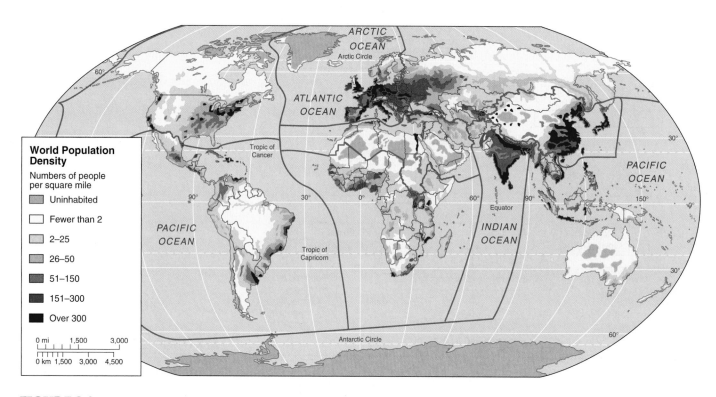

FIGURE 2.4 **World population distribution.** Which world regions have the highest and lowest densities of population? As you read through this chapter, try to explain the differences.

considered as a whole: whether births exceed deaths (natural change); whether immigration exceeds emigration; and the overall balance between natural change and migration.

Natural Change Elements

- The crude **birth rate** is the number of live births per 1,000 habitants per year in a given population. It is related closely to the **total fertility rate**—the average number of births per woman in her lifetime. Total fertility rates of 6 to 7 are typical of many poorer countries, whereas wealthier countries have rates of 2 or below.

- The crude **death rate** is the number of deaths per 1,000 inhabitants per year in a given population. It is often broken down into age groups. **Infant mortality** (deaths per 1,000 live births in the first year of life) and child mortality (deaths per 1,000 live births in the first five years of life) are examples. Infant mortality rates below 10 (10 or fewer infants die per 1,000 live births) in wealthier countries compare with those above 100 in many poorer countries. The crude birth and deaths rates are labeled "crude" because they do not account for age and gender structure of a population and are best used with other indicators.

- The crude birth rate minus the crude death rate equals the rate of natural population increase or decrease. The process of **demographic transition** is summed up in a theoretical model of stages that shows changes over time based on the experience of Western countries (Figure 2.5). The model relates birth and death rates to social and economic circumstances.

Changing circumstances include the spatial reorientation of many people from rural to urban areas, the transformation of economic activity from an agriculturally based economy to a postindustrial one, and the role of women in society.

The Migration Factor

Migration is the long-term movement of people into or out of a place. If immigration to a country or region exceeds emigration and there is natural increase, population will grow. Immigration is a major source of population increase in the United States, as natural increase is slow. In the 1990s, net immigration made up half of U.S. population growth.

Periods of globalization link to major migration of people. In the late 1800s, as trade and European colonization spread, and as refugees from political persecution could escape to new lands, millions of people moved mainly from Europe to the Americas and Australia. In the late 1900s new migration occurred. By 2002, 150 million people lived outside the country of their birth, just 2.5 percent of the world population, but often playing important roles in their host countries. For example, moderately and low-skilled workers moved from South and Southeast Asia to the Persian Gulf oil countries and Europe, sending their paychecks home. Migrants to urban centers such as London, Paris, Washington, and Toronto are dramatically changing the human geography of these areas. Educated men and women from poorer countries around the world moved to the United States, Europe, and Japan for better-paying jobs. At the same time, large numbers of refugees from

Stage 1: High birth rate and death rate: preindustrial society	Stage 2: Death rate falls, birth rate high. Early industrial society: rapid population increase	Stage 3: Rapid population increase giving way to lower birth rates in developed industrial society	Stage 4: Low birth rate and death rate: slow population growth

	2004 Birth rate*	2004 Death rate*
Guinea	43	16
Kenya	38	15
Jamaica	20	7
Japan	9	8
USA	14	8
Uruguay	16	9

*per 1,000 of population per year

FIGURE 2.5 Demographic transition. (a) The diagram shows a view of population change, related to social and economic conditions. In Stage 1 birth and death rates are both high, so population increase is low. In Stage 2 the death rate falls, but not the birth rate, giving high rates of natural increase. Eventually, the birth rate falls and population growth reaches a plateau. A Stage 5 might reflect the experience of some countries where death rates are now greater than birth rates, reducing the total population. (b) Bathing an infant in the high Himalaya of Nepal, where birth rates in villages remain high amid persistent poverty. (a) **Source:** 2004 data from Population Data Sheet, Population Reference Bureau; (b) **Photo:** © David Zurick.

war-torn countries in Africa moved into neighboring countries, often leading to new tensions and problems of accommodating them. European countries became targets for refugees from Afghanistan, Iraq, Palestine, and Kurdish areas.

Overall Population Change

- When natural change is combined with migration changes, an overall population increase of 1 percent will lead to a **population doubling time** of 70 years. An increase of 2 percent means a doubling in 35 years; one of 3 percent means a doubling in 23 years. Wealthier countries today commonly have below 0.5 percent population increase, while poorer countries have rates of 2 to 3 percent. Higher rates of population increase place pressures on economic

resources. Countries with high emigration, low birth rates, or high death rates may have population losses.

- The composition and history of a country's population is often summarized in an age–sex diagram, also termed a "population pyramid" (Figure 2.6). Migration into the country or baby booms show up as expansions in particular age and gender groups; deaths in major wars may be reflected in a narrowing of specific cohorts. If there is an expectation of a long life, the older age groups will have more members.

- Future population projections are notorious for being subject to unexpected events. For example, the growing threat of HIV/AIDS (see "Point–Counterpoint: HIV/AIDS," page 30), or other deadly diseases, may lower future population totals; new baby booms may raise them.

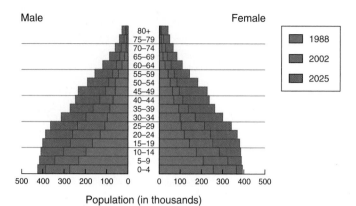

(a) Papua New Guinea

FIGURE 2.6 **Age–sex diagrams (population pyramids).** Diagrams for three years are overlaid or set side-by-side to show changes. In each case, the bars represent a five-year age group (male and female). Total numbers of people are used, rather than percentages of each age group, to allow comparisons over time and place. (a) Papua New Guinea. This shows a typical poorer country with large numbers of young people and fewer old—with increasing numbers in middle age groups by 2025. The progressive increases allow the three years to be superimposed. (b) United States, a typical wealthier country with a more even spread of numbers in each age group and a baby boom (1950–1965) moving upward through the age groups. The three years are separated for clarity. These diagrams occur in each subregion of Chapters 3 through 11. **Source:** U.S. Census Bureau, International Data Bank.

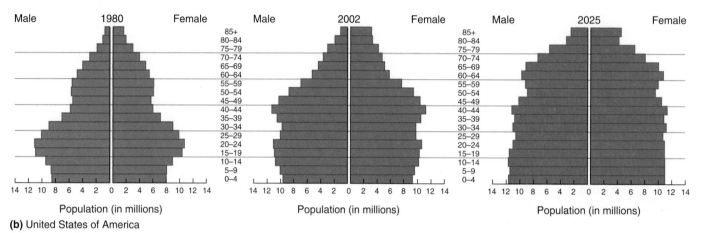

(b) United States of America

How Many People Can Earth Support?

Rising population totals heighten concerns about whether Earth and its resources can support them. In 1798 the English economist Thomas Malthus predicted that world population growth would exceed that of food production, leading to widespread famine. His prediction was not fulfilled because industrial processes raised agricultural productivity enough to feed the increased population. The debate about whether food production can keep up with population increases continues today. The ability of world regions and the countries in them to support projected growth in population is not easy to determine. A set of questions like the following illustrates the difficulties:

Economic questions

What level of well-being is expected?

What levels of technology will be used in growing food, manufacturing goods, and providing services?

Cultural questions

How will average family sizes change?

What support will be provided for young and old?

Are people willing to adopt new lifestyles that might include vegetarian diets (which are more resource efficient), bicycling to work, and spending more tax money on schools and health care?

Can people be forced into adopting new lifestyles, in part by the pressures of globalization?

Political questions

What sort of political system might resolve conflicts among and within countries?

Will organized violence continue to waste human lives and resources?

How will domestic and international trade arrangements work?

Natural environment questions

Do people consider it important enough to maintain a clean environment with conserved wilderness areas that they will alter their demands for cheap and plentiful food?

How much natural-hazard risk can people accept?

What changes will global warming make?

How long will any predictions last, given uncertainties over the usage of such resources as water and fish stocks?

The answers to these questions vary according to where a person lives. If all countries consumed resources at the rate of the United States, the world would already be overpopulated. As it is, the United States, with 5 percent of the world's population, consumes nearly half of the world's oil and large proportions of other resources. A cartoon at the time of the 1992 Rio de Janeiro Environmental Summit showed Uncle Sam telling representatives of poorer countries, "It's a deal. You continue to overpopulate the world, while we squander the natural resources."

We all have choices. Do we belong to the "bigger pie" school, which proposes expanding production through applying more technology in such areas as genetically modified foods, additives, plastics, synthetics, and alternative fuels? Or to the "fewer forks" school, which emphasizes environmental considerations and the slowing, stopping, or reversing of population growth? Or to the "better manners" school, which highlights cultural values as a source of improving the terms on which people interact? Our lifestyle choices concerning transportation, food, and clothing, among many others, will impact future patterns of human and physical geography.

Issues of Political Freedom

Political geography is the study of how governments and political movements (such as nongovernmental organizations, labor unions, and political parties) influence the human and physical geography of the world and its regions. Cultural phenomena, such as religion, and physical features, such as the distribution of fresh water, influence governments and political movements. Self-governing **countries** are the basic political units: within its borders, a country's government is assumed to have political control, or sovereignty, over the country's inhabitants and resources. Each country is ideally recognized by other countries. The numbers of self-governing countries increased from 62 in 1914 to 74 in 1946 and 193 today. However, not all units shown on maps are recognized as independent countries. Some are colonies, while others, such as Taiwan, which the People's Republic of China claims as its territory, lack recognition from other countries.

Country governments promote and protect their peoples in world affairs. Countries tax their citizens to provide public services, including military capabilities, and encourage economic and social welfare. On average during the 1900s, the proportion of a country's wealth taken and used by its government increased from under 10 to over 40 percent and continues to rise. Countries often have systems of regional, state, or local government that carry out some of the governmental responsibilities at different geographic levels. Country governments may also join other country governments in mutual trading or defense agreements. In world regional geography, countries provide the main subunits of study within the world regions.

Nations and Nationalism

A country is rarely, if ever, the same as a nation. A **nation** is an "imagined community," or politicized **ethnic group,** in which people believe themselves to share common cultural features, usually linked to a specific area of land. The cultural features may be language, religion, or other characteristics such as a shared history. For example, the French and Japanese both see themselves as nations. In the country known as the United Kingdom, the English, Scottish, Welsh, and Northern Irish consider themselves distinctive nations. Each fosters separatist groups, and they often enter separate teams in world sporting events. The Scots, Welsh, and Northern Irish—but not the English—now have separate home parliaments with some limited powers, although their people still elect members of the United Kingdom's parliament. Thus a nation may not have control over a country; it may not have its own sovereign government.

Nationalism is the desire of a nation to have its own self-governing country (sometimes called a "state"). The idea was basic to the formation of countries in Europe since around A.D. 1800. As countries finalized their boundaries and constituent areas, the increased levels of communication made possible through printed books and newspapers, the telegraph, telecommunications, radio, and television were used to bolster nationalism. Universal education supported nationalist themes through selective views of history glorifying the national experience. The idea of a "nation–state" emerged—that any nation should have its own state (or country). However, Americans and others with federal governments (see the next section) have to be clear about the use of the term "state." Perhaps "province" is a better term for divisions of a country that do not have full sovereignty, but it is used only in Canada (not India, China, Russia, Nigeria, or Brazil). Despite these reflections and caveats, the term "nation–state" is used widely in the social sciences to denote a country ideally occupied by a people called a nation.

In a country or state, stronger nations often dominate other peoples. Thus the country of France includes culturally distinct ethnic groups: the Bretons, Basques, and Alsatians. The Basque nation seeks to establish its own country across the French–Spanish border, but the Bretons and Alsatians are more content to remain part of the country of France. Germany emerged as a country in the later 1800s, when a group of smaller states united under Prussia's leadership based on an efficient army and a nationalist linguistic banner. In the 1900s, Germany used the idea of uniting separated German-speaking minorities in neighboring countries as an excuse for talk of national supremacy. This led to world wars, when Germany discovered limits to the expansionism its neighboring countries would tolerate.

Indigenous peoples are the first inhabitants of any given area. Many indigenous groups could not resist the spread of mainly European colonialism throughout the world from the 1500s to the 1900s. Their numbers often declined, or they were exterminated as the colonizer took over their lands. Throughout the world, many indigenous groups remain today, generally as minority populations within individual countries. Forcibly assimilated or excluded from decision making, they often evolved their own cultural and political aspirations as distinct "nations." Those that still exist are sometimes regarded as more "primitive" stages of human development and denied human rights and development opportunities. Such indigenous groups compete with other interest groups—as when the Native Americans initially disputed land ownership with European colonists or recently won major rights as the "First Nation" in Canada.

Governments

Some countries have a **unitary government** structure, administering all parts from the center for all aspects of government. Other countries, including most of the world's largest (Russia,

HIV/AIDS

HIV/AIDS is a major threat to world health and especially to millions of people in poorer countries, where 90 percent of infections occur. First recognized in wealthier countries, HIV/AIDS is now a major plague in Southern Africa and is being recognized in the rest of the poorer world. HIV (human immunodeficiency virus) gives rise to AIDS (acquired immunodeficiency syndrome). People contract HIV through unprotected sexual contact with HIV carriers or through contact with HIV-contaminated blood or body fluids (but current medical research suggests not by other contacts with HIV carriers). HIV infections can be passed from mother to baby. Patients become prone to many other sexually transmitted diseases and to other serious illnesses such as tuberculosis (TB). Medical treatments available are complex and expensive, needing close monitoring (Box Figure 1). They do not cure HIV but can prolong life. By 2003, 40 million people had contracted HIV and 7 million had died of AIDS.

For geographers, one of the major issues arising from a study of the HIV/AIDS pandemic (a disease that has a long-term presence around the world) is the difference between materially wealthy and poor countries. There are different experiences and attitudes in wealthy and poor countries.

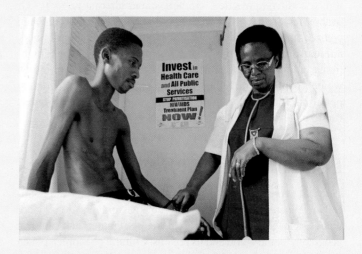

Box Figure 1 Treatment of HIV/AIDS. An HIV-positive man visits a clinic in Cape Town, South Africa. After being close to death, he received antiretroviral drugs in a trial project and is now well enough to run his own small business. **Photo:** © Gideon Mendel/Corbis Images.

Materially wealthy country	Materially poor country
Retroviral drugs are having a significant effect in delaying the onset of AIDS, giving a sense that the disease is on the wane.	The drugs are designed in materially wealthy countries for the strains of HIV that are common there, and those produced by multinational pharmaceutical companies are too expensive for wider use in poor countries. However, Brazilian and Indian producers sell them more cheaply, and the MNCs reduced the price when faced with the terrible outcome of no action.
The greater investment in health care facilities makes the vital monitoring centers widely available.	Monitoring centers are few and far between. Access is being improved in middle-income countries such as Botswana and South Africa, but not in the poorest countries.
HIV rates are low, but are rising again after a period of careful control. The behavior of sexually active males is tending toward less protection, while some of the drugs are having less effect.	HIV rates of infection and deaths from AIDS are high, although recent surveys show that previous estimates on less evidence were often too high.
Men have been affected more than women because of the connection with homosexual activities.	Women are affected more than men. Male circumcision helps prevent infection; women are vulnerable to rape; older male sexual partners pick up infections and pass them on; mobile miners and military personnel are most likely to be infected from prostitutes and pass on infection to marriage partners and the resultant children.
Education programs are effective, together with provision of syringes for drug users, although few countries impose HIV/AIDS tests.	Some countries did not take the threat seriously until the later 1990s, by which time HIV/AIDS was a major cause of death and social disruption. People avoid voluntary involvement because of stigma and social penalties. Botswana, with the highest rate of infection, reduced such high-profile testing to a routine action during doctor visits.
HIV/AIDS is not regarded as a major socioeconomic threat, although more significance is given to the worldwide situation. Cash invested in HIV/AIDS programs increased by 20 times from 1996 to 2003 (although UNAIDS wants twice that in 2005 and even more by 2007).	The high incidence could lead to economic collapse in South Africa, which has more cases than any other country (5.3 million HIV-positive citizens out of a total population of 45 million). More countries are devising national plans to combat the disease. However, much of the funding available has strings attached (e.g., the U.S. provision is not available to family planning programs that "promote" abortions).
Most countries are now open in reporting cases of HIV/AIDS.	Many countries resist full reporting. There is particular difficulty in many Arab countries, where activities contributing to the spread of HIV/AIDS are illegal or not admitted and so there is little detailed monitoring. It is clear that there is a high incidence among sex workers and drug users.

India, Brazil, Nigeria, the United States, and Canada), have a **federal government** structure, dividing the authority for various activities between a central government and partitions called states or provinces. In the United States, for example, the federal government has responsibility for external relations, defense, and interstate relations, while the states have authority for education, local roads, and physical planning.

Governments differ in their influence on internal affairs. Dictatorships and governments that control production from the center intervene much more than democratic governments in economic activities from production to distribution. They may use their powers to coerce and oppress people in such actions as forced labor or ethnic cleansing. Examples are the Soviet Union regime from the 1920s to 1991, the Holocaust against the Jews under the Nazi Germans in World War II, the Serbian treatment of other ethnic groups in the former Yugoslavia, Chinese actions under Mao Zedong and since, and the actions of African dictators such as Idi Amin and Milton Obote in Uganda.

During the 1900s, however, all governments increased their level of intervention in regulating and promoting economic activity. Increased levels of taxation for spending on education, health care, defense, and technology, together with incentives for foreign investment and the manufacturing of export products, strengthened the role of country governments as global interconnections multiplied.

Government functions are concentrated in **capital cities,** where the head of state lives and the administrative and government offices are situated. Many capital cities are the largest cities in the country, like London (United Kingdom), Tokyo (Japan), and Nairobi (Kenya). In many federal countries, the establishment of new capital cities avoided jealousies among existing cities. Washington, D.C., Brasilia (Brazil; Figure 2.7), Abuja (Nigeria), and Canberra (Australia) are examples.

FIGURE 2.7 Brasília, the planned capital city of Brazil. The central avenue lined by public buildings with the Parliament at the far end. The planners added residential wings on each side. Affluent residential districts surround the artificial lake in the distance. Many poorer workers live in linked satellite towns. **Photos:** © Stone/Getty Images.

Political Worlds and Global Governance

Global Governance

No global government exists, and any vision of a worldwide government remains a long way off. However, the term **governance** is increasingly used to include bodies that seek to legislate for and regulate aspects of human activities that are outside the powers of sovereign countries. An evolving global governance complex links countries, international institutions such as the United Nations, and public and private networks of transnational agencies. Much of this complex forms an extension of country government activities into the wider world, but there are also networks that function across country borders, some with little consultation of governments. The process has been called "the stretching of politics."

The United Nations is the largest and most influential institution of global governance, but many others have important functions (Figure 2.8). Countries pay dues to the United Nations, and these are used in its various programs and specialized agencies. Apart from the Security Council and the groups of peacekeeping military that are drawn from member countries, the United Nations does not have specific programs in the security field. Its greatest strengths are in coordinating aspects of human welfare, economic development, and care for the environment. Worldwide organizations linked to the United Nations are designed to promote the economic development of poorer countries (International Monetary Fund, World Bank) or to liberalize trade among countries (World Trade Organization). Although the United Nations' members include almost all the world's countries, it fails to prevent all civil wars, nuclear testing, or drug, weapons, and slave trafficking. The difficulties in obtaining agreement from all or a majority of U.N. members resulted in the United States, supported by other wealthier countries, intervening in the affairs of Kosovo, Afghanistan, and Iraq in the last decade.

Nongovernmental Organizations

Increasingly, nongovernment organizations such as aid bodies have assumed responsibilities for governmentlike activities, including disaster relief. **Nongovernmental organizations (NGOs)** might also be called "private voluntary organizations" or "civil society organizations." They include any group of people engaging in collective action in a noncommercial, nonviolent manner that is not on behalf of a government. Many NGOs have local or country bases, but the largest engage in international activities. Many international NGOs are contracted to supply aid by governments and international agencies irrespective of country borders. Well-known NGOs, such as the International Red Cross, Oxfam, Save the Children, Greenpeace, and *Médécins sans Frontières* (Doctors without Borders), are better known than many smaller countries. NGOs such as Amnesty International and Greenpeace assume advocacy roles, campaigning against human rights and environmental abuses respectively. However, when some NGOs oppose aid agencies, they

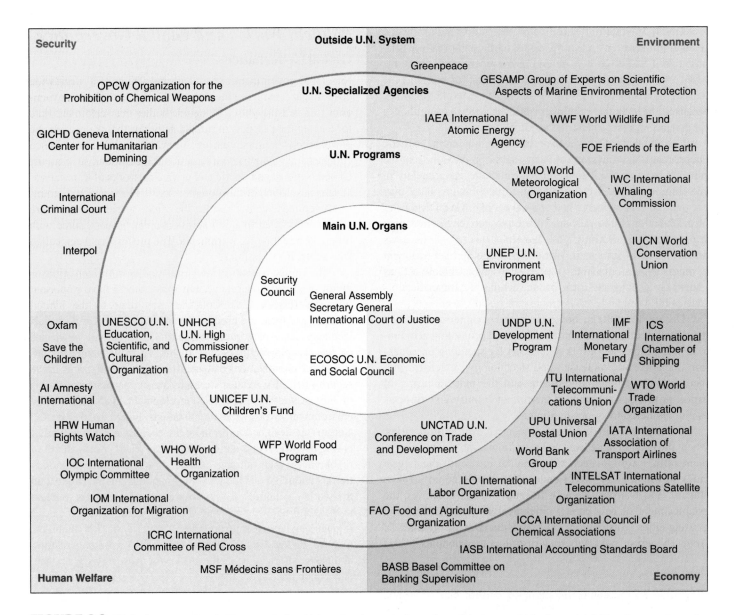

FIGURE 2.8 Global government. The complexity of international organizations; those listed are part of a larger total. The diagram places the organizations in relation to their positions inside and outside the United Nations. Those around the edges are separate international organizations. Countries relate to the complexity of the range of organizations in different ways; small countries find it difficult to send representatives to all bodies.

may hinder projects to bring improved conditions to materially poor people. Their use of the media and politicians may ignore the needs and wishes of those they profess to protect.

Many NGOs work in fields similar to those of the United Nations, which gives a consultation status to NGOs at three spatial levels. "General status" is held by the largest NGOs with global influence and extensive memberships; "special status" is given to NGOs with regional or specialist functions; "roster status" is for NGOs with small memberships but highly specialist roles. NGOs working or consulting with the United Nations rose from under 500 in 1970 to over 2,000 today.

Country Groupings for Trade or Defense

Countries make agreements with other countries to foster security through common trading and defense interests. However, closer political integration among groups of countries is difficult to achieve. No example exists of sovereign countries joining each other and remaining together in a federated country.

Governments have a major influence on the conduct of world trade. They may encourage their people to export goods or may control certain imports by charging taxes, or tariffs, on them. During the later 1900s, there was a widespread political will among countries to free world trade from barriers. The General Agreement on Tariffs and Trade (GATT) was established in 1948 to encourage countries to lower their tariffs. After 1993, the World Trade Organization took over the role of preventing discrimination among trading partners, but its rules appear to many to support the wealthier countries.

Most progress on liberalizing trade has been made at the world regional level in free-trade areas, the members of which impose common tariff rates on imports. The largest trading group at present is the European Union (EU—see Chapter 3). The European Union recently added some of the former Communist countries of East Central Europe. The North American Free Trade Agreement (NAFTA—see Chapter 11) may extend to much of Latin America (see Chapter 10). Even more ambitious is the prospect of the Asia–Pacific Economic Cooperation Forum (APEC; see Chapters 5, 6, 10, and 11) that has pledged itself to free internal trade among its wealthier countries by 2010 and among other countries by 2020. Such groupings of regional interests are considered in each chapter of this text.

The Cold War period generated defense agreements on both sides. The North Atlantic Treaty Organization (NATO) linked North America and western Europe in a common response to a perceived military threat from the Soviet Union (see Chapter 3). The Soviet Union established the Warsaw Pact to unite the countries of eastern Europe in the Soviet defense bloc.

In the 1990s, the end of the Cold War led to shifting emphases among regional groupings of countries. The Warsaw Pact ended after 1991, and NATO extended its agreements to some of the former Soviet bloc countries, despite resistance from Russia. From the late 1990s, the European Union became a more significant political force and developed its own military capability despite objections from the United States and Russia. The Association of South East Asian Nations (ASEAN) began with political objectives during the Cold War (opposing Communist countries such as Vietnam), but afterward shifted to increasingly economic objectives (and admitted Vietnam as a member).

Issues of Economic Inequality

World wealth and poverty are the subjects of economics—how resources are utilized and scarce goods are produced, distributed, and consumed. **Economic geographers** study these aspects and the spatial patterns that result. They study the impacts of peoples and their decisions on the distribution of resource usage, the flows of capital, and the human production of material wealth and poverty. The spatial distribution of wealth depicted in the World Bank map (Figure 2.9) demonstrates the pattern of economic inequality. Perched near the top of the global economic pyramid, some Americans see

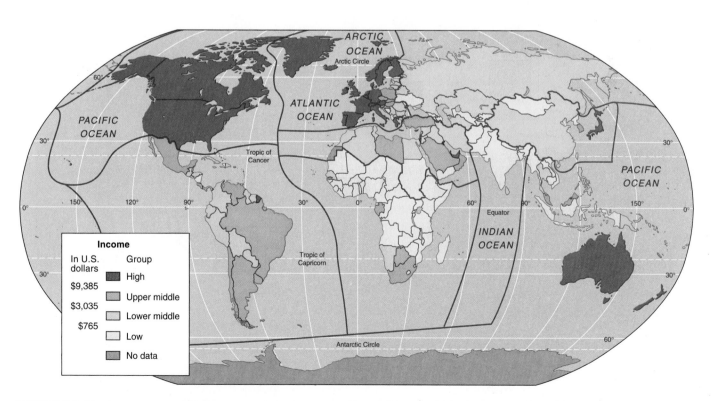

FIGURE 2.9 Major income groups of countries. A World Bank division based on GNI per capita for each country. How do the four categories relate to the major world regions? **Source:** Data from *World Bank Atlas,* World Bank, 2002.

the issues of poverty and deprivation as quite distant. However, the huge numbers of poor people constitute the greatest economic problem facing our world today.

In contrast to the large numbers of poor people in the world, there are relatively few extremely wealthy people. It was estimated that in 2001, the world had 7.2 million people (0.0012 percent of the world total population)—up from 5.2 million in 1997. They owned investable assets of more than $1 million and controlled one-third of the world's wealth. The 1997 estimate listed 425 billionaires, of whom 274 were in the United States. From 1997 to 2000, the numbers of millionaires rose sharply in the United States and Europe, less rapidly in Asia, and hardly at all in Latin America, the Arab world, or Africa. From 2000, however, many of those enriched by high-tech-related and overvalued stocks lost considerable wealth, showing how volatile the system can be.

Measuring Wealth and Poverty

In attempting to give more precise meanings to material wealth and poverty, specific indicators form a common basis for comparing and understanding differences among groups of people.

Ownership of consumer goods is a vivid indicator of differences in material wealth among countries (Figure 2.10). However, poor people's luxuries, such as clean drinking water,

food, clothing, and shelter, are often wealthier people's normal expectations. We have placed consumer goods alongside access to clean water and energy sources to add a further dimension.

Economic development is commonly measured by two statistics of income that are widely reported. **Gross domestic product** (GDP) is the total value of goods and services produced within a country in a year. Gross national product (GNP), now called **gross national income** (GNI), adds the role of foreign transactions to GDP. Per capita figures of the country's total annual income are averages of GDP or GNI per head of the population, not personal incomes.

The divisions shown on the World Bank map (see Figure 2.9) are based on GNI per capita, with countries divided into four income groups: low, lower middle, upper middle, and high.

GDP and GNI values are based on local currencies and converted to U.S. dollars at official exchange rates. Official exchange rates, however, may not reflect the comparable costs of living in a country. The **purchasing power parity (PPP)** estimates of GNI and GDP are more faithful comparisons of living costs among countries. Because prices in India, for example, are much lower for equivalent items you might buy in the United States, US $440 will buy as much in India as US $2,230 does in the United States. To illustrate this idea, *The Economist* devised a "Big Mac index" based on exchange rates against the U.S. dollar. In January 2004, the burgers that sold for an average of $2.80 in the United States

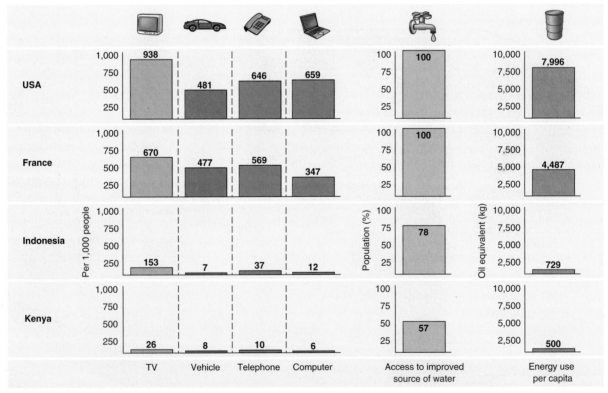

FIGURE 2.10 Consumer goods, water access, and energy use. The ownership of consumer goods is shown as the number of goods per thousand people (e.g., 938 people per thousand have a TV in the United States). Access to clean water is given as a percentage of the population. Energy use is kilograms of oil equivalent per capita. How do these items demonstrate contrasts in affluence among the United States, other materially wealthy countries (France), middle-income countries (Indonesia) and poor countries (Kenya)? This type of diagram occurs in each of the regional chapters, 3–11, enabling comparisons. **Source:** Data for 2002 from *World Development Indicators.* World Bank, 2004.

would cost over $5.00 in Switzerland, $3.45 in the United Kingdom, $2.47 in Japan, $1.97 in South Africa, $1.76 in Brazil, $1.42 in Russia, and $1.23 in China. Countries with high incomes and high living costs have a lower PPP estimate of income than the GDP or GNI based on exchange rates; poorer countries often have higher estimates. For example, in 2001 Switzerland had GNI per capita of $38,330 but a GNI per capita PPP estimate of $30,970; Mexico had comparable values of $5,530 and $8,240. Figure 2.11 shows that, in the wealthiest countries, over 60 percent of world wealth is produced by only 20 percent of the world's population. In the poorest countries 60 percent of the population produces only 20 percent of world wealth.

The United Nations' **human development index (HDI)** is a broader estimate of human well-being, incorporating statistics calculated from life expectancy, education attainment, and health, as well as income. Poorer countries investing heavily in education and health care, such as Costa Rica and Sri Lanka, provide a better quality of life for their people and have a higher HDI than GDP (merely based on a country's income) rank. By contrast, many of the oil-rich Persian Gulf countries have high income rankings based on oil exports but lower HDI rankings because of poor provision of schooling, especially for girls, and health care—although both are improving.

As the emphasis in thinking about economies shifted toward the needs of the poorest people, the United Nations introduced the **human poverty index (HPI).** Linked to the HDI, this is a composite index based on factors that strike balances between individual material poverty and public provision for such needs, and between relevant and available data. For example, the percentage of people expected to die before age 40 indicates health vulnerability (Figure 2.12); the percentage of illiterate adults indicates restrictions on entry to better jobs and full community life; and a combination of percentages of people without access to health care or safe water and of malnourished children under age 5 indicates a lack of decent living standards. The HPI records the proportions of populations affected by such deprivations. Values range from around 10 percent in Cuba, Chile, and Costa Rica to over 50 percent in many African countries and Cambodia in Southeast Asia. Contrasts also occur within countries: in China, the coastal regions have HPI values of 18 percent, while areas in the remote interior have values of 44 percent.

Economic Worlds

The workings of economic processes and strategies for economic organization are often related to political systems that have produced a number of geographic outcomes.

Raw Materials, Manufacturing, Services, and Outsourcing

There are three major groups of industries concerned with the production and distribution of goods.

- The **primary production** of raw materials from natural sources includes minerals, oil, gas, timber, and fish. Farm products come from domesticated plants and animals subject to soil and climate conditions. The industries producing these goods provide livelihoods for many people, with the highest proportions in materially poor countries (Table 2.1).

- The modern materially wealthy countries achieved their prominence through the **secondary production** or manufacturing and construction. Extra value and profit came from using raw materials to produce clothes, furniture, food and drink products, pharmaceuticals, railroads, engines, trucks, cars, airplanes, consumer electrical goods, and many other products. The cost of the raw materials was a relatively small factor; even when combined with the costs of building factories and equipping them with machinery, the wages of factory workers, and the cost of getting the products to those who wanted to buy them, the value of the final goods brought greater profits than primary products. The expansion of manufacturing after 1800 created huge industrial areas dominated by towns, factories, and railroad transportation in Europe and North America. As railroads and larger merchant ships supported demand, sales grew and products became more complex. In 1913 Henry Ford built his factory to produce Model-T Ford cars at Dearborn, Michigan. The assembly of many components into the finished cars was backed by the production of most of the components on the Dearborn site. The next phase was to outsource the component manufacturing to other companies in the local area and later across the United States and world. Today, for example, Boeing aircraft are assembled in Seattle and Airbus aircraft in Toulouse, France, but wings and tailplanes for both are

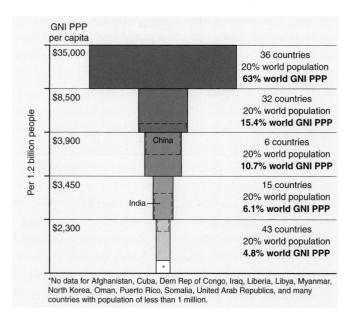

*No data for Afghanistan, Cuba, Dem Rep of Congo, Iraq, Liberia, Libya, Myanmar, North Korea, Oman, Puerto Rico, Somalia, United Arab Republics, and many countries with population of less than 1 million.

FIGURE 2.11 Distribution of world incomes. The 2000 world population was divided into five groups of 1.2 billion people each. Purchasing Power Parity Gross National Income (GNI PPP) was the basis of comparison. The United States is in the top group and itself has 5 percent of the world population and 23 percent of world GNI PPP. Each country's position is shown in the regional chapters, 3–11. **Source:** Data for 2000 from *World Development Indicators,* World Bank, 2002.

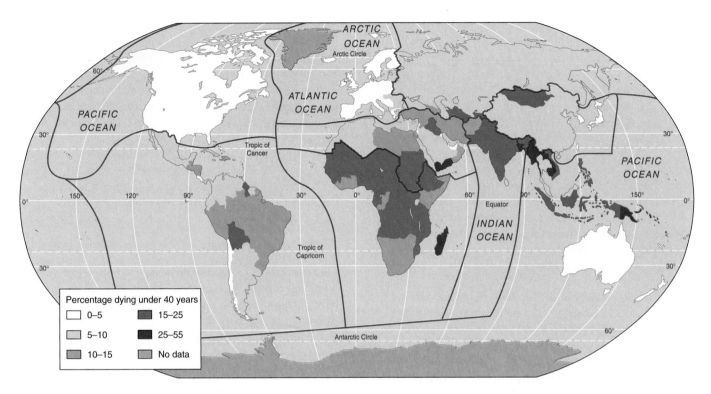

FIGURE 2.12 Early death. This map depicts the percentage of people not reaching 40 years of age for each country. How do the figures relate to major world regions? **Source:** Data for 1990 from *United Nations Human Development Report,* United Nations, 1997.

made elsewhere in the United States and Europe, while parts of the hydraulic system for the landing gear and the electronics systems are sourced across the world. From the 1990s there was an increase in the rate at which manufacturing operations that could be carried out by low-cost labor moved to materially poorer countries in Asia and Latin America. These movements began with textiles and clothing products and then involved electronic goods.

- The service industries of the **tertiary production** sector grew on the backs of the manufacturing companies and from government decisions to provide health services, education, and a wide range of other services. This sector also includes retail and wholesale trade, finance and legal industries, business services, the media, and information technology. Major manufacturing companies diversified into service provision, partly to enhance the sales of their own products. During the

Table 2.1 World Region Comparisons: Economic Sectors

The role of each sector is assessed by using the value added in each as a proportion of each country's gross domestic product and averaged for the region.

World Region	Economic Sectors Value Added as Percentage of Gross Domestic Product		
	Agriculture, etc. (Primary)	Industry (Secondary)	Services (Tertiary)
Europe	5	28	67
Russia and Neighboring Countries	23	31	46
East Asia	8	36	55
Southeast Asia and South Pacific	22	33	45
South Asia	29	24	47
Northern Africa and Southwestern Asia	16	34	50
Africa South of the Sahara	32	25	43
Latin America	12	30	57
North America	2	29	69

second half of the 1900s, all of these areas experienced a huge growth of employment in the materially wealthy urban–industrial countries, with increasing emphasis on high-end business services (lawyers, accountants, IT specialists). The trend was further displayed in separation of a **quaternary production** sector of information-based services (legal, financial, media, Internet). However, as these areas became more complex, the headquarters' offices of many large corporations first outsourced routine paperwork to "office factories" and then sent many aspects to low-cost countries, encouraged and enabled by IT, falling telecommunications costs, and low wages. The growth of the call center industry is one outgrowth of this trend, but higher-level business services, from logging insurance claims to making awards, are also moving abroad. At present, this is particularly important in India.

Free-Market Capitalism

From the 1990s, following the end of the Cold War and the collapse of the Soviet Union's Communist political–economic system, the **free-market,** or **capitalist, system** dominated the world. The free-market system entails the economic decision-making capacities of individual purchasers of goods, who may choose what they want from a range of products. Those who provide for these wants—and sometimes create them through advertising—invest financial capital with the aim of making profits. They "buy" labor and machinery to produce salable goods at the lowest cost. Competition among small firms, large corporations, and countries is an essential feature of the system.

This system marked Western countries for over 200 years and involves the private and corporate organization of investment, production, and marketing. The expansion of trade among resource and market areas further strengthened the Western countries' economies and their multinational corporations. Western countries produced and marketed sophisticated goods at high prices and bought low-priced raw materials and cheap products from the materially poorer countries.

Free-market capitalists face the challenge that affects all economic systems. Fallible humans invest, run companies, and generally perform roles to the best of their ability. Sometimes the investments produce profits, but not always, as the "dot.com" losses in the early 2000s illustrated. Sometimes managers take advantage of weaknesses in the system by fixing prices with their competitors or by dishonest accounting. Even in countries with well-regulated economies, major corporations such as Enron and Kmart may crash and create personal catastrophes for employees, suppliers, and customers.

In theory, governments intervene in free-market economies mainly to regulate the terms of trade and ensure fairness among producers. In practice, government-based decisions, and those of the World Trade Organization, on what trade should happen and what is fair are not necessarily framed or carried out in the public interest. Political considerations favoring some groups of people influence legislative actions. For example, the goal of borders open to trade may be contradicted by placing protective taxes on imported goods. Furthermore, the governments of many wealthier countries provide social services and infrastructure (roads, airports, harbors, water supplies, waste disposal) that give businesses and people in those countries many cost advantages compared to the poorer countries. As a result of such government intervention, capitalist countries become less free-market.

Central Planning

The basis of economic opposition to the free-market Western countries during the Cold War was the Communist centrally planned economic system adopted by the former Soviet Union, its satellite countries, the People's Republic of China, and linked countries such as Cuba. This system places planning and decision-making responsibilities in the central government, on the grounds that the whole country's interests come first and the central ministries know what is best for the people. They plan the production of goods considered essential—whatever the cost and whether or not the goods meet consumer demands. Central governments provide welcome medical care and education and develop strong military defenses.

Those in command of centralized policymaking, however, often made large-scale mistakes, handicapped even more than in the free markets by a lack of information or by personal bias or interest. Many leaders feared to change past policies, even if inefficient or oppressive, while regional bureaucrats often obeyed central commands despite knowing the policies would fail. Overproduction of some goods and underproduction of others led to these countries failing to produce the consumer goods available in most Western free-market countries. Incomes for most families remained modest, while members of the Communist Party hierarchy became relatively wealthy elites.

In the 1980s, comparisons with materially wealthy Western countries via growing global information exchanges fueled dissatisfaction in the Soviet Union. In 1991 the Soviet Union broke up, causing the collapse of economic relationships with the former countries of the Soviet bloc in eastern Europe and others worldwide. These countries entered the global free-market capitalist economic system as separate entities. Already, from 1978, the People's Republic of China had increased trade with other countries and encouraged investment from them. This brought China high levels of economic growth that continued into the 2000s. However, most former Communist countries—and those aligned with them—encountered a traumatic transition to the totally different free-market, now global, economic system.

The Global Economy

There are increasing trends toward global exchanges and institutions.

Global Economic Organizations

The global economic system that emerged after 1990 is marked by institutions of economic governance. The World Bank and International Monetary Fund (IMF), both based in the United

States, lend to materially poor countries. Such loans often demand openness to external investment and foreign goods along with reductions of government bureaucracy and related jobs. Private commercial banks and nongovernmental aid agencies tend to follow the same guidelines for assigning priorities to funding projects in poorer countries. Following the demands places pressures on internal social programs as the government bureaucracy is reduced.

The World Trade Organization's (WTO) role is to ease trade among countries by negotiating reductions in import and export duties. Its basic role is to spread the economic benefits of trade that are slowed by duties. Currently there are 140 member countries of the WTO, each of whom has the power to veto decisions of the organization. However, the WTO garners criticism by appearing to favor the world's more materially wealthier countries in permitting them to discriminate against imports, such as agricultural products and textile manufactures, from impoverished countries. Powerful members of the WTO, such as the United States and the European Union, use the organization as an arena to play out their disagreements concerning agricultural markets in the Caribbean, genetically modified foods, and U.S. concerns over European beef.

Multinational Corporations

Multinational corporations (MNCs) make goods or provide services for profit in several countries but direct operations from a headquarters in one country. The term "transnational corporation" (TNC) is often used instead to refer to corporations that are no longer rooted in a single country. The greater ease of travel and telecommunications contacts, together with the Internet transfer of information, encouraged MNCs to expand in numbers and operations in the later 1900s. They provide a major force in globalization trends.

Multinational corporations place production facilities in countries outside their homelands to take advantage of markets, cheaper labor, land, and energy and sometimes of less stringent worker safety and environmental laws. For example, auto manufacturers spread the manufacture of components across several countries to ensure supplies during labor strikes and to react to local needs (Figure 2.13). The earliest multinational corporations were of U.S. origin, but those of European, Japanese, and South Korean origin are increasingly significant. Of the top 100 MNCs with the most assets outside their home country in the early 2000s, 50 were based in Europe, 27 in the United

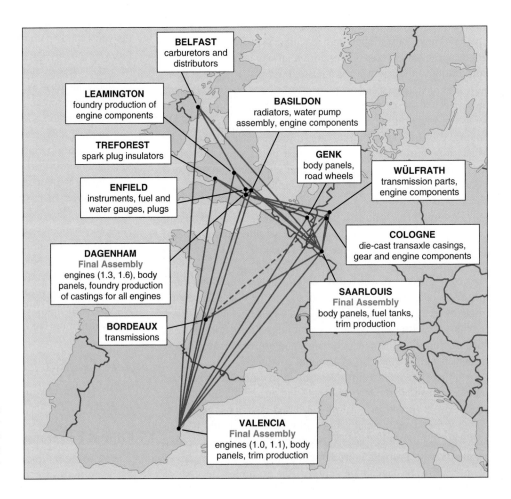

FIGURE 2.13 Multinational corporation's International linkages. The European distribution of factories includes makers of parts and final assembly locations for the Ford Fiesta in the late 1980s. **Source:** From *Global Shift,* 4th edition by Peter Dicken. © 2003 Guilford Press. Reprinted by permission.

BELFAST carburetors and distributors

LEAMINGTON foundry production of engine components

BASILDON radiators, water pump assembly, engine components

TREFOREST spark plug insulators

GENK body panels, road wheels

WÜLFRATH transmission parts, engine components

ENFIELD instruments, fuel and water gauges, plugs

COLOGNE die-cast transaxle casings, gear and engine components

DAGENHAM **Final Assembly** engines (1.3, 1.6), body panels, foundry production of castings for all engines

SAARLOUIS **Final Assembly** body panels, fuel tanks, trim production

BORDEAUX transmissions

VALENCIA **Final Assembly** engines (1.0, 1.1), body panels, trim production

States, and 17 in Japan. They produce a huge range of brands sold worldwide, including Coca-Cola, PepsiCo, Ford, General Motors, Volkswagen, Mercedes, Exxon, Shell, Toyota, Sharp, Samsung, Kellogg, Nestlé, Hyundai, and IBM. By the early 2000s, multinational corporations accounted for 40 percent of all international movements of goods.

Multinational corporations are not only manufacturers. MNCs in service (tertiary sector) industries spread from the 1970s, and, by the early 2000s, over 40 percent of foreign direct investments to countries were directed at them. These services include tourism and travel, data processing, advertising, market research, banking, and insurance. Some manufacturing corporations, such as the Ford Motor Company and General Motors, diversified into financial loans and credit cards.

MNCs wield considerable power in the countries where they operate. Some MNCs act as uncaring monolithic institutions without concern for the best interests of the people they employ in either home or adopted countries. However, other MNCs transfer wealth and technology to poorer countries, provide jobs where none existed in rural areas, and pay better wages and provide better employee benefits and prospects than local companies.

Some local groups resist globally connected companies. When outside corporations force local people to work in dangerous conditions, destroy their environment, or provide little compensation and local development in return for extracting resources, local groups may respond in armed rebellions. The resultant closed mines (as in Papua New Guinea) and disruptions of oil production (as in coastal Nigeria) are in no one's interest.

Global Financial Services

The expansion of financial services to the global scale in the later 1900s both resulted from global economic activity and enabled its expansion.

After 1970, the breakdown of the system of fixed exchange rates led to more frequent flows among currencies. In the oil crises of the mid- and late 1970s, producers raised prices in U.S. dollars, which accumulated in foreign banks. The banks created new financial markets, lending large sums to materially poorer countries, such as Brazil, to develop roads and power dams. In the early 1980s, the United States ran a huge budget deficit financed by borrowing in dollars. This also raised interest rates.

While money flows in the 1970s and early 1980s were mainly in loans from materially wealthier to poorer countries, rising interest rates prevented the poorer countries from repaying the debts. During the late 1980s and 1990s, foreign investment was nearly all supplied by and used in wealthier countries such as the United States, Japan, and the countries of Western Europe (Figure 2.14).

Countries such as Taiwan and South Korea avoided the debt problem of many poorer countries by forcing their own people to save and invest at home. By the early 1990s, growing trade surpluses in Japan, South Korea, Taiwan, and Hong Kong added to

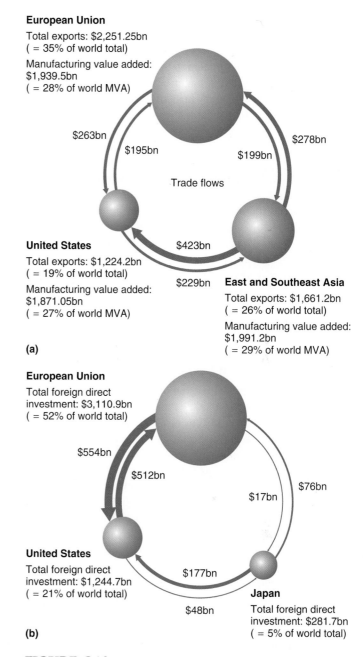

FIGURE 2.14 Concentrations of manufacturing, trade, and foreign direct investment in the late 1990s. The United States, European Union countries, and eastern Asia (including Japan) dominate the world in these important economic areas: 84 percent of world value added in manufacturing, 80 percent of world export value, and 78 percent of FDI (Japan only for eastern Asia). (a) Trade and value added in manufacturing. (b) Foreign direct investment. **Source:** From *Global Shift*, 4th edition by Peter Dicken. © 2003 Guilford Press. Reprinted by permission.

the funds available for new global investment. Banks and other financial houses expanded to service this international money explosion, beginning with American banks but followed by others in Europe, Japan, and the Arab countries. Those in New York, Tokyo, and London trade around the clock.

Danger signs grew with the increasingly open trading. Financial markets dealt more and more in equities (stock and shares) and risky forward contracts based on expectations of commodity production in relation to world prices. Many Asian funds were overinvested in major building projects around the world. Many early "dot.com" businesses did not repay invested capital and failed. Further pressure on both wealthy and poor countries occurred as corporate mergers increased through buyouts of firms followed by the closure of subsidiaries, workforces being laid off, and assets sold.

In the late 1990s, the financial flows slowed for a while when the loans, particularly from Japanese and South Korean banks for construction projects, could not be repaid. In 1997 and 1998, several Asian countries faced economic, political, and social crises as a result; funds from international agencies used to shore up their economies were not available for investment elsewhere, causing countries such as Brazil and Russia to suffer economic slowdowns. Although the Asian countries' economies returned to growth by the early 2000s, many social and political impacts remained. The global economic system became associated with uncertainty.

Global Information Services

In the 1990s, the rapid expansion of Internet-, telephone-, and computer-linked services fueled the growth of information services. E-commerce (electronic commerce) is the trade-based sector of such services. The largest volumes of e-commerce transactions are business-to-business ("B2B"). Large corporations, such as General Motors, work with their suppliers over the World Wide Web. Business-to-consumer ("B2C") facilities include retail sales, bidding (such as for airline tickets), and auctioning. Success in the initial stages of this area was varied; few companies made rapid trading profits, and many went out of business after initial high share valuations. Although some new car sales, for example, took place over the Internet, most people with access to the Internet in the United States used it as a source of information before going to their local auto outlet to buy.

Just as some multinational corporations moved manufacturing facilities to places with lower labor costs, others moved information handling to such places. In 1983, for example, American Airlines established Caribbean Data Services in Bridgetown, Barbados, to process the paperwork related to its tickets and boarding passes. It became the largest single employer in Barbados. In Montego Bay, Jamaica, the Jamaica Digiport Interstate Communications System links clients in Canada, the United States, and the United Kingdom. U.S. insurance companies process claims in Ireland. India, with its large population of English speakers, is growing as one of the world's major call centers for multinational corporations. Some Indian companies even train their staff to respond in American accents.

Global City–Regions

The multiplication and growth of multinational corporations, international financial institutions, dense networks of telecommunications, information-processing facilities, and international airline routes, together with the rising significance of quality business services, placed a new focus on some of the world's largest cities. New York, for example, is the center of a region with 18 million people and has an economic product greater than countries such as Canada or Brazil; its businesses receive 40 percent of their revenues from foreign sources. Foreign banks with New York offices rose from 47 in 1970 to over 200 in the early 2000s; over half of the U.S. law firms with overseas business are based in New York.

Such cities have major impacts on the places immediately surrounding them as well as on cross-border links to other countries. Their geographic scale of size and influence merits the term **global city–regions.** They have concentrations of high-salaried people, high-end technological and business services, specialized workplaces, expensive hotels and homes, major sports stadia, and concert halls. They have high-rise office and apartment blocks and a wide range of arts and sporting facilities. At the same time, their corporations employ increasing numbers of foreign experts and a growing underclass of poorly paid support workers, often migrants from materially poorer countries.

One approach to identifying and classifying global city–regions focuses on the importance of four categories of global corporate services (accountancy, advertising, banking, and law). Prime centers in all categories are New York, London, Paris, and Tokyo, closely followed by Chicago, Los Angeles, Frankfurt, Milan, Hong Kong, and Singapore. Figure 2.15 demonstrates the uneven distribution of these cities—the "control centers" of the global economy. The "top 10" global cities are in the United States, western Europe, and East Asia.

Regional Emphases

Although the free-market capitalist economic system prevailed globally after 1991, its geographic influences and benefits were not evenly distributed. Instead of a single path toward a global economy, distinctive regional variants of capitalist economies developed. The "Asian Way" builds on family linkages connected to government–business liaisons, rather than on the independently verified banking and legal systems that are basic to free-market capitalist economies in Europe and the United States (see Chapters 5 and 6). The "European Way" (Chapter 3) makes much of providing social welfare to support those who are not able to benefit from or exploit the capitalist system. Countries in Latin America moved out of self-sufficient and inwardly focused economic systems and into the world system with some pain (see Chapter 10). The former Communist countries of the Soviet bloc struggled to construct new economic identities (see Chapters 3 and 4). And on a much smaller scale, groups of peoples living in such isolated areas as the Amazon rain forest and Papua New Guinea, as well as increasing numbers in African rural areas, engaged little, if at all, with the global economy. Some local voices were not heard.

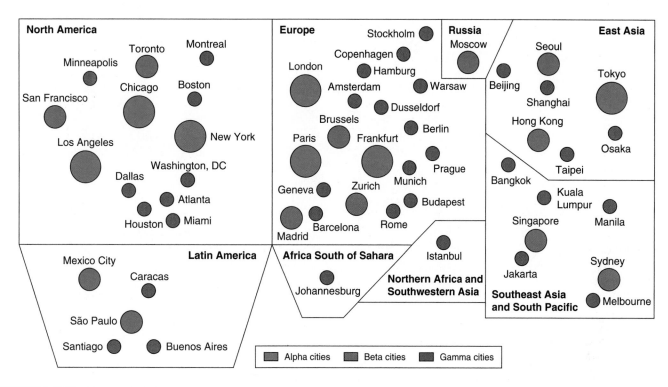

FIGURE 2.15 Global city–regions. Major world cities and their immediate regions are the centers of global economic activities. This threefold classification is based on four categories of international business services: advertising, accounting, banking, and law. Each of the four is given a 1–3 value for a maximum score of 12; values of 10–12 are "alpha," 7–9 are "beta," and 4–6 are "gamma." **Source:** Globalization World Cities Study Group and Network, University of Loughborough, U.K.

Issues of Cultural Freedom and Discrimination

Cultural geography is the study of spatial variations in cultural features, such as material traits, social structures, and belief systems. The culture of a group of people is basic to how they create and re-create the regional distinctiveness that in turn affects their lives and those of their descendants. Some aspects of cultural identity, such as the Muslim religion, operate worldwide; others arise from nationalistic pressures within a country or stem from local concerns. For example, a broader European culture is differentiated among and within countries, as in the northern Flemish and southern Walloon (French) Belgians.

The most important thing to understand about culture is that it is learned behavior. Culture consists of a combination of traditions and behavior practices that are transmitted from generation to generation as well as adaptations, variations, and new ideas or innovations more recently acquired or accepted by various groups of people. Cultural identification is not mutually exclusive. One may be part of many different culture groups at the same time. There is no biological or genetic predisposition to any element of culture.

Languages

A **language** is a means of communication among people, including speech, writing, and signing. It grows out of historic experiences and traditions and often provides a shared identity

for a cultural group. Some groups make a point of using their language to enhance their identity. For example, the French-speaking Québecois people in Québec, Canada, and the Welsh speakers in the United Kingdom increased their focus on language to achieve greater political recognition.

Language is an important factor in geographic diversity. Regional and internal country variations occur. In India, for example, each state designates its own official language(s). In northern India, many local languages give way to Hindi and the state-designated languages, while in southern India colonially imposed English becomes more significant as a common language alongside the multitude of state and local languages.

Thousands of languages are spoken around the world, many by small groups of people in isolated environments such as South America's Amazon River basin. There are twelve dominant languages each with over 100 million people speaking them. Six of these (English, French, Spanish, Russian, Arabic, and Mandarin Chinese) are official languages of the United Nations chosen at the end of World War II by the victorious allies.

Related languages can be grouped in families (Figure 2.16). For example, the Indo-European family includes most of the languages of South Asia (such as Hindi), the Slavic languages of eastern Europe and Russia (like Polish and Russian), and the languages of southern, western, and northern Europe. Some European languages, especially Spanish, Portuguese, English, French, and Dutch, spread worldwide with

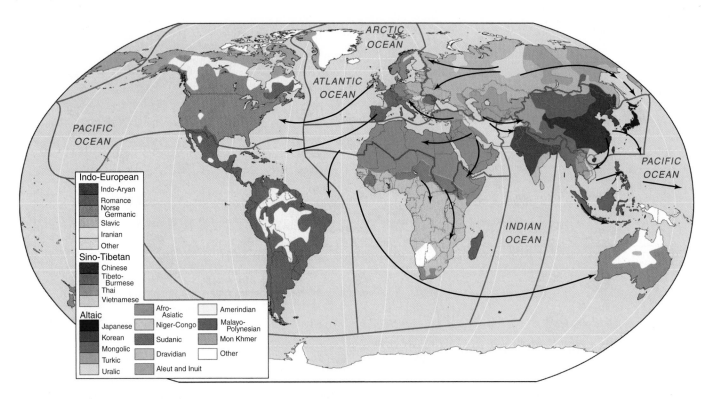

FIGURE 2.16 Cultural geography: world language families. How do the language families relate to major world regions? The arrows show some diffusion routes.

colonization from around A.D. 1450, dominating the Americas and the South Pacific and becoming important in many parts of Africa and Asia.

Other major groups include

- The Sino-Tibetan language family: languages of China (such as Mandarin in the north) and southeast Asia.

- The Altaic family: languages of Central Asia that diffused westward to Hungary and Finland and eastward to Japan and Korea.

- The Afro-Asiatic family, which dominates northern Africa and the Arabian peninsula and includes languages like Arabic and Hebrew.

- The Niger-Congo family: many languages spoken in Africa South of the Sahara, like Shona (Zimbabwe).

- The Malayo-Polynesian family, occurring through Malaysia, Indonesia, the Philippines, and the South Pacific.

- Smaller family groups: the pre-European languages of the Americas (Aleut, Inuit, Amerind) and of Cambodia (Mon Khmer).

Globalization marginalizes and extinguishes some languages and strengthens others. English is increasingly used in communications among scientists throughout the world, in international air traffic control, in computer software, and in the entertainment industry, where films and TV programs in English are dominant. It remains a common language in ex-colonies of Britain where there are rivalries among local languages. In the early 2000s, English met growing competition from Japanese, German, and Chinese to be the dominant computer language.

Religions

Religion also generates strong group loyalties, exclusive attitudes, and geographic variations. Each **religion** is an organized set of practices that professes to explain our existence and purpose on Earth. Some also project a system of values and faith in and worship of a divine being or beings. Regional emphases, such as social and legal practices, and visual features, such as building designs, often reflect religious allegiance (Figure 2.17). Religions play a significant role in transferring cultural values and practices from one generation to the next.

Major World Religions

The religions claiming the largest numbers of adherents are Christianity, Islam, Hinduism, and Buddhism: they each claim around one-fifth of the world's population. Judaism has a smaller number of adherents but widespread influence. Christianity, Islam, and Buddhism are religions that can be joined by anyone in any country and actively seek to extend their membership: they are **universalizing,** or global, **religions.** Hinduism and Judaism are mainly a matter of birth, closely tied to family and region: they are **ethnic religions.** People practicing ethnic religions do not actively seek converts because they see their religions as appropriate only for

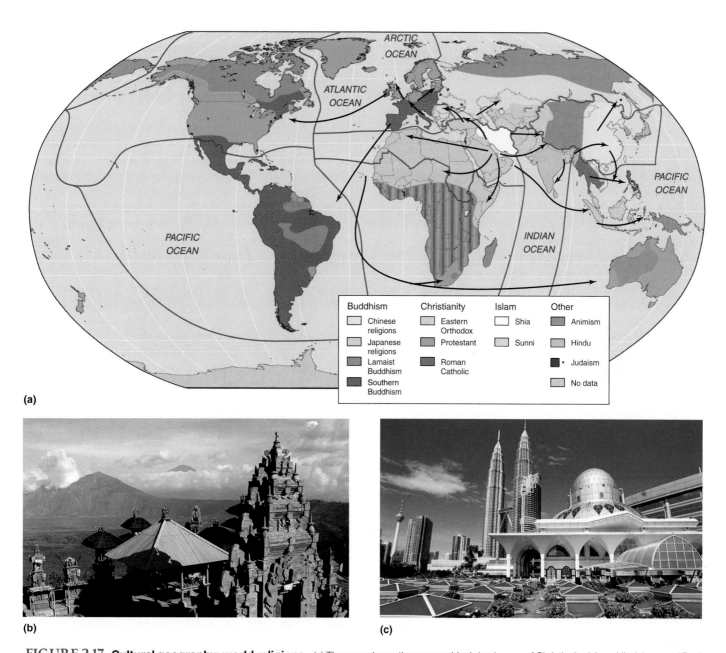

(a)

(b)

(c)

FIGURE 2.17 Cultural geography: world religions. (a) The map shows the geographical dominance of Christianity, Islam, Hinduism, and Buddhism. They also have the highest numbers of adherents. The Jewish religion is spread around the world, mainly in cities. Older and local religious practices remain in localized areas. The map shows the most important religions. Many places have mixtures of allegiance to different religions—sometimes working together, at others in conflict. (b)–(c) Symbols of the main religions: (b) Hindu Temple, Bali, Indonesia. (c) Asy-Syakirin Mosque with the Petronas Towers, Kuala Lumpur, Malaysia: religious and secular symbols. **Photos:** (b) © Paul A. Souders/Corbis; (c) © Tibor Bognár/Corbis.

priate only for their own ethnic groups. Although India and Israel are officially secular countries, there are movements to proclaim India as a Hindu country, while religious Jews seek to extend their control in Israel. Buddhism, often combined with older local religious elements, is the main religion of East Asia, while Hinduism is the predominant religion of South Asia.

Judaism, Christianity, and Islam, the world's most influential monotheistic religions, all originated in the same world region, and all consider the city of Jerusalem to be very sacred.

Christianity and Islam were created from the already existing structure of Judaism. All three believe in the same God. Jews and Muslims both look back to "Father" Abraham, and Muslims recognize Abraham, Moses, and Jesus as prophets. Jesus of Nazareth grew up as a Jew, and Christians use the same sacred writings as Jews in their Old Testament. Connections among other Asiatic religions include the fact that Buddha was Hindu, and Buddhism is accepted alongside local religions in China and Japan.

Christians believe Jesus of Nazareth was the Christ, the Messiah the Jews waited for (and still await because Jews do not believe Jesus was the Messiah). Christianity is the main religion of Europe and former European colonies in the Americas, Africa South of the Sahara, and the South Pacific.

Muslims also reject the idea that Jesus was the Christ or Messiah, but acknowledge him as a prophet of God. Muslims follow Muhammad and practice Islam. Islam is dominant throughout the Arab world of Northern Africa and Southwestern Asia (see Chapter 8) and extends its influence eastward to Central Asia and through Pakistan and Bangladesh to Indonesia (the most populous Muslim country). The major Islamic groups are the Sunnites (about 80 percent of the total) and the Shiites. Shiites form a majority of the population in Iran and Iraq.

The Jewish faith has a different spatial distribution than other major religions. Jews are more dispersed. In A.D. 73, Roman armies destroyed Jerusalem and forced Jews to migrate to all corners of the known world. Thus dispersed, the Jewish **diaspora** formed close-knit exclusive groups. Numerous episodes of oppression against Jewish minority communities occurred. The most extreme anti-Jewish action took place during the Nazi holocaust in Europe in the early 1940s. In 1948, the country of Israel was established as a common homeland for many Jews (see Chapter 8). Today most Jews live in the United States or Israel.

Major religions originated in specific world regions, although many spread out from their earlier centers. Internal divisions are also geographically significant. In Christianity, for example, the rift between Roman Catholicism and the Orthodox churches set apart the cultures of western and eastern Europe. Protestantism grew out of Roman Catholicism in northern Europe after A.D. 1500. Overseas explorations from the 1400s led to Christian missionary expansion outside Europe. Today some 60 percent of world Christians are Roman Catholics, 25 percent belong to Protestant groups, and 10 percent are Eastern Orthodox.

Religion and Society

Religious adherence often determines responses to issues affecting society. Many Roman Catholic and Muslim leaders, for example, officially resist population policies aimed at reducing births—although individual families increasingly make their own choices. Most religions demonstrate care for the natural environment, although the practice of societies has often not matched these beliefs.

Religious differences between peoples may result in conflicts over social, economic, or political ends in disregard (or ignorance) of the views and needs of other groups. One group's actions may then violate or offend another group and result in conflict. Such conflicts often arise from exclusive claims that are fanned for their own ends by those with political power. Christians and Jews, Muslims and Hindus, and Christians and Muslims fight each other. Conflicts extend between Catholics and Orthodox (former Yugoslavia) or Protestant (Northern Ireland) Christians and between Shia and Sunni Muslims. The

Muslim Arabized north of Africa clashes with the partly Christianized black African south, especially in countries such as Nigeria, Chad, and Sudan.

Recent events refute suggestions that economic globalization and an increasing secularization based on scientific explanations of phenomena will reduce religious influences. Although attendance at religious ceremonies and observance of religious rites are decreasing in some countries, especially in the materially wealthier West, science does not provide an alternative system of values or guides to the meaning of life. Today both Western and non-Western societies question the significance of scientific and merely rational approaches as prescriptions for living. As some question the impacts of the scientific basis of modernization, others react against the potentially homogenizing impacts of globalization.

It is likely that religion will continue to influence culture, perhaps in new ways. In the United States, religious groups increasingly get involved in politics, arguing that the country is getting less religious and needs such intervention. From the 1990s, the resurgence of Christian, Hindu, Jewish, Islamic, and Buddhist extremists in some countries is one sign that religious motivations are again significant after the Cold War period of largely political conflicts. In Indonesia, Muslims and Christians fight for political control of some of the many islands. Events in the early 2000s highlighted the growing hatred of Islamic extremists for Western values.

Race, Class, and Gender

Status—and the ability to effect or suppress changes—varies greatly throughout world regions. In many cultures, status is inherited through the family, as in the caste groups of India or the aristocracies of Europe. In other cultures, status may be achieved by creating wealth, political allegiance, media prominence, or sporting performance. While many of the world's wealthier countries increasingly base status on achievement, some people have a greater chance of achieving higher status than others by virtue of their birth into wealthy families, their gender, their upbringing, or their education. Throughout the world, race, class, and gender influence access to position in society.

Race and Ethnicity

An ethnic group is a culture group. Its members identify a common origin (real or imaginary) and are set apart by religion, language, national origin, or physical attributes. Such groups may be called tribes, clans, or segregated minorities.

The concept of **race** is assumed by many to be based on essential biological differences, but characteristics such as skin color, eye shape, or hair type vary as much within identified "races" as between them. Although race and racism are often at the center of human conflicts, the most basic human biological features, DNA and blood type, demonstrate little variation across the human species, which is a single reproducing group. However, cultural status is given to body features as a means of defining what are essentially ethnic "them–us" differences.

Examples of cultural racism include South Africa, where the minority white population operated a supremacist apartheid policy until 1993, separating whites, blacks, and other "colored" peoples into segregated neighborhoods. The country is now reversing this policy (see Chapter 9). In the United States, most African Americans, a minority ethnic group identified mostly by skin color, still struggle against discrimination from people of European origin. Even in Brazil, where European, African, Native American, and Asian peoples mix freely, most wealthy people are of European origin.

In many countries, ethnicity forms the basis of opposing political groups. This is particularly true of African, Latin American, and many Asian countries. Often tribal differences were highlighted and used by colonial powers, playing one group off against another. Since independence, however, the new rulers took up one-party rule and increased the perceived differences by rewarding those in their own ethnic group and penalizing others. This process led to armed conflicts in such places as Rwanda and Burundi, Bosnia-Herzegovina, Kosovo, Afghanistan, Angola, and Sri Lanka. Saddam Hussein in Iraq favored the Sunni Muslims against the dominant Shia Muslims.

Class Distinctions

Class arises from a stratification of society that is imposed through combinations of religious, economic, and social criteria. Wealth, education, and perceived birthright are common bases for class distinctions in all countries. In the United Kingdom, the royal family is linked to a hierarchy of hereditary dukes, earls, and knights. In India, the caste system and, increasingly, education define position in society. In Communist countries, the expectation of a classless society is contradicted by the status and privileges given to members of the Communist Party. Most Americans identify general class differences on the basis of material possessions and appearances. Their classes are based on economics: a lower-middle class of factory and shop workers, an upper-middle class of managers and professionals, and a group of exceptionally wealthy financiers, property owners, and sports and media stars.

Gender Inequalities

Gender—the cultural implications of one's sex—is also responsible for differences and inequalities of opportunity within and among societies that affect geographic differences. Males dominate most societies and have a history of denying full rights to women. Although major changes occurred in the 1900s, particularly in extending voting franchises to women, some countries still deny women the human rights defined by the United Nations.

In the late 1990s, the Taliban in Afghanistan became notorious for taking women out of education, confining them to the home, and selling some of them into slavery. In Africa, female genital mutilation, or "cutting," continues in some societies. In Iran and northern Nigeria, women—but not men—can be stoned to death for adultery. Illiteracy among women is still much higher than among men in most of Africa and Asia

because more men than women have been educated. Few European countries allowed women to vote in elections until the 1900s, with Switzerland delaying this until 1971 and one of its cantons until 1990. Some demographers and human rights organizations argue there may be as many as 100 million fewer women than men in the world due in large part to cultural preferences and practices favoring male children.

Women, even in the world's wealthier countries, commonly receive lower wages than men for the same job and constitute a minority of doctors, engineers, corporate executives, and elected politicians. Some jobs such as nursing, secretarial work, elementary school teaching, and shop clerking have been widely regarded as "women's work" and often have lower status. Further, women commonly perform a major share of home management and child care, which affects their career prospects. In some cases, marriage breakups leave women at a financial and social disadvantage.

The *United Nations Human Development Report* for 1995 included a gender-related development index (GDI) for the first time. The GDI focuses on the same criteria as the HDI (see page 35) but reflects inequalities between men and women. While 177 countries were reported for real GDP (PPP$) per capita and HDI values, only 144 were reported for GDI in the 2004 report (reflecting figures for 2002).

Some aspects of women's inequality are gradually being tackled. The high rates of female illiteracy in poorer countries have major implications for future population resource issues. Better education gives women confidence, enables them to take jobs, improves their self-esteem, and increases their role in decision making. It also impacts such matters as how many children are wanted in a family because their education is perceived as vital but is expensive. In many places, male attitudes and cultural traditions still prevent women from controlling how many children they have. It is clear that many women prefer smaller families, but fears raised in earlier periods of higher infant and child mortality still encourage larger families.

Despite their low status in most societies, women now play major roles in the expanding world economy. During the 1970s and 1980s, the proportion of women in manufacturing jobs in the United States and Canada grew as employment opportunities shifted in their favor. For example, female employment in the electronics industry expanded, but men were laid off when "male" jobs were taken over by machine tools. Expanding numbers of jobs in services such as retailing, health care, teaching, banking, and tourism further increased female employment. In materially poor countries, women comprise around 80 percent of the labor force in export-oriented electronics, apparel, and textile industries.

Wealthier countries, including the industrializing countries of East Asia, experienced greater household prosperity through dual incomes as females entered the paid labor force. Women's employment prospects in Africa and Latin America, however, deteriorated as the result of measures imposed to solve the debt crises in those countries. In Russia, the wide involvement of female labor under the former Soviet Union's Communist regime gave way in the 1990s to a greater focus on male employment.

Cultures and Regions

Cultural differences bring character and identity to the world's major regions, countries, and smaller regions within them. Many cultural differences can be traced back to early human history and the development of agricultural technology, religion, and language in small regions with specific environmental conditions. Regions that are termed "culture hearths" were centers from which techniques, useful materials, and social norms diffused to other regions (see Chapter 1).

Current world geography includes the imprint of many newer cultural overlays on older ones. Regions are constantly being re-created. For example, remnants of the Maya and Aztec empires, as well as of Spanish colonization, still exist in Mexico. The Mayan remains in Yucatán are being revealed from beneath the forest, while modern urban–industrial buildings are replacing Aztecan and Spanish relics around Mexico City. In southwest Asia, the Muslim culture removed most non-Muslim relics, but some, such as the ruins of medieval Crusader castles, still exist. In Communist China, the palaces of former imperial dignitaries occupy large areas of Beijing.

Cultural Fault Lines

Cultural fault lines exist between culture regions at different geographic scales. Tensions along these lines may lead to conflict. One prominent cultural fault line lies between the Muslim and Christian cultural regions from southern Europe through southern Russia. Although the current disputes are not merely religious, the antagonisms between Muslim and Christian cultures along a line through Bosnia, Kosovo, Macedonia, Bulgaria, Armenia, and the Russian republics such as Chechnya bordering on the Caucasus Mountains led to conflicts in the 1990s.

Measuring Cultural Globalization

It is more difficult to measure cultural aspects of globalization than the index described in Chapter 1 (see Figure 1.9). One approach is based on a country's exports and imports of books, periodicals, and newspapers (Figure 2.18). Other possible measures include movies or TV programs, but the data available are not sufficient. Of the 20 countries included in Figure 2.18, some, but not all, with the highest index are multicultural or are English-speaking. Poverty and low rates of literacy are the main barriers to cultural globalization.

Test Your Understanding
2A

Summary

Issues of people and land focus on population distribution and how populations grow or decline. The number of people Earth could support is subject to complex factors.

Political freedom depends on how countries are governed, the role of nationalism in their policies, and attitudes toward indigenous peoples. Countries where political power is concentrated in the hands of a few rulers often adopt harsh policies toward their peoples.

In the 1990s, economic systems were dominated by the free market. Inequality among peoples focuses on the continuing material poverty of so many and the huge material wealth of a few. The global economy spreads world trade through multinational corporations and involves global movements of capital and information. NGOs have local, countrywide, and global roles. Some are well known for bringing aid to needy places across country boundaries and for pursuing human rights and environmental issues. Global city–regions are hubs of international economic activities.

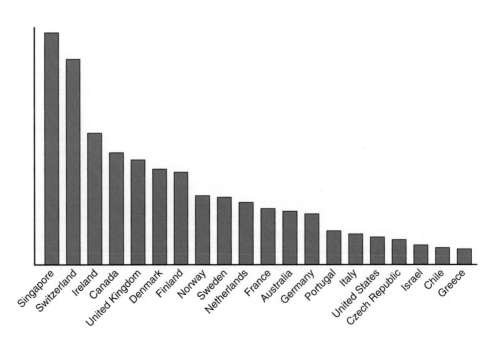

Figure 2.18 Cultural globalization. A classification based on each country's exports of books and magazines. **Source:** Randolph Kluver, Wayne Ru, in *Foreign Policy*, March/April 2004, p.54. Reprinted by permission of *Foreign Policy* via Copyright Clearance Center.

Issues of cultural freedom and discrimination are increasingly significant in world geography. Languages unite or separate peoples and are a major feature of nationalist claims. Religions are fundamental to world regional divisions and foster both togetherness and conflicts. Race, ethnicity, class, and gender affect status in society, and such issues underlie many local conflicts. Cultural differences trace back to early human history in distinctive cultural hearths.

Questions to Think About

2A.1 Compare major political systems. How do they affect economic policies?

2A.2 How is globalization affecting—and affected by—the roles played by multinational corporations, nongovernmental organizations, and intergovernmental organizations (like the EU or the UN)?

2A.3 In what ways do languages and religions define cultural groups? What effect may differences have on human rights?

2A.4 How are gender issues determined by cultural expectations?

Key Terms

population density
physiological density
population–resource ratio
urban area
rural area
demography
birth rate
total fertility rate
death rate
infant mortality
demographic transition
migration
population doubling time
political geography
country
nation
ethnic group
nationalism
indigenous people
unitary government
federal government
capital city
governance
nongovernmental organization (NGO)

economic geographer
gross domestic product
gross national income
purchasing power parity
human development index
human poverty index
primary production
secondary production
tertiary production
quaternary production
free-market capitalist system
multinational corporation (MNC)
global city–region
cultural geography
language
religion
universalizing religion
ethnic religion
diaspora
race
class
gender
cultural fault line

Natural Environmental Issues

Increased global connections have generated greater awareness and concern over the impacts of demographic, political, economic, and cultural pressures on the natural environment. New international networks of scientists and environmental activists try to raise or reduce concerns. However, despite world conferences on these issues, few countries take major measures to control environmental problems.

Physical geography is the study of natural environments and their world distribution. In world regional geography, major physical features, such as oceans, seas, mountain ranges, and rivers, often form boundaries between countries and world regions. For many centuries, human activities relied on natural resources, such as the growing season, water availability, soil types, and mineral-bearing rocks. These influenced the locations of people. Such influences were particularly important before intensive urbanization created widespread new human-engineered environments. Today human beings increasingly intervene in and influence the ways in which natural processes function.

Natural environmental issues raise questions about whether human activities lower or improve the long-term, sustainable productivity expected from natural resources. Damaging environmental quality may threaten human futures. Each chapter in this text discusses the natural environment of the major world region being considered and raises specific concerns about its management.

Earth's **natural environment** is a dynamic system of interacting events that combine in various ways to produce regional differences from high forest-clothed mountains to dried-up desert areas. Natural processes include:

- The workings of the atmospheric and oceanic circulations (atmosphere and hydrosphere) to produce weather and longer-term changes in climatic environments.

- The workings of Earth's interior that cause huge sections of the crust to collide with each other, producing earthquakes, volcanoes, and mountain systems (lithosphere).

- The interactions of atmosphere, hydrosphere, and lithosphere, causing rain, glacier ice, wind, and ocean waves and currents to produce landforms such as hills, valleys, and beaches. Such Earth surface environments are the stages for human activities.

- The actions of living organisms—plants and animals—respond to and modify the local climate, landforms, and soils (biosphere) in ecosystems.

Climatic Environments

The **climate** of a place is the long-term atmospheric condition that makes it more or less habitable. It is determined by the transfers of heat and moisture through the atmosphere and oceans, and by their interaction with the continental surfaces. The transfers are powered by energy from the sun.

Energy Transfers

Earth's atmosphere filters out aspects of solar energy, including ultraviolet radiation, x-rays, and gamma rays that harm living organisms. Mostly visible light rays reach Earth's surface. Absorption of these rays causes rock, soil, and ocean water to be heated and to radiate heat upward. When the heat rays are absorbed in the lower atmosphere by water vapor and carbon gases, they raise the temperature of the air. This is known as the **greenhouse effect,** which is a natural process in Earth's atmosphere. Because the sun is more directly overhead for more of the year in tropical regions, its heating impact on the atmosphere is greatest there (Figure 2.19). Tropical areas have an excess of incoming energy over that lost back to space (Figure 2.20). The polar regions, however, have a deficit of energy; in winter, they have several months of almost complete darkness, losing energy to space.

The tropical excesses and polar deficits are compensated by flows of air and ocean water between the two regions. Tropical oceans become huge reservoirs of heat moved poleward by ocean currents to heat the atmosphere of middle latitudes. The air and water cooled in higher latitudes return to the tropics, where they are reheated. This system makes human habitation possible into high-latitude regions.

Earth's Daily Rotation and Annual Solar Orbit

Earth rotates on its axis once a day and revolves in orbit around the sun once a year. The former creates day and night, and the lat-

ter seasonal changes. Because of Earth's axial tilt, the sun is directly overhead at noon at the Tropic of Cancer (Northern Hemisphere summer) between June 19 and 23 and at the Tropic of Capricorn (Southern Hemisphere summer) between December 19 and 23. The seasonal progression of the overhead sun north and south of the equator brings summers of warmer weather and long polar days to each hemisphere, while the winter hemisphere with low sun angles has cooler weather and long polar nights.

Earth's rotation affects air and water movements—winds and ocean currents—across the surface. The effect bends winds to form circulating weather systems, including cyclones (counterclockwise wind circulation in the Northern Hemisphere, clockwise in the Southern Hemisphere) and anticyclones (clockwise circulation in the Northern Hemisphere, counterclockwise in the Southern Hemisphere). It increases away from the equator toward the poles.

Water Transfers

Oceans are major sources of water that evaporates into the atmosphere, condenses into clouds, and produces rain and snow. Parts of the world with the highest rainfall lie near the equator (Figure 2.21), where warm, humid airstreams collide,

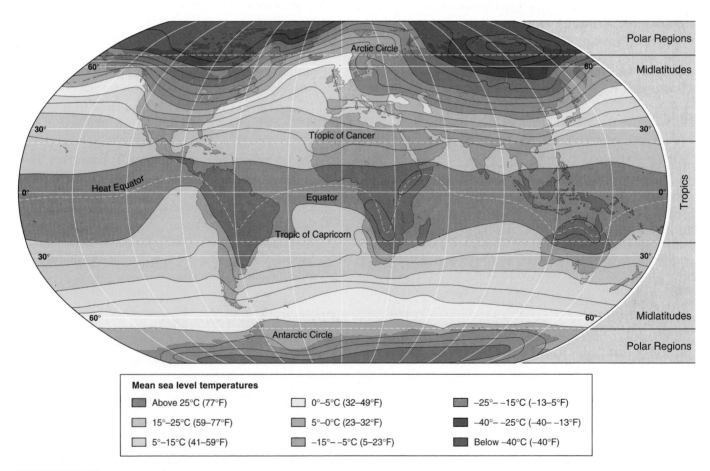

Mean sea level temperatures

Above 25°C (77°F)	0°–5°C (32–49°F)	−25°– −15°C (−13–5°F)
15°–25°C (59–77°F)	5°–0°C (23–32°F)	−40°– −25°C (−40– −13°F)
5°–15°C (41–59°F)	−15°– −5°C (5–23°F)	Below −40°C (−40°F)

FIGURE 2.19 Temperatures at ground level. Isotherms (joining places of equal temperature) for January, during the Northern Hemisphere winter. The heat equator connects points of highest temperature at each meridian of longitude. Compare the Northern and Southern Hemispheres in terms of the extent of very cold temperatures and the position of the warmest band of temperatures. What effect on air temperatures do the oceans have?

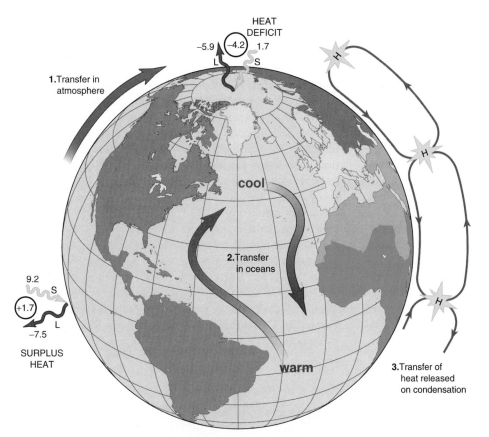

FIGURE 2.20 Global heat transfers. In the tropics, heat losses by long-wave (L) radiation from Earth are less than incoming shortwave (S) solar radiation; at the poles, the losses exceed the gains. Three main mechanisms transfer heat from low toward high latitudes, balancing the excess and deficit: winds in the atmosphere, ocean currents, and movements of humid air (heat is trapped during evaporation and released when clouds condense). Flows of warm air and water toward the poles are paralleled by flows of cold air and water toward the equator.

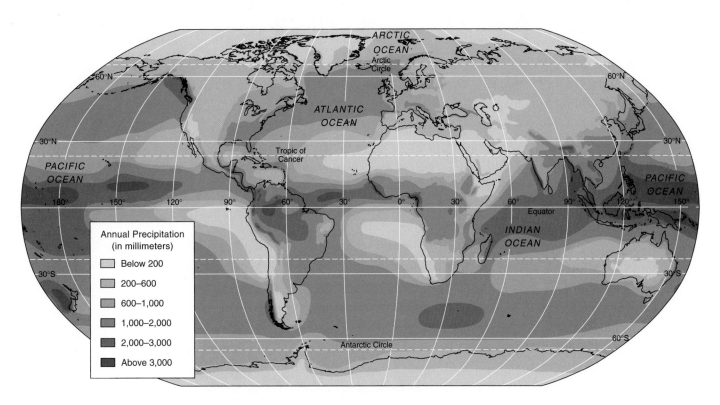

FIGURE 2.21 World precipitation (rain, hail, snow). Locate the main areas of low and high precipitation. Areas with less than 200 mm of precipitation in a year are classed as arid; those with more than 2,000 mm are nearly all in the tropical ocean areas. Warm air (which can hold more moisture than cold air) from the northern and southern tropics converges at the equator, rises, and forms clouds that precipitate rain.

forcing the air to rise and produce frequent rainstorms. Another zone of high rainfall totals is on the ocean-facing west coasts of midlatitude continents, especially where high mountains (as in Canada and Chile) add to the lift caused by the meeting of warm tropical and cold polar air in cyclonic systems. Areas between, including large sectors of oceans, have little, or seasonal, rain.

World Climatic Environments

The receipt and redistribution of solar energy and the circulation of water from oceans to continents vary in their effects around the world, giving rise to specific weather systems and climatic regions (Figure 2.22).

Tropical climates experience high temperatures throughout the year and have short winters, if any. The main climatic variations in the tropics are seasonal differences of rainfall, as can be seen by referring to Africa. The main tropical climates have a north–south distribution, from the equatorial climate with rain at all seasons, through wet–dry seasonal climates, to very dry climates. Places close to the equator have rain in all months of the year (Af), although a few months may be drier than the rest. Places farther from the equator but still within the tropics have a marked alternation of dry and wet seasons (Aw). Eventually, as distance from the equator increases, the dry season becomes so long that annual water shortages occur, and the climate becomes arid, without appreciable rainfall (BWh). The seasonal wet–dry contrasts are greatest in the monsoon climates of South Asia (Am).

Tropical climates have distinctive weather systems. The frequent rains near the equator, where the effect of Earth's rotation is least and circulating weather systems are rare, come from massive thundercloud developments (Figure 2.23). Tropical cyclones (also called hurricanes and typhoons) cause loss of life and property (Figure 2.24) in places poleward of 10° latitude where circulating systems form.

Midlatitude climates have marked seasonal temperature contrasts between summer and winter. Such contrasts are greatest in the centers of the North American and Eurasian continents and least where winds blowing from the oceans moderate winter cold and summer warmth over the west-facing coasts of Europe and the Americas. Most precipitation falls on hills and mountains facing west near these coasts and declines inland to

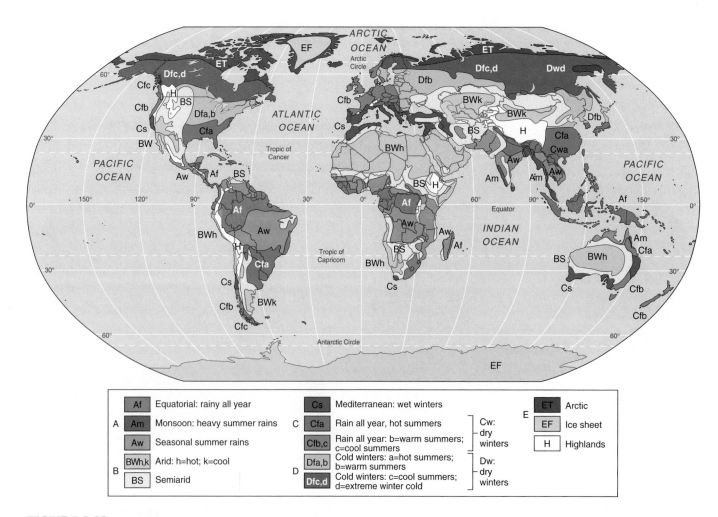

FIGURE 2.22 World climate types. Included are letters referring to the Köppen classification based on the climate characteristics of natural vegetation zones. **Source:** M. Bradshaw, R. Weaver, *Foundations in Physical Geography,* WCB, 1995.

FIGURE 2.24 **Hurricane Katrina over the Gulf of Mexico on August 28, 2005.** It passed over southern Florida (on the right) and narrowly missed Cuba and Mexico (foreground) before turning north toward the Mississippi River delta and New Orleans. Several hundred kilometers across, this hurricane shows the typical cloud-free eye of descending air in the center, surrounded by strong winds and swirling clouds. Once over the Gulf of Mexico, it picked up energy from the warm surface waters, intensifying the winds and surge of ocean water. **Photo:** NOAA.

FIGURE 2.23 **Tropical weather systems: thunderstorm groupings over central Africa.** Seen from a space shuttle, several cloud tops amalgamate in fibrous masses of ice that spread outward and downwind. **Photo:** NASA.

the point where it may become insufficient to support vegetation or agriculture. Midlatitude climates have a west–east distribution, with mild and moist west coasts (Cfb, Cs), continental interiors (BS, Dfa, Dfb, Dw), and east coasts with summer–winter temperature contrasts (Cfa, Dfa). Midlatitude weather systems have winds circulating around low atmospheric pressures in frontal cyclones—which bring rain and high winds—and calmer high-pressure anticyclones. The marine west coast climates in North America and Europe benefit from the ocean currents that warm the air above. On the margins with tropical climates, seasonal alternations with arid climates produce long, dry summers. This regime is typical of lands around the Mediterranean Sea and named after that association. In the large northern continents of Asia and North America, interiors are far from the oceans. They experience drier conditions as well as the greatest summer–winter temperature ranges. In higher latitudes, the humid continental winter seasons have a lasting snow cover.

Polar climates are extremely cold throughout the year. Winter conditions dominate the Arctic climates (ET, EF), although short summer spells may melt some of the snow and ice. Truly polar climates are frozen all year. Polar regions have no weather systems of their own but are dominated by dense, cold air that sinks and flows outward to deepen the winters of

midlatitude regions. Midlatitude cyclones sometimes invade polar regions, bringing high winds and precipitation. Although expeditions to explore Antarctica and the Arctic Ocean take advantage of the long local summers, temperatures remain low. In winter, there is almost total darkness for several months in polar regions. Such conditions support only sparse human settlement.

Global Climate Change

Climates change over time—by small amounts over shorter periods and larger over longer periods. There is a cyclic progression that repeats warmer and cooler phases, resulting from changes in Earth's solar orbit shape and the planet's axis angle relative to solar rays. However, human actions may compound the impacts at particularly sensitive moments.

An intensive freeze, the most recent part of the Pleistocene Ice Age, lasted for most of the last 100,000 years, ending around 10,000 years ago with a period of warming, ice melting, and a sea surface rise of around 100 m (300 ft.) to its present level. During the freeze, huge ice sheets dominated the northern parts of North America and Europe, and sea levels were lowered around the world. After the ice cover retreated, smaller fluctuations of climate brought the warmest conditions around 5,000 years ago. The "Little Ice Age," from approximately A.D. 1430 to 1850, caused upland glaciers to advance several kilometers down valleys and cultivation to retreat from higher areas in midlatitude countries. From the early 1800s, climatic warming resulted in a reversal of those trends.

A present concern of **global warming** is that the carbon gases humans pour into the atmosphere by burning coal, oil, and gasoline will add to the gases that trap solar energy in the lower atmosphere and may enhance the greenhouse effect and natural climatic warming. If ice sheet margins melt and the ocean level rises a meter or so within the next 50 years, the low-lying coasts of coral islands, wetlands, and port areas will be at risk. Such changes would have massive impacts on the high proportion of people who live close to the ocean. The precise trends and impacts on specific regions are difficult to predict, however, because interactions among natural processes are so complex. Even where predictable, the huge scale of size and energy involved in such events as the advances and retreats of ice sheets places them outside the realm of human abilities.

Shaping Earth's Surface Environments

Earth's surface is 71 percent covered by ocean water and only 29 percent occupied by the continents on which people live. The natural environment of continental surfaces forms the height and slope of the land, or its **relief** in features such as mountains, valleys, and plains. The interaction of internal Earth-building forces with external Earth-molding influences, includes weather and the action of the sea (Figure 2.25).

FIGURE 2.25 **Internal and external Earth forces mold the landscape.** The gash across the Carrizo Plain in southern California caused by the San Andreas Fault splitting a section of Earth's crustal rocks. The San Andreas Fault occurs where two plates move against each other, with the western side moving northward. Rain and rivers etched out the line of weakness. Further movement along the fault shifted and offset the line of stream valleys on either side. **Photo:** © Kevin Schafer/USGS/Tom Stack & Associates.

Earth Interior Forces

Earth is a multilayered planet with a hot molten core. Its internal heat provides the energy that forces large blocks of surface rock, thousands of kilometers across and around 100 km (65 mi.) thick, known as **tectonic plates,** to crash into each other or move apart (Figure 2.26a). There are six major and several minor plates (Figure 2.26b). Earthquakes and volcanic outbursts of molten rock from beneath Earth's surface concentrate along plate boundaries, signaling where plates collide or move apart. Where plates move horizontally against each other, transform faults such as the San Andreas Fault in California create earthquakes, but few have volcanic eruptions.

The plate movements cause the opening and closing of ocean basins and the raising of mountain systems. Where plates move apart, or diverge, fissures open and rock erupts as molten lava, adding to the edge of a plate where it solidifies. Such divergent plate margins include the Mid-Atlantic Ridge, where the eruption of molten lava builds Iceland. Plate collisions occur along convergent margins and often force one plate upward to form mountain systems such as the Andes Mountains of South America. The plate that is forced beneath the raised plate is said to be subducted. The subducted solid rock melts under high temperatures generated by burial and friction. The molten rock produced rises toward the surface under pressure, and erupts to create volcanic mountains and piles of lava such as those forming the Columbia Plateau in the northwestern United States.

Surface Changes

Once land emerges above sea level, attacks by the atmosphere and ocean waves etch the details of surface relief. Changes in the temperature and chemical composition of the atmosphere, together with water from rain and snowmelt, react with the rocks exposed at the surface and dissolve them or break them into fragments. Small particles are acted on by chemicals in the water. Such changes are called **weathering.** The broken and dissolved rock material forms the mineral basis for soil. On steep slopes, weathered material moves downhill under the influence of gravity. Such movement may be rapid in slides, flows of mud, or avalanches, or slow in local heaving and downslope creep of the surface. The mobile fraction of this broken rock material often falls into rivers or onto glaciers and is moved toward the ocean.

The concentrated flows of water or ice and rock particles in rivers and glaciers gouge valleys in the rocks—a process called **erosion.** Glaciers formed of ice move slowly, helped by meltwater lubrication in the summer. When the flows reach a lake or the ocean, they stop and the rock particles drop to the lake or ocean floor in the process of **deposition.** Wind blows fine (dust-size) rock particles long distances, while sand-sized particles are blown around deserts or across beaches to form dunes. Along the coast (Figure 2.27), sea waves and tides fashion eroded cliffs and deposited beach features, often moving the rock particles supplied by river, glacier, or wind.

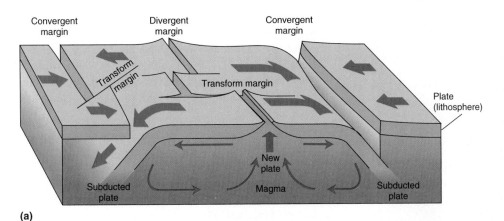

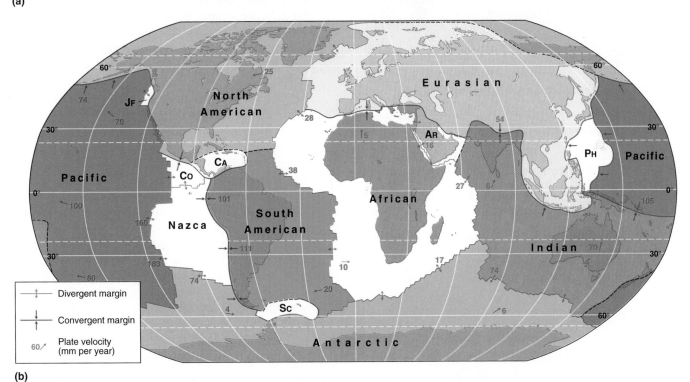

FIGURE 2.26 Plate tectonics, continents, ocean basins, and mountain ranges. (a) The main features of plates and plate margins. Green arrows show plate movement directions; blue arrows depict possible internal Earth convection currents driving movements in the top 100 km of Earth's interior. (b) World map of major and minor plates. The minor plates include Nazca, Cocos (Co), Caribbean (Ca), Juan de Fuca (Jf), Arabian (Ar), Philippine (Ph), and Scotia (Sc). **Source:** Data from NASA.

FIGURE 2.27 Eroding the land. Ocean waves attack the cliffs and form beaches in Oregon. Inland the valleys are caused by a combination of river action and slope processes. **Photo:** © Corbis RF.

Physical and Human Landscapes

In a particular region, the relief features of the land surface are determined by the combination of these internal and external forces, together with modifications added by human occupation, as Figure 2.28 shows. The landscape outcome depends on whether the region contains a plate margin, which climatic elements fashioned the surface, how long the natural forces operated without catastrophic changes, and how people affected the surface processes.

Environments of Plants, Animals, and Soils in Ecosystems

Plants and animals are sustained by a combination of energy from the sun's rays, water circulating from oceans to the continents and back again, and nutrient chemicals in the soils. Most

FIGURE 2.28 Landscape evolution. Western Scotland. The rocks in this northwestern part of Britain were formed over 400 million years ago and then crushed and raised into mountain ranges by colliding plates as an earlier Atlantic Ocean closed. Rivers wore down the mountains. The Atlantic Ocean opened again some 50 million years ago, raising this area with others along its margins. Within the last million years, the Ice Age covered the higher parts with thick ice that moved outward, forming deep valleys such as the one in this photo. On melting, the ice poured water back into the ocean, raising its level and drowning the lower part of this valley as a sea loch. Rivers eroded the hills, depositing deltas along the edge of the deep water. People altered the rates of natural processes by clearing woodland or erecting buildings. **Photo:** © Michael Bradshaw.

plants can capture and store the sun's energy in chemical form, combining with the mineral nutrients drawn from soil to produce the foods that animals require. The atmospheric processes that break up surface rocks, the actions of other organisms, and the water flowing through them make the nutrients available to living organisms.

Ecosystems and Biomes

Plants and animals live in communities, in which they share the physical characteristics of heat, light, water availability, and nutrients. An **ecosystem** is the total environment of such a community and its physical conditions. Ecosystems exist at all geographic scales but for the purposes of this text are discussed in relation to the largest scale, or **biome.** There are five main types of biome—forest, grassland, desert, polar, and ocean. Those on land reflect the major climatic environments, although the natural vegetation and soils of most biome areas have been greatly modified by human activities.

- **Forest biomes** exist in climatic environments where sufficient water is available. Tropical types range from the dense rain forest of equatorial regions, which is characterized by the crowding of great numbers of plant and animal species, to more open forests with fewer species in areas

having marked dry seasons. Midlatitude forests include the deciduous forests, typical of the eastern United States, in which the leaves are shed in winter, and the evergreen forests typical of Canada and western North America. Most midlatitude forests and large areas of tropical forest have been cut for timber or cleared for farming. Some have been replanted.

- **Grassland biomes** occur where longer dry seasons or annual burning by humans restrict tree growth. In the tropics, savanna grasslands are characteristic of much of the seasonal climatic regions, as in Africa with its wide range of large animals from lions to elephants (Figure 2.29). Midlatitude grasslands once occurred in the prairies of North America, the steppes of southern Russia, the pampas of Argentina, and the veldt of Southern Africa. These environments are now dominated by grain farming and grazing.

  With its low shrub vegetation, the Arctic tundra occurs where long cold seasons and strong winds prevent tree growth. It does not qualify as grassland but has some similar characteristics and supports grazing animals.

- **Desert biomes** occur in lands dominated by aridity (where evaporation exceeds water supply) and thus support little or no vegetation. Any plants that survive in such conditions have special means for storing water, as in the cactus, or short-flowering, seed-producing regimes. The largest desert area is the zone extending from the Sahara in northern Africa through southwestern Asia to Pakistan. Other deserts occur in central Australia, southwestern North America, and coastal Peru and northern Chile. In midlatitudes, deserts occur far from the oceans in Central Asia, or in the shadow of mountains as in Patagonia (southern Argentina, South America), and in some areas of western North America.

FIGURE 2.29 World biome types: savanna grassland. Tropical savanna grassland, eastern Africa. The grasses grow in areas of moderate rainfall, supporting large numbers of grazing animals, such as zebras and wildebeest. These become food for lions and other carnivores at the end of the food chain. **Photo:** © Jill Wilson.

- **Polar biomes.** In polar lands, ice cover and low inputs of solar energy provide little sustenance for living organisms. Coastal areas, however, support a variety of animals through the offshore marine resources.

- **Ocean biomes** are differentiated by water temperatures and nutrient availabilities that produce zones of more or less abundant life in tropical, midlatitude, and polar zones. Microscopic plankton plants take advantage of sunlight in the surface waters and form the basis of living systems in the oceans. The richest biomes with the most profuse organisms occur where rising cold water brings nutrients to the surface, as along the west coasts of Africa and North and South America. Fisheries are marked by smaller numbers of a wide range of different species in the tropics and greater numbers of fewer species in midlatitude oceans.

Soils

Soil fertility, together with climatic heat and moisture conditions, governs whether a region will produce good crops, support livestock farming, or have little farming potential. **Soils** form as broken rock matter interacts with weather, plants, and animals. The rock materials supply or withhold nutrients. Water from rain and snowmelt makes any nutrients present available to plants; decaying plant and animal matter releases the nutrients back to the soil in mineral form.

In the United States, for example, the soils of the eastern half developed beneath deciduous forest, in which the annual leaf fall and ample rainfall returned nutrients to fertile brown soils. Farther north, the long winters restricted growth to evergreen forests, and the pine-needle litter made the soils acidic, reducing their nutrients and making them difficult to cultivate. In the U.S. South, rainwater combined with higher temperatures increased the rates of chemical reactions and removed most nutrients from areas of sandy soils through which water passes easily. In the Midwest, development beneath prairie grassland produced the black soils that proved the most fertile and resilient of all. Farther west, the presence of mountains and arid areas resulted in the formation of few well-developed soils: the steep slopes encouraged erosion; low rainfall prevented soil cohesion; and high rates of evaporation caused the accumulation of alkaline salts in the upper layers (salinization).

Human Impacts on Natural Environments

Natural environments operate largely outside human controls, being powered by energy from the sun or Earth's interior. Locally, however, specific human activities change rates of erosion, remove natural vegetation cover, or emit pollutants into the atmosphere and hydrosphere.

Early in human history (see Chapter 1), most livelihoods depended on the local characteristics of weather, rocks and minerals, landforms, water supply, vegetation, animals, and soils. Even at this early stage of low densities of population, human activities locally changed the workings of the natural environment. For example, burning of vegetation in dry seasons around forest margins encouraged the extension of grassland, and the increase of grazing animals—such as the bison in North America that provided meat and hides for Native Americans. Hunting caused the extinction of large mammals such as the mammoth.

Farming and Erosion

The first farmers settled the lighter soils where there was not much vegetation to clear. From around 1000 B.C., new iron implements made it possible to fell trees on a larger scale and extend farmland into heavy clay soil areas. Both phases of woodland clearance increased soil erosion in uplands and deposition in lowlands, together with changes in the species composition of animals and plants. In Asia and elsewhere, the construction of terraces reduced erosion but extended the cultivated area to steep slopes at the expense of forest cover.

Desertification

Desertification is the destruction of the productive capacity of an area of land. It can occur following natural shifts of climate or because of human activities. As a result, deserts, such as the Sahara, expand in area. The southern margin of the Sahara experienced desertification from the 1970s after commercial agriculture had extended into less humid areas and years of drought followed (see Chapter 9). The deaths of cattle and vegetation, and lack of irrigation water, caused many communities to move away from their traditional lands.

Humid areas also experience desertification where vegetation is removed or the soil is stripped away. Environmental contamination, as when mining wastes foul the land or nuclear or other toxic wastes leak, also makes land unusable for many years.

Industrial Revolution

The industrial revolutions after 1750 intensified the scale of human intervention in natural environments (Figure 2.30). The rate of soil erosion from deforested slopes and plowed fields increased. It affected areas of the world that were opened up for agriculture in the 1800s and 1900s by settlers from the colonizing countries who aimed to produce commercial crops for sale to the materially wealthier countries. City expansion produced surfaces of tile, brick, concrete, and asphalt that caused rain to run off the surface more rapidly, adding to the numbers of floods downstream. Pollution of water and air became a health problem in many industrial cities. Demands for fish by a rising world population led to overfishing of the oceans; increasing demand for timber and land led to the cutting of tropical rain forest. Both overfishing and forest cutting remain major aspects of resource depletion.

Atmospheric Pollution

Environmental problems built up with the increase of the materials entering the atmosphere from industrial processes. Factories in the industrial countries poured their wastes into the

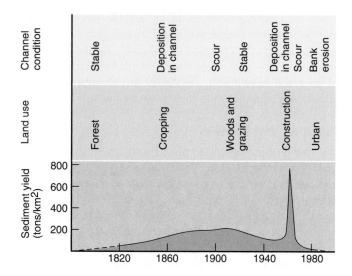

FIGURE 2.30 Human impacts on the environment. A watershed near Washington, D.C., which has been subject to land use changes over the last 200 years. Describe how clearance for farming, land abandonment, and urban expansion affected the sediment flow and hence changed rates of erosion by the rivers of the area. **Source:** Reprinted from "A Cycle of Sedimentation and Erosion in Urban River Channels" by M. G. Wolman from *Geografiska Annaler,* 1967, 49A:385–395, by permission of Scandinavian University Press.

atmosphere and rivers. The carbon gases emitted by burning coal and gasoline exceeded the amounts that natural systems could absorb , and the carbon dioxide content of the atmosphere rose, enhancing the greenhouse effect (see Geography at Work: China's Landscapes and Global Change, page 57).

The growth of the "ozone hole" over Antarctica posed another potential problem. Ozone is a gas that forms a protective shield in the upper atmosphere, intercepting incoming harmful ultraviolet rays from the sun. Depletion of the ozone layer increases ultraviolet ray penetration and the risk of skin cancer. In the 1980s, it became clear that chlorine gases, including the human-made chlorofluorocarbons used in refrigeration systems, gradually permeate upward and destroy ozone by chemical reactions. The reactions are most intense during the polar winters over Antarctica: the October measurements (at the end of Antarctic winter) show the greatest depth of ozone depletion, forming a hole in the ozone layer. People living in southern Chile and Argentina, Australia, and New Zealand take steps to mitigate the danger. Elsewhere around the globe, the ozone layer thinned slowly. Governments agreed to end the use of the ozone-depleting gases, and the policy is reducing ozone destruction, although the potential danger will last for several decades.

While global warming and ozone depletion may have worldwide effects, acid rain affects areas up to several hundred kilometers downwind of major urban–industrial areas. In particular, sulfur and nitrogen gases from power stations and vehicle exhaust react with sunlight in the atmosphere and return to the ground as acids. Soils and lakes that are close to the pollution sources and are already somewhat low in plant nutrients suffer first.

The effects of acid rain may be felt in countries beyond the origin of the pollutants. Canadians claim that emissions from Ohio River Valley power plants in the United States affect their eastern forests and lakes; Scandinavians complain about the emissions from western European industrial areas. The phenomenon is a growing menace downwind of new industrial areas in developing countries. However, something of the complexity of human interactions with natural process was demonstrated in the early 2000s by the recovery of European forests, thought to have been affected by acid rain, as global warming brought higher temperatures, more rain, and greener trees.

Resources and Hazards

Natural resources and hazards are distributed unevenly around the world and have important influences on regional geographic differences. Their study unites aspects of physical geography with cultural perceptions of what is useful or presents difficulties to human populations.

Natural Resources

Natural resources are naturally occurring materials that human societies identify as resources and use to maintain their living systems and built environment. People recognize the availability of resources, their technological and cultural usefulness, and their economic viability.

However, resources valuable to one society or technology are not always rated highly by others. For example, Stone Age peoples used flint and other hard rocks that flaked with sharp edges to make tools and weapons, but such rocks have few uses today. The clay mineral bauxite was ignored until it was found that refining it produced the strong, lightweight metal aluminum. Among energy resources, emphasis shifted from wood to wind, running water, coal, oil, gas, and nuclear fuels.

Natural resources include fertile soils, water, and minerals in the rocks. **Renewable resources** are naturally replenished. The best example is solar energy, which provides a constant stream of light and heat to Earth. Water is a renewable resource that is recycled from ocean to atmosphere and back to the ground and oceans. Many countries use less than 20 percent of the water falling on their territory, returning much of that to natural systems after use. All renewable resources are, however, ultimately finite in quantity and quality. They may be overused. For example, reaching the limits of water supply affects irrigation-based development in arid countries, such as along the Nile River valley (see Chapter 8) and in the western United States (see Chapter 11).

Other resources are nonrenewable: extraction exhausts them. **Nonrenewable resources** include the fuels and metallic minerals available in rocks. Technological advances or new and increased demands, however, assist in finding new sources, extracting sources that were once thought to be uneconomical, and recycling metal products. Such technologies extend the lifetime usefulness of nonrenewable resources.

CHINA'S LANDSCAPES AND GLOBAL CHANGE

Over the past 50 years, population growth combined with economic development and industrial technologies such as fossil fuels and synthetic nitrogen fertilizers have been driving unprecedented ecological changes across China's ancient agricultural landscapes. Current research by Erle Ellis and his team of collaborators in China is demonstrating that the environmental impacts of these changes extend far beyond the region's 2 million square kilometers and 800 million people, impacting both global climate and stratospheric ozone. By a multiscale approach integrating regional data on land cover, terrain, and climate with higher-resolution local measurements from historical aerial photographs, high-resolution satellite imagery, household surveys, interviews with village elders, and field measurements of soils and vegetation, Ellis and his collaborators are now measuring these changes with unique precision at five rural field sites selected across environmentally distinct regions of China (Box Figure 1).

In the late 1980s, while finishing his dissertation on the physiology of crop yield at Cornell University, Ellis became convinced that sustainable agricultural development required a more holistic understanding of long-term agricultural productivity. He set out to investigate the ecology of the world's most productive and long-sustained agricultural systems. China stood out as one of the few places where agricultural systems that had sustained high yields continuously for centuries might still be available for study.

In the spring of 1990, Ellis joined a sustainable agriculture delegation to China and was immediately inspired to begin research there. In the absence of postdoctoral opportunities for foreign researchers, he accepted an English teaching position at Nanjing Agricultural University, and spent the year there learning Mandarin, familiarizing himself with China, and searching for the right project and collaborators. Returning to the United States a year later, he applied for and received a (U.S.) National Science Foundation Environmental Biology postdoctoral fellowship to investigate the nitrogen cycles of traditional versus industrial villages in China.

In 1993 Ellis began field research, believing that a traditional agricultural village might still exist in the Tai Lake Region—the area around Tai Lake bounded by Shanghai, Hangzhou, and Changzhou, long one of the most productive agricultural regions of the world and known in China as the "land of fish and rice," the Chinese equivalent of the "land of milk and honey." He discovered that the region's traditional agriculture had been transformed since the 1970s into one of the most input-intensive agricultural systems in the world and adapted his study to a comparison of the historic (circa 1930) versus contemporary (1995) state of the same village in Wujin County.

Given that nitrogen limits the productivity of most ecosystems, Ellis had hypothesized that the secret to sustaining high long-term productivity was to overcome nitrogen limitation by planting nitrogen-fixing legume "green manure" crops and by efficient nitrogen

Box Figure 1. Carrying out the humification study in the field.

To read more about this study, refer to Ellis, E.C., "Long-term Ecological Changes in the Densely Populated Rural Landscapes of China," in R.S. DeFries et al., eds., *Ecosystems and Land Use Change*, American Geophysical Union, Washington, D.C. (2004).

recycling. Historical records confirmed that the region's traditional farmers sustained rice yields for more than 800 years at levels over 50 percent higher than possible without fertilizers. They accomplished this by means of external sources of nitrogen fertility, including sediments harvested from canals and purchased oilcakes (residues from pressing oil from imported soybeans and rapeseed). Nitrogen fixation and recycling proved to be only minor sources of fertility.

His initial hypothesis disproved, Ellis made another discovery. Because of the ready availability of synthetic nitrogen, sediment was now left to fill in village canals and had increased soil nitrogen concentration over time, producing a dramatic net increase in nitrogen storage across the region's rural landscapes. Potentially, this "sink effect" had global implications, helping to explain the "missing sink" for atmospheric carbon dioxide that continues to complicate our understanding of anthropogenic climate change. (Carbon storage is tightly linked to nitrogen storage in soils.)

Ellis's research was pushed in its current direction by the observation that fine-scale local changes in landscape management could have major regional and global consequences that were observable only by detailed site-based measurements. In 2004, in the final year of a five-year NSF project investigating the global impacts of long-term ecological changes across China's densely populated rural landscapes, his work yielded surprising results, including long-term increases in tree cover that parallel major increases in population density, with simultaneous increases in impervious surface areas that rival those of urban areas.

Natural Hazards

The natural environment poses difficulties and challenges for human settlement in **natural hazards** such as volcanic eruptions, earthquakes, hurricanes and other storms, river and coastal floods, and coastal erosion. Hazards interrupt human activities to a greater or lesser extent, although they seldom deter humans from settling or developing a region if its resources are attractive. For example, people are drawn to living in California or the major cities of Japan despite the likely occurrence of earthquakes. Similarly, people continue to live and work in areas prone to hurricane damage or river flooding.

In areas such as the Mississippi River valley in the United States and the lower Rhine River valley in the Netherlands, protective walls are designed to cope with all but the worst river floods. The highest levels of flooding, as in 1993 along the Mississippi and in 1995 along the lower Rhine, may overtop these walls.

Hazards cause loss of life and destruction of property, but the costs of protection against hazards are also high. Most protection is provided in wealthier countries, where hazards cause the least loss of life but the greatest damage to property. Many poorer countries have few resources available to construct protective measures against natural hazards and often suffer major losses of life after floods, hurricanes, or earthquakes. Comparisons between the few people killed when hurricanes strike the United States and the high number killed in the earthquake and resultant tsunami of Southeast Asia and around the Indian Ocean in December 2004 illustrate this point.

Rio and Kyoto

Concern about global warming has resulted in international conferences aimed at bringing about consensus and collective action to lower the emission of greenhouse gases and other harmful pollutants. Such international cooperation is necessary because air and water are not confined within countries' boundaries. Pollution emitted from one country frequently moves into other countries.

The United Nations Conference on Environment and Development in Rio de Janeiro, Brazil, in 1992, is now commonly known as the "Rio Earth Summit." The Rio conference produced a number of conventions, namely the UN Framework Convention on Climate Change (UNFCCC), the Convention on Biological Diversity (CBD), and the United Nations Convention to Combat Desertification (UNCCD). The UNFCCC established an overall policy framework for addressing climate change and laid the foundation for combating global warming. It went into effect in March 1994 after 50 countries ratified it.

The Rio Earth Summit was followed by another meeting in Kyoto, Japan, in 1997. The resulting **Kyoto Protocol** to the United Nations Framework Convention on Climate Change (UNFCCC) was adopted on the December 11th, 1997, and addressed specific targets for reducing the six primary greenhouse gases: carbon dioxide (CO_2), methane (CH_4), nitrous oxide (N_2O), hydrofluorocarbons (HFCs), perfluorocarbons (PFCs), and sulphur hexafluoride (SF_6). A list of industrialized countries, primarily in Europe but also including the United States, Canada, Japan, Australia, and New Zealand, is known as Annex I countries. According to the Kyoto Protocol, these countries need to reduce their greenhouse gas emissions so that their average yearly emissions for the years 2008 to 2012 are 5 percent less than their emissions in 1990. For the Kyoto Protocol to go into force, two conditions must be met. First, 55 countries must ratify the protocol. Second, the list of at least 55 countries ratifying the protocol must include industrialized countries from Annex I that likewise account for at least 55 percent of the carbon dioxide (CO_2) emissions of those countries

listed in Annex I. The highest emitters are the United States and the Russian Federation, respectively accounting for 36.1 and 17.4 percent of the list's CO_2 emissions. The other countries each emit less than 10 percent, with many below 1 percent.

By the end of 2004, 125 countries had ratified the Kyoto Protocol, easily satisfying the protocol's first requirement. However, most of these countries were developing countries, not listed in Annex I. Most European countries, along with Japan, Canada, and New Zealand, ratified the treaty. As the United States signaled its intentions not to ratify the treaty, the Russian Federation became the key country. After many years it decided to ratify the protocol in November 2004. With the Russian Federation's share of 17.4 percent, the percentage of total emissions for Annex I countries ratifying the protocol amounted to 61.79 percent, well above the minimum 55 percent required for the protocol to go into effect. The Kyoto Protocol went into effect February 16, 2005, thus beginning a new phase of international relations on greenhouse gas emissions and concern for the environment.

Test Your Understanding
2B

Summary

Environmental issues are an outcome of interaction between human beings and the natural world. The linked air and water circulation of the atmosphere and oceans, stirred and circulated by solar energy, sets off a series of weather events that add up to climatic environments. These can be divided into tropical, midlatitude, and polar variations. Climate change is normal in the natural environment, but human impacts are now powerful enough to influence such changes.

Earth's surface features are produced by the interaction between internal forces that produce volcanoes, earthquakes, and mountain ranges and climatic forces of temperature changes, wind, water, and ice flow. Plants, animals, and soils are parts of ecosystems fueled by solar energy and with nutrient circulations linked to broken rock materials and atmospheric gases. Forest, grassland, desert, polar, and ocean biomes are distinctive ecosystems.

Human interactions with the natural world convert forest and grassland into farmland but also pollute air, water, and soil. Humans designate natural resources and combat natural hazards.

Questions to Think About

2B.1 How do (a) natural environments affect human actions and (b) human activities affect natural environments? Refer to recent events to illustrate your answer.

2B.2 What causes natural environments to change?

2B.3 What are some examples of renewable and nonrenewable resources?

2B.4 Analyze a recent natural hazard in terms of its causes, effects, and prospects for future control.

Key Terms

natural environment	biome
climate	forest biome
greenhouse effect	grassland biome
tropical climates	desert biome
midlatitude climates	polar biome
polar climates	ocean biome
global warming	soils
relief	desertification
tectonic plates	natural resource
weathering	renewable resource
erosion	nonrenewable resource
deposition	natural hazard
ecosystem	Kyoto Protocol

World Regions, Human Development, and Human Rights

Dividing the world into regions leads geographers to consider how some places experience different living conditions than others. The answers to this investigation vary from place to place and depend on such factors as

- The complex interactions of people with the natural environment.
- Considerations of the historic growth of population numbers and cultural expressions.
- The evolution of political systems.
- The growth of economic output.
- How different views of human rights affect those living in different parts of the world.

In a globalizing world, the issues arising from these studies become more obvious and urgent. In this chapter we have highlighted a series of geographically related analyses of population, resources, political freedoms, wealth and poverty, cultural freedoms, and environmental concerns. They all affect human development and human rights.

Human Development

Throughout history, the differences among places led some countries that were well endowed with natural resources and strong leadership to assume positions of dominance and superiority. For example, the Chinese long believed they were the only civilized people and all others were barbarians. They spread their culture widely in East Asia, but held back from wider geographic control from the 1400s. During the next centuries, European countries claimed they were taking their form of civilization and modern ways to parts of the world that they

often colonized. Such assumptions increased differences among places. The histories of today's materially poorer regions, especially their experiences of European colonialism since A.D. 1500, are basic to an understanding of geographic differences among countries. In the later 1900s, the United States and Soviet Union contested world economic, military, and political dominance, each believing that their systems were superior. After 1991, the breakup of the Soviet Union left the United States as the only "superpower" (Figure 2.31).

Despite the trends toward globalization, the world remains full of differences and inequalities. The material wealth of many Americans contrasts with the poverty of some parts of that country and the extreme material poverty of millions in Africa and Asia. Such differences led to studies of how some regions and countries move ahead and others fall behind—a matter of more or less **development. Human development** is the process of enhancing human capabilities and improving life quality through such aspects as education, health care, and access to adequate clean water and energy supplies. Although this definition refers to all levels of well-being, most concern is given to the possibilities of helping materially poor in "less developed" or "underdeveloped" countries and regions to catch up with the wealthier countries. **Sustainable human development** involves economic growth that does not deplete renewable resources for the future. It thus links to both human and natural resources, drawing together studies of human and physical geography.

Issues of Human Development

The United Nations Human Development Program and recent World Bank publications focus on the eradication of material poverty. In 2000, the United Nations and other global organizations formulated the Millennium Development Goals, to be achieved by 2010–2015 (Table 2.2).

It is encouraging that the last 50 years saw major reductions of income poverty in large parts of the world, improvements in human development indicators—particularly in health

FIGURE 2.31 U.S. military might. The armament in planes and missiles on a single U.S. aircraft carrier is greater than that of most countries. **Photo:** © Corbis RF.

Table 2.2   Millennium Development Goals.

In 2000, many target dates were set between 2010 and 2020, but lack of progress to 2005 suggests that the goals will take much longer to achieve.

Goal 1: Eradicate extreme poverty and hunger.

Goal 2: Achieve universal primary education.

Goal 3: Promote gender equality and empower women.

Goal 4: Reduce by two-thirds, between 1990 and 2015, the under-five mortality rate.

Goal 5: Improve maternal health.

Goal 6: Combat HIV/AIDS, malaria, and other diseases.

Goal 7: Ensure environmental sustainability.

Goal 8: Develop a global partnership for development.

and education—and the wider spread of law and fair administration of justice. However, it remains challenging that so many people in the world remain materially poor.

A study of approaches to development includes not only the theories about how it has taken or may take place and an understanding of strategies adopted to make it happen, but also insight into the ideologies that decide the purposes of development. Several phases of ideas can be described.

"Developed Countries Thought They Knew Best"

In the 1950s and 1960s, the combination of many former colonial countries becoming independent and the growth of the Cold War (perceived at the time as **First World** versus **Second World**) rivalries formed the background to a preoccupation of some economists with development issues. Western economists assumed that the materially wealthy Western countries understood how economic growth worked. The Soviet and Chinese Communist regimes offered different inducements. Both had interests in promoting their ideas in the new countries. The **Third World** was born as an alternative way, but the Third World countries could not avoid pressures to align themselves to the First or Second World. Most of its countries remained poor despite these external influences.

Two main approaches were adopted. First, it was suggested that the change from traditional (underdeveloped) to modern (developed) would take place through a transition from agriculture-based societies to industrial production and mass consumption of the manufactured goods: from primary through secondary to tertiary and quaternary production. However, the world's materially wealthier countries are home to the most profitable sectors and have more diversified economies than the poorer countries (see Table 2.1). This brings a greater prominence in the new, high-value manufacturing industries such as building airplanes, electronics, and pharmaceuticals, as well as in the high-profile business services.

In the second approach a movement from rural to urban living brought improved access to waged jobs, education, health care, and other services. Urban centers have become

home to three-fourths of Western country populations and to increasing numbers in the Soviet Union, but to only one-third or less in poorer countries.

The free-market capitalist countries and the centrally planned Communist countries both encouraged **modernization** in developing countries, but neither had overwhelming success. Outcomes were patchy, resulting in geographically uneven development among and within countries. The formula worked in a few countries, particularly those, such as Japan and South Korea that had U.S. help in war recovery investment. Other Asian countries, such as Malaysia, Thailand, and Singapore, also invested in manufacturing and service industries and experienced rapid economic growth. Larger countries, such as Mexico, Brazil, India, and China, possessed internal resources and a potentially large market for home-produced goods, but economic growth focused on a few urban areas. The smallest countries often lacked both resources and large markets. Most developing countries established only a few, often low-technology, industries making goods for internal consumption.

"History Provides the Key"

In the 1960s and 1970s, descriptive studies of historical experiences, mainly focused on events before independence, shifted some emphasis to understanding how colonialism contributed to underdevelopment and dependency of poorer countries on the colonial powers. Such studies provided generalized patterns (or models) of the processes involved in colonization and integration with Western countries. They were, however, descriptions of experiences, rather than strategies for later growth.

One set of studies found that most materially wealthier places got wealthier, while few poorer places experienced material improvements. As worldwide trade developed from the late 1400s, many European merchants enhanced their own positions at the expense of other parts of the world. They removed valued commodities from around the world to build their own wealth. This trend was continued from the 1800s as they traded low-value raw materials from abroad with their own high-value manufactured goods. Such processes resulted in the recognition of a set of materially wealthier core countries and another set of dependent peripheral countries.

A second set of studies focused on the transportation relationships between the expanding merchant-based societies of Europe and the "new" lands that were settled, appropriated, and developed as a result of expansionist activities from the late 1400s. Figure 2.32 is one such model of this type.

"Third World Answers Back"

By the 1960s and 1970s attempts to transfer the Western prescriptions for economic growth were not bringing economic development to all countries. Concerns of those living in Third World countries went wider than the economic lag. Resentment of previous colonial domination, combined with ideas from the apparently successful Communist countries at the time, resulted in the conviction that the materially poorer countries had been unjustly made dependent on Western countries. A chain of dependence was

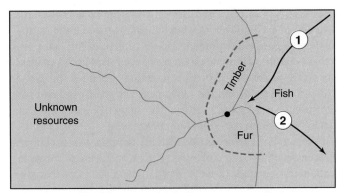

New land, ocean transport

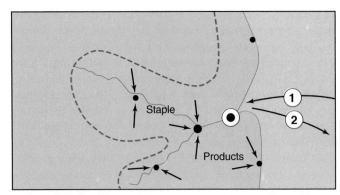

River transport

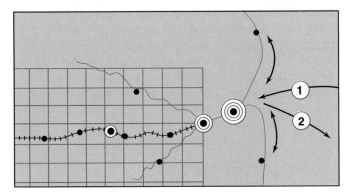

Rail transport

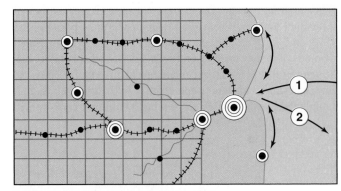

Road/rail transport

FIGURE 2.32 Colonial settlement patterns. The sequence of stages provides a general picture of how places such as Australia and Argentina grew with settlement from Europe. Other former colonies did not go through all the stages. At first, imports from the home country (1) are paid for by exports (2). Early settlement relates to river transportation; and later to railroads, which establish new patterns of movements and economic development. City growth is shown by numbers of circles.

identified, starting with rural peasant farmers linked into local market towns, international ports, and centers of global trade. A wish to be independent, self-sufficient, and released from involvement in the West-dominated world economy placed an emphasis on home-based manufacturing (**import substitution**), local trade barriers, restrictions on foreign corporations, and the formation of trading groups of countries with similar concerns. By the 1990s, however, it became clear that isolation from the world economy might have some internal cultural and political value, but it did not lead toward equality with the Western countries.

"Bottom-Up, Not Top-Down"

From the 1970s, greater emphasis was placed on self-reliance and the local characteristics of places rather than on overall prescriptions that ignored spatial variations. The idea of a single path to development that worked in a straight line of cause and effect was rejected in favor of more flexible approaches. Geographic differences increased in significance.

Rural development became a focus for grassroots and populist pressures to reestablish local communities and strengthen their abilities to maintain themselves and combat external pressures. Basic needs programs to increase the availability of food, clothing, and housing, however, achieved little, and separation of communi-

ties from urban influences often left them in a time warp. In some countries, such as China, Cuba, and Tanzania, socialist influences vowed to link culture, history, and local institutions to an integrated communal approach mobilizing human and natural resources. A focus on environmental consciousness encouraged sustainable production. Another strategy freed women from virtual slavery in many places, adding to an increased emphasis on human rights and justice. Such ideas, however, often proved difficult to maintain in the face of external economic and political pressures to produce goods for export and to welcome foreign direct investments.

In many poor countries, there are few jobs in the **formal economy:** these pay salaries that are taxed. Few people own title deeds to their home and the land on which it stands. The majority of people have to gain income as best they can in the **informal economy.** They cannot use their homes or land as collateral for loans, and it has been shown that this failure to extend formal property rights to most of the population is a major cause of poverty and a lack of enterprise opportunities.

Alternatives to Modernization

In the 1990s, the uneven and often unsatisfactory results of attempts to modernize on Western patterns led to new approaches. The experience of Asian countries that achieved economic growth through export-based manufacturing led major global agencies, such as the World Bank, United Nations, and many nongovernmental organizations, to insist that countries applying for aid should export manufactured goods and allow foreign investment—a process known as **structural adjustment.** Such policies were linked to downsizing internal bureaucracies and central government planning. Copying the Asian way, however, proved no easier than copying Western experiences.

Alongside such structural adjustment, other new approaches were based on local cultures. Some looked back to precolonial conditions. Other groups of people set up microcredit banks. They built on the experience of the Grameen ("Village") Bank in Bangladesh, which grew out of a program of small individual loans by Professor Mohammad Yunus beginning in 1974 and was formally established in the mid 1980s. Small amounts of money enabled craft workers and others to establish businesses that helped them emerge from poverty. By 2001, the Bangladeshi Grameen Bank had 1,170 branches across Bangladesh, with 11,500 staff serving 2.4 million borrowers in

over 40,000 villages. Each working day, the bank collected an average of $1.5 million in weekly installments. Of the borrowers, 94 percent were women, and 98 percent of the loans were repaid on time. These methods are now applied in 58 countries, including materially wealthier countries such as the United States and Canada. As the movement has grown and extended its influence, so have its critics, who suggest, for example, that such loans add to the oppression of women.

However, it again proved difficult to promote local cultural and environmental distinctiveness in an age of mass communication and tourism. In the 1990s, a new division of the world emerged from technological innovation and transfer (Figure 2.33). Around 15 percent of Earth's population, living in Canada and the United States, western and northern Europe, Italy, Japan, South Korea, Taiwan, and Australia, is responsible for nearly all technological innovations and their applications. A further 50 percent of the world's population is able to adopt some technologies in production and consumption. The remaining 35 percent of the world's population lies outside these two zones, is "technologically disconnected," and often is

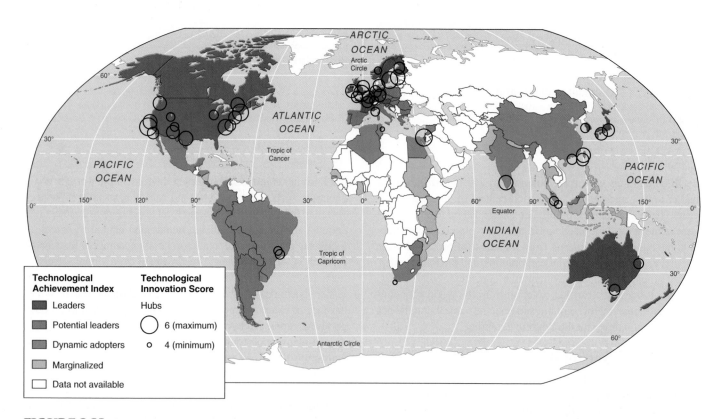

FIGURE 2.33 Global hubs of technological innovation. In 2000, *Wired* magazine consulted local sources in government, industry, and the media to find the locations that matter most in the new digital geography. Each was rated from 1 to 4 in four categories, giving maximum scores of 16 and minimum of 4: the ability of area universities and research facilities to train skilled workers or develop new technologies, the presence of established companies and multinational corporations to provide expertise and economic stability, the population's entrepreneurial drive to start new ventures, and the availability of venture capital to ensure that the ideas make it to market. Forty-six locations were identified as technology hubs, shown on the map as black circles. The hubs with the highest scores include: Silicon Valley (California), Boston (Massachusetts), Stockholm (Sweden), Israel, Durham-Raleigh-Chapel Hill (North Carolina), and London (U.K.) . Ten others in the top 40 occur in the United States, 12 in Europe, and 8 in East and Southeast Asia. **Source:** *United Nations Human Development Report,* 2001.

caught in a trap of poverty, disease, low agricultural productivity, and environmental decline: these people need technological solutions they cannot afford.

Development and Globalization

People conscious of globalizing trends question whether they will override whatever underdeveloped countries try to do for themselves. Friction of distance has ceased to determine the economic costs of locating production facilities around the world, and low labor costs have increased in significance. Fewer controls on trade and money movement allow MNCs to choose their places of investment worldwide. Better communications by satellite TV and the Internet enable wider global information transfer, including the marketing of global corporation products. The speed and flow of transactions among financial markets in New York, Tokyo, and London expanded to incorporate regional global city–regions in countries experiencing economic growth (see Figure 2.15).

Some media and pressure groups imply that everywhere is becoming similar, based on Western society and culture. A more realistic view is that globalization does not result in uniformity, but its greater flexibility, openness, and complexity lead to new differences among places and the spatial inequalities of uneven development. Localization and fragmentation result from local interpretations of global forces.

This is not a new condition. Globalization grew out of the European expansion from the late 1400s. Rival European countries imposed political and economic power differences, together with cultural shifts, on people in the colonized regions. The processes spread at uneven rates and in varied directions, with the overall result of increasing differences among people and places. After independence, many former colonial countries attempted industrialization to emphasize their postcolonial freedom and desire for self-sufficiency: they could choose to enter the world economic system, or not. This process proved difficult to achieve. Western corporation branch plants and free-trade zones were established in many materially poorer countries, taking advantage of low labor costs to make electronics, textiles, furniture, and leather goods. In the 1990s, data processing also moved to low labor-cost locations.

The combination of growing multinational corporation influences and the increasing focus of economic activity in global city–regions highlight new geographic patterns of flow among people, finances, goods, and ideas. Any global convergence of economy and culture is paralleled by the divergence of dispersed manufacturing around the world. Despite wider availability, Internet access also highlights the differences among wealthier and poorer places: the New York–New Jersey urban area has more telephone lines than the whole of Africa.

Similar trends of convergence and divergence affect cultural transfers. The predominance of Western tastes in consumer preferences, food outlets, and mass media offerings is most noticeable in the major cities and among the urban elites. As the better-off in materially poor countries adopt Western lifestyles, however, they often combine them with local elements, as demonstrated in popular music, movies, and fast-food outlets. Such developments, by diversifying the cultural environment, often set off a return influence by immigrants in the Western world.

The world's leading industry of the early 2000s, tourism, provides one of the most significant links between development and globalization. Closely related to conspicuous consumption in materially wealthier countries, tourist venues based on jet aircraft and ocean liner access enhance the differences among places. From the late 1960s, visitors from wealthier countries have traveled more and spent more on their vacation trips. Many underdeveloped countries have adopted tourism as a strategy for modernization. Although tourism has provided positive effects such as greater understanding of conditions in other parts of the world, its negative side is often a low local return. Food and trinkets brought in have replaced local products. International corporations build the hotels and the cruise ships. Historical reconstructions for tourists have subdued the unpleasant aspects and guilt arising from, for example, slavery in favor of inaccurate stereotypes.

Responsible Development

By 2004, "responsible growth" was a term replacing earlier versions. Arising from the Johannesburg World Summit on Sustainable Development held in August 2002, a consensus emerged directing world leaders to new development paths that build on the UN Millennium Development Goals (see Table 2.2) and link economic growth, environmental sustainability, and social equity. In this view, directed at the initial period to 2015 and on toward 2050, poverty reduction is not an end in itself, but is a precondition for peaceful coexistence and ecological survival.

Issues of Human Rights

Defining Human Rights

In the late 1700s, the concept of **human rights** emerged in part from revolutions in Europe and the United States that emphasized, for example, "Liberté, Égalité, Fraternité" in France. This concept was enshrined in the U.S. Bill of Rights. Stressed again after World War II by the United Nation's Declaration of Human Rights (Table 2.3), international agreements and actions often refer to them. However, few countries fully implemented the UN list, and some groups claim that the imposition of human rights legislation impinges on their traditional rights.

Whereas pressure for human development emerged from mainly economic considerations that were later broadened cultural, political, and environmental concerns, pur human rights has often been a special objective of l

Table 2.3   United Nations Declaration of Human Rights

Freedom from discrimination because of gender, race, ethnicity, national origin, or religion

Freedom from want and a decent standard of living

Freedom to develop and realize one's human potential

Freedom from fear of threats to personal security in arbitrary arrest or violence

Freedom from injustice

Freedom of thought and speech to participate in decision making and forming associations

Freedom for decent work without exploitation

philosophers, and political pressure groups. During the Cold War, political rhetoric reduced human rights to a propaganda weapon, with the West emphasizing civil and political rights, while the Communist world focused on economic and social rights. Today human rights activists, particularly in the West, claim all people should live under social arrangements that protect them from the worst abuses and deprivations.

The United States and western Europe have long argued for **political rights** (the right to vote and participate in one's own government). Others, including the former Communist governments, argue for **social rights** (the right to have a job and earn a living with basic material standards). People in materially poorer countries argue for **cultural rights** (the right to protect one's cultural traditions, often in response to Western ideals and practices). An example of cultural rights involves Western medicine patents, where people in other countries have to pay royalties for a medicine they had discovered hundreds of years before the institution of patenting systems.

In practice, there is considerable debate as to what should be included or is feasible as human rights. At times, human rights issues provide "red herring" arguments. For example, the world's wealthier countries often argue against trading with poorer countries because their factories do not observe human rights by employing children or ignoring environmental protection. Such arguments can be seen as a disguise for protectionist policies designed to secure low-wage, low-skill jobs in the wealthier countries. After 9/11, there was much criticism in Western countries of the Taliban government in Afghanistan and other strict Muslim regimes oppressing women and carrying out amputation punishments for minor crimes. Yet Muslims claim that their strict rules protect women, while many Western governments allow men the freedom to mistreat and degrade women. Different cultural definitions of human rights contribute to differences among people's expectations.

Agreement on the nature of human rights and justice is not global. Muslim countries in particular debate the largely secular basis of the United Nations list that conflicts with their religious beliefs and fierce legal provisions. Saudi Arabia refuses to attend UN conferences on such topics on the grounds that it has a God-given right to gender discrimination.

Human Rights and Justice Systems

Most of the human rights listed in Table 2.3 relate to personal and group justice and are potentially subject to legislation. Threats to personal security, injustices, and discrimination by gender, race, ethnicity, national origin, religion, or age are problems that supporters of human rights wish to end. Government policies and courts of justice legislate for and enforce some rights. In some cases, one person's right is another's restriction.

The death penalty is the ultimate act against a human's right to life. Some Asian countries impose it on drug smugglers; some U.S. states impose a death penalty on murderers. Slavery is another restriction of basic human rights and was mostly abandoned in the 1800s, although cases still emerge in countries such as Sudan and Côte d'Ivoire. Long-term imprisonment for political views was common in the Soviet Union and under the South African apartheid regime until the early 1990s. It continues in China, Myanmar, and many other countries. Amnesty International is an NGO that monitors such actions. Imprisonment for long periods without accusation or trial occurs in many countries and in 2001 was extended in the United States and United Kingdom as part of antiterrorism policies. Most justice systems are based within countries, but, at the global and world regional levels, new courts are taking up international issues, particularly of human rights and war crimes.

Human Development and Human Rights

Although having different emphases and emerging from different sources, the two strands of human development and human rights reinforce each other. Human rights add to the development agenda by drawing attention to those who are accountable for respecting rights and adding social justice to economic principles. This shifts priorities toward the most deprived and excluded. At the same time, human development brings a long-term perspective to fulfilling the rights by assessing the workings of socioeconomic contexts and institutional constraints. It highlights the resources and policies needed to overcome the remaining gaps.

In the early 2000s, geographic differences continue in many areas of discrimination, poverty, personal insecurity, injustice, and abuses of free speech. For example, internal armed conflicts affect many parts of the world, holding back human development, abusing human rights, and creating increased numbers of dispossessed refugees. The moves toward democracy in Africa and eastern Europe with multiparty elections brought some advances in human development and human rights. These trends, however, also led to new conflicts in some countries over previously suppressed ethnic demands.

The Rest of the Book

Chapters 3 through 11 each discuss a major world region. For comparisons among them, each chapter has a similar structure and uses similar maps and diagrams that have been introduced in this chapter.

- At the start, there is a statement about the region's distinctiveness in the world.

- A section on cultural history outlines the origins of present human geographic patterns.

- A study of the natural environment focuses on climates, landforms, vegetation, and soils, leading to discussion of environmental issues and problems.

- There is an account of the global forces affecting the region and how places in the region deal with them.

- Each region is then divided into subregions for more detailed consideration of the political, ethnic, and economic aspects.

- In each regional study there are boxes that focus on specific localities, highlight concerns that stimulate debates, and illustrate the research being carried out by geographers.

Our final Chapter 12 seeks to bring together some conclusions arising from the world regional studies in the global and local contexts. We trust that this will show how geographers are realistically facing our present and future worlds.

Test Your Understanding
2C

Summary

Human development and human rights are linked. Much was achieved in improving the well-being of peoples during the 1900s, but many needs remain. Although many attempts have been made to understand how human development may occur, the geographic situations are too varied to make it possible to devise a single way ahead. Changing emphases on the country and global trends lead to a better understanding of the top-down and bottom-up issues. Human rights provide a facet that highlights shortfalls in development.

Questions to Think About

2C.1   How have the varied elements contributing to development emerged from the experiences of different countries?

2C.2   How important is it that one-third of the world's population is poor? What can be done about it?

Key Terms

development	formal economy
human development	informal economy
sustainable human development	structural adjustment
First World	human rights
Second World	political rights
Third World	social rights
modernization	cultural rights
import substitution	

Chapter 3

Europe

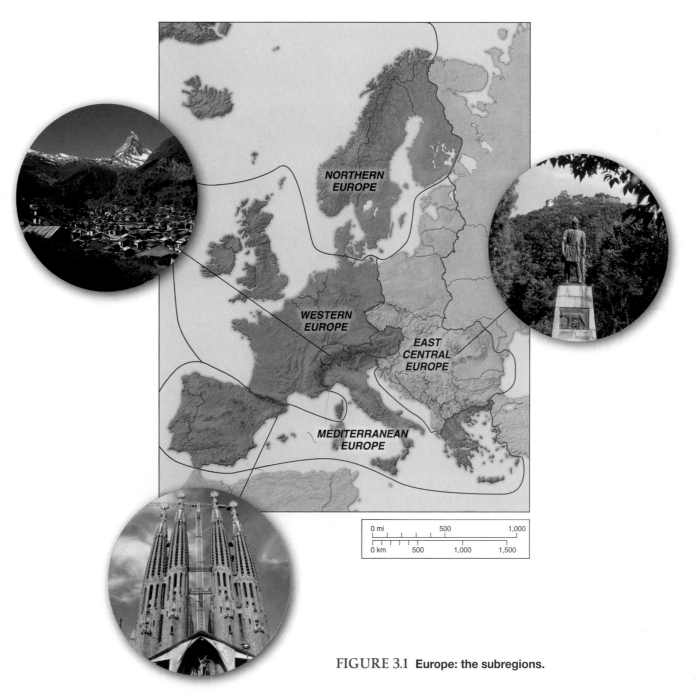

FIGURE 3.1 **Europe: the subregions.**

Themes in This Chapter

Beginning in the 1400s, when the Portuguese began sailing the coast of Africa and Columbus landed in the Americas, Europeans launched an age of discovery and set into motion a new wave of globalization that continues to reverberate around the world today. The drive to accumulate material wealth led to periods of intense competition and frequently war between European powers. In the aftermath of World War II, a new spirit of cooperation led to greater economic and political integration and the rise of the European Union. Europeans also became major proponents of human rights and human development around the world. Europe remains culturally diverse, but now its population is aging and presenting new economic challenges. Peaceful and prosperous, Europe attracts immigrants and asylum seekers from around the world. The themes are:

Europe sets the new global order.

Europe's cultural diversity.

Temperate, midlatitude natural environments.

Setting a new global standard in the modern era.

Four subregions:

- Western Europe: Europe's industrial heartland.
- Northern Europe: high living quality in harsh environments.
- Mediterranean Europe: Europe's oldest civilizations.
- East Central Europe: former Communist countries adopting capitalist practices.

Point–Counterpoint: The European Union

Personal View: Bosnia-Herzegovina

Geography at Work: Business Development

European Influences

Many great ideas and inventions originated in other world regions, often long before Europeans made any discoveries. Nevertheless, Europe (Figure 3.1 and Table 3.1) is a hearth for many contemporary global ideas and practices. Some of these include democracy, Christianity, colonialism, imperialism, capitalism, the Englightenment, nationalism, fascism, socialism, communism, and genocide. As stated, not all of these are wholly European. Christianity, for example, began in Southwestern Asia; but for centuries the religion was solely in Europe, where it took on a European character before spreading around the world. Europeans also made such scientific discoveries as the Earth being spherical and revolving around the sun, the laws of motion and gravity, the theory of evolution, genetic science, the causes of disease, radioactivity, and the theory of relativity, European inventions include movable type for the printing press, as well as the microscope, telescope, steam engine, railroad, internal combustion engine, automobile, radio, orbital satellite, and digital computer. The Industrial

Revolution also began in Europe, radically altering Europe's economy and society and then making similar dramatic changes in other world regions.

Europeans used their technologies to explore and gain more material wealth from abroad. They also used their technology to transform Europe's natural environments (Figures 3.2 and 3.3). Farms and market towns replaced forests, and in time, factories and cities replaced many farms. Rivers were straightened for navigation and dammed to prevent flooding and provide power. The Earth was mined for coal and ore and then drilled for oil. All these activities improved the material standard of living but profoundly affected Earth's ecosystems.

(a)

(b)

FIGURE 3.2 Diverse transformed landscapes of Europe.
(a) Urban Leiden, low-lying Netherlands; (b) Rural Monterchi in hilly Tuscany, Italy. **Photos:** (a) © Mike Camille; (b) © Alasdair Drysdale.

FIGURE 3.3 **Europe: physical features, countries, and capital cities.**

Table 3.1 Summary Information of Europe's Subregions: people and economy. Data by subregion, area, population, income (GNI PPP—Gross National Income Purchasing Power Parity), urbanization, Human Development Index (HDI), Human Poverty Index (HPI).

Subregions	Land Area (km²)	Population (millions)		GNI PPP 2003		Index 2002	
		Mid 2005 Total	Est. 2025 Total	Total (US$ billions)	Per Capita (US$)	HDI (rank of 177)	HPI (% population)
Western Europe	1,421,726	249.1	258.5	6,927	32,958	12.0	11.8
Northern Europe	1,257,150	24.5	26.0	726	30,475	8.0	7.8
Mediterranean Europe	1,030,430	121.8	121.9	2,844	21,670	22.8	11.3
East Central Europe	1,341,140	126.6	119.1	1,247	13,767	47.6	
Totals or averages*	5,050,446	522.0	525.5	11,744	24,717	22.6	10.3

*Note: GNI—Gross National Income; PPP—Purchasing Power Parity

Source: World Population Data Sheet, Population Reference Bureau (2005), World Development Indicators, World Bank (2005), Human Development Report, United Nations (2004), and Encarta Microsoft (2005).

For better or worse, over the last 500 years, Europeans used their local ideas to lay the foundation for the modern global economy, structure the world's political system, and spread their culture around the world. Europeans frequently forced their ways on other cultures as they took control of lands and peoples around the world. Though dominating, they integrated local ideas, technologies, and faiths from cultures in Asia, Africa, and the Americas into their own cultures. After the trauma of World War II, Europeans became less aggressive in their interactions with other cultures, even becoming major proponents of human rights and human development. European cultures continue to interact with and react to cultures in other areas of the world, illustrating that globalization is not destroying geographic differences but is often resulting in different and alternative cultural practices. Indeed, though Europeans are now cooperating both economically and politically, Europe is as diverse as ever.

Europe once was thought to be its own continent, but it is now known that Europe is really a peninsula of the Eurasian continent. Much of Europe is as far north as Canada. London and Paris, for example, lie at latitudes much farther to the north than either Montréal or Québec City. However, Europe has a long indented coastline, leaving few places within it far from the moderating effects of a sea or ocean. With westerly winds and the warm currents from the Gulf Stream, Europe's climates are very moderate for its northerly latitude.

Diversity, Conflict, and Technological Innovation

Current attempts to unite European countries into a whole occur against a history of divisions and divergences. Politically, Europe is a very fragmented world region, with many small and medium-sized countries that reflect its varied history of conflict and cooperation. The diversity of Europe began centuries ago, when numerous peoples migrated into the world region, bringing with them new cultural practices.

Migrations of Peoples

The Greeks, Romans, and Celts were the first to inhabit Europe, followed by Germanic, Slavic, and other peoples.

Cultural Groups: Greeks, Celts, Romans

Modern **Greeks** inhabit only the southeastern corner of Europe. Their ancestors, however, were firmly established in southern Europe by 1000 B.C. The Greek city–states had differing forms of government, but some established democracy, later adopted elsewhere in Europe and the world. Early Greeks also contributed to the sciences and humanities. A number of their ideas still influence the way many people view the world.

At the same time that Greek civilization emerged in southeastern Europe, **Celtic** peoples (**Celts**) occupied large areas of central and Western Europe. They gave their names to such places as Bohemia, Gaul, the Alps, and the Rhine. Celtic culture began to wane as the Romans and then later peoples took over many of their lands. Remnants of their culture can be found in the language and culture of the modern Welsh, Scots, Irish, and Bretons of western France (Figure 3.4a).

Around the 100s B.C., Rome eclipsed the Greek city–states. The **Romans** created a large empire that stretched from the eastern Mediterranean to the British Isles. Crucial to keeping their empire together, the Romans built an extensive road system, remnants of which are found today. Many of their army camps grew into large cities, such as London and Paris. Aspects of Roman culture were passed down through the ages. The Romans spoke Latin, which persisted in the lands that the Romans occupied longest: Gaul (France), Hispania (Spain and Portugal), the Italian peninsula, and Dacia (Romania). Without political unity after the fall of Rome, Latin evolved into the distinct languages of French, Spanish, Portuguese, Italian, and Romanian (see Figure 3.4a). The Romans adopted Christianity as the empire's official religion in A.D. 381. In time, Christianity spread well beyond the former borders of the Roman Empire (Figure 3.4b), but the center of the Roman Catholic Church, the branch of Christianity with the most members, is still located in Rome.

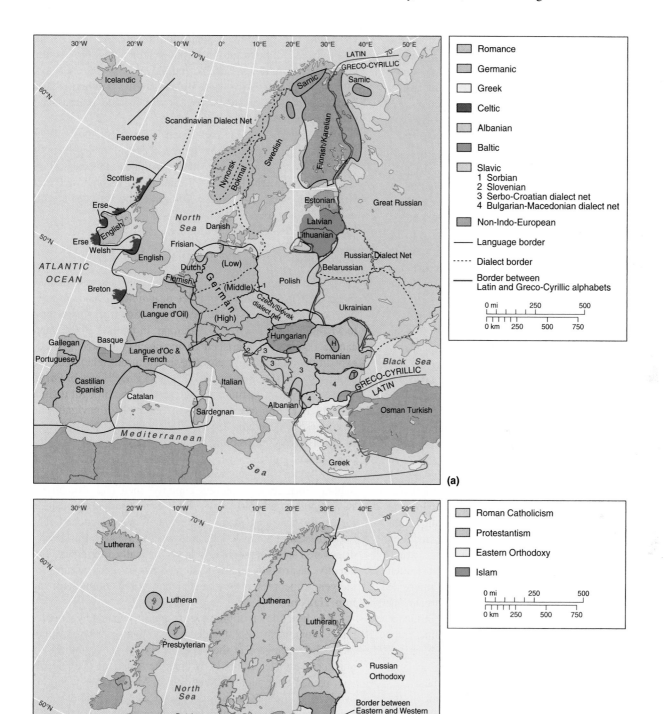

FIGURE 3.4 (a) Spatial distributions of languages and selected dialects of modern Europe. The Latin alphabet, such as that used for English, is used by languages to the west of red line, and generally in Roman Catholic and Protestant Europe. The Greco-Cyrillic alphabets are generally used by Eastern Orthodox Christians to the east of the line.

(b) Spatial distributions of religious groups in modern Europe. Are there any spatial correlations between language and religious groups? **Source:** Jordon-Bychkov, *The European Culture Area,* 4th ed. Reprinted by permission of Rowman & Littlefield.

Jews lived in Europe since Roman times, first along the Mediterranean and then in small groups everywhere else, especially East Central Europe. Jews contributed greatly to European culture. Persecution culminated in the Nazi Holocaust in the 1930s and 1940s that sharply reduced their numbers.

Cultural Groups: Germanic and Slavic Peoples

During Roman times, **Germanic peoples** arrived from the east conquering whatever Celtic lands the Romans had not taken, namely the areas just north of the Danube and east of the Rhine. These tribes continually threatened the Roman Empire, sacking Rome itself for the first time in A.D. 410. By the end of the 400s, Gaul was taken over by the Franks, eventually to be renamed for them (France). The Burgundians lent their name to a province (Burgundy) that was eventually absorbed into France. The Visigoths and Lombards moved into the Italian peninsula. The latter name is found in the modern Italian provincial name of Lombardy. The Angles and Saxons moved into the British Isles, pushing the Celtic peoples farther into the fringes of Europe. Even today, the English are considered Anglo-Saxons.

Other Germanic tribes moved north into Scandinavia. By the A.D. 800s, they developed a distinct **Viking** culture. The Vikings were a seafaring and martial people who experienced overpopulation in their home areas. During this time of warm climate, they sailed northward, colonizing the Faroe Islands, Iceland, and southwestern Greenland, and probably reached North America by the A.D. 900s. From the south, the area of modern Germany, the Vikings felt the pressure of Christianizing forces. In response, they invaded the British Isles and northern France and sacked wealthy monasteries. To the southeast, they influenced the lands around the Baltic Sea, and they played a formative part in founding Kievan Rus', which became one basis for the future Russia. Although they are often caricatured as violent pirates, Vikings mostly settled alongside the local inhabitants of areas that they invaded, organizing wide-ranging trade across northern Europe. Their influence ended after the 1200s, when cooling climate, volcanic eruptions in Iceland, and political factions caused the Greenland colonies to die out and halved the population of Iceland. Plagues reduced the numbers of people in their homelands.

Germanic culture is still prevalent today. Though the Franks, Burgundians, and Lombards adopted the Romance languages of the Roman provinces they conquered, other Germanic peoples, like the Vikings, maintained their Germanic languages through the centuries and are clearly seen on the map today (see Figure 3.4a). Germans, Austrians, Dutch, and the Scandinavians (such as Danes, Norwegians, and Swedes) are the most numerous of today's Germanic peoples. The Germanic peoples also converted to Christianity and later became the driving force behind the creation of the branch of Christianity known as Protestantism (see Figure 3.4b).

The last of the major groups to migrate into Europe were the **Slavs,** who began arriving in the A.D. 400s. During the next few centuries, Slavs pushed as far west as the Elbe River in the middle of modern Germany and as far south as the Adriatic coast and into the Balkan peninsula, threatening the Greeks. The Slavs are divided into three major groups: western, southern, and eastern. Poles, Czechs, and Slovaks are western Slavs. Slovenes, Croats, Serbs, and Bulgarians are southern Slavs. The eastern Slavs are Russians, Ukrainians, and Belarussians (see Chapter 4).

The groups thus far mentioned account for most of the modern European peoples and represent considerable diversity. Yet a few other groups migrated into Europe at this early stage and left their mark. Latvians, Lithuanians, Estonians, Finns, Hungarians, Albanians, Roma (Gypsies), and Basques are the most well known.

The Rise of European Global Power

Our global economy is fundamentally free market, or capitalist, in nature. **Capitalism,** the practice of individuals and corporations owning businesses and keeping profits, traces its origins to Mediterranean and Western Europe. In the late 1400s, mercantile capitalism flourished as merchants invested in trade expeditions that brought profit in the form of precious metals (gold and silver). It began in 1418 when Prince Henry of Portugal established an institute at Sagres, where he brought together scholars to improve and teach the methods of navigation to Portuguese sea captains. He hoped that better skills would lead to the discovery of a sailing route around Africa and on to the Spice Islands in the east. In 1441, a ship finally sailed far enough south to reach wetter parts of Africa south of the Sahara. Portuguese explorers captured men and women for slavery and found gold. News of this event sparked enthusiasm for exploration, and many new voyages were launched. The Portuguese and Spanish, later followed by Western Europeans, discovered and conquered new lands, radically altering and frequently destroying many local economies and cultures of indigenous peoples as they began a new era of **colonialism** and **imperialism.**

The best known of the new voyages was made by Christopher Columbus in 1492. Funded by the Spanish crown, Columbus sailed westward in the belief he would get to India by a quicker route. He did not know that the Americas (supposedly later named after another Italian sailor, Amerigo Vespucci) lay in his path. At the same time, Vasco da Gama led the Portuguese explorations around the southern tip of Africa to India. Portugal and Spain both gained huge wealth from trading with and colonizing the Americas, Africa, and parts of Asia. The French, Dutch, and British followed them in the 1500s and 1600s. Wars in Europe and declining home economic bases reduced the roles of Spain and Portugal, whose rulers spent their New World wealth on armies to maintain their positions in Europe. Home-based merchant wealth shifted power toward the northWestern European countries. The Dutch emerged as a maritime power in the 1600s, establishing colonies in the East and West Indies but losing territory in North America to Britain in the 1660s. Over the following century, Britain won a competition with France for supremacy in North America and India, setting the basis for its worldwide empire.

European exploration not only resulted in the spread of European culture and practices. Europeans experienced many things in their travels and brought much back to Europe that they incorporated into their cultures. European diet is one aspect of culture that changed dramatically because of exploration. For example, what would the Irish, German, and Polish cultures be without the potato? Yet the Europeans knew nothing of this tuber until they traveled to the Americas. The potato, along with maize (corn), is significant because it has much greater yields than the grains grown in Europe until that time. Subsequently, the potato contributed to population growth that fueled further exploration, migration, and even the Industrial Revolution. Through this exploration, the Europeans learned about cotton, tobacco, tea, cocoa, and a wide range of other products that they commonly use today. From their encounters with the Chinese, they learned about such technologies as the compass, gunpowder, paper, and printing.

The Europeans were able to explore rapidly and build trading networks by exploiting Arab, Indian, and Chinese ports and links.

Industrial Revolution

The growing overseas trade and merchant wealth of the countries of northwestern Europe led to increasing demands for manufactured goods, stimulating a series of technological innovations and organizational changes. The resources of cottage weavers and blacksmith forges were hard pressed to fill the expanding markets for cloth and metal goods. From the mid-1700s, machinery, at first driven by waterpower, increased productivity in the metal and textile industries. The concentration of machines in factories that were soon powered by steam from coal-burning furnaces required growing amounts of capital and numbers of workers. This led to the expansion of urban–industrial centers on or near coalfields (Figure 3.5) and to systems of

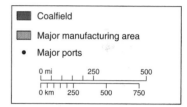

FIGURE 3.5 Europe: major manufacturing areas at the start of the 1900s. Compare their distribution with the late 1900s distribution of population (see Figure 3.23). Note: The contemporary political boundaries are drawn for reference.

banking and investment. Rivers, canals, and the sea were the initial forms of transportation used to assemble raw materials for the new industries and to distribute their products. Huge port facilities grew in the estuaries of major European rivers. During the 1800s, railroads increasingly gained in prominence, connecting ports and coalfield industries.

Beginning in Great Britain, the **Industrial Revolution** spread across the English Channel to the Netherlands, Belgium, northern France, and the western areas of Germany. In the late 1800s and 1900s, the Industrial Revolution diffused further to Central and Eastern Europe and to other areas of the world. Factories for metal smelting and fashioning, textile machinery, steam engines, and chemical refining were located in areas with plentiful coal resources. The European empires did not transfer their new manufacturing technologies to their colonies. Instead, they used their colonies to produce the raw materials needed in Western European factories. Cotton, wool, indigo, tobacco, and foodstuffs are a few examples. Forced to supply Western Europe with raw materials and having Europe as the only source of finished products, the colonies were pushed into economic dependence. Local economies and traditional ways of life were brought to an end as peoples in the colonies had to change their lives radically to produce exports for Europe.

Modern Countries: Nation–States

The current system of international relations and the character of peoples' identities began in Europe. Beginning shortly after A.D. 1000, the identities of Europeans changed as people began to shift their loyalties from more local feudal leaders to their emerging countries. An important step came with the Treaty of Westphalia in 1648 that marked the end of the Thirty Years' War, a bitter religious war between Roman Catholic and Protestant leaders. The treaty laid down rules for religious toleration and ended the arbitrary behavior of monarchs by establishing a legal system of international relations. Monarchs retained considerable power but had to explain to their people why their actions were in the best interests of their country rather than in their own personal interests. Though the rules of the Treaty of Westphalia have been modified, the current system of international relations is still referred to as the "Westphalian system."

The Westphalian system led to the development of **nation–states,** first in Europe and then in much of the rest of the world. As noted in Chapter 2, a nation is an "imagined community" of people who believe themselves to share common cultural characteristics; **states,** often called "countries," are politically organized territories with independent governments. Europeans developed the idea of the nation and combined it with that of the state to form the **nation–state ideal,** which is the belief that each nation should be free to govern itself and can do so only if it has its own state (country). Thus nations become linked with states. However, the nation–state ideal is just that, an ideal and not reality in most cases. The French nation, for example, believes that the French state should be solely inhabited by the French nation. However, France contains peoples such as the Basques, who regard themselves as a separate nation. Since the age of nationalism beginning in the late 1700s, many dominant European nations imposed their national cultures on other nations living within their borders in an attempt to make the nation–state ideal a reality. In response, many minority peoples resisted, and some—like the Basques—have tried to establish their own nation–states. The nation-building projects of dominant peoples who try to homogenize their states and of minority peoples who try to declare their own nation–states resulted in much violence, even war.

To use the terms nation, state, and nation–state correctly, remember that nations are peoples and states are countries. Thus, for example, the French are a nation and France is a state. Because France is inhabited mainly by the French nation, it is a specific kind of state known as a nation–state—that is, a state inhabited by or intended for a single people. It is incorrect to refer to France as a nation because France is not a people. However, France can be called a country, a synonym for state. Indeed, some prefer the term country instead of nation–state because few nation–states truly fit the definition of the term.

The grouping of peoples into nations was brought about not only by the Westphalian system but also by technological innovations such as the printing press, gunpowder, and modern militaries. The scientific ideas of the Enlightenment contributed greatly, too. After Gutenberg invented movable type for the printing press in 1447, books became common. One no longer had to be wealthy to obtain books. As literacy rose in conjunction with Enlightenment ideas of individuality, freedom, and rational thought, common people saw fewer differences between themselves and their rulers. By the 1700s, many questioned the privileged positions of the aristocracy, turning their loyalties instead to one another and their countries.

As literacy spread, governments had to choose language standards. Regional dialects broke down as people throughout their countries conformed to the same rules of spelling and writing. Until this time, the dialects of any language were so distinct that people could hardly communicate with one another, though they spoke the same language. Moreover, dialects frequently changed every few miles. The printing press and language standardization made it possible for larger groups of people to communicate with one another. In many countries, a Bible translation was the main means of spreading the new language standard.

Gunpowder, too, helped bind people together. It led to the development of rifles and cannons that required soldiers to drill together constantly so that they could work as a team. The individuality of medieval knighthood was replaced by modern armies of soldiers who dressed and acted alike.

Though the new technologies and Enlightenment ideas changed society, people were reluctant to challenge the supposedly divine authority of monarchs. After people in the 13 British colonies in North America rebelled against the British monarch and defeated the British military, people in Europe saw that their monarchs were not invincible, protected by the hand of God. The American Revolution soon inspired the French Revolution. Napoléon Bonaparte fed the new nationalist zeal in France, formed a new national army, and defeated the imperial forces of Europe. It required the British and Prussian national armies to defeat Napoléon's national army. Though Napoléon was defeated, he demonstrated that the nation–state, with a population that saw itself as one people and supportive of the state that represented them, was an efficient form of government. The death knell was rung for empires, city–states, and the like, though many hung on for another hundred years before being replaced by nation–states.

Nationalism and World Wars

As nationalism grew in Europe, it was fanned by the competitive nature of capitalism. Economic competition turned into nationalist competition. Before long, the armies that Western Europeans created to conquer and colonize the rest of the world were turned toward one another. In 1914 war erupted between the European powers, later to be known as World War I. Though Germany and Austria-Hungary were decisively defeated in 1918, the trouble was not over. The war costs and protectionism caused European economies to slump in the postwar 1920s and 1930s, bringing hardship to millions of individuals and families. Discontent and resentment grew in the defeated countries. The nationalist competition became more bitter and fed the more extreme but opposing ideologies of fascism and communism, two other concepts Europe gave to the world. A European war engulfed the rest of the world once again between 1939 and 1945, known as World War II. The intolerant side of nationalism under fascism led to the extermination of millions of people of specific groups such as Jews and Roma (Gypsies), a phenomenon called **genocide.**

Test Your Understanding
3A

Summary

Europe set into motion many of the global processes that we experience today. Although individual countries are now less powerful than they were early in the 1900s, many European cultural characteristics are part of other world cultures.

The current human geography of Europe grew out of varied influences, including the movements of Mediterranean, Germanic, Slavic, and Asian peoples. Europe developed global trading links from the 1400s and dominated the Industrial Revolution of the 1700s and 1800s. Europe also developed new political ideas concerning governance.

Questions to Think About

3A.1   What were Europe's historic contributions to the development of the world's present economic order? How did they affect Europe Itself?

3A.2   What were Europe's historic contributions to the development of the world's present political order? How did they affect Europe itself?

3A.3   Which factors promoted European migration and colonization from the 1600s to the 1800s?

3A.4   What led to the development of nations and the nation–states?

Key Terms

Greeks	colonialism
Celts	imperialism
Romans	Industrial Revolution
Germanic peoples	nation–state
Vikings	state
Slavs	nation–state ideal
capitalism	genocide

Natural Environment

The geography of Europe is marked by closeness to the ocean and rapid changes in physical landscape over short distances, factors that influenced human actions over time. The natural environment, particularly the temperate climates, the small scale of geologic provinces, and the long, indented coastline, contributes to the abrupt geographic differences from place to place.

Midlatitude West Coast Climates

Europe's north–south extent is slightly longer than its east–west extent. This is reflected in its range of climatic environments (Figure 3.6). Most are **midlatitude west coast climates** ranging from the icy northern coasts of Norway near the Arctic Circle and the summer midnight sun to the southern Mediterranean warmth of Portugal. No part of Europe is more than 500 km (320 mi) from the coast, allowing mild, humid oceanic atmospheric influences brought by westerly winds to affect the whole region.

The far north areas of Norway, Sweden, and Finland have polar climates, with long, very cold winters and months of snow cover. The coldest weather is somewhat ameliorated along the coasts of these northern lands with air warmed by the North Atlantic Drift current, which brings warm ocean waters across the North Atlantic from the Gulf Stream. Most of the western coastal countries from Norway to Portugal have mild winters and warm summers with precipitation throughout the year. Much of their warmth and humidity comes from the winds that bring North Atlantic maritime air.

In the far south along the Mediterranean coastlands, there is a seasonal contrast in which the midlatitude belt of cyclones moves southward in winter, bringing rain and wind, while the hot, dry air from the Sahara to the south creates droughtlike conditions in the summer. This southern variant is known as a **mediterranean climate.**

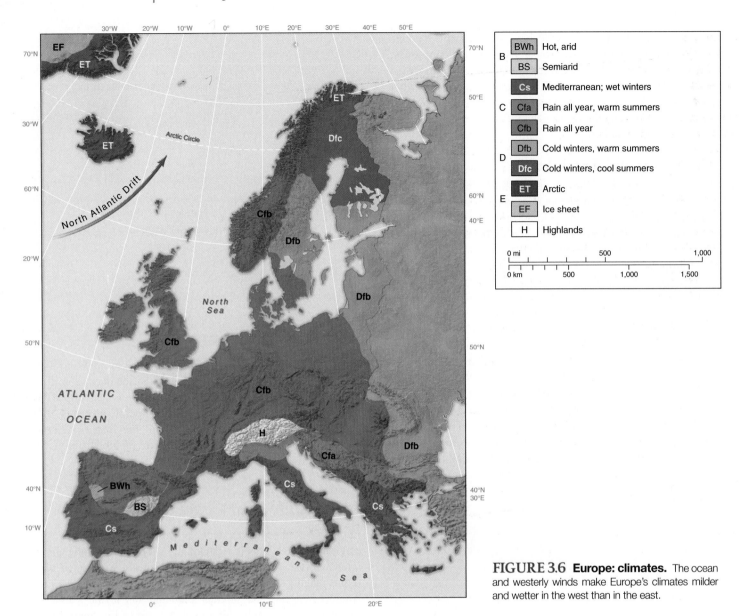

FIGURE 3.6 Europe: climates. The ocean and westerly winds make Europe's climates milder and wetter in the west than in the east.

In the **midlatitude continental interior climate** of Central and Eastern Europe, winters can be severe as cold winds bring freezing temperatures from Russia and the northern countries. Summers are warmer than on the coasts and are marked by thunderstorms.

Climate changes affected Europe during the period of human occupation from the later part of the Pleistocene Ice Age. As climatic conditions warmed, the ice sheets, which had extended as far south as the Thames River valley in England and the plains of northern Germany and Poland, melted and retreated northward. Climates throughout Europe got warmer until some 5,000 years ago, when they reached their warmest point. After that time, conditions fluctuated with warmer periods in Roman times, from A.D. 900 to 1200, and since 1850 were punctuated by colder periods. The cooling in the 1300s had much to do with the end of Greenland colonies and with

forcing farmers in Europe to lower the upper level of cultivation on hills. Since the mid-1800s, the retreating glaciers in alpine valleys (Figure 3.7) influenced the tourist industry and the generation of hydroelectricity. The retreating glaciers and decreasing extent of snow cover reduced the winter sports season but increased the summer holiday warmth. Inlets to hydroelectricity projects are now sited higher up the valleys, increasing the elevation (and energy) of water falling on the electricity generating turbines.

Although lands around the Baltic Sea continue to rise gradually above sea level in response to the postglacial melting of the ice sheet, the weight of which caused that area to subside, most of Europe is concerned about the potential impacts of global warming. Rising sea levels would affect the extensive low-lying parts of Europe, including the many areas of former coastal wetland that have been reclaimed an

(a)

(b)

FIGURE 3.7 Europe: changing climate. A hundred years has made a difference in Zermatt, Switzerland. (a) The glacier shown on the left side of the 1880 painting disappeared from the modern photo (b), having retreated over a kilometer up the valley, leaving bare rock and icemelt deposits exposed on the valley floor. Locate the church in both views and compare other evidence of the 100-year differences between the two pictures. How might the changing climate have affected tourism? **Photos:** © Photo Klopfenstein-Adelboden.

intensively populated. The Netherlands has particular worries, but attention has also focused on the plight of Venice in northern Italy. Venice, a medieval trading city, is one of Europe's greatest architectural treasures and tourist attractions, but it is gradually subsiding (Figure 3.8). Increasingly frequent high tides weaken its foundations. In 1990 St. Mark's Square was flooded 9 times, and in 2000 it was flooded 90 times. In 1966 a south wind raised the tide level by 2 meters (6 ft.), covering St. Mark's Square with oil-polluted water that left stains on all the buildings. As Venice's buildings became less secure, the city's population fell from 175,000 in 1950 to 80,000 in 1990. The industrial development of the lagoon that backs Venice is a major part of the problem because its industries and oil refineries pollute the waters and make protection of Venice difficult in the face of demands for deepwater channel access.

FIGURE 3.8 Venice. A medieval port city that accumulated great wealth on its unique site in a lagoon. Venice is prone to flooding, as seen here. **Photo:** © Michael Bradshaw.

Geologic Variety

Within its relatively small area, Europe includes almost the world's entire range of geologic features. There are ancient shield areas around the Baltic Sea, the uplands of central Europe, the young folded mountains of the Alps, and the extensive plains in countries around the North and Baltic Seas (see Figure 3.3). Volcanoes and earthquakes are active along the Mediterranean Sea.

The Mediterranean Sea is the remnant of a larger ocean that occupied the area between Africa and Europe but closed as the two continents clashed along a convergent tectonic plate margin (Figure 3.9). Within this zone, the young folded mountains of the Alps form the highest ranges, with many peaks rising above 4,000 m (13,000 ft.). The highest point is Mount Blanc (4,807 m, 15,771 ft.), found on the French–Italian border. Farther east, the ranges are not so high in Austria, where few peaks exceed 3,000 m (10,000 ft.).

The high Alps are part of a series of ranges that includes the Sierra Nevada in southern Spain; the Pyrenees between France and Spain; the Apennines that form the Italian peninsula; the coastal ranges of the Dinaric Alps in Croatia, Bosnia, Macedonia, and Albania; and the Pindus in the Greek Peninsula. The curve of the Carpathian Mountains in Slovakia and northern Romania, continued in the Balkan Mountains of Bulgaria, forms a further extension. Such ranges create the dominantly mountainous environment of alpine Europe and the hilly environments of Mediterranean Europe and the southern Balkans. Lowland areas in these mountainous areas are restricted to river plains and deltas formed by the deposition of rock fragments and particles worn by erosion from the uplands.

To the north of the young folded mountains, most of Europe is lowland, with extensive areas less than 300 m (1,000 ft.) above sea level. The lowlands dominate northern Germany, Poland, and the Baltic countries—part of the North European Plain that continues eastward into Russia. They are formed of more recent layers of rock, covered in the north by glacial deposits.

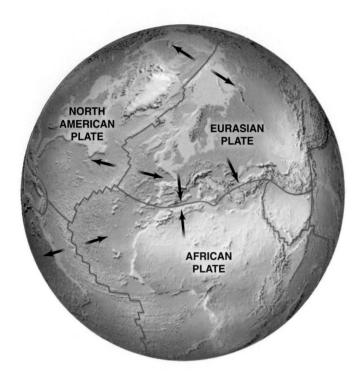

FIGURE 3.9 Europe and plate tectonics boundaries (red). The relationship of major relief features to plate margins. The southern margin of the continent is part of a convergent plate boundary that was responsible for raising the Alps and other mountain ranges along the Mediterranean Sea. The west-facing coasts result from the opening of the Atlantic Ocean with a divergent boundary in the Mid-Atlantic Ridge that surfaces with volcanic activity in Iceland.

The lowlands are framed by hilly areas of older rocks that once formed mountain ranges but were worn down by erosion and then raised again by faulting. These include much of Spain and Portugal (the Meseta), France's Massif Central and Brittany areas, the Rhine Highlands of southern Germany, the Bohemian Massif of the Czech Republic, the uplands of western and northern Britain, and those of Norway. As the Atlantic Ocean opened along a divergent plate margin, uplift occurred along continental margins, together with volcanic activity. Rifting extended through the North Sea area, providing a downfaulted block of rocks that became oil reservoirs.

Long Coastlines and Navigable Rivers

Europe is marked by peninsulas such as Scandinavia (Norway and Sweden), Jutland (Denmark), and Brittany (France) in the north, and Iberia (Spain and Portugal), Italy, and Greece in the south. The British Isles formed another peninsula until waves cut the Strait of Dover a few thousand years ago. Arms of the ocean, such as the Baltic, North, Mediterranean, Adriatic, and Aegean Seas, reach far inland. Smaller islands in the Baltic and Mediterranean Seas add to the length of coastline that encouraged many groups to engage in trade and develop ship technology at a time in history when water transportation was easier

than land transportation. Many **estuaries,** where rivers meet the seas, facilitated port building. Skills and technology of that phase were available to expand mercantile capitalism around the globe between A.D. 1450 and 1750.

On land, connections between places and world trade routes were eased by the existence of major river valleys. The Rhine and Elbe Rivers in Germany and the Danube River flowing from Germany through Austria and the Balkans were particularly important. The Rhone, Seine, and Loire Rivers in France, the Thames River in England, the Vistula River in Poland, and the Po River in Italy were also significant in movements of people and goods from early times. River valleys large and small also provided good soils. Sites for towns at river crossings attracted and concentrated the settlement of growing populations. Modern investments in the Euroports at the mouths of the Rhine and Rhone Rivers are important for maintaining Europe's continuing role in world trade.

The Danube River is the longest in this region but is less used for transportation than the Rhine because it flows through less economically developed countries. In the last 50 years, efforts to coordinate the management of Danube waters for transportation and hydroelectricity proved difficult. The Danube flows through more countries (eight) than any other major world river, and the Cold War limited traffic eastward from Austria.

The Rhine is Europe's second longest river and the world's busiest waterway (Figure 3.10), used annually by some 10,000 ships that carry 250 million tons of cargo. The watershed includes significant parts of four countries (Switzerland, Germany, France, and the Netherlands). The river waters come from melting snow and ice in Switzerland and from tributaries (Aar, Neckar, Main, and Mosel). The Rhine has major roles in transportation, industrial water supply, and, in the Swiss sector, the generation of hydroelectricity.

Some of the main efforts at managing the Rhine waterway were devoted to making navigation possible for large barges from the international port of Rotterdam at its mouth up to Basel, Switzerland. The river was canalized to that point. As canalization proceeded, rivers were straightened, producing faster flow, channel bed erosion in some sections, and silt deposition in others. Channel deposition partly filled some sectors and caused flooding. Higher levees were constructed to protect from flooding the lands on either side that were used more and more intensively. At the Rhine mouth in the Netherlands, greater efforts were put into keeping out the sea than to maintaining the levees, which were breached in places during the high river levels of early 1995.

The transportation uses of the Rhine River are linked to the growing industrialization of its watershed since the late 1800s. The coalfield and steelmaking areas of the Ruhr and Saar are now less productive and polluting, but some 20 percent of the world's chemical industry output occurs along the river with major centers around Basel, Mannheim, and the Ruhr area. Pollution was at its worst in the 1970s; since then, an international

nutrients through an annual leaf fall raised the quality of soils. The most fertile soils developed on river deposits and on the loess (windblown glacial debris) covering the common limestone rocks in the northern plains. Fir and pine trees growing on sandy soils and on the thin soils of uplands tended to lower the quality of their soils by acidification.

Because clearing light woodland was easier, initial settlement favored thinly vegetated uplands on limestone, sandstone, and some granite rocks during the period of warming climate to around 5,000 years ago. Also favored were the easily cultivated **loess** soils that mantled the southern edge of the North European Plain from Poland to northern France and southern Britain. They provided an easy route of diffusion for farming technology originating around the Mediterranean Sea, via the Danube River valley.

After the introduction of Iron Age tools, human groups rapidly cut into the denser forest on the lowlands. By the time of the Roman occupation of France and lowland Britain, a large proportion of the forest on the lighter lime-rich soils had been cut and some inroads made to the denser forest on heavier soils. Further expansion of the cultivated area onto the clay soils occurred during the Middle Ages.

The maritime emphasis of Europe extends to its exploitation of fish resources in the Mediterranean, Baltic, and North Seas and the North Atlantic Ocean. Fishing, related ports, and ships grew in significance in the later medieval period. In the 1900s, overfishing led to decreasing supplies and made sea fishing a source of contention and conflict among European countries.

Natural and Human Resources

Europe contains the natural mineral resources that formed the basis of technological revolutions from the Bronze Age (3500–1000 B.C.) (tin and copper) to modern industrial revolutions (especially coal and iron). Its well-watered lands and moderate temperatures fostered tree cover and a large proportion of fertile soils that provided the basis for supporting relatively high densities of people at each successive stage of history.

By late medieval times, water channels were used widely to generate power in mills and furnaces. Later, the presence of upland areas with high levels of precipitation and snow or ice storage made it possible for engineers to develop hydroelectricity resources there. As populations grew, the development of technologies to support economic growth expanded human resources, making it more likely that the new ideas would spawn further changes.

The rivers and extensive coastline encouraged trade and the exchange of ideas across the region. In the 1800s and 1900s, the human landscapes and historic buildings became resources that made Europe the world's major center of tourism. At the same time, some industrial resources that had been important declined in significance. Coal was replaced by oil and natural gas because of high costs of underground mining and the environmental impacts of burning coal.

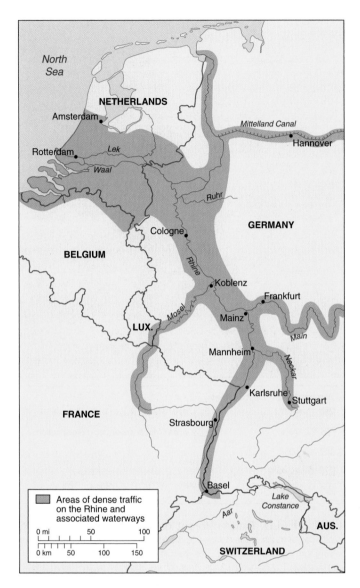

FIGURE 3.10 Europe: Rhine River and associated waterways. Compare the impacts of the industrial areas along the river's length. Rotterdam is Europe's number one port because of its location at the point where maximum Rhine River traffic meets the ocean, requiring change from barge to containers or bulk cargo ship. New canals increased the amount of trade passing through Rotterdam. **Source:** Reprinted and modified by permission from *Financial Times,* February 2, 1995.

agreement has been reached to reduce the problem, helped by improved water treatment technology. Concerns remain over nondegradable chemicals and metals that are still at high levels in the river.

Forests, Fertile Soils, and Marine Resources

Between 15,000 and 10,000 years ago, Europe was recolonized by forest as the ice sheets retreated northward. Human occupation began in a forested environment in which the circulation of

Environmental Issues

Europe's large populations and their development of technologies for exploiting environmental resources have had major impacts on the natural environment. Today, the wealthier countries are particularly concerned to maintain environmental quality, pouring billions of dollars into such measures.

From Forests to Farms

As soon as humans cut the forest, soil washed down hillsides more rapidly and contributed to the building of lowland river plains and the growth of deltas at their mouths. In the Middle Ages, for example, a combination of growing populations, rigid political systems, close grazing by sheep and goats, and climate change caused intense soil erosion on the hills of the Mediterranean peninsulas, together with the downstream extension of river plains into coastal deltas. Hilly areas of southern Italy and Greece were denuded of their soils, leaving rocky outcrops. In much of Europe outside the alpine area, however, the slopes are less steep, and cultivation methods were adopted that maintained the soils and their productivity over many centuries. In the 1800s, competition from cheap grain imported from newly opened and settled lands in North and South America resulted in once-plowed lands in Europe being sown with grass for livestock production, further reducing soil erosion and helping to maintain soil quality.

Impacts of Industrialization

The Industrial Revolution and the spread of factory-concentrated production led to widespread pollution of the rivers and air. Many rivers lost their fish stocks. Occasional major pollution incidents still kill fish in major rivers such as the Rhine. Great efforts are being made to improve the quality of river water and the European Union sets rising standards to be attained by specific dates. In the Thames River of England, the reduction of pollution was so successful that fish stocks revived in the 1990s after decades of absence.

The winter smoke fogs (smog) that blighted the major industrial and urban areas of Europe in the 1950s were controlled by legislated reductions in coal burning in some countries. More recent air pollution occurs where high densities of road traffic pour sulfur, nitrogen oxides, and carbon particles into the air. Such emissions react with sunlight to lower air quality. Under meteorological conditions of slow-moving air, pollutants may accumulate to dangerous levels in broad valleys.

Emissions of sulfur compounds from thermal power stations, especially those burning coal, create **acid deposition**—of dry particles near the source or of wet "acid rain" farther downwind (Figure 3.11). Acid deposition occurs around all the main industrial areas of Europe. It affects areas of coniferous forest, thin soils, and shallow lakes in the Alps and Scandinavia. The highest levels of acid pollution were in southern Poland close to the poorly regulated coal mines of the Communist era.

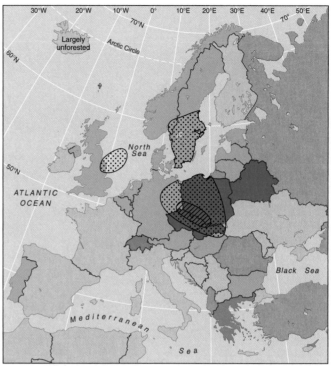

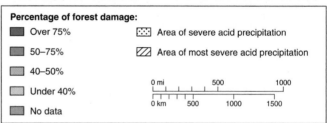

Percentage of forest damage:

- ■ Over 75%
- ■ 50–75%
- ■ 40–50%
- □ Under 40%
- ▨ No data

- ⬚ Area of severe acid precipitation
- ▨ Area of most severe acid precipitation

FIGURE 3.11 Europe: forest damage as of 1998. Westerly winds cause much of the damage to the east of major industrial areas. Most severe damage is in the Black Triangle. Compare to Figure 3.5. **Source:** Jordon-Bychkov, *The European Culture Area*, 4th ed. Reprinted by permission of Rowman & Littlefield.

Awareness of environmental problems led to legislation and institutions to monitor and fight air and water pollution, especially in non-Communist Europe. Although road congestion increased with more cars and trucks, new cars in the 1990s emitted 93 percent less carbon dioxide and 85 percent less hydrocarbons and nitrogen gases than in 1970. Lighter vehicles, electronic engine management, more economical engines, and the retiring of older cars will continue to reduce atmospheric pollution.

East Central Europe

Environmental degradation was a legacy of Communist governments in East Central Europe. Manufacturing industries did not have to adopt procedures to reduce air and water pollution. Industrialized areas were badly affected by the burning of low-quality coal that emitted high proportions of sulfur and particles into the atmosphere (Figure 3.12). Chemical works

(a)

(b)

FIGURE 3.12 **Environmental degradation in East Central Europe.** (a) Kraków-Nowa Huta, Poland. Sędzimir Steel Mill, formerly the Lenin Steel Mill. Communist governments rarely required any kind of pollution controls, and little is different for this factory today. The cemetery in the foreground is a telling commentary. (b) Pollution levels dropped in many areas of East Central Europe with the end of Communism in 1991, after many inefficient factories shut down because they could not compete in the global market. UN funds were provided to shut down this heavily polluting factory in Copşa Mica, Transylvania, Romania. Considered the most polluted place in Europe, local people called it "black town" because everything was black from the air and sky to laundry on clothes' lines and children's faces. Shown a few years after operations ceased, the sky is now blue and the grass green again. **Photos:** (a) © Jerzy Jemiolo; (b) © George W. White.

emissions polluted the rivers. When compared to standards of the Environmental Protection Agency in the United States, levels of chemical compounds found in the air and water were hundreds, sometimes thousands of times higher than what is recommended as safe. One of the worst areas lay along the Czech, Polish, and East German borders. Known as the **Black Triangle,** this area witnessed the death of as much as 90 percent of forests from coal mining and other heavy industries in some locales.

Global Environmental Action

The negative environmental effects of industrialization have motivated Europeans to engage in global efforts to reduce pollution. Representatives from European governments were active participants in the Rio Earth Summit in 1992 and the meeting in Kyoto, Japan, in 1997 (see Chapter 2, p. 58). European countries comprise most of the countries on Annex I of the Kyoto Protocol. Collectively, they account for more than 32 percent of carbon dioxide (CO_2) emissions on the list, meaning that their ratification was key to gaining the number of countries totaling the 55 percent of emissions necessary for the protocol to go into effect. Though their listing on Annex I means that they are industrial countries primarily responsible for the emission of greenhouse gases, European countries were willing to ratify the Kyoto Protocol and reduce their emissions. Indeed, Romania was the first Annex I country to ratify the protocol

(32nd country overall to ratify). When Iceland ratified the protocol in 2002, it was the 55th country overall to ratify the protocol, thus satisfying the first requirement for the protocol to go into effect. European enthusiasm for the Kyoto Protocol led to a joint statement of the European Union and Japan (74th to ratify and a key country on Annex I) calling for other countries to ratify the protocol. European countries were also instrumental in motivating the Russian Federation to ratify, which it did in November 2004, likewise making the protocol go into effect by satisfying the protocol's second requirement that Annex I countries representing 55 percent of the carbon dioxide (CO_2) emissions of the Annex I list ratify the protocol.

As countries went through the process of ratifying the Kyoto Protocol, European countries began taking effective action to lower their carbon dioxide (CO_2) emissions. By 1999, Germany and the United Kingdom were the most successful of European Union countries. Germany succeeded in lowering its emissions by almost 16 percent of its obligated 21 percent. The United Kingdom achieved 8.4 of its 12.5 percent obligation. Overall, the European Union lowered its carbon dioxide emissions by 1.4 percent, with an 8 percent target. Most successful were East Central European countries, which lowered their emissions from 27 to 68 percent, though their targets were 6 to 8 percent. Success was primarily achieved following the end of Communist governments beginning in 1989 and the subsequent economic slowdown that shut many factories down.

Mediterranean Sea

The growth of population in the 1900s, combined with crowding into coastal locations, industrialization, and the great increase in tourism, led to pollution of the Mediterranean Sea—an important issue for many prospective tourists (Figure 3.13). The Mediterranean Sea is also one of the world's major shipping lanes and has the world's highest level of oil pollution. It is an almost closed sea with hardly any connections to oceans where water mixing dilutes pollution. Confined within the Mediterranean, the chemicals and other nutrients from pollution decay and thereby deplete oxygen, killing sea life and creating large algae blooms that thrive in the anaerobic conditions. The Adriatic Sea has some particularly large algae blooms. The most polluted European areas of the sea are those near urban–industrial areas, such as Barcelona, Marseilles, Genoa, Naples, and Athens.

In Spain, for example, the coastal population increased from 12 percent of the national total in 1900 to 35 percent in the 1990s, when the annual influx of tourists added several million visitors. Although the Mediterranean countries of Europe have populations that are not increasing in total, tourism and the movement to the coastal areas continue to grow. Tourists arrive mainly in July and August, when water is short and local sewage systems find it hard to cope.

The North African and Asian Mediterranean countries—which also pollute the largely enclosed sea—are experiencing major population increases (see Chapter 8). For example, the cities along the African coast are likely to double their populations by 2025. Tourism around the Mediterranean Sea is expected to grow from around 100 million local and foreign visitors in the early 1990s to between 170 and 340 million.

Following a 1975 UN conference, the governments around the Mediterranean Sea tackled their pollution problem through the Mediterranean Action Plan. Sewage treatment improved in the wealthier countries such as France, but the poorer countries of North Africa cannot afford the necessary investment. Even if pollution is reduced as countries get wealthier, the coastline and its delicate ecosystem are changed irrevocably when coastal wetlands are reclaimed and built over.

Waste Management

One of the major environmental concerns in Europe is the disposal of rising quantities of trash and industrial waste. Since 1950, the amount of waste per head doubled, although it is still less than half that in the United States. Europe, however, has more people on less land. Industries produce increasing quantities of hazardous wastes, ranging from poisons to waste oils, heavy metals, and radioactive substances, some of which remain toxic for long periods.

Waste management is becoming a growth industry, requiring capital and high technology. As environmental legislation increases, large-scale management corporations will be required. At present, most trash is still buried in landfill sites, the management of which is being improved to exclude gases that might cause combustion or to tap them for use in power generation. Incineration is increasing to reduce the bulk of waste, but it still leaves ashes and emits polluting gases. Recycling is increasingly popular, but the processes are often costly. Recent legislation in Germany obliged companies to take back the packaging used in their products, but the outcome overloaded the German recycling system, creating large "mountains" of paper, glass, and plastic, together with exports of waste to other countries, using up their own capacity.

FIGURE 3.13 Marbella, Spain. Crowded beach on the Mediterranean coast welcomes millions of tourists, particularly from Northern Europe, during the summer months. **Photo:** © Alasdair Drysdale.

Test Your Understanding
3B

Summary

Europe has a mainly coastal environment with midlatitude climates that include cold polar varieties in the north, all-year mild and humid conditions in the west, more continental extreme conditions toward the east, and summer droughts around the Mediterranean Sea.

The physical landscapes of Europe include alpine, lowland, hilly plateau, glaciated, and river plain areas. Lowland and better soils predominate between the northern and alpine mountains. The rocks contain energy and metallic mineral resources. Easy access to ocean transportation was an important factor in Europe's economic growth.

The postglacial forest vegetation that covered the region was largely removed from the areas of better soil and replaced by cultivation and urban–industrial areas. Soil erosion, water pollution, and air pollution resulted from intensive use of the natural environment.

Questions to Think About

3B.1   How does Europe's climate change from north to south? From east to west?

3B.2   Which of Europe's rivers are longest and most navigable?

3B.3   What impact has industrialization had on Europe's natural environment?

3B.4   How supportive are European countries of the Kyoto Protocol?

Key Terms

midlatitude west coast climate

mediterranean climate

midlatitude continental interior climate

estuary

loess

acid deposition

Black Triangle

Global Changes and Local Responses

Europe after 1945

After World War II (1939–1945), Europeans seriously reevaluated their role in the world and their relationships with one another. The war was so devastating that even the winners suffered destruction and huge financial, political, and cultural losses. In Western Europe, the U.K., France, and the Netherlands were confronted with independence movements in their colonies at a time of weakness. Commonly fueled by the ideologies of nationalism and communism that originally came from Europe, four centuries of building colonial empires on which "the sun would never set" ended.

At the same time, the politically and economically weak European countries faced the United States and the Soviet Union, which emerged as new world powers. At the end of World War II, the Soviet Union showed its strength when the Red Army moved into most of the East Central European countries and fostered the establishment of **Communism.** The Communists believed that capitalists used their riches to manipulate their governments in order to protect, even increase, their privileged positions in society and keep the majority of society, especially the working classes, powerless and in relative poverty. Instead, Communists argued for **democratic centralism:** the belief that the Communist Party, the political party of the working class, was the only true representative of the people and, therefore, the only party with the right to govern. To keep capitalists and others from taking advantage of the people, communists also believed in **state socialism:** governance by the Communist Party, actively running the political, social, and economic activities of the people. The state owned all the businesses and decided what was produced. The capitalist practice of competing companies producing similar products was seen as unnecessary. Rather, large corporations owned by the state made each product. The state, not the free market of consumers, decided what needed to be produced through a **planned economy.** At the same time, the other new superpower, the United States, supported the revival of the countries of Western Europe, where the Marshall Plan injected huge sums to assist economic recovery and where the threat of Soviet expansionism was met by the establishment of the North Atlantic Treaty Organization (NATO).

With Soviet Communism firmly established in East Central Europe by the early 1950s, many Europeans felt that their nationalist notions and capitalist practices were threatened. Europeans in countries about to lose their colonies knew that their countries were too small to compete individually with the Soviet Union and United States. Moreover, the two world wars revealed the ugly sides of nationalist political competition, economic protectionism, and capitalist competition. Competition could lead to failure as well as success. On the other hand, cooperation in a non-Communist form was thought to result in everyone's success.

NATO and the European Union

Immediate cooperation came about in non-Communist (Western, Northern, and Mediterranean) Europe with the **North Atlantic Treaty Organization (NATO)** in 1949 (Figure 3.14). NATO included the United States, which was seen as an ally against the Soviet military threat then dominating East Central Europe. After the breakup of the Soviet Union in 1991, many questioned the need for NATO. Others feared that Russia would eventually become a formidable power again, though the Soviet Union no longer existed and Russia was weak. Having emerged from more than 40 years of Soviet domination, East Central European countries were eager to join NATO. Russian objections and the cost of expansion delayed NATO expansion. NATO accepted Poland, the Czech Republic, and Hungary as new members and is considering other countries. In 2002 NATO formed a partnership with Russia, the country that NATO was created to defend against! Post–Cold War NATO is actually more focused on resolving or policing disputes within the expanded Europe and its immediate neighbors—as in Bosnia, Kosovo, and Macedonia.

To compete successfully again in the world economy over the long run, Europeans in the non-Communist countries created the European Economic Community (EEC), known today as the **European Union (EU)** (Figure 3.15). In 1949, not long after the end of World War II, Belgium, the Netherlands, and Luxembourg joined together in the Benelux customs union. In 1952 the Benelux countries joined France and West Germany to form the European Coal and Steel Community (ECSC). In 1957 the five ECSC countries plus Italy signed the Treaty of Rome, establishing the European Economic Community (EEC). The EEC was expected to create a common market in which goods, capital, people, and services moved freely between countries. The European Commission became the executive arm of the community

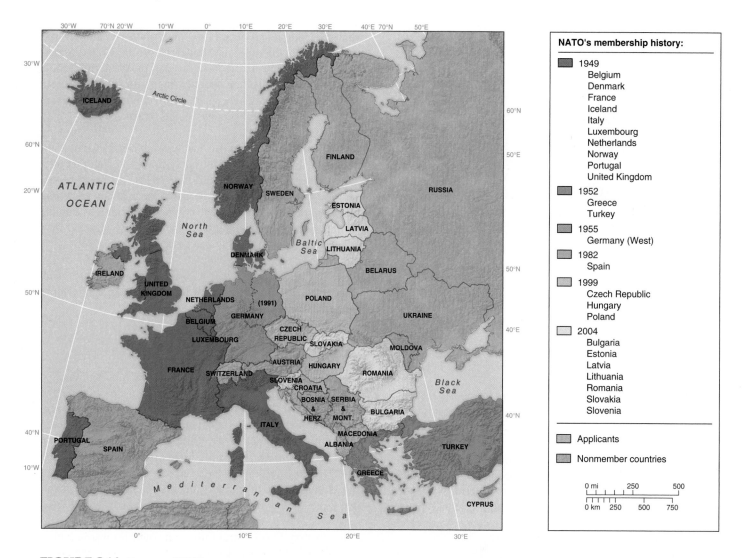

FIGURE 3.14 Europe: NATO. The North Atlantic Treaty Organization, showing countries that are members and those that are applying for membership. The United States and Canada are also original members. In 2002, Russia formed a special partnership with NATO.

and was based in Brussels, Belgium. In 1967 the EEC changed its name to the European Community (EC) to emphasize the move from economic toward political goals. Denmark, the Republic of Ireland, and the U.K. became members in 1973, Greece in 1981, and Portugal and Spain in 1986. By then the European Parliament, with elected members from all these countries, was created and located in Strasbourg, France. Other EU organs are headquartered in countries such as Luxembourg (Figure 3.16).

The Single European Act, signed in 1986 when Spain and Portugal joined, set out the steps needed to implement a single market. The Treaty of Maastricht (1991) attempted to set a timetable for monetary and political union and changed the name of the organization to European Union (EU) in 1993. Austria, Sweden, and Finland joined in 1995. In 2004, 10 countries of East Central Europe, many formerly part of the Soviet bloc, joined the EU in the organization's largest expansion to

date. Bulgaria and Romania hope to join the EU in 2007. Turkey's application has long been under consideration, and Croatia applied for membership in 2003. Many Europeans see the EU as good, while many others do not (see Point–Counterpoint: The European Union, p. 88).

The European Union represents yet another idea emanating from Europe: **supranationalism.** Supranationalism is the idea that differing nations can cooperate so closely for their mutual benefit that they can share the same government, economy (including currency), social policies, and military. During the Cold War Communism tried to offer a form of supranationalism, but after 1990 most Europeans abandoned the Communist experiment, leaving those in the EU as the primary advocates of supranationalism. Members of the EU are still working out the details of their cooperation, but what they have accomplished is remarkable, considering that the more predominant nationalist idea, subscribed to by most of

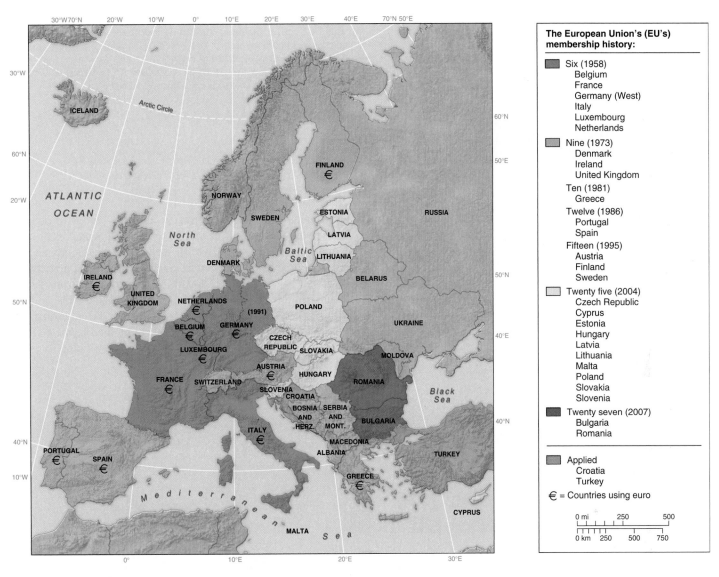

FIGURE 3.15 European Union, its growth to date.

The European Union's (EU's) membership history:

- Six (1958)
  - Belgium
  - France
  - Germany (West)
  - Italy
  - Luxembourg
  - Netherlands

- Nine (1973)
  - Denmark
  - Ireland
  - United Kingdom
  - Ten (1981)
    - Greece
  - Twelve (1986)
    - Portugal
    - Spain
  - Fifteen (1995)
    - Austria
    - Finland
    - Sweden

- Twenty five (2004)
  - Czech Republic
  - Cyprus
  - Estonia
  - Hungary
  - Latvia
  - Lithuania
  - Malta
  - Poland
  - Slovakia
  - Slovenia

- Twenty seven (2007)
  - Bulgaria
  - Romania

- Applied
  - Croatia
  - Turkey

€ = Countries using euro

FIGURE 3.16 In Luxembourg, the European Union flag (blue with yellow stars) proudly flies with seemingly equal importance as the national flag. The historic central part of the once strongly fortified Luxembourg City looms in the background. **Photo:** © David C. Johnson.

the world, holds that such cooperation is impossible between nations. Nationalism may still preclude total political union in Europe.

It remains to be seen if supranationalism will work. The combination of NATO security in the Cold War and linked economic policies enabled Europe, by the later 1900s, to regain a major place within the global economic and political system. For example, European countries have many of the highest GDPs per capita in the world (Figures 3.17 and 3.18).

Devolution within European Countries

As Europe moves toward greater economic and possibly political integration, it is simultaneously experiencing **devolution:** local peoples desiring less rule from their national

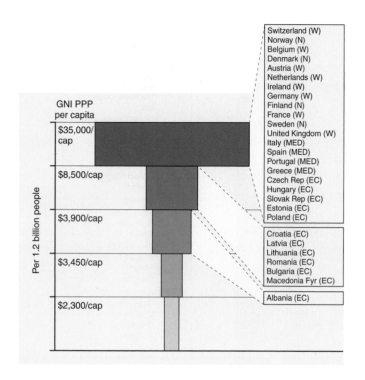

GNI PPP
per capita

Per 1.2 billion people

$35,000/cap

$8,500/cap

$3,900/cap

$3,450/cap

$2,300/cap

Switzerland (W)
Norway (N)
Belgium (W)
Denmark (N)
Austria (W)
Netherlands (W)
Ireland (W)
Germany (W)
Finland (N)
France (W)
Sweden (N)
United Kingdom (W)
Italy (MED)
Spain (MED)
Portugal (MED)
Greece (MED)
Czech Rep (EC)
Hungary (EC)
Slovak Rep (EC)
Estonia (EC)
Poland (EC)

Croatia (EC)
Latvia (EC)
Lithuania (EC)
Romania (EC)
Bulgaria (EC)
Macedonia Fyr (EC)

Albania (EC)

FIGURE 3.17 Europe: national incomes compared. The countries are listed in the order of their GNI PPP per capita. Compare the relative material wealth of the countries in Western (W) and Northern (N) Europe with those in Mediterranean (MED) and East Central (EC) Europe. **Source:** Data (for 2000) from *World Development Indicators,* World Bank, 2002.

governments and seeking greater authority in governing themselves. Devolution occurs in the United Kingdom, for example, where Scots and Welsh seek greater autonomy and recently obtained their own parliaments with limited powers. Estonia's, Latvia's, and Lithuania's independence from the Soviet Union and the breakup of Czechoslovakia and Yugoslavia were other forms of devolution. Devolution is also seen in the bloody campaigns fought by the Basque peoples straddling the Spain–France border and the Catholic, in Northern Ireland.

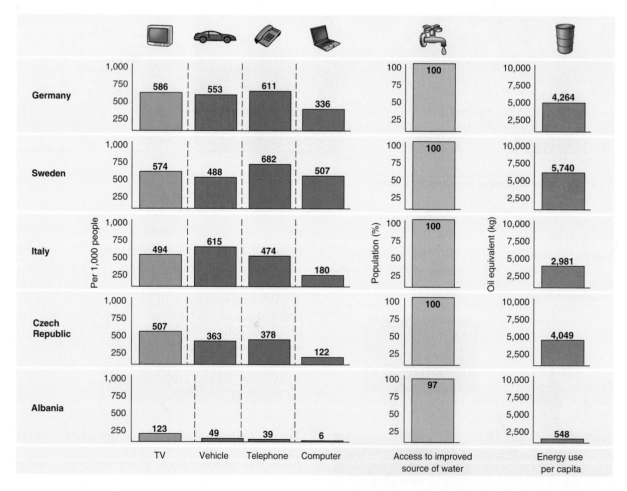

FIGURE 3.18 Europe: ownership of consumer goods. Note the diversity within Europe. Summarize the differences among the countries by subregion. **Source:** Data (for 2002) from *World Development Indicators,* World Bank, 2004.

Devolution pressures are not coming only from ethnic and national minorities. They are also coming from the people of provinces that straddle neighboring countries. For example, Strasbourg is the major city in the mid–Rhine Valley, providing goods and services that the smaller cities do not. Though Strasbourg is in France, it is on the German border, making it a city that is closer to many Germans than German cities of a similar size. Thus many Germans obtain goods and services in Strasbourg. High unemployment rates on the French side of the Rhine River and high-paying jobs in neighboring Germany lead to many French desiring to work in Germany. In the past, differing government policies of bordering countries—customs, border guards, different currencies—made it difficult for citizens to travel routinely to neighboring countries only a few miles away. Nevertheless, interaction was desired, and peoples of border regions have pressured their national governments to make daily commutes easier. To do this, people living in border areas have formed **Euroregions,** which straddle country boundaries. The people of Euroregions work to make transboundary movement easier within their Euroregions. For example, border controls between France and Germany no longer exist, and passes to museums on the French side of the Rhine boundary are valid in German and Swiss museums on the other side of the Rhine. Euroregional organizations also lobby their respective national governments for policies to facilitate movement. Euroregions are appearing in large numbers, now straddling almost every country boundary (Figure 3.19).

Euroregions are challenging national governments everywhere in Europe, and with support from the EU. In addition to the rise of Euroregions, many provinces of countries are exerting their authority. Almost all the provinces in Germany, for example, have offices in Brussels, where they lobby the EU directly for funds. The interests of Euroregions and countries' provinces are not the same as each other's or countries' interests. These devolutionary examples, along with the supranational examples from the previous section (such as the EU), show that the power of nation–states is being challenged from above and below. Indeed, these examples illustrate that many local voices are supporting supranational organizations like the EU with the intent of weakening nation–states to strengthen local governance. The nation–state has been a powerful force for the last 200 years, especially in Europe where it began. However, new kinds of governments and new relationships between levels of government are clearly being worked out.

Population Patterns

Dynamics

European countries are notable for their zero population growth rates (Figure 3.20). Europe's population is decreasing in most Mediterranean and East Central European countries and experiencing only small increases in most Western and Northern European countries (Table 3.2). Rates of natural population change were between -0.6 and 0.3 percent in 2000. Throughout Europe, total fertility rates declined from as high as three births per female in 1965 to commonly fewer than two in 2000. Italy

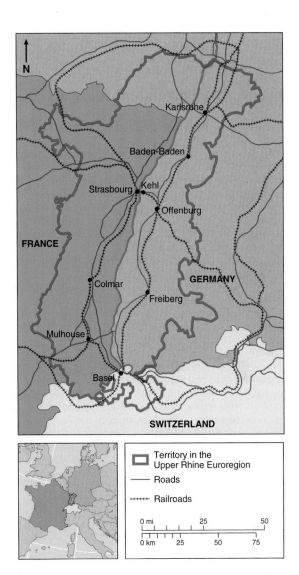

FIGURE 3.19 **The Upper Rhine Euroregion.** Notice the close proximity of French, German, and Swiss cities.

and Spain had the lowest total fertility rate (1.2) in the world. Though these two countries and Portugal are predominantly Roman Catholic, the church's opposition to birth control is clearly having little effect. In Italy, few babies are born outside marriage, but the access of women to careers, young people continuing to live with parents, and the end of pressures to have children tend to defer marriage and reduce the numbers of children. Western Europe has also experienced a greater number of divorces and later marriages, and increased numbers of widows resulting from the longer life expectancy for women (80 years) created more and smaller households so that more housing units were required. The weak economies of the former Communist countries of East Central Europe also have dampened population growth in that subregion.

In the demographic transition, European countries approach or are in the fourth stage (Figure 3.20), where birth rates and death rates are almost equal. The 2002 age–sex diagrams (Figure 3.21) clearly show where these countries are

THE EUROPEAN UNION

The EU is now a major factor in changing the human geography of Europe. Yet many issues cause continuing debate (Box Figure 1), including questions asking whether the European Union is a good idea and whether it will succeed.

Those in favor look at the situation of European countries in the world and note that countries like the United States, Japan, China, and Russia are much larger than any one European country. This makes it difficult for any single European country to compete with these other countries. An economic and political union gives the Europeans more clout in the global economy and world politics. For example, the two largest economies in the world, the United States and Japan, have 297 and 128 million people respectively (2005 data). The largest EU country is Germany (with the world's third largest economy), and it has a population of only 83 million. However, the combined population of the 25 members exceeds 450 million people and has considerably more economic and political clout.

Internally, union has other advantages as well. Free movement of labor, capital, and goods strengthens businesses and allows great opportunities to buy the best products at the lowest prices. To facilitate such movements, 12 of the 15 EU members in 2002 adopted a common currency known as the euro. It replaced the traditional national currencies such as the German mark, the French franc, and the Italian lira. Imagine if all 50 states in the United States each had their own currencies. Imagine then that if you wanted to buy a product from a state other than yours, travel to another state, even for a few hours, or take a trip through several states, you would have to exchange your state's currency for those of other states. In the process, you would have to pay fees for making the exchanges. You probably would not travel to or purchase products from other states as you do now, using the dollar. It would also frequently cost you more to restrict your actions to your own state. This was the situation in the 12 EU countries before they adopted the euro.

As mentioned, not all Europeans favor integration or believe that the EU will work over the long run. For integration to succeed, the member countries have to adopt similar or the same laws and economic policies. For the euro to work, for example, all member countries must limit their spending and keep their annual budget deficits within 3 percent of their GDPs. To do so, many countries may be unable to pay for their social programs and stimulate their economies as they see fit. They would have to give up a lot of what they value and suffer through economic slumps of high unemployment for the benefit of the other member countries. If they choose to break the rules and spend, then they devalue the euro and damage the economies of the other member countries. Many Europeans oppose the euro because of the restrictions that come with it. The United Kingdom, Denmark, and Sweden have not adopted the euro and continue to use their own national currencies.

Many Europeans are opposed to more than just the euro. Integration involves the adoption of common laws and the removal of barriers, including border controls between member countries. States will not be able to stop the entry of foreigners, whether from other member countries or from abroad as they enter through other member countries. This means that these foreigners will take local jobs, demand cultural rights, and generally be a visible foreign presence. It also means that countries will not be able to stop the surge of cheap foreign goods and possibly contaminated food, driving businesses into bankruptcy and creating unemployment. In short, the EU is seen by many as the loss of national sovereignty and the erosion of national identity.

Germany and France, along with the Benelux countries and Italy, have typically been great supporters of the EU. Germany alone contributes 30 percent of the EU's budget. The smaller countries of Ireland, Portugal, and Greece are great proponents of the EU because they receive much more from it than they pay into it. The United Kingdom, Denmark, and Sweden tend to be most opposed to the EU's integration policies. It is even said that the United Kingdom joined the organization to slow down integration and prevent the emergence of a monolith on its doorstep that would be stronger than it.

(a)

(b)

Box Figure 1 In December 2000, leaders of the EU's 15 member countries negotiated a treaty that allowed the admission of 10 new member countries, mostly from East Central Europe. The Nice Treaty, named after the French city on the Rivera where it was negotiated, had to be ratified by each and all of the 15 member countries. Fourteen had done so by the end of 2001 but the people of Ireland rejected the Treaty in June of that year. Determined to have the Nice Treaty ratified, the Irish government scheduled a second election in October 2002. Though the Nice Treaty was ratified by Irish voters in the second election, the campaign leading up to the election was bitter. **Photos:** (a) © John Cogill/AP/Wide World Photos; (b) David Crausby/Alamy Images.

(Continued)

One of the biggest concerns of the current EU members involves enlargement. Thirteen countries applied for membership in the late 1990s (see Figure 3.15 and Box Figure 1). Most became members in 2004, though Romania and Bulgaria were told to wait until 2007 and negotiations with Turkey creep along. In the meantime, Croatia applied for membership in 2003. The inclusion of these countries could make the EU a more powerful force than it is today, but it could also delay or thwart further integration. Some EU countries have been working together for more than 40 years and are ready to move farther down the road of integration. Current members debate whether it is best to wait for new members to catch up or to allow those ready to move ahead on a new issue to do so. As new members have joined, the EU seems to be adopting the multiple path of continued integration. The adoption of the euro with three members opting not to participate is one example of this tendency.

The other problem with EU enlargement is that the 10 new countries are mostly much poorer economically than the 15 older members and may be a financial drain. The 10 East Central European countries have 53 regions similar to provinces. In 1998, before they became members, 41 out of the 53 regions were below 50 percent of the EU's GDP. Only the regions around Prague and Bratislava had per capita GDPs close to the EU average. These two regions aside, EU regional development funds clearly need to be redirected from Ireland, Portugal, Greece, and southern Italy to the East Central European countries. Germany feels that it already contributes a lot and would have to contribute more. France is worried about protecting its agricultural sector. In addition, worker productivity and wages in the East Central European countries were only about 40 percent of the EU average. With freedom of movement, current members fear that unemployment will rise in their countries as East Central European workers would willingly work for less. Thus even the most supportive countries of the EU moved to postpone and restrict the conditions of enlargement in the hopes that the economics of the applicant countries will improve before membership takes full effect.

POINT

Economic union pools together the resources of member countries and thereby strengthens the economies of every member country.

Economic union allows for the free flow of capital and labor, permitting capitalist tendencies to strengthen members' economies.

The euro, or common currency, further facilitates the movement of capital and labor by doing away with the costs of converting currencies.

Political union increases influence in regional and global politics because it combines the political and military strength of member countries.

Political union results in common laws and standards for individuals and the environment in member countries. It makes for better social and natural environments.

COUNTERPOINT

Economic union undermines the ability of member governments to make economic decisions that are in their national best interests.

The free flow of capital and labor undermines attempts by member governments to protect their national economies.

The euro forces member governments to have monetary and budgetary policies that may harm their national economies.

Political union forces member countries to adopt foreign policies that go against their national interests (such as forcing some, like Ireland, to give up their neutrality).

Common laws and standards among member countries undermine national needs and traditions, often watering down social and environmental laws.

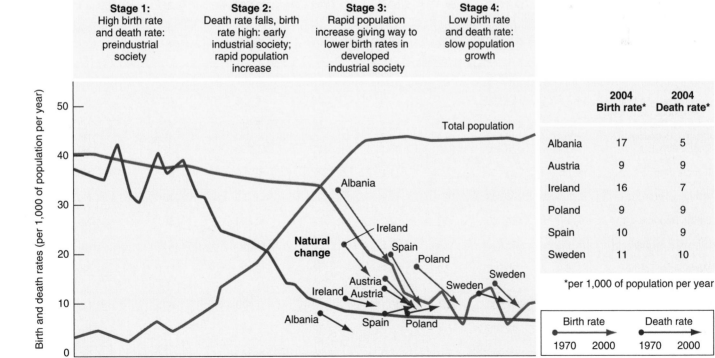

FIGURE 3.20 Europe: demographic transitions. Birth rates and death rates are almost equal in most countries. What does this signify?

Source: *2004 Population Data Sheet,* Population Reference Bureau.

Table 3.2   European Countries. Data by country, area, population, income (GNI PPP—Gross National Income Purchasing Power Parity), Human Development Index (HDI), Human Poverty Index (HPI).

Country	Capital City	Land Area (km²) Total	Population (millions) Mid-2004 Total	Population (millions) Est. 2005 Total	GNI PPP 2003 Total (US$ billions)	GNI PPP 2003 (US$) Per Capita	Index 2002 HDI Rank of 177	Index 2002 HPI % Population
Western Europe								
Belgium, Kingdom of	Brussels	33,100	10.4	10.8	299	28,930	6	12.4
French Republic	Paris	551,500	60.0	63.4	1,640	27,460	16	10.8
Germany, Federal Republic of	Berlin	356,910	82.6	82.0	2,267	27,460	19	10.3
Ireland, Republic of	Dublin	70,280	4.1	4.5	120	30,450	10	15.3
Luxembourg, Grand Duchy of	Luxembourg City	2,586	0.5	0.6	24	54,430	15	10.5
Netherlands, Kingdom of	The Hague	37,330	16.3	17.4	464	28,600	5	8.2
United Kingdom of Great Britain and North Ireland	London	244,880	59.7	64.0	1,639	27,650	12	14.8
Austria, Republic of	Vienna	83,850	8.1	8.4	239	39,610	14	No data
Swiss Confederation	Bern	41,290	7.4	7.4	235	32,030	11	No data
Northern Europe								
Denmark, Kingdom of	Copenhagen	43,090	5.4	5.4	168	31,213	17	9.1
Finland, Republic of	Helsinki	338,100	5.2	5.3	141	27,100	13	8.4
Iceland, Republic of	Reykjavik	103,000	0.3	0.3	9	30,140	7	No data
Norway, Kingdom of	Oslo	323,000	4.6	5.1	170	37,300	1	7.1
Sweden, Kingdom of	Stockholm	449,960	9.0	9.9	238	26,620	2	6.5
Mediterranean Europe								
Helenic Republic (Greece)	Athens	131,990	11.0	10.4	213	19,920	24	No data
Italian Republic	Rome	301,270	57.8	57.6	1,543	26,760	21	11.6
Portuguese Republic	Lisbon	92,390	10.5	10.4	183	17,980	26	No data
Spain, Kingdom of	Madrid	504,780	42.5	43.5	905	22,020	20	11.0
East Central Europe								
Albania, Republic of	Tirana	28,750	3.2	3.7	15	5,530	65	No data
Bosnia-Herzegovina, Republic of	Sarajevo	51,130	3.9	3.9	26	6,320	66	No data
Bulgaria, Republic of	Sofia	110,910	7.8	6.5	60	7,610	56	No data
Croatia, Republic of	Zagreb	56,540	4.4	4.3	48	10,710	48	No data
Czech Republic	Prague	78,860	10.2	10.1	160	15,650	32	No data
Estonia, Republic of	Tallinn	45,100	1.3	1.2	17	12,480	36	No data
Hungary, Republic of	Budapest	93,030	10.1	8.9	139	13,780	38	No data
Latvia, Republic of	Riga	64,500	2.3	2.2	24	10,130	50	No data
Lithuania, Republic of	Vilnius	65,200	3.4	3.5	38	11,090	41	No data
Macedonia, Former Yugoslav Republic of	Skopje	25,710	2.0	2.2	14	6,720	60	No data
Poland, Republic of	Warsaw	312,680	38.2	36.6	437	11,450	37	No data
Romania	Bucharest	237,500	21.7	18.1	159	7,140	69	No data
Slovak Republic	Bratislava	49,010	5.4	5.2	72	13,420	42	No data
Slovenia, Republic of	Ljubljiana	20,050	2.0	2.0	38	19,240	27	No data
Yugoslavia, Federal Republic of (Serbia-Montenegro)	Belgrade	102,170	10.7	10.7	No data	No data	No data	No data

Source: World Population Data Sheet, Population Reference Bureau (2005), World Development Indicators, World Bank (2005), Human Development Report, United Nations (2004), and Encarta Microsoft (2005).

Country	Ethnic Group Percentages	Languages: O-Official N-National	Religion Percentages
Western Europe			
Belgium, Kingdom of	Flemish 55%, Walloon 33%	Flemish (Dutch) (O) 56%, French (O) 32%	Roman Catholic 81%, Protestant 19 %
French Republic	Celtic/Latin with Teutonic, Slavic, Nordic, African, Indochinese, Basque	French (O), with dialects	Roman Catholic 81%, Muslim 5%, Protestant 2%
Germany, Federal Republic of	German 92%, Turkish 2%	German (O), English, Russian	Protestant 35%, Roman Catholic 34%, Muslim 2%
Ireland, Republic of	Celtic, English	English (O) mainly used, Irish (O) (Gaelic) on west coast.	Roman Catholic 93%, Church of Ireland (Anglican) 3%
Luxembourg, Grand Duchy of	Celtic/French/German 75%, some Portuguese, Italian	Luxembourgisch, German, French	Roman Catholic 97%
Netherlands, Kingdom of	Dutch 96%, Moroccan, Indonesian, Turkish	Dutch (O), Frisian, English, German	Roman Catholic 34%, Protestant 25%, Muslim 3%
United Kingdom of Great Britain and North Ireland	White 94%, south Asian, black, other 6%	English (O), Welsh, Scottish Gaelic	Anglican 47%, Roman Catholic 16%, other Protestant 2%, Muslim 2%
Austria, Republic of	Germans 99%	German (O), other SE European	Roman Catholic 78%, Protestant 6%
Swiss Confederation	Germans 65%, French 18%, Italians 10%	German (O), French (O), Italian (O)	Roman Catholic 46%, Protestant 40%
Northern Europe			
Denmark, Kingdom of	Danish, Inuit (Eskimo), Faroese, Greenlander	Danish (O), Faroese	Lutheran 87%
Finland, Republic of	Finn 93%, Swede 6%	Finnish (O), Swedish (O)	Lutheran 89%
Iceland, Republic of	Norwegian/Celtic descendants	Icelandic (O), Danish, English	Lutheran 96%
Norway, Kingdom of	Germanic (Nordic, alpine, Baltic)	Norwegian (O), English	Lutheran 86%
Sweden, Kingdom of	Swedes, Finns, Danes, Norwegians, Greeks, Turks	Swedish (O), English	Lutheran 94%
Mediterranean Europe			
Helenic Republic (Greece)	Greeks 98%	Greek (O), Turkish	Greek Orthodox 98%
Italian Republic	Italian, Sicilian, Sardinian, German, French	Italian (O), German, French, Slovenian	Roman Catholic 98%
Portuguese Republic	Mediterranean	Portuguese (O), English, French	Roman Catholic 94%
Spain, Kingdom of	Spanish 74%, Catalan 16%, Basque 2%	Castilian Spanish 74%, Catalan 16%	Roman Catholic 97%
East Central Europe			
Albania, Republic of	Albanian 95%, Greek 3%	Albanian (O), Greek	Muslim 70%, Greek Orthodox 20%, Roman Catholic 10%
Bosnia-Herzegovina, Republic of	Muslim 40%, Serb 38%, Croat 22%	Bosnian (O), Serbian (O), Croatian (O)	Muslim 40%, Orthodox 31%, Roman Catholic 15%, Protestant 4%
Bulgaria, Republic of	Bulgarian 85%, Turkish 9%	Bulgarian (O), ethnic languages	Bulgarian Orthodox 85%, Muslim 13%
Croatia, Republic of	Croat 78%, Serb 12%	Croatian (O) 99%	Roman Catholic 77%, Orthodox 11%
Czech Republic	Czech 81%, Moravian 13%, Slovak 3%	Czech (O), Slovak, German	Roman Catholic 39%, none 40%, Protestant 5%
Estonia, Republic of	Estonian 64%, Russian 29%	Estonian (O), Latvian, Russian	Lutheran, Estonian Orthodox, Russian Orthodox
Hungary, Republic of	Hungarian (Magyar) 90%, Roma (Gypsy) 4%	Hungarian 98%	Roman Catholic 63%, Calvinist 20%, Lutheran 5%
Latvia, Republic of	Latvian 55%, Russian 32%	Latvian (O), Russian	Lutheran, Roman Catholic, Russian Orthodox
Lithuania, Republic of	Lithuanian 80%, Russian 8%, Polish 8%	Lithuanian (O), Russian	Roman Catholic (most), Lutheran, other Protestant
Macedonia, Former Yugoslav Republic of	Slav 65%, Albanian 21%	Slavic 70%, Albanian 21%	Orthodox 67%, Muslim 30%
Poland, Republic of	Polish 98%	Polish (O), English, German	Roman Catholic 95%
Romania	Romanian 89%, Hungarian 7%, Roma (Gypsy) 2%	Romanian (O), Hungarian	Romanian Orthodox 87%, Protestant 6%, Roman Catholic 5%
Slovak Republic	Slovak 86%, Hungarian 11%, Roma (Gypsy) 2%	Slovak (O), Hungarian	Roman Catholic 60%, Protestant 8%, Orthodox 4%
Slovenia, Republic of	Slovene 88%, Croat 3%, Serb 2%	Slovenian (O), Bosnian, Croatian, Serbian	Roman Catholic 83%
Yugoslavia, Federal Republic of (Serbia-Montenegro)	Serb 62%, Albanian 17%, Montenegrin 5%	Serbo-Croat 99%	Serbian Orthodox 65%, Muslim 19%, Roman Catholic

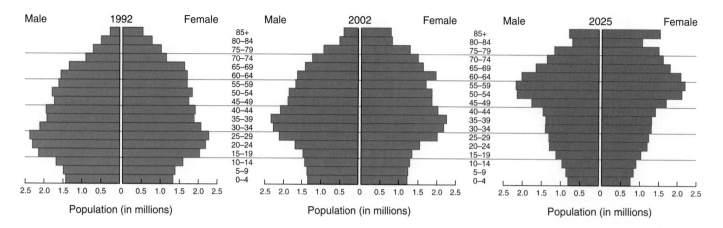

FIGURE 3.21 Age–sex diagram of Italy. The narrowing base of each successive pyramid is indicative of countries with declining birth and total fertility rates. **Source:** U.S. Census Bureau. International Data Bank.

in the demographic transition. The exceptions to population decline are countries (Albania, Bosnia-Herzegovina, Macedonia) with significant Muslim populations. The traditional way of life still practiced in these countries is just as likely an explanation for population growth as religious belief. Europe's population is clearly aging, with those over 65 years making up increasing percentages of the total population (Figure 3.22). Consequently, the burden on the welfare systems is likewise increasing and resulting in higher taxes and the reduction of some welfare programs, such as health and education, after years of wide-ranging coverage. Many of these countries have had to cut back on the services they provide.

Urban–Rural Shifts

Western Europe has high densities of population, except for its marginal highlands (Figure 3.23). The highest densities are in the urbanized industrial belt that runs southeastward from central Britain, through northern France, Belgium, and the Netherlands, and into Germany. Moderate population densities occur across the French farmlands, eastern and southwestern England, parts of southern Germany, and northern Switzerland. The lowest densities are in the uplands of Scotland, parts of Ireland, the high Alps and Pyrenees in southern France, Switzerland, and Austria.

In Northern Europe, most people live toward the southern part of the subregion. Although (without Greenland and Iceland) they cover an area that is almost as large as Western Europe, these countries, apart from Denmark, have low densities of population because of difficult climates and terrains. (Compare Figures 3.3 and 3.6 with Figure 3.23.)

The distribution of population in the Mediterranean countries of Europe reflects the hilly and mountainous nature of the terrain. Most people live in the lower parts of major river valleys and along the coasts. The Po River valley of northern Italy is the largest area of high population density, with others along

the Portuguese coast and eastern Spain. The Alps, Apennines (Italy), Greek mountains, and the Pyrenees and parts of central Spain have very low densities.

East Central Europe has a more even population distribution than other European subregions. The main concentrations of people are in the urban–industrial areas on either side of the borders between the Czech and Slovak Republics and southern Poland. Farther south, the Danube River valley has the main population centers in the cities of Belgrade (Serbia and Montenegro), Budapest (Hungary), and Bucharest (Romania). The predominance of farming elsewhere is reflected in moderate population densities, with low densities on the poor soils near the Baltic Sea and mountainous areas such as the Carpathians and the Dinaric Alps.

Urbanization and Urban Landscapes

European countries are among the most highly urbanized countries in the world (see Table 3.2). The high proportions reflect the economic focus on urban-based manufacturing and service industries. Europe has few extremely large cities because functions are often spread among several cities (Table 3.3). Paris (France), with around 10 million people, and London (U.K.), with over 7 million, are by far the largest urban centers.

The geographic characteristics of the city landscapes of Western Europe result from centuries of making and remaking built environments (Figure 3.24). A number of Western European cities contain relics of historic cultures from Roman times onward. In medieval times, a network of towns grew with walled defenses and market and administrative functions. Internal spatial differentiation by class was not strong.

With the Industrial Revolution of the 1800s, cities grew tremendously; land use differentiated into central business districts of shops and offices, industrial areas of large factories, worker housing, and suburbs in which the growing management

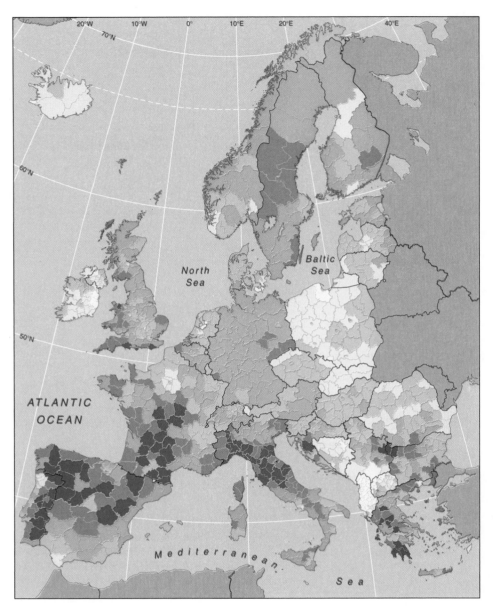

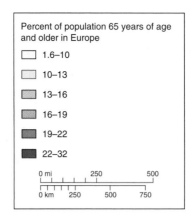

Percent of population 65 years of age
and older in Europe

☐ 1.6–10

☐ 10–13

☐ 13–16

☐ 16–19

☐ 19–22

☐ 22–32

0 mi   250   500

0 km   250   500   750

**FIGURE 3.22 Europe's spatial
patterns of aging.** Notice that the older
areas tend to be both in the peripheral areas
of countries and the peripheral areas of
Europe as a whole. Why do you suppose
that most of East Central Europe has few
areas with a significant population over age
65? Does it have to do with birth rates,
death rates, and/or emigration/immigration?
Source: Kazimierz J. Zaniewski, Department of
Geography and Urban Planning, University of Wis-
consin-Oshkosh.

classes lived. In the early 1900s the development of public
transportation in cities led to further differentiation of land use
and the building of suburban housing linked by streetcar or bus
to the city centers, where retailing, commercial, and manufacturing activities concentrated.

During World War II, bombing and ground fighting reduced
parts of many European cities to rubble, destroying older areas,
including medieval and industrial buildings. The period from
1945 to 1970 was one of rehabilitation, expansion, and restructuring of cities. Nearly all cities expanded their functions and
populations. City centers were rebuilt with utilitarian buildings
(Figure 3.25a), while large tracts of public housing catered to
lower-income groups. New towns were built at a distance from
the largest cities, separated by "green belts" of rural land in
which new building was seldom allowed. A major result was

decentralization—a movement away from the previous focus on
the central business district and toward more economic activity
in the extensive suburbs built since 1945.

From the 1970s, there was disillusionment over the outcome
of postwar reconstruction and the bleak nature of low-cost public
housing. Large public housing tracts, especially where high-rise
buildings were common, proved unsuccessful and often became
centers of unemployment and crime. A significant move began
toward public and private cooperation, particularly in efforts to
make city centers and old waterfronts more livable (Figure
3.25b). Areas with buildings of historic interest were declared
conservation areas, but the finances for pursuing such a policy
now came from private investors who saw the potential returns
from encouraging tourism and in building new accommodations
for those wishing to move back near the city center.

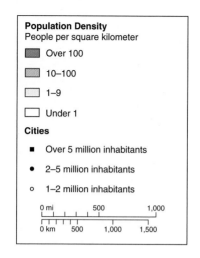

FIGURE 3.23 Europe: population distribution. Explain the heaviest and lightest concentrations of people in each subregion. **Source:** Data from *New Oxford School Atlas*, p. 98, Oxford University Press, UK, 1990.

Older couples and younger professionals moved to neighborhoods near city centers. They bought and renovated poor and dilapidated housing, increasing the value of the housing stock and raising neighborhoods from low income to higher income. This process is known as **gentrification.** Allowing only pedestrian traffic in central areas and arcades of specialty shops became features of city centers, together with the congregation of high-order services requiring support from people living in a wider area—theaters, opera houses, cinemas, and clubs. Areas around the city center that had become derelict, such as old railroad yards, factories, workshops, and worker housing, became sites for the new facilities.

In the 1980s relaxed planning restrictions led to a further dispersal of economic activity from city centers to the suburbs and beyond. Manufacturing had already moved out from cramped inner city sites to suburban industrial parks. Now retail and office facilities followed to suburban warehouse shopping, hypermarkets, shopping malls, and office parks. Nevertheless, densities of European cities are still much higher than in the United States, and suburbanization is much less pronounced.

In Northern Europe, cold climates and lack of farm land are linked to high percentages of urbanization. Population totals are small and the major cities are not very large—although they include 25–33 percent of the national population (in comparison, Paris and London have 17 percent and 12 percent, respectively). The main cities of Northern Europe were little changed by warfare, and their centers continue to be dominated by government and commercial buildings built in the 1800s and early 1900s (Figure 3.26).

Table 3.3 Populations of European Cities (in millions)

Western Europe	2003	2015*
Paris, France	9.8	10.0
London, U.K.	7.6	7.6
Lyon, France	1.4	1.5
Marseilles-Aix-en-Provence, France	1.4	1.4
Lille, France	1.0	1.1
Birmingham, U.K.	2.2	2.2
Leeds, U.K.	1.4	1.4
Manchester, U.K.	2.2	2.2
Tyneside (Newcastle), U.K.	1.0	1.1
Berlin, Germany	3.3	3.3
Essen (Rhein-Ruhr North), Germany	6.6	6.6
Düsseldorf (Rhein-Ruhr Middle), Germany	3.3	3.3
Cologne (Rhein-Ruhr South), Germany	3.1	3.1
Hamburg, Germany	2.7	2.7
Munich, Germany	2.3	2.3
Stuttgart, Germany	2.7	2.7
Frankfurt (Rhein-Main), Germany	3.7	3.7
Mannheim (Rhein-Neckar), Germany	1.6	1.6
Bielefeld, Germany	1.3	1.3
Hannover, Germany	1.3	1.3
Nuremburg, Germany	1.2	1.2
Aachen, Germany	1.1	1.1
Rotterdam, The Netherlands	1.1	1.2
Amsterdam, The Netherlands	1.1	1.2
Dublin, Republic of Ireland	1.0	1.1
Vienna, Austria	2.2	2.2
Northern Europe		
Copenhagen, Denmark	1.1	1.1
Stockholm, Sweden	1.7	1.8
Helsinki, Finland	1.1	1.1
Mediterranean Europe		
Lisbon, Portugal	2.0	2.1
Porto, Portugal	1.3	1.4
Madrid, Spain	5.1	5.3
Barcelona, Spain	4.4	4.5
Rome, Italy	2.7	2.6
Milan, Italy	4.1	4.0
Naples, Italy	2.9	2.9
Turin, Italy	1.2	1.2
Athens, Greece	3.2	3.3
East Central Europe		
Warsaw, Poland	2.2	2.2
Katowice, Poland	3.0	3.0
Prague, Czech Republic	1.2	1.2
Budapest, Hungary	1.7	1.7
Belgrade, Serbia and Montenegro	1.1	1.1
Bucharest, Romania	1.9	1.8
Sofia, Bulgaria	1.1	1.0

* = estimated

Source: United Nations, Department of Economic and Social Affairs, Population Division, Urban Agglomerations 2003.

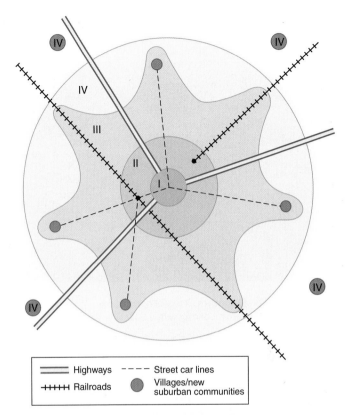

FIGURE 3.24 Generalized morphology of non-Communist European cities. (I) Medieval core. (II) Early growth area later penetrated by railroads and the location of early industrial factories and worker housing. (III) Star-shaped pattern created by suburbanization along street car lines in the early 1900s. (IV) Post–World War II suburbanization with automobiles.

FIGURE 3.25 Europe: urban landscape in Plymouth, England. Modern changes in the city center. The main shopping area was totally rebuilt after World War II bombing and remains the largest retail district in the city despite out-of-town supermarket and warehouse shopping developments from the 1980s. Vehicles were excluded from some streets in this shopping area in the early 1990s to encourage pedestrian traffic. **Photo:** © Michael Bradshaw.

FIGURE 3.26 Denmark. Copenhagen's urban landscape illustrates the combination of modernity with a Scandinavian flavor. **Photo:** © Loren W. Linholm.

The Mediterranean countries are less urbanized than countries in either Western or Northern Europe, but include some of Europe's largest cities (see Table 3.3). Many Mediterranean cities have ancient Greek, Roman, and Moorish historic cores surrounded by medieval, 1800s, and modern industrial and residential developments. Although their historic sectors gain much attention from tourists, Mediterranean cities are also often ports, industrial centers, administrative centers, or magnets for beach-based tourism. The huge new tourist cities of southeastern Spain, the French Riviera, eastern Italy, and southern Greece have many new urban forms blended with their old historical features.

Urbanization levels in East Central Europe range from a low of 42 percent in Albania to a high of 75 percent in the Czech Republic. Still, no cities are of great size. East Central European cities lack the high-rise office buildings containing commercial and professional service facilities and apartments that dominate the central parts of large cities in Western Europe. Many are dominated by obsolete, polluting, and rusting industrial facilities built in the 1960s and 1970s. The lack of major world cities in East Central Europe arises from the small size of the countries, the late arrival of industrialization, and the focus on manufacturing growth linked to medium-sized industrial town expansion during Soviet times. The city landscapes of East Central Europe suffered great war damage and were subject to post-1945 industrialization and the construction of monolithic buildings (Figure 3.27a), but they often retain older townscapes (Figure 3.27b).

The largest cities in the southern part of East Central Europe were often built as expressions of the Austro-Hungarian Empire presence. Budapest, the capital of Hungary, combines two cities on opposite banks of the Danube River and is a government center with industrial suburbs. Its central areas include spacious 1800s buildings. Bucharest, the capital of Romania, had much of its central area built to resemble Paris. Romania's Communist leader, Nicholae Ceauşescu, destroyed Bucharest's urban landscape and put up his own grandiose buildings to bolster his image.

FIGURE 3.27 East Central European cities. (a) Warsaw, Poland: The 38-story Palace of Culture was built with funds from the Soviet Union in 1952–1955. The building, which houses several scientific and cultural institutions, theaters, and Congress Hall, was built on the World War II ruins of the city. (b) Czech Republic: Having escaped much of the destruction of World War II, Prague is one of the best-preserved cities in East Central Europe and subsequently a major tourist destination. Photo shows tourists on the famous Charles Bridge with Prague's castle in the background. **Photos:** (a) © Jerzy Jemiolo; (b) © Emily A. White.

(a)

(b)

Cities in the Balkans often contain historic centers, such as the port of Dubrovnik (Croatia). An earthquake destroyed much of Skopje, the capital of Macedonia, in the 1960s. Cities such as Sarajevo, the capital of Bosnia, expanded after 1950 with central apartment blocks, suburban industries, and single-family homes, but suffered destruction in the early 1990s civil war.

Women, Power, and Social Position

Compared to other areas of the world, women in Europe generally have a high degree of political power and a large share of the economic wealth. At a basic level, equality can be measured by life expectancy. For the region, men typically live 75 years on average, while women live to 81. Power often begins with literacy and education. In contrast to many other areas of the world, women's literacy levels in Europe are equal to men's, at just under 100 percent. Frequently a higher percentage of women receive university degrees. Norway ranks highest with 67 percent of all university degrees received by women. Bosnia-Herzegovina, Portugal, Iceland, Poland, and Estonia closely follow, each having greater than 60 percent. In the fields of education, humanities and arts, social and behavioral sciences, and medicine, women commonly receive 60 to 70 percent of the degrees. On the other hand, women receive only 15 to 20 percent of the engineering degrees, though in Bulgaria and Romania more than 43 percent of engineering degrees are earned by women. The overall geographical pattern shows that women in Northern and East Central Europe are highly educated. In the latter subregion, Communism, which advocated that women were equal to men, clearly had an influence on East Central European society, though those countries are no longer Communist.

A more direct indicator of women's power is their representation in government. Northern European countries ranked among the highest in the world in 2004, with Sweden placing second with women holding over 40 percent of the seats in the lowest house in parliament. The geographical pattern was varied in the other subregions. Women generally held a high percentage of parliamentary seats in Western Europe, though Ireland and France were low. Women had poor representation in Mediterranean Europe, but Spain was notably high. The percentages of parliamentary seats held in East Central Europe spanned the spectrum despite generally high education levels among women. Overall, women had greater political representation in the vast majority of European countries than women in the United States.

Changes in Europe's Economic Geography

Two world wars and intervening economic depressions shattered the economies of European countries and their economic relationships with the rest of the world. After 1945, greater governmental intervention restored **productive capacity** (the amount of goods a country's businesses can produce) and now ensures education, health care, unemployment benefits, and pensions for all. In non-Communist Europe, older, "heavy" **(producer goods)** industries, such as steelmaking, heavy engineering, and chemicals located on coalfields, were replaced in value of output and employment by motor vehicles, consumer goods, and light engineering products. The availability of electricity spread, so that new products could be manufactured wherever there was plentiful semiskilled labor, often in the larger cities that were also large markets for the consumer goods. This process led to greater material wealth in the cities compared to the older industrial and rural areas. The numbers employed in primary and secondary industries gave way to service industries—based in offices rather than factories. This further increased urbanization. The polluted old industrial centers declined, and unemployment in them rose.

In the second half of the 1900s, new industrial areas grew in places more suited to the needs of developing technologies and industries, though long-established production continued in the older areas. The major investments in factories, housing, human skills, and infrastructure made it too costly in financial and human terms to undertake sudden locational shifts of existing production facilities. The advantages of producing goods in an area that has a trained labor force, assembly and distribution systems of transportation, and support by financial and other services, build up and reinforce the original locational advantages as **agglomeration economies.** Keeping production in an area, although its costs may be higher than those in possible competing areas, creates **geographic inertia.** Only when a substantial change in costs occurs, new products emerge, or the demand for the original product is reduced do new areas develop and older manufacturing centers decline.

Furthermore, **deindustrialization** occurred when the numbers of jobs in manufacturing fell rapidly and factories became derelict in older industrial areas. A decline of 20 percent in European manufacturing jobs between 1970 and 1985 was more than balanced by a 40 percent rise in tertiary sector jobs. However, the skills of blue-collar miners and factory production-line workers were seldom convertible into the new white-collar office, hospital, or classroom jobs that were taken by younger and better-educated people. Those workers unable to retrain for the new jobs faced long-term unemployment. Older production workers, female workers, and young people with a poor education entering the labor force were particularly at risk.

Planning and Privatization

In non-Communist Europe after 1950, the problems of declining old manufacturing areas in particular became political issues and led to additional programs of social welfare, regional policies for siting new industries in the older industrial areas, and retraining programs. Governments attempted to redress the differences in unemployment between new and old industrial areas by forcing manufacturers to locate new factories in materially poorer, rather than wealthier, areas of their country. State industries were located in such areas despite higher operating costs. Italy, France, and Britain introduced such policies. Investment in southern Italy—the Mezzogiorno—is a major example of government-directed industrial location.

Beginning in the 1970s, EU regional policy took over from national policies aimed at redistributing employment opportunities. The European Regional Development Fund was established in 1975. There was a net flow of funds to the margins of EU countries, to their older industrial areas, and especially to the newer (and materially poorer) member countries in Mediterranean Europe. The main emphasis of EU regional policy was on granting loans to create jobs directly or indirectly. The loans were used mainly for infrastructure projects, especially roads, telecommunications, water supplies, and waste disposal. Job creation outcomes had modest success, with a few thousand new jobs created. Nevertheless, the EU continually modifies its policies for regional development funds to help the economically poorer areas of Europe. For the years 2000 to 2006, the EU allocates funds according to the following list of priorities (Figure 3.28): (1) development of the most disadvantaged regions (Objective 1), (2) the conversion of regions facing structural difficulties (Objective 2), (3) interregional cooperation (Interreg III), (4) the sustainable development of urban areas in crisis (Urban II), (5) the development of innovative strategies to support regional competitiveness.

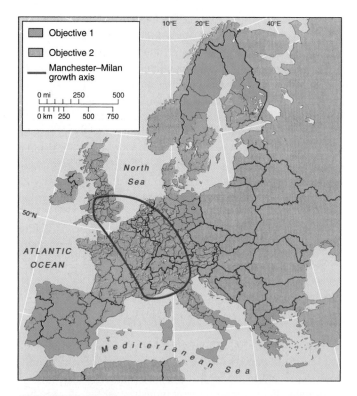

FIGURE 3.28 European Union structural funds, 2000–2006, areas eligible under Objectives 1 and 2. All of Ireland and Portugal, along with most of Scotland, qualified for structural funds until the current round. Note the locations of the most disadvantaged areas. Former East Germany is among them. **Source:** Copyright © European Communities, 1995–2002.

Regional development funds have improved the local living situation in the economically depressed areas of the EU. However, such policies produced a modest effect at great cost and, by focusing on internal issues, often made the countries uncompetitive in world markets. However, in Ireland in particular, many multinational corporations were attracted, and the wages they paid increased personal well-being.

A major shift in Western Europe in the 1980s and early 1990s moved these countries away from a relatively stable economic and social order in which those less able to compete were looked after by government provision of industrial protection and welfare. The other side of this stability had been that the countries of Western Europe began to price their products too high for world markets, resulting in reduced sales and thus income. Management then decided to lay workers off rather than increase income by lowering the prices of their products to boost sales volume. Moreover, increasing welfare costs put country budgets into deficit and raised internal tax demands. To combat this economic problem, many Western European countries, led by the United Kingdom, moved toward practices followed in the United States, where private enterprise is encouraged, but potential high rewards are balanced by high levels of risk, making for a less stable human environment. During the 1980s Britain led the way in privatizing state enterprises, many of which suffered financial losses and were a drain on tax income. The privatized concerns, including steelmaking and electricity, gas, and water utilities, commonly made considerable profits after this change, especially by cutting their labor forces, leading initially to more unemployment. However, by the early 2000s, unemployment rates decreased in countries where the most extensive privatization plans initially had caused the high unemployment rates. The United Kingdom, for example, had a low unemployment rate of approximately 5 percent while countries like France, Germany, and Italy—which had modest privatization programs—had employment rates around 10 percent.

Agriculture

Centuries ago, peoples entering Europe settled land, cleared the forests, and established the two- and three-field system. Under feudal conditions, farmers divided their land into thirds and rotated what they grew on each third every year, leaving one-third fallow to regenerate. In areas of poor soil, farmers alternated their crops between two fields. Wheat was the most common crop, but rye, oats, and barley were cultivated. Livestock played an important role, with much of the grain going to fatten the livestock. Following the plagues of the 1300s, feudal obligations gave way to individual ownership. Most people were farmers who owned small plots of land that primarily fed their own families but frequently produced a small surplus that could be sold for profit.

From the mid-1700s, the Industrial Revolution transformed agriculture. Scientific crop rotation, which involved the planting of soil-enriching crops such as clover and turnips every four years, made it unnecessary to leave half or one-third of the land fallow every year. The use of tractors, fertilizers, and pesticides required fewer people to work larger farms. Small farms of a few hectares were taken over by more successful farmers, resulting in fewer but much larger farms, a trend known as **concentration.** Many people moved from the countryside to the cities, where they found jobs. Tractors, fertilizers, and pesticides also resulted in greater agricultural productivity per hectare, known as **intensification.** Today agriculture still occupies almost as much land as before (Figure 3.29) but provides only 2 to 7 percent of total employment.

Modern agriculture motivated farmers to switch from producing a variety of crops for their families to single crops that brought high profit, such as sugar beets or oil seeds. This is known as **specialization.**

Concentration, intensification, and specialization radically changed agriculture and the rural way of life over the last hundred years. In contrast to the past, most modern farms are large, and the farmers who run them require management skills. Farms also depend on industries that produce seeds, fertilizers, pesticides, and machinery to produce agricultural goods. They rely on food processing plants and marketing agencies to get their products to the consumer. The term **agribusiness** was coined to describe how commercially oriented farming has become and the close links that farming has with other industries. Thus, while fewer Europeans work on farms, agriculture is still a very significant sector of European economies.

Though agricultural production increases in Europe, the number of farmers continues to decrease. Farmers worry that their declining numbers will result in less political power and

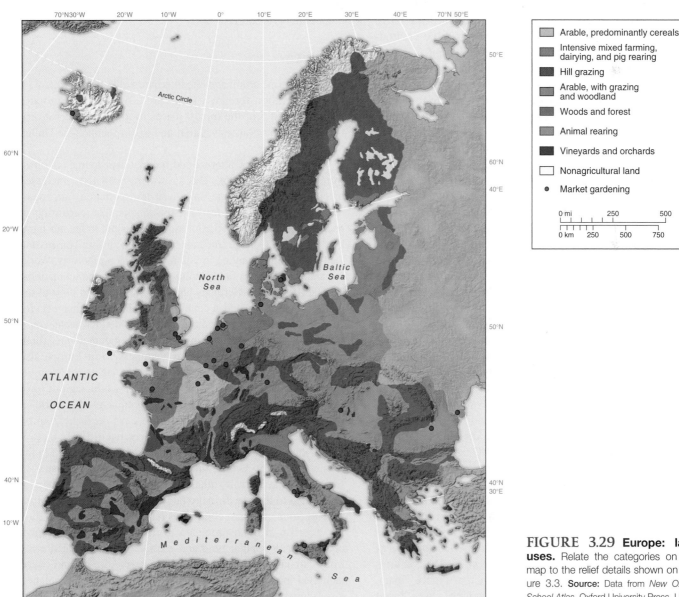

FIGURE 3.29 Europe: land uses. Relate the categories on this map to the relief details shown on Figure 3.3. **Source:** Data from *New Oxford School Atlas*, Oxford University Press, UK.

will endanger the future of farming. European governments, however, have been sensitive to the need to protect a traditional way of life and maintain a stable food supply. The EU Common Agricultural Policy (CAP) at first provided a set of price supports that made farming almost risk-free at a time of advancing technology. Output increased rapidly in the 1960s and 1970s, but costs to consumers and taxpayers rose too. Providing agricultural subsidies to farmers is the largest expense of the EU's funds. In the most productive farming areas of southeastern England, northeastern France, northern Belgium, and the Netherlands, larger farms had access to more capital and adapted easily to new methods. Mountains of grain and butter, and lakes of milk and wine, resulted. For example, the cereal crop rose by 40 percent between 1975 and 1985, generating a surplus and huge storage costs.

In the hilly regions, where marginal livestock farming replaced crops, EU policies supported farmers for social and environmental reasons as much as economic ones. Governments feared rural depopulation and soil erosion if the farms were abandoned. Governmental support helped the environment and slowed depopulation, but did not reverse it.

During the 1980s, the CAP changed to reduce the huge surpluses of produce in storage as a result of price guarantees to farmers. Agriculture was then linked more closely to market prices but an immediate lowering of guaranteed prices would have been devastating to farmers and rural areas. More gradual reductions were engineered through downward revisions of price thresholds and milk quotas, and by forcing overproducers to pay part of the storage costs. It was also recognized that less production would reduce stress on the natural environment.

In 1988 further measures encouraged major cutbacks in production. Older farmers were helped to retire early by being given pensions in exchange for ceasing production of surplus capacity crops. Arable land was retired from production in exchange for grants to farmers who would continue to look after the land but not produce certain crops on it. Livestock farmers were encouraged to produce less from the same area, a process known as **extensification.**

By the late 1990s, the results of the attempts to reduce farm output in the EU countries were not clear. Setting aside land does not always reduce productivity in proportion to the area set aside. Farmers tended to take the subsidies to retire their less productive land, resulting in only minor decreases of overall production. Continued surpluses and the growing global economy in the 2000s led to greater complaints from other countries such as the United States that EU agricultural subsidies gave European farmers unfair advantages in international trade. To further reduce agricultural surpluses, the EU implemented further CAP reforms. The EU decided to "decouple" subsidies from production, meaning that farmers no longer received more subsidies for producing more. Instead, the EU began paying a flat amount that was linked to such concerns as rural development and environmental protection. The countries that benefited greatly from CAP funds (such as France, Spain, Ireland, and Portugal) prevented "decoupling" from being

implemented in all cases. However, if production does not decrease, pressure to apply the decoupling principle more universally is likely to increase.

As European farmers faced the loss of long-term subsidies and increased global competition in the 2000s, they were hit by continued urban expansion and the relaxation of planning restrictions on the conversion of farmland. The feeding of inappropriate animal matter to cattle in the late 1990s led to the spread of bovine spongiform encephalopathy (BSE)—a disease similar to Downes cow syndrome in the United States and generally known as "mad cow disease." An outbreak of foot-and-mouth disease in the U.K. in 2001 led to the slaughter of thousands of livestock to stop the spread of the disease overseas. Though the disease was halted, the outbreak cost millions and further damaged the reputation of the British meat industry. The EU prevented the U.K. from exporting its beef until 2002, but farmers in other countries were affected by a loss of confidence among meat buyers and the reoccurrence of the disease more widely in Europe.

Tourism

Western Europe dominates the international tourist market with over 162 million foreign visitors in 2001. In 2001 European countries held 5 of the top 10 places in terms of international tourist receipts: France was first and Spain second (the United States was third).

The numbers of internal and international tourists doubled from 1970 to the 2000s. This growth resulted from increased incomes and leisure time, improved transportation (air, roads, rail, ferries) and new lifestyle expectations. The expansion of package tourism made foreign travel accessible to greater numbers of people. College students, for example, buy rail passes and visit major cities, staying in youth hostels along the way. Working and middle-class families use campgrounds and recreational vehicle (RV) parks. Germany and the U.K. provided most of the tourists in other European countries, which are also magnets for American and Japanese visitors. Although Mediterranean countries were the main receiving areas, the mountainous, rural, coastal, and historic urban areas within Western Europe also increased their tourist trade. While tourism is increasingly significant in Europe, tourist industries in other world regions have developed more rapidly. Europe still dominates the world in tourism, though its share of tourism receipts declined from 64 percent of the world total in 1975 to 49 percent in 1997.

In the mid-1990s, tourism generated one job in eight in the whole European Union, making it the largest industry in Europe. Unlike in many other industries, the growth in tourism has not led to the rise of many large international tourist corporations. Smaller tourist companies remain the norm.

Tourism is often promoted to help regional development because many popular rural and coastal areas are located away from urban–industrial economic growth areas and need increased local economic activity to prevent further population loss. Tourist development supplements local household incomes, often improves the facilities available in an area, and adds to amenities that attract other forms of employment. Much tourism employment, however, is seasonal and poorly paid. The unskilled and low-paid nature of many jobs means they are taken by migrants and thus have less impact on the local economy than anticipated.

Global City–Regions

One way that Europe plays a leading role in the global economic system is through its cities, where multinational corporations site factories and where sophisticated service industries bring important global roles. Of the global cities identified in Figure 2.15, Europe has by far the most with well over one-third. These European cities are no longer at the top of the list of the most populous cities in the world, but they are among the most globally connected cities. London and Paris, along with only New York and Tokyo, are the most important cities for accounting, advertising, banking, and law. When we think of cities such as London and Paris, we should not simply consider their businesses as only serving Londoners and Parisians, respectively. The businesses within them serve people all over the world, whether in the Americas, Africa, or Asia (Figure 3.30). Many millions of dollars flow in and out of these cities from and to other countries around the world. After London and Paris, cities such as Frankfurt, Milan, Zürich, Brussels, and Madrid rank high in global importance.

Test Your Understanding
3C

Summary

World Wars I and II had a devastating impact on Europe's political and economic systems. Following World War II, Europeans were forced to rethink their relationships with one another and the wider world. Europe was divided into a democratic/capitalist West and a Communist East. The West formed the North Atlantic Treaty Organization (NATO) and the European Economic Community (now, the EU). The East adopted Communism and was tied to the Soviet Union (see Chapter 4). Beginning in the early 1990s, East Central European countries abandoned Communism and applied to join NATO and the EU. All across Europe, devolutionary pressures increased.

Europe generally has low population growth rates and fertility rates, and most European countries are near or in stage four of the demographic transition. The population in Europe is densest in the industrialized areas of Western Europe. Densities are lowest in the cold, mountainous areas, especially in the north, and more evenly distributed across East Central Europe. Europe is highly urbanized. Women have comparatively high levels of political and economic power.

Economically, Europe experienced shifting trading links and new locations for manufacturing. European countries still maintain a high standard of living. Europe's cities, particularly those in Western Europe, continue to play a major role in the global economic system.

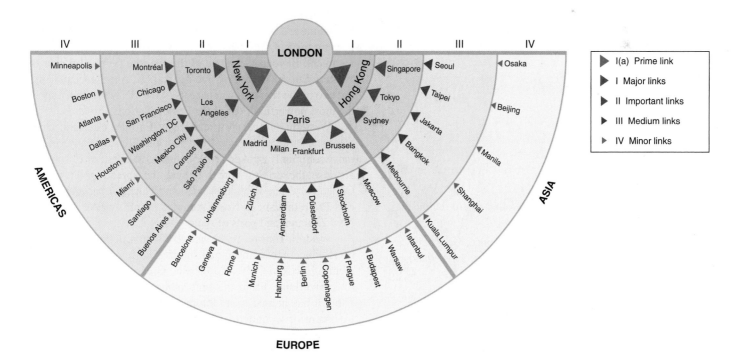

FIGURE 3.30 London. As one of the four most globally connected cities, London is a major center of accounting, advertising, banking, and law. Why do you think Hong Kong is a major link with London when Tokyo ranks higher as a globally connected city (see Figure 2.15)? **Source:** Taylor, P.J., D.R.F. Walker, and J.V. Beaverstock, *Globalization and World Cities Study Group and New Research Bulletin 6, Firms and Their Global Service Networks.*

Agriculture was modernized so that the region that was a net importer is now a net exporter of food. The manufacturing industries modernized in type and levels of productivity. Service industries, especially financial services and tourism, are the main employers, contributing increasingly to the continuing affluence of the region.

Questions to Think About

3C.1 What kind of new supranationalist organizations did Europeans create after World War II, and how are they supranationalist in their functioning?

3C.2 What are the trends in population growth rates in Europe? What are the causes of these trends?

3C.3 What happened to urban landscapes in Europe in the decades after World War II?

3C.4 What are the spatial patterns of economic development in the European Union? That is, which areas are more economically developed, and which areas are less economically developed?

3C.5 What are some of the problems that European agriculture has faced in recent decades?

Key Terms

Communism
democratic centralism
state socialism
planned economy
North Atlantic Treaty Organization (NATO)
European Union (EU)
supranationalism
devolution
Euroregions
gentrification

productive capacity
producer goods
agglomeration economies
geographic inertia
deindustrialization
concentration
intensification
specialization
agribusiness
extensification

Subregions

Though Europe is smaller than many other world regions, it is marked by distinct subregions:

- *Western Europe:* Austria, Belgium, France, Germany, Luxembourg, Netherlands, Republic of Ireland, Switzerland, and United Kingdom.

- *Northern Europe:* Denmark, Faeroe Islands, Finland, Greenland, Iceland, Norway, and Sweden.

- *Mediterranean Europe:* Greece, Italy, Portugal, and Spain.

- *East Central Europe:* Albania, Bosnia-Herzegovina, Bulgaria, Croatia, Czech Republic, Estonia, Macedonia, Hungary, Latvia, Lithuania, Poland, Romania, Serbia and Montenegro, Slovakia, and Slovenia.

Western Europe

The countries of Western Europe include Austria, Belgium, France, Germany, Ireland, Luxembourg, the Netherlands, Switzerland, and the United Kingdom (U.K. formed of England, Scotland, Wales, and Northern Ireland) (Figure 3.31). This subregion contains three of the four largest populations in Europe (see Table 3.2).

Western European countries have been most significant in creating Europe's image as a global leader. The United Kingdom, France, and the Netherlands were three of Europe's most powerful colonial powers. Along with Germany and Belgium, they were also among the first countries to experience the Industrial Revolution. Colonialism and industrialization gave Western Europe the ability to establish many of the global trade flows that are still in effect today. Though Western European countries are no longer major colonial powers, they still import raw materials from countries that were their former colonies and then sell finished products back to them. The economies of Germany, France, and the U.K. rank, respectively, as the third, fourth, and fifth largest in the world. These three countries are members of the Group of Eight (G8), an informal organization representing the world's most materially wealthy countries. (The other countries in the G8 are the United States, Canada, Japan, Italy, and Russia.) The human environment of this subregion ranks high in health, education, and income standards, with all countries ranking within the top 20 on the human development index.

The European Union helped many Western European countries maintain and even increase their economic and political standing in the world. At the same time, these countries helped the EU become the powerful force that it is today. They account for 5 of the 6 founding members and 7 of the 25 current members. Beyond the scope of the EU, the political power of Western Europe is underscored by the U.K. and France occupying two of the five permanent seats of the UN Security Council, the most powerful organ of the United Nations.

Western Europeans spread many European cultural characteristics around the world. English and French are two of the most commonly spoken world languages. Western Europeans are also the driving forces behind the spread of Protestant Christianity. The British, French, and Dutch in particular, carried Protestantism to North America, South Africa, Australia, New Zealand, and parts of Asia.

Western Europe is the hearth for international organizations concerned with human rights. The International Movement of the Red Cross and the Red Crescent traces its origins to Switzerland. From the time of its Geneva Convention in 1864, it has helped to set the modern standard for the ethical treatment of wounded soldiers and civilians during wartime. The International Court of Justice, located in The Hague, Netherlands, initially began in 1899. At the time, it was designed to find peaceful resolutions to conflicts between countries. Since then, it has expanded its scope to include many human rights issues. For example, it is involved in

FIGURE 3.31 Western Europe: countries of the subregion. The western margins include Scotland, Ireland, Wales, southwest England, and Brittany in northwest France.

assessing the responsibilities of those accused of atrocities in the horrific Balkan wars of the 1990s. Oxfam (short for the Oxford Committee for Famine Relief) began in Oxford, England, in 1942 to help starving people in Greece who were caught between Nazi occupation and an Allied blockade. Since that first mission, Oxfam has undertaken numerous operations around the world to bring food and medical supplies to those suffering from war and similar causes. Amnesty International began in London in 1961 and works today to free individuals around the world who are imprisoned for expressing their opinions. Belgium's genocide law, which was enacted in the 1990s, is one of the most far-reaching in the world. It allows non-Belgians to file complaints for crimes committed anywhere in the world. When lawsuits were filed against American leaders for the deaths of school-

children in the 1991 Gulf War, the United States pressured the Belgium government into amending its genocide law, threatening to move NATO headquarters from Brussels. The Belgium government amended the law to refer cases to the home countries of the accused if such countries are democratic.

Countries

Located primarily along the maritime edges of Europe, Western European countries exerted their independence early in history. Coupled with technological innovation, such independence allowed Western Europe to dominate international relations and the global economy from the 1600s to the early 2000s. Strong local voices within Western Europe led to the development of nationalism and the nation–states seen on the map today.

France and the United Kingdom

France is one of Europe's older and most powerful and influential countries (Figure 3.32). Beginning with the Capet family in the 900s and its small family holdings around Paris, the country grew over the centuries, becoming a nation–state soon after the French Revolution in 1789. The French Revolutionary expression of "liberty, equality, and fraternity" inspired other revolutions in Europe. The French bring a strong sense of individuality to their nationalism, and France has a very centralized form of government with most decisions made in Paris. In 1982 the central government created 22 provinces called "regions" and granted them considerable autonomy in raising taxes and spending funds on local projects to satisfy the desire for more local decision making.

The United Kingdom (U.K.) adopted its name in 1801 and included the whole of Ireland until 1922. England, Wales, and Scotland are district entities that together with Northern Ireland comprise the United Kingdom of Great Britain and Northern Ireland. The United Kingdom has a political geography that reflects strong local voices within the country. The rise of Scottish and Welsh nationalism over the last few decades resulted in greater autonomy for both.

The Northern Ireland situation is complex. In 1922, independent-minded Irish succeeded in removing all but six counties of Ireland from the United Kingdom. The freed countries became the Irish Free State. Still part of the British Commonwealth, the Irish Free State had to swear allegiance to the British monarch until 1949, when it achieved its independence and adopted the name "Republic of Ireland." The six counties of the north remained part of the United Kingdom after 1922 as Northern Ireland. The violence in Northern Ireland often is depicted as a religious conflict between Roman Catholics and Protestants, but it is really a conflict between Irish nationalists and Unionists, also called Loyalists. Irish nationalists, feeling oppressed by the U.K. government, seek to unite the six counties of Northern Ireland with the Irish republic to the south. Because most Irish nationalists are Catholic and have the goal of uniting Northern Ireland with a Catholic country and because Unionists/Loyalists tend to be Protestant and want to maintain a union with a Protestant country, the issue is often described as a religious conflict. As the conflict became progressively violent, even ambivalent Catholics sought protection of other Catholics and Protestants sought refuge with other Protestants. The conflict has led to religiously segregated communities, though almost as many violent acts are committed by Catholics against Catholics and Protestants against Protestants as across religious lines. Since the late 1990s, the U.K. and Irish governments have tried to reduce violence and produce a combined government for Northern Ireland (like the devolved assemblies in Scotland and Wales). However, longstanding attitudes have slowed progress of disarmament and trust between the groups.

The global connections that France and the U.K. had with the rest of the world changed when these countries withdrew from most of their colonies in the 1950s and 1960s. Though these two countries no longer directly control other peoples and their lands, they both are very involved with their former colonies. The U.K. maintains ties, for example, through the British Commonwealth. In 1945 France created a currency known as the CFA franc (franc of the French Colonies of Africa) for use in its western and central African colonies. These colonies are now independent countries, but the CFA franc still exists and ties their economies closely to France. France also maintains the French Foreign Legion, a military unit of non-French citizens that fights for French interests. France and Canada are the main supporters of the Agency for Francophony, an organization that promotes the French language and culture around the world.

The Low Countries

The Netherlands, Belgium, and Luxembourg lie along the deltas of the Rhine and Meuse (Maas) Rivers and were once all called "The Netherlands," meaning "Low Countries," The modern country of the Netherlands has one-quarter of its land below sea level. Dutch ingenuity and hard work reclaimed much of the country from the sea, mostly in the form of *polders* (Figure 3.33). In the past, windmills helped to pump the land dry, and more than 1,000 are still in working order today, though diesel and electric engines do the pumping.

The people of the Netherlands voiced their desire for independence from Catholic Spain during the Protestant Reformation in the 1500s. The seven northwest provinces, calling themselves the United Provinces of the Netherlands, under the leadership of the Holland province, fought for their independence and achieved it in 1648, the year of the Treaty of Westphalia and the end of the Thirty Years' War. In 1815 the southern provinces were joined with the United Provinces to

FIGURE 3.32 Paris, France. The Arc de Triomphe de l'Etoile stands on the hill of Chaillot and serves as the center of twelve radiating avenues, one of which is the Champs Elysées. In 1806, Napoleon conceived of a triumphal arch in the spirit of ancient imperial Rome that he could dedicate to the glory of his armies. Designed by Jean François Thérè Chalgrin (1739–1811), it was completed in 1836. Now a symbol of French patriotism, a huge French flag is hung from the ceiling of the arch on national holidays. **Photo:** © Yann Arthus-Bertrand/Corbis.

FIGURE 3.33 Land reclamation in the Netherlands. With one quarter of the country's land below sea level, the Dutch have been building dikes to keep the seawater from inundating their land for more than a thousand years. The term polder is used to describe land enclosed by dikes. **Source:** From Hoffman, *A Geography of Europe: Problems and Prospects.* Copyright © 1983. This material is used by permission of John Wiley & Sons, Inc.

the north, but almost two centuries of separation made union impossible. Most of the southern provinces declared their independence from the Netherlands in 1830 and became the country of Belgium. The province of Luxembourg lost most of its territory to Belgium at that time, but what remained became the country of Luxembourg in 1868.

From the mid-1600s, the Netherlands became a colonial power in the East and West Indies, and Belgium also later had colonies, including Belgian Congo. Colonialism helped to make Dutch cities such as Amsterdam and Rotterdam great trading centers, the latter still being the largest port in the world. Over half of the Netherland's GDP comes from international trade. These countries tried to remain neutral in recent wars but were usually overrun because they lay in the path of easy access from Germany to France. After World War II, Belgium, the Netherlands, and Luxembourg worked together as **Benelux** (**Bel**gium, the **Ne**therlands, and **Lux**embourg). Belgium's capital, Brussels, serves as the seat for the European Commission and headquarters for NATO. The Hague, the Netherlands, is the seat of the World Court, and Luxembourg is the location of many EU agencies.

Alpine Countries

The three alpine countries of Austria, Switzerland, and Liechtenstein are centered on the high mountains of the Alps with their glaciated valleys. Austria and Liechtenstein began as dynasties, though Austria grew into one of Europe's largest empires under the Hapsburgs. Modern Austria represents most of the German-speaking regions of that former empire, which disintegrated at the end of World War I. Vienna's grandiose buildings are a clear reminder of earlier, more regal times. Allied agreements after World War II kept Austria from joining the European Economic Community, but in 1995, not long after the breakup of the Soviet Union, Austria joined the European Union.

Switzerland is a country of strong local voices. Its people felt strongly that they should not be incorporated into any empire, especially the Hapsburg empire. In 1291 three communes in the Alps swore to help defend one another from outside aggression. The Swiss Confederation is now composed of 26 cantons that each have considerable local decision-making powers compared to the federal government. This internal political geography unites a country with a complex geography of languages and religions. Over two-thrids of the country is German-speaking, but French dominates in the west, Italian in the south, and Romansh in the southeast. Roman Catholicism and Protestantism are both common in the country. For external relations, Switzerland has a long history of neutrality in wars and declined to join the EU, though Swiss citizens approved a referendum in 2000 allowing for cooperation with the EU concerning such matters as the free movement of people, agricultural trade, and transportation linkages. The Swiss also maintain a mission to the EU in Brussels, and in 2005 the EU opened an office in Bern, Switzerland.

Germany

Germany is one of the younger nation–states in Western Europe, having come into existence only in 1871. Since then its boundaries have changed. After World War II, the Allies could not agree on a common form of government, so they divided Germany into occupation zones that became a Communist country (German Democratic Republic, East Germany) and a democratic capitalist one (Federal Republic of Germany, West Germany). On October 3, 1990, the two countries reunified (Figure 3.34). With a federal structure like that of the United States, Germany is composed of 16 provinces called *länder*. Most, such as Bavaria and Saxony, are based on historical territories and craft their own laws and policies concerning such issues as education, transportation, and social welfare.

People: Ethnicity and Culture
Immigrant Workforces

World War II destroyed many Western European cities, particularly industrial cities. The advent of the airplane, especially the bomber, resulted in many civilian deaths and the destruction of vast areas far from the front lines. Much rebuilding had to be done, but many young men of prime working age had been killed during the war. German industry in particular needed workers and looked to southern Europe: Portugal, Spain, Italy, Yugoslavia, and Greece. When not enough workers were found in these places, Turkey became another source. Over time, Turks came to represent the largest immigrant workforce in Germany (Figure 3.35).

The U.K., France, and the Netherlands did not have to search for labor. As decolonization progressed in the 1950s and 1960s, tens of thousands of people preferred to migrate to the former ruling countries. Thus the U.K. received many migrants from the Caribbean, Pakistan, and India, while France accepted large numbers from Algeria and other African countries, as well as Southeast Asia, especially Vietnam. Thousands from Indonesia migrated to the Netherlands.

These foreign workers were frequently granted only limited rights. They obtained the legal right to work but in some cases not citizenship or all the legal rights enjoyed by full citizens. The Germans coined a term for these foreign workers: *gastarbeiter* or **guest worker.** Though allowed to stay as long as work was available, decades if necessary, the fundamental principle was that they would always be foreigners and would eventually go back to their respective homelands. Though more than 7 million immigrants, of which 2 million are Turks, live in Germany, few have more than guest worker status. With so many immigrants, Germany changed its citizenship law in 1999, which previously required German blood, allowing many more immigrants to become German citizens. Of the

FIGURE 3.34 Berlin: changing landscapes. Located near the center of the city, Potsdamer Platz was a thriving business district before World War II. (a) During the Cold War, it was immediately behind the Berlin Wall (foreground) and became a landscape of tank traps, mines, guard dogs, and patrolling soldiers with machine guns. (b) In 1991, with new apartment buildings, a repaved street, and beginnings of a reopened subway station. Use the mound of Hitler's bunker on the left, the East German transmission tower, the church dome, and the tall, almost windowless concrete building as reference points. **Photos:** © George W. White.

(a)

(b)

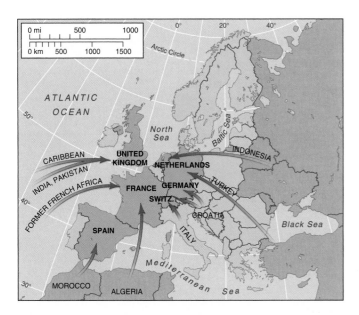

FIGURE 3.35 Europe: sources of guest workers, 1970 to 1990s. The sources vary from the former colonies of the United Kingdom, France, and the Netherlands to the countries around the Mediterranean Sea. What contributions have such workers made to the sending and receiving countries?

growth rates among Western Europeans dropped considerably. The current population decline has economic consequences. The number of retired people is growing while the number of workers is declining (see Figures 3.22 and 3.28). The declining working generation is increasingly having difficulties supporting pensions for the growing retired generation. Industry also has difficulty finding workers and frequently seeks foreign workers, though this may be unpopular among many citizens.

Refugees

The relatively high level of economic health and political freedom found in Western Europe has been attractive to political refugees and asylum seekers from other areas of Europe and the world. With very liberal asylum laws, Germany attracts and receives the largest numbers of refugees, but asylum seekers are also an issue in France and the U.K. The wars in the Balkans in the 1990s increased the number of refugees, particularly from Bosnia-Herzegovina. Germany accepted by far the largest share, with over 300,000. Large numbers of refugees in Western Europe come from countries such as Sri Lanka, Turkey, Ghana, Iran, Iraq, Afghanistan, Somalia, Ethiopia, Democratic Republic of Congo (Zaire), Cambodia, and Vietnam. Together with guest workers, they contribute to an ethnic diversity in Western Europe that is much greater than usually acknowledged.

Economic Development

Western Europe remains the economic heart of the region, with 60 percent of all manufacturing jobs in the region and 75 percent of the research and development that devises and applies new technology. Germany in particular has the third largest economy in the world and the largest in Europe. It is often referred to as the "motor" of Europe and is responsible for much of Western Europe's economic strength. Total German exports almost meet the combined exports of France and the U.K., and its central position helps make Germany the major trading partner for almost 20 European countries. By contrast, France and the U.K. each are the main trading partners for only three to four other European countries.

Sophisticated Manufacturing Industries

Two of Western Europe's major industries are automobiles and airplanes. The automobile industry includes foreign MNCs and formerly nationalized companies, such as Renault in France, that are now mostly privatized. General Motors and Ford Motor Company of the United States opened their European plants in the 1950s and built a complex set of interrelationships throughout Europe among assembly plants, independent suppliers of components, and specialist factories producing engines and transmission systems (see Figure 2.13). The Japanese carmakers Nissan, Toyota, and Honda all built assembly plants in the U.K. to gain a foothold in the EU market. Locally owned firms, such as Peugeot-Citroën (France) and BMW

4 million foreign-born workers in France, just over 1 million have French citizenship. The United Kingdom granted citizenship to immigrants from its former colonies but tightly controls further immigration.

By the 1970s Western Europe was rebuilt, but slumping economies hurt further by an international oil crisis meant that less labor was needed. As unemployment rose, many Western Europeans blamed the migrants for taking their jobs. Though many guest workers lost their jobs, most did not leave. Unemployment benefits allowed them to maintain a standard of living that was higher than in their home countries. Moreover, many had raised families in their host countries, and their children and grandchildren had no desire to relocate to the countries of their parents' birth. Though many guest workers thought fondly of their ancestral homelands, their children viewed them as foreign lands. In the 1980s the German government offered Turkish guest workers cash to go back to Turkey if they took their entire families with them. As many unemployed Turkish men considered the offer, it was not uncommon for their children to run away in fear that they would have to go live in what they felt was a Third World country where they could not speak the language.

As Western European economies improved again in the late 1980s, hostility toward foreign workers subsided for a time. However, following the end of Communism in other areas of Europe in 1989, a flood of new people looking for asylum and work came from East Central Europe. Many Western Europeans became hostile toward Eastern Europeans but even more so against non-Europeans. At the same time, population

(Germany), also have factories in countries outside Europe, including China, Brazil, and the United States. Engineering expertise, especially in the U.K., is responsible for many developments in Formula I racing cars.

The automobile industry finds it difficult to expand further. Germany produces around 4 million cars per year, France around 3 million, and the United Kingdom and Belgium around 1 million each. The total is well above the numbers that can be sold. In the 1980s, only Germany expanded the number of cars produced as overproduction became common—a situation made worse by the opening of Japanese-owned factories in the U.K. Automation and just-in-time delivery of components were introduced to cut costs, causing tens of thousands of layoffs in the assembly and component factories. They placed a greater emphasis on close geographic links—including location and transportation—between the carmakers and their suppliers.

In 1998 the merger of Daimler-Benz (Mercedes) and Chrysler suggested a new approach toward world markets that involves some of the largest automobile makers in different countries working together. Renault took over Nissan (Japan) in 2002, and Volkswagen owns Seat (Spain) and Skoda (Czech Republic). European carmakers also are active in the global spread of automobile plants in Brazil, India, and China.

The aerospace industry is another key sector. Aeronautical research and manufacturing are well developed in Western Europe. Defense was a reason for the support of this industry, but commercial production receives more attention. The small size of Western European countries makes it difficult for them to compete commercially with a country as large as the United States and its airline manufacturers such as Boeing. To compete, a group of Western European airplane manufacturers formed the Airbus consortium in the late 1960s. In the same spirit of the European Union, Airbus pools the resources of several countries. Companies in France, Germany, the U.K., and Spain all manufacture various components that are then shipped to assembly plants in either Toulouse or Hamburg (Figure 3.36). Toulouse is the bigger assembly plant and headquarters for Airbus. The first airplane rolled off the assembly line in 1972, and Airbus soon captured 10 percent of market share. Growth has been steady and rapid. Today Airbus employs 44,000 and from 1999 captured the biggest share of new passenger jet orders worldwide.

Competition between Airbus and Boeing is now fierce. With commercial airline traffic at a high and likewise airport congestion, the two companies are staking their market positions on two different scenarios of the future. Airbus believes that continued congestion will result in greater difficulties for airlines in obtaining gates at major airports such as London's Heathrow, Tokyo's Narita, and New York's John F. Kennedy. Therefore, Airbus launched the A380, a plane that has 50 percent more floor space and at least one-third more seats than Boeing's 747. The plane flies greater distances, and versions offer shower facilities, a piano bar, an exercise room, and a barbershop. The plane was scheduled to make its first commercial flight in 2006. In contrast, Boeing believes that greater airport

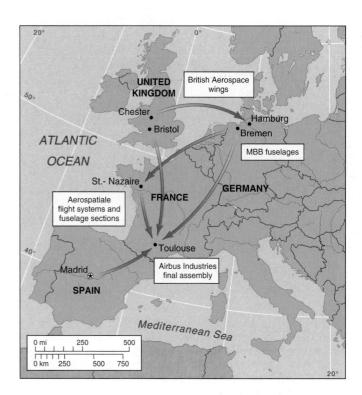

FIGURE 3.36 Western Europe: aerospace industries. Cooperative European manufacture of Airbus aircraft. The high levels of capital inputs and technological complexity required for constructing passenger aircraft make it impossible for the aerospace industry in one country of Europe to support the whole process. What are the political and economic implications of these movements?

congestion will lead to both new airport construction and the expansion of current facilities, calling for more midsize and faster jets. Thus Boeing is developing a modestly sized but very fast subsonic cruiser. It is difficult to predict the future, but European cooperation in aeronautics may make Western Europe the world's leader in the commercial airline industry.

Energy Sources

Domestic coal was the main source of energy until the 1950s: as late as 1952, it provided 90 percent of energy needs in the U.K. and 92 percent in Germany. After 1955 home-produced coal declined rapidly as environmental legislation made burning it more costly and as cheaper imported oil, mainly from Southwest Asia, took over in residential, industrial, petrochemical, and transportation uses. By 1972, oil supplied 65 percent of the subregion's energy needs, while coal was reduced to 22 percent, mostly used to generate electricity.

Oil price shocks in the 1970s raised the oil price 10 times. This gave a boost to exploration and production in the North Sea basin (Figure 3.37), extending northward from the discoveries of natural gas in the Netherlands in 1963 to the sea area between the U.K. and Norway. Early estimates of reserves and production were raised again and again: the Netherlands gas fields have already extracted twice the original reserve estimate and nearly

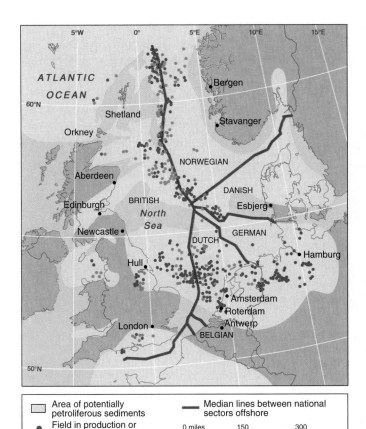

FIGURE 3.37 North Sea oil and natural gas fields (1995).
Source: Pinder, *The New Europe.* Copyright © 1998. Reprinted by permission of John Wiley & Sons, Ltd.

as much again remains available in the rocks. In the North Sea oil and gas fields, production continued to rise in the 1990s, and exploration shifted west of the northern British Isles.

Oil prices in the mid-1980s fell in response to new discoveries that created competition and conservation by users. Natural gas became the cheapest and cleanest fuel for electricity generators. By discovering its own oil and gas fields, reducing the growth in its demand, and adopting new technologies, Western Europe obtained good prices and a greater security of delivery from competing producers of oil and gas, particularly those in Northern Africa and Southwest Asia. The vast natural gas resources of Russia were brought by pipeline into Germany, France, and Italy. In the future, this source of supply is likely to expand because Russia needs to export vast quantities to pay for the consumer, capital, and producer goods that it imports (see Chapter 4).

Of other sources of energy, nuclear-powered electricity generation is the most important, especially in France and Belgium (77 and 57 percent of the national totals, respectively, in 1996). It would have been more important elsewhere if early promises of low cost and safety had materialized and the Chernobyl nuclear power plant explosion in Ukraine in 1986 had

not occurred. Hydroelectricity is important in alpine Europe. There is some development of wind power in northern Germany and the United Kingdom and of tidal power in northern France, but these and other alternative energy sources provide a tiny fraction of current needs.

Service Industries

While manufacturing in Western Europe ventured into new sectors, it fell behind the world leaders in the United States and Japan. However, the development of the service sector almost rivaled that in the United States. In the late 1990s, employment in the service sector made up over 65 percent of the total workforce in most Western European countries.

Service jobs grew in importance as the countries increased their populations, became richer, and instituted strong social welfare programs. Jobs in retailing, wholesaling, education, health care, and government employment relate closely to numbers of people and population distribution. The providers range from private corporations, such as food and drink retailing, to state-controlled institutions in health care and education. Such services support and depend on the manufacturing sectors of the economy but play an indirect role in production. For example, better education and health care may provide a more skilled and adaptable workforce.

The main growth areas in the service sector are in producer services and tourism. **Producer services** are involved in the output of goods and services, including market research, advertising, accounting, legal, banking, and insurance. They serve other businesses rather than consumers directly. Greater use of computers and information technology increased **productivity,** which is frequently measured by the amount of product generated or work completed per hour of labor. Yet despite the increase in productivity, producer service jobs have increased by a quarter to a third in the countries of Western Europe since the 1970s and employ around 10 percent of the labor force. Producer services are related closely to global commercial developments and concentrate in the centers of major cities, where agglomeration economies are significant. Agglomeration economies, as stated earlier, involve the clustering of businesses in a location, often near governmental agencies, to produce savings from the sharing of infrastructure, labor pools, market access, and reduced transportation costs. London, Paris, Amsterdam, Frankfurt (Figure 3.38), and Munich have major shares of these industries. While some functions, such as back-office routine processes, are decentralized in suburban and small-town "paper factories"—often in landscaped office parks—major city centers retain the high-order functions in which face-to-face personal contact is important. In the early 2000s, many back-office and call center jobs went to cheaper locations in Ireland and overseas to India and the Caribbean.

Mergers and takeovers after the mid-1980s led to a smaller number of very large producer service firms, each broadening the range of services. Banking, insurance, property services, and accountancy were combined. Centrally located specialist firms took over provincial firms to extend their influence

FIGURE 3.38 Western Europe: financial services. Frankfurt, Germany, is one of Europe's financial centers. Can you distinguish between the preserved medieval buildings and the newer office complexes? **Photo:** © Ray Juno/Corbis.

nationwide and internationally. Although financial firms dealing with huge sums of money are prone to occasional scandals, the level of control in Western Europe is high enough to attract many transactions of world financial significance.

Northern Europe

Northern Europe is sometimes called "Norden" and extends from the North European Plain to beyond the Arctic Circle and encompasses the Scandinavian and Jutland peninsulas, Finland, and several islands in the Baltic Sea and North Atlantic Ocean. It includes the countries of Denmark, Finland, Norway, and Sweden, together with the former colonies of Denmark, Greenland, the Faroe Islands, and Iceland (Figure 3.39). The subregion is sparsely populated but rich in natural resources.

From the 800s through 1200s, the subregion's Viking inhabitants extended their control to the British Isles, northern France, Iceland, Greenland, the Baltic area, and Ukraine. Later, Sweden, as an imperial state, controlled much of Northern Europe and areas that are now in Russia, the Baltics, northern Poland, and Germany at various times until the early 1800s. Since then, Sweden and Denmark gave up imperial ambitions and often take positions of neutrality in international relations.

Northern European countries are great advocates for human rights. Sweden is home to the Nobel Institute, though a committee of the Norwegian parliament awards the Nobel Peace Prize. Finland is known for the International Helsinki Federation for Human Rights. Compared to the rest of the world, women in these countries hold the highest percentage of elected and nonelected positions in government. Some would argue that the high level of women's empowerment has likewise contributed to the subregion's considerable economic prosperity and some of the world's highest GDP per capita and HDI figures (see Figure 3.17 and Table 3.2; see "Geography at Work: Business Development," p. 112).

Countries

Throughout much of their histories, the peoples of Northern Europe remained mostly free from outside rule and thus were able to express their local voices in self-governance. Within the subregion, the Danes and Swedes frequently controlled the others, as well as people and lands outside Northern Europe. Norway, Finland, and Iceland have enjoyed independence only in the last hundred years.

Sweden has been the largest and politically strongest Northern European country through history, with only Denmark as a rival. At its height from 1610 to 1718, "the Great Power period," Sweden controlled the areas now known as Finland and the Baltics, and northern areas of Poland and Germany. The Swedish army was able to defeat Danish, German, and Russian forces, often simultaneously, before it was permanently weakened in 1721. Sweden lost most of its possessions but retained much of Finland. In 1809 Sweden lost Finland and in 1812 acquired its current boundaries; but a few years later and until 1905, Sweden gained considerable control over Norway.

Though Sweden has not engaged in a war since 1812, the Swedish government cooperated with Nazi Germany by providing it with ball bearings necessary for weapons and by allowing German troops to be transported by rail across Sweden to Norway and Finland. In contrast, Swedes such as

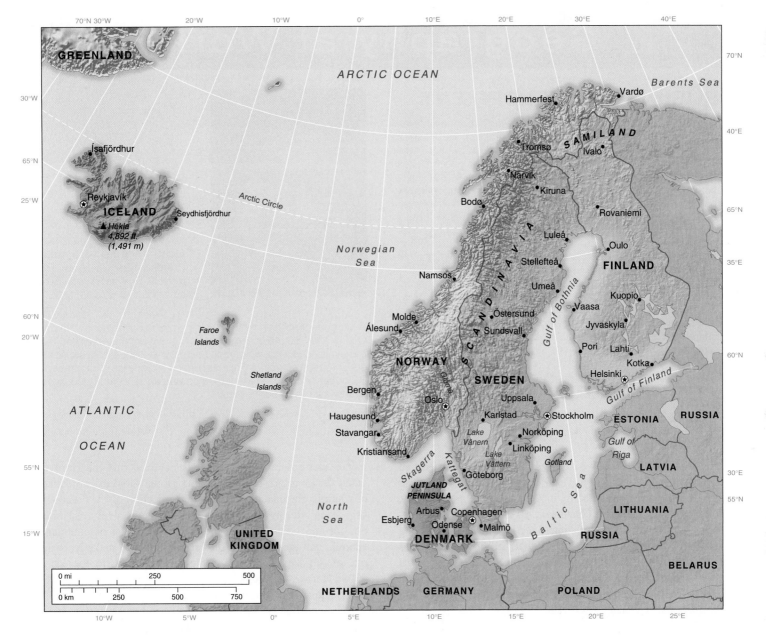

FIGURE 3.39 Northern Europe: the countries, cities, and physical features. Norway and Sweden comprise the Scandinavian Peninsula; the Jutland Peninsula is part of Denmark.

Raoul Wallenberg rescued thousands of Jews from the Nazis' Holocaust. Not long after the end of the Cold War, Sweden joined the EU in 1995. It decided, along with Denmark and the U.K., however, not to adopt the euro as its currency. Since the end of the Cold War, Sweden has also invested heavily in the Baltic countries.

Denmark is one of the smallest countries in Europe but has had strong global connections since early times. Claiming the oldest capital city and flag, the Danes also claim that their queen, Her Majesty Queen Margrethe II, has the oldest royal lineage in the world, dating from Viking King Gorm in the early 900s. During the Middle Ages, Denmark was a major power in Northern Europe. The Danes willingly joined the EC

(now EU) in 1973, but they have resisted adopting many EU standards because they have usually been lower than their own. Danes even threatened to unravel the EU in the early 1990s when they failed in 1992 to ratify the Maastricht Treaty, which called for greater political and economic union among the member countries. In a second vote in 1993 the Danes narrowly ratified the treaty, but more recently refused to adopt the euro as their currency.

Denmark retains close ties with Iceland, the Faroe Islands, and Greenland, which were all part of it at one time. The populations of the latter two are only 46,000 and 52,000, respectively. Iceland gained some autonomy in 1918 but retained strong links with Denmark until World War II, when U.S. and U.K. forces

BUSINESS DEVELOPMENT

Geography is very useful to individuals working in business development For example, Anne Hunderi, who received her master's degree in geography at the University of Oregon, now works for the Hordaland county council in Norway. Hordaland county includes the city of Bergen, Norway's second largest city. Since 1997, Anne has worked in various departments within the county but works primarily on developing the tourism industry and helping small businesses obtain start-up loans. Much of Anne's help goes to specific groups like immigrants, young people, and women. For example, Anne has helped catering businesses and tailors to establish themselves. One interesting endeavor that Anne has helped to get its start is "Fresh," a small business that makes soaps and beauty products. Anne even used the resources of her position to open up her own bed and breakfast.

Anne believes that geography has great breadth and allows individuals with geographic knowledge to bridge the human and natural worlds. Geography gives individuals the flexibility to become involved in a wide variety of concerns such as planning, transportation, and environmentally related issues. Though Anne now works on local projects, the flexibility of geography also allows Anne to work on international issues. For example, before working

Box Figure 1: Anne Hunderi. Using geography to promote local and international business development.

in her current position, Anne was the project coordinator for Hordaland in a US $6 million European Union project that identified the best ways to extend microcredit (small business loans) in six Western European countries. She has also promoted Hordaland tourism in Scotland, coordinated study trips abroad, and arranged trade missions for local businesses.

occupied it after Germany conquered Denmark. Iceland declared full independence in 1944. The Faroe Islands and Greenland achieved a degree of autonomy in 1948 and 1979, respectively, though both remain part of Denmark. All three places are inhibited by physical difficulties of climate and land and rely on fishing the surrounding waters. In the late 1990s. fish stocks declined as other countries increased their take. Competition for use of the surrounding waters led Iceland and the Faroe Islands to extend and defend their 200-mile limits against European fishers until international agreement was obtained.

Norway is a mountainous country with limited farmland primarily around Oslo and Trondheim. During the Ice Age, glaciers excavated deep valleys as they cut their way to the sea. When the ice melted, the sea rose and flooded inlets that are known as **fjords.** At the inland end of these fjords, small patches of good agricultural land formed and were called "Viks." The people who inhabited them were called "Vikings." The Norwegians are proud of their Viking heritage and their seafaring ways. Modern Norway was controlled either by Denmark or Sweden and did not achieve self-governance until 1814. Even then, the Swedish king remained the king of Norway until 1905, meaning that independence was not fully achieved until 1906, when the Norwegians elected their own king. Since then Norwegians have guarded their independence carefully. Norway is a member of NATO, but Norwegians rejected referendums to join the EU. Until the 1970s, Norway's economy was based on fishing and the smelting of

metals using hydroelectricity. It is believed that the Norwegians' desire to protect their fishing grounds and their practice of whaling played a major role in the rejection of EU membership, even after new wealth came with the discovery of North Sea oil and natural gas in the 1970s. The oil was abundant enough, and the home market small enough, to make Norway one of the world's largest crude oil exporters. In the early 2000s Norway built pipelines to carry natural gas to the U.K. as the latter's gas supplies depleted.

Finland was within the Swedish kingdom from the 1100s to 1809, when it became a Russian possession. Finnish independence was declared in 1917, but Stalin's armies attacked Finland in 1939 with the intent of reincorporating Finland into the Soviet Union. The Finns fought off the attack but sought a closer alliance with Nazi Germany to protect their country's independence. As the Red Army rolled back Hitler's forces, it launched new campaigns against Finland. Vastly outnumbered, the Finns sued for peace, surrendering strips of land adjacent to the Soviet Union. They agreed not to join any anti-Soviet political organization, including NATO and the EU. The Soviets were particularly concerned about Finnish broadcasting, which could be received in nearby Estonia, a Soviet republic where the language was so closely related to Finnish that Finnish was easily understood. After the dissolution of the Soviet Union in 1991, Finland was freed from its forced neutrality, quickly joined the EU, and established close links with the Baltic countries, especially Estonia.

People: Ethnicity and Culture

Culturally, the Scandinavians (Swedes, Danes, Norwegians) are Germanic peoples, specifically the northern branch, and descendants of Vikings. Their languages and histories are closely related (see Figure 3.4a). The inhabitants of the Faroe Islands, Iceland, and Greenland are descendants of early Scandinavian settlers and likewise have similar cultures, though Greenland also is inhabited by small numbers of Inuits. In contrast, the Finns and Sami (commonly but derogatorily called Lapps) are Finno-Ugric peoples whose languages, related to Hungarian, are not Indo-European like most of the other languages of Europe and bear no resemblance to the Germanic languages of the Scandinavians. The Sami practice a traditional nomadic lifestyle of herding reindeer. They are found in the far northern reaches of Norway, Sweden, Finland, and adjacent areas of Russia. Since the mid-1990s the Sami have cashed in on a "Santa Claus" industry that flies in children and their families during the winter season.

Evangelical Lutheran Christianity is the major religion for northern Europeans (see Figure 3.4b). Officially, 90 percent or more of the population are Lutherans in the four major countries and in Iceland. This Protestant variant of Christianity influenced the lives of the people, inducing very serious and community-conscious attitudes toward work and social life. In recent years, the combination of affluence and materialism broke many of these strong cultural links and loosened the control exercised by the churches.

Economic Development

The economies of the countries of Northern Europe relied on primary products until development of manufacturing and service industries in the 1900s. Denmark is a major agricultural country; 75 percent of Denmark's lowland area on the Jutland peninsula and on the islands between the peninsula and Sweden is farmed with an emphasis on dairy and livestock products (see Figure 3.29). Denmark also commonly has the largest fish catches in the EU and engages in world shipping. Sweden has agriculture in the south, while the north has significant timber and one of the world's major iron-mining industries. The Swedish sawmill industry is Europe's largest and accounts for about 10 percent of the world's exports. Sweden was also the world's third largest exporter of pulp and paper in the early 1990s. Finland is another major wood-producing country, and Norway has fishing and shipping industries. The discovery of major oil and gas reserves beneath the North Sea brought new wealth to Norway from the 1970s (see Figure 3.37).

In terms of industry, Denmark is known around the world for Tubourg beer and the toy company Lego, though Denmark has many other industries that manufacture furniture, handicrafts, high-tech medical goods, automatic cooling and heating devices, stereo equipment (Bang & Olufsen), and sensitive measuring instruments. The Finnish company Nokia is world renowned for its mobile phones. Finland itself has more mobile phones per capita than any other country, with 65 cellular phones per 100 inhabitants. Finland's other important industries are glassware, metal, machinery, and shipbuilding.

Sweden is the largest and most industrialized Northern European country. In addition to forestry, its biggest industries are in engineering, iron and steel, chemicals, and services. Engineering products account for the largest share of Sweden's manufacturing industry. Swedish engineering companies such as SKF, ABB, and Ericsson and inventions such as the ball bearing gave Sweden a good worldwide reputation in this sector. Volvo and Saab are well-known automotive companies but Saab also produces commercial and military airplanes. Many of Sweden's traditional mechanical manufacturing companies diversified into the electronics industry. ABB is the largest producer of industrial robots in Europe. In telecommunications, Ericsson sells digital exchanges and phone systems. The Swedish chemical industry originally produced matches and explosives, but paint and plastics grew in importance after World War II. In the last few decades, pharmaceuticals became important with companies such as Astra and Pharmacia & Upjohn. The largest number of people in Sweden are employed in the service sector: financial, educational, and medical services. Much of the employment is in the public sector, but high costs led to increasing privatization.

The four largest Scandinavian countries have some of the highest GDP per capita figures in the world. In recent decades, Sweden maintained its income level, while Finland (new industries such as Nokia mobile phones), Norway (oil and natural gas), and Denmark (high-tech industries) increased their affluence. The small, stable populations made it possible to develop societies in which poverty is limited, as reflected in the ownership of consumer goods (see Figure 3.18).

Test Your Understanding
3D

Summary

Western Europe contains many of Europe's former colonial and imperial powers, and it is where the Industrial Revolution began in the 1700s. Countries of this subregion are still powerful today and rank high on the human development index.

Three of the four largest countries by population in Europe are in Western Europe (France, Germany, and the United Kingdom). The population of Western Europe is static in overall numbers and aging. Decreasing numbers of people in the working age groups support increasing numbers of retired people. In the 1960s and 1970s, immigration was encouraged to bring in guest workers.

The manufacturing industries modernized in type and levels of productivity. Service industries, especially financial and tourism aspects, are the main employers, contributing increasingly to the continuing affluence of the subregion.

Northern Europe consists of affluent countries with small populations. Denmark and Sweden were once great empires, but in recent times Northern Europe is known for its neutrality and concern for human rights. The economies of these countries were based on primary products from farm, mine, forest, and ocean, but now income is from manufacturing and services.

Questions to Think About

3D.1   How have Western and Northern Europeans contributed to the struggle to achieve human rights around the world?

3D.2   Why do so many difficulties confront the movement toward peace in Northern Ireland?

3D.3   From where around the world are migrant workers and refugees coming to Western and Northern Europe? Which specific European countries are they going to?

3D.4   What are Western and Northern Europe's important industries?

3D.5   What factors account for the economic well-being of the Northern European countries?

Key Terms

Benelux   productivity
guest worker   fjords
producer services

Mediterranean Europe

Mediterranean Europe consists of four large countries—Portugal, Spain, Italy, and Greece—and five small ones: Andorra, Monaco, Vatican City, San Marino, and Malta (see Figure 3.40 and Table 3.2). Gibraltar remains a British colony, although discussions between the U.K. and Spain continue over its future. Portugal is not strictly a "Mediterranean" country, because its coast faces the Atlantic Ocean, but it is part of the Iberian Peninsula with Spain, and its history and culture are closely linked to that of its neighbor. Southern France is also oriented to the Mediterranean Sea, but its position within France and, therefore, governance from Paris makes it more conveniently included in Western Europe.

From ancient Greek and Roman ideas to those of the Italian Renaissance, Mediterranean Europe played a major role in directing the early course of Western civilization. In the 1400s, Portuguese and Spanish exploration launched the Age of Discovery and soon marked the beginning of European colonization of other lands and peoples. With colonization, the Portuguese and Spanish transplanted their languages and Roman Catholicism around the world. By the Industrial Revolution of the 1800s, however, Mediterranean Europe had lost much of its political power, and Greece was part of the Ottoman Empire. In the 1900s, differences grew among the Mediterranean countries. Italy became the most industrialized and Spain followed, while Portugal and Greece still rely on agriculture, fishing merchant marines, and tourism for much of their overseas income.

FIGURE 3.40 Mediterranean Europe: the countries, cities, and physical features. The countries occupy peninsulas extending southward into the Mediterranean Sea.

Countries

Portugal is one of the oldest countries in Europe, tracing its roots to the year 1143; but it did not attain its current territory until the 1400s, when the Moors from north Africa were finally expelled from the Iberian Peninsula. Portugal's boundaries also have been the most stable in Europe. Spain, too, traces its roots to the 1100s through the Kingdom of Castile, but did not attain its current control of most of Iberia until the 1500s.

With the riches gained from the sea voyages of the 1400s, Portugal and Spain emerged as the first European imperial powers. Competition between them led the pope to divide the world between them by the Treaty of Tordesillas (1494), designating a line of longitude 370 leagues (approximately 1,500 km or 1,000 mi.) west of the Azores as the eastern boundary of Spanish influence and the western boundary of Portuguese influence. While Portugal colonized the coast of what became modern Brazil (see Chapter 10) and parts of Africa and Asia, Spain established its rule in the rest of Latin America and some Pacific Ocean islands.

The subsequent rise of France, the Netherlands, and the U.K. as colonial powers sharply curbed the global influence of Portugal and Spain. By the early 1800s, both countries lost their Latin American colonies, although they maintained African colonies until the 1970s (see Chapter 9). Two dictators, Antonio Salazar, who came to power in Portugal in 1926, and Francisco Franco, who took control of Spain in 1938, kept these two countries inward looking. Following their deaths in the 1970s, Portugal and Spain became more democratic and outward looking, granted their remaining colonies independence, and joined the European Community (now EU) in 1986.

Spain is much larger than Portugal, but great devolutionary forces are working against it. The Basques, descendants of a people who lived in Europe before the arrival of the Indo-Europeans, seek independence. They live in northern Spain around the city of Bilbao and also in southwestern France. Some use violence to achieve their goal. The Catalans on the northeast coast, where Barcelona (Figure 3.41) is their main city, also press for greater autonomy. Catalans (17 percent of the Spanish population) frequently prefer to fly their own flag and speak their own language that is close to Spanish but still distinctive. The Galicians, north of Portugal, have been less vocal but are culturally different than Castilian Spaniards, having more in common with Portuguese culture.

Though Greece and Italy are hearths for some of Europe's oldest civilizations, the modern nation–states of Greece and Italy came into existence only in modern times. Greece emerged as a nation–state in 1832 but did not take its current shape until the early 1900s when Ottoman control ended. Italy is the youngest of the major Mediterranean nation–states, coming into existence in 1861 and experiencing some boundary changes as late as the 1940s. Regionalism within Italy is very strong, with most Italians considering themselves Sicilians, Tuscans, Venetians, and so on first and Italians second. A strong north–south

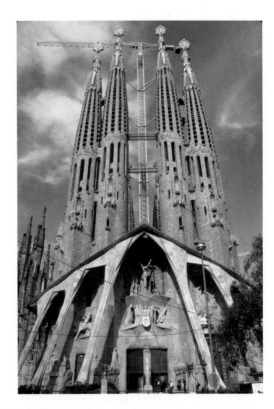

FIGURE 3.41 Barcelona, Spain. Barcelona is world renowned for its Temple de la Sagrada Familia. Begun in 1882 and designed by the famous Catalan architect Antoni Gaudi (1852–1926), this church is far from complete. **Photo:** © Loren W. Linholm.

differentiation is part of this regionalism. The south is still very rural, Roman Catholic, and under the control of familial organizations such as the Mafia. The Mafia's main reputation is as a crime organization, but it is also a way of doing business through family connections. In contrast, the north is very urban, modern, industrialized, and materially wealthy. The Communists of northern Italy, centered in the industrial towns, long gave Italy the distinction of having the largest Communist Party in non-Communist Europe. The Communist Party's influence waned in the 1990s as the Northern League had tremendous growth. The Northern League is a political party and movement that seeks greater autonomy for northern Italy. Unlike the Communists, who draw their support from factory workers, the Northern League is made up of young entrepreneurs engaged in activities tied to the modern global economy.

People: Ethnicity and Culture

The Portuguese, Spanish, and Italians all speak languages in the Romance (also Latinic) branch of the Indo-European language family (see Figure 3.4a) and are Roman Catholic (see Figure 3.4b), having inherited both characteristics from the Romans. Though the Greeks were also in the Roman Empire, their culture predates the Romans and remained distinctive during and after Roman control. The Greeks are Eastern Orthodox Christians, as are many people in East Central Europe, Russia, and its neighboring countries. As a language, Greek has its own branch within the Indo-European family.

In Italy, there are only tiny areas in the northeast (German) and northwest (French) where Italian is not the dominant language. The modern differences between northern and southern Italy grew out of the medieval occupation of the region south of Rome by Muslims and its later domination by feudal systems under French and Spanish kings. The northern city–states maintained trading links and developed industrial output at an earlier date. It was not until 1861 that Italy achieved unification.

Economic Development

The four main countries of Mediterranean Europe remained as peasant-farming countries with feudal or fragmented types of social organization while Northern Europe industrialized in the 1800s. Industrialization began in the late 1800s in northern Italy and in the Catalonian region around Barcelona in northeastern Spain. Most modernization occurred after World War II and the incorporation of the Mediterranean countries in the EU. Industrialization in Greece and Portugal has been very modest. The two countries remain among the poorest of the EU countries and receive large sums of regional development funds (see Figure 3.28).

Italy's GDP is almost twice the total of the other Mediterranean Europe countries and is close to that of the U.K. As one of the world's largest economies, Italy is a member of the Group of Eight (G8). The country's economic power comes mainly from northern-based and high-tech industries. The Po River valley between the Alps and the Apennines is the largest center of manufacturing in Mediterranean Europe. For example, Fiat manufactures automobiles in Turin. Venice, at the mouth of the Po River, is famous for its glass manufacturing. Milan, the largest city in northern Italy, is a major center of financial and other service industries as well as a producer of a diverse group of manufactured goods including tractors, domestic electronic goods, china, fashion, and pharmaceuticals. Milan and its surrounding towns produce nearly one-third of the Italian GDP and form one of the major growth areas of Europe.

Agriculture

Agriculture is very important to the Mediterranean countries. The warm dry summers and cool winters that seldom drop below freezing allow the Mediterranean countries to produce crops that are difficult to grow in the other subregions of Europe such as olives, table and wine grapes, citrus fruits, figs, and specialized cereal grains for pasta. In addition, Portugal produces most of the world's cork for wine bottles, obtained from the bark of the cork oak (Figure 3.42). Furthermore, specialization has led to **market gardening,** the commercial production of these crops.

FIGURE 3.42 A cork oak forest in Portugal. Bottle corks are made from the bark of these trees. Little processing is required. Once the bark is stripped from the tree, bottle corks are cut. Numbers on the trees refer to a calendar year and indicate when the bark can be stripped from the tree again. **Photo:** © Alexander B. Murphy.

Tourism

The Mediterranean's warm climate, sunny beaches, and historic centers attract tourists, making tourism a major industry. Each year, Portugal and Greece each receive 12–13 million tourists. Italy and Spain typically each receive 40–50 million tourists per year. As a result, English and German are commonly understood in the tourist areas of these countries. In Portugal, the Algarve (along the southern coast) and the Madeira islands are popular. In Greece, many tourists are found on the islands or in the capital, Athens, Barcelona, Madrid, the cities of southern Spain, and especially the beach resorts of the southeast and Balearic Islands account for most tourist visits to Spain. In Italy most tourists go to the historic centers of Rome, Florence, Venice, and Pompeii (near Naples) (Figure 3.43 and Figure 3.8). The east coast beaches, Sicily, many smaller towns with their art treasures, and the winter

FIGURE 3.43 Mediterranean Europe: Italian tourist attractions. Florence has its cathedrals and art galleries by the Arno River. **Photo:** © Michael Bradshaw.

sports and lake resorts in the Alps attract many others. Venice has become so popular with international tourists that the city has many more tourists than residents. An increasing problem for Venice is rising sea level. While global warming is debated, sea-level rise has resulted in seawater breaching canals and flooding streets more and more every year (see the "Natural Environment" section in this chapter).

East Central Europe

Modern East Central Europe is made up of the Baltics (Estonia, Latvia, and Lithuania), Poland, the Czech and Slovak Republics, Hungary, and what is often called the Balkans: Slovenia, Croatia, Bosnia-Herzegovina, Serbia and Montenegro, Macedonia, Albania, Romania, and Bulgaria (Figure 3.44 and Table 3.2). These countries are grouped together because Communist forms of government and economies were imposed on them shortly after World War Il ended in 1945 (see the section on "Global Changes and Local Responses" in this chapter). Most of these countries were directly controlled by the Soviet Union and called Soviet satellite states. The Baltic countries were incorporated into the Soviet Union from 1945. Though Soviet domination ended from 1989, these countries share common experiences in moving from Communism to more democratic forms of government and capitalist economies.

Countries

The countries of East Central Europe emerged as nation–states in the late 1900s and early 2000s. From the beginning of the nationalist idea in late 1700s to the end of World War I in 1918, most of East Central Europe was dominated by four great empires—the Russian, German, Austro-Hungarian, and Ottoman empires (Figure 3.45a). World War I started in the subregion—specifically in Sarajevo, Bosnia-Herzegovina—when a Serb nationalist assassinated Archduke Ferdinand of the Austro-Hungarian Empire. In the aftermath of the war, diplomats met in Paris in 1919 to create a lasting peace by redrawing Europe's boundaries. Sir Halford J. MacKinder, a British geographer and Member of Parliament, had long argued that future peace could only be ensured by separating the German and Russian empires. He proposed the creation of a zone of small countries for this purpose. Woodrow Wilson argued for the right of "national self-determination" for the peoples of Europe. With both MacKinder's and Wilson's complementary ideas, a series of new or modified countries was created from Finland to Greece (Figure 3.45b). It was at this time that Yugoslavia and Czechoslovakia came into existence. With them, the term "Eastern Europe" came into use.

After World War II, Stalin's Red Army moved into most of Eastern Europe. Through Stalin's manipulations, including boundary changes, especially affecting Poland (Figure 3.45c), their governments and economies adopted Communist forms. Winston Churchill said that an "Iron Curtain" had fallen across Europe. Politically, Europe was divided into East and West during the Cold War. The term "Central Europe" disappeared with

FIGURE 3.44 Central Europe: countries, cities. This subregion has been a buffer zone between the major powers of Germany and Russia for centuries, and the countries are in transition from a period of domination by the Soviet Union that ended in 1991 to subsequent orientation westward to the rest of Europe.

the Nazis, who used it as a synonym for the empire they wished to create. "Eastern Europe" then referred to the Communist part of Europe between the Soviet Union and Western Europe. The East Germans were added while the Baltic countries were excluded because they were incorporated into the Soviet Union. Finland and Greece were likewise excluded from the concept of Eastern Europe because they did not adopt communism.

By the dissolution of the Soviet Union in 1991, many of the peoples of this subregion resurrected the old concept of Central Europe and emphasized that they were Central Europeans and not Eastern Europeans, a term that had taken on very negative connotations. The peoples of the Baltic countries also readopted the term "Central Europe" for themselves. Though many of these countries were part of an older Central Europe, they represented the portion east of the Germans. Thus we have the term "East Central Europe." During the Cold War, the other

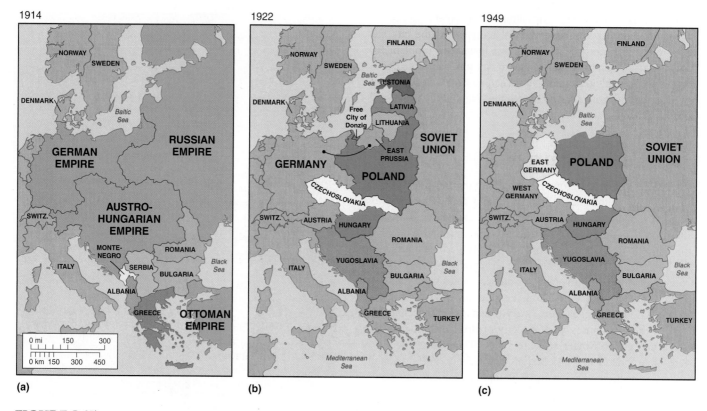

FIGURE 3.45 **The boundaries of East Central European countries at three different dates in the 1900s.** Compare to the contemporary map of the subregion in Figure 3.44. What are the locations of countries such as Poland, Latvia, and Yugoslavia at each of these times? **Source:** From Demko & Wood, *Reordering the World,* 1st ed.

Central Europeans—the Germans, Austrians, and Swiss—became so Western oriented that they are rarely labeled Central Europeans anymore, or even "West Central Europeans."

Devolutionary processes occurred in East Central Europe after the end of Soviet domination from 1989. The declarations of independence from the Soviet Union by the Baltic republics are one example. Czechoslovakia divided into separate Czech and Slovak Republics in 1993. In these cases, the movements toward independence advanced peacefully. The Czechoslovak situation was referred to as the "Velvet Divorce," coined from the "Velvet Revolution" of 1989 that peacefully ended Communism in that country.

"Yugoslavia"

Devolutionary processes also occurred in Yugoslavia but were not so peaceful as the country broke apart. The Yugoslav case is important because it illustrates the ugly aspects of fervent nationalism, and it changed international laws concerning human rights. The Yugoslav situation also took on international significance as it involved other European countries and the United States, including NATO, as well as Russia. As of 2002, Yugoslavia ceased to exist as a country. The country's last two republics abandoned the name and now call themselves Serbia and Montenegro. Slovenia, Croatia, Bosnia-Herzegovina, and the Former Yugoslav Republic of Macedonia (FYROM) were republics of Yugoslavia but seceded in 1991 and 1992. For some republics such as Slovenia, independence was achieved relatively quickly and without much bloodshed. For others such as Bosnia-Herzegovina, the opposite was the case (Table 3.4).

Prior to World War I and the creation of a Yugoslav state, Serbia and Montenegro were independent countries. Both, however, achieved independence only in 1878 and obtained much of their territories during the Balkan wars of 1912–1913. The Slovenes and Croats had lost their independence in the Middle Ages, eventually finding themselves divided among a number of provinces within the Austro-Hungarian Empire before World War I.

In the late 1800s Serbia's leaders planned to annex to Serbia the territories of their South Slavic brethren in Austria-Hungary. Serb nationalists were upset when Austria-Hungary occupied Bosnia-Herzegovina in 1878 and outraged when Austria-Hungary formally annexed the territory in 1908. Not long afterward, a Serb nationalist sought revenge by assassinating Austria-Hungary's Archduke Ferdinand when he visited Sarajevo in 1914. The episode began World War I.

Yugoslavia came into being after World War I. At the time, it was called the Kingdom of the Serbs, Croats, and Slovenes (Figure 3.46a), reflecting the fact that the new country incorporated a

Table 3.4 Time Line of Yugoslavia

1878	Serbia and Montenegro become independent; Austria-Hungary occupies Bosnia-Herzegovina.
1908	Austria-Hungary formally annexes Bosnia-Herzegovina.
1912–1913	First and Second Balkan Wars. Serbia, Bulgaria, and Greece annex portions of Macedonia.
1914	Serbian nationalist assassinates Austrian Archduke Ferdinand in Sarajevo. Austria-Hungary declares war on Serbia. World War I begins.
1919	Paris Peace Conference. The Kingdom of the Slovenes, Croats, and Serbs is created and includes Montenegro, Bosnia-Herzegovina, and Vojvodina.
1929	The Kingdom of the Slovenes, Croats, and Serbs becomes "Yugoslavia."
1941	Yugoslavia is dismantled and occupied by Axis powers during World War II.
1945	Josip Tito, leader of the Partisans, who are Communists, takes control of a re-created Yugoslavia. Six internal republics are created.
1980	Tito dies and nationalists begin expressing themselves. Slobodan Milosević eventually catapults to power by fanning ethnic hatreds.
1991	Slovenia and Croatia declare independence; bloodshed ensues as the Yugoslav military tries to prevent secession and Serbs in Krajina of Croatia declare their independence from Croatia.
1992	Bosnia-Herzegovina and Macedonia declare independence. Bloodshed ensues as many Serbs resist the independence movement.
1995	Croatian army retakes Krajina and forcibly expels Serb population. Dayton Peace Accord ends open hostilities in Bosnia-Herzegovina and divides between a Muslim-Croat federation and a Serb republic.
1999	Serb military and paramilitary units terrorize ethnic Albanians in the Serb province of Kosovo. NATO begins intensive bombing campaign of Yugoslavia. Millions of ethnic Albanians flee to Albania and Macedonia. Success of the air campaign leads to the stationing of NATO and Russian ground forces in Kosovo.
2000	Slobodan Milosević loses a national election and is no longer the leader of Yugoslavia.
2001	New Yugoslav government arrests Slobodan Milosević and sends him to the International Court of Justice in The Hague, the Netherlands, where he is the first leader of a country indicted for genocide and crimes against humanity.
2002	Yugoslavia changes its name to Serbia and Montenegro.

number of nations. In 1929 the name "Yugoslavia" was adopted. "Yugoslavia" translates from the Slavic languages into "the land of the South Slavs."

The relationships of South Slavs are close and overlapping; identities are complex and intertwined. The Slovenes and Croats are Roman Catholic, while the Serbs, Macedonians, and Bulgarians—also South Slavs—are Eastern Orthodox Christians. Other South Slavs are Muslims like many Bosnians. Many Bosnians claim that a Bosnian can also be Roman Catholic or Eastern Orthodox. Despite having differing faiths, all South Slavs speak closely related languages with the Serbs and Croats essentially speaking the same language—Serbo-Croatian—though many Serbs now say they speak Serbian and many Croats say they speak Croatian.

Many South Slavs wanted to be free of Austria-Hungary even before the war began, but they also had differing goals for freedom. Some wanted independence, but not having independent governments of their own, they lacked fully recognized diplomatic corps and had little voice at the Paris Peace Conference in 1919. On the other hand, the Serbs had a diplomatic corps that spoke for the South Slavs. The Western powers were aware that many Slovenes and Croats wanted their own independent countries, yet they felt that these countries would be too small to be economically viable. Thus they supported the Serb proposal to unite the South Slavic lands of Austria-Hungary with Serbia and Montenegro to create a single country.

Many South Slavs had serious disagreements over the meaning and purpose of the new country. Were they to form a common national identity or remain distinct? Should the country have a strong centralized government with everyone obeying the same laws, or should the provinces exercise their autonomy? While such debates continued, Serbia's capital, Belgrade, became the new country's capital, the Serbian king the new country's king, and Serbian symbols, such as the flag, the new country's symbols. Resentment toward the Serbs quickly grew and worsened as Serb nationalists used their power to insist that the Serb way become everyone else's. During World War II, the country disintegrated as some supported the Nazi Germans and others resisted them.

Following the war, the Partisans, who were Communist under the leadership of Josip Tito, gained control of the country. Tito's parents were Slovene and Croat, yet as a Communist he did not believe in nationalism, so he suppressed any expressions of national pride. Instead, Tito promoted the "Yugoslav" idea while not insisting on any one group's characteristics for Yugoslav characteristics. He generally allowed everyone to continue with their differing cultural practices but insisted that they all consider themselves Yugoslavs. Tito also set up the internal political geography of the country by creating the six republics that we see today (Figure 3.46b), with Serbia having the autonomous regions (later called provinces) of the Vojvodina and Kosovo. Each republic was named after and dominated by one national group except for Bosnia-Herzegovina (see Figure 3.47 and "Personal View: Bosnia-Herzegovina," p. 122).

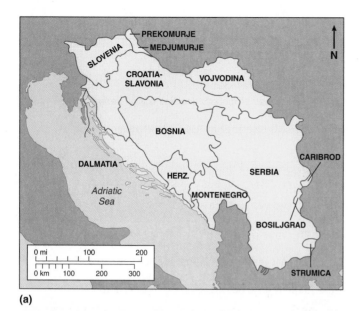

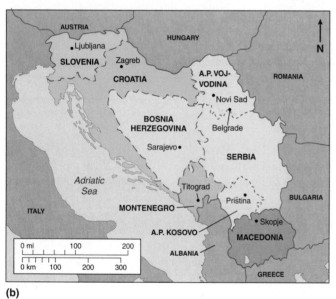

FIGURE   3.46 Yugoslavia. (a) The territorial formation of Yugoslavia in 1919: (b) Yugoslavia's administrative boundaries under Tito (1945–1980). The boundaries lasted until 1991. **Source:** (a) from Stephen Clissold (ed.), *A Short History of Yugoslavia.* Copyright © Cambridge University Press, 1966; (b) From Fred Singleton, *A Short History of Yugoslavia Peoples.* Copyright © Cambridge University Press, 1985.

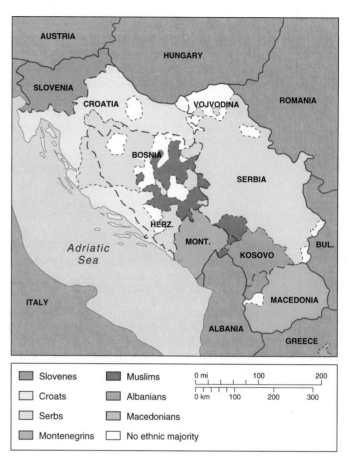

FIGURE 3.47 Importance of ethnic differences: the former country of Yugoslavia in the Balkans of Europe. After 1991, Yugoslavia broke into five independent countries, although Serbia and Montenegro continued to call themselves Yugoslavia until 2002. The independence movements resulted in war in Croatia and Bosnia-Herzegovina. Kosovo (Serbia) experienced violence in the late 1990s, bringing U.S. and NATO involvement in 1999. The colors show areas where particular ethnic groups comprise over 50 percent of the population.

The nationalists of various groups were unhappy with Tito's policies and practices, but Tito kept them in check with his charisma and heavy hand. After Tito died in 1980, the politics within Yugoslavia changed. Nationalism and old resentments resurfaced. Slovenes and Croats felt that the Serbs had forced them to become more Serbian. Slovenia and Croatia were the wealthier republics, and their people had strong feelings against the profit from their hard work going to subsidize the poorer republics. In 1991 Slovenia and Croatia declared independence, and the Yugoslav army attempted to prevent the secession movements. For Slovenia, the fight was brief. Within a short time, Belgrade accepted Slovenia's independence. The situation in Croatia was bloodier. Twelve percent of Croatia's population was Serb, most of whom lived along Bosnia's border in a region known as Krajina ("the borderland") and in eastern Croatia. The Belgrade government, dominated by Serbs, used the Yugoslav army, also dominated by Serbs, to help the Serbs of Croatia establish their own republic. Although Croatia achieved independence, it could not control a significant piece of its territory.

Yugoslavia without Slovenia and Croatia, not to mention Macedonia, which was making plans for secession, left Bosnia-Herzegovina in a Yugoslavia more dominated by Serbs than ever before. Before long, Bosnia-Herzegovina also began moving toward secession. Bosnia-Herzegovina's ethnic complexity, however, made secession a difficult issue. In the 1991 census, 44 percent of the population was Muslim, 31 percent Serb, 17 percent Croat, and 8 percent of other groups. Many from all the groups supported the independence of their republic. Serb nationalists were the greatest opponents, and with support from the government in Belgrade and the Yugoslav army, they began an armed fight to keep Bosnia-Herzegovina within Yugoslavia. Eventually, Croat nationalists turned against government forces with the goal of incorporating territories of the republic into Croatia. Without a neighboring country to support them and under attack from two directions, the Muslims remained loyal to Bosnia-Herzegovina. Serb and Croat nationalists claimed that Muslims were oppressing them in an attempt to make Bosnia a Muslim state. Ironically, Serb and Croat nationalists helped to make Bosnia-Herzegovina more Muslim as they tried to separate its non-Muslim territories.

The international community proposed a number of partition plans to end the conflict. However, the geographic distribution of groups within the republic was complex. Though some groups were associated with particular areas, other patterns existed. For example, Muslims tended to live in the cities, while Serbs were in the surrounding countryside. In 1995 the United States became involved and brokered the Dayton Peace Accord. Though recognizing the integrity of the republic, it temporarily divided the republic into two "entities." The north and east came under the Republika Srpska (Serb Republic) with the capital in Banja Luka. The rest became part of a Muslim-Croat Federation with the capital in Sarajevo. The situation is peaceful, but tensions and animosities remain, as do UN peacekeepers. At the same time, Bosnia-Herzegovina's economy is in shambles. Unemployment has been as high as 70 percent in many communities, and the country has relied heavily on foreign aid. Bosnia's territory is used to funnel prostitutes and illegal immigrants into and across Europe.

Macedonia was fortunate to escape bloodshed when it negotiated the withdrawal of Yugoslav troops in April 1992. Because war broke out at the same time in Bosnia-Herzegovina, the Macedonians may have benefited from a decision in Belgrade not to engage the Yugoslav army in two conflicts at the same time. Nevertheless, Macedonia was still in a precarious position.

Macedonia is an old region, having existed in ancient times. Alexander the Great, for example, was from Macedonia. Compared to the modern republic, historic Macedonia stretches beyond the boundaries of the former Yugoslav republic into western Bulgaria and northern Greece. Serbia, Bulgaria, and Greece fought the Balkan wars of 1912–1913 to gain control over the territory. Though Serbs, Bulgarians, and Greek nationalists all claimed Macedonia on the grounds that Macedonians were really members of their own respective nations, all three agreed that Macedonians were not their own distinct nation. The situation changed when Tito recognized the Macedonians, created the Macedonian republic, and recognized the Macedonian language. The Bulgarian and Greek governments still do not recognize the Macedonians as a separate people. Fearing that the Macedonian government would claim historic Macedonia, the Greek government blocked international recognition of the country as long as it insisted on calling itself Macedonia

until a compromise was found for the country's official name: the Former Yugoslav Republic of Macedonia (FYROM). This name prevents the country's government from claiming historic lands in either Greece or Bulgaria.

Though Belgrade allowed Macedonia (FYROM) to secede without a military struggle, it also left Macedonia landlocked and surrounded by hostile countries. In addition, Macedonia was the former Yugoslavia's poorest republic. Twenty percent of its GDP still comes from agriculture, and unemployment is around 30 percent. Moreover, 23 percent of Macedonia's population is Albanian. In 2001 Albanian nationalists began an armed insurgency. NATO forces are helping local peoples to achieve and maintain peace, as they also are in Bosnia-Herzegovina.

Together, Serbia and Montenegro contain almost half of former Yugoslavia's population with around 11 million inhabitants. With 62 percent of the population, Serbs are clearly the majority. Montenegro (meaning "black mountains") is by far the smaller of the two republics in the new federation, with Montenegrins representing only 5 percent of the country's population. Montenegrins are very closely related to the Serbs. Some simply call them "mountain Serbs." Some Montenegrins have been unhappy with Serb policies and seek independence for their republic. Comprising almost 17 percent of the population and found mostly in Kosovo in the south where they are more than 90 percent of that province's population, Albanians are another sizable minority. A Hungarian minority lives mostly in Vojvodina in the north of Serbia.

Though many of the nations/republics of Yugoslavia seceded from the country because they feared or were tired of Serb nationalism, the Serbs had a different view of the situation. Like majority groups in other countries, many Serbs believed that their wealth was squandered by Yugoslavia's minorities. They saw their monies invested in the poor areas, many of which were Muslim and did not benefit the Serbs. Additional fears were fanned by the fact that these minorities had higher population growth rates. Serbs also felt that they were persecuted by Slovene and Croat nationalists. Slobodan Milosević catapulted to power by playing on these fears. He used the Yugoslav army and Serbian paramilitary units to "protect" Serbs. After losing these wars and bringing ruin to the Yugoslav economy, Milosević began a war against the Albanians of Kosovo. NATO bombing stopped the persecution, then NATO troops were stationed in the province. Lost wars, Kosovo occupied by foreign troops, and a severely damaged economy led Serbs to oust Milosević in elections at the end of 2000. Not long after, Milosević was taken into custody and charged with crimes against humanity and genocide at the International War Crimes Tribunal for the Former Yugoslavia (ITCY) in The Hague, the Netherlands. Many Serbs supported Milosević because he was supposedly protecting Serbs throughout the former Yugoslavia. Until recently, only a few had begun to realize the atrocities that he committed in pursuing these goals. Perhaps they will learn more as Milosević's trial unfolds.

BOSNIA-HERZEGOVINA

The wars that ensued in Yugoslavia after the country began disintegrating in 1991 were particularly bloody in Bosnia-Herzegovina, lasting from early 1992 until the end of 1995, when the Dayton Peace Accord was brokered (Box Figure 1). Bosnia-Herzegovina was different from other republics in that it was not named after an ethnic group that inhabited it. Instead, it was named partly after a river and partly after a medieval territory. Bosnia derives its name from the Bosna River. Herzegovina stems from the German word herzog, which means "duke." Therefore, Herzegovina literally means "dukedom." Bosnia-Herzegovina most likely kept its historical name because no single group occupied the territory to justifiably change its name for that group (see Figure 3.49). For a long time, Bosnia-Herzegovina's multiethnicity worked well. The people of the republic were even able to showcase their ethnic harmony to the world when they hosted the 1984 Winter Olympics in Sarajevo.

The war destroyed Bosnia's ethnic harmony and caused many to flee and seek refuge elsewhere. Miro Paunović and his family are one such example. Miro was born in Mostar and lived there for most of his life until the beginning of the war in 1992. Mostar is the main city of Herzegovina and is internationally known for its stone bridge. Referred to locally as the "Old Bridge" (*stari most*) (Box Figure 2), it was built by the Ottoman Turks in 1561 and came to symbolize ethnic harmony within all of Bosnia-Herzegovina. Sadly,

the bridge was destroyed by the Croatian army on November 9, 1993, the anniversary of *Kristalnacht* (the night when Nazis attacked Jews and Jewish businesses throughout Germany in 1938). Miro has fond memories of the bridge and will never forget how people would travel from all over Yugoslavia every summer to compete in a bridge diving contest. Miro was really happy when the bridge was rebuilt in 2004.

Miro also misses the old town center where people would come to shop and meet at cafes to socialize. In terms of its site, Miro says that Mostar is not unlike Cumberland, Maryland, where he now lives. Both cities are crowded down along rivers that wind their way through mountains. Mostar's river is the Neretva, which flows south into the Adriatic Sea, some 50 miles away. The short distance to the sea made it easy for Miro and his family to travel to the beaches of the Adriatic, often for just one day. For vacations, Miro's family would typically spend a couple of weeks on the coast, usually traveling to the famous cities of Dubrovnik and Split. Mostar's proximity to the coast means that both the summers and winters are relatively mild. Skiing is popular in this mountainous country, but those interested in doing so must travel further inland, to the north and east, to places such as Sarajevo before they can find enough snow to ski.

Mostar is a regional center for southeastern Bosnia-Herzegovina. Its factories produce a variety of goods, including aluminum, military airplanes, car and truck parts, and clothing. The mountainous character of Bosnia-Herzegovina means that the country produces far less agriculturally than many of the other Balkan countries. Nevertheless, some agricultural goods are produced. The area around Mostar is known for wine grapes, nectarines, and plums. Not surprisingly, Mostar has a winery and a distillery that makes a hard alcohol called *Loza*.

When Miro went to school as a young boy in Mostar, the educational system was somewhat different than what his children are now experiencing in the United States. All children went to elementary school, which lasted until the eighth grade. Foreign languages were taught in Miro's school beginning in the fifth grade. English, German, Russian, and French were the choices. Miro selected English, probably the most popular foreign language at the time. From the ninth through twelfth grades, young people went to specialized high schools, of which Mostar had five. The *Gymnasia* was for those bound for university. Another school emphasized automotive training, while a third focused on working with industrial machines. A fourth specialized in food technology, and the fifth centered its curriculum around economics. According to Miro, this latter school went through the greatest transformation in recent years. During Communism the school emphasized the teachings of Marx and Engels but has since changed to the modern principles of capitalism. Miro states that school was quite rigorous in Mostar. He considers what his children are now learning in school in the United States and notes that he learned the same information when he was two to three years younger than American students.

Miro went to the automotive school and received a diploma in auto mechanics. Like all young males at the time in Yugoslavia, Miro had to complete a year of military service before pursuing his career. When he returned to Mostar after his military service, Miro immediately accepted a job in a bar before ever looking for a job as an auto mechanic. He had the position for about three years when the war broke out in April 1992. Like

Box Figure 1 Map of Croatia and Bosnia.

Box Figure 2 Mostar, Herzegovina. The "Old Bridge" (*stari most*) was built by the Ottoman Turks in 1561 and came to symbolize ethnic harmony within all of Bosnia-Herzegovina. The bridge was destroyed by the Croatian army on November 9, 1993, but with international support is being rebuilt. **Photo:** © Ronald Wixman.

most people in Bosnia-Herzegovina, Miro did not fit easily into any of the ethnic categories of Serb, Croat, or Muslim. Usually people were identified according to their father's status. In Miro's case, his father was a Serb from Serbia, not Bosnia-Herzegovina. Miro's father had come to live in Mostar many years before while serving in the Yugoslav army. While in Mostar, he met and married Miro's mother, a Muslim, and stayed to raise a family. Like many Muslims in Bosnia-Herzegovina, Miro's mother was nonpracticing. With Mostar also having a large Croat population, it meant that Miro did not force himself into any one category, nor did he have to for most of his life. As far as he was concerned, he was a Yugoslav. He, like everyone else, had the right to travel anywhere throughout Yugoslavia and associate with whomever he pleased regardless of religion.

When the war began in 1992, his family found itself in a difficult situation. Serbs were a minority in Mostar. Moreover, Miro's parents were not categorized together in the same group, although it should be noted that Miro's parents were divorced for a few years by this time. The Yugoslav army offered Serbs where they were minorities in Bosnia-Herzegovina safe passage to Serbia proper. Miro and his sister, Maja, were airlifted to Zaječar in Serbia to join their father, who had moved back to his hometown following his divorce from Miro's mother in 1988. Miro's mother, Melva, stayed in Mostar until some of her family became victims in the war. Her niece Maja, who had the same name as her daughter, was wounded. Maja lost her leg when the shell from a bomb landed in front of her apartment building in Mostar. The war prevented Maja from receiving proper medical care, and her leg had to be crudely amputated. Maja then fell victim to a bone infection. Thanks to the help from Veterans for Peace, Maja was taken to a hospital in Cumberland, Maryland, for proper medical treatment. By the time Maja was ready for transport to the United States, her

father was also wounded and could not accompany his daughter. Melva undertook the journey with her. Maja's father was able to join his daughter later.

After Miro's mother was settled in Cumberland, she contacted Miro and his sister in Serbia and asked them to join her, their cousin, and their uncle in Cumberland. By this time, Miro had met a woman named Nataša and married her. They had a son named Milan. Nevertheless, Miro welcomed the opportunity to go to America, so he and his new family moved with his sister to Cumberland in 1995.

Miro first obtained a construction job that required him to drive 70 miles to work every day. Even though Miro never had to commute so far for work in Bosnia, he was happy to be working and living in a peaceful place. After some time and with a lot of hard work, Miro and his mother were able to purchase a house in Cumberland. Miro's mother and his sister lived in the lower half of the house, while Miro lived with his young family in the upper part. Miro's wife, Nataša, had another child, named Nina. Miro was happy to have his relations in close proximity as is the custom in Bosnia-Herzegovina. In many ways, Miro's new family was as complex as the one that he was born into. Miro had Bosnian citizenship while his wife and son were citizens of Serbia and Montenegro. Miro's youngest child, Nina, was born an American. In time, however, Miro's entire family became Americans.

After a little more than a year, Miro took another job in a small electronic shop. It paid more but it was also 70 miles away, not far from his previous construction job. Miro worked there for about a year before he got a job as a truck mechanic in Cumberland. It pays less than the electronics job but it is the career that Miro trained for while in Bosnia-Herzegovina, and Miro really likes working within a couple miles of home. It allows him to arrive home about the same time his children return home from school. Milan and Nina are now in elementary school and are bilingual. Milan would like to be a computer technician, an astronaut, or a car designer when he grows up. Nina wants to be an artist or zookeeper, perhaps both. Miro's sister Maja pursued and received a degree in education at Frostburg State University, married, and moved to Toronto, Canada. Miro's mother, Melva, received an occupational therapy degree at the local community college and moved to Baltimore to work at a hospital. Miro's wife, Nataša, began to take courses at Frostburg State University in the fall of 2004 but had not chosen a major yet. Miro says that he probably will pursue additional community college work in mechanics when Nataša finishes her education and begins to work.

Miro's father visited Miro's family in 2002. Retired, he found it difficult to live in the United States where the culture was foreign to him and the lack of well-developed public transportation made it difficult for him to get around. After six months, he returned to Zaječar, Serbia, where Miro's younger brother, Miloš, also lives. Nataša's parents also came to live with Miro and his family in 2002. Though English has been difficult for them, Nataša's father, Dragoslav, was able to obtain a job in an Italian bakery while Nataša's mother, Miroslava, was hired at a motel. Within two years, they were able to save enough money to put a down payment on their own house.

In addition to the family members already mentioned, three other families from Mostar live in Cumberland. One is that of Miro's cousin, and the second is that of Miro's uncle. The husband of the third family was one of Miro's friends in Mostar. Finding work, learning English, and adjusting to American culture has been difficult at times, but these families from Bosnia-Herzegovina and Serbia are overcoming these challenges and creating new lives for themselves in the United States.

People: Ethnicity and Culture

Culture

Most of the peoples of East Central Europe are Slavs. Poles, Czechs, Slovaks, and Sorbs are Western Slavs. Slovenes, Croats, Serbs, Bulgarians, and Macedonians are South Slavs. Their languages are all closely related, even to those of the East Slavs, such as the Russians (see Figure 3.4a). As Roman Catholics, the West Slavs, Slovenes, and Croats use the Latin alphabet (see Figure 3.4b). The other South Slavs—the Serbs, Macedonians, and Bulgarians—use the Cyrillic alphabet because they are Eastern Orthodox Christians. Cyrillic is a modified version of the Greek alphabet that the Orthodox monks Cyril and Methodus specifically created to convert the Slavs to Christianity. Though all the Slavs are related to their fellow Slavs the Russians, the Eastern Orthodox Slavs of East Central Europe have a closer connection to the Russians, who are also Eastern Orthodox Christians. They frequently align themselves politically with the Russians, especially the Bulgarians.

Romanian stands out as a language because it is a Romance language. With Italian as the closest related language, Romanian is the only Romance language spoken in the subregion (see Figure 3.4a). Romanians are primarily Eastern Orthodox, but their language and Western orientation led them to adopt the Latin alphabet. Romanians see themselves as descendants of Dacians, a people who lived 2,000 years ago with an empire centered in Transylvania, and of Romans. Roman ancestry gives Romanians a sense of good pedigree (Figure 3.48).

FIGURE 3.48 Deva, Transylvania, Romania. This statue of a Dacian soldier atop a pedestal with the shewolf nursing Romulus and Remus, the symbol of Rome, illustrates the Romanian belief that their nation descended from a Daco-Roman mix. Located in front of a Hungarian castle, this statue declares the legitimacy of Romanian control over the ethnically mixed territory of Transylvania. **Photo:** © Emily A. White.

Lithuanian and Latvian are also together in a category. Under the influence of the Poles, the Lithuanians became Roman Catholic. The Latvians accepted Protestantism from Swedes and Germans. The Estonians, however, do not speak an Indo-European language but rather a Finno-Ugric language like Finnish and Hungarian. The Hungarians are primarily Roman Catholics, but many in the eastern part of Hungary are Protestants. Albanian is an Indo-European language but stands alone in its own category. With 70 percent of its population Muslims, Albania is the most Muslim country in Europe, though a significant number of Albanians are Eastern Orthodox and Roman Catholic.

East Central Europe was a thriving center for Jewish and Roma (Gypsy) culture, but both groups experienced intense discrimination throughout history. The Nazi Holocaust in particular resulted in the extermination of more than 6 million Jews, not to mention many from other groups such as Slavs and Roma. Europeans who are not Roma have historically discriminated against the Roma because the Roma have comparatively darker skin and, like Jews, have non-Christian beliefs and practices. Their migratory lifestyles do not fit the settled, land-owning ways of most people. Roma still are persecuted intensely and are forced to live as an underclass.

Ethnic Tensions

The breakup of Yugoslavia in the 1990s is Europe's most recent major tragedy involving war and massive human rights violations. Because many of the ethnic groups in the conflict had differing religions, many international diplomats and commentators, as well as governments, tended to label the groups according to religion and even viewed the conflict as a religious one. They rejected recognizing such groups as the Bosnians because Bosnians did not claim to be of a single faith but rather of any faith found in Bosnia-Herzegovina. In doing so, they ignored the fact that many Serbs, Croats, and Muslims fought together to prevent Bosnia-Herzegovina from being carved up by Serb and Croat nationalists.

The fact that many groups of differing faiths have joined together during conflicts, such as the one in Bosnia-Herzegovina, illustrates that conflicts in East Central Europe are caused by more factors than simple religious differences. Historical, political, and economic geography have played great roles. In Yugoslavia, for example, Slobodan Milosević used ethnic politics to gain power. He used thugs in the region of Kosovo to attack Albanians living there. When Albanians fought back, it aroused the fears of many Serbs in Yugoslavia that this ethnic minority, which was economically poor and had high birth rates, was persecuting Serbs and destroying the Serb heartland. Milosević vowed to protect Serbs and Serb culture by calling for the crackdown on all minorities in Yugoslavia. Serb nationalists enthusiastically supported Milosević. In response to Milosević's policies, however, Yugoslavia's ethnic minorities moved toward secession.

The history of Yugoslavia's political geography made it easy for a politician like Milosević to pit the various ethnic groups against one another. Yugoslavia was a country that was created from territories of the Austro-Hungarian and Ottoman empires. Both of these empires had completely different political and economic traditions and orientations. In Bosnia-Herzegovina, for example, Serbs historically tended to live in the rural areas, and Muslims tended to live in the urban areas. The rural Serbs were comparatively poorer and less educated than the urban Muslims. These rural Serbs resented the urban Muslims, and this resentment expressed itself during the war in the 1990s. Though the two groups each had their own religion, the differing rural and urban traditions played a greater role in the conflict.

Ethnic tensions exist in other countries as well, and the study of geography helps us understand them. Albanians, for example, live beyond the borders of Albania, in the neighboring Yugoslav province of Kosovo, where they account for 95 percent of the population, and in adjacent areas of Macedonia, where they are 23 percent of the population (see Figure 3.47). Hungarians also live in the countries surrounding Hungary. Large numbers of Russians are found in the Baltic countries. The major problem is a disjunction between the distribution of nations and country boundaries. Many nations live in territories that extend beyond the countries designated for them and are seen as minorities in other groups' countries. As a minority, they are a threat to the dominant groups. For example, if Hungarians live in southern Slovakia, the Vojovodina, Transylvania, and other places, should not these territories be given to Hungary? If Albanians live in Kosovo and western Macedonia, should not these territories be awarded to Albania? This solution would certainly bring nations and countries into greater geographical alignment.

The problem with giving minorities their own countries is that it would not end the existence of minorities within countries. As minorities become majorities, new minorities come to the forefront or into being. If Kosovo became an Albanian country, it would have Serb minorities. A similar situation already occurred with the cases of the Hungarians and Russians. When Slovakia, for example, was part of the Hungarian kingdom, Slovaks were the minority and Hungarians the majority. When Slovakia was detached from Hungary, Slovaks became the majority in a new Slovakia, but the Hungarians living there became a minority. After 1991, Russians found themselves as minorities in new countries (Estonia, Latvia, Lithuania) that once had been part of the Russian-dominated Soviet Union (see Chapter 4). Because it is not possible to give every ethnic group its own country without creating new minorities, many international leaders do not support the breaking up of countries.

Some argue that minorities should leave a country to avoid conflicts. Many minorities, however, occupied their land as long as or even longer than the majority. In many cases and as shown in the examples already discussed, boundaries were once drawn differently so that minorities actually formed a majority with others of their group. For example, the boundaries of Hungary once included the Hungarians of Slovakia, Transylvania, and other areas in a Hungary that was much bigger than it is today. Tensions between Hungarians and their neighbors are not just over ethnic differences. Conflict arises over the issue of whether Hungary should be able to regain control of its lost territories. Hungarians who want Hungary to reclaim its lost territories are known as irredentists. **Irredentism** is the desire to gain control over lost territories or territories perceived to belong rightfully to one's group. With boundaries having changed so frequently in East Central European history, irredentism is a major issue and a source of much conflict, more so than conflict arising from simple cultural differences among groups.

Current majority groups, however, rarely want to lose any of their territories, though other groups may live in them. Territories often have great meaning for nations. For example, though Kosovo is more than 90 percent Albanian, Kosovo is the cradle of Serb civilization. It contains the great monasteries and other cultural artifacts of the Serb nation. Many Serbs feel that losing control of Kosovo would result in the loss of much of their culture. A dilemma then emerges. The nation–state ideal implies that a nation may only possess territory occupied by its members. Yet some territories are so important that nations will not give them up, even if their nation is not a majority in them.

To conform to the nation–state ideal, some nations feel that they must eradicate other groups from their territories to justify exercising control over their territories, though the practice certainly violates both the human rights of the victims and international law. Eradicating people from a territory because they are ethnically different is known as ethnic cleansing. Serb nationalists coined the term to refer to their own actions during the war in Bosnia-Herzegovina in the 1990s. They attempted to legitimize their claim to the republic by making the territory purely Serbian. Ethnic cleansing is now a term used to refer to similar acts perpetuated in other places around the world, not only currently but also in history. The Holocaust, for example, which refers to the Nazis' attempt to exterminate Jews and other groups during World War II, is now referred to as **ethnic cleansing,** though the term did not exist at the time.

Ethnic cleansing takes three forms: assimilation, expulsion, and extermination. All three forms were employed in the former Yugoslavia. Bosnia-Herzegovina became infamous for not only concentration camps but also rape camps. Rape is certainly used to terrorize, but it frequently had an added purpose in that war. Despite differing political and economic traditions and differences between individuals, Serbs, Croats, and Muslims were racially and linguistically similar, even sharing many cultural characteristics. The only obvious cultural difference between the groups was religious belief. Extremists thus saw religious conversion as a means of assimilation. Because Balkan peoples typically follow the tradition of raising a child in the father's faith, rape was seen as a means of conversion and thus assimilation. After being raped, many women were told that they were then obligated to raise their children in the rapist's faith, making the child a member of the rapist's nation. Many women were held in the camps so that they could not obtain abortions. Genocide is the attempt to exterminate an entire ethnic group. The attempt by Serb nationalists to end Islam in Bosnia-Herzegovina by exterminating Muslim men and impregnating Muslim women with "Serb" children was an example of **genocidal rape.**

Widespread atrocities, including rape, in Bosnia-Herzegov-ina led to changes in international law. To deal with the atrocities, the World Court in The Hague set up a special war crimes tribu-nal, known as the International Criminal Tribunal for the former Yugoslavia (ICTY). The tribunal tried numerous cases and con-tinues to do so. Previously, international law viewed rape as an unfortunate by-product of war, but the tribunal set the precedent for recognizing that rape is used as a weapon in war and deserves greater punishment than before. Rape is now officially recog-nized as a war crime. In one case, the tribunal convicted a man for the crime of not preventing a rape.

Economic Development

After the Industrial Revolution began in England in the late 1700s, it took considerable time for it to spread to East Central Europe. The outside empires that dominated most of East Central Europe through the 1800s and up to 1918 treated their East Central Euro-pean domains as colonies, preferring to extract resources and agri-cultural products rather than invest in industrialization. The Czech and Polish lands were somewhat exceptional in that a number of cities had factories. Hungary and Slovenia also developed indus-try. Nevertheless, a large number of people in the subregion still work in agriculture. Albania, for example, ranks as one of the highest with 60 percent of its labor force working in agriculture.

After World War II, Communist economic policies were imposed on most countries in East Central Europe. The Com-munists tried to improve the standard of living by encouraging industrialization, particularly in areas with coalfields, and by providing jobs for everyone, whether it made good economic sense or not. During Communist times, farming efficiency improved as industrialization provided tractors to replace horse-drawn plows. Great strides also were made in technolo-gy, education, health care, and welfare. Literacy increased dra-matically and infant mortality plummeted. One of the difficulties in the transition to capitalism was that people in the former Communist countries were not accustomed to paying for health care and higher education. Unemployment was also a new and shocking experience. For many people, capitalism was a step backward from Communism.

In terms of trade, Soviet Communists preferred not to be ensnared by capitalist practices, which they considered to be corrupting. Therefore, they tried to restrict trade among fellow Communist countries. Stalin set up the Council for Mutual Economic Assistance (CMEA or COMECON) to compete with the European Economic Community (EEC) in Western Europe and even prevent East Central European countries from attempting to join the EEC. COMECON linked the subregions' economies with one another and with the Soviet Union. Stalin also insisted on specialization, with the northern countries focusing on industry and the southern ones producing agricul-tural products. Such specialization forced greater dependency on the Soviet system and prevented any country from realign-ing itself with the West. Cheap oil and natural gas from the Soviet Union also increased dependence but left a legacy of contaminated soil and water.

After the end of Soviet control from 1989, the countries of East Central Europe experienced economic crises as they reori-ented themselves from incorporation in the Soviet economic system to the world economic system (see Chapter 4). Exchanges with one another and the former Soviet states greatly declined. Countries were challenged with breaking up business monopolies owned by the state, providing productive jobs, and cleaning up the environment. The difficulties in moving to capi-talism led to an immediate drop in GDP, but growing trade with Germany in particular and new economic policies caused the GDP to rise again (Figure 3.49). Personal income, which was low during Communist times, remained low and resulted in very low consumer goods ownership compared to the rest of Europe.

The Communist system focused on reorienting the agricul-tural economies of East Central Europe to industrial economies. While the Communists created factory jobs, similar factory jobs in the West were replaced by service jobs as the world economy continued to develop. In 1997 the service sector accounted for 60 to 70 percent—more in a few countries—of GDP in Western Europe. In contrast, the service sector in East Central Europe accounted for only 50 to 60 percent of GDP on average and was well below those levels a few years earlier. Albania at 23 percent and Romania at 36 percent represented the lowest. Their low rankings illustrate that they have had the greatest difficulties changing their economies and improving the standards of living of their people.

Poland, the Czech Republic, Hungary, Slovenia, and the Baltic countries have had the greatest successes in adopting capi-talist practices. These countries were also the most industrialized

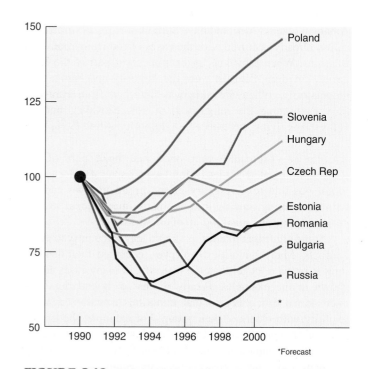

FIGURE 3.49 GDP growth in East Central Europe after the end of Communism. Notice the initial rapid decline in GDP.
Source: Copyright © The Economist Newspaper Group, Inc. Reprinted with permis-sion. Further reproduction prohibited. www.economist.com.

before and during Communism and have had stronger traditions of democracy. Significantly, they also were closest to countries of the European Union, making them easy beneficiaries of trade and investment from countries in the other subregions of Europe, especially from Germany. For example, 70 percent of Poland's trade was with the EU in 1996.

One of the biggest success stories comes from the Czech Republic. One example is the Czech automotive company Skoda. It was like the automotive companies in the other Communist countries in that it produced poor-quality, two-cylinder vehicles. Volkswagen, however, purchased Skoda, and with German management and huge sums of cash, the company has become the most successful formerly Communist company anywhere. Skoda employs 4 percent of the Czech workforce and accounts for 14 percent of Czech exports. Eighty percent of Skoda's vehicles are exported, including to Germany with its competitive auto market.

The Czechs could have great success in earning money through beer exports because they are originators of great beers. For example, Pilsner beers come from the town of Plzeň, and Budweiser traces its roots back to the town of Budweis (now České Budějovice). However, the American company Anheuser-Busch uses its financial might, with which it obtained the Budweiser trademark, to keep the Czechs from exporting Czech Budweiser, though the Czech beer was brewed hundreds of years before the American brand existed.

Hungary exerted its economic independence during Communist times. For example, it allowed private businesses that employed fewer than five people. The mixture of large-scale Communist economic policies with small-scale capitalist practices led to the term "goulash Communism," a term derived from the traditional Hungarian stew, which is a mixture of vegetables and meat. Capitalist practices were minimal, but they were greater than in other Soviet bloc countries and gave Hungary a head start in the transformation process. Though a modest-sized country, Hungary captured over one-third of all 1990s foreign direct investment in the former Communist countries, including the former Soviet Union. In 1989, 65 percent of Hungary's trade was with COMECON. By 1998, 80 percent of Hungary's trade was with the EU.

Before Yugoslavia was engulfed in war beginning in 1991, it had one of the strongest economies and highest standards of living of Communist Europe. Not a Soviet satellite, Yugoslavia pursued its own course. It employed the Communist idea of centralized planning, but compared to the Communist countries in the Soviet sphere, it also allowed a more genuine practice of the Communist belief that workers should manage their companies. Subsequently, productivity was high in Yugoslavia. Travel to non-Communist countries was not as restricted as in the Soviet sphere. As a result, thousands of Yugoslavs, especially Slovenes and Croats, sought work as guest workers in countries such as Germany. These workers sent millions of dollars back to family members in Yugoslavia.

Slovenia prospered more since independence in 1991 because it further developed its economic contacts with Germany and other EU countries. Slovenes now have the highest standard of living and the most modern economy of the former Communist

countries. Croatia may have been just as prosperous, but independence in 1991 was followed by the devastation of war and little foreign investment. Now that the wars are over, Croatia is rebuilding and the economy is growing. Serbia and Montenegro's economy faces great difficulties. Economic boycotts during the wars and NATO bombings devastated the country's economy. Though the war is over, the country is receiving little foreign investment. Macedonia escaped most of the ravages of war, but its landlocked position among unfriendly neighbors and its distant location from the wealthier countries of Europe attract little foreign investment to this largely agricultural country.

Test Your Understanding
3E

Summary

Mediterranean Europe comprises the countries of three peninsulas—Portugal and Spain, Italy, and Greece. These countries represent some of the oldest civilizations in Europe. Apart from northern Italy, which is as prosperous as parts of Western Europe, these countries lagged far behind their more affluent neighbors until their incorporation in the European Community (now EU) in the 1980s. They are now developing manufacturing and service industries, and all rely heavily on income from tourism.

East Central Europe consists of three small Baltic countries (Estonia, Latvia, and Lithuania), Poland and Hungary with their plains, and the hilly areas of the Czech and Slovak Republics, all with mixtures of farming and manufacturing. Farther to the south in the subregion are the former Yugoslavia (Slovenia, Croatia, Bosnia-Herzegovina, Serbia and Montenegro, and Macedonia), Romania, Bulgaria, and Albania, some of which have experienced varying degrees of civil strife from full-scale war in Bosnia and Croatia to ethnic clashes in the other countries, including that in Kosovo.

The countries of East Central Europe struggled through the transition from Communism to democracy and capitalism, with many of them recently having become members of the European Union. In recent years, many of them experienced economic growth as they established economic trade with the rest of Europe. Others continue to struggle.

Questions to Think About

3E.1   How did Mediterranean Europe begin European colonialism?

3E.2   What attracts tourists to Mediterranean countries, and what are some of the environmental impacts of tourism? (See also the "Natural Environment" section in this chapter.)

3E.3   How does geography help us to understand ethnic tensions in East Central Europe?

3E.4   Can you explain the greater economic growth in Poland, the Czech Republic, Hungary, and Slovenia in the 1990s compared to that in the other East Central European countries?

Key Terms

market gardening	ethnic cleansing
irredentism	genocidal rape

Chapter 4

Russia and Neighboring Countries

FIGURE 4.1 Russia and Neighboring Countries:
The subregions.

THE SOUTHERN CAUCASUS

THE SLAVIC COUNTRIES

CENTRAL ASIA

| 0 mi | 500 | 1,000 |
| 0 km | 500 | 1,000 | 1,500 |

Themes in This Chapter

Russia and Neighboring Countries is rich in natural resources but have harsh natural environments that make difficult living conditions. Much of the region's population, as well as its political center and economic activities, are located in the western part of the region. Though the region is influenced by both European and Asian cultures, languages, and religions, the Slavic peoples dominate numerically, politically, and economically. Indeed, Russians have been the biggest empire builders in modern history, and the Russian Federation is still the largest country in the region. The themes are:

Countries redefining their relationships with the world and with one another.

The settlement of peoples and the rise of Russian political power.

A natural environment of mid- and high-latitude climates and plentiful resources.

New global and local relationships.

Three subregions:

- The Slavic Countries: majority culture inhabiting the largest area.
- The Southern Caucasus: old cultures in a mountainous environment.
- Central Asia: five Muslim countries in challenging environments.

Point–Counterpoint: Russia: Still a World Power?

Geography at Work: Wine Industry

Personal View: Russia

New Relationships

Russia (formally called the Russian Federation) and its Neighbor Countries comprise 12 countries that can be divided into three subregions (Figure 4.1 and Table 4.1). Together with the Baltic states of Estonia, Latvia, and Lithuania (see Chapter 3), they once formed the Union of Soviet Socialist Republics (USSR), more commonly called the Soviet Union. The Soviet Union disintegrated in 1991, but today's countries that comprise the region continue to share a common political and economic legacy of the Soviet system. Indeed, 11 of the new countries that emerged from the Soviet Union's collapse formed the **Commonwealth of Independent States (CIS)** in 1991. Georgia joined in 1993; thus all 12 countries discussed in this chapter are members of the CIS. The three Baltic republics (Estonia, Latvia, and Lithuania) decided to seek greater cooperation with European organizations.

The CIS is an organization of cooperation between independent countries, not a single country itself. It is an economic union that recognizes the necessity of maintaining the trade flows established during Soviet times. Member countries meet to discuss and work out economic needs and policies. Beyond economics, the entire membership has little in common, particularly in terms of culture and natural environment (Figure 4.2). The Slavic countries (Russia, Ukraine, and Belarus, including neighboring Moldova), the Southern Caucasus (Georgia, Armenia, and Azerbaijan), and Central Asia (Kazakhstan, Turkmenistan, Kyrgyzstan, Tajikistan, and Uzbekistan) are all very different subregions.

In time, these three subregions will likely integrate themselves into other world regions. Some researchers already include some or all of the Slavic countries with Europe and Central Asian countries with Northern Africa and Southwestern Asia (see Chapter 8), while China is actively forming links with countries along its border. In reality, this process is far from complete, and it is premature to redraw the boundaries of the world regions so radically. Despite these countries' differing cultures and orientations, their common Soviet legacy has influenced their cultural development in particular ways and presents them with similar recurring economic and political problems. These countries have found it to their mutual benefit to continue working together. Because Russia remains the dominant political and economic force of the region, we find it appropriate to call the region "Russia and Neighboring Countries."

Stretching from the eastern edges of Europe to the Pacific Ocean, Russia and Neighboring Countries is the world's largest region in terms of land area. However, with much of

Table 4.1 Summary Information on the Subregions of Russia and Neighboring Countries. Data by subregion, area, population, income (GNI PPP—Gross National Income Purchasing Power Parity), urbanization, Human Development Index (HDI), Human Poverty Index (HPI).

| Subregions | Land Area (km²) | Population (millions) | | GNI PPP 2003 | | % Urban | Index 2002 | |
		Mid 2005 Total	Est. 2025 Total	Total (US$ billions)	Per Capita (US$)	2005*	HDI (rank of 177)*	HPI (% population)*
Slavic Countries	17,920,400	204.1	185.3	1,607	5,523	64.5	71.0	no data
Southern Caucasus	186,100	15.9	17.2	53	3,230	56.0	90.0	no data
Central Asia	3,994,400	58.7	72.3	179	3,286	40.6	99.4	no data
Totals or averages*	22,100,900	278.7	274.8	1,839	4,013	53.7	86.8	

Source: World Population Data Sheet, Population Reference Bureau (2005), World Development Indicators, World Bank (2005), Human Development Report, and United Nations (2004).

(a)

(b)

(c)

FIGURE 4.2. **Russia and Neighboring Countries: Diverse landscapes and people.** (a) New church, Kharbarovsk. (b) Georgian boys dance group "Preserve Our Culture." (c) Russian village, south Siberia. **Photos:** © Ronald Wixman.

its land either in the frozen arctic environment of the north or the hot deserts of the south, the region is the least populated on Earth. Great distances (Figure 4.3), river systems that flow toward the edges of the region such as to the Arctic Ocean, and extreme climates of hot and cold that make it difficult to build rail and road have prevented governments from integrating the region with highly developed transportation and communication linkages. The region is inhabited by over 100 ethnic groups (Figure 4.4), so that many of its peoples have as much in common with people of neighboring regions as with one another.

Crossroads, Imperialism, and Cultural Diversity

Russia and Neighboring Countries accounts for more than four times the Earth's land surface as the European world region but are divided into only 12 countries, compared to Europe, which has more than 30. The Russian Federation is larger than all the other countries in the region put together—nearly twice the size of the United States. The smallest country, Armenia, is only the size of Maryland and Delaware combined.

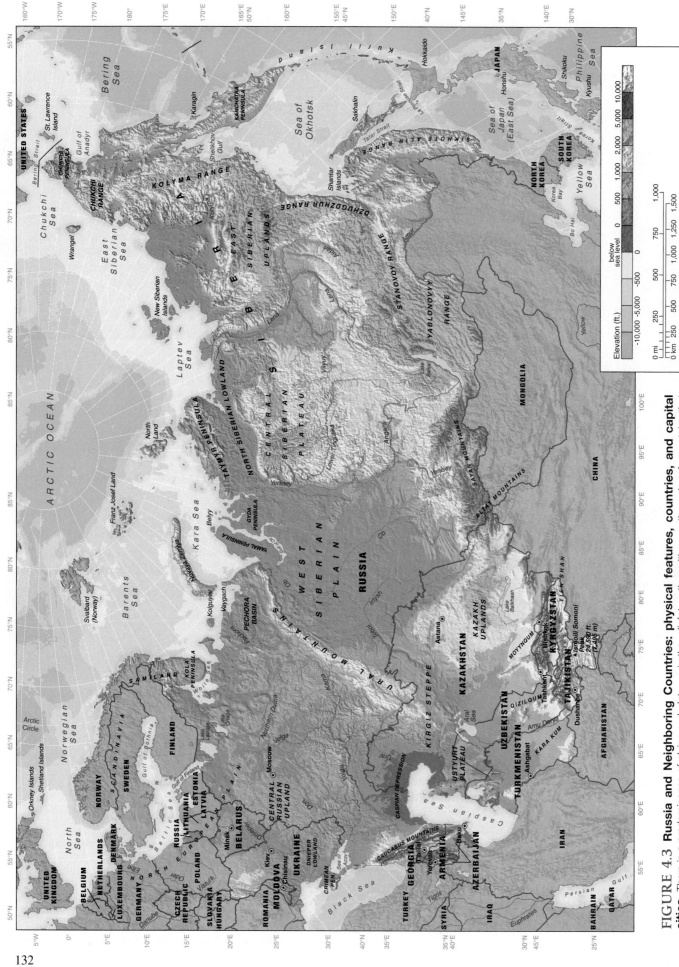

FIGURE 4.3 Russia and Neighboring Countries: physical features, countries, and capital cities. There is a predominance of plains and plateaus in the relief, together with a southern rim of mountains that resulted from clashing tectonic plates.

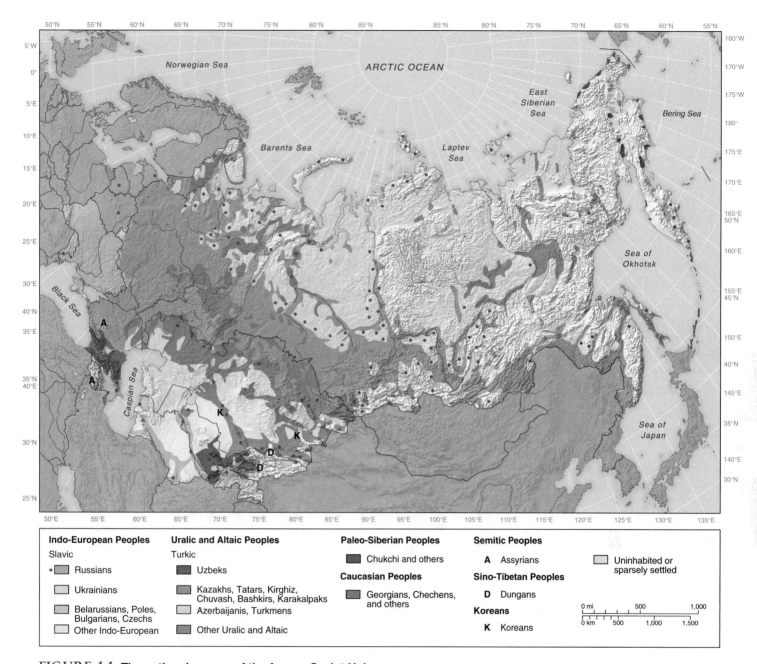

FIGURE 4.4 **The national groups of the former Soviet Union.** Note how the Russians spread across southern Siberia.

These differences reflect the cultural and political history of the region. The Russian Empire was one of Europe's great empires in modern times. Unlike the European empires such as France and the United Kingdom, which conquered lands overseas, the Russian Empire conquered peoples and land adjacent to it, extending its borders as it annexed new territories. As early as the mid-1800s, the Russian Empire had become very multinational, with Russians barely constituting a majority of their empire's population. Russia, however, retained control over most of its conquered lands. Siberia—a huge, resource-rich, yet sparsely populated area—represents most of the land that the Russians conquered in history. The other countries that declared independence from the Soviet Union in 1991, though

much smaller, represent old cultures. Along with Russia, many of them lay along the crossroads of ancient migration and trade routes and never existed as separate countries.

Over the last two millennia, three major invading influences were Christianity, penetrating the region from the southwest; Islam, entering from the south; and Mongol culture, sweeping in from the east. Many tribal and ethnic groups moved in from all directions, especially the east. The semiarid steppe grasslands of southern Russia and Ukraine served as the main route for those who invaded Europe from the east. Some invaders settled for a time in the steppes before moving on to Europe, while others decided to remain permanently. Central Asia lay along the Great Silk Road (Figure 4.5), the

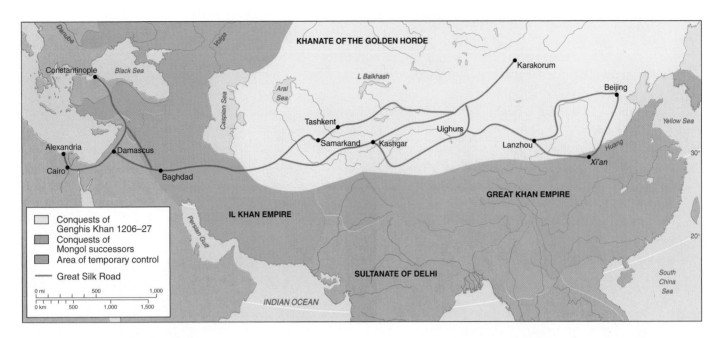

FIGURE 4.5 The Great Silk Road. Note how Central Asia, namely with its cities of Tashkent, Samarkand, and Kashgar, serves as an important link in this old trading route, which existed in several variants. From *Cultural Atlas of Russia* by Robin Milner-Gulland. Reprinted by permission of the author.

pathway that brought many goods and ideas from East and South Asia that eventually helped to build the European empires. The Caucasus Mountains separated the ancient civilizations of Southwest Asia and the cultures of the North European Plain. Trade routes between the Viking homelands in Scandinavia and the Byzantine Empire ran through the Slavic areas located on the North European Plain. Centuries of migration and imperial conquests created a region that is culturally very complex today.

Eastern Slavs

As early as 1500 B.C., a Proto-Slavic people occupied land between the Vistula and Dnieper Rivers. From A.D. 400 to 900, they divided into three main branches. The western and southern branches settled in East Central Europe (see Chapter 3). The eastern branch moved north and east, developing its own character as it came into contact and exchanged ideas and practices with the Baltic, Finno-Ugric, and Viking peoples. Eventually, the Eastern Slavs subdivided further into Russians, Belarussians, and Ukrainians (see Figure 3.4a).

Rus

From the 800s, Vikings (called *Varangians* in Russian) moved into the North European Plain along the Volga River. They established control and began trading the amber, beeswax, fur, and precious metals of their newly acquired lands with Byzantium, the Orthodox Christian realm to the south, with its capital at Constantinople. The Vikings ruled as overlords by establishing their own dynasties or by marrying into and taking over

local dynasties. As Eastern Slavs mingled with Vikings, Balts, and Finno-Ugric peoples, *Rus* came into being. The term *Rus* applied to both the people and their land.

The most important of the earliest Rus principalities were Novgorod and Kiev. Novgorod was founded first, but Kiev grew more quickly in power, exacting tribute from the other Rus principalities. Kievan Rus reached its zenith in the 900s and 1000s. In 988, Prince Vladimir brought Eastern Orthodox Christianity to Rus, tying the Eastern Slavs to Constantinople and leading to the adoption of the Cyrillic alphabet, which was derived from the Greek alphabet. Over time, Kievan Rus could not maintain its authority over the other principalities of Rus. Each began to exert its individual authority. Novgorod and a newer principality called Vladimir-Suzdal, with its capital at Vladimir, were two of the most notable (Figure 4.6).

In the 1200s, the Mongols (or Tatars) swept in from the east and attacked the principalities of Rus. Those in the south suffered most. Kiev was sacked in 1240. Novgorod escaped invasion, but it had to pay tribute to the Mongols. The Mongols established the Khanate of the Golden Horde, a state that stretched from the steppes to central Asia and Siberia. It included the southern Rus principalities and extracted tribute from the others. The Mongol influence on the development of Russian culture was very strong. Centuries after the fall of the Mongols, Napoléon said, "Scratch a Russian, and you will wound a Tatar."

Muscovy

Moscow emerged in the principality of Vladimir-Suzdal on the trade route between the Baltic and Black Seas. It was an insignificant trading post at the time of the Mongol invasions,

▨ Kievan Rus 900	── Boundaries around 1100
▨ Land gained by 1054	0 mi 250 500
▨ Area temporarily paying tribute	0 km 250 500 750

FIGURE 4.6 The Rus principalities c. 900–1100. Kiev and Novgorod were initially the most powerful cities; the rise of Moscow dates from the 14th century, after the Tatar invasion. From *Cultural Atlas of Russia* by Robin Milner-Gulland. Reprinted by permission of the author.

but its isolated, forested location offered some protection from the Mongols, and it soon grew in importance. When the southern Rus principalities came under control of the Mongols, the Orthodox Christian metropolitan (provincial primate) of Kiev and all Rus moved north from Kiev to Vladimir in 1299, increasing the authority of Vladimir-Suzdal. In the early 1300s the prince of Vladimir-Suzdal cooperated with the Mongols by collecting tributes from the other Rus principalities on the Mongols' behalf. Around 1318 the prince married the sister of the khan of the Golden Horde and received the title of "Grand Prince." In 1327, the metropolitan of the Orthodox Christian church moved from Vladimir to Moscow. During the 1400s, the grand princes of Moscow conquered the other Rus principalities. The consolidated territories governed from Moscow came to be known as Muscovy.

By 1480 Muscovy was so powerful and Mongol authority so weak that Grand Prince Ivan III (the Great) simply stopped paying tribute to the Mongols. Ivan III married Sophie Paleologue, the niece of the last Byzantine emperor. After Constantinople fell

to the Ottomans in 1453, Ivan was seen as the true inheritor of the Christian realm. He adopted the title "czar," derived from the Latin *caesar* for emperor. With the succession of Ivan IV (the Terrible), the patriarch of the Eastern Orthodox Church in Constantinople accepted Ivan's title of czar. Moscow claimed to be the "Third Rome," successor to Constantinople and Rome. Even today, Russians are familiar with the concept that Moscow is the "Third Rome."

The Russian Empire

Muscovy grew in power and influence. In 1613 Mikhail Romanov ascended the throne. The Romanov dynasty lasted until the last czar was deposed in 1917. From 1480, Muscovy's territory continued to expand (Figure 4.7), especially to the east, into Siberia and Central Asia, but also to the west, bringing the Russians into contact with Central and Western European culture. The greatest territorial expansions were made under Peter the Great (1682–1725), Catherine the Great (1762–1796), and Alexander I (1801–1825).

Peter expanded Muscovy's boundaries to the Baltic Sea, where he founded the city of St. Petersburg and made it his country's new capital in 1712 in an attempt to Westernize his country. Peter built the first Russian navy, reorganized the military and government along European lines, and founded modern institutions devoted to the sciences and arts. Muscovy became the Russian Empire. Catherine continued to Westernize Russia in the late 1700s. As well as expanding farther east into Siberia, she also conquered lands to the west, incorporating Poland, and to the south, gaining all of Ukraine, where she founded the Black Sea port of Odessa. She established Russian influence in the Balkans. By the end of Catherine's reign, Russia was clearly one of Europe's great powers.

Alexander I was considered the "Savior of Europe" for defeating Napoléon's attack on Moscow in 1812. The victory allowed Alexander to play a major role in redrawing Europe's boundaries at the Congress of Vienna in 1815. Russia's boundaries moved farther west with the acquisition of Finland, more of Poland, and Bessarabia (now Moldova). Around the same time, Russia expanded into the Caucasus, taking the area around Baku. Russia also took more interest in Alaska.

By the early 1900s, the Russian Empire achieved its greatest extent, stretching from Europe to East Asia and the Pacific Ocean. Russia was a world power, but its strength was derived from its sheer size. Internally, it had many problems. Politically, the czars remained in total control, continuing to claim to speak in the name of God, but absolute power inhibited the creation of an efficient government. Technologically, Russia was decades behind Western Europe. The technology of the Industrial Revolution developed very slowly in Russia, leaving Russia farther and farther behind as time progressed. Most Russians remained feudal peasants farming with horses or oxen. At the same time, Russia had expanded its boundaries to include more than 100 different peoples. After the rise of nationalism in Europe in the mid-1800s, Russia's nationalities began clamoring for independence. This situation worsened as Russians themselves became

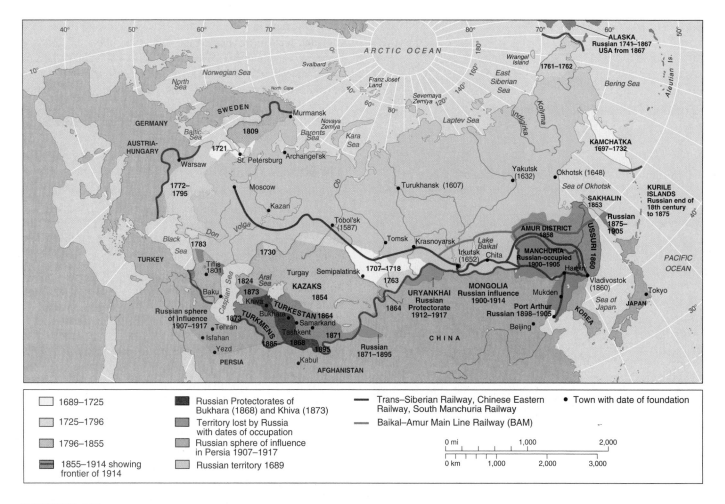

FIGURE 4.7 Russian Federation: history of growth. The expansion of the Russian Empire from the 1600s to the early 1900s. The Trans-Siberian Railway was built through territories that were regarded as safely Russian. It was rerouted after territory was lost. The Baikal–Amur Main Line Railway (BAM) was begun in 1974 and opened in 1989. It opened up more of the vast mineral and forest resources of eastern Siberia and the Russian far east to greater exploitation.

nationalistic and began suppressing non-Russians. They developed **Russification** policies to force Russia's minorities to become more Russian ("Russified").

The Soviet Union

Competing for Control

World War I (1914–1918) exerted great stress on the Russian Empire. By 1917, food shortages and a huge death toll combined with long-standing opposition to czarist rule. Revolutionaries deposed Czar Nicholas II and set up a provisional government that proved weak. A number of national groups staked their claim to territory and declared independence for Finland, the three Baltic countries (Estonia, Latvia, and Lithuania), Belarus, Ukraine, the countries of the Caucasus (Georgia, Armenia, and Azerbaijan), and Central Asia (at the time called Turkestan). The Polish lands of Russia joined with the Polish lands of Germany and Austria-Hungary and an independent Poland was declared. Bessarabia (now Moldova) was annexed by Romania.

By late 1917 the Bolsheviks, a group of Communists, also known as "Reds," overthrew the provisional government. Civil war ensued when anti-Communists, called "Whites," tried to dislodge the Bolsheviks from power. By 1922 the Reds gained the upper hand, expelled their enemies, and reclaimed Belarus, Ukraine, the countries in the Caucasus, and Central Asia.

In 1922, under the leadership of Vladimir I. Lenin, the Bolsheviks established the Union of Soviet Socialist Republics (USSR), commonly called the Soviet Union. This new country succeeded the Russian Empire, but the Bolsheviks changed the old system of government and economy. They abolished the monarchy and government by a privileged few and replaced them with "soviets." Soviets were workers' and soldiers' councils that drew their members from the common citizens. Economically, the Bolsheviks despised capitalism, viewing it as a system that allowed a select few—the bourgeoisie (middle class)—to grow wealthy by exploiting the masses—the proletariat (the industrial working class). The Bolsheviks reversed the results of capitalism by putting the workers in charge of factories and businesses. They sought to create a "dictatorship of the proletariat" and, thereby, a "workers' paradise." Religion was seen as a tool of the oppressors because the clergy sanctioned the rule of political leaders and the capitalist exploiters who oppressed the common people. Consequently, the Bolsheviks attacked churches and mosques, dynamiting many and turning others into scientific centers such as planetariums to show the error of religious belief.

Five-Year Plans

Lenin died in 1924, and within a few years Joseph Stalin came to power and radically changed government policy. Stalin's heavy-handed form of governing became known as Stalinism, and some debate whether it was truly a form of Communism. Stalin believed it necessary to transform the Soviet Union's economy forcefully into the communist ideal of a "workers' paradise" based on industrialization. This was also necessary to overcome the technology advantages of the West. However, the Soviet economy was largely agricultural, and farmers could not relate to the urban–industrial ideology of Communism. Therefore, Stalin implemented the first **five-year plan** in 1928.

The five-year plan called for collectivization and industrialization. Under collectivization, small family farms were merged to create large farms with thousands of acres. Unlike the small farms, large farms were better designed to use modern farm machinery, then under production in the new factories. The government became the owner of the collectives and farmers became employees. With collectivization, farmers became more like factory workers, even living in tightly packed housing like urban factory workers. Thus they were more sympathetic to the Communist way.

The plan called for government ownership of all industries. In what is known as the **command economy,** the government set quotas favoring heavy industry over production of consumer goods. The five-year plans also established **central planning.** In contrast to capitalist economies, supply, demand, or profit did not dictate what would be produced. With central planning, the government decided how many goods and services were needed by society, almost without cost considerations. Rapid, forced industrialization prevented the Soviet Union from experiencing the world economic depression of the 1930s as the government kept investing in the economy and providing jobs.

World War II

In 1939 Stalin entered a nonaggression pact with Hitler, allowing the Soviet Union to take control of territories that formerly belonged to the Russian Empire: the Baltic states, eastern Poland, and Bessarabia (now Moldova). Finland was also a target, but the Finnish forces successfully defended their country from the Red Army.

On June 22, 1941, Hitler launched Operation Barbarossa, the invasion of the Soviet Union. The campaign was initially devastating. The Soviet Union survived largely because of its sheer size. A vast amount of land was lost and millions of people were killed, but even larger areas were left unconquered. The Soviets moved war production farther east, out of the range of the Nazi military, including its air force. With help from the United States and the United Kingdom, production increased. Finally, the harsh Russian winter stopped the German Nazi armies. In the meantime, Stalin moderated many of his harsh policies, including the persecution of the Russian Orthodox Church. After a fierce battle at Stalingrad in 1943, Soviet forces began rolling back the Nazi invaders and began winning what the Soviets called the "Great Patriotic War." By May 1945 the Red Army swept through East Central Europe and occupied much of Germany, including the capital, Berlin.

Its victory in World War II allowed the Soviet Union to annex the Baltic countries and Moldova, former territories of the Russian Empire they also marched into in 1940, and new territory in East Central Europe never before governed by the Russians (East Prussia, the northern half of which is now Kaliningrad, and areas of Poland). Victory also allowed Stalin to establish and support Communist governments in East Central European countries. To counter NATO and the Marshall Plan (economic aid from the United States) and later the European Economic Community, Stalin created the Warsaw Pact and Council for Mutual Economic Assistance (CMEA, also known as COMECON).

Test Your Understanding — 4A

Summary

In 1991, 15 new countries replaced the Soviet Union as Communism ended in this world region. Since then these new countries have worked to establish new relationships with one another and the world. One example is seen in the Commonwealth of Independent States (CIS). During Soviet times, most of the republics were inward-looking and very dependent on Russia. Old trade links are difficult to change, but most of the new countries seek to break that dependence on the Russian Federation and to establish stronger ties with other countries, including the United States. Though the Russian Federation prospers by the old dependencies, it too seeks new relationships with countries of other world regions.

The present distribution of peoples and cultures can be traced back to migrations up to around A.D. 1000. The Orthodox form of Christianity, the invasions of Mongol hordes and Ottoman Turks, and the conversions to Islam along the region's southern borders had major influences on subregional cultures. Much of the region remained under feudal and agricultural conditions until the early 1900s.

Russia expanded its territories from the 1600s and began to modernize its productive capacity along Western lines. The Bolshevik Revolution of 1917 led to Communist control of the former Russian Empire as the Soviet Union. After World War II, Soviet influence extended over countries in East Central Europe to form the Soviet bloc with economic and defensive ties.

Questions to Think About

4A.1   What is the relationship between the countries of the Commonwealth of Independent States (CIS)? Do they act as one country? Why or why not?

4A.2   How did the Rus begin as a people, and which Rus principalities began the Russian Empire?

4A.3   How did the Soviets design the economy of the Soviet Union?

Key Terms

Commonwealth of Independent
   States (CIS)
Russification
five-year plans

command economy
central planning

Natural Environment

The variety of natural environments in Russia and its Neighboring Countries reflects their huge land area. The continental interior climates, often with long, harsh winters; the vast plains; and the massive areas covered by a variety of natural vegetation types and soils provide a distinctive stage on which the development of human geographies occurred.

Midlatitude Continental Interior Climates

Russia and its Neighboring Countries nearly all lie north of 40°N, and most of their area is north of 50°N. As such, these countries lie as far north as Alaska and Canada. The region contains places on Earth that are farthest from oceans and their moderating effects. They become extremely cold during winter and hot during summer, a situation described by the term **continentality.** The warmest climates of the region are around the Black and Caspian Seas and in the Central Asian countries (Figure 4.8).

The largest proportion of these lands is more than 500 km (320 mi.) from the ocean, and many parts are over 2,000 km (1,250 mi.) from the ocean. With greater distance inland, the extremes of summer and winter temperatures increase. The greatest contrast is in eastern Siberia, where large areas have January temperatures below –30°C (–22°F) and July averages of 12° to 16°C (56° to 60°F) (annual differences of 45°C or 80°F). In western Russia, winter temperatures average –5° to –10°C (15° to 25°F), and summer temperatures are 15° to 20°C (59° to 68°F, a difference of 25°C or 50°F). With greater distance northward, the climate gets colder and the winters longer; northern Russia lies north of the Arctic Circle. In the far east of Siberia, proximity to the Pacific Ocean results in more humid conditions, although the winters remain long and very cold as Arctic winds sweep out from the continent's interior.

Greater distance inland reduces precipitation. Few parts of the region have over 80 cm (32 in.) a year. Much has moderate precipitation (40–80 cm, 16–22 in.) falling mainly in summer but producing a long-lasting snow and ice cover during the winter. Parts of southern Central Asia are arid because of the high evaporation rates in warmer latitudes and distance from rain-bearing air masses. They rely on snowmelt from the mountains along their southern borders. Parts of the far east are subhumid. Low precipitation and low temperatures make eastern Siberia inhospitable.

Southern Mountain Wall

The southern boundary of the region is marked by mountain systems. They were pushed up along the convergence of tectonic plates carrying the continent of Eurasia on the north and the landmasses of Africa, Arabia, and India to the south (Figure 4.9). The Caucasus Mountains, between the Black and Caspian Seas, are part of this line of mountain ranges, which extends through the Elburz Mountains of northern Iran to the Tien Shan and the Pamir Mountains along the southern borders of the Central Asian countries. These mountain ranges rise to over 7,400 m (24,000 ft.). They are snowcapped and in spring provide considerable meltwater for streams flowing through the dry areas of southern Russia and Central Asia. In the far east, the East Siberian Uplands and the volcanic peaks on the Kamchatka Peninsula parallel the Pacific coast and its tectonic plate boundary in areas that have few people. The Earth movements not only raised high mountain ranges but also produced deep basins such as those filled by the Black Sea, Caspian Sea, and Lake Baikal.

Plateaus, Plains, and Major River Valleys

Plains and low plateaus dominate most of the landscapes of this region. The North European Plain widens eastward from Poland into Belarus, Ukraine, and European Russia until it ends against the Ural Mountains that mark the line between Europe and Asia. Plains around the northern shores of the Black, Caspian, and Aral Seas extend into the nearly level, vast West Siberian Plain. Farther east, the relief becomes more hilly in the Central Siberian Plateau on ancient mineral-bearing rocks. Such extensive plains, interrupted by low hills, eased the invasions from eastern Asia and central Europe into Russia in medieval times and the later Russian imperial eastward expansion.

Across these mainly low-lying landscapes flow some of the world's longest rivers. In the west, the Don River system flows into the Black Sea and the Volga River into the Caspian Sea. The Amu Darya and Syr Darya Rivers of Central Asia flow into the Aral Sea. The Ob, Yenisey, and Lena are the longest rivers of all and flow from the southern mountains of Central Asia and eastern Siberia northward to the Arctic Ocean.

While the rivers in western Russia are used intensively for transportation, generating hydroelectricity, and as sources of industrial and domestic water, those farther east, with the

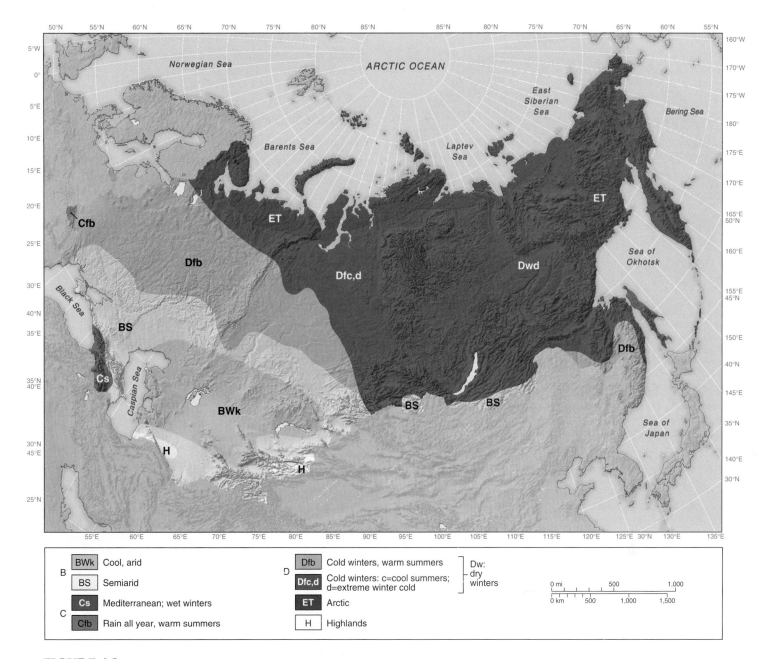

FIGURE 4.8 Russia and Neighboring Countries: climates. The dominant midlatitude continental interior climatic environments are characterized by harsh winter conditions, aridity in the south, and Arctic conditions in the north.

exception of the Angara River, are less exploited. This is partly because the areas through which they flow are sparsely settled, but also because the rivers flow northward, are frozen for much of the year, and even in midsummer have frozen mouths along the Arctic Ocean that cause meltwater to back up and flood vast areas of wetland on either bank.

Desert, Grassland, Forest, and Tundra

The natural vegetation, much altered by human occupation, is heavily influenced by its latitudinal location and relationship to climatic factors of temperature and water availability. Hot

desert in the south changes northward to steppe grassland, to deciduous and coniferous forest, and to tundra on the shores of the Arctic Ocean. These latitudinal distributions are interrupted only by mountain ranges. (Figure 4.10a and b):

The deserts of the area east of the Caspian Sea contain few patches of grass and oasis vegetation and were once occupied by nomadic peoples. Trade routes via the region's oases provided long-established links between Europe and China.

North of the desert, the **steppe grasslands** grow on fertile **black earth soils** (chernozems), which are similar to the soils of the North American prairies. The steppes extend from southern Poland eastward through Ukraine into Kazakhstan, forming

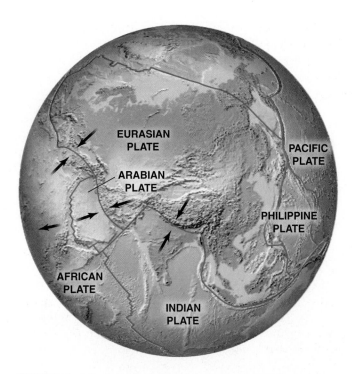

FIGURE 4.9 Russia and Neighboring Countries: plate tectonics. The map of tectonic plates and their margins shows that the Eurasian plate clashes with other continental plates from the Caucasus to Central Asia; it clashes with the oceanic Pacific plate in the east.

difficult, since removal of the vegetation cover causes rapid melting of the permafrost and collapse of the underlying soil layers as the released water flows away.

Natural Resources

Russian imperial and Soviet political, military, and economic expansions were driven and enabled by the use of plentiful natural resources. Russia became a superpower in the mid-1900s, when political power was linked to steel and fuel output. The rocks of this region contain huge quantities of a wide variety of mineral resources. Metallic minerals include iron and gold as well as many minerals that are in short supply elsewhere in the world. There are also diamonds and other valuable products. Many of these resources occur in the ancient rocks of the Siberian plateaus or have been washed by erosion into the interven-

FIGURE 4.10 Russia and Neighboring Countries: natural vegetation and soils. (a) The section X–Y across the map illustrates the types of vegetation and soils in a north–south succession. (b) Map of natural vegetation and its links to soil types. **Source:** Data from Archie Brown et al. (eds.), *The Cambridge Encyclopedia of Russia and the Former Soviet Union*, 1994, Cambridge University Press. Reprinted with permission of Cambridge University Press.

one of the world's major arable regions. In the Middle Ages, the steppe grasslands provided the grazing lands through which wave after wave of invaders moved westward into Europe.

Farther north, trees can grow where evaporation rates decrease, and a zone of wooded steppe gives way to deciduous forests, linked to fertile **brown earth soils.** The trees have been largely cut to extend farmland.

Northward and eastward—where temperatures and precipitation decline—the deciduous forest gives way to forests dominated by hardy birch trees and evergreen pine, fir, and spruce. This **northern coniferous forest,** or **taiga,** dominates vast areas of land from Moscow northward and across most of Siberia. Pine and fir trees in the west give way eastward to larch. The taiga forest occupies areas with very long winters and poor glacial soils. The underlying poor **podzol soils** have a gray sandy layer under a surface of slowly decaying leaves and above a layer where iron minerals accumulate, often becoming very hard and impeding drainage.

Around the shores of the Arctic Ocean and extending southward into the plateaus farther east is an area where trees will not grow. It is covered by grasses and low-growing shrubs. Such **tundra** vegetation is underlain by permanently frozen ground, or **permafrost,** that also extends southward beneath the taiga forest. The continuous and broken permafrost areas cover most of central and eastern Siberia and are up to 3,000 m (10,000 ft.) thick in parts of eastern Siberia. Farming is

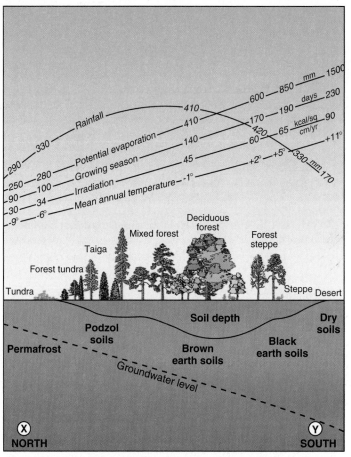

(a)

ing areas of sedimentary rocks in the northern, sparsely inhabited areas. Fuels include coal, oil, natural gas, and uranium. The Caspian Basin and Siberian lowland the contain substantial oil reserves, perhaps as much oil as the Persian Gulf area (see Chapter 8).

The countries of this region continue to be world leaders in producing fuels and other minerals and their export assumes great significance in the face of economic losses from the transitions in manufacturing and other economic activities. The main problem for the CIS countries today is to build the requisite oil and gas pipelines to world ports for both Caspian Sea and Siberian oil. Both cases pose many political, economic, and environmental problems. By the early 2000s, it appeared that Caspian Sea oil could be transported via a combination of routes through southern Russia and to the coast in Georgia, although alternative routes through Iran and Turkey are emerging (Figure 4.11). Also, in the early 2000s, the Russian government entered into negotiations with both Japan and China concerning pipelines that could be built from Angarsk near Lake Baikal to either Nakhodka on the coast where it

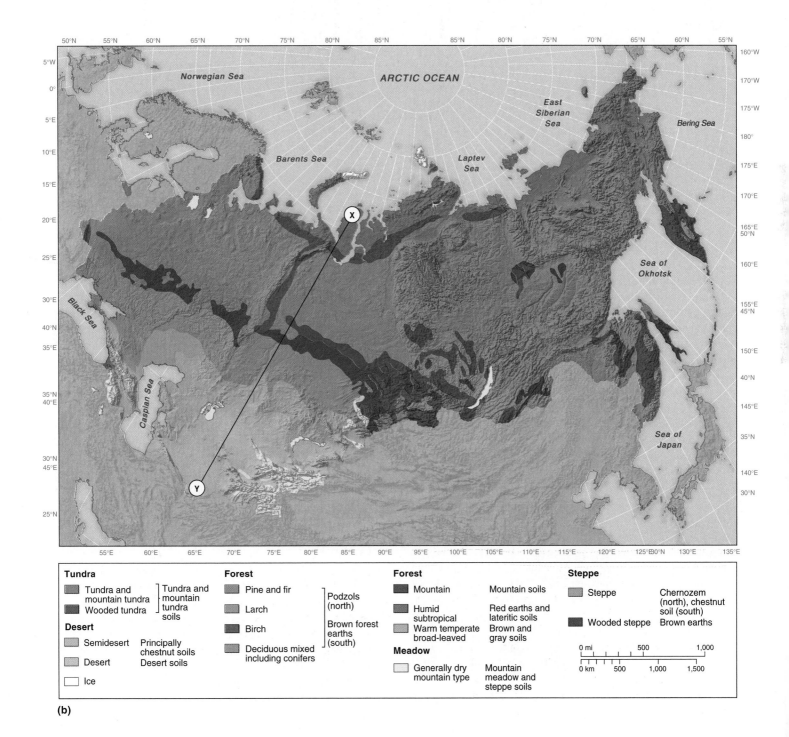

(b)

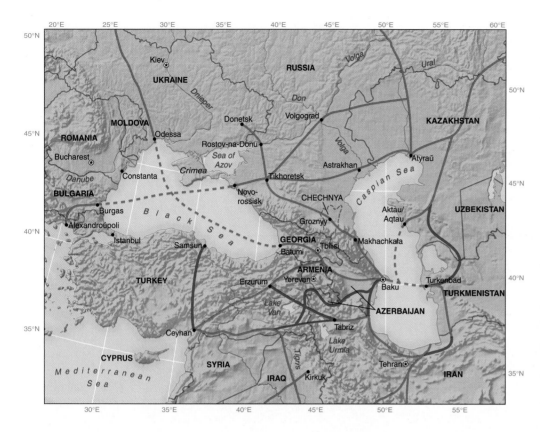

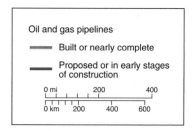

FIGURE 4.11 Russia and Neighboring Countries: pipeline outlets for Caspian Sea oil and gas. Russia wants oil and gas exports from the countries of the Southern Caucasus and Central Asia to flow through its Black Sea port of Novorossisk via the Groznyy (Chechnya) pipelines. The countries of the Southern Caucasus and Central Asia and their international oil company partners are examining the routes through Turkey and Ceyhan on the Mediterranean Sea or through northern Iran. How does this example illustrate the interdependence of countries in the world economic system and the importance of individual countries and their actions?

could export to Japan, or to Daqing China (Figure 4.12). In late 2004 the Russian government decided to enter a cooperative agreement to pump oil to Japan. China, however, is desperate for oil and natural gas for its rapidly growing economy. Alternative deals may yet result in the construction of branch pipelines to China.

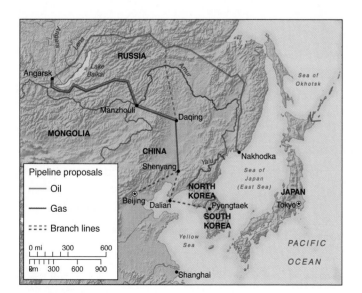

FIGURE 4.12 Russia. New oil and gas pipelines proposed for the Russian far east and East Asia.

The taiga across northern Russia forms the world's largest forest area, covering 7.5 million km^2 (2.9 million mi.^{2}; compared to 5.5 million km^2, or 2.12 million mi.^{2}, of rain forest in Brazil and 4.5 million km^2, or 1.7 million mi.^{2}, of related coniferous forest in Canada). It provides a huge reserve of biodiversity and is important in world climate change, taking up one-third as much carbon dioxide as the tropical rain forest absorbs over a similar period. Owned by the state, this resource has been well regulated and little reduced by logging or fires. State planning located all the Soviet Union paper and pulp mills west of the Urals, making it costly to transport the wood from Siberia. Although large corporations began to cut more of these forests from the 1990s and potential tax revenues from such cutting make it an attractive prospect for Russia, the central bureaucracy continues to slow the process.

Environmental Problems

Human activities caused serious environmental damage in many parts of the region (Figure 4.13). For centuries, inhabitants of the region thought that the vast quantities of resources in their lands were inexhaustible. Later Communism preached that nature should be transformed to serve human needs. Rapid industrialization, as exemplified by Stalin's five-year plans, led to the massive exploitation and extraction of minerals and the construction of enormous factories, often built quickly with little thought to environmental protection.

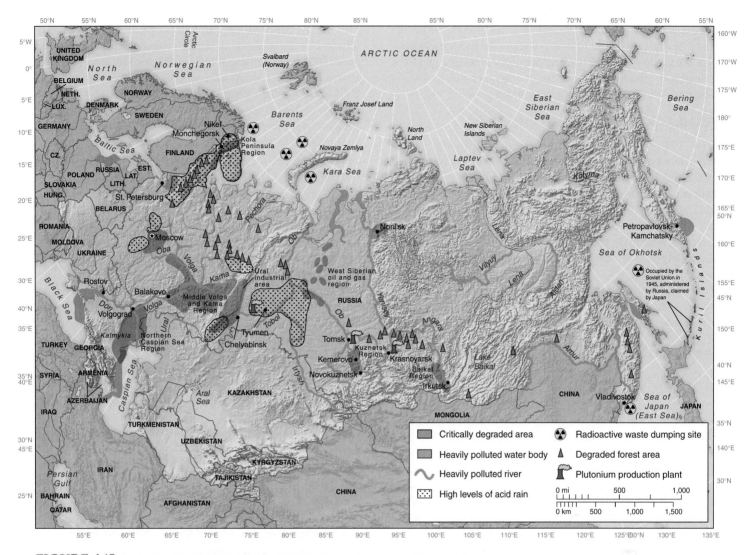

FIGURE 4.13 Russia and Neighboring Countries. Areas of environmental degradation.

The difficulties of developing the natural resources within this world region and the priorities of production over environmental protection led to increasing environmental degradation of the tundra, forest, and desert areas. The emphasis on heavy industry (especially chemicals and steel), the nuclear industry, faulty storage of toxic rocket fuel and oil, testing of weapons, and excavation of metallic minerals laid waste to huge areas, many of which remain unreported and unreclaimed. A few examples illustrate the range of environmental problems facing Russia and Neighboring Countries:

Oil Pollution

One of the major legacies of the Soviet Union is frequent oil pipeline breaks and leakages. Hundreds occur every year. Some of the resulting oil spills cover hundreds of square kilometers. Inefficient construction practices are the major cause. In 2004, an oil spill in the Samara region was caused by an incision carelessly made to illegally siphon oil.

Sumgait, north of Baku in oil-rich Azerbaijan, is one of the world's most polluted places. Once one of the main suppliers of petrochemicals to the Soviet market, its 30 or so industrial plants are now a huge wasteland of broken concrete, rusting factories, and abandoned railroad tracks. The air smells of chlorine and sulfur. The town is another legacy of environmental costs resulting from Soviet central planning.

Pollution at Norilsk

The city of Norilsk in western Siberia has the most polluted environment in Russia. The region contains some of the world's richest deposits of mineral ore: 35 percent of the world's nickel, 10 percent of the world's copper, significant cobalt reserves, and 40 percent of the platinum group of metals. The Norilsk Metallurgical Combine releases millions of tons of pollutants into the atmosphere each year. In 1991, for example, the amount was nearly 2.4 million tons, with the sulfur dioxide content at 40 to 50 times legal limits. The stinking acidic haze blackens snow, kills trees for miles around, and poisons the river. Many locals, most of whom work in the factories, believe that the noxious fumes inoculate them against disease, but local physicians report high incidents of respiratory illnesses and shortened life expectancy, as low as 50 years.

Nuclear Pollution

The 1986 Chernobyl nuclear reactor explosion, which became a symbol for such disasters, affected most of the area immediately around it north of Kiev, but the fallout was worst to the north in Belarus (Figure 4.14). Nuclear dumps on islands, peninsulas, and in seas, especially along the Arctic Ocean, have also released radioactive pollution. Well before Chernobyl, the "Kyshtym incident" in 1957 in the Urals exposed almost 500,000 people to harmful doses of nuclear radiation. Coupled with other industrial activities, nuclear contamination puts the Urals high on the list of the most polluted areas of the CIS. By 2004 Chernobyl was offered as an international tourist venue.

Aral Sea Contraction

One of the greatest environmental disasters in the world occurred in the lands east of the Aral Sea (Figure 4.15). Although Central Asia is arid, snow on the high mountains to the east melts in spring, providing water to such rivers as the Amu Darya and Syr Darya that flow into the Aral Sea. This water provided the basis of local irrigation farming and small urban settlements through history, but the Soviet Union adopted ambitious plans to use the water on irrigated cotton farms inside and outside the main river basin. Supported by growth-oriented bureaucrats, who often acted against the advice of government scientists, the project extracted so much water that the rivers stopped flowing into the Aral Sea. The sea is now less than half its previous size, all transportation on it has ceased. Glaciers in the mountains are in retreat, groundwater levels have fallen accompanied by ground subsidence, and dust storms now affect areas that had seldom experienced them. In 2005 an $86 million grant from the World Bank provided a 13 km (8 mi.) dam to regulate the flow of the Syr Darya. The dam is allowing the northern Aral to rise 42 m (138 ft.), bringing it within 13 km (8 mi.) of the town of Aralsk.

Threatened Fisheries

The Black Sea, which is surrounded by six countries, is also under threat from environmental degradation. A proposal by the Ukrainian government to build a large oil terminal and refinery near Odessa threatened local holiday resort beaches and offshore locations that support up to 70 percent of the sea's fish.

The practice of putting industry before environment occurs in the other five countries bordering the Black Sea (Russia, Georgia, Turkey, Bulgaria, and Romania) and in the 17 countries whose rivers drain into the sea. Fish stocks and plant and animal life in the Black Sea are killed by oil, toxic waste, ships' ballast water, and nutrients from fertilized fields. Only five of the 26 fish species caught in the 1960s were still found in 1992, when the total fish catch of the six rim countries dropped to 10 percent of their 1986 haul. Offshore drilling for oil and gas takes place near the Crimean Peninsula, threatening further environmental damage.

Global Environmental Politics

As a heavily industrialized part of the world, Russia and its Neighboring Countries account for a notable percentage of the world's environmental pollution. Indeed, on the list of Annex I

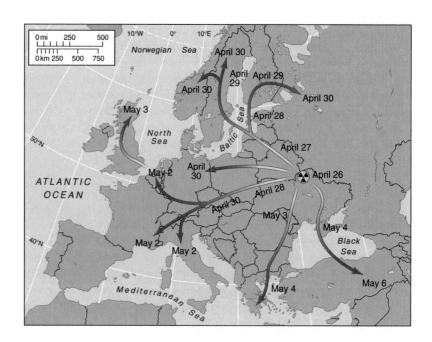

FIGURE 4.14 Russia and Neighboring Countries: impact of a nuclear disaster. The areas affected by the Chernobyl nuclear power plant explosion that released clouds of radioactive gases in 1986. The various winds at the time carried the radioactive gases in numerous directions over the days that followed, as seen on the map. In Ukraine, over 1.5 million people were affected, 10 percent of whom were severely radiated. All 600,000 members of the cleanup team suffered minor illnesses. Local death rates will increase because the evacuation of the inhabitants after the explosion was delayed. Farther afield, the radiation was detected over Scandinavia and even in the United Kingdom.

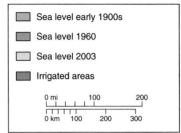

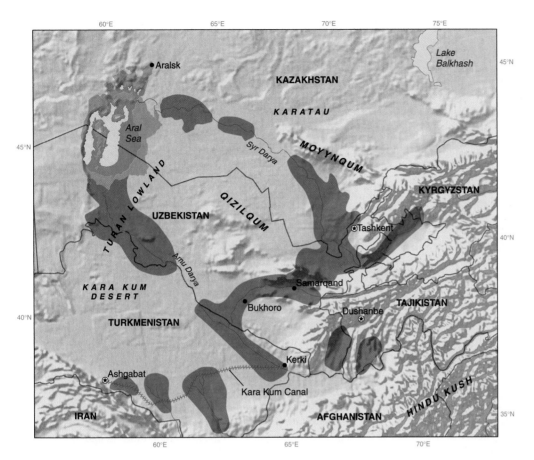

FIGURE 4.15 Central Asia: Aral Sea environmental disaster. The Aral Sea, Kazakhstan, and its surroundings. The sea is a basin of inland drainage in an arid region. It is supplied by meltwater from the mountains to the southeast. Irrigation projects such as those supplied by the Kara Kum canal have been steadily diverting water from the Aral Sea.

countries of the Kyoto Protocol, the Russian Federation solidly ranks as the second greatest carbon dioxide (CO_2) emitter with 17.4 percent of such emissions, well behind the United States at 36.1 percent and well ahead of third-ranked Japan at 8.5 percent. With such a large percentage of CO_2 emissions, the Russian Federation played a key role in determining whether the Kyoto Protocol would receive the required ratification of Annex I countries representing 55 percent of CO_2 emissions in order to go into effect. With highly inefficient and polluting industries, the Russian Federation may have had little incentive to sign a protocol to require its industries to pollute less. On the other hand, the economic downturn leading to the shutdown of inefficient factories in the 1990s meant that the Russian Federation was meeting Kyoto targets without effort. European diplomats heavily lobbied the Russian government to ratify the Kyoto Protocol in return for European economic investment, particularly in environmentally friendly industrial technologies. In early 2001 the Russian government indicated that it would ratify the Kyoto Protocol. The U.S. government, however, was against the protocol and sent representatives to Moscow every time the Russian government signaled its intention to move forward with the ratification process. Repeated lobbying by the American and European governments continued until the Russian government decided to ratify the Kyoto Protocol in December 2004. The Russian Federation's ratification fulfilled the requirement for Kyoto to go into effect 60 days later on February 16, 2005.

Though the Russian Federation was the key country in the world region for the implementation of the Kyoto Protocol, all the other countries of the region, except Belarus, Kazakhstan, and Tajikistan, signaled their support by already having ratified the protocol before the Russian Federation.

Test Your Understanding
4B

Summary

Russia and Neighboring Countries has midlatitude continental interior climates ranging from arid through humid to Arctic conditions and marked by extreme seasonal fluctuations of temperature. Although mainly a world region of plains, low plateaus, and wide river valleys, its southern margins are marked by high mountain ranges along major junctions of tectonic plates.

The natural vegetation and soils follow the climatic pattern with a south-to-north sequence of desert, steppe grassland, deciduous forest, northern coniferous forest, and tundra.

Natural resources are of vital importance to this world region's countries. The rocks of the region contain some of the world's most significant reserves of fuels and metallic ores. Human activities, particularly in the Communist era, led to major environmental problems. The Russian Federation became key to the implementation of the Kyoto Protocol.

Questions to Think About

4B.1   How are the climate and natural vegetation of this region related to each other?

4B.2   How would you characterize the environmental problems of this world region?

4B.3   Why do you think the Russian Federation decided to ratify the Kyoto Protocol even though it was lobbied not to by the U.S. government?

Key Terms

continentality

steppe grasslands

black earth soils

brown earth soils

northern coniferous forest (taiga)

podzol soils

tundra

permafrost

Global Changes and Local Responses

In the decades following World War II, Communism further transformed the political and economic systems of the Soviet Union. With the desire to spread Communist ideology around the world, the Soviet Union led a struggle to compete against the capitalist countries of the West, particularly the United States. The world's two superpowers initiated a "Cold War" as they attempted to undermine each other's systems. To do so, they supported foreign governments, movements, and insurgencies that represented their respective ideologies in other countries, primarily in poorer countries. By supplying arms, they turned local and regional conflicts into international conflicts that sometimes resulted in tremendous death and destruction.

Both superpowers were determined to prove the superiority of their respective systems. In 1957 the Soviet space program launched *Sputnik,* the first artificial Earth satellite, and set a number of other records in outer space. The United States geared up its own space program, setting records such as landing the first person on the moon. At the same time both countries competed in an arms race, building thousands of nuclear weapons, though the competition never developed into a hot war.

In its antagonism to capitalism, the Soviet Union and its allies closed off their economies and societies from the capitalist world. Refusing to get caught up in the inequities of capitalism, the Soviet Union and its satellite countries in East Central Europe sought to become economically self-sufficient. The closed Soviet society meant that individuals in the capitalist West knew less and less about life in the Soviet Union as time progressed. From the outside, the perspective of most Americans was that the Soviet Union was a powerful and hostile competitor, one that must be as strong as the United States because it financed Communist movements around the world and could build sophisticated nuclear weapons and rocket-base delivery systems. Americans and others in the West did not see that the Soviet system had calcified and become extremely inefficient. Inequities also grew within the Soviet Union, especially in relation to ethnic groups. Moreover, support of Communist movements worldwide and the arms race were serious economic drains on the country.

Communism at an Economic Standstill

When Stalin died in 1953, his series of five-year plans had helped the Soviet Union to partially catch up to the West, despite the serious setbacks of World War II. The West itself, however, continued to develop economically so that the Soviet Union was still behind despite spectacular scientific breakthroughs such as the space program. By the end of the 1950s, the Soviet Union had caught up only to the West of the 1920s, the time when Stalin began the five-year plans. Industry in the West now worked with new materials such as plastics and other synthetics and used new fuels such as petroleum and natural gas. The Soviet Union relied on materials such as metal and wood and almost exclusively used coal as a fuel. Realizing this change, Soviet planners implemented a shift in material and fuel use.

However, capitalist economies changed in other ways as well. Consumers in the West were now more demanding. For example, in the early 1920s Americans were forced to buy black Model Ts because Ford would not produce any other color. Lagging sales in the late 1920s forced Ford to produce Model Ts in other colors. By the 1960s Americans demanded many choices in the products they bought. American businesses responded, but this required industrial restructuring. Huge factories churning out massive quantities of the same item were replaced by smaller production facilities that were geographically spread out and capable of retooling to respond to constantly changing consumer demand. Production flexibility became the key to success. The Stalinist Soviet economy was so rigid that Soviet planners could not restructure, and consumers had little voice to change the system.

The weaknesses of the economic system only compounded over time. For example, the military accounted for a large portion of the country's economy and saw little need to be efficient. With the government owning all businesses and an absence of competition, managers saw no need to use fuel wisely or search for fuel alternatives in a country rich in natural resources. The communist guarantee of a job for everyone meant that labor costs were high though wages were low, and no attempts were made to increase productivity by updating machinery and computers. Finally, the practice of central planning meant that government bureaucrats, not supply and demand, determined what was produced. The bureaucrats proved highly inefficient. As resources became scarcer because they could not be adequately exploited or delivered, the various ministries of the government hoarded them, attempting to become self-sufficient because other agencies could not supply them. This practice created further redundancies, shortages, and squandering of resources. Bureaucrats played it safe by locating new enterprises where they could best obtain supplies. In most cases, the locations were big cities. This led to excessive migration to cities, creating congestion, high living costs, and environmental problems. Ironically, though government bureaucrats helped ruin the Soviet economy, they became the unhappiest segment of society and greatly wanted change. As the privileged within the Soviet Union, they were displeased to see their standard of living drop below that of the poorer groups in the materially wealthy capitalist countries.

Perestroika *and* Glasnost

Mikhail Gorbachev became the leader of the Soviet Union in 1985. He immediately set out to reform his country's economic and political systems. In doing so he highlighted two concepts: *perestroika* (economic restructuring) and **glasnost** (informational openness). For *perestroika,* Gorbachev believed that it was necessary to divorce economics from politics, allow more local control, and introduce free-market practices. Such policies went against Communist ideals and established interests. The bureaucrats fought Gorbachev and his policies, creating great political turmoil.

As political battles raged, Gorbachev's economic policies ran opposite of what had been practiced since the 1920s. A crisis ensued as the economy attempted to reverse direction. Companies, having existed in a noncompetitive environment, now had to meet their own costs and find their own customers. No longer able to waste resources, they had to cut costs. To generate income, they cut production and raised the prices of their goods, but they soon found that they could not afford to buy supplies from one another. Not only were goods very expensive, but cash was also in short supply. For example, the Soviet Union previously sent machinery to Cuba and in return received sugar. With free-market reforms, Soviet factories wanted cash, not sugar. The same problem emerged between companies within the Soviet Union and the COMECON trading bloc of East Central Europe. Furthermore, the lack of a capitalist banking and financial system meant that cash could not flow easily. Poor transportation and communication systems only exacerbated the situation. Production declined. To cut costs, companies laid off workers, increasing unemployment. The Soviet economy continued to spiral downward with frequent labor strikes and rampant crime.

Gorbachev also stressed the concept of *glasnost,* the policy of providing government information to citizens. *Glasnost* was intended to have a positive effect on the country by empowering citizens with knowledge. Freedom of information allowed citizens to learn about corruption, government abuse, forced labor, and many of the other problems of the Soviet government. Non-Russians at last could vent their anger at the Soviet government for its Russification policies. In contrast, Russians complained that the Soviet government suppressed Russian culture, especially the Russian Orthodox Church, and distributed Russia's resources to the country's minorities. To avoid the wrath of citizens, many politicians echoed their anger, championed local and regional causes, and turned against the central government in Moscow. To curry favor with the citizenry, local and regional politicians helped to hoard resources for their people, further contributing to the country's economic crisis. Before long, government leaders helped their republics move toward independence. Even Boris Yeltsin, the elected leader of the Russian Republic in 1990, called for Russia's independence from the Soviet Union.

When Mikhail Gorbachev became the leader of the Soviet Union in 1985, he believed in Communism and tried to preserve the Soviet Union by implementing reforms. Ironically, in 1991, he became the Soviet Union's last leader as his policies led to the unraveling of the country. Gorbachev tried one last time to keep the country together with a new union treaty that relinquished much of Moscow's political power to the republics. On the eve of its signing on August 19, 1991, dissatisfied conservative Communists who wanted to restore the old system attempted to seize power in Moscow while Gorbachev was away. Massive public demonstrations supported Gorbachev. When Boris Yeltsin supported the demonstrators and protected them from a military crackdown, the coup failed. The coup attempt only accelerated the dissolution of the Soviet Union. Led by Lithuania, the Soviet republics all declared independence by December. While Gorbachev's policies led to the rapid demise of his country, his reforms paved the way for further changes, without which the country would probably have disintegrated anyway, perhaps more violently.

After the demise of the Soviet Union in 1991, many people around the world wondered if Russia would remain a world power (see Point–Counterpoint: Russia: Still a World Power? on page 148) because it did not have the economic strength of a world power. By 2000 Russia produced only 0.08 percent of the world GDP. The GDPs per capita within Russia and the other CIS countries were far below those of capitalist countries such as those in Europe and the United States (Figure 4.16). Almost 14 percent of Russia's people officially live in poverty, earning less than $31 per month. The ownership of consumer goods reflects this modest income (Figure 4.17). TV set ownership is high, as television provided mass communication from the government to the people during the Soviet era.

Human Rights

In imperial Russia, the czars killed or sent their political opponents to Siberia, far from the political center of the empire. After the Bolsheviks came to power in 1917, they continued the practice of banishing their opponents, often harshly. In 1919, for example, they had several hundred thousand Don Cossacks killed. In 1921, they established the first extermination camp for the opponents of their regime. Most camps, however, were designed primarily for slave labor but had extremely high death tolls.

Under Joseph Stalin, the camps grew dramatically in size and number. In 1928 they had approximately 30,000 prisoners. By 1930 the numbers grew to 600,000, when Stalin created the Main Directorate for Corrective Labor Camps. The acronym for

RUSSIA: STILL A WORLD POWER?

As Communism lost its grip on East Central Europe after the fall of the Berlin Wall in 1989 and then fell with the dissolution of the Soviet Union in 1991, Russians were confronted with profound change. On the positive side, it promised political and economic reform, with greater freedom of expression and a better standard of living. On the negative side, it coincided with the loss of much of their country, the open expression of great anti-Russian feelings from people within their world region, and the questioning and potential end of Russia as a world power. These experiences were a tremendous blow to the prestige of the Russian people.

When the 14 non-Russian republics declared their independence from the Soviet Union in 1991, it seemed that the peoples of these republics simply were expressing their right of self-determination. The Russians, however, believed that the Soviet republics were Russian lands regardless of who lived in them. They saw the Soviet Union as rightfully theirs because it was created from the Russian Empire, lands that they struggled for and acquired over the course of centuries. Ukrainian and Belarussian independence particularly is difficult for Russians to accept because much of the land and peoples of these republics were part of *Rus*—the original Russian heartland and people. Though these Soviet republics declared independence, many Russians do not see these 14 new countries as foreign. Instead, they see them as part of their "Near Abroad" and feel that they have an exclusive voice in both the internal and international relations of these countries.

As noted in the "Population Distribution and Patterns" section (pages 153–154), millions of Russians live in the "Near Abroad." The adjustment from majority to minority status has been difficult for Russians, both in and out of Russia. The policies of these independent republics have caused millions of Russians to migrate to Russia.

Countries of the "Near Abroad" also are crucial to Russia's status as a world power. The Baltic republics housed key military installations. The Crimean Peninsula is a major Russian naval facility. It is no wonder that the Russian government tried to hold onto the Crimean Peninsula and its naval fleet after Ukraine declared independence. It would take decades and billions of dollars of investment for Russia to re-create a comparable naval facility along the Russian coast of the Black Sea. Kazakhstan was the center of the Soviet space program. The Russians see the Soviet space program as their accomplishment, one that only Americans have matched. Control over the oil in the Caucasus and Central Asia is also key to Russia's role as a world power.

Coping with the loss of the "Near Abroad" is compounded by struggles by non-Russians within the Russian Federation for independence. Chechnya is the most vexing of them all. As the Soviet Union disintegrated in 1991, the Chechens moved toward independence from the Russian Federation and declared it in 1994. The Russian military responded by supporting a Chechen rebel group that sought to overthrow the Chechen government and keep Chechnya within Russia. The military campaign was not as quick as the Russian government had hoped. Military operations continue, though intervals of negotiations bring relative peace. The campaign, however, has been brutal. By early 1996, an estimated 40,000 to 100,000 people, mostly civilians, had been killed in a republic of just 1 million. Street-to-street fighting and bombings by the Russian air force have made Chechnya's cities uninhabitable. Accused of gross human rights violations, Russia's international reputation has suffered from the war in Chechnya. However, Russians fear that if they grant Chechen independence, then Russia's other minorities will also move toward independence. In addition, Chechnya, especially Groznyy, is a major transit point for oil leaving the Caspian Sea region (see Figure 4.11). Keeping the oil flowing through this pipeline is crucial for Russia. If proposed pipelines through Georgia and Iran succeed in taking business away from the Russian pipeline, Russia will lose billions in income, along with ability to influence international affairs.

Russia's ability to influence world affairs diminished in other ways in the 1990s. Soldiers in Russia's military are vastly underpaid, unprepared, and demoralized from their experience in Chechnya (Box Figure 1a), which followed the earlier defeat in Afghanistan in the 1980s. Though Russia is a nuclear power, the United States, viewing itself as the sole victor of the Cold War, acted unilaterally in international affairs. At the same time, the North Atlantic Treaty Organization (NATO) expanded into East Central Europe to include countries formerly dominated by the Soviet Union. For members and potential members of NATO, the expansion was a defensive move. For Russians, NATO's moves were provocative, particularly because they were made without Russian consent. NATO was a Cold War institution created to defend Western Europe from Soviet Communism. Because Soviet Communism no longer existed and Russia was very weak, NATO expansion suggested that the organization was not defensive but offensive. It played on Russian fears that NATO was nothing more than another force, not unlike Napoléon or the German armies of the two World Wars, to menace Russia.

By the late 1990s Russians felt ostracized by the United States and Western European countries. Closer relations with China led to the signing of a friendship treaty in 2001. The treaty is a significant agreement because it sets aside decades of tension caused mostly by border disputes. The treaty allows for cross-border trade. For example, Russia will send oil and military supplies to China. Both China and Russia agree to crack down on Islamic fundamentalists who straddle their borders and threaten the territorial integrity of their countries. In short, a Russo–Chinese alliance is a signal to the rest of the world that Russia is still a world power.

Vladimir Putin, the leader of Russia, and the events of September 11, 2001, have changed Russia's relationship with much of the world. Putin has not complained much about American international diplomacy, though it offends long-held Russian interests. Instead, he has been more conciliatory in hopes of gaining U.S. cooperation and increasing Russia's international influence. A warming of relations between the United States and Russia began and then increased after the events of September 11, 2001. Islamic fundamentalism in Afghanistan was a concern that led to the Soviet invasion of that country in the 1980s. The U.S. attempt to suppress Islamic fundamentalism parallels Russian concerns. The U.S. criticism of the Russian military campaigns in Chechnya turned to sympathy. This occurred when al-Qaeda fighters taking refuge in Chechnya aided the Chechen cause. Russian leaders were able to describe the war in Chechnya as another front in the war on terrorism. Russia was also able to press the Georgian government to do something about Chechen rebels hiding in the Pankisi Gorge. Russians are sensitive about foreign activity in a "Near Abroad" country such as Georgia, but with Russian–American cooperation in the war on terrorism, Russians voiced little concern about the United States dispatching elite military forces to Georgia to take care of a problem that vexed Russia.

Russia also played a crucial role in determining the fate of the Kyoto Protocol. The United States opposed the protocol but needed Russian opposition to keep the protocol from going into effect. However, European countries and Japan supported the protocol

Box Figure 1 (a) Russian soldier walks past wreckage of a Mi-8 helicopter south of Grozny, Chechnya, March 23, 2005. The Mi-8 helicopter that belonged to Russian Interior Ministry troops crashed while preparing to land after suffering a technical malfunction. (b) Russian paratroopers march through Red Square at a military parade during Victory Day celebrations in Moscow, May 9, 2003. **Photos:** (a) © Musa Sadulayer/AP/Wide World Photos; (b) © Ivan Sekretarev/AP/Wide World Photos.

and needed Russian support for it to go into force. Finding itself the crucial decision maker, the Russian government delayed its decision, allowing each side opportunities to sway it with offers. By the end of 2004, the Russian government decided to join with European countries and Japan and ratify the Kyoto Protocol.

Since 2000 Russia has become a more full-fledged member of the informal organization representing the world's wealthiest countries, once known as the Group of Seven (G7) but, with Russia, increasingly known as the G8. In 2002 NATO gave Russia a voice in NATO affairs, though Russia is not a full member. Russia also has offered to supply the United States and its allies with oil. Showing a preference to buy oil from Russia rather than from many of the countries of Southwestern Asia, such as Iran, the United States has become more supportive of the Russian oil pipeline from the Caspian Sea. With an improving economy, more effective political leadership, and a changed climate of international affairs, the Russian Federation has the opportunity to exert itself as a world power (Box Figure 1b).

Debate whether Russia is still a world power.

POINT	COUNTERPOINT
Russia lost possession of 14 republics (now independent countries) and with them sizable populations and resources.	Russia's heartland remains together, and Russia is the world's largest country, with a sizable population and considerable resources.
Russia's international power is undermined by the presence of 25 million Russians who now find themselves as ethnic minorities in other countries. Russia must be careful not to offend these other countries and thereby endanger the Russians who live in them.	Russians living as ethnic minorities in other countries increase Russia's political influence in those countries because they can vote and hold political office. Russia can also pressure those countries on behalf of Russians living in them. The Russian minorities also strengthen Russia's trading links with these countries.
The Russian government has been ineffective in controlling independence movements in its own republics. It has been condemned for human rights violations in Chechnya. Its military is underpaid, unprepared, and demoralized. Russia was unable to prevent the expansion of NATO.	Russia gained sympathy from the United States for its "war on terrorism" in Chechnya. Russian soldiers are very patriotic, as in the past, and Russia is a nuclear power. With peacekeeping troops in the former Yugoslavia, Russia influences events in that area of the world. Russia also now has a voice in NATO's affairs.
With rampant corruption, Russia's economy is weak and has floundered since the end of Communism in 1991. Russia is unable to capitalize on selling its vast resources on the world market. Its control of Caspian Sea oil is questionable. Russia has little influence in global economic organizations.	Russia's economy is steadily improving, and the government is fighting corruption. Russia has signed agreements to supply oil to other countries and is able to control the flow of the oil from the Caspian Sea. Russia has joined the Group of Seven (G7), making the organization the Group of Eight (G8).

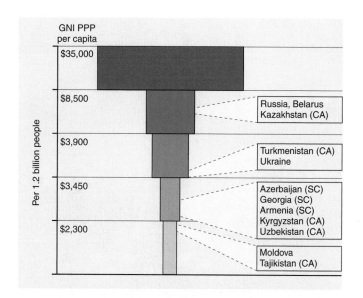

FIGURE 4.16 Russia and Neighboring Countries: national incomes compared. The countries are listed in the order of their GNI PPP per capita. CA = Central Asia, SC = Southern Caucasus. **Source:** Data (for 2000) from *World Development Indicators*, World Bank, 2002.

the Russian name, ***Glavnoe upravlenie ispravitel no-trudovykh lagerei*** ("Main Directorate for Corrective Labor Camps") is ***gulag.*** Many of the *gulag* camps were located in Siberia (Figure 4.18). The first dramatic increase in the number of slave laborers was caused by Stalin's program of collectivization in the agricultural sector. Many of the prisoners were the *kulaks,* successful farmers who engaged in capitalist practices Stalin

labeled as "rich." Not allowed to join the collectives after their land was taken away from them, the *kulaks* fell into dire poverty and then were sent to the *gulag.*

Stalin also saw the *gulag* as a useful tool for economic development. Slave labor was essentially free, costing only a few bowls of thin soup and a couple of slices of bread a day for each laborer. Slave labor could be used to complete dangerous work that no free worker would agree to do. For example, in 1931 Stalin directed slave labor to build a 227-km (141-mi.) canal across northern Russia from the White Sea to the Baltic Sea. Using pickaxes, shovels, and wheelbarrows, more than 100,000 workers completed the canal in 1933. The cost was tens of thousands of lives. Stalin then used slave labor from the *gulag* to complete such projects as the Moscow–Volga Canal and the Baikal–Amur main railroad line. Slave labor was also used to build hydroelectric stations, thousands of kilometers of roads, and industrial complexes in isolated locations in Siberia (such as the previously mentioned complex in Norilsk) and northern Russia.

The *gulag* was filled with millions more after Stalin began a series of purges in 1934, known as the Great Terror. The purges sent millions of Communist Party members, military officers, scientists, intellectuals, artists, and common citizens arbitrarily accused of treason to the *gulag.* Their free labor further fueled the Soviet economy and helped in the war effort after Hitler declared war on the Soviet Union in 1941. Engineers in the *gulag* system designed the advanced Soviet airplanes and tanks produced during the latter years of the war. Laborers in the *gulag* assembled these weapons. After the war, Stalin sent millions more to the *gulag.* Many were German prisoners of war. Others were accused of collaborating with the

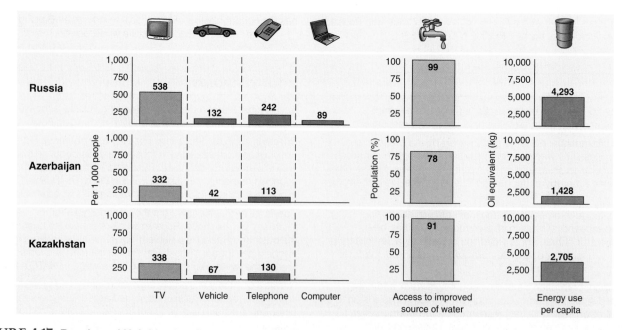

FIGURE 4.17 Russia and Neighboring Countries: ownership of consumer goods, access to clean water, and energy usage. What is the significance of relatively high television ownership? **Source:** Data (for 2002) from *World Development Indicators*, World Bank, 2004.

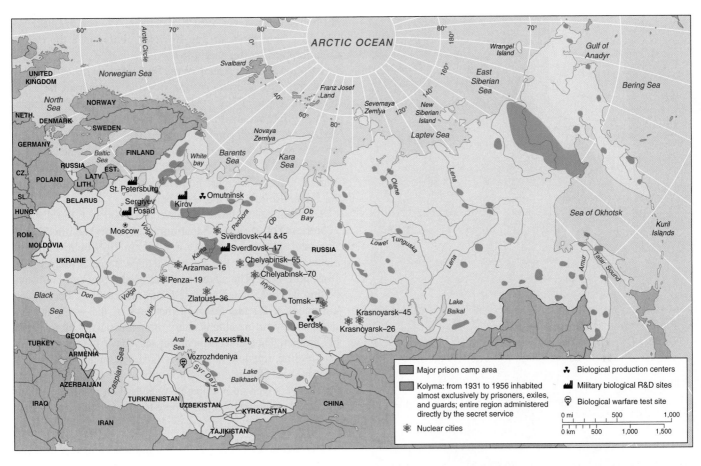

FIGURE 4.18 Gulag camps and Soviet secret cities. Shown on this map are *gulag* camps (slave labor camps) where prisoners were housed and their labor used for economic output. For example, the prisoners in the Kolyma area mined gold. Also shown are Soviet "secret cities," built to isolate special types of military research and with access to them limited; they now have little, if any, role. **Source:** Map from *The Unquiet Ghost* by Adam Hochschild. Copyright © 1994 by Adam Hochschild. By Meridien Mapping, Oakland, California. Reprinted by permission of Georges Borchardt, Inc.

enemy, including entire ethnic groups within the Soviet Union. Russians, Ukrainians, and East Central Europeans who lived under German occupation during the war were seen as contaminated with capitalist ideas and were also sent to the *gulag.*

Many *gulag* prisoners were forced to work in coal, copper, and gold mines. The Kolyma gold mining region in far eastern Siberia was one of the most treacherous *gulag* camps. Kolyma was isolated and known for having the coldest recorded temperatures on Earth, sometimes as low as –98°F (–72°C). Thirty percent of the prisoners died in their first year at Kolyma. Very few survived a second year. As many as 3 million perished between 1937 and 1953 in Kolyma alone. About 2 pounds of gold were produced for every person who died.

After Stalin's death in 1953, the number of people sent to the *gulag* dramatically decreased but did not stop. It is difficult to know exactly how many perished in the *gulag,* although estimates are in the tens of millions. People outside the Soviet Union heard of the Soviet *gulag,* but few facts were known until Alexander Solzhenitsyn wrote about it in two books, *One Day in the Life of Ivan Denisovich* and *The Gulag Archipelago.* Solzhenitsyn had personal experience in

the *gulag.* After writing a letter containing critical remarks about Stalin, Solzhenitsyn was sent to the *gulag* from 1945 to 1953. Solzhenitsyn's later literary efforts earned him the Nobel Prize for literature in 1970. After writing *The Gulag Archipelago,* Solzhenitsyn was sent into exile in 1974. He returned to Russia after the fall of Communism in 1991 (Figure 4.19).

Andrei Sakharov was another major figure who struggled for human rights. A physicist, Sakharov was known as "father of the Soviet hydrogen bomb." He was also concerned about the radioactive hazards of nuclear testing. After he published an essay titled "Reflection on Progress, Coexistence and Intellectual Freedom" in the *New York Times* in 1968, Sakharov was fired from the Soviet weapons program. He continued arguing for peace and human rights and won the Nobel Peace Prize in 1975. In 1980, not long after he criticized his country's invasion of Afghanistan, Soviet authorities sent Sakharov to internal exile in Gorky, 250 miles east of Moscow. After Gorbachev came to power, Sakharov was freed in 1986 and later invited to help in governmental reforms before his death in 1989.

FIGURE 4.19 Since the end of Communism in 1991, people have been freer to express their opinions about historical events. New civil war (1917–1922) monument in Kharbarovsk commemorating both the Reds (Communists) and Whites (anti-Communists). Rather than vilifying one side, both sides are depicted as patriotic despite their opposing beliefs. **Photo:** © Ronald Wixman.

Since the end of Communism, human rights abuses have decreased but not ended. The system of police practices and prisons that violated human rights did not completely disappear, particularly not in many Central Asian countries. Court systems need to be reformed. Defendants sit in jail before their cases are heard, and low pay makes many judges prone to accepting bribes. Juries are rare. Judges in Russia act with the advice of two citizens who are popularly called "nodders" because they always agree with the judges. Judges frequently serve as prosecuting attorney as well as judge. Not surprisingly,

the conviction rate has been 99.6 percent. The war in Chechnya caused a new series of human rights abuses (see the Point–Counterpoint box on page 148).

Following the breakup of the Soviet Union in 1991, democratic reforms allowed groups working for human rights to blossom, especially in the Slavic countries. For example, Amnesty International now has as many as 40 chapters across the region. Though some human rights groups are outgrowths of international organizations, some are unique to the concerns of the region's people. The Committee of Soldiers' Mothers of Russia, "Mother's Right" Foundation, and the Moscow Center for Prison Reform are a few examples.

Human rights, however, have worsened in Central Asian countries. For example, Saparmurad Niyazov has ruled Turkmenistan since independence in 1991. In 1999 he had himself elected president for life and built a huge statue of himself cast from pure gold. The statue revolves so that the sun is always behind it. Niyazov calls himself Turkmenbashi ("father of the Turkmens"). He has renamed the days of the week and the months of year, naming January after himself and April after his mother. Though Turkmenistan may have the most repressive regime, freedom of the press is almost nonexistent across Central Asia. Political opposition groups are weak, and any of their leaders who do not operate outside their respective countries are mostly likely in jail, where torture is common.

Women's Roles

Communist ideology professed that women were equal to men, and women achieved this equality following the revolution in 1917, before they were allowed to vote in the United States. During Soviet times, women moved into positions traditionally held mostly by men in other societies (like the United States), such as jobs in government, economics, medicine, and engineering. In some professions women formed the majority. Most physicians, for example, were women. During World War II, some of the most decorated fighter pilots and infantry were women. Though Communism gave women great freedom in their career choices, it also idealized the factory worker, not the physician or engineer. Consequently, women were given the freedom to move into professions that lost much of their prestige compared to the same professions in the capitalist world. However, women also were encouraged to do industrial and construction work, holding more than 60 percent of the construction jobs.

Women achieved equality in their careers but usually not at home. After women completed a day's work at their jobs, many arrived home to undertake the traditional responsibilities of housework and child care. Men's attitudes and behaviors in regard to their wives did not change much during Communism. As a result, between home and work, women ended up working long hours. As the region has transformed from Communism to capitalism, the roles of women have changed. The idea that women are equal to men still exists, although many women lost their jobs or had jobs that were not paid for long periods. Full and meaningful equality has yet to be realized.

Population Distribution and Patterns

In the Russian Federation, the greatest concentration of people is in western Russia (Figure 4.20). Higher population densities continue eastward along the Trans-Siberian Railway to the southern end of Lake Baikal and Vladivostok on the Pacific coast. The extensive areas of mountains and desert in the south and of permafrost-ridden lands in the north and east contain few people. Ukraine, Belarus, and Moldova cluster around the most densely populated western Russia. They have moderate population densities that are evenly spread, with greater concentrations of people around industrial cities.

The Caucasus Mountains create uneven distributions of people in Georgia, Armenia, and Azerbaijan, with few people in the mountains and most in the plains. The distribution of population in Central Asia is marked by contrasts between areas of few people in the arid and mountainous parts and areas of higher densities in the irrigated lowlands. The main concentrations of people are all separated from other world economic centers by high mountains and the difficulties of having to travel or send goods through Russia.

During Russian imperial and Soviet times, the government in Moscow maintained control of their vast country by encouraging Russians to move to other republics and the minority peoples, sometimes against their will, to move great distances from their territories. Consequently, when the 15 republics of the Soviet Union became independent countries in 1991, many people found themselves outside their designated countries. Twenty-five million Russians, or 17 percent of the Russian population of the Soviet Union, live in the "Near Abroad" (Figure 4.21). High concentrations

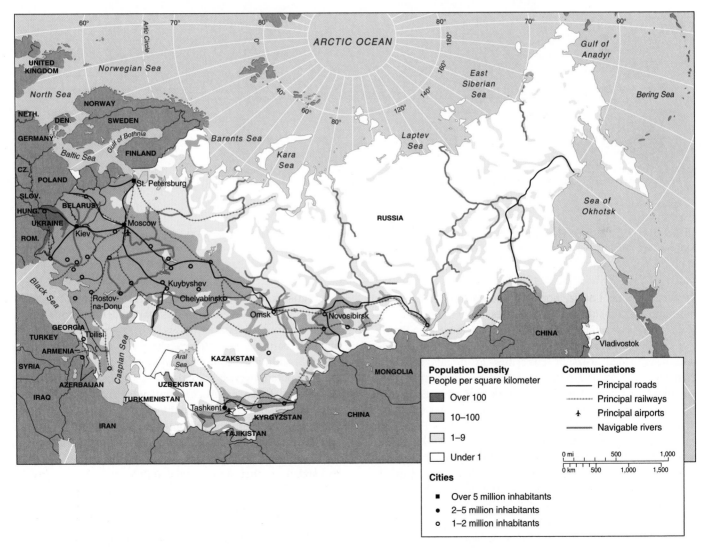

FIGURE 4.20 Russia and Neighboring Countries: distribution of population. For Russia, comment on the location of the highest and lowest densities of population. How do these reflect climatic conditions, initial historic superiority, the line of the Trans-Siberian Railway, and other factors? Compare this map with Figures 4.6, 4.7, and 4.8 **Source:** Data from *New Oxford School Atlas,* Oxford University Press, UK.

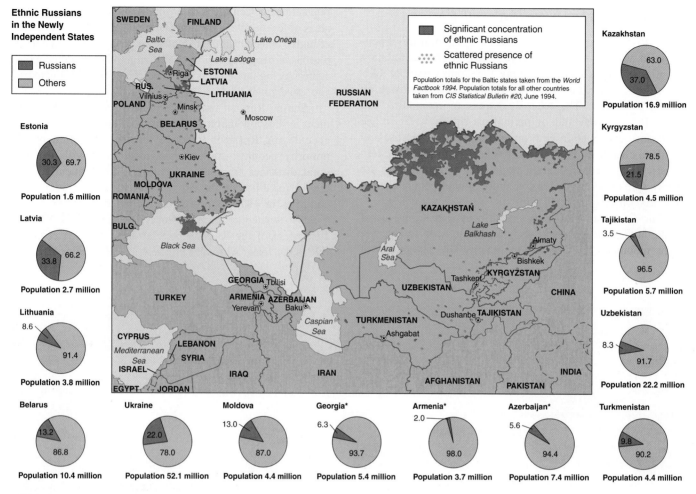

FIGURE 4.21 Ethnic Russians as minorities in neighboring countries. In which countries are Russians the largest minorities? Do Russians outside of Russia enhance or hinder the political power of Russia in the region?

live in Estonia, Latvia, northern and eastern Ukraine, and northern Kazakhstan. In many locales in the Crimean Peninsula (part of Ukraine), Russians constitute more than 90 percent of the population. Many of the new countries see their long-established Russian minorities as threats and have insisted that these Russians assimilate (learn the native language and way of life) or go back to Russia. By 2001 it was estimated that as many as 5 million Russians emigrated from the independent republics to the Russian Federation (Figure 4.22). In June 2003, the leader of Turkmenistan ended dual citizenship for Russians. Russians choosing Russian citizenship desperately tried to sell their property quickly because non-Turkmen were not allowed to own property. However, with only a few weeks' notice, many Russians had to sell at very low prices.

In the Southern Caucasus, most Armenians and Azerbaijanis do not live in their respective countries. Of the 6.3 million Armenians in 2005, only 3.0 million lived in Armenia. Many lived in Azerbaijan and Georgia. Of the 19 million

Azerbaijanis, 8.4 million lived in Azerbaijan. Most lived in neighboring Iran. Azerbaijan was 90 percent Azerbaijani, but 90 other groups lived there as well. In contrast to Armenians and Azerbaijanis, most Georgians lived in Georgia, which had 4.5 million inhabitants. However, 30 percent of Georgia's population was composed of other groups such as Armenians, Russians, Azerbaijanis, Ossetians, and Abkhaz. The peoples of the Central Asian countries also lived in one another's countries. For example, the boundaries of Kyrgyzstan, Tajikistan, and Uzbekistan weave through the fertile and densely settled Fergana Valley, leaving many Kyrgyz, Tajiks, and Uzbeks in their neighbors' countries.

Population Dynamics

One of the greatest concerns for the Slavic countries is population decline. Overall, these countries may decline by 20 million by 2025 (Table 4.2). Growth rates were typically from 0.0 to −0.7 percent in 2001. Fertility rates were around 1.2, far below

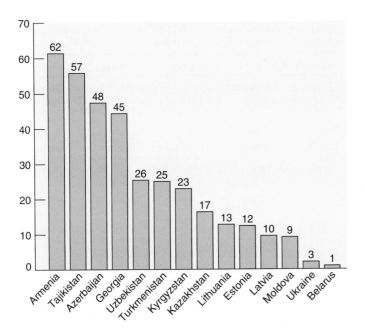

FIGURE 4.22 Migration of ethnic Russians from the former Soviet republics from 1989 to 1998. Each number represents the percentage of ethnic Russians who have emigrated from that country's total population of ethnic Russians.

the fertility rate of 2.1 needed to maintain a population at its current number. The age–sex diagram for Russia (Figure 4.23) shows an uneven pattern of age groups as a consequence of World War II and a baby boom from 1950 to 1970, when large families were encouraged by the state. Women dominate the over-60 age groups to a greater extent than in other parts of the world as a result of male deaths in World War II and Stalin's persecutions. In 2004 the birth rates were 10 per 1,000 of the population and falling, while the death rates were 17 per 1,000 and rising (Figure 4.24), yielding one of the fastest rates of natural decrease in the world that year.

For the Russian Federation, such figures hide variations within this huge country. In some parts, deaths outnumber births by two to one. The most significant geographic variation is between the western regions of the Russian Federation and the more recently settled regions of the north and Siberia. In western Russia, birth rates are low, death rates high, and the population is aging quickly. Yet the total numbers fall less rapidly because other Russians are immigrating to western Russia from the former Soviet republics and from the north and Siberia. Emigration is particularly strong from the areas where indigenous peoples have gained control over local government.

Economic decline has been the major cause of population decline in the Slavic countries and the Southern Caucasus. Lack of government subsidies for industries in the north and Siberia is one factor. Lack of similar subsidies in the older urban areas, especially the industrial cities, and the inability of industries to adapt to market economies led to high unemployment. Many who are still working have not been paid for months or years.

Abortion rates are high, a legacy of Soviet times when abortion became a common means of birth control. Health care has also suffered greatly with economic decline. Hospitals have frequently run out of basic supplies such as drugs and anesthetics. Economic hardship also has increased alcoholism, particularly among men. For many hospitals, 90 percent of emergency cases have been alcohol-related. The entire situation has given women little incentive to have children.

Though Moldova is also transforming its economic system from Communism to a market economy, its population is not declining. Unlike the neighboring Slavic countries, Moldova is overwhelmingly agricultural, not industrial. The traditional large farming family has not been greatly disrupted by the transition. However, growth rates are not nearly as high as they once were. Thus the transition has had some effect on the population of Moldova. Like Moldova in its subregion, Azerbaijan stands out as a country of positive population growth in its subregion of the Southern Caucasus. The total fertility rate, which is above 2.0, more than offsets emigration.

While most of the Slavic countries and those in the Southern Caucasus have stable or declining populations, the five countries of Central Asia anticipate growth from the 2005 total of around 59 million people to 72 million by 2025. Natural increases remained below 2 percent in the countries with large Muslim majorities: Kazakhstan (50 percent non-Muslims) had the lowest increases (0.5 percent per year), followed by Kyrgyzstan with 70 percent Muslims (1.3 percent per year). The age–sex diagram for Kazakhstan (see Figure 4.23b) contrasts with those of the other areas of this world region. Death rates are very low—a recipe for increasing population. The fertility rates fell from around 6 to around 3 between 1970 and 2000. Such rates contrast with the population decline in the rest of the former Soviet domain.

The Economy of the Land: Agriculture

The Slavic countries' northerly latitudes and distant locations from moderating maritime influences result in low moisture. Low-temperature climates in the north and high-temperature ones in the south make farming difficult almost everywhere (Figure 4.25). Even in the best locations, the late arrival of spring or the early appearance of winter can ruin a year's crops. The most productive farming areas are in the west of Russia, Ukraine, and Moldova in the former steppe and deciduous forest areas (see Geography at Work: Wine Industry on page 160). Ukraine's vast plains of black soil produce abundant crops of wheat and make the country agriculturally self-sufficient. Moldova also has low relief and good soils like western Ukraine. Fruits, vegetables, wine, and tobacco are the main farm products. Belarus, however, is a country with low relief but mostly poor soils. Having been shaped by continental glaciation, the southern part of the country is covered by the Pripyat Marshes and the northern part by glacial moraines. The sandy glacial soils are good for

Table 4.2 Russia and Neighboring Countries. Data by country, area, population, income (GNI PPP—Gross National Income Purchasing Power Parity), urbanization, Human Development Index (HDI), Human Poverty Index (HPI).

Country	Capital City	Land Area (km²)	Population (millions)		GNI PPP 2003		% Urban	Index 2002	
			Mid 2005 Total	Est. 2025 Total	Total (US$ billions)	Per Capita (US$)	2005	HDI (rank of 177)	HPI (% population)
Slavic Countries									
Russian Federation	Moscow	17,075,400	143.0	130.2	1,279	8,920	73	57	No data
Belarus	Minsk	207,600	9.8	9.4	59	6,010	72	62	No data
Moldova	Chisinau	33,700	4.2	4.0	7	1,750	45	113	No data
Ukraine	Kiev	603,700	47.1	41.7	262	5,410	68	70	No data
Southern Caucasus									
Armenia	Yerevan	29,800	3.0	3.3	12	3,770	65	82	No data
Azerbaijan	Baku	86,600	8.4	9.7	28	3,380	51	91	No data
Georgia	Tbilisi	69,700	4.5	4.2	13	2,540	52	97	No data
Central Asia									
Kazakhstan	Akmola	2,717,300	15.1	15.9	92	6,170	57	78	No data
Kyrgyzstan	Bishkek	198,500	5.2	6.7	8	1,660	35	110	No data
Tajikistan	Dushanbe	143,100	6.8	9.2	7	1,040	27	116	No data
Turkmenistan	Ashkhabad	488,100	5.2	6.6	28	5,840	47	86	No data
Uzbekistan	Tashkent	447,400	26.4	33.9	44	1,720	37	107	No data

Source: World Population Data Sheet, Population Reference Bureau (2005), World Development Indicators, World Bank (2005), Human Development Report, United Nations (2004), and Encarta Microsoft (2005).

potatoes. Toward the north in the Russian Federation, around Moscow, arable land gives way to livestock farming. East of the Urals, farming is restricted to areas that have sufficient water and length of summer growing season. The warm climates of the Southern Caucasus make it possible to grow citrus fruits, tea, tobacco, cotton, and rice.

Soviet leader Nikita Khrushchev sought to increase agricultural production in the 1950s with his **Virgin Lands Campaign** to make the Soviet Union self-sufficient and independent of the global economy. This campaign promoted farming in lands where it had never been done before. Many of these lands had poor soil quality, and they did not have enough water or heat to grow crops. Much of the land in the Virgin Lands Campaign was in the semidesert and desert areas of Central Asia, especially in the Kazakh Republic. Some of the land was also in the adjacent dry areas of the Russian Republic, much of it stretching across southern Siberia. With the importation of water through irri-

gation systems, it was believed that these dry areas would become productive. From 1917 to 1987, irrigated land in the Soviet Union increased from 3.5 to 20 million hectares. At the same time, large drainage networks were built in the lands north of Moscow to dry the waterlogged soils. From 1956 to 1987, drained land increased from 8.6 to 19.4 million hectares.

The Virgin Lands Campaign proved a huge failure over the long run. Production increased, but many problems resulted. Massive amounts of water were diverted from the Caspian and Aral Seas, wreaking havoc with their ecosystems (see Figure 4.15). Pouring water on desert soils led only to soil salinization. Drainage projects were no more successful. Farm production fell. Eventually the Soviet government stopped pushing for the expansion of the Virgin Lands Campaign. Over the ensuing decades much of the land was slowly abandoned, though some of the better lands are still farmed.

Country	Ethnic Groups Percentages	Languages O=Official N=National	Religion Percentages
Slavic Countries			
Russian Federation	Russian 82%, Tatar 4%, Ukraine 3%	Russian (O), Tatar, etc.	Russian Orthodox, Muslim
Belarus	Belarussian 78%, Russian 13%, Polish 4%	Belarussian (O), Russian (O)	Eastern Orthodox 80%, Roman Catholic 18%
Moldova	Moldovan/Romanian 65%, Ukrainian 14%, Russian 13%	Moldovan (= Romanian) (O), Ukrainian, Russian	Eastern Orthodox 99%
Ukraine	Ukrainian 73%, Russian 22%	Ukrainian (O), Russian, Romanian, Polish	Ukrainian Orthodox, Roman Catholic
Southern Caucasus			
Armenia	Armenian 93%, Azeri 3%, Russian 2%	Armenian (O) 96%, Russian 2%	Armenian Orthodox 94%
Azerbaijan	Azeri 90%, Daghestani 3%, Russian 2%	Azeri (O) 89%, Russian 3%, Armenian 2%	Muslim 93%, Eastern Orthodox 5%
Georgia	Georgian 70%, Armenian 8%, Russian 6%, Azeri 6%	Georgian (O), Russian	Georgian Orthodox 65%, Muslim 11%, other Orthodox 18%
Central Asia			
Kazakhstan	Kazakh 46%, Russian 35%, Ukraine 5%, German 3%	Kazakh (Qazaq, O), Russian	Muslim 47%, Russian Orthodox 44%
Kyrgyzstan	Kirghiz 57%, Russian 18%, Uzbek 14%	Kirghiz (O), Russian	Muslim 70%, Russian Orthodox 20%
Tajikistan	Tajik 65%, Uzbek 25%, Russian 3%	Tajik (O), Dari, Russian, Uzbek	Sunni Muslim 80%, Shiite Muslim 5%
Turkmenistan	Turkmen 77%, Uzbek 9%, Russian 7%	Turkmen (O) 72%, Russian 12%, Uzbek 9%	Muslim 85%, Eastern Orthodox 11%
Uzbekistan	Uzbek 80%, Russian 6%, Tajik 5%	Uzbek (O) 74%, Russian 14%	Muslim (mostly Sunni) 88%, Eastern Orthodox 9%

Mikhail Gorbachev attempted agricultural reforms in 1988 when he permitted individuals and cooperatives to lease land privately from the government for farming. Within a few years the government allowed greater privatization. By the end of 1992 all state farms became the property of their members. At the same time, however, the Soviet Union began to disintegrate. Like all other sectors of the economy, the agricultural sector suffered and production fell. For example, the number of cattle fell from 60 million to 36 million from 1985 to 1997. Similarly, pigs fell from 39 million to 20 million and sheep from 65 million to 24 million over the same period.

By 1996 private farms accounted for only 2 percent of agricultural production in the Russian Federation. Few people own private farms because the infrastructure and economic system do not support them. Many have opted to stay on the collectives, though they are now owners. The collectives still accounted for 52 percent of agricultural production in 1996.

Interestingly, much production came from private plots on the collectives. For decades, Soviet authorities allowed individuals to cultivate small plots on the large collectives for their personal use. The total acreage on these personal plots was minuscule compared to the rest of the country's farmland. However, their production was comparatively very high. In 1970 they accounted for 31 percent of production. Their significance increased during the economic chaos of the 1990s: by 1996 they accounted for 46 percent of production. This method of privately using collectively owned land was more productive than private farms. Individual incentive mixed with collectively owned land and farm machinery that was better integrated into the economic system proved more effective than the private farms that lacked machinery and needed to be run as individual businesses.

Despite the economic hardship of transition, Russia, Ukraine, and Kazakhstan supplied approximately 11 percent of the world's traded wheat in 2004.

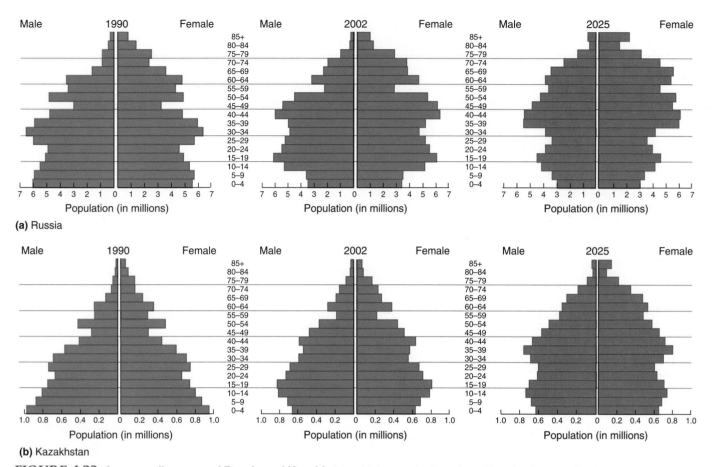

(a) Russia

(b) Kazakhstan

FIGURE 4.23 Age–sex diagrams of Russia and Kazakhstan. (a) Account for the variety of irregular shapes in Russia's pyramids. Do they correspond to historical events experienced by Russia? (b) How do Kazakhstan's pyramids compare to Russia's? What explains the differences? **Source:** U.S. Census Bureau, International Data Bank.

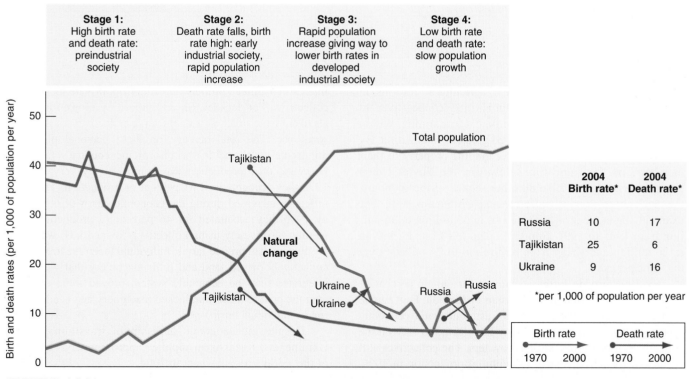

FIGURE 4.24 Russia and Neighboring Countries: demographic transition. Some death rates in the mid-1990s were higher than the birth rates; some countries resembled the world's less developed countries. **Source:** 2004 Update: Population Data Sheet, Population Reference Bureau.

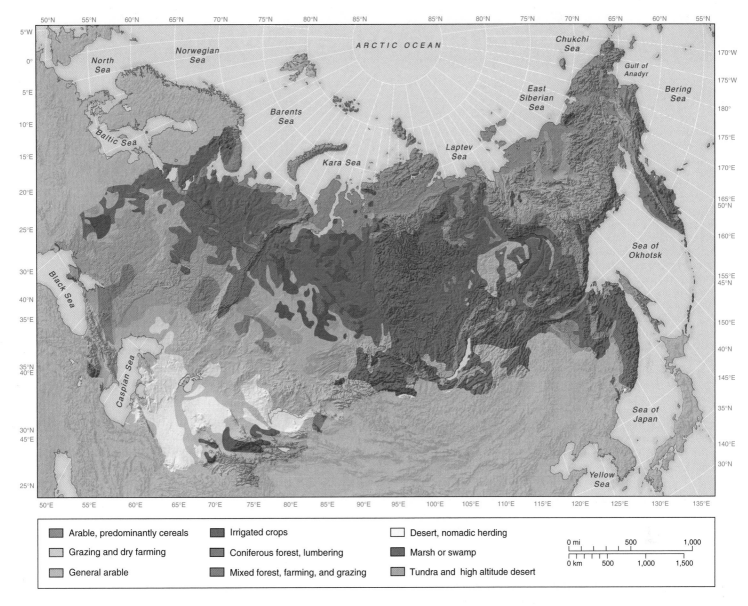

FIGURE 4.25 Russia and Neighboring Countries: major land uses. Relate the different types of farming to the natural regions as designated by climate, natural vegetation, and soils. **Source:** Data from *New Oxford School Atlas,* Oxford University Press, UK.

Legend:
- Arable, predominantly cereals
- Grazing and dry farming
- General arable
- Irrigated crops
- Coniferous forest, lumbering
- Mixed forest, farming, and grazing
- Desert, nomadic herding
- Marsh or swamp
- Tundra and high altitude desert

Urban Patterns and Linkages

Urbanization

The highest rates of urbanization in this world region are in areas that emphasize urban–industrial development. Thus the Russian Federation, Ukraine, and Belarus are the most urbanized, while Moldova and most of the Central Asian countries with their traditional ways of life are among the least urbanized. The Russian Federation has the greatest number of cities and the largest cities of the region (Table 4.3).

Urban Landscapes

In 1917, at the time of the Bolshevik Revolution, 17 percent of Russians lived in mainly small, provincial towns and cities. These prerevolutionary Russian towns were often more like overgrown villages, except for those such as Moscow and St. Petersburg that had grand designs of buildings (Figure 4.26). In the Southern Caucasus and Central Asia, the cities, which lie on ancient and medieval trade routes, are very old with architecturally beautiful structures that attract tourists.

Joseph Stalin's five-year plans emphasized intensive centralization and rapid industrialization. Industrial centers expanded and new specialist resources centers emerged, often in new towns in remote regions. Urbanization increased, from 48 percent of the region's population by 1959 to more than 70 percent in the 1990s. Stalinist urbanization produced distinctive, standardized cities throughout the Soviet Union. Housing for the workers consisted of numerous poor quality, high-rise, concrete buildings with few shopping opportunities in their vicinity because Soviet planners did not foresee the need for much shopping.

WINE INDUSTRY

Robert W. Hutton (Box Figure 1), retired subject cataloguer for the Library of Congress, finds geography important to understanding the wine industry. In his cataloguing activities, Robert had a lot of books on Russian wine, but the wine experts he worked with knew little of winemaking in Russia. Having a bachelor's degree in Russian from Haverford College and a master's in geography from Columbia University, Robert found his niche with so many books on Russian wine. Many of the books in Russian contained a lot on information on the former Soviet republics, especially in the warmer southern republics stretching from Moldova to Kazakhstan. In 1989 Robert was able to travel to Georgia, his first trip to the region. A second trip was to Moldova in 1993, and then one to Ukraine and Crimea in 1998. In each case Robert was able to talk to the local winemakers in technical Russian winemaking language. Such cultural understanding gave a good impression of Americans.

Robert finds that each wine is unique to the area where it is grown, and geography's concern with both the physical and human worlds takes into account all the factors explaining where and how wine is made. The term *terroir* refers to the physical geography of winemaking—that is, the concern with such factors as climate and soil. However, economics and cultural attitudes are just as important in understanding the wine industry. Whether during czarist or Communist times, Robert found that the concern seemed to be with quantity over quality. Since the end of Communism in 1991, winemakers have become more interested in producing better wines, but they find that their customers from the region prefer high alcohol content over flavor.

Box Figure 1 Robert Hutton: retired subject cataloguer for the Library of Congress.

Robert is currently working with a winemaker from the region who is writing a comprehensive book on winemaking in Ukraine. It will be the first book written on Ukrainian wines since the end of Soviet times, and the English edition, which Robert is helping to produce, will be the first such book in English.

One major trend since the breakup of the Soviet Union in 1991 is the building of suburbs, particularly around Moscow. The congestion and pollution of inner-city streets by the growing use of cars and trucks, together with the increase of urban crime, caused richer families to move out. The mayor of Moscow contributed to the movement by ejecting families from apartment blocks designated for renovation and commercial sale.

A new focus of the market economy was not only the construction of new private housing but also the conspicuous appearance of suburban shopping malls with supermarkets, clothing chains, electronic stores, movie theaters, and foreign-owned hotels (Figure 4.27). In the 1990s most were built in Moscow, followed by St. Petersburg. By 2003, 28 percent of Russia's retail sales were in Moscow, though only 2–4 percent of Russians shopped in modern facilities. As a result of the fierce competition in Moscow, investors began searching for new cities, usually regional capitals with more than 1 million inhabitants. Kazan, Samara, Yekaterinburg, Krasnodar, and Kaluga have emerged as new cities of investment. For example, Marriot Renaissance opened a hotel in Samara in late 2003, and Ikea has chosen Kazan to build a mall in which to locate its first store in Russia.

Secret Cities

In addition to the cities marked on official Soviet maps, a number of "secret cities" were revealed after 1991. Some have over 100,000 people. These cities were linked to the nuclear industry, biologic warfare research, or missile and weapon design (see Figure 4.18). Built in the Stalinist monolithic style, such cities developed as a combination of scientific research institutes and labor camps; each is surrounded by a zone of cleared land and electrified fences up to 30 km (20 mi.) wide. The cities grew out of the early days of secret nuclear bomb making. Scientists who were enticed to work in them had better living conditions than other citizens. Many such cities were duplicated at great cost in case the United States struck first in a nuclear war. In some, the factories, reactors, and even housing were built underground.

Assembling scientists secretly in specialized centers was successful, turning the Soviet Union into a military superpower, albeit at the expense of consumer needs, long-term economic growth, and environmental damage. There is a concern about the level of environmental damage caused by the dumping of wastes and unreported nuclear accidents on a scale greater than the Chernobyl accident. Leaks from the biological weapons

Table 4.3 Cities of Russia and Neighboring Countries

Slavic Countries	Population 2003 (in millions)	Population 2015 (in millions)
Moscow, Russia	10.5	10.9
St. Petersburg, Russia	5.3	5.2
Kharkov, Ukraine	1.5	1.4
Kiev, Ukraine	2.6	2.6
Minsk, Belarus	1.7	1.7
Nizhniy Novgorod, Russia	1.3	1.2
Novosibirsk, Russia	1.4	1.3
Yekaterinburg (Ekaterinburg), Russia	1.3	1.2
Samara, Russia	1.2	1.1
Omsk, Russia	1.1	1.0
Chelyabinsk, Russia	1.1	1.0
Volgagrad, Russia	1.0	1.0
Ufa, Russia	1.0	1.0
Rostov-na-Donu, Russia	1.1	1.1
Kazan, Russia	1.1	1.1
Dnepropetrovsk, Ukraine	1.1	1.0
Donetsk, Ukraine	1.0	0.9
Odessa, Ukraine	1.0	1.0
The Southern Caucasus		
Baku, Azerbaijan	1.8	2.0
Tbilisi, Georgia	1.1	1.8
Yerevan, Armenia	1.1	1.0
Central Asia		
Tashkent, Uzbekistan	2.2	2.3
Almaty, Kazakhstan	1.1	1.1

Source: United Nations, Urban Agglomerations 2003 (2003).

FIGURE 4.26 Russia. The Cathedral of the Intercession, commonly known as St. Basil's Cathedral, is located on Red Square, the center of Moscow. Ivan IV (the Terrible) had the cathedral built to commemorate Russia's victory over the Mongols. Originally intended as eight smaller, clustered cathedrals to represent Russia's eight victorious battles over the Mongols, the result became a single cathedral with eight domes and chapels signifying the victories. **Photo:** © Ronald Wixman.

centers gave rise to anthrax outbreaks near Yekaterinburg. The liquid fuel used in Soviet rockets is supertoxic and highly carcinogenic, and the reduction of rocket weapon stocks poses questions about its disposal.

Entry to these cities is still restricted, but life for the people in them has changed. It is difficult for the cities to get new work apart from contracts to carry out the more dangerous types of nuclear experiments. Privileges and status vanished as salaries shrank and were often not paid after 1991. People stay in the cities, however, since they see them as safe havens from the reported crime and instability elsewhere during the period of transition. New scientists, however, do not volunteer to move there.

Global City–Regions

Russia and Neighboring Countries tries to better its economy by engaging in the global economy, but its attempts are inhibited by the small role that its cities play as global cities. Of the global cities identified in Figure 2.15, Russia and Neighboring Countries has only one—Moscow—and it does not rank high on the

FIGURE 4.27 Moscow: IKEA. In recent years, European- and American-style super shopping centers have been appearing in the region. Notice how this IKEA has its name written in Cyrillic (ИКЕА) as well as Latin. As elsewhere, this shopping center, which is along the northwestern beltway (MKAD), depends upon and promotes greater automobile usage. **Photo:** © M. Blinnikov.

list. Almaty (Kazakhstan), Kiev (Ukraine), St. Petersburg (Russia), and Tashkent (Uzbekistan) offer some global city services, but this world region has very few global cities for so many urban areas. The lack of global connectedness is the direct result of the Communist legacy. Globalization is linked closely to modern capitalism, and Communism precluded the development of the types of accounting, advertising, banking, and law that are so characteristic of globalization. Refusing to engage in global capitalist competition, the Soviet government likewise focused inward on self-sufficiency, and engaged in trade only with fellow Communist countries. Consequently, earlier global cities of this world region lost many of their global connections. Since the end of Communism in 1991, the capital cities, especially Moscow but also cities such as St. Petersburg, are developing accounting, advertising, banking, and law firms and have received the greatest amounts of international investment. In time they will likely move up in the global city rankings.

Questions to Think About

4C.1   Do you think that Mikhail Gorbachev's policies destroyed the Soviet Union, or was the country destined to disintegrate anyway? Explain.

4C.2   What kind of human rights abuses occurred during Soviet times?

4C.3   Do you believe that women are more equal to men in the Slavic countries than women in your own country? Explain your answer.

4C.4   What are the population growth trends of the region's countries, and what causes these trends?

Key Terms

perestroika	*gulag*
glasnost	Virgin Lands Campaign

Test Your Understanding
4C

Summary

For a number of years after World War II, the economy of the Soviet Union prospered, but by the late 1960s, it began to languish. Mikhail Gorbachev attempted reforms in the late 1980s with *perestroika* and *glasnost*. The reforms were partially successful, but the Soviet Union disintegrated. *Glasnost* revealed many human rights abuses during Soviet times.

The people of this world region continue to wrestle with new and old human rights abuses. The full rights of women have been recognized since early Communist times. However, individual women suffer from discrimination at times.

Population is not evenly distributed across this world region. Most people live on the fertile soils of the North European Plain west of the Urals. Populations are low in the cold climates of the north, especially in Siberia and in the mountains and deserts of the south. In the 1990s the populations of the Slavic countries began to fall as death rates exceeded low birth rates, resulting in slowly decreasing populations.

Ukraine and Moldova have some of the world's best farmland, and the warm climates of the Southern Caucasus make it possible to grow citrus fruits, tea, tobacco, cotton, and rice. However, much of the region's lands are too hot, cold, dry, or wet for agriculture. During Communist times, the government tried to increase the amount of farmable land with the Virgin Lands Campaign. The economic transition in the 1990s brought hardship for the agricultural sector as farmers had difficulty transitioning to a market economy.

Urbanization increased dramatically during Communist times but was planned by the government and linked to industrialization. The transition to market economies has led to suburban growth and emergence of shopping malls. The inward orientation of Soviet Communism kept the world region from developing many global cities. Moscow ranks highest, and it and other cities are likely to rise in the global city rankings as Russia and its neighbors establish new political and economic links with the rest of the world.

Subregions

Russia and Neighboring Countries can be divided into subregions that share many similar characteristics:

- *The Slavic countries:* the Russian Federation, Ukraine, and Belarus, including neighboring Moldova.
- *The Southern Caucasus:* Georgia, Armenia, and Azerbaijan.
- *Central Asia:* Kazakhstan, Turkmenistan, Kyrgyzstan, Tajikistan, and Uzbekistan.

The Slavic Countries

The Slavic countries of this subregion are the Russian Federation, Ukraine, and Belarus. Moldova also is included because many Slavs live there, and it is closely tied to the Slavic countries (Figure 4.28). The Russian Federation is by far the largest in land area of any country in the subregion and in the world. It is nearly twice as large as Canada, the United States, or China. In 2005 it had 77 percent of the CIS area and 51 percent of the CIS population.

Though the Russian Federation and the other countries of the subregion are experiencing economic hardship as their economies move from Communism to capitalism, the Russian Federation still exerts considerable power. The Russian Federation remains a nuclear power and continues to hold one of the five permanent seats of the UN Security Council, the most powerful organ of the United Nations. Because it contains substantial portions of the world's natural resources, the Russian Federation has considerable economic potential. The leaders of the Russian Federation have pressed to have their country added to the Group of Seven (G7), an informal organization representing the world's wealthiest countries. Though no official rules for membership exist, Russian representatives have been regularly invited to meetings, leading many to refer to the organization as the Group of Eight (G8).

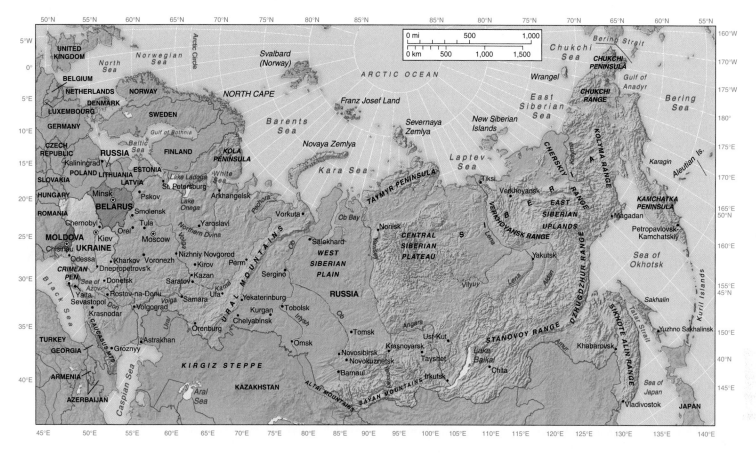

FIGURE 4.28 Slavic Countries: major cities and physical features. Note the distribution of the major cities and their concentration in the western part of the subregion.

Countries

The Slavic countries seen on the map today became independent only in 1991 with the boundaries they had as Soviet republics. This situation is also true for the Russian Federation, though the Russians controlled the Soviet Union and its predecessor, the Russian Empire. Prior to 1991, Ukraine was independent only for a brief period after World War I until it became part of the Soviet Union in 1922. Before that it was part of the Russian Empire. Belarus was never independent before 1991 and was usually part of either the Russian Empire or the Polish–Lithuanian Kingdom. Of the former Soviet republics, Belarus is the most closely tied to the Russian Federation. Since 1991, Belarussians have considered creating a Russian–Belarussian Federation.

Moldova, too, was never independent and has been a distinct territory for fewer than 200 years. It was part of the Romanian province of Moldavia until the Russian Empire annexed it in the 1800s and named it Bessarabia. Romania annexed the territory after World War I, but the Soviet Union annexed it again after World War II. Many Romanians and Moldovans hoped to unite their two countries after the dissolution of the Soviet Union in 1991, but the Russian military stationed in the country prevented this. Worried about a union of Moldova and Romania, Russians and Ukrainians living in Moldova declared their own republic in the Transnistria region. Other ethnic minorities have made similar proclamations. The government has not been able to suppress these independence movements completely.

Culture: Ethnicity

Native Moldovans are closely akin to Romanians and are not related to the Slavs (see Table 4.2). Russians, Ukrainians, and Belarussians are Eastern Slavs, offshoots of the broader Rus people and culture that emerged in the 800s and 900s when the Rus adopted Eastern Orthodox Christianity. All three Slavic groups are closely related as seen, for example, in the name *Belarussian,* which translates as "White Russian." Differences between them developed after Poles, Lithuanians, and Austrians ruled the western lands of the Rus from the 1300s to the 1800s. During this time, Ukrainians and Belarussians developed separate identities from Russians. Polish, Lithuanian, and Austrian influences are seen in the fact that nearly 20 percent of Belarus's population is Roman Catholic. A number of Poles and Lithuanians live in Belarus today after Poland's boundary was relocated in 1945.

The name *Ukraine,* meaning "borderland," illustrates history's shifting boundaries and influences. To the Russians, Ukraine is Russia's borderland; but to many Ukrainians, who have stronger ties with Europe than the Russians, Ukraine is Europe's borderland. Russian ties with Ukrainians and Belarussians were reestablished in the early 1800s when the Russian Empire extended west. Russification began soon afterward and was particularly strong during Soviet times. Today 78 percent of Belarus's population is Belarussian and 13 percent Russian, but 63 percent of the population regularly speaks Russian and not Belarussian. Not surprisingly, Belarus's leaders have expressed interest in uniting Belarus with the Russian Federation. In Ukraine, Russians, living primarily in the industrial areas of eastern Ukraine, account for 22 percent of Ukraine's population. Ukrainians, however, fear Russian domination and cultivate their ties with the West.

Economic Development

Soviet economic policies tied the Slavic countries closely together, but many Ukrainians, Belarussians, and Moldovans prefer greater independence for their countries. They seek new trading relationships with other countries, but old connections are hard to break and new relationships difficult to form. For example, during Soviet times, eastern Ukraine was the most important iron and steelmaking region of the Soviet Union. Eastern Ukraine can provide for Ukraine's economic independence today, but the area's industries depend on imports of oil and natural gas, primarily from the Russian Federation, to meet 85 percent of their energy needs. Belarus also has much heavy industry that can provide economic self-sufficiency, but with more than half of its trade with the Russian Federation, it too depends heavily on Russia. Moldova likewise still trades more with the Russian Federation than with Romania, though many Moldovans desire closer ties to Romania.

The Slavic countries developed heavy industry during Soviet times but did not keep equipment up to date (Figure 4.29). Since 1991, the Slavic countries have worked to update their factories, attract foreign investment, engage in the global economy, and develop more service industries.

The Russian Federation

The Russian Federation is a large country with considerable geographic diversity. Its political and economic geography requires special treatment. The Russian Federation is the modern political state representing the land known as Russia. To many Westerners, Russia is a mysterious land, hidden in cold, dark forests on the eastern and northern fringes of Europe. Europeans have regularly included the Russian heartland within Europe but at the same time have considered Russians too "Asiatic" to be European. Europeans struggled for centuries to understand Russia, exemplified by Winston Churchill's remark that "Russia is a riddle wrapped in a mystery inside an enigma." Depending on their relationship with Europe and their desire to be within Europe, Russians themselves have frequently alternated between emphasizing their European qualities and their wider role spanning eastern Europe and northern Asia.

FIGURE 4.29 Kerch, Crimea. Following the end of Communism in 1991, outdated, inefficient factories like this one were abandoned. In the 2000s they have been seen as sources of raw material. Thus they, like this one, have been stripped of their bricks, steel, and other materials, which were then sold. **Photo:** © Ronald Wixman.

In the 1990s and early 2000s Russians still saw their country as a world power (see the Point–Counterpoint box on page 148), but economic decline, marked by a 60 percent drop in GDP, challenged that desire. As the Russian Federation struggled to maintain its influence in world politics, its relationships with the former Soviet republics changed considerably. While external relationships evolved in new directions, Russia's internal political geography and its economic and social relationships were all dramatically altered.

Political Divisions

The internal political geography of the Russian Federation includes a mixture of political units. To a large extent the political geography of the country was inherited from the Soviet system, though some changes were made after 1991. Notably, the Kremlin, the center of government in Moscow, does not have as much power over the political units as before.

The 89 political units fall into two categories: administrative and autonomous. The administrative units consist of 6 federal territories (*krays*), 49 regions (*oblasts*), and 2 federal cities (Moscow and St. Petersburg). Much like the states, counties, and municipalities of the United States, they were created to administer the large country. In contrast, the autonomous units, consisting of 21 republics, 1 autonomous region (*oblast*), and 10 autonomous districts (*okrugs*) (Figure 4.30), are able to craft many of their own laws and govern themselves somewhat

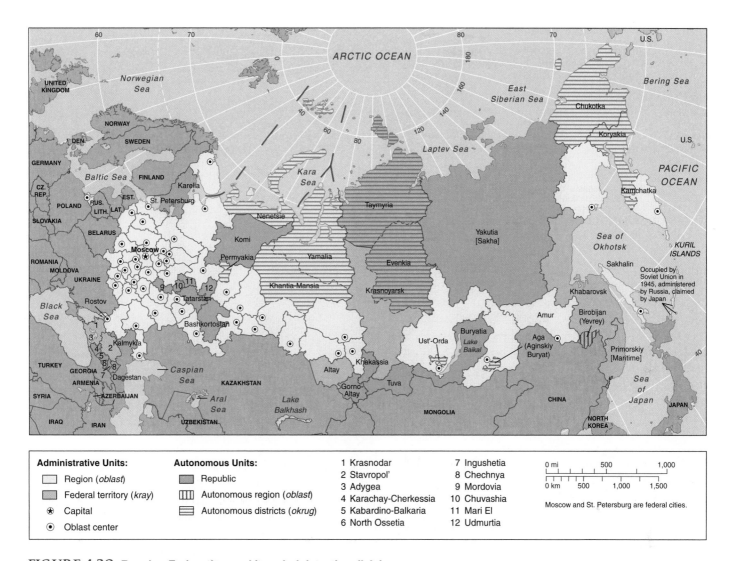

FIGURE 4.30 Russian Federation and its administrative divisions. Notice the uneven allocation of power.

differently than in the rest of the Russian Federation. The resource-rich republics tend to exercise the greatest authority over their own governance.

The autonomous territories were established by the Soviet Union to reflect the presence of ethnic minorities such as the Tatars and Sakha (Yakuts). Soviet law protected minority languages, religions, and cultures. However, only 52 percent of the Russian Federation's approximately 30 million non-Russians live today in the autonomous territories. As a way of controlling their vast country, the Soviets drew boundaries for the republics that deliberately left many members of ethnic groups outside their intended territories and included many Russians within them (Figure 4.31). Also, many recognized nationalities did not receive republic status. Of the 90 numerically significant groups recognized in the 1989 census, only 35 had a homeland. In addition to the previously mentioned policies of Russification, boundary drawing was clearly a means that the Soviets used to divide and dilute non-Russian groups. The

situation is particularly true in the North Caucasus, where the native peoples have most fiercely resisted Russian rule through history. For example, the Karbardians were grouped together with the Balkars, though they had more in common with their neighbors the Cherkessians.

Kalmykia illustrates some of the complex factors that produced the current ethnic geography of the Russian Federation. The Kalmyks were Buddhists pushed out of their ancestral lands in western China by Han expansion in the A.D. 1500s. They reached the steppes west of the Volga River mouth in 1608, where Peter the Great later gave them a kingdom, or *khanate*. From the mid-1700s Russians and Germans took some of this land, causing a large group of Kalmyks to attempt to return eastward in 1771. Those who returned lost their towns and were resettled on collective farms under Stalin, who then accused them of siding with Germany during World War II and scattered them across Siberia. Allowed back in 1957, the Kalmyks are not a majority in their own republic but claim political power. They

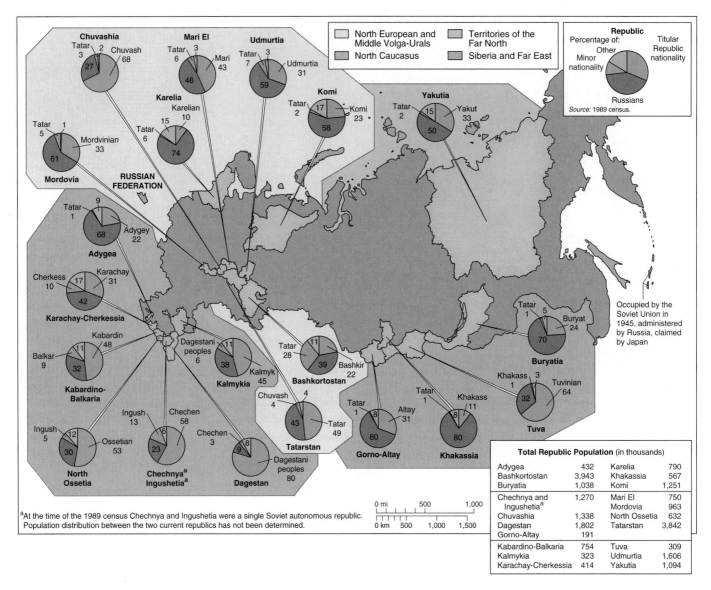

FIGURE 4.31 The ethnic composition of the Russian Federation's autonomous republics. Pie charts show the percentages of each ethnic group.

rebuilt Buddhist temples, and their old Mongol script is being revived in schools. Soviet agricultural policy, however, turned over half of the land into desert through overgrazing and poor irrigation projects and left the republic in poverty.

The autonomous political units can be summarized into four groups:

North European and Middle Volga–Urals: Both areas have been integrated into the Russian state since the 1500s. The peoples of the North European territories mostly practice Eastern Orthodox Christianity. Their territories are largely in the boreal forests, which are agriculturally poor and have low population densities. The economies are resource-oriented, producing electro-energy, wood products, and ferrous metals. The Middle Volga-Urals group represents the meeting ground between Finno-Ugrian, Turkic, and Slavic peoples. The large numbers

of Russians are Orthodox Christians, but many of the indigenous peoples are Muslims. Located in the mixed forest and forest–steppe region, these territories have economies that vary considerably, from traditional agriculture to extractive industries and modern manufacturing.

North Caucasus: Several autonomous territories occupy the foothills on the northern side of the Caucasus Mountains. Some extend into the steppe. Most of the peoples are Muslims, but Christians live in the area, too. The Russians were not able to exert full control over the area until the mid-1800s. Ethnically and linguistically, the area is one of the most complex places in the world. The various peoples, however, share certain economic and cultural similarities that facilitate a "mountaineer" identity. Their shared opposition to Russian rule has fostered cooperation with one another. The Chechens in particular have resisted Russian rule. Their attempts to make Chechnya independent of Russia since 1991 have resulted in bloody military conflicts between themselves and the Russian army.

Siberia and the Far East: These autonomous territories stretch east, with many along the Chinese and Mongolian borders. They contain Turkic or Mongolian peoples (Figure 4.32). Many are Buddhists and practiced nomadic herding until they were incorporated into the Soviet Union. They often keep their traditional ways of life amid large pockets of Russians who engage in extractive industries, and raw material and food processing. Tuva was annexed only as recently as 1944. The 1993 Tuvan constitution claims complete sovereignty for the republic and the right of secession, though this policy violates the constitution of the Russian Federation. Violence directed at Russians caused many Russians to leave.

Stalin created the Jewish (Yevrey) Autonomous Oblast in 1934 and designated it as a homeland for Soviet Jews. It was hoped that it would be a counterattraction to Palestine (now Israel), but its remote location and harsh environment made it unattractive. In 1989 only 9,000 Jews lived there. Eighty percent of the population was Russian.

Territories of the Far North: More than 30 different ethnic groups live in the far north, many numbering only in the thousands. Seven of the groups have *okrugs* (autonomous districts) and only one, the Yakuts, have their own republic, meaning that the majority of peoples lack a designated homeland. The territories are rich in diamonds, gold, iron ore, timber, and other natural resources. During Soviet times, a flood of Russians seeking to exploit these resources reduced the percentages of the indigenous peoples to less than 16 percent in the *okrugs*. The Russians concentrated in towns associated with industry. Pipelines, other modern infrastructure, and the resulting pollution made it difficult for the indigenous peoples to practice their traditional ways of life, which include hunting, fishing, and reindeer herding. The Association of Peoples of the North formed in 1989 to voice the concerns of the indigenous peoples.

Heartland and Hinterland in Russia

Another basis of regional differences in Russia are **heartland** and **hinterland.** The heartland lies west of the Urals and includes many of the original territories of Rus and then Muscovy. It contains the greatest concentration of Russian people and accounts for much of the country's economic and political activity. It is also known as the Russian homeland. The Moscow and St. Petersburg urban regions, the Volga River valley, and the Urals contribute to the heartland's prominence. The Moscow region, approximately 400 km (250 mi.) square, is home to 50 million people and was the focus of Soviet central planning and transportation routes linking the entire country. Local manufacturing includes vehicle, textile, and metallurgical industries. St. Petersburg, a major Baltic port north of Moscow, is a smaller manufacturing center but still produces around 10 percent of the total Russian output, including shipbuilding, metal goods, and textiles.

Southeast of Moscow, the Volga River is lined by a series of industrial cities that use the river, linked since the 1950s by a canal outlet to the Black Sea. Around 25 million people live in the Volga River region, which was developed for manufacturing during and after World War II—at a distance from advancing German armies and helped by the discovery of major local oil and natural gas fields. Manufactures include specialized engineering and the Togliatti car plant built by Fiat of Italy.

East of the Volga River basin, the Urals contain metal ores. Like the Volga region, the southern Urals were developed during and after World War II, principally as a metals center. Oil and natural gas fields are to the south and east.

The Russian Federation's hinterlands—dependent and tributary to the heartland—include Kaliningrad *oblast* along the Baltic Sea, the Arctic regions around Murmansk, the mining areas east of the Urals, resource-rich Siberia, and the Pacific region inland from Vladivostok. Large expanses of the hinterland are virtually empty of people and economic activity and remain inaccessible to development. Siberia forms a huge area that can be divided into the more developed southern margins along the line of the Trans-Siberian Railway and the northern lands.

Siberia is an essential part of Russia, making up three-fourths of its land and providing a large proportion of its raw materials—a common feature of hinterland regions. In 1990 Siberia produced 73 percent of Russia's oil, 90 percent of its natural gas, 61 percent of its coal, all of its diamonds, and 30

FIGURE 4.32 Russia. Buryat singers at Datsan Temple (southeast of Irkutsk. **Photo:** © Ronald Wixman.

Personal View

RUSSIA

Irkutsk is a rather special city in Siberia. It lies five time zones and 88 hours by Trans-Siberian Railway east of Moscow and another three time zones and 72 hours on the train from Vladisvostok on the Pacific Ocean. It was one of the earliest Russian settlements in Siberia, dating from 1652, when trappers and traders established a wooden fort at a crossing of the Angara River just downstream from Lake Baikal. Today it is home to over 700,000 people and has a much greater variety of buildings than other cities along the railway, ranging from the wooden houses with carved window frames and painted shutters that give it a villagelike appearance to the monolithic former Communist Party headquarters (Box Figure 1). Its people have a strong pride in their city and its cultural history, having produced many poets and artists. The expulsion of Europeanized groups from St. Petersburg to Siberia early in the 1900s

added to the ethnic range that includes Slavic Russians and Asian groups, such as the Buddhist Buryats living on the Russian side of the Mongolian border and the Yakuts from northern Siberia, who speak a Turkic-derived language.

A major stop and junction on the Trans-Siberian Railway, Irkutsk connects southward into Mongolia and China by rail and northward into Siberia by bus and boat in the summer. It also has good air links to Moscow and other Russian cities, as well as places in northern China, Mongolia, and Japan. Chinese traders regularly appear selling a range of cheap goods. Telephone communications are good, with calls abroad often being clearer and cheaper than local or internal Russian calls. Fax, e-mail, and other telecommunications facilities are available, although many people still use telegrams. The post office remains under bureaucratic control, with letters having to be sent separately from parcels and many forms needing to be filled out.

(a)

(b)

Box Figure 1 **Irkutsk, Siberia.** (a) Unlike many other Russian cities, Irkutsk has many of its prerevolutionary (pre-1917) buildings. (b) Apartment buildings of the Communist period line the Angara River, where people recreate in summer. **Photos:** © Ronald Wixman.

percent of its timber and electricity. It also has a growing fishing industry on the Pacific coast. Farmed lands are mainly in the west of Siberia. Siberia remains a region of promise for enterprising Russians, often based on romantic aspirations such as taming the wilderness or making a fortune.

Along the southern strip of Siberian Russia are a number of Soviet-planned industrial centers that are separated by large tracts of sparsely settled land. In the Kuzbas region around Kuznetz, 2,000 km (1,200 mi.) east of the Urals, a major coalfield was initially developed to supply coal to the steel manufacturing centers of the Urals. This led to local industrialization, linked to the discovery of iron ore and the return transport of bauxite from the Urals. Aluminum, steel, and products using

them are major products. The world's largest aluminum plants at Krasnoyarsk, Bratsk, and Sayansk provided metal for the Soviet aircraft and missile industries, but these demands disappeared in the 1990s, and the output of aluminum fell by 70 percent from 1991 to 1993. Privatization of the aluminum smelters after 1994 involved foreign capital and led to a major increase in exports.

Farther east, centers close to Lake Baikal, including Irkutsk (see Personal View: Russia on page 168), form a narrow belt of industry using hydroelectricity generated in streams flowing to the Yenisey and Lena River systems. Mining and lumbering are also important in isolated localities, while larger cities provide services to extensive areas in northern Siberia where population and mining activities are scattered.

Irkutsk is an industrial and commercial city. Its domestic electricity comes from a local hydroelectric dam. Such power is also used in a nearby aluminum factory, although the future of this and other manufacturing facilities based on metal and engineering products is in doubt. Irkutsk is also a center for coal mining and oil production in the surrounding area.

The city center shops increased their range of wares rapidly in the mid- and late-1990s: a bakery may also sell cameras, while a pharmacy may sell shoes. The department stores are more like indoor markets, with stalls competing to sell the same products. Fruit and vegetables are commonly sold in street or sidewalk booths. Some sports shops sell Western goods, but the prices are too high for ordinary Russians.

The Siberian climate brings varied hardships to the people of Irkutsk. Its winters are long and very cold. The first snow flurries may come in September, but by October, the ground is covered by snow that lasts until late March, although it is only 15 cm (6 in.) deep and little more falls until spring. Temperatures at their worst plunge to −40°C (−40°F) for a few days and never rise above freezing during winter; the January mean temperature is −21°C (−5°F). In summer, daytime temperatures reach 20°–25°C (70°–75°F), and the thick winter ice on Lake Baikal melts. There is little rain, but dry, dust-laden winds and mosquitoes make life uncomfortable. Although they are used to these conditions, local people need to wear thick fur coats and hats in winter and prefer to live in the centrally heated apartment blocks with running water rather than in the older wooden houses.

Although the city-edge apartment blocks are favored for living, they are often difficult to locate because of irregular numbering. The city authorities and large factories built them for their workers, but during the 1990s, the occupiers assumed ownership. Utilities are monopolies and own the household appliances as part of a contract to supply unmetered electricity or gas. Without meters, there is no check on the amounts used, and people tend to be wasteful until supply cuts occur. Water is supplied centrally, as is central heating by water that is pumped around the whole city; warm water is turned on in October and off in April, and there is no other control of this process.

Irkutsk draws its university students mainly from the local area. There are residence halls, but most students live at home. Science, engineering, medical, and language courses, often linked to teacher training, are most popular, but there are also cultural, theater, musical, and historical studies courses. Students get into courses on the basis of their preuniversity qualifications, and they have to pass annual tests to progress and qualify for the small monthly maintenance grant. Tuition is free. Personal achievement measured in tests or longer examinations is a problem because Russian students are used to working together and may talk to each other across the exam room.

Students have fewer clubs and societies than their counterparts in Western universities, but they enjoy hockey and basketball in the winter and soccer in the summer. Irkutsk won a national hockey competition in 1998 and has an open-air stadium in the middle of the city (where those watching sometimes endure temperatures of −30°C). Russian hockey is played on larger, more open ice-covered fields and has less personal contact than the hockey played in the United States and Canada. Volleyball is popular as an informal sport that requires little equipment. Theater performances, especially of music or the ballet, are inexpensive—a dollar or so for a ticket. But Russians generally prefer entertaining at home to going out to be entertained, and there are few night clubs or discos. Birthday parties especially are major social events.

The university halls provide a diet of basic food, often of low quality based on soups and meat snacks. Men living in apartments struggle to feed themselves because cooking is regarded as a woman's job, as are clothes washing and housecleaning. Some students will work part-time to gain income, but this is not common. Resources in the university are poor because of low funding, and staff and students occasionally strike over pay and conditions. Professors are state employees whose pay is often delayed or withheld; when they retire, their pensions may not be paid for many months.

The issue of low or no pay is important in Irkutsk. People manage to survive, however, by growing their own produce at out-of-town *dachas*. These are small wooden houses, often near Lake Baikal, on the grounds of which potatoes and other vegetables are grown; they offer some escape from the polluted air of Irkutsk.

Many Russian students entertain strong hopes of moving to a Western country. Their idealistic notions of living elsewhere are fueled by preparations that involve immersing themselves in Western clothes, music, competitive values, and language (English, German, and Japanese courses are popular). Factors such as continuing military service—which all young people have to enter for two years, usually after completing their studies—and the lack of local job opportunities also make them want to emigrate. Even if university graduates obtain a good job in Russia, the pay is poor and they will not be able to live independently with their own apartments. Many continue to live with parents, even after they marry.

People living in Irkutsk saw many changes in the 1990s. Shops now contain a greater range of foods and consumer goods, although prices for the most desirable items remain too high for most people. The job outlook is not hopeful, and housing is generally of poor quality compared to Western norms. Easy access to the surrounding countryside, cheap musical events, and the great strength of deep personal relationships—as opposed to the mere acquaintanceships so common in the individualistic West—remain the strengths of Russian society.

The far east, flanking the southern Pacific coast of Russia, has ports such as Vladivostok, and metal industries along the Amur River. In the 2000s Japanese corporations began investing in the region's mines. The large population in neighboring China, coupled with a new Russian law in 2003 that allows foreigners to lease land in Russia for 49 years, has attracted many Chinese farmers to this part of Russia.

Soviet centralized planning often ignored the geography of resources and sited production facilities for political, rather than economic, reasons. Special defense-related factories, for example, were built in remote locations where it was expensive to maintain them. Workers were paid more to work in these faraway facilities, and the cities that grew up around them received many subsidies. The Russian Federation stopped giving financial support to these poorly located factories and cities. Without subsidies, many factories could not compete economically and had to shut down. In turn, unemployed people emigrated to the Russian heartland. Subsequently, regional differences and local lifestyles are now greater within the Russian Federation.

The realities of such factors as closeness to consumer markets or ports with world trading connections are causing geographic shifts in regional production patterns. Moscow, for example, has succeeded the best in the transition to capitalism. As much as 20 to 25 percent of Moscow's population is middle class, the highest percentage in the Russian Federation. In other areas, less than 10 percent of the population has attained middle-class income.

Foreign Investment

In the 1990s corruption, poor infrastructure, and an unwieldy bureaucracy prevented much foreign investment from flowing into Russia (Figure 4.33). The situation is slowly changing, and Russia is now receiving greater amounts of foreign investment. Most investment has come from foreign corporations joining Russian groups to exploit Russia's vast mineral wealth, namely oil, natural gas, and metal ores. The Russian automotive industry has also received investment.

The oil and natural gas reserves are of particular interest to industrialized countries and the main potential source of foreign investment (Figure 4.34—compare with Figures 4.11 and 4.12). Western oil corporations that hoped to be placed in charge of oil exploration and marketing found that the Russian government expected them to enter joint ventures with companies such as Lukos oil and Gazprom, which produces and distributes natural gas. Some 34 percent of the world's gas reserves are controlled by Gazprom, which supplies 20 percent of Europe's demands.

Through the 1990s, Japanese companies did not invest in Russia, primarily because the Russian government continued to hold onto portions of the Sakhalin Islands that it took from Japan at the end of World War II. However, by the mid-2000s, Japanese firms began investing in the abundant natural resources of Russia's far east. Mitsui and Company and Mitsubishi Corporation purchased 45 percent of Sakhalin Energy Investment Company and engaged in a $10 billion project to obtain liquefied natural gas for Japan.

Russia's mineral wealth extends to a wide variety of metal ores. At present, multinational mining corporations are interested in developing gold mining, as in the world's largest gold deposit at Sukhoi Rog in Siberia. Russia also produces one-fourth of the world's diamonds.

One of the most promising areas for foreign investment is in automobile manufacturing. Russia had only 13 million private autos, with an average age of over 10 years in the late 1990s. The market is growing, however, and all cars made are sold. In 1998, in one of the biggest foreign investments to date, the Gorky firm in Nizhniy Novgorod entered a joint agreement with Fiat of Italy to use idle defense factories for component manufacturing. General Motors also negotiated similar arrangements with the largest Russian auto producer, Avtovaz.

Trade

Since 1991, trade patterns for the Russian Federation have changed dramatically. During Soviet times, most trade was with the republics that would become independent and with the COMECON partners in East Central Europe. In 1990, for example, 70 percent of exports and 47 percent of imports were with the republics that became members of CIS. In 1996 exports and imports to these countries dropped to 21 and 30 percent, respectively. Trade with the former COMECON countries declined similarly. In contrast, trade with the European Union (EU) and the United States increased rapidly. The EU accounts for almost 40 percent of Russia's non-CIS exports and 54.3 percent of its imports. The United States and Germany are the Russian Federation's most important trading partners; Italy and the Netherlands played prominent roles, too. The Russian Federation is primarily exporting raw materials such as oil, natural gas, wood products, and metals, and importing finished products such as machinery and equipment. The situation reflects Russia's abundant stocks of natural resources and antiquated industries. If current trends continue, Europe will obtain 94 percent of its oil and 81 percent of its natural gas from Russia by 2030. Poland, Slovakia, and Hungary already receive 100 percent of their oil from Russia, and the Baltic states, Slovakia, and Romania currently obtain 100 percent of their natural gas from Russia. The potential for growth in trade between the EU and Russia is great and may eventually equal EU trade with the United States, at present several times greater than that with Russia.

Science, Sports, and Society

Communism strove to excel in science and sports and to improve life in society as never before seen in human history. Science and sport achievement was also a way to show that Communism was superior to any other political–economic system. Scientists were pampered in terms of incomes

Country	Total Foreign Investment in Billions of U.S. Dollars	Investment per Resident of the Resident Country
United States	$316.5	$1,149
Czech Republic	3.7	370
Poland*	9.6	246
Brazil	30.5	175
Hungary	1.7	170
China*	39.0	31
Russian Federation	2.7	19
Ukraine	0.6	12
Indonesia*	2.1	9
Pakistan*	0.4	3

Source: *Economist* Intelligence Unit
*Estimate

FIGURE 4.33 **Foreign direct investment inflows: dollars per person, 2000.** The Russian Federation receives comparatively little foreign direct investment in terms of total dollars and in terms of dollars per Russian. For example, the Czech Republic, which has a population of 10 million, receives more money than the Russian Federation with a population of 144 million.

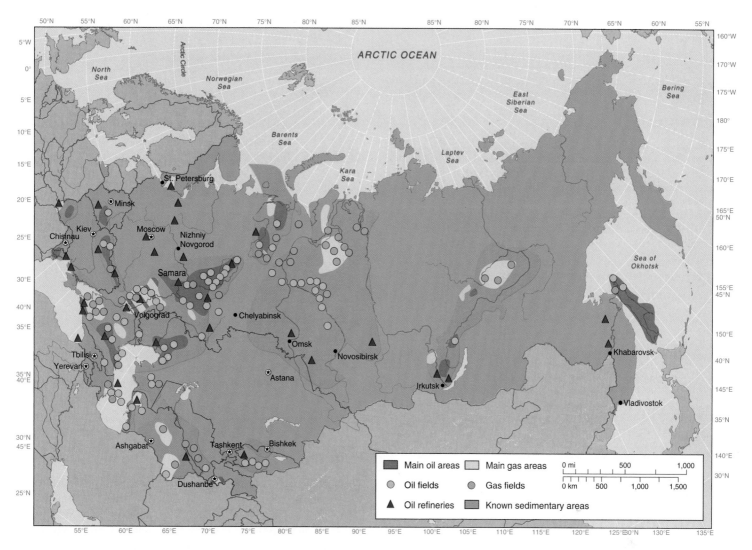

FIGURE 4.34 Russia and Neighboring Countries: oil and gas fields. Assess the availability of oil and natural gas to Russia, the other CIS countries, and the potential for developments in Siberia.

and equipment and concentrated in new cities established for their work—some of which remained secret because they were dedicated to military purposes (see Figure 4.19). Akademgorodsk, outside Novosibirsk in the center of Russia, has around 30 research institutions. The Budker Institute for Nuclear Physics, for example, focused on abstract research on fundamental particle physics. As already stated, the Soviets invested heavily in a space program as well to best the United States.

Like scientists, athletes were pampered and given the best sport facilities in which to train. Athletes were not expected to hold jobs, only train for sports. The Soviet Union sent its athletes to international events, especially the Olympics, where the Soviet Union usually won the most medals of any country.

Everyone in society was guaranteed a free university education and a job. Health care was provided without cost to everyone. Maternity leave was generous. The theater and arts

received generous support. The average citizen could go to a play, see a ballet, or visit a museum for a low cost. The goal of the Soviet government was to improve society to the degree that all citizens received all their basic needs and much more, regardless of their social class. The Soviet government claimed to have eliminated unemployment and poverty. Though the Soviet system eventually failed, it brought about great political, economic, and social transformation in its 74 years of existence.

Following the end of Communism in 1991, public institutions and welfare support systems were the main losers as a result of the political and economic changes. Many things that Russians took for granted, such as a job and free university education, were no longer available to many. After losing support for their efforts, many scientists and athletes migrated to other countries. Not surprisingly, some Russians still support communist ideals.

Test Your Understanding
4D

Summary

The Slavic countries share a similar history and culture. The Russian Federation is larger than the others combined, and it is difficult for the others to break their dependence on the Russian Federation. The Russian Federation reorganized its internal political geography to give greater power to local voices, for both ethnic minorities and Russians. A heartland and hinterland distinction is significant within the Russian Federation. Economic restructuring continues and new trading links develop as the Slavic countries compete in the global economy.

Questions to Think About

4D.1   What are the cultural similarities and differences of the peoples of the Slavic countries?

4D.2   How is the Russian Federation's internal political geography structured?

4D.3   What new trade links are the Slavic countries developing with one another and the rest of the world?

Key Words

heartland   hinterland

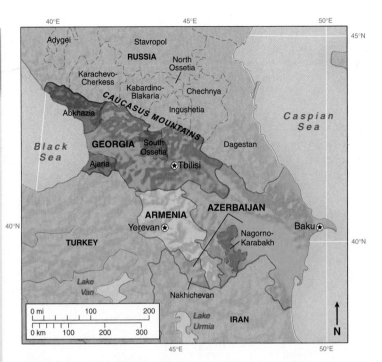

FIGURE 4.35 The Southern Caucasus: countries, cities, and major physical features. Darkened areas within countries represent areas that have acted autonomously, often without the consent of their respective central governments.

The Southern Caucasus

Georgia, Armenia, and Azerbaijan straddle the Caucasus Mountains and are frequently called the *Transcaucasus,* meaning "across the Caucasus" (Figure 4.35 and Table 4.2). This term reflects a Russian ethnocentric view of the region: these countries are on the other side of the Caucasus Mountains from Russia and were once Russian colonies. The more neutral term "Southern Caucasus" is used to refer to these countries, and "Northern Caucasus" is applied to the part of the Caucasus in the Russian Federation.

Armenia and Georgia are both mountainous with many peaks rising above 5,000 m (15,000 ft.). Azerbaijan is mountainous in the west along the borders of Georgia and Armenia, but the eastern areas of the country, formed by the Caspian Sea coast, are flat with areas below sea level. The origins of the word *Azerbaijan* are not clear, but one version derives from Persian words that mean "land of fire." "Land of fire" could refer to the surface oil deposits that burned naturally in the past or to the oil fires in Zoroastrian temples that once dominated the region. Zoroastrianism no longer exists, but it is a religious forerunner to Christianity and Islam, tracing its origins back to Azerbaijan. Zoroastrians believed that the Earth would be consumed in fire following judgment day. Interestingly, this belief developed in an area of the world where the Earth is oil-soaked and burns easily.

Countries

Though the three countries of the Southern Caucasus are relatively new, the peoples and political relationships extend far back into history. Armenians, for example, trace their ancestors back to 6000 B.C. In the 100s B.C. the Armenian empire controlled most of the Southern Caucasus and stretched across what is now northern Iran, Iraq, Syria, and eastern Turkey. Modern Armenia is small compared to its predecessors and does not even include a majority of all Armenians. Important historical places, such as the medieval capital Ani and Mount Ararat, the historically accepted landing place of Noah's ark, are now in neighboring Turkey. This situation illustrates that the modern Armenian state encompasses only the eastern portion of historic Armenia.

Georgians have lived in the Southern Caucasus almost as long as Armenians and likewise built empires that stretched across the subregion and beyond, reaching their height of power from the A.D 1000s to 1200s. It was around this time that the Azerbaijanis (also known as Azeris) emerged as a people in the subregion.

Independence and self-rule are features of the distant past. Over the last 800 years, the peoples of the Southern Caucasus were ruled from the outside. The most notable foreign rulers were the Russians, who took control in the 1800s and continued exercising their power until the Soviet Union dissolved in 1991. As former Soviet republics and current members of the CIS, Armenia, Georgia, and Azerbaijan struggle with establishing a viable political and economic existence. Located between Russia, Iran, and Turkey, these countries reflect the differing cultures and political views of their larger neighbors. Past outside political control and continuing outside influence create tension and conflict between the countries of the Southern Caucasus.

Culture

Though the Southern Caucasus is a small area of the world, the subregion is culturally very diverse (Figure 4.36), partly because it lies at the historic contact zone between the Turkish, Persian, and Russian empires. Many languages are spoken in the subregion. They are of different language families, so most have little in common. The Georgian language is in the Caucasian language family and has a unique alphabet. Armenian is Indo-European but stands alone in its own branch and has a distinct alphabet of 38 letters, derived mostly from Greek. Azerbaijani is in the Ural-Altaic language family (see Figure 2.16).

The Georgians (see Figure 4.2b) and Armenians are both Christian, and both adopted Christianity early in history, in the A.D. 300s. The Armenians claim their country was the first in the world to adopt Christianity officially. The Armenian Apostolic Church has been independent since the Middle Ages and expresses a unique view of Christianity. The Georgian Church is associated with the Eastern Orthodox Christian churches.

Arabs introduced Islam in the 600s and 700s into Azerbaijan. In the 1500s the Shia branch of Islam came to the country and now dominates, making Azerbaijan and Iran the only two countries in the world where Shia Islam is practiced by the majority of the population and controls the government. Despite their close ties with Iran, Soviet secular policies influenced Azerbaijanis greatly. For example, Azerbaijani Muslims, unlike those in Iran, drink wine, and women are not veiled or segregated (Figure 4.37). In 1991 the Azerbaijani government also went against the wishes of Iran and adopted a modified Latin alphabet for Azerbaijani instead of the Arabic alphabet—the original alphabet of the Qu'ran and the alphabet used in many Muslim countries. The Latin alphabet is customarily used in Roman Catholic and Protestant countries but also in nearby Turkey, where the language is similar to Azeri.

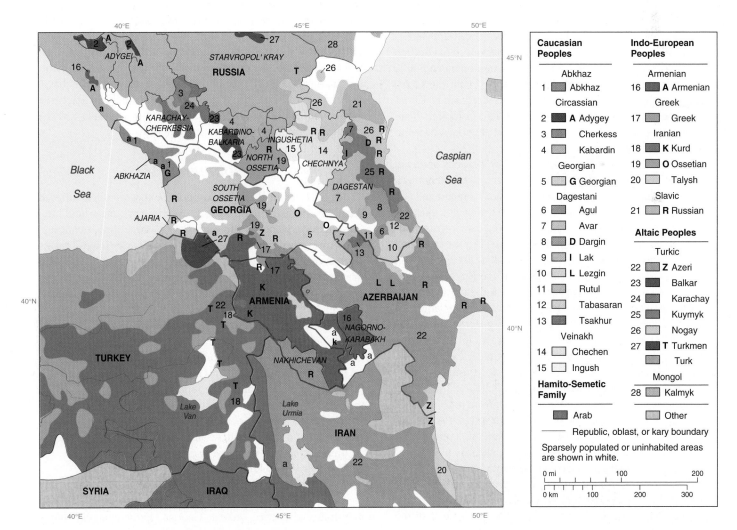

FIGURE 4.36 Ethnolinguistic groups in the Caucasus region. Compare the distribution of groups to the location of political boundaries. Do these comparisons help to explain conflict in the subregion?

FIGURE 4.37 Southern Caucasus. Azerbaijani children. **Photo:** © Ronald Wixman.

Ethnic Peace and Conflict

History has given rise to differing cultural combinations in the Caucasus. Some ethnic groups are closely related and others are not. Many live in peace, while others are locked in conflict. For example, the Adjarians, who have an autonomous republic in Georgia (see Figure 4.35), live peacefully with the Georgians, though they are Muslims and the Georgians are Christians. Despite differing religions, the Adjarians are indistinguishable from Georgians, and most Adjarians consider themselves Georgians.

Relations among other groups have not been so peaceful. The Ossetians and Abkhaz both have their own autonomous republics within Georgia, but they distrust the Georgians and feel no loyalty to the Georgian state. Both groups sought independence for their republics in 1991 and 1992, but the Georgian military intervened and great bloodshed resulted. The Russian army helped to maintain a cease-fire in South Ossetia. Though Russian military personnel aided the Abkhaz cause, the Russian government worked with the UN to establish a cease-fire in Abkhazia.

The persecution of Armenians has had a lasting effect on this part of the world. In 1895 the Ottoman government massacred 300,000 Armenians within its realm. Again in 1915, during World War I, the Ottoman government tortured, exterminated, and deported its Armenian population, claiming that the Armenians were a threat. Somewhere between 600,000 and 2 million Armenians were exterminated out of a prewar population of about 3 million in what can be referred to as the "Armenian genocide." Many Armenians became refugees, migrating across their traditional homeland or leaving it altogether. By 1917 fewer than 200,000 Armenians remained in Turkey.

It was not just this one period that inflicted a toll on Armenians. Over 1,000 years, foreign invaders wreaked havoc numerous times and scattered the population. Today over half of the Armenian population lives outside of Armenia in a diaspora. About half of the diaspora community lives in other CIS countries. The other half lives in communities from India across to

Southwestern Asia, Europe, and North America, with a sizable number in the United States. Interestingly, many Armenians in the diaspora speak the western dialects of Armenian, associated with eastern Turkey, formerly in the Ottoman Empire. Armenians in Armenia speak the eastern dialects. Nevertheless, despite dialectical differences, the Armenian diaspora has close ties with Armenia. Many émigrés now serve in the Armenian government, and the Armenian government considers Armenians abroad to be members of the Armenian nation.

The largest conflict since the dissolution of the Soviet Union in 1991 involved Armenia and Azerbaijan. In 1924 the Soviet government created an autonomous territory within Azerbaijan known as Nagorno-Karabakh. It was 94.4 percent Armenian. By 1979 Armenians represented only 76 percent of the region's population. Armenians began to fear their loss of numbers and objected to Azerbaijani laws that restricted the development of the Armenian language and culture. Clashes between the Armenians of Nagorno-Karabakh and Azerbaijanis began in the 1960s and developed into war by 1992. Armenian forces of Nagorno-Karabakh seized most of the territory and advanced westward to link their territory with Armenia. Afterward they moved into Azerbaijan proper, but success brought condemnation. The Turkish and Iranian governments warned the Armenians to cease hostilities. Finally, peace talks sponsored by the UN, Russia, Iran, and a number of other countries met with success, and the shooting war ended in 1994. In addition to Nagorno-Karabakh, Armenian forces continue to control approximately 20 percent of Azerbaijan. An official agreement on the governance and the political status of territories has not emerged.

Economic Development

The economies of Georgia, Armenia, and Azerbaijan suffered greatly from the ethnic conflicts, even in Armenia, where little fighting took place. In the early 2000s, these three countries' economies steadily improved. All three countries have climates warmer than those of the other former Soviet republics and can produce agricultural products not available to the north. Thus, during Soviet times, for example, Georgia supplied over 90 percent of the Soviet Union's tea and citrus fruits. Armenia supplied fruits, especially grapes, and Azerbaijan produced tobacco, cotton, and rice.

While encouraging agricultural production, Moscow limited industrial development in Georgia, Armenia, and Azerbaijan. These policies made these republics heavily dependent on the other Soviet republics, especially Russia, for markets in which to sell their agricultural goods and sources for their industrial goods. Such policies also created a situation in which all three countries still have relatively low ownership levels of consumer goods (see Figure 4.17).

Since 1991 these countries have worked to reduce this dependence on Russia by greatly altering and increasing their industrial and service sectors. Georgia, located on the sunny, warm, eastern shores of the Black Sea, has great tourist potential. Azerbaijan, located on the Caspian Sea, will become one of the world's leading oil producers if it can exploit its oilfields fully.

American oil companies are developing new fields, but continued investment depends on the subregion's political stability. In addition to problems with the Azerbaijani government, foreign investors are concerned about shipping the oil through existing and proposed pipelines (see Figure 4.11). The Russian Federation tries to make Azerbaijan use the existing oil pipeline via Groznyy (Chechnya) to the Russian port of Novorissisk. This is an example of the Russian Federation's continuing attempts to control events in the countries that were once part of the Soviet Union. To avoid continued Russian control, Georgia, Armenia, and Azerbaijan are developing trading relationships with other countries, most notably with Iran, Turkey, the United States, and those in Europe.

Central Asia

The former Central Asian republics of the Soviet Union now form five independent countries—Kazakhstan, Tajikistan, Uzbekistan, Turkmenistan, and Kyrgyzstan (Figure 4.38). In 2001 Kazakhstan was the largest of the five countries in area and ninth in the world, but Uzbekistan had the most people (see Table 4.2). These countries have similar landlocked situations, arid or semiarid climates, and Muslim faith of many of their peoples. The subregion's fertile river valleys of the Amu Darya and Syr Darya are one of the cradles of human civilization. They played important roles in the trading of goods and ideas from the time of the earliest civilizations in Mesopotamia, China, and the Indus River valley. Straddling old trade routes between east and west, most notably the Great Silk Road (see Figure 4.5), great cities such as Bukhoro (Bukhara) and Samarqand (Samarkand) (Figure 4.39) emerged. During the 700s and 800s, Bukhoro became one of the leading centers of learning, culture, and art in the Muslim world. Its grandness rivaled the other Muslim cultural centers of Córdoba (Spain), Baghdad (Iraq), and Cairo (Egypt). Some of Islam's greatest historians, geographers, astronomers, and other scientists came from the area. In the late 1300s, Timur (Tamerlane) emerged as the dominant leader in Central Asia and conquered lands far to the west, south, and east. Within his vast empire, Samarqand served as his capital. A new flowering of culture began as numerous scholars and artisans came to reside there. Timur's grandson was one of the world's first great astronomers. Literature flourished, and great religious structures and palaces were built.

Despite a history of great cultural, political, and economic power, Central Asia is often ignored because it appears to be isolated from world trade and has little political or economic power in current world affairs. However, Central Asian countries occupy strategic geopolitical positions with overland access to Russia, Afghanistan, China, Iran, and Pakistan. When the United States overthrew the Taliban as part of the attempt to root out Osama bin Laden and al-Qaeda in 2001, Uzbekistan and Tajikistan played large roles in the military and humanitarian campaign with their airbases and entry points to Afghanistan.

Countries

Central Asia's five countries were once part of a land called Turkestan. Turkestan was not a single, consolidated country but a loose confederation of tribes. Beginning in the 1800s, the

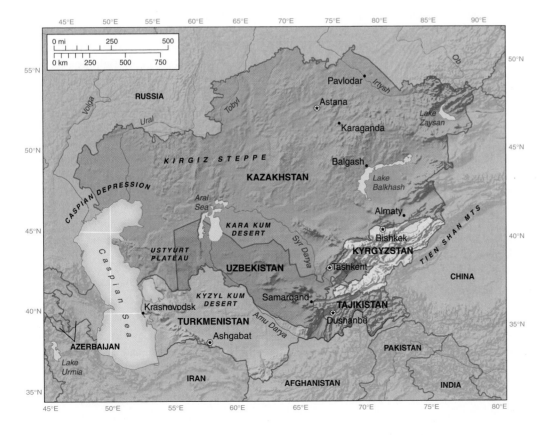

FIGURE 4.38 Central Asia: countries, cities, and major physical features. These countries have a mountainous southern border. Under Russian rule, they were carved out of the former Turkestan area. Only Almaty and Tashkent have populations of over a million people.

FIGURE 4.39 Central Asia: Samarqand, Uzbekistan. Historic Muslim buildings highlight the importance of Islam in Central Asia. Madrasah (theological seminary) buildings of the 1500s and 1600s border Registan Square. Typical features of Islamic art include arches, tilework, domes, and minarets. **Photo:** © Ronald Wixman.

Russian Empire expanded forcibly into Turkestan and took firm control of it by the end of the century. Remnant territories of the Central Asian peoples were annexed by neighboring countries. For example, some Uzbeks and Tajiks found themselves in Afghanistan. In the 1920s, not long after the creation of the Soviet Union, Soviet authorities drew boundaries for new republics in Central Asia. These republics became the countries of Central Asia that appeared on the world map after the Soviet Union broke up in 1991.

Accounting for more than 40 percent of the population of the five Central Asian countries and the only one that borders all the others, Uzbekistan is the dominant country in Central Asia. The fact that large numbers of Uzbeks live just across the borders in Kazakhstan, Kyrgystan, and Tajikistan emboldens Uzbekistan's government to declare its right to intervene in its neighbors' affairs to protect Uzbeks living outside its borders. On the constructive side, Uzbekistan has entered economic agreements with its neighbors and has taken the lead in solving the environmental problems associated with the Aral Sea. During Soviet times, Moscow created the Muslim Board of Central Asia and headquartered it in Tashkent. Tashkent likewise became the center for the religious training of Muslim clerics as the Soviets closed similar Islamic institutions in the other republics. Consequently, Uzbekistan has had a lead in Islamic affairs in the post-Soviet period.

Though Central Asian countries willingly joined the CIS, they are concerned with Russian domination and seek to diminish Russian influence within their countries and reverse past Russification policies. For example, the Russian and then Soviet government settled many Russians in northern Kazakhstan during the 1800s and then again during the Virgin Lands Campaign of the 1950s and 1960s (see Figure 4.21). Soviet industrialization also integrated the northern part of the country with the rest of the Soviet Union. To diminish the power of the large Russian presence now in Kazakhstan resulting from these earlier policies, the Kazak government moved the capital of the country from Almaty (Alma-Ata) in the Kazak-dominated southeast to Akmola in the Russian-dominated areas of the north on June 10, 1998. Akmola, meaning "white tomb," was renamed Astana, meaning "capital." The government claimed that Almaty had outgrown its location and that Astana (Akmola) was a more central location within the country and thus a better place for the seat of government. The move also means that more Kazaks will likely move to the Russian areas. Feeling unwelcome under many of Kazakhstan's governmental policies, many Russians are leaving the country. Similarly, Turkmenistan's government ended dual citizenship for Russians in June 2003, causing many Russians to emigrate.

Culture: Traditional and Modern

Modern Central Asians are settled, but many of their ancestors were nomadic, and their cultures still reflect the traditional ways of life of their ancestors. For example, in traditional Kazakh culture, it is customary to ask about the well-being of someone's livestock before inquiring about the person's health and that of his or her family. The dwelling of these nomadic peoples was the yurt, a circular tent consisting of a willow wood frame covered in wool felt. An opening at the top allows smoke to exit from the fire used for cooking and heating. Modern Central Asians no longer live in yurts, but they use them as decorative motifs for building or erect them in their yards and sleep in them during the summer. The yurt is an important symbol of national identity and appears on the national flag of Kyrgyzstan.

Carpet making and horse breeding are important national traditions for the Turkmens. Five traditional carpet designs are incorporated into Turkmenistan's national flag. Akhalteke, a breed of horses well adapted to the desert, is the breed of national significance, appearing as the central figure in Turkmenistan's national emblem. Many Turkmen still own at least one akhalteke.

The Uzbeks, who prospered greatly from the Great Silk Road (see Figure 4.5), wear clothing made with fine fabrics, color, and ornamentation (Figure 4.40a). All were expensive in earlier times, and the ability to incorporate as much as possible in one's dress showed one's wealth. This earlier culture practice is evident today. On any given day, it is still common to see individuals elegantly dressed, though they have no special function to attend.

Modern Central Asians are by no means homogeneous, as each group has a unique history of ethnic development, though many of them are closely related:

- Kazakhs emerged as a distinct people in the 1400s from a mixture of Turkic and Mongolian nomads of central Asia.
- The ancestors of the Kyrgyz probably originated in Mongolia (Figure 4.40b). After the 800s, they mixed with Turkic tribes from the south and west and adopted their current name, Krygyz, which means "40 clans" in the Turkic languages. The 40 clans are represented on the national

(a)

(b)

FIGURE 4.40 Central Asia. (a) Uzbek woman in front of suzane embroidery. (b) Kyrgyz man drinking kumys (alcoholic drink made of fermented horse milk). **Photos:** © Ronald Wixman.

flag with a sun that has 40 rays. During czarist times, both Kazakhs and Kyrgyz were called Kyrgyz, illustrating that the languages of the two peoples are closely related.

- Turkmens trace their ancestors back to Oghuz tribes that inhabited Mongolia and southern Siberia around Lake Baikal. In the 700s these tribes migrated into Central Asia and assimilated Turkic and Persian tribes, giving rise to the Turkmen. "Turkmen" probably means "pure Turk" or "most Turklike of the Turks."

- Uzbeks are a Turkic people who moved into the lands now known as Uzbekistan in the 1500s. They are closely related to Turkmens but also have close ties with Tajiks.

- Tajiks probably acquired their name from an old Arab tribe. The Tajik language is a Persian language and was not distinguished from Persian (Farsi) until the Soviets designated Tajik as a unique language. At the same time, Tajiks had not differentiated themselves from Uzbeks, though the Uzbeks are a Turkic people. Both groups commonly spoke each other's languages.

During late Russian imperial and Soviet times, Central Asians were subjected to Russification policies. Soviet policies were more intense, forcing the Central Asians to adopt the Cyrillic alphabet of the Russian language for their native languages. After 1991 many Central Asian countries, with the exception of Kazakhstan, returned to using the Latin alphabet (used for English and Turkish), began expunging Russian words from their languages, and downgraded the status of the Russian language. The policies were problematic because many Central Asians speak only Russian. Moreover, Russians emigrated, and the Russian government was offended and applied political and economic pressure on these countries. By the late 1990s many Central Asian governments curtailed their attacks on the Russian language, and many of them designated Russian as the second official language of their countries. However, in the early 2000s Saparmurad Niyazov, the leader and virtual dictator of Turkmenistan, introduced a unique Turkmen alphabet to replace Cyrillic.

Most Central Asians are Muslims. Those living closest to the Islamic heartland, such as the Uzbeks, converted to Islam as early as the 600s, soon after the rise of Islam. Those farther north, such as the Kazaks and Kyrgyz, converted to the religion only in the last 200 years. Kyrgyz also mix Islam with their traditional belief in totemism, the religious idea that humans have spiritual kinship with particular animals (such as reindeer, camels, and bears). The sun, moon, and stars also have great religious importance for the Kyrgyz.

Within the Russian Empire and later in the Soviet Union, Islamic beliefs were suppressed by deportations, the breakup of nomadic lifestyles by collectivization, and the immigration and resettlement of Russians. Since independence in 1991, the Muslim majorities played significant roles in the political transition of these countries. Some of the countries look to links with other Muslim countries such as Iran, Iraq, Turkey, Kuwait, and Saudi Arabia. These countries finance the building of mosques and other Islamic institutions in central Asia.

Though Central Asians have a strong faith in Islam, Communism influenced the development of belief and practice. For example, Islamic political groups are discouraged, even banned in some countries. Traditional practices like veil wearing are rare. Women are generally viewed equal to men.

People: Ethnic Conflict

The ebb and flow of history created an ethnically complex subregion with significant minority populations in most countries (see Figure 4.21). The ethnic mosaic and many of the tensions that exist between groups can be attributed to Soviet policies. In the 1920s the Soviets drew the republics' boundaries in a way to help the Communist government in Moscow keep control over Central Asia. For example, the Soviets drew the boundaries of Kyrgystan, Tajikistan, and Uzbekistan so that they arbitrarily zigzag through the fertile and densely settled Fergana Valley, leaving many Kyrgyz, Tajiks, and Uzbeks in their neighbors' countries. These boundaries pit Central Asians against one another, formerly weakening their resistance to Moscow's rule and even resulting in the people looking to Moscow to maintain peace.

Soviet policies also resettled many Russians in the subregion. The presence of Russians was another means for the Soviet government to control Central Asia. During World War II Stalin built many factories in Central Asia because the subregion was out of reach of the Nazi military. He moved Russians and other peoples from across the Soviet Union to work in these factories. As part of this effort, he also shifted many Central Asians to each other's republics and relocated ethnic groups he did not trust to Central Asia.

Soviet policies eventually split Central Asian societies in two. The groups for whom republics were named (titular groups) tended to maintain traditional ways of life in rural areas, while the imported ethnic minorities, particularly Russians, became the industrial, business, and civil servant class and dominated the urban areas. For example, in Tajikistan's capital city, Dushanbe, only 39 percent of the population was Tajik in 1989. Many central Asians view the Russians within their countries with resentment and suspicion—as unwanted reminders of Soviet rule. Moreover, while many central Asians speak Russian, few Russians bother to learn Central Asian languages. Central Asians also resent the privileged economic positions that Russians continue to hold within their countries. New anti-Russian policies and a general anti-Russian attitude of many Central Asians have caused many Russians to emigrate since 1991.

The relationships between Central Asian governments and ethnic groups have not been positive either. For example, anger at the ethnic minorities in Uzbekistan has led to violence and the emigration of these minorities. Tajik authorities have criticized neighboring Uzbekistan for not recognizing the uniqueness of Tajik culture and forcing many Tajiks in Uzbekistan to register as Uzbeks. At the same time, violence against Uzbeks in neighboring countries caused many Uzbeks to migrate to Uzbekistan. After a civil war in Tajikistan in the 1990s, thousands of Tajiks moved north into Kyrgyzstan in an attempt to escape the conflict, but this only intensified ethnic tensions in Kyrgyzstan.

Economic Development

During the Russian imperial period and especially under Soviet domination, the economic output of these five countries was redirected to supplying the needs of Russia. When the Civil War in the United States in the 1860s prevented Russian industry from receiving its cotton supplies, the Russian government saw a great opportunity in Central Asia to create its own supply of cotton. The Soviets later continued developing the cotton industry but also imposed other forms of economic development on the subregion. Industrialization was based on producing iron and steel and tractors, while farming focused on growing irrigated cotton. The extraction of coal, iron, chromium, oil, and natural gas turned the subregion into a colony producing raw materials for Russian factories.

Following independence in 1991, GDPs dropped considerably and are generally below their past levels though they are improving (Figure 4.41). Soviet infrastructure still keeps Central Asia dependent on the Russian Federation. Pipelines and transportation lines are of Soviet specifications and run primarily to the Russian Federation. Central Asians are leery of Russia taking advantage of them but do not want to depend on China to the east. Goods could travel through Azerbaijan, Armenia, and Afghanistan, but all three of these countries are very unstable. Turkmenistan is working on a rail line to Iran. Western countries, however, are uneasy about shipping goods through Iran and may not trade with Central Asia if Iran is the only choice for transit. Thus Central Asia remains heavily dependent on the Russian Federation. Such trade benefits the Russian Federation but not necessarily Central Asia. The Russian Federation buys cheap raw materials from Central Asia and sends back more expensive finished products. Kazakhstan and Turkmenistan need a route to transport oil and natural gas to Western customers and are frustrated by Russia's failure to support new

Post-Soviet Dip

GDP, local currency at constant prices
1990 = 100

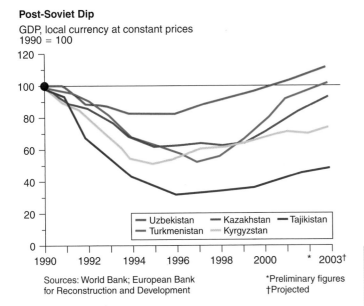

Sources: World Bank; European Bank
for Reconstruction and Development

*Preliminary figures
†Projected

FIGURE 4.41 GDP growth in Central Asia after the end of Communism. Notice that many of the economies have not fully recovered. **Source:** "Post-soviet dip" from *The Economist*, July 26, 2003. Copyright © The Economist Newspaper Group, Inc. Reprinted with permission. Further reproduction prohibited. www.economist.com

pipelines (see Figure 4.11). The Russian Federation has no incentive to help these two countries in this matter because it also sells oil and natural gas to the West. Kazakhstan has barged oil across the Caspian Sea and then through Iran.

Because all five countries produce similar commodities (oil, natural gas, and cotton), rivalry rather than cooperation is most common, especially in attempts to attract foreign trade. Kyrgyzstan and Tajikistan, however, do not have large fuel reserves like the other three. Consequently, the other three countries have shut off fuel as a political lever against their neighbors.

The issue of water resources also encourages rivalry. Water is scarce in this dry area of the world but important for agriculture and power generation. Disputes exist over the allocation of water that flows in rivers that pass through a number of the countries. Major rivers originate in the mountains of Kyrgyzstan and Tajikistan and flow down to the lowlands of Turkmenistan, Uzbekistan, and Kazakhstan (see Figure 4.15). Each country frequently complains that the others withdraw an unfair share of the water from the rivers. Serious tensions have arisen between Kyrgyzstan and Uzbekistan over water in the Fergana Valley, where agricultural reform and land privatization programs are endangered by the water disputes.

Test Your Understanding
4E

Summary

The countries of the Southern Caucasus and Central Asia are among the world's most complex in terms of ethnic differences and conflicts. They are tied economically to the Russian Federation but seek new relationships with other countries of the world. Their world position at the heart of Asia makes them strategically significant.

Questions to Think About

4E.1 What are the cultural similarities and differences among the peoples within the countries of the Southern Caucasus and Central Asia?

4E.2 What are the actual and potential sources of political conflict in the Caucasus and Central Asian countries?

4E.3 How do the physical and economic geographies of the Central Asian countries differ?

Chapter 5

East Asia

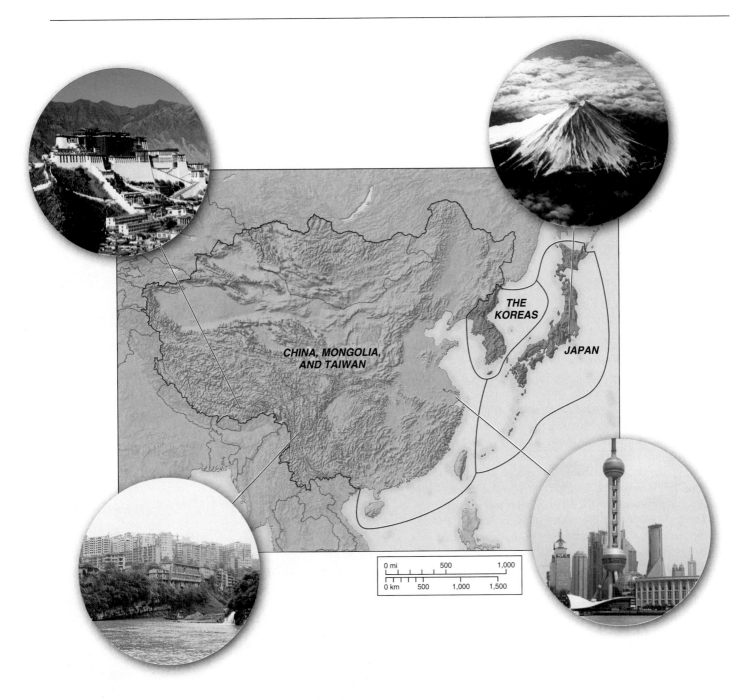

FIGURE 5.1 **East Asia: subregions and typical views.**

THE KOREAS

CHINA, MONGOLIA, AND TAIWAN

JAPAN

0 mi   500   1,000

0 km   500   1,000   1,500

Themes in This Chapter

East Asia poses increasing challenges—political, economic, and cultural—to the leading world regions in North America and Europe. Based on a similar historic cultural heritage, Japan, China, South Korea, and Taiwan developed successful economies in the second half of the 1900s. Faced with conflicting European political–economic systems of capitalism and communism, countries developed their own responses and their own global involvements. North Korea did not, and remained isolated under its dictator, while Mongolia is physically isolated on the Chinese–Russian border. Major themes include:

Economic growth and global roles in the late 1900s.

Past clashes of internal and external cultures contributing to present geographic differences.

Natural environmental variety.

Global influences to and from East Asia.

Three subregions:

- Japan, the region's economic leader into the 2000s.
- The Koreas, North and South: contrasting responses to change.
- China, Mongolia, and Taiwan: different economic, political, and cultural forces.

Point–Counterpoint: Population Policies in China
Personal View: Beijing Tourist Guide
Geography at Work: Frontier Forces

East Asian Miracle

The reemergence of East Asia (Figure 5.1) as a region with world political, economic, and cultural impacts was a major event of the later 1900s. In the 1990s it was called the "East Asian Miracle." Today East Asia, led by Japan, China, South Korea, and Taiwan, refocuses global economic growth and political power on the countries situated around the Pacific Ocean. East Asia not only interacts with external globalization trends but also contributes to those trends (Figure 5.2). In the early 2000s the world region was home to 24 percent of the world's population and produced 20 percent of economic output from 9 percent of the land area. It is a third force in the global economy alongside the United States and Europe.

The region's present character builds on millennia of cultural and technological development and their interactions with the natural environment (Figure 5.3). From ancient times, the Chinese instigated many of the most significant human advances from technology to art. Until well into the 1900s, they insisted that all other people were "barbarians" in comparison to their own "civilized" ways; the Chinese saw themselves as the "Middle Kingdom"—at the center of the world. The Japanese had similar attitudes to outsiders. In studying this region, therefore, we must understand the traditional cultures, the political and economic changes of the last century, and the responses to Western global political, economic, and cultural dominance.

182

From Poverty and Defeat to Renewed Eminence

After World War II, poverty, political disruption, and cultural confusion were the rule in East Asia. In 1945 defeated Japan was economically devastated and hated in the region for its wartime acts, while the other countries struggled to recover from Japanese invasions. The Koreas gained independence following decades of Japanese occupation and exploitation. Splitting Korea into North and South led to the 1950s Korean War, which destroyed much of both countries. In China, the warring Nationalist and Communist

FIGURE 5.2 China: central Shanghai. The low-rise old town surrounded by new apartments, offices, and hotels. Old meets new. Western lifestyles, clothing, and mobility meet Chinese traditions and government. International tourists flock to the city. **Photo:** © Michael Bradshaw.

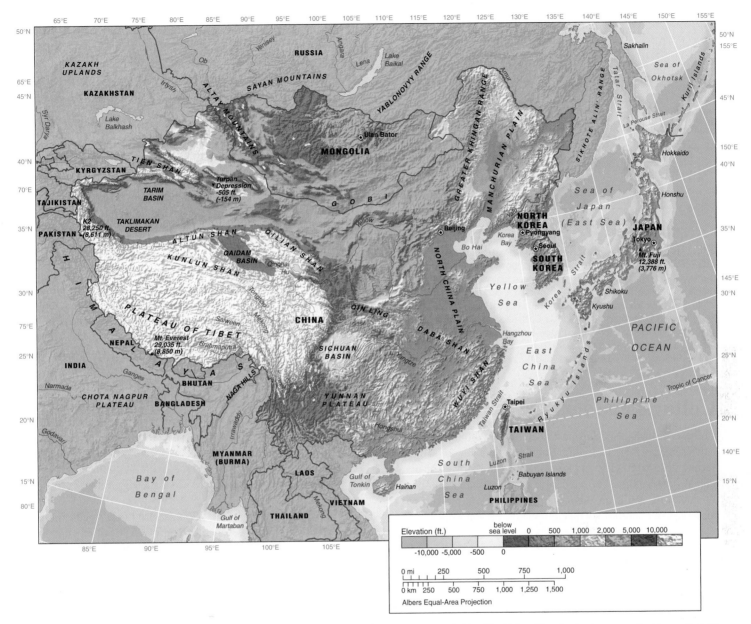

FIGURE 5.3 East Asia. Mountains, lowland plains, major rivers, countries, and capital cities. The Yellow River is also known as the Huang He and the Yangtze River as the Chang Jiang.

factions resumed their prewar conflict. When the Communists prevailed in 1949, the remnant of the Nationalist army fled to Taiwan. By 1960, Japan still had a gross national income per capita that was only one-eighth that of the United States, while other East Asian countries had GNIs per capita equal to or lower than those of African countries at the time.

Over the next 40 years, East Asian countries experienced widely different fortunes. Japan's export-led growth raised its gross domestic product one hundredfold between 1960 and 2000 at current dollar values, making it the world's second richest country. Behind Japan, the newly industrializing South Korea, Hong Kong (before and after it became part of China in 1997), and Taiwan increased their incomes six- to sevenfold from 1980

to 2000. Meanwhile, over nearly 30 years from 1949, the People's Republic of China under Mao Zedong built national cohesion at tremendous cost in terms of lives and changes of policy. From the 1980s, however, China's economic growth increased its total GNI from $202 billion in 1980 to $1,065 billion in 2000. By 2002 it had overtaken Japan in PPP GNI to take second place in the world and by 2004 became the third world trading country (Figure 5.4). However, neither North Korea with its declining economy nor Mongolia with its tiny population and economy experienced such growth. To explain these changes affecting the geography of East Asia, it is necessary to examine the history of the human occupation of the region, highlighting the interactions of people and environment.

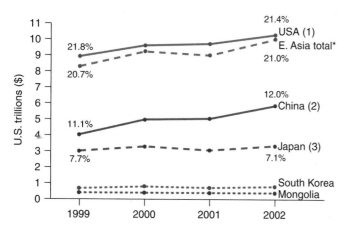

FIGURE 5.4 East Asia: an increasing proportion of world GDP PPP. The percentages given are of the world total GDP PPP for each year. The figures in parentheses are the country's world position. This graph shows how in total East Asia is closing economically on the United States; no data are available for North Korea or Taiwan, and those would bring equality or a slight advantage to East Asia. Second, it shows how China's economy is gaining ground while Japan's falls back. Third, the East Asian countries showed more of a dip in 2001 than the United States. **Source:** Data from Human Development Reports (United Nations) 2001–2004.

Cultural and Political Influences

Chinese, Koreans, and Japanese share many cultural characteristics (Figure 5.5), including Buddhism, the use of Chinese script, close family life and kinship links, and a focus on communal organization. The modern business phenomenon of the "Asian way," built on this common culture, is marked by a hierarchy of relationships that define how people work and live with each other: ruler–subject (boss–employee), father–son, husband–wife, older person–younger person. In each, the former is regarded as protecting the latter, while the latter respects and obeys the former. There is a parallel emphasis on maintaining poise and proper respect ("face") in public. While the

similarities are important and arise from common cultural histories, differences emerged between the Chinese, Japanese, Mongols, and Koreans.

Chinese Empires

East Asian cultures developed largely through the interactions of past Chinese civilizations and empires with surrounding territories. Established Chinese empires added land or lost it to rivals. At times, internal anarchy fragmented the area controlled by Chinese emperors. Each phase left its marks on peoples and landscapes.

Through periods of both unity and disunity, Chinese people produced a series of technological and cultural achievements, including the invention of paper, printing, the compass, the crossbow, paddleboats, clocks, and gunpowder. The Chinese developed a distinctive written script and styles of paintings (Figure 5.6), pottery, and building designs. They took significant steps in astronomy, mathematics, the understanding of weather and minerals, and engineering skills such as deep mining. Many of their achievements were taken up in Western Europe at a later stage; for example, Chinese paintings influenced French painter Claude Monet in the late 1800s.

Foundations of Chinese Cultures

By 2000 B.C. the first Chinese civilization developed in the Huang He (Yellow River) valley (Figure 5.7), based on wealth from agricultural surpluses and craft skills. It controlled much of the lower Huang He and lower Chang Jiang (Yangtze River) basins.

The Zhou dynasty (1122–256 B.C.) used Iron Age technology, irrigation, and deeper plowing of the soil to support economic growth. Improved transportation opened wider

FIGURE 5.6 Chinese art. Sorting of the cocoons from the *Book of the Silk Industry,* Qing dynasty, early 1800s. **Photo:** © Giraudon/Art Resource, NY.

FIGURE 5.5 East Asia: cultural features. A temple beside Dongting Lake, lower Chang Jiang, China. People come to pay homage to the Buddha and/or local deities, and it is a tourist stop. **Photo:** © Michael Bradshaw.

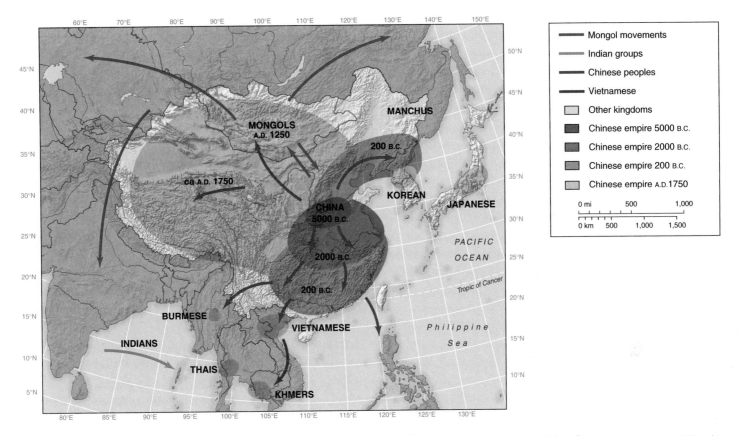

FIGURE 5.7 East Asia: the expansion of the Chinese empires and the location of adjacent kingdoms. The Mongol and Manchu kingdoms were the only external ones that expanded to take over all of China.

geographic trading areas but the prosperous Chinese lands attracted attacks from outside. During the political and social upheavals of this period, **Confucius** (Kong Fuzi), an administrator, called for better-trained and more able managers to focus on orderly conduct and proper relations in all social contexts. Hundreds of years later, knowledge of Confucian principles made appointments to public office possible through a system of merit-based written exams. His significance as an educator led to the adoption of his birthday in September as National Teachers' Day.

Daoism, the teaching of Laozi, disdained the Confucian hierarchical system for organizing society, preferring a return to local, village-based communities with little government interference. Daoism is based on the apparently opposing concepts of yin and yang (including the "dark side" and "sunny side" of a hill, negative and positive, male and female, evil and good, Heaven and Earth). Yin is female, linked to the lightweight, a perception that affects Chinese gender values. In fact, yin and yang are complementary, interdependent principles operating in space and time, emblems of the harmonious interplay or balance of all pairs of opposites in the universe (Figure 5.8). Yin and yang are seen as two "breaths" (*qi,* or cosmic energy, pronounced "chee"). Every person's duty is to strengthen, control, and expand the *qi* received at birth.

Around 220 B.C., the Qin dynasty (spelled "Ch'in" in older forms of Romanization), after which China is named, imposed strict laws to weld the separate feudal states into a centralized and culturally uniform empire with a standardized written script. It sited its capital near modern Xi'an. Armies extended the empire southward into the Xi Jiang (West River) basin and westward as far as Lanzhou. The high cost of completing the Great Wall (Figure 5.9) caused internal resentment.

After Southeast Asian Buddhism had been rejected earlier in south China, Buddhist monks from India reached northern China in the early years of the first millennium A.D. Adding a spiritual dimension to materialistic Confucianism, Buddhism also brought a more systematic framework to Daoist philosophies. After 600, these links resulted in a period of literary and artistic brilliance, religious toleration, and foreign trade. A literate Chinese elite, however, rejected Buddhism as disruptive.

Over the next three centuries, the Tang dynasty spread Chinese influence into the northeast of modern China, Korea, Japan, and northern Vietnam. Zen Buddhism, a form that is centered in meditation and moments of insight in the course of mundane activities, arrived in Korea and Japan in the early 1200s.

Lamaistic Buddhism was established in Tibet from the A.D. 600s, combining Indian Buddhism with Tibetan rituals. The Dalai Lama leaders had both religious and secular roles. The term *Dalai*

FIGURE 5.8 Yang and yin. The Pudong area across the Huang Pu from central Shanghai. Some believe that the TV tower completed in 1998 is too much (male) yang and that later buildings around it are more rounded and modest (female) yin to give balance. **Photo:** © Michael Bradshaw.

FIGURE 5.9 China: Great Wall. Built in stages and connected around 200 B.C., the Great Wall was 2,400 km (1,500 mi.) long. It was surprisingly effective in keeping out aggressors for many centuries. Today it is a major tourist attraction. Badaling is the most popular center for visitors, having a new fast highway link to Beijing that makes a half-day visit possible. **Photo:** © Michael Bradshaw.

taxes imposed on them, combined with crop failures and famines, opened the way for rebellion and the establishment of the Ming dynasty in 1368.

Later Empires

Renewed Chinese expansion occurred under the Ming dynasty (1368–1644), during which the economy, literature, and education advanced. Chinese naval power under Admiral Cheng-Ho (1405–1433) ventured beyond India to the Persian Gulf, Red Sea, eastern Africa, and Madagascar. Although there was trade, the Chinese did not conquer new lands and withdrew from wider links.

The succeeding Qing (Manchu) dynasty ruled from 1644 to 1911. By the mid-1700s it had expanded the Chinese realm to its farthest limits in the northeast, Mongolia, Xinjiang (the northwest), Tibet, Burma (Myanmar), and Taiwan. The expansion once again brought Lamaistic Buddhists and Muslims into western parts of China. In 1820 China was the world's wealthiest empire, producing nearly one-third of world GDP. After that time, however, poor internal government faced by Western influences and conflicts led to a decline of political and economic power.

Japanese Isolationism

The most powerful of the kingdoms, or empires, in the lands surrounding China developed in Japan. Ancient traditions place Japan's foundation in 660 B.C. The Ainu, a non-Chinese people, were among the earliest inhabitants but retreated northward as other groups invaded through Korea.

("ocean of wisdom") was conferred by a Mongol ruler in the 1200s, and Lamaism had a continuing influence on Chinese dynasties that wished to establish political control over Tibet.

Mongol Invasions

In the 1200s the Mongols invaded China. Kublai Khan, the Mongol leader, established his capital in northern China at Beijing. He and his descendants ruled as Chinese emperors, driving the center of native Chinese intellectual and economic development southward. They also ruled much of modern Russia. Many Mongols intermarried with Chinese. In the 1300s, growing Chinese resentment at the new rulers, disruptions, and

The principle of **Shinto** encapsulated Japanese traditional values in a national religion built on animism, ancient myths, and customs. It is a **pantheistic religion** and involves emperor worship and Japanese superiority. It became an instrument of political control. In Shintoism, both living and nonliving objects possess spirits, with Mount Fuji and other mountains being given special reverence. The Japanese crowded into the lowlands, avoiding mountains (in a combination of reverence to nature gods and fear of volcanic eruptions and earthquakes). Japanese Buddhism tolerates Shintoism, and both are followed by most of the population. Chinese Daoism and Confucianism also influenced social practices.

Japanese emperors resided at Kyoto, retiring from public life and delegating administration of the country to leading families of court nobles. From the 1100s, these feudal lords took over the imperial administration, and armed samurai warriors under a shogun (military leader) acted over the heads of the powerless court. The shoguns dominated Japan and largely maintained its isolation from the rest of the world.

When the last shogun resigned in 1867, the Meiji emperor regained his position as titular head of government. The royal capital moved to the shogun center in Edo, later called Tokyo ("eastern capital"). However, in the 1889 constitution, the emperor became a figurehead and political power was taken by leading figures in the parliament.

Korean Origins

The area that is now North Korea was the center of the first Korean kingdom, around which the peninsula was unified in the A.D. 600s. After the Mongol invasion and retreat, Confucian principles of government were adopted. The Yi dynasty, founded in Seoul in A.D. 1392, ruled the kingdom of Korea until 1910, although it became subject to the Chinese in 1644. Close proximity to Japan attracted aggressive interest from the growing military power. In 1894–1895, as rebels opposed the Korean government, the newly strengthened Japanese forces overran Korea, defeating both Chinese and Russian attempts to aid the Koreans. Japan annexed Korea in 1910.

Responses to European Intrusions

The Portuguese led European expeditions to the region at the start of the 1500s but were soon followed by the Dutch, Spanish, French, and British. Although European colonists never occupied China, Korea, or Japan, the arrival of traders, missionaries, and military forces affected them all. Their partial incorporation into Western global economic and cultural systems began.

China Resists Colonization

In 1557 the Portuguese established Macau as the first trading port on the Chinese coast. China resisted other attempts to set up trade or political links until the global economic system expanded during the 1800s.

The British East India Company raised demand for Chinese tea, exported through the southern port of Guangzhou (Canton). At first the company paid for tea in silver but replaced silver with opium from India as the trade increased. In 1796, to preserve its privileged trading position at Guangzhou, the company bowed to the Qing emperor's ban on importing opium to China. However, the company relied for an increasing share of its income on its Indian monopoly over the production and distribution of opium, and it sold the drug in open auctions in India. Private merchants then took it to the Chinese coast. By 1819, the opium trade, stimulated by lower prices, increased so much that it exceeded the value of tea exports and required payment in Chinese silver, draining those resources. When a Chinese official reasserted the policy of forbidding opium imports, the Opium War of 1841–1842 forced the Qing government to sign the Treaty of Nanjing. This treaty established a framework for continuing British trade with China and the opening of other ports. The port of Hong Kong was built on a surrendered island at the entrance to the Zhu Jiang (Pearl River) estuary. The Kowloon Peninsula opposite Hong Kong Island was ceded to Britain in 1852, and in 1898 China leased the New Territories to Britain for 99 years.

Further local wars during the 1800s ceded rights of Chinese trade and residence through specified ports to Britain, France, and the United States. The increasing external control of its economy affected the Chinese deeply. While colonial status was unthinkable for the Chinese, the trading concessions through eastern ports weakened the Qing emperors' authority and internal political structure.

Disrupted Chinese Republic

In 1912 a republic replaced the enfeebled Qing dynasty. Sun Yat-sen, who had experienced some years of exile for rebellion in the 1890s, and his Kuomintang party (Guomindang, or the Chinese Nationalist Party) claimed overall political power. Attempts to reunify China under a common government in the 1920s and 1930s, however, contended with strife among local warlords, a Communist rebellion, and Japanese attacks. The main Japanese invasion of China began with a 1933 expansion out of the northeast that it had occupied earlier in the 1900s, followed by a general declaration of war in 1937. The Kuomintang and Chinese Communists united in resistance.

Japanese Aggression

After 1867, the new government of Japan concluded that it should take major actions to prevent world domination by Western countries. The Meiji leaders followed the Western world's path to international political power through rapid industrialization. At the end of the 1800s, its new military power enabled Japan to conquer Taiwan and Korea and to win a war against Russia that enabled it to take Manchuria in northeastern China.

Modernization in Japan built on economic stability and government encouragement. Manufacturing provided the basis of modern Japanese growth. New factories built in the coastal zone from Tokyo to Osaka made iron and steel, ships, construction engineering products, and textiles. The economy responded to external trading pressures as well as a national desire to be a military power.

In the early 1900s Japanese governments became more democratic, and social groups established political parties and labor unions. In the period up to World War II, industrial expansion was based on government–business links, cheap labor, large numbers of small businesses, and improved infrastructure. Overseas, Japan became known for its cheap goods, often of mediocre quality. The 1930s economic depression—when other countries raised tariff charges against cheap Japanese imports—led to an internal pact between Japanese military leaders and the elite families, the *zaibatsu,* who owned the biggest industries. The military leaders took over political power and shifted industrial production to naval ships, air force planes, and other military equipment. This phase culminated in World War II.

After invading China, the Japanese military–industrial war machine, invoking the past glories of the shogun leaders and overriding the civil government, occupied much of Southeast Asia and many Pacific islands. Japan's antagonism toward the United States came to a head with the surprise attack on the U.S. naval base at Pearl Harbor, Hawaii, in December 1941. The Pacific war that followed was a major part of World War II and ended in the 1945 defeat of Japan with the destruction of much of Japan's industrial capability by U.S. bombing and postwar dismantling.

During World War II, Japanese military leaders promoted themselves as a master race. The rest of East Asia was made wary of Japanese political imperialism by the harsh manner in which the Japanese occupied other countries. Prisoners of war and civilians in occupied countries were used as forced labor, and some women were made sex slaves for soldiers. After 1945, groups in the formerly occupied countries pressed for redress, and in the 1990s the Japanese government apologized to them.

Natural Environments

The natural environments of East Asia produce contrasts of mountains and lowlands, arid and well-watered lands, desert and forest. Large numbers of people crowd into the small proportions of river lowlands. People and their ways of life are much affected by natural events, and themselves have major impacts on the physical environment by modifying natural features and processes.

Subtropical and Midlatitude Climates

East Asia has a variety of climatic environments (Figure 5.10). In southern China, a subtropical **monsoon climatic environment** has summer southeasterly winds that are drawn into central Asia as the continent heats up and rising air occurs in areas of low pressure. These winds bring moisture from the tropical

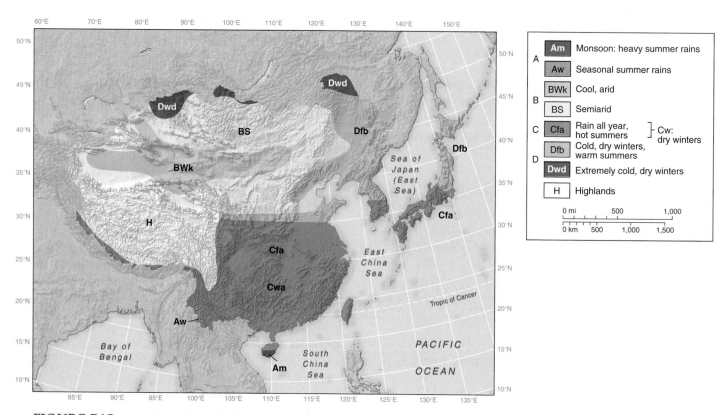

FIGURE 5.10 **East Asia: climate regions.** Note the contrasts within the region.

oceans and heavy rains. In mid- to late summer the heating of the Pacific Ocean makes the region subject to typhoons (the Asian equivalent of hurricanes) with strong winds and rain. In winter the winds blow outward from the high pressure over the cold continent and are cooler and drier.

Farther north, in coastal China, the Koreas, and Japan, the **midlatitude east coast climatic environments** bring summer rains and drier winters. From south to north, conditions change as in the eastern parts of North America from Cuba and Florida to New England and Newfoundland. Although summers have periods as warm as those farther south, icy winds blowing from Siberia make winters much colder. Japan receives more winter precipitation than the continental countries after the winds blowing from the continental interior pick up moisture on crossing the Sea of Japan and precipitate snow on the west-facing mountains.

The western parts of China and the whole country of Mongolia have arid or semiarid climatic environments because of their distance from the ocean: humid ocean air loses its moisture by raining before reaching the interior. Winters are extremely cold. In the westernmost parts of China, high altitudes in the mountains and Tibetan plateau cause even greater summer–winter and daily ranges of temperature.

Mountains and Major Rivers

Mountain systems and relatively small areas of lowland (see Figure 5.3) form the diverse relief of East Asia. High mountains extend eastward from the Himalayan Mountain ranges and the Tibetan Plateau, occupying over one-third of China. On the border with Nepal and India, the Tibetan Himalayas are the world's highest mountains, including Mount Everest (8,850 m; 29,035 ft.) and nearly 50 other peaks that rise over 7,500 m (25,000 ft.). Although the high proportion of mountains and destructive events suggest a difficult environment for human occupation, the region supports some of the highest densities of population in the world.

The volcanic and earthquake-prone islands of Japan are subject to major shocks and eruptions every decade or so as tectonic plates clash (Figure 5.11). In southern China, some of the most distinctive hilly landscapes are in areas of limestone rock, where the almost vertically sided hills are riddled with caves in what is known as a karst landscape.

In summer, the meltwater from the high mountain snowfields in the west combines with intense rains toward the coasts, supplying high flows in some of the world's most active rivers. The three major rivers of China are, from north to south, the Huang He (Yellow River), the Chang Jiang (Yangtze or Long River), and the Xi Jiang (West River)—with its wide lower section, the Zhu Jiang (Pearl River). Inland, these major rivers carve deep valleys with steep slopes and boulder-strewn streambeds. The rivers carry large quantities of water toward the sea, together with eroded mud, sand, and rock fragments. On their way to the coast, they drop their load of sand, silt, and clay to form wide plains and deltas. The tidal range and wave activity are both low around these coasts, allowing the large river-borne loads of silt and clay to build deltas at river mouths.

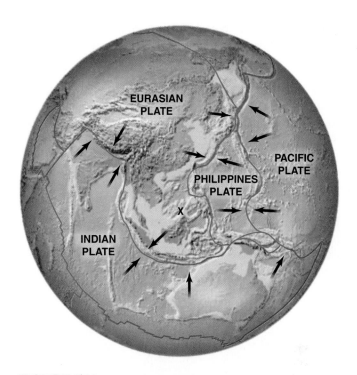

FIGURE 5.11 East Asia: plate margins. Relate the plate margins to the formation of the Himalayan ranges in the west of the region and the volcanic and earthquake activity in Japan in the east. 'X' marks the position of the Spratly Islands.

Forests, Grasslands, and Desert

The range of climatic and relief environments in East Asia produces a variety of natural ecosystems, although human land uses replaced or modified the forests, grasslands, and soils. In southern China, extensive areas are dominated by a great variety of species, including teak forest. Farther north, midlatitude deciduous and evergreen forests were largely cleared from the lower and more cultivable areas but remain on steep slopes in northeast China, the Koreas, Taiwan, and Japan. In northwest China and Mongolia, the increasing aridity causes the forests to give way to grassland and desert. The Gobi Desert occupies much of Mongolia and northernmost China.

Country Boundaries

Although most country borders in East Asia follow physical features such as mountain ranges or rivers, their locations reflect cultural and colonial history. Boundary disputes, including those between China and Vietnam and China and India, remain central to the political concerns of countries. Mongolia acts as a **buffer state,** helping to reduce direct conflict across the long boundary between China and Russia.

A dispute where there are no physical boundaries concerns the ownership of the Spratly Islands in the South China Sea (approximately 8°N, 124°E). All the surrounding countries, including Taiwan, but mainly China and Vietnam, claim the islands. The sea forms a major shipping route for 70 percent of

Japan's oil imports, so pressures exist to keep it international. However, in the late 1990s, the discovery of oil in rocks beneath the seas surrounding the Spratlys coincided with disappointing results from Chinese offshore exploration farther north. Both China and Vietnam issued exploration licenses to international oil corporations in adjacent sections of the area. By 2005, China and Japan were also disputing offshore areas between them that promised new finds of oil.

Natural Resources

The major natural resources of East Asia are surface water flows, fertile soils, and minerals in the rocks. The mineral resources of East Asia (Figure 5.12) include coal, oil and gas, iron, gold, and precious stones. Coal deposits are widespread in China, while major deposits of oil and natural gas occur in western China and in offshore locations that are still being explored. Japan has few mineral resources and relies on imported raw materials for its industries.

The initial Chinese civilization near Xi'an was based on millet, growing wheat on the fine wind-blown soils (loess) there and wet rice farther south. This made it possible to feed large numbers of people and maintain high population densities. The annual floods renewed fertile alluvial soils along the lower valley floors and enabled abundant crops to be harvested.

Environmental Problems

Despite high population densities in some parts, the mountains and arid areas that make up half of the region's area attracted few people.

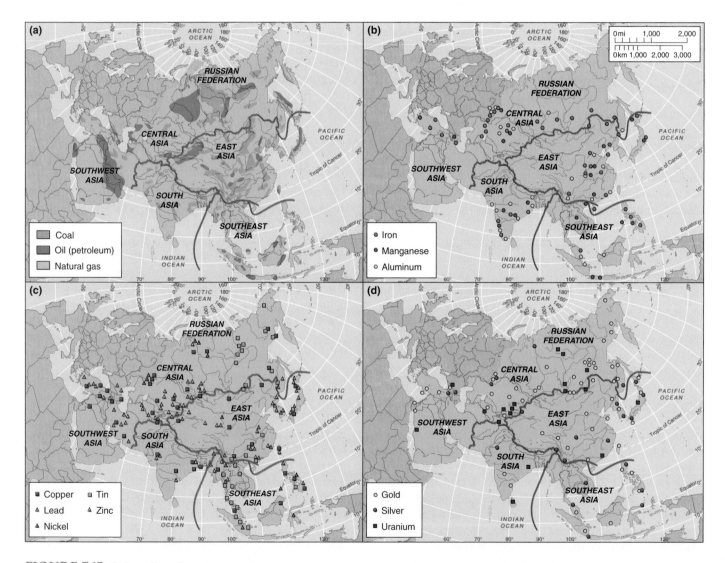

FIGURE 5.12 Asia: mineral resources. Compare the resources of East Asia with other parts of the continent. (a) Fuels: coal, oil (petroleum), natural gas. (b) Metals: iron, manganese ores, bauxite (aluminum ore). (c) Metals: copper, lead, nickel, tin, zinc. (d) Precious metals: gold, silver, uranium.

Natural Hazards

As elsewhere in the world, short-term major natural events, such as earthquakes, volcanic eruptions, floods, and typhoons (Asian hurricanes) present hazards to human occupation but do not discourage high population densities. Lying along a convergent plate boundary makes Japan subject to major earthquakes, such as the one in 1995 that destroyed much of Kobe, and occasional volcanic eruptions. Geologic activity also occurs some distance from the plate boundaries. In 1976 a massive earthquake hit the coal-mining area around Tangshan (near Tianjin) in northeastern China, killing 242,000 people. Rehabilitation, however, took only a few years.

Flooding is an annual hazard on the low-lying major river plains. Chinese flood hazards attract publicity because of the large numbers of people affected when the Huang He or Chang Jiang floods, but similar hazards also occur on a local scale in other river basins and in neighboring countries. The worst floods occur when precipitation and high meltwater flows together enter rivers in the summer months. The 1991 floods— the worst in China during the 1900s—affected both the middle basin and the provinces at the river's mouth.

Human Impacts and Responses

Human activities may cause or enhance natural disasters. Human actions often make the flooding worse. The 1981, 1991, and 2002 floods along the Chang Jiang followed abnormal precipitation, although excessive deforestation in the upper parts of the river basin increased the rapid runoff. River damming for hydroelectricity, transportation, and irrigation, as in the massive Three Gorges project on the Yangtze River, raises river flow in some places and reduces it elsewhere. Urban areas and mining pollute water with toxic chemicals and waste. Deforestation resulting from expanding agriculture and cutting forests for lumber often creates soil erosion, downstream floods, and silting. The silt that is swept out to sea blankets and destroys offshore fisheries and coral reefs.

In the mid-1990s, 180 million people in the more northerly Huang He basin suffered the opposite problem—too little water. Building dams in the headwater areas reduced overall flow downstream, while mid-basin irrigation works diverted waters that had been the mainstay of the lower basin. Successive years of drought from 1987 reduced flow in the lower basin to less than half its previous level. By the early 2000s the Beijing metropolitan region suffered from water shortages and declining well water levels.

An example of environmental degradation occurred when China plowed the formerly grassed semiarid areas of the northwest. From 1948 to 1978, soil erosion by wind and water removed over 700,000 hectares (1.7 million acres) per year from productive activities. Government agencies responded by planting a "Great Green Wall" of trees 400–1,700 km (250–1,000 mi.) wide. Dust storm days in Beijing fell from 30 per year in the 1970s to 12 by the late 1980s, indicating a degree of success, but the incomplete program ended in the 1980s, after which Beijing experienced an increase in dust storms.

Forest planting to protect landscapes became a feature of many parts of China from the 1980s. The 1981 flooding in Sichuan highlighted the effects of poor land use management, including excessive tree felling. Major tree plantings, especially in shelter belts, reduced wind erosion. In a massive drive from 1999 to 2004, some 8 million hectares (20 million acres) of farmland in the fragile upland west were converted to woodland, and another 11 million hectares (25 million acres) of bare hills were planted with trees. The reduction in output from the lost farmland was largely compensated by increased yields on the better remaining farmland.

Diseases

The countries of East Asia are notable for their improvements in health care provision in the later 1900s. This had major impacts on reducing death rates and infant mortality rates throughout the region.

However, HIV/AIDS is an increasing threat in China. It was recognized as a national problem in 2001, when health officials in China admitted to a "very serious epidemic." At the end of 2000 the United Nations AIDS Office estimated 1 million HIV sufferers in China and predicted that there might be 20 million by 2010. One of the sources of HIV is in Henan Province, where, from the late 1980s, several counties made an industry of blood collection. The opportunities to earn the equivalent of US $5 per unit at government and army collection stations attracted many people, but failures to follow sterilization guidelines resulted in the transmission of the virus. Local officials covered up the problem, and AIDS sufferers reported harassment by police when applying for help from health agencies. From 2001, Beijing authorities spent heavily on education programs to clean up the blood collection process, although they could not afford life-prolonging medications for HIV/AIDS sufferers. Other sources of HIV/AIDS infection included intravenous drug use (about 50 percent of cases) and sexual transmission.

In the early 2000s China became subject to a virulent form of influenza, the Sudden Acute Respiratory Syndrome (SARS), which spread widely to other parts of the world after not being admitted until a number of travelers caught it and returned to their home countries. Around half of those infected died; but it was brought under control, and subsequent outbreaks were short-lived.

Pollution

As industrial activities increase, more factories emit fumes, more vehicles are used, and farming is made more productive by the use of fertilizers, pesticides, and mechanization. In areas of greater population density and modern industrialization, levels of air and water pollution are high. Large cities regularly exceed the standard levels for concentrations of carbon monoxide, sulfur dioxide, suspended particles, and lead in the atmosphere. Five Chinese cities—Beijing, Shanghai, Shenyang, Xi'an, and Guangzhou—are among the 10 most polluted cities in the world. This is partly because China depends on coal for 80 percent of its energy needs. Small factories and power plants that burn coal inefficiently produce one-third of the particulates and sulfur dioxide, while domestic users produce another one-third. Acid rain from the particulates and other emissions affects downwind areas, with some Chinese pollution traveling as far as the United States. Cities in all the East Asian countries experience rising pollution from motor vehicles, meeting it by measures that reduce emissions, and slowing the addition of new vehicles.

Water pollution is particularly serious around the expanding major cities, where raw sewage and untreated industrial effluent are common along rivers and in offshore fisheries. In the 1960s high mercury levels from industrial processes poisoned fish and killed over a thousand Japanese people who ate them, but this problem has been solved by new measures and tests. In northeast China, heavy metals from the metallurgical industries around Shenyang polluted local drinking and crop water. It caused a surge of birth defects and a drop in life expectancy of 10 years below the national average of 70 years. While access to safe drinking water is increasing in China, water pollution from untreated urban wastes and rural fertilizers caused Shanghai to move its drinking water sources upstream in the mid-1990s. Hotels in China provide bottled water for guests to use in cleaning teeth. Polluted water and acid rain both reduce crop productivity and make it essential to cook all vegetables by boiling or stir-frying at higher temperatures than the boiling point.

Environmental Policies

Although the governments of East Asian countries were late in adopting environmental policies, they are now setting aside wildlife sanctuaries, establishing forest management procedures, and considering the overall control of water resources. Some countries passed environmental legislation to reduce pollution of air and water but had difficulties enforcing the standards. To combat air pollution, Shanghai restricts the allocation of new motor vehicle licenses and has phased out small motorized bicycles. In China, the government raised coal prices to reduce the inefficient use of this polluting fuel, but it fears that acid rain will spread to new areas in the time it takes to implement the policy. The problem of companies corrupting government officials for short-term gains slows progress in environmental quality in many East Asian countries.

Japan, with so many people and factories on relatively small islands, made the greatest strides in environmental control after the worst years of the 1960s. Tokyo traffic police wearing smog masks became a symbol of pollution, but legislation improved air quality. One of the major environmental problems Japan faces today, when virtually all its dumping sites are filled, is the disposal of trash, which increased by a third in the 1980s and 1990s. As a result, Japan is becoming a world leader in developing new technologies of recycling materials and burning of trash, freeing up landfill areas.

Test Your Understanding
5A

Summary

In 2004 East Asia had 24 percent of the world's population and produced just over 20 percent of the world's economic output. The region includes countries that are among the world's materially wealthiest to its poorest. The "East Asian Miracle" had major impacts on Japan, South Korea, Taiwan, and China.

Chinese culture dominated the region's history, but smaller kingdoms around the margins established separate national entities that formed the basis of modern countries. Buddhism accommodated to Chinese and Japanese religious ideas, while Muslim influences affected the northwestern parts of the region. All the countries in this region resisted European colonization.

The varied physical environments of East Asia include contrasts of rainfall and drought, the world's highest mountain systems, major river basins, and hazards such as typhoons, floods, earthquakes, and volcanoes. The huge concentrations of people and industry and high rates of recent economic development raise concerns of environmental pollution and degradation.

Questions to Think About

5A.1   To what extent can East Asia be defined as the "Chinese realm"?

5A.2   What influences worked against the total Chinese or European domination of East Asia before 1950? How did such interactions affect later developments?

5A.3   Does the wide range of natural environments in East Asia contradict the idea of its character as a major world region? Explain.

Key Terms

Confucius (Kong Fuzi)

Daoism

Shinto

pantheistic religions

monsoon climatic environment

midlatitude east coast climatic environment

buffer state

Globalization and East Asia

Most countries of East Asia have a major involvement in the global economic system. They cluster toward the top of the world range of gross national income (Figure 5.13), ownership of consumer goods, access to clean water, and energy usage (Figure 5.14). For U.S. traders, Japan is an established major competitor, while the People's Republic of China, South Korea, and Taiwan are the three main emerging big markets. However, North Korea and Mongolia remain largely outside the world economy.

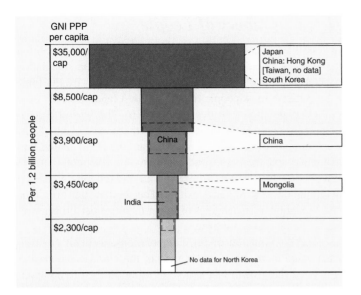

FIGURE 5.13 East Asia: country average incomes compared. The countries are listed in the order of their GNI PPP per capita. **Source:** Data (for 2000) from *World Development Indicators,* World Bank, 2002.

evolution of the Asia–Pacific Economic Cooperation Forum (APEC) reflects the global trading links from East Asia to Southeast Asia, the South Pacific, and the Americas.

Although the growing economies are highlights in this region, major political, demographic, and cultural changes accompanied an increasing involvement in world events and global trends. Greater population mobility effected economic and cultural exchanges, but also facilitated the easy transmission of diseases.

East Asian Governmental Contrasts

Other than North Korea, the East Asian countries emerged from inward-looking attitudes after 1945. The U.S. postwar reconstruction involvements in Japan, South Korea, and Taiwan included the installation of democratic governments. Apart from periods of military control in South Korea, and those of virtually single-party government in the other countries, the emphasis lasted. After its establishment in 1949 (Figure 5.15), the People's Republic of China was governed from Beijing by the Communist Party. It emphasized national unity and made farmers landholders, but also made horrendous mistakes that held back the country. Since 1978, China's adoption of an "Open Door" policy resulted in massive changes in the lives of the Chinese people, but did not disturb the continuity of the governing elite. By contrast, Mongolia is a tiny country between Russia and China that follows policies it believes will placate its neighbors, although the end of the Soviet regime in 1991 led to greater

The recent economic growth of these countries was at first based on exporting products to American and European markets. By the 1990s it was also bolstered by trade among countries within this region and Southeast Asia. Asian internal trade is now as great as external trade, showing that the future lies with both intraregional trade and a global scope. The

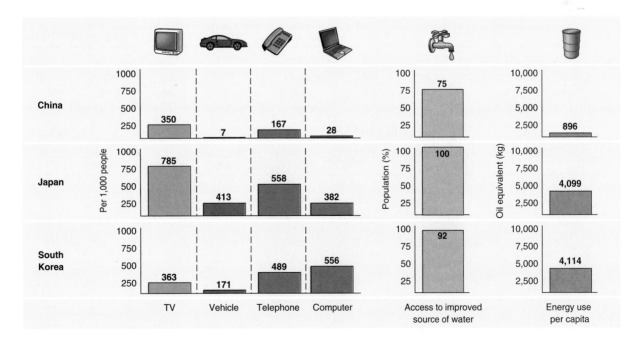

FIGURE 5.14 East Asia: ownership of consumer goods, access to clean water, and use of energy. How do these figures relate to economic growth? **Source:** Data (for 2002) from *World Development Indicators,* World Bank 2004.

FIGURE 5.15 China: The Tiananmen with the gateway into the Imperial Forbidden City. The portrait of Mao Zedong over the gateway marks his 1949 declaration of the People's Republic of China from the balcony above. **Photo:** © Michael Bradshaw.

political freedom. North Korea remains in the thrall of an inward-looking ruler who continues to lead the country into poverty as an international pariah.

Huge Numbers of People

In 2005 East Asia was home to 1.5 billion people, 24 percent of the world's population. Despite an emphasis on population control measures, this could rise to 1.7 billion by 2025 (Table 5.1). Because most people live on the best land, there are contrasts in population densities (Figure 5.16). The densest population areas are along the eastern and southern parts of Japan, the western parts of the Koreas and Taiwan, the north central part of China, and the coasts and major river valleys of southern China. However, half of China and all of Mongolia are desert or mountain with lower densities. The western half of China and Mongolia have densities of fewer than one person per km², except along major routeways.

The combination of huge populations, difficult environments, and increasing involvement in the world economy has generated the growth of massive cities and urban living environments (Table 5.2). Urbanization is a major feature of the region, but it is expressed in geographic differences.

Japan's Urban Concentrations

Japan is a densely but unevenly populated country, in which urban–rural contrasts have become more pronounced. In the short period from 1920 to the early 2000s, Japan changed from a largely rural country, in which industrial towns housed 25 percent of the population, to one in which almost 80 percent

Table 5.1 East Asia: Data by subregion, country, area, population, income (GNI PPP—Gross National Income Purchasing Power Parity), urbanization, Human Development Index (HDI), Human Poverty Index (HPI).

Subregions	Land Area (km²)	Population (millions) Mid 2005 Total	Est. 2025 Total	GNI PPP 2003 Total (US$ billions)	Per Capita* (US$)	% Urban 2005*	Index 2002 HDI (rank of 177)*	HPI (% population)*
Japan	377,800	127.7	121.1	3,641	28,820	78	9	11.1
South and North Korea	219,560	71.2	75.6	(SK)859	(SK)17,930	70	(SK)28	
China, Mongolia, Taiwan	11,200,260	1,335.9	1,510.9	6,635	11,867	68	78	16.2
Total or Average*	11,797,620	1,534.8	1,707.6	9,881	19,623.3	72.3	43.5	13.7

Country	Capital City	Land Area (km²)	Population (millions) Mid 2005 Total	Est. 2025 Total	GNI PPP 2003 Total (US$ billions)	Per Capita (US$)	% Urban 2005	Index 2002 HDI (rank of 177)	HPI (% population)
Japan	Tokyo	377,800	127.7	121.1	3,641	28,620	79	9	11.1
Korea, Republic of (South)	Seoul	99,020	48.3	49.8	859	17,930	80	28	no data
Korea, People's Dem Republic (North)	Pyongyang	120,540	22.9	25.8	no data	no data	60	no data	no data
China, People's Republic of	Beijing	9,596,960	1,303.7	1,476.0	6,435	4,990	37	94	13.2
Hong Kong (to China, 1997)		1,040	6.9	7.9	196	28,810	100	23	no data
Mongolia	Ulan Bator	1,566,500	2.6	3.4	4	1,800	57	117	19.1
Taiwan	Taipei	35,760	22.7	23.6	no data	no data	76	no data	no data

Source: World Population Data Sheet, Population Reference Bureau (2005), World Development Indicators, World Bank (2005), Human Development Report, United Nations (2004), and Encarta Microsoft (2005).

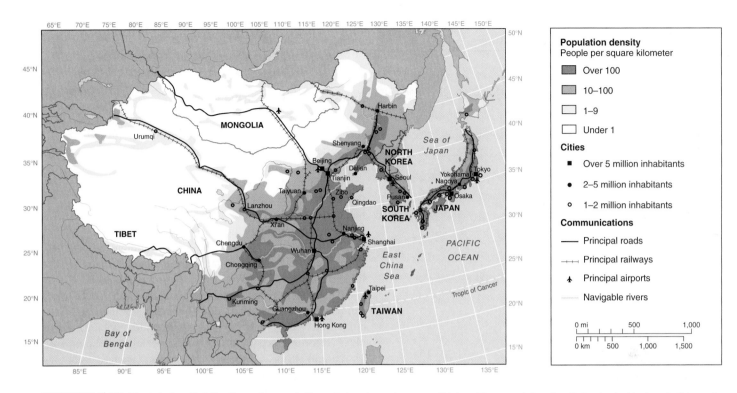

FIGURE 5.16 East Asia: distribution of population. Relate the main areas of high and low population density to physical factors. Is this a satisfactory basis for explaining the differences? **Source:** Data from *New Oxford School Atlas*, Oxford University Press, UK.

Country	Ethnic Groups %	Languages O=Official N=National	Religion %
Japan	Japanese 99%, some Koreans, indigenous	Japanese (O)	Shinto/Buddhism 85%
Korea, Republic of (South)	Korean	Korean (O), English	Christian 26%, Buddhist 23%, None 49%
Korea, People's Dem Republic (North)	Korean	Korean	(Around 60% non-religious), Buddhism, Confucianism
China, People's Republic of	Han Chinese 92%, Zhuang, Mongolian, Tibetan, Uygur	Mandarin (N), Yue (S), dialects	(Officially atheist) Buddhist, Confucian, Daoist, Muslim, Christian
Hong Kong (to China, 1997)	Chinese, Europeans	English, Chinese	Buddhism, etc., Christian
Mongolia	Mongolian 90%, Kazak, Chinese, Russian	Khalkha Mongol 90%, Turkic, Russian	Tibetan Buddhist, some Muslims
Taiwan	Taiwanese 84%, mainland Chinese 14%	Mandarin Chinese, local dialects	Buddhist 50%, Confucian/Daoist 25%, Christian 5%

Table 5.2 East Asia: Major Urban Centers

City	Country	2003 Population (millions)	2015 Population (millions)
Beijing	China	10.8	11.1
Changchun	China	3.0	3.6
Chengdu	China	3.4	3.9
Chongqing	China	4.8	5.8
Dalian	China	2.7	2.9
Guangzhou	China	3.9	3.9
Harbin	China	2.9	2.9
Hong Kong	China	7.0	7.9
Jinan	China	2.6	2.9
Nanjing	China	2.8	3.0
Qingdao	China	2.4	2.7
Shanghai	China	12.8	12.7
Shenyang	China	4.9	5.2
Taipei	China	2.5	2.4
Taiyuan	China	2.5	2.8
Tianjin	China	9.3	9.9
Wuhan	China	5.7	8.0
Xi'an	China	3.2	3.6
Zibo	China	2.7	3.0
Inchon	South Korea	2.6	2.8
Pusan	South Korea	3.6	3.4
Pyongyang	North Korea	3.2	3.5
Seoul	South Korea	9.7	9.2
Taegu	South Korea	2.5	2.5
Nagoya	Japan	3.2	3.3
Osaka-Kobe	Japan	11.2	11.4
Tokyo	Japan	35.0	36.2

Source: United Nations Urban Agglomerations 2003 (2003).

now live in cities. At the end of World War II half of the Japanese population was still rural, but further industrialization brought rapid urbanization. From 1945 to the mid-1970s, jobs in the manufacturing and service sectors multiplied, and migration to urban areas increased rapidly. Today the main concentrations of people are on the small areas of lowland, and 70 percent of the total population lives in the Pacific coastlands from Tokyo to Osaka. This urbanized zone is often called the "Tokaido **megalopolis**"—a series of almost continuous metropolitan centers with urban functions that exchange flows of people and goods with the surrounding areas.

After the mid-1970s, improving transportation encouraged people to move out of central cities that were being stifled by congestion and pollution. This was more a process of **suburbanization,** when people move to the outskirts of existing cities, than of **counterurbanization,** when they move away from metropolitan centers to small towns in rural settings. As urban populations grew, the more remote and environmentally difficult areas in the mountains and along parts of the western

coasts of Japan suffered depopulation. Such regions often have up to 25 percent of their population over 65 years old, and government grants are available to help people stay or settle there.

Japanese City Landscapes

All Japanese cities exhibit the crowding of buildings due to the shortage of available land. However, detailed study shows that they reflect irregular historic growth followed by rapid modern changes and expansion.

Castle towns, ports, or centers on main routes to religious shrines form the historic cores of metropolitan areas. Many modern Japanese central cities retain older street patterns, although the buildings are recent. Only the sturdiest ancient buildings survive. Most of the older buildings in Japanese towns were built of wood and other nondurable materials that fires, earthquakes, and World War II bombing destroyed.

The modern cities contain sectors for commerce, industry, and residences. After 1970, downtown areas of many Japanese cities were redeveloped with high-rise buildings but retain an older sector that acts as a tourist and specialist shopping area. The central business districts of the largest Japanese cities contain areas specifically devoted to hotels, department stores, financial businesses, and national or local government offices. High-rise offices, hotels, and apartment blocks are a response to high land costs, as are the underground developments in central Tokyo that include restaurants and car parking lots. Infilling of the Tokyo Bay is yet another symptom of the desire and need for more land. Smaller business districts occur in suburban areas of the major cities. Out-of-town shopping began to take hold in the 1990s.

Japanese industrial development is strongly localized, with heavy industry in coastal locations, often on reclaimed land because of the needs for space and proximity to ocean transportation links to raw material suppliers and markets. Such zones shut off city centers from the waterfronts. Light industry is more widely distributed and intermixed with housing. Older firms retain inner-city sites, taking advantage of the availability of labor, linkages with other firms, and market outlets. Local governments encourage the establishment of light industrial estates on the city margins, including science parks for high-tech industries.

Housing often consists of postwar apartment blocks near the city center, giving way to suburban bedroom areas. Poor, cheaper downtown housing units are refurbished to maintain the labor force in those areas. Transport routes determine the main expansion of suburban housing.

Urban Focus in the Koreas

In South Korea, the urban population increased from 25 percent in 1950 to 50 percent in 1980 and 80 percent in 2005. The main population and industrial facilities are in the northwest, centered on the capital, Seoul, Inch'on, and Taejŏn and along the southern coast, including Pusan, Taegu, and Ulsan. Almost totally destroyed in the 1950s fighting, Seoul was rebuilt rapidly (Figure 5.17).

were market and administrative centers, often with walls that restricted expansion. The exceptions were some port cities, of which Shanghai became the largest (Figure 5.18).

The largest cities in China (see Table 5.2) include the capital Beijing, Shanghai, Tianjin, and Chongqing, all of which are treated as political provinces. Shenyang, Wuhan, Guangzhou, Changchun, Chengdu, Xi'an, and Harbin each grew from around 1 million or fewer people in 1950 to multimillion cities today. It is likely that further growth of the major cities will take Beijing, Shanghai, and Tianjin to around 15 million people each by 2015, when China will have over 100 cities with populations of more than 1 million people.

Policies toward urban growth fluctuated from the mid-1900s. The Communist rule from 1949 alternately favored urbanization and movements back to the countryside. In the Mao Zedong era, the wish to encourage modernization through industrialization led to rapid urbanization from 1949 to 1960. In this period, the urban population more than doubled, from 49 million to 109 million, representing a growth from 9 to 16 percent of the total population. Part of this urbanization resulted from industrialization around previously administrative cities such as Xi'an and Nanjing. Another part came from the development of new inland centers such as Baotou, the iron and steel center in Inner Mongolia.

From the 1960s to mid-1970s, urban population was controlled by the rural–urban registration system (*hukou*), established in the 1950s to prevent rural-born people from becoming legal residents of cities by denying them services such as schooling, formal-sector jobs, and housing. During the Cultural Revolution, over 20 million young urbanites, as well as many others, were moved to rural areas to be reeducated in the virtues of rural life. Little investment was made in urban housing or transport infrastructure as any available capital went into expanding heavy industry. By 1978 the Chinese urban population made up around 13 percent of the total, but living conditions in urban areas had declined.

FIGURE 5.17 East Asia: Seoul, capital of South Korea.
The Chongno tower and other modern buildings dominate the center of the city, which was largely destroyed in the 1950s war against North Korea. The volume of road traffic reflects the country's economic growth. **Photo:** © Jose Fuste Roga / Corbis.

FIGURE 5.18 China: The Bund, Shanghai. This waterfront area was part of the 1800s port development by Europeans. More recently a raised walkway has been built that separates the Huang Pu River from the busy roadway behind, and new hotels and other high-rise blocks tower over the older buildings. **Photo:** © Michael Bradshaw.

North Korea's population is also concentrated around its western coasts in the capital city, Pyongyang, and Namp'o. Countrywide, fewer than 60 percent live in towns. Although possessing greater mineral resources than South Korea, particularly coal and metallic ores, North Korea has not developed a major manufacturing capability that draws more people to the towns.

Urban versus Rural Population in China

In China, the long-term population contrast between environmentally productive and difficult areas is being overlaid by rapid urbanization. Until the mid-1900s, most Chinese cities

From 1978 to the early 2000s, reforms placed an emphasis on urban jobs, higher incomes, and better access to services. They led to a rapid rise of urban population numbers, although the *hukou* system still operates in theory. The official urban population rose from 13 to 37 percent of the Chinese population. However, newly urbanizing areas are often still classified as rural, including the dispersed, small-town urbanization in the coastal provinces, the expanding Beijing–Tianjin area, and the corridors along the lower Huang He, Chang Jiang, and Xi Jiang. These areas raise the total urban population toward half the Chinese total. Chinese people are increasingly making their own decisions and moving to large urban areas. Many Chinese cities still retain relics of older cultures. Former palaces and temples occupy large sections of central Beijing (Figure 5.19). In many cities, modern roads, squares, and buildings replaced the old city walls and cramped housing. Monolithic and utilitarian buildings typical of centrally planned countries mark areas of Chinese cities built after 1949. The periods of urban neglect and the post-1978 renewal of urban and industrial construction are identifiable in city zones. The production sector and urban infrastructure compete for urban investments. Close control on living accommodations, however, reduced the incidence of **shantytowns**—housing built informally, often without permission, on "spare" land.

From the late 1970s, new urban-based industries recruited labor in the countryside on temporary contracts; such workers now comprise up to 10 percent of urban inhabitants.

The industrial area in southern China's Guangdong province inland of Hong Kong exemplifies the present rapid urbanization. Factories and huge apartment blocks sprang up in the 1990s to form cities of several million people around Shenzhen. Many new urbanites do not bother to acquire temporary residence certificates. In 2004 Shenzhen reregistered the rural population in its special economic zone, giving them better pension, health care, and welfare benefits. The growing middle-income class in China is almost exclusively urban, working in administration or as managers and technicians. Their numbers more than trebled from 1980 to the early 2000s. The Chinese government believes that 200 million people in this group will be able to afford private cars and own their housing by the early 2000s. To date it seems that, although the government gives them little recognition, this group prefers the current political security that brings economic well-being to initiating disruptive changes that might bring greater democracy.

Over the same period, increased investment in housing and urban infrastructure tripled the housing space built each year, mainly in large apartment blocks. Higher standards of building, sewage services, transportation, and domestic water and gas supplies improved the lot of urban dwellers. The huge new apartment and condominium blocks, however, encroach on farmland that is particularly valuable in a country with so many people and insufficient farmland to grow the necessary food. A debate continues over the relative merits of expanding the

FIGURE 5.19 Expansion of built-up Beijing, 1949–2000. The imperial Capital/Palace is in the center of the old city, built over a period of 500 years. In 1949 the urban area was a compact walled city with outlying villages. The dark lines are modern ring roads. The lands close to the old city were built on first, and suburban developments later pushed outward. **Source:** From *Beijing* by Sit, Figure 5.7, p. 134. Copyright © 1995. Reprinted by permission of John Wiley & Sons, Ltd.

Built-up area as of 1949–50
Added built-up area 1951–85
Built-up area 2000
Reserved green space
Rural land uses

0 mi   2.5   5.0
0 km   2.5   5.0   7.5

largest industrial cities or placing more resources in widespread small and medium-sized towns that are integrated with the surrounding rural areas.

Global City–Regions in East Asia

The growth of global and intraregional trade in East Asia led to the emergence of global city–regions (see Figure 2.15) as essential nodes linking this region to the rest of the world through business and political exchanges. Tokyo, Japan, is one of the world's top four global city–regions—with New York, London, and Paris—while Hong Kong and Seoul (South Korea) are in the second rank. Osaka (Japan), Taipei (Taiwan), and Beijing and Shanghai (China) are major cities where there is growth in international business services.

Population Pressures in Japan and the Koreas

All the major East Asian countries face the pressures of increasing populations in limiting environments, and have taken measures designed to slow population growth. Japan's total population increased from around 90 million in 1960 to 128 million in 2005. The combination of current low fertility (1.3) and population growth rates (natural increase of 0.2 percent) leads to estimates that the total will peak around 128 million in 2010 and fall to 121 million by 2025. Infant mortality at 3.4 per 1,000 live births is the world's lowest, and life expectancy at 81 years is the highest.

In 1948 concern over providing for its large and rising population after World War II led Japan to pass the Eugenics Act, allowing abortions and instigating a series of family planning drives. As the birth rate fell over 50 years from 34 to 9 per 1,000, the death rate fell from 14 to 8 per 1,000. Japan is in the final stage of demographic transition (Figure 5.20). The age–sex diagram for Japan (Figure 5.21a) has bulges and troughs. There were more births in the postwar and late 1960s periods. Between those dates, there were fewer births as the economy grew; after 1970 birth control and affluence combined to lower fertility.

Changes in the population's structure and distribution have important economic and social implications. In particular, Japan's population is aging. In 1970, 7 percent of the population was over age 65, rising to 19 percent by 2005. Meanwhile, smaller nuclear families and accommodations in suburban houses became common, reducing the traditional care of older people by eldest sons. The shift to smaller households caused the total number of households to increase from 23 million in 1960, when they averaged 4.5 persons per household, to over 40 million in the 1990s, with 3 persons each. The aging of the population increases pressures on pension provision, medical costs, and leisure provision for the retired. It also adds to firms' wage costs as more employees reach senior positions. Japanese people are now worried that there will not be enough labor for their industries or funds to support the increasing numbers of elderly people. Couples are encouraged to have more children.

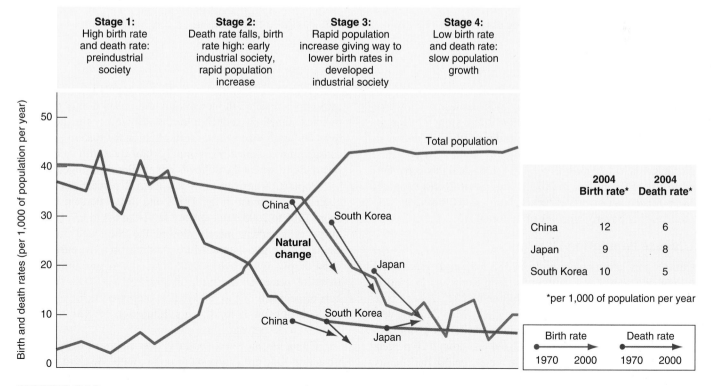

FIGURE 5.20 East Asia: demographic transition. Compare the positions of East Asian countries in this transition with each other and with countries in other major world regions. **Source:** *2004 Update: Population Data Sheet,* Population Reference Bureau.

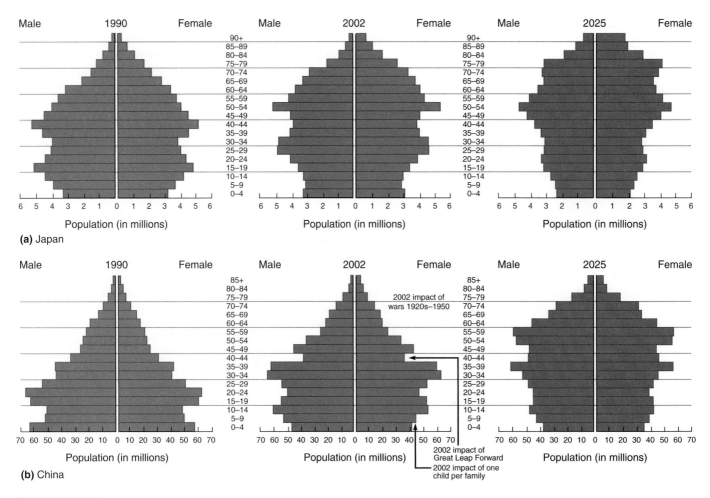

FIGURE 5.21 East Asia: age–sex diagrams in 1990, 2002, and 2025. (a) Japan. (b) China. **Source:** U.S. Census Bureau, International Data Bank.

South Korea's population of 48 million people in 2005 is likely to rise slowly to 50 million by 2025 because of a low rate of population increase. Life expectancy rose from 47 years in 1955 to over 70 years today. As in Japan, the population in older age groups will increase markedly in the next 20 years. North Korea had 22 million people in 2004, and this is predicted to rise to 25 million in 2025. While South Korea's modest rise is related to growing affluence, North Korea's has a context of increasing poverty and lack of opportunity.

Chinese Population Dynamics

China's population increased from 583 million in 1953 to 1.3 billion (including the 7 million in Hong Kong and Macau) in 2005 and may grow to nearly 1.5 billion by 2025. A huge population was at first seen as an advantage by Mao Zedong, who attempted to mobilize the great numbers of Chinese people as a "cohesive force." When it was realized that around 20 percent of its annual income increase would merely enable the country to keep up with the basic needs of the extra people, China made several attempts to restrict family size (see the Point–Counterpoint: Population Policies in China page 202).

From the 1970s, the large Chinese population and growth projections resulted in increasingly draconian family planning policies aimed at reducing births. However, the policy of reduced family size gained greater hold in the eastern areas and coastal cities than in the essentially rural west with its minority populations, where the policy is not applied so stringently.

Population growth rates fell from 2 percent per year in 1965 to 1.4 percent in the 1980s and to 0.6 percent in 2005. From 1965 to 2005, total fertility fell from 6.4 to 1.7. Even this major achievement merely slowed the continued growth of China's huge population. More children survived early childhood despite fewer babies being born, as infant mortality fell from 200 per 1,000 live births in 1945 to 32 in 2005. The numbers of people living to old age also increased as life expectancy doubled from 35 to 70 years. Education cut adult illiteracy from 80 to 19 percent.

China has moved toward the later stages of demographic transition. One projection showed that, with one child per family, China would reach a maximum population of 1.25 billion in 2000 and would then decline to under 500 million by 2070. With two children, a maximum of over 1.5 billion would be reached in 2050 and subsequent decline would be slow. With three children per family, the population could rise to over 4 billion by 2080.

The age–sex diagram for China demonstrates the changing history of population growth. For example, the lower numbers in the 40- to 44-year age group in 2002 reflected the impact of the major population disaster linked to the Great Leap Forward. The diagram also highlights the changing proportions of men and women. There are more women than men in the older age groups but fewer in the younger groups as a result of the one-child policy: girl babies are often aborted or killed to save the family "space" for a boy child. The resulting gender imbalance could pose major social problems for China. Demographers predict that by 2025, tens of millions of marriage-age men will not be able to find wives. The shortage of women is already reflected in the demand that produces an increased trafficking of women and girls from inland China and the neighboring countries of Burma and North Korea for prostitution and marriage.

The future of China's population is linked to potentially political issues such as the problem of caring for the elderly as they grow in numbers. Traditionally, extended families looked after older people, but urbanization often breaks close family ties. In the face of small government provisions for old age and the closure of many state-owned firms that provided pensions, Chinese families are striving to increase their savings for later years.

Overseas Chinese

The numbers of people leaving China represents a tiny proportion of the population. Official international migration out of China in the early 2000s was up to 400,000 persons per year, with 100,000 of those moving to the United States. For example, people in Changle county receive a total of around $100 million each year in remittances from relatives working abroad. Nearly 80 percent of the population of Houyu village in Changle county now lives in and around New York. Chinese contract laborers on short-term arrangements dominate construction sites around the Pacific Ocean. However, no figures exist on Chinese migrants smuggled illegally into countries such as the United States, Japan, Europe, and Siberian Russia; estimates are as high as 200,000 per year. For example, it is known that organized human smuggling rings operate from Fujian province.

The current migrations follow centuries of emigration that made groups of Chinese people—the overseas Chinese— prominent contributors to most East and Southeast Asian countries. Most of the overseas Chinese come from the coastal areas from Shanghai to Guangdong and base their business interests on strong family and clan links. Although their numbers are relatively small compared to the total Chinese population—21 million people in Southeast Asia and 21 million in Taiwan—their influence is much greater. Chinese capital and entrepreneurs dominate the private-sector economies in Indonesia, Malaysia, Singapore, the Philippines, and Thailand. For example, in the early 2000s they comprised 3.5 percent of the Indonesian population but controlled around 75 percent of the listed companies.

Although some anti-Chinese discrimination continues in countries such as Indonesia, Malaysia, Thailand, and Vietnam, their role in economic development makes the Chinese difficult to exclude. Overseas Chinese financiers, particularly in Taiwan and Hong Kong, are also major contributors of foreign investment to new developments within China. The **transnational Chinese economy** is based on these linkages and has increasing links in both personnel and finances to Chinese-based business activities in the United States, western Canada, and the rest of the world.

In the Economic Forefront: Japan, South Korea, and Taiwan

In the late 1900s Japan became the world's second largest economy, with 8 of the 10 largest global banks and huge capital surpluses from selling its products abroad. Major Japanese industries invested in the United States, Europe, and Southeast Asian countries to manufacture goods for their domestic markets. They then invested in China. Today Japanese companies produce goods in China and southeast Asia for their brands selling worldwide. South Korea and Taiwan also built economies based on exporting to the rest of the world. From the mid 1980s China joined these growing economies, and by the early 2000s its GNI PPP was double that of Japan.

East Asian Multinationals

Japanese, South Korean, and Taiwanese exports of manufactures and foreign investments led to the growth of huge multinational corporations. From Japan, automakers such as Honda, Toyota-Lexus, Nissan, and Mitsubishi, together with electronics and electrical goods makers such as Sony, Sharp, Fujitsu, Panasonic, and Matsushita, built worldwide brands manufactured in many countries. From South Korea, multinationals such as Hyundai, Samsung, Daewoo, and Lucky Goldstar (LG) penetrated markets around the world, while Taiwan's smaller companies dominated aspects of IT products.

Many Japanese and South Korean employees from managers to technical staff worked abroad in subsidiary companies of the multinationals. In the 1980s and 1990s around 50,000 Japanese were involved in such situations, and the South Korean numbers increased from 150,000 to 250,000. Short business trips abroad grew from 600,000 to over 2 million in Japan and from just over 100,000 to nearly 1 million in South Korea. Government employees in overseas embassies also doubled during this period. These figures are similar to those in Europe and North America, demonstrating the growing role of these two East Asian countries in the global economic system.

The Example of Toyota

The Japanese company Toyota is the world's third largest automotive group in sales after General Motors and Ford (Figure 5.22). In the early 2000s its capital resources were double those of either of the other two. Toyota's experiences illustrate some

POPULATION POLICIES IN CHINA

Chinese population change since 1950 was marked by major shifts in policies and attitudes. Although the Maoist government watchword of the early 1950s was "strength in numbers," the census of 1953 registered a total population of 583 million, and the first birth control campaign was launched in 1956 in urban areas. Its impact was minimal, however, in comparison with the 25–30 million who died as a result of the Great Leap Forward beginning in 1958. Birth rates increased until China launched another program of birth control in the early 1960s, based on delayed marriages and a wider distribution of birth control pills. It had some effect but was then overtaken by the disrupting Cultural Revolution, which swept aside many of the government-established groups to administer family planning. Once again, birth rates and population totals increased rapidly.

When an influential nuclear scientist, Song Jian, attended a population symposium in 1979, he made some rough calculations of Chinese population growth on the basis of the poor 1964 census results and formulas used to predict missile trajectories. Asked to flesh out these predictions in relation to whether families would have one, two, or three children, he worked with four other scientists in the first computer-based forecast in China. They predicted that a rigorously implemented one-child policy would keep China's population to 1 billion by 2000, falling to 700 million by 2050 with increases in female sterilization, abortion, and other birth control methods. The projections for two- and three-child families forecast continuing increases in the future.

In 1979 Deng Xiaoping approved the one-child policy before further debate, and this key plank of Chinese reforms was implemented without any law being passed by the National People's Congress. Initially a "temporary measure," it has lasted for over 20 years with its complex regulations, harsh punishments, and 80,000 full-time family planning workers to enforce them (Box Figure 1). Despite the new economic freedoms granted to peasant farmers, the policy subordinated individual interests to those of the country; local officials could monitor the most private aspects of individuals' lives. By 2005 there were over 100 million one-child families, accounting for 30 percent of all families.

Previously in the 1970s, some one-child policies had been implemented by urban authorities in Shanghai and other cities. Urban dwellers living in state units could be tracked. But from 1979, the countrywide one-child policy was directed at rural areas where the communes (and their controls) were being disbanded, food production was increasing, and extra labor was needed.

Box Figure 1 Chinese population trends. Family planning poster in Chengdu. The single child depicted is a girl: the one-child policy collides with a cultural preference for boys. What is the purpose of the English? **Photo:** Alasdair Drysdale.

of the ways in which Asian multinational corporations (MNCs) grew to worldwide prominence in the later 1900s and some of the challenges they face in the 2000s. Its policies remained successful as it expanded exports and manufacturing abroad. As globalization becomes more pervasive, Toyota may have to allow control of parts of its activities to occur outside Japan.

Kitchiri Toyota first researched gasoline engines in 1930, switching the family textile business into cars. Family directors still sit on the Toyota board. Although Toyota's headquarters are in Tokyo, its Takaoka base is a sprawling collection of 35-year-old factory buildings amid the rice fields of Aichi province to the southwest. Forty years ago, the city of Kolomo was so impressed

(Continued)

The policy included "sticks" and "carrots." Families complying with the policy received certificates that enabled them to claim free health care for the mother during pregnancy and delivery, free health care for one child, free education for one child, promotions and better pay for parents of one child, free family planning supplies and operations, and sometimes free vacations. However, families not complying lost free health care and free education for the first child, received lower pay or lost jobs, were excluded from social clubs and organizations, and came under pressure and were even forced to have abortions.

The policy was not implemented evenly, however. In some areas, the houses of those resisting the set quotas were burned, and women were imprisoned and subjected to late abortions, with some babies even being killed as they were being born. Crowded urban areas often had stricter enforcement than western rural areas. However, the southern industrializing area in Guangzhou was not subjected to the policy in order to attract foreign investment to an area of plentiful cheap labor. In Xinjiang province in the far west, the minority people were allowed two and three children. In some rural areas, the peasants rioted, killing local party officials and family planning workers until the policy was softened in 1984 and parents were allowed to have second children under certain circumstances. Elsewhere, in rural areas in particular, baby girls were killed, and the act was glossed over before officials knew about it, creating imbalances between males and females in those areas.

The policy did not meet its target; the 2000 population was 1.27 billion and continuing to increase instead of slowing. Even that figure is debatable because peasants mislead officials and officials misreport statistics. In Hubei province, when county officials had insufficient emergency relief food, they discovered that the total population was

10 percent higher than the census figure, with most families having three children, because local officials had reported numbers to match the official quota. Occasional events bring to light families with up to six children. In 2002 a pharmaceutical company revealed that it produced 25 million vaccines for infants, although the estimated baby population was only 20 million, indicating a higher birth rate.

Would China have been better off without this draconian policy? Comparisons with India suggest it might have been. In India, which has only moderately successful policies on family planning, fertility dropped, and the country claims it has already avoided 230 million extra births and will stabilize its population in 2040, the same year that China hopes for this result. A 1986 study in two counties of Shanxi province allowed peasants to have two well-spaced children. By 1996 the growth rate had fallen, fewer third children were born, fewer abortions were carried out, and less female infanticide occurred. Less coercion went with better relations between the peasants and the authorities. This is an important factor in any country.

The policy continues and is justified by the China Population and Information Center (website: www.cpirc.org.cn) as being effective in some parts of the country where it is accompanied by better application of the policies, more consultation, and better advice. Even in the areas of greatest compliance, such as Beijing, only two-thirds of couples held single-child certificates. The center claimed, however, that "family planning has become a social habit."

However, fears of a population age imbalance in an aging society led cities such as Shanghai to waive restrictions over having a second child. Beijing, however, did not follow Shanghai's example. In 2004 there were 134 million Chinese aged over 60 years, half of that age group in Asia. In 1970 there were six children to every old person in China; today there are two elderly persons to every child.

POINT

The one-child policy results in fewer births that place claims on resources.

It brings slower population increase and a more balanced population structure in the early stages.

It allows better planning for urban expansion, housing, education, job availability, and transport and utility infrastructure.

It provides control by the government and local officials in the interests of the whole population.

It brings a rapid realization of the importance of family planning in the face of imminent overpopulation disaster. Other policies had failed.

It is important to maintain this process.

COUNTERPOINT

Family planning is most successful when parents act on their own initiative, perhaps guided by information.

It upsets the population balance between sexes (abortion and murder of baby girls) and age groups (in the later stages when there are small numbers of young people and increasing numbers of older people).

It ignores other factors in population growth, including migration, the effect of natural disasters (as during the Great Leap Forward) and diseases, and the impact of political chaos (as in the Cultural Revolution).

It often requires draconian methods of control that violate human rights.

It ignores cultural and economic factors that favor large families.

The one-child family is at risk and fragile in placing all hopes on a single child: most families would consider two children ideal. Only children are often unduly pampered and obese.

with the company's local investment that it renamed itself Toyota City. The Toyota City factories produce 700,000 vehicles a year and provide the model "mother factory" for overseas plants.

At first, Toyota copied—and improved—others' innovations, such as Ford's moving production assembly line. Costs were tightly controlled and profits plowed back. The company continued to

be financially flexible and lean. At Takaoka, Toyota developed *jidoka*—automation with human intelligence, combining high quality, short delivery times, and low-cost production. Combined with *kaisen*—continuous improvement—and just-in-time delivery of components, it led to strong demand for the reliable Toyota vehicles. Such methods were adopted by MNCs worldwide.

FIGURE 5.22 Japan: Toyota cars. The new Toyota Prius hybrid sedan comes off the production line at Tsutsumi plant, Toyota City, in October 2003. Demonstrating the company's technological lead, the Prius was soon in great demand, at six times the monthly target of 3,000 cars. **Photo:** © Toyota Motor Corp/Handout/Reuters/Corbis.

In the early 2000s Toyota developed CCC21 (Construction of Cost Competitiveness for the 21st Century), a program aimed at cutting billions of dollars from its costs. The company is reviewing its design, manufacturing, and other costs with the objective of better using equipment and human resources. It will not, however, fire any of its 36,000 workers at Toyota City but will reduce the 4,000 temporary workers. Toyota also negotiates with component suppliers to reduce costs by up to 30 percent. Toyota holds an average 25 percent interest in most of its component makers. Denso, its chief supplier (the fourth largest component maker in the world), employs 85,000 people and is the chief of over 300 suppliers. Such policies and links illustrate a typical Japanese conservative attitude, which builds in linkages committed to lifetime employment and long-term agreements. However, Toyota's competitors, such as Nissan and Mitsubishi, are breaking apart their supplier chains to achieve cost savings, and this will eventually cause Toyota to rethink its policies.

After 1980, its adaptability enabled Toyota to change from a Japanese auto exporter into a global operator with 41 manufacturing subsidiaries in 25 countries. It has strong market shares in the United States and Europe, where Toyota's conservative Japanese outlook also applies. It does not merge with rivals, it does not have foreign managers among its Tokyo vice presidents, and it has never sent a female executive overseas. The company emphasizes the importance of cash reserves and profitability.

Toyota's successful strategy began with making mid-priced cars for the growing numbers of middle-class Japanese, and it still has 42 percent of its domestic market. The large home sales are a major strength in light of international currency fluctuations.

When the company expanded overseas, local production within the United States and Europe minimized problems of currency exchange-rate shifts. Toyota also targeted local demands, producing light trucks for the U.S. market that is the source of two-thirds of its current profits. Europeans bought smaller cars. The Lexus luxury brand sells in both regions.

To keep up with market trends, Toyota launches a new model every month, requiring huge investments that are questioned by the increasing numbers of foreign shareholders. The realization that even such large single companies cannot satisfy all the world's market areas led to Toyota joining an alliance with Peugeot-Citroën of France to market small cars in Eastern Europe.

Recent Entry: China

Mao Zedong tried to make China dependent on its own resources. He attempted to produce a total revolution against traditions and the rest of the world. He almost destroyed the country. However, after Mao's death in 1976 the Chinese government turned around and integrated the country with the global economic system. China is becoming a major world power. It is the world's most populous country, and between 1980 and the early 2000s, its "open door" policies were the basis for economic growth at the fastest sustained rate of any country. This economic growth multiplied the value of its output fivefold and raised 600 million people out of absolute poverty. Chinese factories now make many goods bought in Western stores, from electronics to clothing.

While the surrounding countries remain wary of China's military and new economic strength, many Chinese still see their country as poor and struggling to escape centuries of foreign intervention. The neighboring countries fear a Chinese people who continue to perceive a sphere of influence extending to lands it held during its period of greatest expansion in the Ming dynasty.

China expects treatment that reflects its position as a growing power in the world. Other countries get involved in internal Chinese matters, promoting human rights and democratic institutions. As the People's Republic of China reconsiders its relations with Taiwan, Japan, and the Koreas, as well as other Asian countries such as India, each of these countries reorients its outlook, trade, and policies to take account of China's growing role in the region and world.

Current foreign direct investment in China is fourth in the world, after the United States, Germany, and the United Kingdom, and outstanding among the world's less materially wealthy countries. Most investments in China come from Chinese in Hong Kong and Taiwan and from other Chinese living abroad—but also increasingly from Japan. To justify such economic investment, China's huge supply of low-cost labor includes college graduates and computer engineers who are paid one-tenth as much as their Taiwanese or Japanese counterparts. However, Chinese entrepreneurs themselves face difficulties. Their government is slow to expand the list of companies allowed to go public with share offerings.

Western and Asian multinational corporations trading in China include oil companies, electronics groups, auto manufacturers, and restaurant chains. For example, in 1987 Kentucky Fried Chicken (KFC, Figure 5.23) opened a huge outlet in China next to Tiananmen Square in central Beijing. Despite the 1989 incident in this square, when Chinese army tanks attacked peaceful demonstrators, McDonald's followed in 1992 and by 2001 had nearly 400 outlets throughout China. Both KFC and McDonald's plan to expand further. The first Starbucks in China—serving coffee in a tea-drinking country—opened in 1999, and by 2004 the firm had hundreds of outlets. The most contested site for Starbucks was in a Beijing Forbidden City souvenir shop, with 70 percent of 60,000 people surveyed opposing it. But it continues to serve coffee there. Such symbols of an intruding global economy, however, excite little public antagonism. Even at the height of anti-American demonstrations over the bombing of the Chinese Embassy in Belgrade, Yugoslavia, in 1999, the Chinese who stoned the American diplomatic missions left KFC and McDonald's largely untouched. Many Chinese regard the expansion of multinational food franchises as a positive sign that they are involved in the global economy.

China and the World Trade Organization

China joined the WTO in November 2001. The Chinese leaders expect WTO membership to boost exports and foreign investment, but it will also force its industries to become more competitive in world markets in both quality and price. WTO rules will make China cut import tariffs and allow foreign businesses to compete in its highly protected areas such as telecommunications, insurance, and banking industries.

Such trends raise internal opposition from inward-looking nationalists and the bureaucrats managing the Beijing-based public monopolies. Some Chinese see WTO agreement to Chinese membership as a plot that will enable the United States to gain control of China's economy. They stress the dangers of rising unemployment and fear that foreign multinationals in China will merely replace current government monopolies or strip the assets of firms they take over. Cheaper agricultural imports could mean trouble in rural areas, and openness to integrated global economies could aggravate regional differences within China. Other criticism focuses on the debt and a rising budget deficit caused by increased government spending on housing and other infrastructure construction that external companies require before investing in China.

Neighboring countries also criticize the growing Chinese dominance of international investment and its competitive entry to world markets. Fears of total Chinese economic domination in world markets, however, are unlikely to be realized. Other countries possess different comparative advantages and are developing greater technological expertise. Increased personal wealth inside China expands internal markets for its goods and those from other countries. Perhaps Taiwan will be most affected by China's growth as it invests in production facilities on the mainland but suffers capital and brain drains and a loss of factories to the mainland.

In the immediate future the Chinese government plans to build on its WTO membership by establishing trading agreements with other countries. Major arrangements for the supply of energy and raw materials were agreed in 2004 with Russia, Venezuela, Brazil, and the Central Asian countries, while trade links were agreed with the Southeast Asian countries. In particular, China has entered trade and student exchange programs with European countries following the United States reductions in foreign students.

International Tourism

Increasing personal wealth and the development of hubs for international airline routes resulted in rising foreign tourist travel for South Koreans, Taiwanese, and Japanese. Tourism became an industry that both brought in visitors and made it possible for people to visit other places around the globe. In 2002 Japan attracted over 5 million visitors, while 16 million of its people vacationed abroad—both numbers having risen by 50 percent since 1990. In a country favored by Japanese tourists, direction posts in Switzerland give information in Japanese as well as English and local languages. South Korea received over 5 million tourists in 2002, while 7 million of its people were outbound, almost doubling since 1990.

During the 1990s China's government decided to make its tourist industry one of the world's largest (see Personal View: Beijing Tourist Guide on page 206). By the early 2000s, numbers of international visitors (over 37 million in 2002) made it fifth in the world after France and Spain and close to Italy and the United States. Income from this source increased eightfold from 1990. In addition to the great historic treasures such as the Terracotta Warriors (Xi'an) and the Great Wall (Beijing), China has improved its museums of the country's historic artifacts; it has built a trade in handicraft products including silk carpets and embroidery, lacquer furniture, and jade carvings; and it has many new top-class hotels and first-class roads and airport facilities. Hong Kong remains an important part of this strategy, as will be the staging of the 2008 Olympic Games in

FIGURE 5.23 China: central Beijing. A KFC outlet alongside Chinese shops, together with a BMW and lots of bicycles: a mixture of global economy and local trends. **Photo: © Michael Bradshaw.**

Personal View

BEIJING TOURIST GUIDE

Angela is the English name of a Chinese tour guide (Box Figure 1). She works in an industry that the Chinese government has identified as making a major contribution to its economy. In 2001 China became the world's fifth largest destination for international tourists (after France, Spain, the United States, and Italy), with receipts increasing eightfold from 1990. In 2008 Beijng will host the Olympic Games, and preparations are under way to improve airports and major roads. Chinese people are encouraged to spend their saved money on visiting other parts of their country.

When she was born, Angela's parents lived in northeast China (Heilongjiang Province). Her father came from a Beijing intellectual background (her grandfather was a poet of the contemporary scene) but moved to a rural area in 1964, aged nearly 18 years, when Mao Zedong encouraged the urban young to do so. However, life was not easy there. The farmers made it difficult for the skinny city boy, who at first was not strong enough to cope with the heavy work. Social tensions were common between farmers and newcomers from the cities.

The compensation for her father was meeting her mother, a farmer's daughter who liked the young man who could write and brought an exotic element to her village. They married, and Angela was born. She remained an only child at that stage because her parents could not afford more.

Box Figure 1 Tourist guide Angela with clients. In Tiananmen Square, Beijing. **Photo:** © Michael Bradshaw.

Beijing. Moreover, the Chinese themselves are encouraged to support the home tourist opportunities, while increasing numbers visit other countries. As part of the infrastructure linked to tourism and business travel, in 2005 the Chinese central aircraft purchasing agency ordered over 80 new airliners from Airbus, as well as more Boeing 737s and advance orders for the Boeing 787. These planes are allocated to Air China and the regional airlines within the country.

Will Miracle Growth Continue?

The East Asian economic growth in the later 1900s was termed the "East Asian Miracle" in a 1993 World Bank report that argued that distinctive Asian characteristics motivated and fueled economic growth:

- Successful countries had strong governments that managed stable business environments by such measures as keeping inflation low. Their tax structures distributed some of the growth rewards throughout communities. Domestic savings increased.

- Governments encouraged export competitiveness that led to partial global integration.

- Education programs focused on primary and secondary education for both boys and girls as a priority over prestigious higher education for a few.

By the mid-1990s, the rapid economic growth in some countries began to slow. It was geographically uneven and subject to interruptions. For example, in China, the rising incomes of people along the coast, who have access to global trade, contrasted with the continuing extreme poverty of many inland areas. In the crisis years of 1997 and 1998, economic problems spread from Southeast Asian countries to South Korea and Hong Kong, but had less impact on China or Taiwan. The International Monetary Fund made loans in exchange for an end to protectionist policies that kept Western foreign investments out of East Asia. South Korea recovered its economic impetus within a couple of years, but in that time it also suffered unemployment, corporation failures, and an increase in long-term poverty.

In 2001 a World Bank report, "Re-Thinking the East Asian Miracle," provided alternative explanations centered around a combination of global trends and local attitudes:

- Asian economic growth up to the mid-1990s resulted mainly from increased inputs of capital, labor, machinery, and infrastructure (transportation, power, and the like), rather than from rising productivity (increasing output per employee).

- The advantages of activist government industrial policies and liaisons between bankers and industry with government support appeared less convincing when large corporations used government funds wastefully or corruptly.

In 1980, when Angela was five, her family moved back to Beijing because her grandfather was ill and needed help. Such a move would have been impossible had not another family wished to move from Beijing to a rural area. An exchange was arranged. A further problem was that Angela's birth had been registered in the rural area, but she did not have the ID necessary for her to go to school in Beijing. She was nicknamed "little black." Her father wrote to the Beijing Municipal Security Bureau requesting an ID for Angela, which was granted six months later. Angela could now go to school, although at seven years she was older than others in her class. Her father had given her a start by teaching her some calligraphy and poetry.

On moving to Beijing, Angela's parents were allotted a one-family apartment across the street from one of Beijing's major hospitals. They shared a single water tap with other apartments. Further pressures on Beijing housing in the early 1980s resulted in two more families having to share the single-family apartment. Although some tasks, including cleaning the shared toilet, were supposed to be shared, the new families did not comply, and no one wanted to take responsibility for repairs to windows, for example. The overcrowding, almost constant quarreling, and poor conditions made it difficult for Angela to work at home or invite school friends home. It was not until the 1990s that the family could move to better accommodation. Angela regrets not being able to experience the more communal life of the single-story hutongs with their central courtyard that are a feature of inner Beijing. In these, elderly people mingle with children, and there is a sense of community and looking after one another.

After high school, Angela studied for four years at the Beijing Tourism Institute. She applied to this university and gained admission by passing the National Entrance Examination of Colleges. After gaining her degree, she spent six months as an intern in the state travel industry, guiding one to four people around Beijing and gaining confidence by 1998 to embark on the nationwide tours. Like other guides, she has her own personal approach, but believes that she is able to talk about any aspects of Chinese life that interest her groups. Competence as a guide is more important than political correctness. She and other guides are open about the mistakes, as well as the achievements, of the Mao Zedong era.

Further reform changes from the late 1990s affected Angela's way of life. She works on contract to a tourist company and is paid when working. Previously refused as part of the Chinese way of life, tips have become an important part of the tourist employees' incomes. She has her own pension and health care plans. In 2002 Angela bought her own apartment with modern facilities, and married a lighting engineer in 2003. If they have a child, it can be just the one.

The apartment is close to her parents. In 2003 her father retired from his work on the *Chinese Writers* magazine at age 57, three years before the normal time for office jobs. He receives the state pension of 80 percent of his magazine salary and takes part-time jobs. In 2005, aged 55, her mother retired from her job as a Chinese herbal pharmacist. Coming from a poorly educated rural background, she had to work and study hard for two years of evening classes to achieve the necessary qualification; her first job was in a hospital and her last with a wholesale pharmaceutical company. Angela's father sees the reform changes as being good on the whole for Chinese people; her mother is concerned that Angela has to pay health and pension costs.

Many young people like Angela enjoy the reform freedoms and accept the costs of having to be responsible for their own futures. However, there is also a common openness in discussing whether the reformed and open system is better than the old centrally planned version. Some aspects, such as land ownership and a (reducing) number of industries, are still centrally controlled.

- The policy of encouraging export sales and protecting home markets was a poorer motivator of economic growth than openness to trade in the world of globalization.

- Increasing global links showed the weakness of such social institutions as strong government, family control of businesses from the smallest to the largest corporations, and regulatory agencies without sufficient powers.

- Many of the profits from export earnings went into real estate speculation rather than further investment in productive facilities. Money bound up in such speculation was not available to meet a crisis.

The late 1990s crisis thus dealt a blow to assumptions that the "Asian way" would produce continuing economic expansion. Geographic factors, beginning with the development of distinctive cultures in specific places through history and interactions of people with the natural environment, bring a greater degree of diversity to this region than one simplistic prescription could change in a few years. While the Japanese economy falters, however, the Chinese keeps on expanding as fears of it "overheating" are dispelled.

Global and Local Culture

After centuries of isolation from the rest of the world, East Asian countries are now exchanging their wealth of economic and cultural activities with other parts of the world. Western cultures in particular have become widely pervasive, but local influences respond. In the political and economic realms, capitalism and democracy are dominant, influencing Chinese Communism, albeit with local distinctiveness. In the social, consumer, and artistic realms, Western influences are gaining ground in such areas as sports events, clothing, fast foods and food tastes, and housing equipment expectations. For example, brides in China today often have a Western-style wedding in white and then change into traditional dress for another reception.

Increasingly, however, cultural globalization brings reciprocal changes. The Chinese wish to excel at all they do, especially in front of audiences in other countries. China has long had its Western ballet and orchestral groups and took to soccer and athletics as well as table tennis. In each case, the Chinese contributed to a dominantly Western tradition. In South Korea, young pop stars and record companies adapt Western methods as they work on their links to the Chinese in developing a market among the younger upwardly mobile Chinese. More Western influences appear on CCTV1, the Chinese government monopoly TV station, including children's programs such as the U.S. *Sesame Street* (allowed in as a 1998 "scientific" contribution) and the U.K. *Teletubbies* (*Tianxian Baobao,* "Antenna Babes"). Moreover, the greater personal freedoms in China, together with the growth of international tourism, have brought out a religious sensitivity, a regard for the ancient ways, and revivals of traditional music, acrobatics, and Chinese opera. Such national feelings are encouraged by the Chinese government.

Human Rights

Changes are making human rights an important issue in many East Asian countries, where family responsibilities and strong governments long worked against individual freedoms. Lack of basic freedoms remains the case in North Korea, and China continues to restrict activities by many religious groups. The operation of the police and law courts still makes it difficult for people to obtain justice, but people in China now enjoy many freedoms in such areas as owning property, deciding a marriage partner, deciding where to live, moving about the country, gaining a passport for travel, and speaking one's mind. Although there are elections for some political posts, the central government leaders emerge through a complex process within the Communist Party. And yet there appears to be little opposition that questions the Party's control. Japan, Taiwan, Mongolia, and South Korea have also been through periods of political controls that limit human rights, but are now more open and enable democratic processes in the Western sense.

Test Your Understanding
5B

Summary

East Asia is in the forefront of the global economy, especially Japan, South Korea, Taiwan, and China, which produce many goods sold around the world. They are marked by increasing urbanization that causes even greater population densities but enhances the economic well-being and life expectancy of many. Two-way international tourism and movements of businesspeople emphasize these links, while the Chinese living overseas have important influence in their own countries and invest back in China. North Korea and Mongolia, however, remain poor, largely excluded from a global role by political dogmatism and poor external connections, respectively.

Questions to Think About

5B.1   Compare the economic rises of Japan and China.

5B.2   How do modern economic processes produce intensive urbanization?

5B.3   Suggest ways in which East Asian countries have responded to Western global trends.

Key Terms

megalopolis	shantytown
suburbanization	transnational Chinese economy
counterurbanization	

Subregions

Two powerful countries, Japan and China, dominate East Asia. At present, Japan's strength is economic. China's strength is political–military, cultural, and increasingly economic. Mongolia is linked to China because of its location between China and Russia. Taiwan is also linked to China, which claims possession of the island, but many Taiwanese bitterly oppose moves to reunite them to mainland China. As separate countries, North and South Korea have very different levels of global involvement but may reunite in the future. Here are the subregions:

- *Japan* remains a leader in world economic terms.
- *The Koreas, North and South,* provide contrasts of political and economic development.
- *China—including Hong Kong (Xianggang) since 1997 and Macau since 1999*—is considered with the island of *Taiwan,* which it claims, and the neighboring country, *Mongolia.*

Data for the region and its countries are listed in Table 5.1.

Japan

Japan's present government, economy, and constitution were put in place during the post–World War II U.S. occupation that lasted from 1945 until 1952. While the emperor remains a symbol of national unity, political power rests in networks of politicians, bureaucrats, and business executives. The parliament, or Diet, appoints the prime minister, usually the leader of a majority coalition.

Japan's (Figure 5.24) recovery from defeat and destruction in World War II to become the materially wealthiest country in East Asia with the second highest total annual income in the world constitutes a great economic growth success story. Following postwar reconstruction, economic growth was rapid. From 1960 to 1990, Japan's total GDP grew from 9 to 53 percent of the U.S. total and from 3.5 to 14 percent of the world total. In the 1990s and early 2000s, however, as the U.S. economy surged further ahead, the Japanese economy slowed (see Figure 5.4). Although in 2002 its per capita income (GDP PPP) remained much higher than that of China ($27,380 to $6,790), the total Chinese economic product was 50 percent higher than the Japanese.

Japanese Ethnicity

The isolation from the European influx of traders imposed by the military shogun rulers from 1650 to 1850 had a consolidating and lasting effect on Japan's ethnicity and culture. Throughout Japan's modernization and postwar reconstruction, its people maintained a traditional culture that encouraged attitudes based on loyalty to the emperor, family, and workplace. The devotion of Japanese workers to their employer's interests in exchange for jobs throughout their working lives, health care, and social support provided an alternative platform to

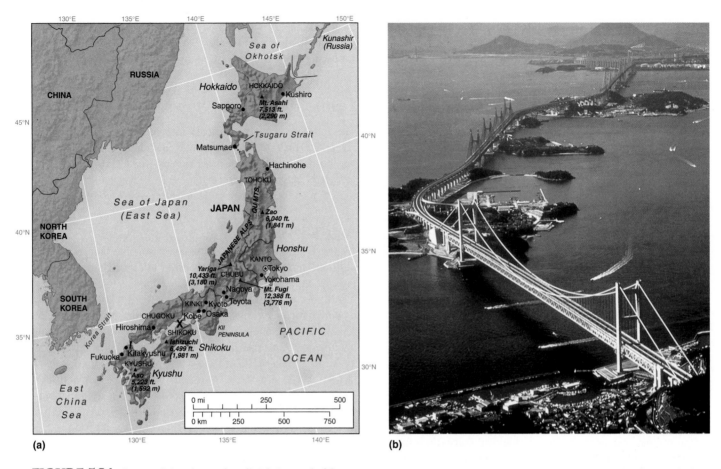

FIGURE 5.24 Japan: islands, major districts, and cities. (a) Assess from the map the significance of the mountainous interior and the long coast for Japanese geography. (b) Many islands are linked by bridges and tunnels. The Seto Great Bridge links the islands of Shikoku and Honshu (x on map). **Photo:** By permission of Japan's Ministry of Foreign Affairs.

Western motivations for competing in the global economic system. In the late 1990s, however, Japanese employers canceled or modified such agreements in the face of pressures on business costs from newly emerging countries where wages and support costs were less.

The main Japanese traditions, from Shintoism to Buddhism and Confucianism, agree on the subordinate status of women. Although women's status is changing with their greater education and involvement in the labor force, they still work for lower wages, and few attain management positions or lifetime employment status.

Three minority ethnic groups remain important. First, the Ainu in Hokkaido retain ethnic and language character despite Japanese dominance since the 1800s. Second, on the largest island of Honshu, the Burakumin, numbering around 1.2 million people, were outcasts and allowed to work only in the lowest occupations until their emancipation in 1871. It was not until 1969, however, that the Japanese government put funds into upgrading their settlements to nationwide standards. Third are foreign nationals, three-fourths of whom are Koreans, moved forcibly to Japan in the early 1900s. Until the 1980s, they found it difficult to gain Japanese citizenship and are still subject to discrimination.

Migration into Japan is restricted, and only 1 percent of the population is of foreign origin. Concern over the ethnic background of foreign workers at a time of demand for labor in the 1970s led Japan to recruit 140,000 workers from the Japanese populations of Peru and Brazil—the *Nikkeijin* (ethnic Japanese born overseas). They look Japanese but are culturally Latin American, spoke little Japanese on arrival, and had different behavior patterns. Japan also allows the immigration of foreign brides for men left behind in rural areas by the migration of women to urban jobs. Such "mail-order brides" come from the Philippines, Vietnam, and Thailand. Seldom speaking Japanese, they take up a life of extreme social loneliness.

Japanese Identity

Defeat in World War II brought huge challenges to the Japanese people and their views of themselves and their country. They were invaded by global (American) economic, political, and cultural interests. A U.S.-style constitution was imposed. A degree of peace and prosperity brought economic growth, based first on major orders during the Korean War (1951–1953) and then on access to the U.S. markets. By the 1980s the

Japanese economy was second only to the United States, and the focus on exporting manufactured goods while resisting imports brought a huge trading surplus and massive bank deposits.

Their prosperity gave Japanese people pride in exporting goods to foreign markets, in taking over foreign businesses including much of Hollywood, and in seeing "just-in-time" production methods adopted widely. However, the 1990 stock market and 1993 land and property crashes renewed the tensions between internal and external forces. Major Japanese companies faced bankruptcy and downsized their management capability. They laid off staff, who had expected a lifetime job and for whom this brought a major loss of face. They adopted U.S. attitudes to employment and the running of corporations. This crisis of identity came when the traditional authority of family and school was also challenged.

Through the 1990s and into the early 2000s, public figures such as authors, popular comic-strip writers, and the mayors of some major cities raised nationalistic issues. These included the use of the rising sun flag, the role of the Self Defence Force (and forbidden wider military development), and the U.S.-mandated postwar constitution. The shift to Western production and staffing policies by major corporations included the takeover of Nissan by French company Renault. Events such as the Gulf War became significant to the sensitivities of Japanese national pride. In 1991, forbidden by its constitution from providing military resources to the international consortium freeing Kuwait after the Iraqi invasion, the Japanese government

offered a huge sum to help pay other countries. Many Japanese were upset when their country was unfairly, but widely, derided for taking the soft option of paying instead of sending soldiers.

Economic Development

The Japanese economy changed massively after World War II. From 1960 to 2002, farming, fishing, and mining—together forming the primary sector of the economy—declined from 33 to 1 percent of output; manufacturing rose to 30 percent before falling to 21 percent; and the service sector rose from 38 to 68 percent.

Agriculture

Japanese farms occupy small areas of suitable land within the hilly and mountainous country, often intermingled with housing and industry (Figure 5.25) that compete for the land and pay higher prices than farmers could afford without subsidies, by which the government maintains agriculture as a significant sector of the economy. After 1945 the Japanese government bought rice at a price that enabled the farmers to mechanize, and tariffs protected them from world market competition. New tastes for Western foods, however, halved rice demand after 1960. The rice that could not be sold formed a stored "mountain" that was too expensive to export. Rice acreage fell as farmers converted land to cultivate vegetables and fruit, reducing the rice store. Rice still costs Japanese consumers six times what Americans pay and 13 times what Thais pay.

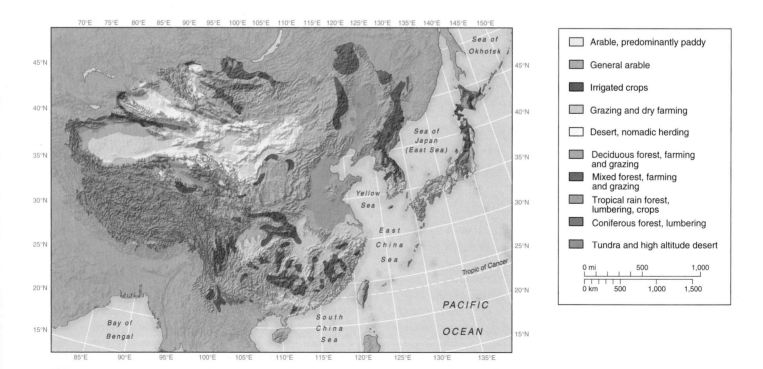

FIGURE 5.25 East Asia: farming and other land uses. Notice the contrast in intensity of production between eastern and western China.
Source: Data from *New Oxford School Atlas*, Oxford University Press, UK.

Although subsidies made farm incomes secure, migration from rural to urban locations for the better-paid jobs, better education opportunities for children, and social life halved the rural population from 1960. After 1970, only 12 percent of farmers worked full-time without off-farm jobs. The remaining farmers were often aging and found it difficult to compete in a world of declining subsidies and rising foreign competition. The size of most farms remained stable, however, because land increased in value and many owners held on to their property as a family investment.

In the early 2000s Japanese farmers faced new shocks. Improved transportation links and refrigeration brought large quantities of Chinese tomatoes, eggplants, onions, and garlic bulbs to Japanese supermarkets. For example, on Tokyo's wholesale markets Chinese shiitake mushrooms sold for less than one-third the price of the Japanese varieties and quickly took a 40 percent share from Japan's 30,000 shiitake growers. Forest products and the bulrushes used for making tatami mats are also under pressure from cheaper Chinese sources.

Manufactures: Reconstruction

After dismantling Japan's World War II industrial base, the United States poured money into the country to rebuild its industries and infrastructure. The Cold War beginning in the late 1940s, the Communist takeover in China, and the Korean War of the early 1950s resulted in American markets opening to Japanese products. The cheapness of raw materials on world markets and its own low-cost, young labor force enabled Japanese manufacturers to undercut their rivals in the world's wealthier countries and gain overseas markets. The Japanese government assisted and advised industry, with the **Ministry of International Trade and Industry (MITI)** encouraging export sales through a worldwide network of market intelligence gathering offices. Following the rebuilding of heavy industries such as steelmaking and chemicals in the 1950s, Japan diversified into shipbuilding, automaking, and light manufactures that were demanded by consumers at home and abroad. In the mid-1950s, when the Suez Canal was closed and Western companies sought cheaper ways of transporting oil, Japan made timely transportation innovations such as building very large oil tankers and bulk carriers for other commodities. Japanese-built bulk tankers and carriers brought home oil from southwestern Asia and coal and iron ore from Australia.

During the 1960s, total output from the Japanese economy increased by 10 percent per year. At this stage, Japanese workers waited for the gains in prosperity as their firms plowed profits back into new developments. It was not until the 1970s that Japan established a favorable trade balance and paid better wages so that its own consumers could buy a range of Japanese-made products. Import restrictions discouraged foreign-made products.

Manufactures: Increasing Costs and "Just in Time"

In the 1970s labor shortages, increased oil prices, a world shipping slump, and reduced demand for some products caused Japanese corporations and their supportive government to rethink policies. For example, as Japan depended on (now more expensive) imported oil for 70 percent of its energy needs, the Japanese government reduced the costs of internal hydroelectricity and nuclear power and subsidized unprofitable coal mines.

By the 1980s the older industries such as steelmaking, shipbuilding, petrochemicals, and cement making suffered from overcapacity on world markets and competition from newly industrializing countries such as South Korea. Factory closures led to social problems resulting from unemployment, particularly around Osaka.

However, responses to the changes took Japanese industries into a new era. Greater investment raised the output of light industries such as those producing cameras and household appliances. After copying others' technologies and designs, Japanese firms became initiators. The increased application of technology brought productivity and international competitiveness gains in industries including electronics, robotics, and new materials. By 2000 Japan had half of the world's industrial robots, four times the number installed in the United States. Careful control of inventories by "just-in-time" deliveries cut warehousing costs in assembly industries and freed more factory space for production. Toyota—a company that led the way in technologies of production—found that the indiscriminate use of robotics and automation was too expensive. The company moved to a system that involved robotics where appropriate but also employed the greater use of people skills in making cars.

Japan's successful exports became known for their quality, reliability, and market-leading technology. In the 1980s other countries built huge deficits of payments to Japan, causing the Japanese currency, the yen, to double in value compared to other international currencies. This made Japanese exports more costly for other countries to buy. Imports became cheaper, however, so that the Japanese spent more on products made abroad as well as those made in Japan. The major boom in retailing was joined by another in the construction of new houses and apartment blocks. Japan turned from a country of producers into one of consumers. As demand for more luxury and prestige goods rose, Japanese manufacturers moved into these fields, competing with imported goods. Japanese firms, especially those involved in electronics or automobile production, formed conglomerates known as *keiretsu*. These industrial groups, built around financial nuclei that usually included a major bank, produced giant trading and manufacturing companies.

Manufactures: Global Investments

From the 1980s, the Japanese corporations with large yen trading surpluses invested more overseas. They first bought out their foreign mineral suppliers and later established Japanese factories overseas. Building large productive units in other countries during the 1980s and 1990s also helped them penetrate markets that had previously placed quotas on the import of, for example, Japanese-made cars. In the late 1980s, after major investments in the United States and Britain to establish 14 Japanese auto plants, Japan switched investment to Southeast

Asia. The economic growth of Thailand, Malaysia, and other countries in Southeast Asia received a major stimulus as Japanese investments poured in.

From the mid-1990s, a further phase of Japanese investment abroad expanded previous investments in cheap-labor, assembly-line factories by moving design and research facilities to Southeast Asian countries. Japan also made increasing investments in China, particularly in the northeastern port of Dalian and other areas close to Japan. By the early 2000s most Japanese multinational corporations such as Sony, Hitachi, Toshiba, NEC, and the automakers Toyota, Honda, Mazda, and Nissan used parts and whole products manufactured and assembled in China. Chinese manufactures, often with Japanese corporate badges, began to flood the Japanese markets at home and overseas. Japan's fragile financial system had another economic factor to cope with. Japan now has a huge trade deficit with China.

Renewed Economic Challenges and Services Growth

By the early 2000s Japan again faced a need to change. It suffered from falling prices and business collapses, its established systems of government–business linkages were undemocratic and increasingly inefficient, and its banking systems came under pressure because of huge debts, particularly resulting from investments in Southeast Asia. While Western nations imitated Japanese production processes and competed against Japanese goods in world markets, the Japanese looked more closely again at Western ways, particularly in relation to financial control.

The huge trade surplus from exports of manufactures resulted in the rapid growth of Japan's financial and business services. In 1986, as well as being the center of government administration, Tokyo became the world's second financial center after New York and in front of London. The additional functions led to further increases of population in the Tokyo region.

Throughout Japan, the increase in service occupations, from education and health care to retailing and tourism, widened the range of employment and contributed to a diversified economy that is not so dependent on the fluctuations of world markets as when manufacturing dominated it. Japan's popular culture ranks high among the country's latest exports. Revenue from royalties and sales of Japanese music, video games, anime, comics (*manga*), films, and fashion rose 300 percent over the last decade, garnering the country $12.5 billion in 2002.

Japan's Regions and Cities

Not all of Japan is equally prosperous. Japanese physical environments—from snowy Hokkaido in the north to subtropical Okinawa in the south and from the volcanic mountain spine to the coastal lowlands—combine with distinctive contrasts in urbanization, manufacturing emphases, and rural ways of life. Japan consists of four main islands: Hokkaido, Honshu, Shikoku, and Kyushu (see Figure 5.24).

Hokkaido

The northernmost island of Hokkaido has over 20 percent of the land but only 5 percent of the Japanese population and produces just 3.8 percent of the total GDP. The island's previous inaccessibility from the rest of Japan was reduced by a direct rail link through the Seikan tunnel under the Tsugaru Strait and better air services. The improved access encourages farmers to grow new breeds of rice and produce dairy products and crops linked to the increasing Japanese demand for Western-style foods. Farm products now exceed the value of Hokkaido's traditional natural resource–based industries (fishing, coal mining, and forestry). The 1972 Winter Olympics at Sapporo—now Japan's fifth-largest city—gave a boost to tourism on the island. The recession of the late 1990s, however, affected this region badly, with many bankruptcies and increasing unemployment.

Honshu

The main island of Honshu has three parts. Northern Honshu (Tohoku) has a relatively sparse population for Japan, but its density is twice that of Hokkaido. There is better transportation infrastructure, although the central mountains make crossings between the two coasts difficult. This region is Japan's leading rice producer, with 25 percent of the total, much on reclaimed coastal lands.

In the early 2000s, the economic problems facing Japan brought particular hardships to northern Honshu. Always a vulnerable area with little manufacturing employment, its main sources of income and jobs in fishing and farming suffered from declining stocks of fish and price competition from overseas suppliers. The main fishing port of Hachinohe had newly opened pinball stores that were full of unemployed fishermen and workers from idle fish-processing plants. The apple orchard owners had to diversify because the price of apples was so low, while Chinese produce undercut the local garlic growers. Public money financed infrastructure projects rather than local industries. Workers in short-term construction jobs had to move elsewhere to continue employment.

Central Honshu includes the Tokyo region (Kanto). It is the heart of Japan in terms of population concentration and economic activity: 31 percent of the population produces 37 percent of Japanese GDP on less than 10 percent of the land. Tokyo, Kawasaki, and Yokohama fuse into a huge urban complex around Tokyo Bay (Figure 5.26). Tokyo itself, having 35 million people in 2003, is the world's most populous metropolitan area. Tokyo is Japan's center of government, finance, and internal and international service industries; Yokohama is Japan's busiest port; and Kawasaki has heavy industries. Oil refineries, steel mills, chemical plants, and power stations line the waterfront and cover large sectors of reclaimed land. The

FIGURE 5.26 Japan: Tokyo–Yokohama–Kawasaki urban area. Sukibayoshi pedestrian crossing, Ginza. **Photo:** © Vol. III/Corbis.

rest of the plain around this megacity still produces much rice, together with fruit, vegetables, poultry, and pigs for the nearby markets. Volcanic mountains around the northern and western margins of the bay area have winter sports facilities, and their hot springs attract other tourists.

The Pacific coastal area south of Tokyo is an industrial corridor of connected lowlands where a line of cities along the "bullet train" (*shinkansen*) and major highway routes urbanized former agricultural land. Nagoya is Japan's third largest city and the center of a textile region that now has automaking (Toyota City), oil refining, petrochemicals, and engineering industries. This region's diversified and growing industries suffered least in the late 1990s recession.

West and north of the Tokyo complex, the land in Chubu is more mountainous. The northern coastlands facing the Japan Sea are another rice-growing district. Mountains separate it from the main economic areas of the country, although highway and rail links penetrate these Japanese "Alps." The 1998 Winter Olympic Games were held at Nagano in this area, which receives winter snow from winds crossing the Sea of Japan. The mountains include national parks, such as that around Mount Fuji, Japan's highest peak (3,776 m, 12,388 ft.), and have hot springs. Chubu leads the country in generating hydroelectricity.

Southern Honshu contains Japan's second most prosperous region, Kinki, which includes Osaka and its immediate hinterland. Three cities dominate the eastern part of this region. Kyoto is Japan's sixth largest city and was the country's capital from A.D. 794 to 1868; it remains a cultural and tourist center and has craft industries such as silk, pottery, and traditional furniture. Osaka, Japan's second largest city, and Kobe are industrial ports on

Osaka Bay. Following an early specialization in textile manufacturing, the area shifted into iron and steel, chemicals, and shipbuilding in the 1920s, adding oil refining, petrochemicals, automaking, and domestic appliances after 1945. Reclaimed land from the Inland Sea became the site of port facilities and heavy industry using imported raw materials. In 1995 the new Kansai airport was opened, built on reclaimed land south of Osaka, but it is the subject of worries about subsidence of the fill materials and the threat of rising sea levels as a result of global warming. This part of southern Honshu became Japan's "rust belt" of declining industries in the 1980s, and it was hit hard in the late 1990s recession. In 2001 the area around Osaka and Kyoto had Japan's second highest unemployment rates of 6 to 7 percent (compared to a national average of 5 percent).

The western peninsula of southern Honshu (Chugoku) has coastal lowlands on either side of a mountainous spine. While a harsh winter climate and poor links with the rest of Japan mark northwest-facing coasts, the southern coast looking to the Inland Sea has a better climate. The latter area responded economically to improved accessibility by train and road after 1945.

Shikoku

Shikoku is a mountainous, small island with only 5 percent of Japan's land. The Pacific side of the island has a warm, moist climate, but mountains isolate it from the northern coasts facing the Inland Sea that are industrialized and irrigated for crops. Major bridges now connect the island with the rest of Japan.

Kyushu

Kyushu, the westernmost island, is connected by tunnel to Honshu. It was the historic point of entry to Japan for Buddhism and early Christianity, new writing script, and trade. Two-thirds of the island's population lives in the northern region around Fukuoka and Kitakyushu, both having long-established metal, chemical, and shipbuilding industries. These began to decline, but in the 1980s, Kyushu attracted automaking and high-tech industries with expanding research and development facilities.

Farming and fishing remain important in the economy. Much of Kyushu is a rice-growing area. It is warm enough for a second crop of vegetables or rushes, while early crops are grown for the large cities farther north. Mandarin oranges, tobacco, and livestock are produced on the hills. There is a growing tourist industry.

Ryukyu Islands

Offshore, the Ryukyu Islands, including Okinawa, stretch for 1,000 km (650 mi.) southward toward Taiwan (see Figure 5.3). Their economy has not developed with the rest of Japan, and they have high unemployment. Many still have American military bases alongside the traditional fishing and agriculture, and a tourist industry is developing. There are pressures for the removal of the U.S. military personnel, although most of the remaining U.S. troops are based on Honshu.

The Koreas

North Korea and South Korea (Figure 5.27a)—officially the People's Democratic Republic of Korea and the Republic of Korea, respectively—occupy the hilly Korean peninsula that extends southward from the Chinese border along the Yalu River. The peninsula almost touches Japan at its southern end. The two Koreas shared a common ethnicity and history to the mid-1900s but after that moved in different directions.

At the end of World War II, the United States and Soviet armies defeated Japan and divided the "freed" North and South Korea at the 38th parallel. The Soviet Union supported North Korea, and the United States supported South Korea. Independent republics were created in 1948. North Korean forces invaded South Korea in 1950. The United Nations sent military under U.S. commanders to support South Korea, and, following destruction over much of the territory as warfare moved back and forth, the two sides agreed on an armistice in 1953. The agreement resulted in the formation of a Demilitarized Zone between the two countries (Figure 5.27b). This strip of "no man's land" is flanked on both sides by heavily militarized areas. Tensions between the countries remain, causing them to maintain a high level of defensive capabilities. North Korea possesses the ability to make nuclear and biochemical warheads.

By the late 1990s there were signs of closer ties developing between the two countries. Presidential visits led to reunions of Korean families separated for 50 years. Talk of reunification of the two countries gathered some momentum, but both countries had reservations about such a move. North Korea did not wish to lose its independence, while South Korea counted the potential cost.

Countries

With huge support from the United States after the Korean War, South Korea became one of the newly emergent urban–industrial countries linked to the West and the global economic system. South Korea's governments ranged from paternalistic democracies to dictatorships, using absolute power to promote personal savings and encourage export industries. By the 1980s and 1990s, such policies led to rising prosperity for many and large trade balances for the country. Freer elections introduced from the mid-1980s led to increases in democratic processes during the 1990s. South Korea's financial sector, together with its mode of industrial organization in large conglomerate firms, or *chaebol,* threatened to weaken much of the country's economy and political stability in the late 1990s, but attracted successful rescue measures.

North Korea maintained a Communist regime under the personal dictatorship of Kim Il Sung and his son Kim Jong-il, who succeeded in 1994. Supported by the Soviet Union and China in the Cold War period, North Korea's isolation from the rest of the world and suspicions of South Korea and the West

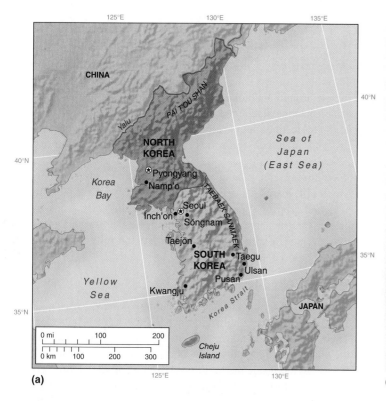

FIGURE 5.27 The Koreas, North and South. (a) After the history of Japanese occupation in the early 1900s, the Koreas prefer the term "East Sea" to "Sea of Japan" but have not convinced others to change the name. (b) The Demilitarized Zone. A bridge crossing the frontier river has been destroyed and trees cut to maintain an open-vision area. **Photo:** © Alasdair Drysdale.

caused it to invest heavily in military capabilities instead of economic development. After the loss of Soviet Union subsidies in 1991, North Korea depended increasingly on international aid to make up for internal economic collapse and famine. Its main bargaining point is its possession of nuclear and biochemical weapons and ballistic missiles to deliver them. North Korean brinkmanship policies attract international aid to prevent military action.

South Korean Economic Development

Reconstruction and Economic Growth

After World War II and the Korean War (1950–1953), South Korea established import-substitution industries that developed manufacturing skills and reduced the need to pay hard currency for foreign goods. The transportation infrastructure built by the Japanese when they ruled Korea in the early 1900s and rebuilt with external funds after the 1950s war helped develop the Korean economy. South Korea built a new series of major highways, and its public transport system is one of the world's best.

The devastating Korean War in the early 1950s took place when three-fourths of the South Korean people still gained a living from farming. In postwar land reforms, the government took over large, inefficiently run estates and encouraged a new group of small landowners to apply imported fertilizers. Farm productivity rose, and South Korea became largely self-sufficient in food. The farming and rural emphasis gave way to manufacturing, services, and urban living. A country of subsistence farmers under largely feudal control was transformed in a generation to the world's main maker of large ships and memory chips, the fifth largest automaker, and the eleventh largest world economy—with greater output than the whole of Africa South of the Sahara. From the early 1960s to the early 1990s, few other countries' economies grew so fast. Exports rose from $33 million in 1960 to $162 billion in 2002. South Korea's total economy (GDP PPP) is now 13 percent of China's and 24 percent of Japan's.

Manufacturing and the Chaebol

The iron and steel, shipbuilding (Figure 5.28), chemicals, automaking, and textiles industries established from the 1950s continue to be important, although in the 1990s there was a swing toward high-tech industry. In this development, the South Korean government supported the family-owned companies that developed into huge conglomerates—the *chaebol*, such as Hyundai, Daewoo, LG (Lucky Goldstar), and Samsung. The *chaebol* became diverse in their products, starting up subsidiaries that often had little chance of success, taking on massive debts, and providing major problems for the country. However, when Halla, the twelfth largest *chaebol*, went bankrupt in 1997 with debts of 20 times the company's assets, Hyundai, the largest, was found to have guaranteed around 15 percent of the debts. This highlighted the fact that the founders of the two *chaebol* were brothers and had strong cross-company links. Family loyalty—part of Korean (and Confucian) culture—was exposed as making problems for running large

FIGURE 5.28 South Korea: Ulsan City shipyard. The Hyundai *chaebol* is a major builder of large ships. Note the space required for assembling the materials and the range of equipment used in construction. **Photo:** © Vander Zwalm Dan/Corbis.

corporate businesses in a free market that demands increasing transparency. The *chaebol* also controlled large parts of the South Korean financial markets, providing an environment for building up debts that were shared through many links.

As an example of *chaebol* history, South Korea was once called the "Republic of Hyundai" after one of its most powerful *chaebol*. After the 1997 crisis exposed its huge debts and damaging grip on the economy, Hyundai was restructured around several independent companies. Hyundai Motor, however, is the only one that makes profits, driven by its U.S. sales. Roh Moo-Hyun, who became president of South Korea in 2003, pledged to make the *chaebol* more accountable. In this he has the support of young reformists, who wish to change the country's conservative, business-driven political system to one that is more liberal.

South Korea was a major investor in modern China in the early 1990s, taking advantage of cheap labor costs in the nearby Chinese province of Shandong. The initial rush of small companies to invest in Chinese locations prepared the way for major South Korean corporations such as Daewoo, Samsung, and Hyundai, which link to hundreds of smaller companies producing their components and products.

In late 1997 indebtedness reached the point where the South Korean currency unit, the won, and its stock exchange collapsed. Devaluation of its currency increased South Korea's indebtedness to other countries. Bank closures, bankruptcies, and unemployment stirred militant trade unions into political action against the stringent conditions proposed by external loan providers. By the early 2000s the South Korean economy had largely recovered, helped by the breakup of some family-owned *chaebol,* and several of the new firms merged with U.S. and European corporations.

Developing Broader Interests

In 2002 the South Korean government signaled its intention to develop the island of Cheju-do (Jeju) off its southern coast. The island's natural environmental potential includes its pleasant climate and attractive scenery, from an extinct volcano to many waterfalls and beaches. It is a traditional honeymoon destination, center of tangerine plantations, and home of professional women divers who pluck seafood from the ocean floor. The project, to be completed by 2010, symbolizes South Koreans' sense of vision, their appreciation of the country's geographic position, and probably their overly optimistic hopes. South Korean officials managing the project make much of their country's central position between Japan and China and the government's intentions to make business friendlier. Cheju is two hours' flying time from Beijing, Tokyo, and Hong Kong. It is proposed to turn the island into an international business and tourism hub akin to Hong Kong and Singapore, with a free-trade area in which English will be a second language. Overseas universities could set up campuses alongside a science and technology park, duty-free shopping malls, and 20 golf courses. The South Korean government is committed to paying half the projected costs of the project, which will require a new airport, better roads, and more hotels. It depends, however, on large investments from the private sector. Moreover, the local population of half a million complains about the lack of consultation and worries about the environmental impacts.

Although mostly noted for its economic growth, South Korea also has an increasing regional contribution in popular music that is both cultural and economic. Young pop stars such as Boa develop alongside other youth culture inputs in movies, TV shows, computer games, and fashion. This trend relates partly to South Korean consumers preferring local products to the invading Western music and movies. It is also part of South Korea's development of knowledge-based service industries. South Korean "K-pop" bands top charts in Taiwan and Cambodia, while Korean TV dramas are highly rated in Vietnam. China, with its huge and growing market for music, is a prime target for record companies, which envisage setting up an academy for discovering young Chinese pop stars. The main problems for a South Korean venture in China are the rogue bands copying songs and the rampant piracy and copying of CDs. The 2002 World Cup soccer finals, cohosted in South Korea and Japan, provided a further platform for wider penetration of both Western and South Korean products.

Economic Stagnation in North Korea

When the Japanese ruled the whole of Korea in the early 1900s, the northern part provided iron, coal, and raw materials for a chemical industry and had plenty of hydroelectricity potential. However, from its creation in 1948, North Korea has been an inward-looking country with an economic policy advertised as isolationist self-reliance. This had a historic precedent when Korea was called the "Hermit Kingdom." In fact, the new country depended on aid and good prices from China and the Soviet Union. These arrangements ceased in 1989. Factories became idle, and people lacked proper nutrition. By the early 2000s the GDP per capita in North Korea was one-eighth of South Korea's. In 2000 the death rate rose 50 percent higher than in 1994, indicating widespread famine and disease. The North Korean economy collapsed, making the country and its government dependent on international aid, of which the United States is the largest contributor. North Korea became the world's largest recipient of famine relief. The best food produced in North Korea feeds the army and party officials, while other people depend on donated food, often of lower quality.

North Korea faces a difficult period. After years of a declining economy, the people, duped by hype from their leaders or cowed by fear, do not appear likely to revolt. But they have few skills or resources to attract external foreign investment. North Korea faces three alternatives. It could continue its path to greater poverty and eventual collapse, be reunified with South Korea—although that country views the prospect with unease after Germany's experience incorporating its eastern area (see Chapter 3)—or follow a path of economic liberalization similar to that of China, take offered aid in exchange for reducing its nuclear program, and encourage investment from foreign countries. Some South Korean multinational corporations are poised to capitalize on low-wage labor and the markets for its goods in North Korea. Many South Korean families with relatives in North Korea wish to help them take advantage of better economic conditions. The North Korean government, however, has avoided following China's lead because its members fear losing power.

Test Your Understanding
5C

Summary

Japan is small, densely populated, and island-based. It became the world's second wealthiest country following a dramatic economic recovery after 1945. Japan's agriculture remains highly protected from foreign competition. In the later 1900s its industrial economy expanded from dependence on importing raw materials and producing goods for export markets. It moved from iron and steel manufacturing to making autos and high-tech goods. In the 1990s the rising value of the yen, caused by selling more exports than buying imports, led to competition in world markets from cheaper sources. This caused a slowing of Japan's economy and its investment in other countries. It also led to a nationalistic mood that is popular with the proud people.

South and North Korea have the same ethnic composition of Korean people but contrast in the outcomes of events over the last 50 years following the destructive war of the early 1950s. South Korea received help with reconstruction after the 1950s war from the United States in particular. The United States opened its markets, and South Korea became one of the world's growing urban–industrial centers. Its economic wealth brought increased standing in the world and in East Asia. North Korea retained its Communist government, backed until 1989 by economic and political links to China and the Soviet Union. In the 1990s North Korea became impoverished and was under suspicion by other countries of developing nuclear and biochemical weapons.

Questions to Think About

5C.1   In tracing the history of economic growth in Japan from the mid-1800s to the present, how would you relate the events to global connections and local voices?

5C.2   What are the main regional differences within Japan?

5C.3   Explain the economic growth of South Korea.

5C.4   Imagine a conversation at an occasional meeting of relatives living in North and South Korea. Include comparisons of living conditions and the roles of each country's government.

Key Terms

Ministry of International Trade and Industry (MITI)

China, Mongolia, and Taiwan

The People's Republic of China (PRC) is the world's most populous country, with 1.3 billion people in 2005, and the third largest by area (Figure 5.29). Hong Kong (Romanized Cantonese for the name, or Xianggang, the pinyin version of Mandarin Chinese) and Macau, previous colonies of the United Kingdom and Portugal, respectively, were returned to PRC control in 1997 (Hong Kong) and at the end of 1999 (Macau). Mongolia (2.6 million in 2005) and Taiwan (22.7 million) are tiny by comparison with the PRC but have distinctive roles in relation to China and the rest of the world. These two countries are discussed together after the section on China.

China

After the end of World War II, China experienced violent political, economic, and cultural shifts. Today it is marked by stable government, but rapid changes in an expanding economy, and a focus on its national strengths and traditional culture in the modern setting.

Political, Economic, and Social Ups and Downs

From 1800 to 1950, China declined from producing 33 percent to 5 percent of world economic output. After 150 years of divisions among competing warlords and ideologies and the disruption of the Japanese invasions in the 1930s, Mao Zedong, the leader and dictator of China from 1949 to his death in 1976, wished to make rapid and massive changes. The People's Republic of China's first aim was to turn a backward, feudally divided agricultural country into a united, advanced, and centralized socialist industrial country. From 1949 to 1959 Mao established central planning, mainly through the collectivization of agriculture in communes. The "Great Leap Forward" of 1959 to 1962 and the "Cultural Revolution" from 1966 to 1976 attempted to change the economy and society in revolutionary ways but caused economic and social devastation. Mao died in 1976. Mao's successor, Deng Xiaoping, brought in reforms that abolished most of the communes and engaged with the global economic system. His saying, "It doesn't matter whether it's a black cat or a white cat, as long as it catches mice," enabled the adoption of more open policies in the economy.

Ethnicity in China

The **Han Chinese** make up 92 percent of the Chinese population. They are of similar appearance to most of the national minorities but derive their ethnicity from the emergence of the common Chinese administrative culture in the A.D. 200s and 300s. Most non-Han groups did not have their own writing systems and adopted the Han Chinese system. The Han culture diffused by conquest and movements of people to administer new areas. More recently, Han people moved into the northeast after being excluded until the late 1800s and, from 1949, into westernmost China.

From 1949, the Chinese government encouraged the greater use of the northern Mandarin language that is spoken in north central China and Beijing alongside the retention of minority languages. Minorities, such as the Mongols in the north, the Tibetans and Uighur in the west, and the more dispersed Manchu and Muslim Hui, were allowed limited jurisdiction over their affairs, particularly the expression of their culture in music and drama, in autonomous regions and counties within the main provinces. Local resentment of the Chinese takeover of these lands, however, causes the eruption from time to time of anti-Chinese demonstrations in Tibet and Xinjiang.

Most minority groups live along the northern borders with Russia and Mongolia and in western China. Central Asian ethnic groups dominate the north and northwest, while Tibetans are the main group in the far west. Tibet was taken under Chinese control again in 1950 , but, following a 1959 rebellion, the PRC imposed Chinese economic, social, and political control. Despite large subsidies, the material condition of Tibetans did not improve after the Chinese occupation, and the Cultural Revolution was a disaster for Tibet with further bloody repression. The former Tibetan Dalai Lama lives outside Tibet, but his son recently succeeded him under an official Chinese move.

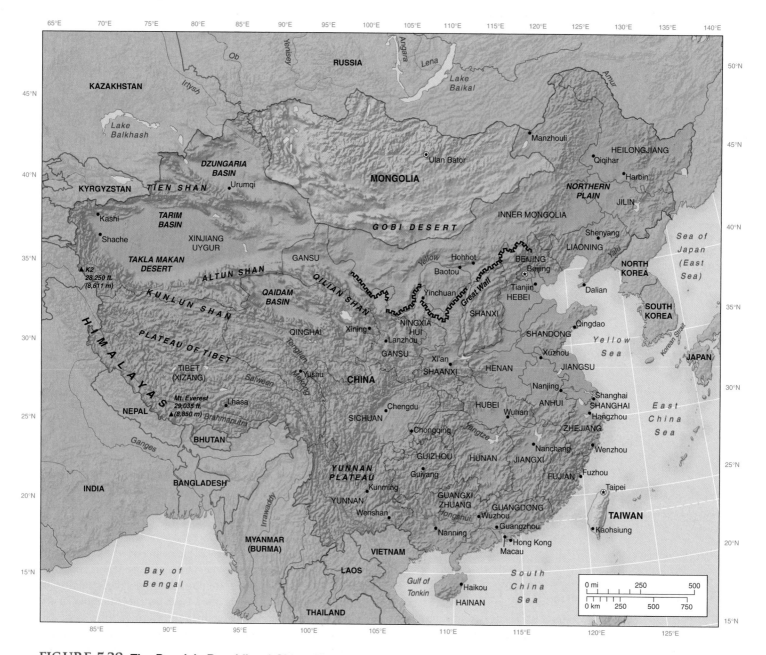

FIGURE 5.29 The People's Republic of China, Mongolia, and Taiwan: major geographic features, including towns and cities, major rivers, and provincial boundaries. China has 23 provinces (capital letters), four cities with provincial status (Beijing, Shanghai, Tianjin, and Chongqing), and five autonomous regions that have large national minorities (Tibet, Inner Mongolia, Guangxi, Ningxia, and Xinjiang). The Yellow River is also known as the Huang He and the Yangtze River as the Chang Jiang.

Xinjiang Province, formerly known as East, or Chinese, Turkestan, is linked culturally to Central Asia (see Chapter 4), but China reasserted its control of Xinjiang in 1949. Han Chinese were resettled there, increasing from 5 to 38 percent of the 18 million people in the province today. The local Uighurs resist the PRC intrusions by occasional terrorist action. Local autonomy is unlikely, however, as China develops relations with Central Asian countries, extending the rail system to Kashgar and the border, and invests in a pipeline to take oil and natural gas from Xinjiang eastward to Shanghai.

About 2 million ethnic Koreans live in northeast China, mostly in the Yanbian Korean Autonomous Prefecture. They are credited with introducing rice paddy farming to this area in the 1800s. However, former languages were replaced by Mandarin Chinese.

Collectivization and Communes

China modeled its initial revolutionary changes on the Soviet Union experience. Large-scale industrialization and the collectivization of agriculture were the first priorities. Two-thirds of available investment went into manufacturing, where output rose rapidly, but hardly any into agriculture.

Although little investment went into improving agriculture, rural social structures were revolutionized. Before 1950 most Chinese were tenants or landless, working tiny parcels of land. The landlords or rich peasants who owned over 70 percent of the cultivated land made up fewer than 10 percent of the population. During the early and mid-1950s, central government took over land ownership and in theory returned control to local people. The process involved several steps:

- In 1952, 300 million landless peasants received their own plots of land, houses, implements, and animals, free from debt.

- By 1954, over half of these peasant farmers joined "mutual aid teams." Each team comprised up to 10 households, and the owners pooled labor, tools, and work animals at planting and harvest. Such teams often became permanent, grouped in larger agricultural producers' cooperatives. By early 1956, over 90 percent of the rural population joined cooperatives.

- Next, land was pooled under central management, with members of the cooperatives receiving income based on their land and labor inputs. In "advanced" cooperatives, the land became the property of the organization apart from tiny garden plots for each family. A typical advanced cooperative included approximately 150 households farming a total of around 400 acres of land. This process, known as **collectivization,** was completed by late 1956.

- Finally, the government established communes. Each **commune** combined agriculture, industry, trade, education, and the formation of a local militia. The commune leadership planned the agricultural year, other economic activities, and social functions. They guaranteed food, clothing, education, housing, and arrangements for weddings and funerals to members. By September 1958, communes included 98 percent of peasant households. Although some communes in favored areas prospered and provided good schooling and health care, few achieved the expected potential economies of scale based on sharing equipment, buying seed and fertilizer in bulk, and combined production agreements. While some managers attended to community welfare, many became unreasoning dictators. The commune experiment achieved a degree of communal working, but local arrangements were often interrupted by political upheavals. Commune life failed to motivate the peasant farmers, for whom political propaganda and punishments never generated the same commitment as individual incentives.

The Great Leap Forward and Cultural Revolution

The **Great Leap Forward** of the late 1950s aimed to increase the rate of industrialization by producing basic industrial products, such as iron and steel. It attempted to integrate the rural and urban areas by setting up small-scale production units in both. Throughout much of China, community efforts devoted to iron smelting, often in small backyard furnaces, caused a neglect of farmwork and worsened the drought-based famine of 1960, in which 25–30 million people died. The plan was a disaster. Both industrial and agricultural production dropped, and China was forced to import wheat from Western countries. Stung by failures and misunderstandings, the Chinese leaders abandoned most Great Leap Forward policies.

After a few years of higher production, the country once again turned inward as Mao Zedong attempted to change the whole basis of Chinese society and its administration. The **Cultural Revolution** of 1966 to 1976 diverted attention from increasing economic productivity to protests against unchanging local authorities and leadership cliques. Mao Zedong's "Little Red Book" summarized his wish for revolutionary actions to purge society of its traditions as well as links to the West. The Red Guards, composed mainly of young adults, carried out his tenets, destroying management structures at all levels of society and trashing cultural relics. Such fear-based tactics kept Mao Zedong in power until his death. His policies, however, swept aside many of the short-lived economic gains of the early 1960s in a massive persecution of the educated classes, technologists, and former leaders.

China Joins the Wider World

After Mao Zedong died in 1976, his former chief deputy, Deng Xiaoping, redirected China's policies. His "open door" policy, based on China accumulating wealth in favored parts of the country to diffuse it to the poorer areas, reversed Mao's movement toward self-sufficiency and isolation from world markets.

Changes began on the farm. In the 1980s, commune-based agriculture gave way to the **household responsibility system,** allowing families to decide which crops to grow. This built on traditional family ties as opposed to Western individualism. At first a local government initiative, central government supported it after 1981 and made it the hub of its reforms. Households could lease collectively owned land for up to 15 years. Small groups entered short-term contracts, leasing land to achieve production quotas set by the government. Farmers could sell any surplus crops or animal products in local markets. By 1984, 99 percent of households participated.

Deng's more open economic policy attracted investment from outside China and generated the world's highest rate of economic growth from the mid-1980s to the early 2000s. Foreign investment brought rapid economic growth to many coastal regions of China. The central Communist government, however, stayed in power, reducing some of the oppressive measures that restrained personal freedoms.

After Deng Xiaoping died in 1997, his successors continued the policies that brought economic growth and involvement in the global economic system. In 1998 further reforms enabled and encouraged families to buy their own housing, leading to a growth of speculative building. The acquisitions of Hong Kong and Macau in the late 1990s highlighted the process of reclaiming long-term Chinese territories.

From the 1970s onwards, many families' material expectations increased. In the 1970s the desired items of the growing middle class were a watch, a bike, and a sewing machine. Through the 1980s they were TVs, washing machines, and refrigerators. In the 1990s attention shifted to sound equipment, microwaves, and air conditioning, and by the 2000s it focused on computers, cars, and apartments. People were encouraged to buy rather than save.

Internal Political Tensions in the 2000s

In the early 2000s China faces increasing tensions between the government's desire to maintain central control over the economy and the growing influence of links to the world's free-market system. In 1991 around 100,000 private businesses employed 1.8 million people; by 2001, 24 million people worked in 1.5 million private businesses with over eight employees and around 30 million in smaller, individual concerns. In that decade, private industry grew from a few percent to 40 percent of total Chinese industrial output. Many publicly owned industries are no longer cost-effective against international competition and face social consequences from shedding workers. Joining the World Trade Organization in late 2001 brought further challenges to Beijing's centralized control over the economy when the government had to cope with rising urban unemployment, rural unrest, and funding shortages resulting from difficulties in collecting taxes.

The Chinese government handles the tensions in various, often contradictory, ways. While educated Chinese now have increased opportunities to shape careers, go abroad, or pursue research interests without party interference, the army and security forces continue to respond firmly to internal dissent and unrest. The justice system remains poor and subject to political pressures that prevent magistrates from acting fairly: for most people outside political and business elites, trials and imprisonment remain arbitrary and often long-term. The government diverts internal pressures for greater democracy by a renewed emphasis on Chinese nationalism through a historical focus on the pre-Communist triumphs of past civilizations and greater publicity and support for Chinese cultural and athletic achievements.

Despite cries for reforms, the likelihood of a popular backlash is diminished by wide contentment with the economic progress and a fear among ordinary people of acting outside the established system. The local and provincial party leaders who continue to block some reforms hang on to their powers—much like their feudal predecessors. On the other hand, experiments of local democracy in rural areas often lose their impact because central ministries impose restrictions on powers of action. Officials continue to demand illegal fees. Many mid-ranking officials understand "political reform" to mean merely streamlining the bureaucracy. Even among Communist Party members, liberals vie with old-guard Maoists and free-market capitalists with leftists. Each group has its own journals and Internet websites. Although more opinions are aired than was possible in 1989, multiparty democracy is not yet even a remote possibility.

In 2002 major changes in the governing council brought the "fourth generation" of leaders led by Hu Jintao (after Mao, 1949–1976, Deng Xiaoping, 1978–1995, and Jiang Zemin, 1995–2003). Having been picked by Deng in their early forties, they are all technocrats with Communist Party and regional experience, used to working together to achieve consensus. All are dedicated to the new reforms related to an outward-looking focus and economic growth. In 2003 their influence led to new personal freedoms.

Economic Development

Attempted Geographic Shifts, 1949–1976

The Communist government of the People's Republic of China tried to expand output rapidly, but the Great Leap Forward and Cultural Revolution interrupted progress. Attempts were also made to distribute economic activities more evenly through the country to increase equality within China and enhance national defense. In 1949 the northeast, together with Shanghai and Tianjin, produced 70 percent of national output. The nearness of the northeast to Japan and the Soviet Union, and the coastal vulnerability of Shanghai, caused the Chinese government to move the production of military goods inland.

The policy of wider diffusion had an ideologic, as well as practical, basis. The Communist Party leadership wished to demonstrate that past locations and concentrations of industry depended on the flawed precepts of the capitalist system and foreign intrusions. Factory locations close to raw material sources had better economic foundations. From the mid-1950s, new manufacturing enterprises moved to regions with local coal, hydroelectricity, oil deposits, or strategic factors. However, many of the factories built in the interior proved hugely expensive to run.

Despite the new thrusts of policy and the powers possessed by the central government to make changes, the main centers of manufacturing production remained in the northeast. Central and western interiors and southern China developed slowly through this period because of political disruptions and poor transportation linkages to new industrial centers.

Changing Directions after 1976

In 1978 China's total economic output was lower than South Korea's or Taiwan's had been in the 1960s. Deng Xiaoping's major changes in outlook and policy began with a new approach to rural life and encouraged investment from foreign countries and corporations with a view to increasing manufactured exports. China's total economy multiplied sixfold within 20 years.

After 2000, China experienced a further revolution based on the use of the Internet and mobile phones. Its telecommunications system was upgraded in the 1990s by a fiber optic grid laid across the country. That made possible an increase in telephone lines from fewer than 10 million in 1990 to 125 million in 2000, when 2 million were being laid each month. Mobile phone users multiplied from 5 million in 1995 to over 300 million in 2004, making China the world's largest user of cellphones. However, the lack of a national payment system for buying goods online, the continued state control and policing of website commerce, and sensitivity about website content delay wider use of the Internet.

Chinese economic development does not conform to the pattern followed in other countries, which began economic growth with the production of raw materials and food for export and moved to simple manufactured goods before more sophisticated ones and a range of service industries. China already produces all types of goods in all economic sectors, from rice, rag diapers, and plastic toys to microchips, and spans the entire value chain.

Farming and Rural Living in the 2000s

Agriculture remains prominent in the Chinese rural economy. Despite farming being given low priority by government and separated from other aspects of the economy through the Mao Zedong years, in 1980 agriculture still employed almost the entire—and growing—rural workforce. Low levels of mechanization demanded high labor inputs (Figure 5.30).

By 2000, only half of the rural workers were in farming. From the early 1980s, commune controls were relaxed, and individuals and groups could plan their own program. The personal involvement, longer-term contracts, and lower quotas to be fulfilled before selling on the open market led to growing confidence and higher output. New rural sources of income through the township and village enterprises (TVEs) took underused rural labor and offered higher wages. Over 130 million employees, making up over 30 percent of the rural workforce, worked in the TVEs. Initially this expansion grew out of collectively owned enterprises, rising from 22 to 36 percent of all Chinese industrial output by the late 1980s. After that, individual and private enterprises became the new driving forces in rural industrial expansion. Millions of people remained in industrializing rural areas instead of moving to the cities. However, many millions more moved out of rural areas, a growing migrant population in the country's cities.

China attempts to feed 21 percent of the world's population from 7 percent of the world's arable area. Farming is both land- and labor-intensive, and output per unit of land is high. Much agricultural land is cropped two or three times per year, and competition is increasing between land-intensive crops—wheat, corn, soybeans, and cotton in the north and rice and sugarcane in the south—and labor-intensive crops that produce greater value per unit of land, such as vegetables and fruit. Patterns of farm production move toward more complex land uses, such as the integrated planting of mulberry, sugarcane, and bananas around fishponds in the Zhu Jiang delta. Growth in the textile industry led to a huge demand for cotton. By 2004, extra planting led to oversupply and falling prices. Meat and milk production continue to be small-scale, "backyard" enterprises.

The more profitable specialist crops, including industrial crops (cotton, soybeans), fruit, and vegetables, together with livestock products, threaten the output of some traditional staples, particularly wheat and rice. While the Chinese government needs to ensure continuing grain supplies, it requires greater diversification and intensity of production. By encouraging grain growers to increase yields by using more imported potash fertilizer, China produced enough grain in the 1990s to put a surplus into storage. However, continuing population increases made greater demands on food in both quantity and quality, forcing China to increase its imports of grain and other foods. These are funded by increased exports of manufactures.

Water Resources

New projects to move water from south to north will succeed only if they are linked to efficient management of the water (Figure 5.31). Work on the South–North Transfer Project started in 2002, but its huge costs arouse controversy inside and outside China: environments will be altered and lives disrupted. The western route needs tunnels; the central route needs a canal or aqueduct 1,240 km (770 mi.) long; the eastern route will follow established water courses that are highly polluted. In the central route, the Danjiangkou reservoir water level will have to be raised 13 m (43 ft.), with major impacts on 200,000 displaced families, deforested hills, and a lowering of water levels below the dam. Even if completed, the transfer project will be only a partial solution to the lack of water in the north. An alternative approach could begin by coordinating and testing water conservation and antipollution measures in local areas, but governments and their engineers prefer major projects because the proposed scale promises big results from single-focused efforts, the bureaucratic administration is easier, and more opportunities exist for media attention.

FIGURE 5.30 China: rice paddy near Yichang, Chang Jiang (Yangtze River) valley. A water buffalo and plow preparing land for planting rice and other paddies already planted. The older house that was used for both humans and animals has been augmented by a new home for the farmer's family—with a similar traditional structure. Although Chinese farmers include the poorest people, many have experienced better incomes in the last 20 years. **Photo:** © Michael Bradshaw.

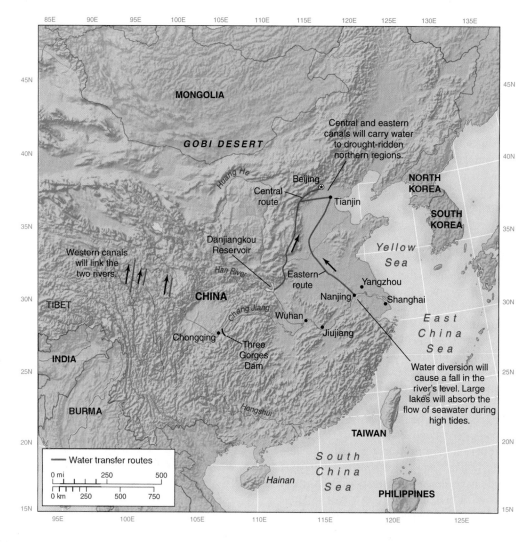

FIGURE 5.31 China: potential water diversions. Northern China and the Huang He basin are increasingly short of water. Three potential diversions have been identified. The western routes are short, but mountainous, requiring tunnels; the central route requires a canal or aqueduct 1,240 km (770 mi.) long direct to Beijing; the eastern route follows existing channels, but their waters are polluted. Compare this map with the climate map, Figure 5.10, and the population map, Figure 5.16, to understand more fully the need for such diversions of water.

Chinese Energy Policy

The current Chinese economic expansion places huge pressures on energy supplies, especially in electrical power. Although China ranks third in the world for producing energy (first in coal, fifth in oil, and sixth in hydroelectricity), it suffers from the high costs of distributing the fuels and electrical power over long distances. Oil and natural gas finds are mainly offshore or in the far west, distant from economic centers; coal is produced in the north and hydroelectricity in the south. The shortfall in power supplies is as high as 20 percent in the expanding southern areas. One of the major foreign investment opportunities is to expand power provision. In the early 2000s, after discussions about the proportion of profits to be allowed to foreign corporations investing in construction projects, building began on the first of around 50 power stations that will use Chinese coal and foreign capital. Although producing only 2 percent of China's

electrical power in 2004, nuclear power stations will increase in significance, especially in southern coastal areas at a distance from coal supplies.

China has the world's greatest hydroelectricity potential, but even if it were all developed, it would supply just 6 percent of the country's needs. The building of the Three Gorges Dam began in 1993. Much of the dam and its ship locks were completed in 2003 (Figure 5.32a, b). The final stage is due in 2009. This project illustrates the environmental and social issues facing major hydroelectricity projects. When complete it will generate enough electricity to save the yearly burning of 45 million tons of polluting high-sulfur coal and will help flood protection and navigation. The dam, however, created a lake 600 km (450 mi.) long below Chongqing, and 1.3 million people are being resettled. Farmers on good valley land are moved to often poorer uplands, and towns that were submerged are rebuilt on higher

(a)

(b)

(c)

FIGURE 5.32 **China: Three Gorges Dam project, Chang Jiang (Yangtze River).** (a) Three Gorges Dam, closed in 2003 and raising the water level behind it to 140 m (440 ft.) above sea level, 75 m above the downstream side. By 2004, 6 of the 18 turbines were installed and generating electricity; completion is expected in 2009, when the level behind the dam will be raised to 175 m (550 ft.) above sea level. (b) Ship locks at the Three Gorges Dam. At present there are three lifts of 25 m (80 ft.), taking around three hours to pass through the locks. An identical sequence enables ships to go in the opposite direction. (c) The impact upstream. The older buildings in the city of Zhongxian (downstream from Chongqing) will be knocked down in time for the 2009 water rise to 175 m, and the new city being built on higher land will be occupied. **Photos:** © Michael Bradshaw.

land (Figure 5.32c). Although questioned by environmentalists and internal advisers, this project is Chinese government policy and is on target for completion.

Coal reserves remain China's chief source of energy. They occur mainly in the northern half of the country, resulting in heavy use of the rail network and massive investment in the railcars that carry it southward. Coal transportation takes up 60 percent of the railroad capacity, and plans are in hand to build coal slurry pipelines from interior mining areas to the coast.

One outcome of the Great Leap Forward and Cultural Revolution was the opening of small local mines run by villages in southern China; these now produce around half of the total output. China's mining economy has two aspects—one with large-scale, capital-intensive, and technologically advanced mines, and another with small-scale, labor-intensive, low-technology mines. The smaller rural mines help with chronic underemployment, but their operation often lacks environmental and safety management. An estimated 10,000 Chinese coal miners die on the job every year, accounting for two-thirds of the world's coal mine deaths. Up to 80 percent of the fatalities occur in the small mines. In 2005 new laws emphasized safety procedures and equipment.

In the late 1990s China's oil industry was reorganized into two major companies, Sinopec and Petrochina, which evolved from two government monopolies. One had been responsible for refining and distribution and the other for exploration, but the new companies do both. They competed first for a large proportion of the 90,000 gas stations that were mostly owned and run by local governments, rural collectives, or individuals. Moving before foreign corporations could get into the market in the wake of China's entry into the WTO, both companies bought up thousands of the best-situated gas stations, sometimes at ridiculously high prices. They now look to joint ventures with foreign companies, headed by BP, Exxon, Mobil, and Shell. The investment presumes more cars and trucks will be on the roads. Although China has fewer cars in proportion to the population than any other country, their numbers increased from 300 to 4,200 per million in the 1990s as incomes rose. China is becoming a major oil importer in a limited world market.

Growth in Chinese Manufacturing

After 1978, the drive to greater efficiency and an open-door policy to encourage foreign investment became more important than self-sufficiency and led to what has been termed the "Great Leap Outward." Today China's manufacturing sector comprises a complex intermarriage of local enterprise, foreign multinational capital, and public ownership.

At first, government policy to relax trade control and encourage economic links to the rest of the world was a response to the need for foreign exchange to buy grain and high-tech equipment. Improving foreign exchange balances stimulated further exports and foreign direct investment. Such policies placed pressures on the publicly owned enterprises that formed the heart of Chinese industry up to 1978. From the 1980s, reduced central government subsidies and the "management responsibility system"—an industrial equivalent to the household responsibility system—placed more decision making in such firms' hands. By the early 2000s, 95 percent of industrial production sold at market prices. The central government, however, now focuses its own investment on only 1 percent of existing state-owned enterprises, allowing others to be merged, taken over by workers, or go bankrupt. By the end of 2004, the remaining "flagship" state-owned enterprises reported increasing profits from transportation, coal products, oil, chemicals, and metals that partly compensated for less profitable enterprises. Requirements to provide job security and social benefits make government-owned enterprises less profitable than private firms. However, loans to failing state-owned enterprises and grandiose socialist housing projects still amount to around 30 percent of GDP.

China has its own multinational corporations and encourages them to generate more world-known brands. One such company is Qingdao Haier, centered in the coastal city of Qingdao in Shandong province. From being a government-owned firm producing shoddy home appliances, it grew by the late 1990s to sell 15 percent of the Chinese washing machines and 33 percent of the refrigerators. It improved quality and listened to customer needs. For example, when Sichuan peasants used a washing machine to rinse soil off potatoes, the machines were adapted to prevent clogging. A small machine for washing a single change of clothing sells well in crowded urban centers. As Haier expands, however, it enters the very competitive production of TVs and pharmaceuticals, looking to overseas markets when financial resources and qualified staff are in short supply.

The construction industry received a huge boost from the reforms that made it possible for people to buy their homes. The construction of high-rise offices and hotels, new highways, new airports, factory buildings, and port facilities impresses the visitor. However, there has been much overbuilding, and there are continuing worries over rising land and house prices.

Increasing Steel Output

As rapid economic growth occurred in southern coastal China, the old industrial area of northeast China became the country's "rust belt" (Figure 5.33). China built its heavy industrial capacity in this area in the 1950s, but few factories and their equipment were updated. Output of steel products continued, but many were unmarketable because of their price or poor quality. In the 1990s, when market forces began to take over from government funding, local coal mines (20 percent of China's total) and oil companies (33 percent of China's total) in the northeast could not be paid for their products by the bankrupt industrial users.

Yet China's steel industry is still the world leader, with increasing production from new coastal and inland mills. The 2003 production was 222 million tons—more than that of Japan (100 tons) or the United States (90 tons)—after almost doubling in 10 years. The steel from the newer Chinese mills competes with U.S. and Japanese products and often undercuts

FIGURE 5.33 China: northeast industrial region. The steelworks, China's largest, at Anshan, Liaoning Province. **Photo:** © Colin Garratt; Milepost 92½/Corbis.

them in price, causing layoffs and hardships in those countries. The Chinese steel industry, however, is fragmented as the result of central government policy of placing a steel mill in each province. It is also inefficient, with many mills incurring losses and overcapacity, and produces mainly basic steel for which there are few markets. China imports better steel from South Korea and Japan for the sheets used in autos and computers.

China's Automaking Industry

Making cars is becoming one of China's major industries. Chinese and foreign firms invested $25 billion from 1994 to 2004, half of that since 2002. Total sales (2 million cars in 2003) are still behind the United States (17 million). Many foreign firms have established factories: Volkswagen has 30 percent of the market based on factories around Shanghai. Not only do major U.S., European, and Japanese automakers assemble cars in China, but their parts suppliers moved into multiple Chinese factories, making China a primary global source. China is now the place to launch and develop new models (Figure 5.34).

The Chinese automaking industry is increasingly competitive, with over 120 companies that range from small firms making crude autos based on old Soviet designs and the availability of cheap parts to the most sophisticated modern vehicles. In the center of this spectrum are firms such as Wanfeng, which began in 1996 by hammering motorcycle wheels, became the world's largest seller of aluminum alloy wheels, and now produces a cheap Jeep Cherokee look-alike from local parts in a new, but basically equipped, Shanghai factory. The largest Chinese parts manufacturer, Wanxiang, employs over 30,000 and increasingly exports its products.

Expansion of Tertiary Industries

Before 1978 there was only a single state bank for all financial operations, but there are now insurance, securities, and wider banking facilities. By the early 2000s over 200 foreign banks were active, many with licenses to carry out business in Chinese currency. In particular, Shanghai's new Pudong area has seen a huge expansion of financial investment with 150 foreign

institutions and the headquarters of 20 world-scale banks, together with the Shanghai Stock Exchange and Futures Exchange.

Growth also occurs in retailing, with supermarkets, shopping malls, home improvement stores, electronics outlets (from computers to cellphones and sound equipment), and auto sales doing well. In wholesale distribution, logistics control is a major area of growth: for example, FedEx has made Shanghai its Asia–Pacific hub and has doubled the flights to and from North America.

The media industry also grows, often with strong links to foreign financing and ideas. State broadcasting (TV and radio) and newspapers are subsidized but often operate at a loss because people see them as propaganda organs with boring content. In 2004 rationalization closed 1,400 state and party newspapers and forced many periodicals to finance themselves. However, China seeks to be a global media power. It is restructuring and focusing the main TV channels, and the state-owned Shanghai media group is expanding fast with foreign partners. Digital TV expanded in 2005. The movie industry is also growing, and going to the movies is the favorite leisure occupation of urbanites.

There is still an impression among Chinese local leaders that only factories will bring income growth, whereas services can reduce living costs, cause less pollution, create more jobs, and have growing markets. To date, the Chinese central government supports transportation systems, the financial sector, and tourism, but identifies needs for raising standards in such areas as accounting, medical care, and education.

Attacking Poverty in China

Many major cities are marked by new demonstrations of conspicuous consumption, but poverty—both absolute and relative—still exists. Up to the late 1970s, rural life remained harsh. In 1978 China estimated that 33 percent of its rural population (260 million people) lived below its defined poverty line. As the initial impacts of 1980s agricultural reforms doubled output and incomes, rural industries created over 100 million new jobs, and by the late 1990s the incidence of rural poverty declined from 33 to 9 percent (from 260 million to 97 million people).

As the government's emphasis in the late 1980s shifted to developing the coastal belt, economic disparities between coastal and interior places widened, so that, in the mid-1990s, the human poverty index for interior provinces was 44 percent, compared to 18 percent in the coastal provinces. Migrations of people from rural areas to cities unprepared for their arrival caused urban food subsidies to rise to five times the funding of rural health, education, and relief programs. Even within the cities, Chinese citizens see a few people becoming very wealthy while others face increasing unemployment as publicly owned facilities close. By 2004 the numbers of poor fell to 3 percent of China's population, but this still included 30 million people. Most of the remaining poor live in areas with few natural resources and poor infrastructure—where improvements

FIGURE 5.34 China: automaking. The General Motors launch of its Hy-Wire car at Badaling by the Great Wall, November 2004. This car features hydrogen fuel cell propulsion and other technology of the future automaking industry. **Photo:** © Andrew Wong/Reuters/Corbis.

will be more difficult. Using another measure, the numbers of hungry people in China fell from 16 to 11 percent over the decade 1990–92 to 2000–02 (194 to 142 million).

Regions and Cities in China

With such a large area, China's boundaries enclose many local regional variations (Figure 5.35). Physically, China is a country of contrasts. Culturally, the regional distributions of languages and dialects, ethnic differences, and historic relics provide a variety of hindrances and resources for human development. Politically, China is divided into 23 provinces and other province-level units, including four great cities (Shanghai, Tianjin, Beijing, and Chongqing) and five autonomous regions with large ethnic minorities (Inner Mongolia, Guangxi, Ningxia, Xinjiang, and Tibet).

Economic differences between wealthier and poorer areas are linked to regional concentrations or absences of resources. In the 1980s the government divided China into three regions that were expected to have different types of development. By the early 2000s, the central government renewed efforts to encourage economic development in the interior. A "Go West" strategy adopted in 1999 aimed to affect half the country's area and 25 percent of its population.

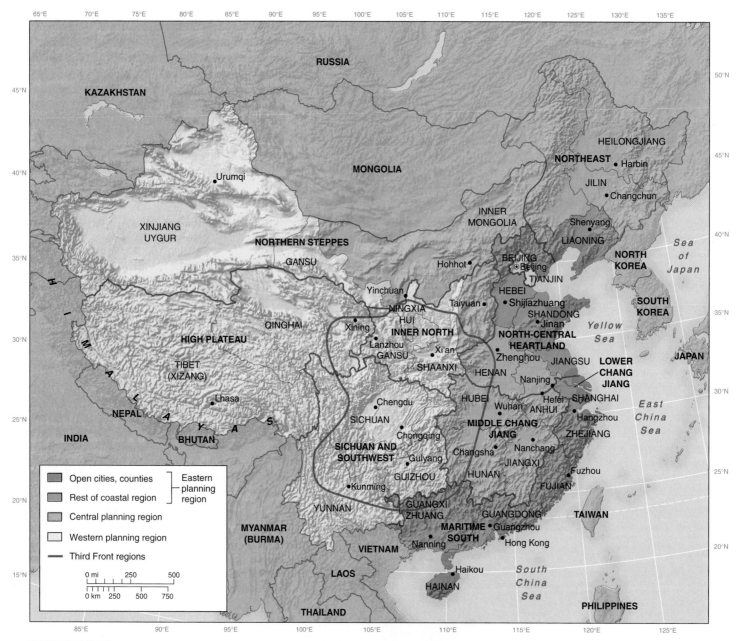

FIGURE 5.35 People's Republic of China: internal geographic and planning regions.

Investment in infrastructure, development of the legal system, and ecological responsibility tried to attract new industries to areas with natural resources and cheap labor. Although the policies had some success, they were often seen as intrusive by local minority groups.

Eastern Coastal Regions

The "Golden Coastline" of the eastern region includes the capital, Beijing, port cities such as Shanghai and Hong Kong, and other economic growth areas such as the Guangdong region in the south. Though this zone is only one province deep along the eastern coast, it has the greatest potential for economic development and the easiest links to the global economic system. Its specialty of export-oriented goods motivates further economic growth.

Four **special economic zones** were established in the southeast in 1979 and a fifth on the island of Hainan in 1988. Fourteen coastal ports were designated as open coastal cities in 1984. Not only do these places have advantages for access to foreign capital and technology, but they also offer cheap local labor and buildings with tax concessions for equipment.

Not all parts of this coastal belt experienced the same rapid economic growth. The southernmost part of the Chinese coastal zone, *the maritime south,* had rapid economic growth after the 1978 relaxation of external trading rules. By 1990, 60 percent of over 12 million township small business enterprises were located in the southern coastal provinces. This area became the Chinese "sun belt." Summer rainfall and mild winter temperatures increase southward, reflected in the crops that range from winter wheat in the north to rice and tea farther south. Two crops of rice are harvested in the longer summers of the far south. Special economic zones established from 1979 encouraged foreign investment and technological innovation in manufacturing, especially around Guangzhou at the head of the Zhu Jiang estuary. The new industries made textiles and apparel, electronics goods, chemicals, and machine tools. A huge area of rapidly expanding production grew around Shenzhen in Guangdong province inland of Hong Kong. Facilities in this area include an oil refinery, nuclear power plant, airport, deepwater shipping berths, and inland superhighway links. Hong Kong and the offshore economy of Taiwan provide links to the global economy.

Although *the northeast* was not much settled by Chinese until the early 1900s, the Japanese then occupied it and developed it into one of the most industrialized regions of China based on the local occurrence of coal and iron. Industrialization of this region continued after the 1949 communist takeover, making it the center of large-scale industry in China. Also after 1949, growing corn, soybeans, and rice extended agriculture farther across the rich black earth. The region produces iron and steel products, machinery, and vehicles, has a major oilfield and is supported by a dense railroad network. Dependence on heavy industry made this region the "rustbelt" of China. Poor management and high costs led to the closing of many state-owned enterprises and impoverished many of the 107

million people in the region, who were dependent for welfare and pensions on their employers. Although in the early 2000s China's leadership made a major point of promising aid to the region as the "elder son of the republic," the aid was limited to bailing out some of the loss-making state-owned enterprises. Few foreign manufacturers were attracted, apart from BMW and Volkswagen autos and South Korean LG's new research center. The port of Dalian gained from being close to Japan.

The north central heartland of China has a long history of human occupation, political leadership of China, and intensive production. Its industrial and administrative cities include Beijing and Tianjin. It comprises the plains, formed largely of river delta deposits, into which the Huang He and Chang Jiang drain. Nearly one-third of the Chinese people live in this region. After 1949, new drainage and irrigation works improved and regulated the flows of the major rivers, but pollution by unregulated industries remains a major problem. The region has an agricultural base of winter wheat and specialist crops. Oil is produced near the mouth of the Huang He.

In the lower Chang Jiang lowlands and central coastlands, *the Shanghai area* is the most advanced and prosperous city of a productive farming region. It is built along the Huang Pu near the mouth of the Chang Jiang. Through the 1980s, the policy to encourage growth in the south affected the Shanghai area adversely. While incomes in Guangdong and Fujian doubled, those in Shanghai saw no growth. From 1991, however, the central government made Shanghai a special economic zone, and it became the center of China's main economic growth area. This region experienced massive redevelopment and building of new cities after years of slow growth. The cities of the lower Chang Jiang region have traditional manufacturing (textiles) mixed with post-1949 heavy industries (steel, shipbuilding, and oil refining), together with huge recent investments in strategic manufacturing industries based on modern technology (autos, electronics, and industrial equipment).

Shanghai is China's main port after Hong Kong. It is the financial capital and a major center of higher education and research and development. Most of its growth is funded by internal Chinese investment, and many of its products are consumed internally rather than exported. Much of the city is being rebuilt with new bridges, better roads, river port and airport facilities, and a subway system.

The new economic zone in Pudong across the Huang Pu from Shanghai (see Figure 5.8) already has 4,000 enterprises and 400,000 workers. Further developments around Suzhou are linked by expressway road. The Shanghai authorities initiated Suzhou New District in 1990 to move industry westward out of its central area. Then the Singapore government, in a publicized demonstration of inter-Chinese working, contracted to build the Suzhou Industrial Park and expand supercity facilities between Shanghai and Suzhou.

Shanghai is the center of China's growing automaking industries, often based on joint ventures with multinational corporations. Volkswagen of Germany started making cars in

Geography at Work

FRONTIER FORCES

Select Disc 2 of the CD-ROM set, *Interactive World Issues of Place and Planet,* included with this text. The section on *China: Frontier Forces* is a study by Chinese geographer Chui Yang and U.S. geographer Joe Glodney of the geography of the arid–humid borderlands around Lanzhou, China. They examine the local physical geography and people's responses to it.

Begin with Part 1/1 and work through the interviews and videos. This study refers to a wide range of the sort of information used commonly by geographers in analyzing situations in different parts of the world.

Box Figure 1 The "Singing Sand Dunes" of the Gobi Desert. Photo: © Vol. III/Corbis.

Shanghai in 1985 and produces one-third of the national total, expecting to double its output in the early 2000s. It uses parts that are 85 percent locally made. A joint venture between China and General Motors is located in Shanghai's Pudong Development Zone. Other vehicle manufacturers are based in the northeast and between Shanghai and Beijing. Brilliance China (luxury vans) and its subsidiary Shenyang Jinbei Passenger Vehicle Company (vans) import engines and components from Japan.

Peugeot and Citroën (France), Chrysler and Ford (U.S.), and Japanese companies also produce cars, trucks, and minivans in China. As in other countries in East Asia, the multinational corporations press the government for greater openness to the world economy to establish their own factories and offices inside the country, but then encourage the government to protect their investment by excluding potential competitors.

Hong Kong

Hong Kong comprises an island (centered on Victoria) and mainland peninsula (Kowloon) that were ceded to the British in the mid-1800s. A more extensive area of mainland, the New Territories, was leased to the United Kingdom for 99 years in 1898. All returned to Chinese rule in 1997. The Basic Law, under which Hong Kong operates within China, guarantees retention of its free-market system for 50 years and sees the city as a complementary financial center to Shanghai.

Hong Kong, the world's busiest port, has an excellent natural harbor with deepwater access and extensive water frontages (Figure 5.36). Goods received from inland China are sent to worldwide destinations, while others collected from the rest of the world are sent into China. These are the features of an **entrepôt.** As the Chinese economy opened up and expanded after 1976, particularly in the special economic zone around Shenzhen, the value of Hong Kong's trade multiplied sevenfold. At

the time of political transfer, Hong Kong was deeply enmeshed with interior China in trade, investment, and personal contacts. In 1996 China accounted for 37 percent of Hong Kong's trade and Hong Kong for 47 percent of China's trade. Over 20,000 trucks crossed the Hong Kong–mainland border each day. As part of China, Hong Kong began with advantages over its internal and external rivals for dealings inside the growing country. New ports opening along the Zhu Jiang delta in the early 2000s relieved the huge pressure on Hong Kong's port facilities.

The population of Hong Kong, nearly 7 million in 2005, is almost totally Chinese. As the main urbanized area became overcrowded from the 1950s, new towns spread into the New Territories. They now accommodate around one-third of the

FIGURE 5.36 People's Republic of China: Hong Kong. View from Victoria Peak on Hong Kong Island to Kowloon across the harbor. Hong Kong has some of the highest densities of population in the world. **Photo:** © Alasdair Drysdale.

total population. The high population density causes half of government expenditure to go for roads and public transportation. In the late 1970s electrification upgraded the railroad route inland to Guangzhou (Canton), and its use increased. Two cross-harbor tunnels and an underground transit system opened in the 1980s. A new Hong Kong International Airport was built offshore north of Lantau Island and was linked by bridges to the mainland. It replaced the older airport in Victoria Harbor that could not be expanded on its limited site.

Hong Kong's growth as a manufacturing center took off in the 1970s, and by 1990 manufacturing provided a third of all employment. The production of electronic goods and scientific equipment increased, but the textiles and clothing industries declined because of foreign competition and rising labor costs in Hong Kong. In the 1980s manufacturing moved out of Hong Kong and expanded just over the border in southern China, with its cheap land and labor costs. Despite the pressure on farm incomes from office and factory jobs, Hong Kong grows nearly a third of its food needs on the small area of nonurbanized land in the New Territories.

In the early 2000s, the shift out of manufacturing as factories moved inland led to service industries accounting for 85 percent of Hong Kong's GDP and 80 percent of employment (from around 50 percent in 1980). Hong Kong is one of the world's major finance centers and the third largest gold market, accounting for up to 15 percent of the world total. Hong Kong provides expertise in finance, trade, and public administration and has an important stock market. It is the regional headquarters for 800 foreign companies.

Tourism became Hong Kong's second major industry with 6.5 million visitors in 1990. Many new hotels catered to this influx of people, which added an extra 10 percent to the population at peak times. After Hong Kong's transfer to China in 1997, tourism continued at a high rate. The numbers of visitors topped 16 million in 2002.

Macau's New Approach to Gaming

Before its 1999 incorporation in China, Macau had become a byword for corruption and crime. Over half of all Asian gaming took place in Macau's casinos, and gaming taxes accounted for 60 percent of its government revenue. The Chinese government, instead of closing down the gambling industry in Macau, is reforming it. With a shortage of tax revenues, China now considers combining gaming in Macau with other tourist activities.

Interior behind the Coastal Regions

The central region is a densely populated interior zone without coastal outlets but with natural resources of minerals, water, soils, and not-too-steep slopes that allow agriculture to flourish. This region specializes in farming and energy production. Such resources provide a basis for local industrialization. Its main problem is the lower prices received for local food and raw materials compared to those paid for manufactured goods from the coastal zone factories.

The inner north region is centered in Shanxi and Shaanxi. Xi'an in Shaanxi was a historic center of Chinese civilization, but this region now forms the northwestern rim of Chinese economic activity. The loess soils of windblown origin are easy to work but also easy to erode, attracting major efforts to overcome soil losses. Spring wheat and millet are basic crops. Local coal mining and hydroelectric potential provide a basis for modern industry, but distance to markets and low levels of technology slow development.

The middle Chang Jiang basin combines wider plains and hilly areas. It also has cold winters and decreasing rainfall westward. Rice, winter wheat, and cotton grow on the plains around industrializing cities linked to local hydroelectricity potential and river transportation. Wuhan is the main industrial center of this region. The encouragement of growing interregional linkages, including the Three Gorges project (see Figure 5.32) along the Chang Jiang basin, which is being called "China's Soaring Dragon," partly relieved the tensions between this region and the east coast.

The Sichuan basin and Yunnan Mountains of inland southwest China were cut off by gorges along the Chang Jiang, but the Three Gorges project makes this area more accessible. The highly cultivated rice-growing Sichuan plain contrasts with the surrounding mountains and plateaus. Mineral resources are underexploited. The isolation of this area, however, stimulated industrialization for local and national defense needs during World War II and as part of China's strategy after 1949.

Deep Interior Regions

The *western region* is sparsely populated, with extensive arid and high mountain environments. The region's future is concerned with animal husbandry and mineral exploitation. Relatively little friction occurs across the long international borders because so few people live near them.

In the northern steppes from Inner Mongolia westward to Xinjiang, plateaus and encircling mountains separate grassy or arid basins. The climate is dry with warm summers and very cold winters. Livestock keeping is the main traditional occupation. Ethnic minorities such as Mongols, Uighurs, and Kazaks live there and frequently demonstrate opposition to the central Chinese government. The proximity to the Russian border led China to develop parts of this area, and extensive mineral deposits will be exploited as the transportation links for oil and gas are established.

The high plateau of Tibet and Qinghai, which is mostly above 3,000 m (10,000 ft.), occupies the southern half of westernmost China. The People's Republic of China reduced the region's isolation and introduced modern industry, but only 6 million people live in a huge area. New buildings in Lhasa, the capital of Tibet, symbolize the Chinese control of the province (Figure 5.37). In 1950 Chinese armies invaded Tibet and, despite subsequent uprisings, maintained rule by oppression.

(a)

(b)

FIGURE 5.37 People's Republic of China: Tibet. Compare the roles and designs of these buildings, old and new. (a) Potala Palace, Lhasa, Tibet, China, former residence of the Dalai Lama. (b) Telecommunications building. Lhasa, Tibet, China, part of modern Chinese control and modernization. **Photos:** (a) © Robert Harding Picture Library Ltd./Alamy; (b) David C. Johnson.

Mongolian Isolation

Mongolia is a landlocked country of mountain ranges, the Gobi Desert, and semiarid grassy steppes akin to the northern steppes region of China (see Figure 5.3). It has a small population and struggles to maintain its independence as a buffer state between China and Russia. The capital, Ulan Bator, is connected by rail to both China and Russia.

After being a Chinese province, Mongolia became independent in 1921 with Soviet Union backing, and a Communist government was installed in 1924. The Soviet Union modified Mongolia's East Asian character, replacing the Mongolian alphabet with Cyrillic script and reorganizing the education system. Mongolia became dependent on Soviet aid and trade. After Russia's military forces were withdrawn in 1992, the Mongolian people strove to earn a living while attempting modernization in a landlocked, semiarid environment. In 2000 the former Communist Party was overwhelmingly reelected on the basis of policies that combine reform, social welfare, and public order.

Mongolian People

Mongolia has a dispersed, low-density population with falling natural increase and life expectancies now in the 60s. Ethnically, the Mongol people dominate Mongolia, although it also has 10 percent who are Chinese, Russians, or Kazakhs.

Just over half of Mongolia's population lives in urban places as the result of the pull of growing industrialization and the push of rural poverty. Mongolia's capital, Ulan Bator, houses nearly one-fourth of the country's population, including those living in large tented (yurt) areas.

Mongolia's Limited Resources

Mongolia has a strategic location in the heart of Asia but restricted economic potential. Many people still gain a living from herding livestock on the semiarid grasslands. Farmers responded to privatization and decontrol of meat prices in the early 1990s by massively increasing their herds. As agricultural cooperatives were split and land abandoned because there was no equipment to cultivate it, crop areas declined.

For the future, Mongolia seeks to increase its output of minerals such as copper, gold, and molybdenum and encourages foreign investment to provide alternatives to the previous Russian links. Private enterprises became the main source of economic growth in the 1990s, with rising production of gold, although a major fall in world copper prices and weaker prices for cashmere woolens slowed growth. Mongolia suffers from a limited range of resources and the problems of making the transition to a market economy in a landlocked position.

Taiwan

Taiwan has 10 times the population on one-fourth the area of Mongolia. It is a well-watered island country, with a mountain "backbone" along its eastern coast. After World War II, Taiwan returned from Japan to China. In 1949 the Kuomintang forces escaping from China took control of Taiwan. The Kuomintang leaders brought their families to join several thousand military personnel who had been stationed on Taiwan to quell anti-Kuomintang riots. Taiwan prospered with high-tech export industries and, after 1980, became a major investor in mainland Chinese industries.

Originally settled by aborigines of Malay–Polynesian origin, the island's population today is nearly all of Chinese origin. The Taiwanese have demographic characteristics similar to those of other materially wealthy countries, including life expectancies of 75 years. Taipei is the capital and largest city of Taiwan, which has nearly three-fourths of its population living in urban areas as a result of the growing industrial and service-centered economy.

While Taiwan maintains its independence, the People's Republic of China insists it remains a part of the mainland country and intends to reincorporate it. The PRC attitude places Taiwan outside the normal niceties experienced by independent countries, such as membership in the United Nations. For 50 years after 1949, the PRC threatened to invade Taiwan and backed off only under U.S. threats.

China's leadership is determined to apply to Taiwan the "one country, two systems" formula adopted for the incorporation of Hong Kong. Almost half the Taiwanese population already supports reunification with China based on a Hong Kong type of provision that preserves Taiwan's prosperity and its role as a major investor in Chinese economic growth. In the early 2000s, "China creep" threatened to replace the threats as Taiwan became more enmeshed in mainland China's economy. China is Taiwan's third largest trading partner, following the United States and Japan. Nearly half of the investment in Shanghai's Pudong development project is Taiwanese. Although Taiwan's government officially limited Taiwanese investments in China, over $60 billion followed indirect routes through Hong Kong, despite few return profits.

Many Taiwanese, however, continue to watch events in Hong Kong before agreeing to become fully part of China again and, in the meantime, continue to strengthen their military capabilities. Supporters of Taiwanese independence argue that it is different from Hong Kong: Taiwan is 170 km (100 mi.) offshore, has no built-in deadline date for return to China, and has a more mature democracy than Hong Kong and much of the rest of East Asia. But pressures grow on Taiwan as many of its best workers and managers move to the mainland, unemployment increases in Taiwan, and the World Trade Organization members asks new questions about the Chinese being able to use the Taiwanese ports and airports freely. In 2005 the first direct flights for airline passengers between Taiwan and mainland China occurred over the New Year holiday when Chinese families get together.

Taiwanese Economy

After 1949, the Kuomintang government encouraged rapid industrialization, rural change, and urbanization. The increasing rice surplus gave way to more diversified farm products, including sugar, tea, vegetables, and fruits. The expansion of urban-based industrial jobs and buildings, plus the mechanization of farming, reduced the farm population from 6 million in 1965 to fewer than 4 million in the 1990s, while an initial agricultural trade surplus gave way to increasing food imports.

Starting in the mid-1960s, export-processing zones focused attention on manufacturing for export. By the 1980s the emphasis switched to high-technology development. Private ownership of firms increased from 52 percent in 1960 to over 80 percent in the 1980s. The government retained control of banks, interest rates, and exchange rates to maintain a consistent financial environment. It cultivated a strong domestic manufacturing industry, at first based on transistor radios, followed in the 1970s by the assembly of Japanese electronic goods, and then promoting links with companies such as Philips (Netherlands) to build its own integrated circuit industry. Taiwan now leads the world in such technologies as integrated circuits, laptop computers, modems, and data communications.

By the 1990s Taiwan had one of the largest world trade surpluses and was able to export capital, particularly to China and Thailand. As is the case with Japan and South Korea, direct investment in Chinese manufacturing contributed to many "Taiwanese" products. As Chinese costs undercut Taiwanese producers, Taiwan continues to invest in new technologies such as biotechnology and new integrated circuit designs. Such investment, however, demands so much capital that it poses major risks for the Taiwanese government, such as overspending and eventually losing in competition with the Chinese. Already, Taiwan's furniture and textile industries have moved to mainland China, leaving increased unemployment, and its electronics industries are under major threat.

Large financial surpluses meant that major financial problems in the late 1990s affected Taiwan less than South Korea and Southeast Asia. Taiwan's other positive features include mainly small corporations and more flexible conditions of operating that enable failing companies to cease trading. Such conditions result in higher productivity compared to the South Korean *chaebol* and Japanese conglomerates.

Test Your Understanding
5D

Summary

China has 20 percent of the world's population. In 1976, still under a Communist regime, it emerged from isolation and instigated a mixture of industrial development, greater agricultural productivity, and greater national unity. It interacted with the global economic system in exporting its products and importing foreign capital.

China's agriculture became more productive after 1976 but is near the limits of areal expansion. The country's industries grow rapidly in southern and coastal China but elsewhere need new plants and infrastructure. China's economy is now a mixture of state-controlled enterprises and foreign multinational and local investment in manufactured goods for export.

Mongolia remains poor and isolated, has a semiarid climate, and was under Soviet Union domination until 1990. After losing Soviet financial aid, its economy struggled with transition. Taiwan, claimed as an integral part of the People's Republic of China, prospered as an independent developer of high-tech products and invested large sums in mainland China.

Questions to Think About

5D.1   How did Chinese government economic policies from 1949 to 1976 and from 1976 to the present differ? What impacts did changes have on China's agricultural, manufacturing, and urban geography?

5D.2   Why do the age–sex diagrams for China and Japan show different patterns? What are the consequences of such population structures for education, labor forces, house-building policies, and the proportion of older people?

5D.3   What roles might Hong Kong and possibly Taiwan play within China?

Key Terms

Han Chinese	Cultural Revolution
collectivization	household responsibility system
commune	special economic zones
Great Leap Forward	entrepôt

Chapter 6

Southeast Asia and South Pacific

FIGURE 6.1 **Southeast Asia and South Pacific: subregions and aspects of the region's geography.** Photo: (left) © Thomas Hutchinson.

The world region of Southeast Asia and the South Pacific is a contemporary convergence of environments and peoples, the influence of empire and economic globalization, and the delicate balance of competition and cooperation. Proximity to water and coastal access dominate the region's physical situation and some countries consist of numerous islands. Culturally, Southeast Asia and the South Pacific region includes ancient Hindu and Buddhist peoples, the world's most populous Islamic country, and areas such as Australia and New Zealand, where Christianity is the dominant religion. Countries and cities in the region compete intensely for market growth and contemporary global connections. At the same time efforts exist to strengthen intergovernmental economic and political relationships among the region's diverse cultural traditions and political ideologies. The main themes are:

The contemporary impact from centuries of global economic exchange.

The physical geographic impact of latitude, altitude, and a strong maritime influence.

A variety of people with Hindu, Buddhist, Muslim, Chinese, and European cultures.

Three subregions (Figure 6.1):

- Southeast Asia: continental and island Asian countries.
- Australia and New Zealand: materially wealthy countries with small populations.
- South Pacific Islands: tiny countries—the idyllic and the poverty-stricken.

Is Antarctica a region?

Geography at Work: The Indian Ocean Tsunami of December 2004

Point–Counterpoint: Singapore

Personal View: Malaysia

A World of Influences

Crossroads in the Sea

Southeast Asia and the South Pacific encompass a wide range of historical influence, religious tradition, linguistic diversity, and economic goals. The region is a geographic crossroads and geographic transition zone from the Indian Ocean to the Pacific Ocean and from the northern latitudes to southern latitudes (Figure 6.2). Varied beliefs and traditions intersect here with contemporary economic forces.

Some of the world's longest-established indigenous groups still live in traditional ways amid expanding modern cultures. Traders from China, India, and the Arab world exchanged goods in the region and influenced local human geographic patterns. Commerce and conquest brought a lasting imprint of Buddhism, Hinduism, and Islam to the region (Figure 6.3).

From the 1500s, Portuguese traders, followed by Spanish, Dutch, British, and French merchants and government-sponsored colonizers, took control of most of the kingdoms, sultanates, and sheikdoms of Southeast Asia. Only Thailand (then known as Siam) succeeded in maintaining its independence. Christian missionaries established new religious allegiances. European-controlled trading companies used regional populations to extract raw materials for export to Europe. The Europeans imported Chinese and Indian laborers, who worked in mines and on plantations and added to today's multicultural populations and sources of diversity and conflict.

The Japanese occupied most of Southeast Asia during World War II. In the postwar years, Thailand, Malaysia, Indonesia, and the Philippines resisted the diffusion of communism from the north that overtook Vietnam, Laos, and Cambodia. The United States feared the spread of communism would overtake bordering countries one after another like a row of dominos toppling the next in line. This fear was known as the **domino theory** and was used to support the U.S. policy of **containment,** where the objective was to stop or contain the spread of communism. In 1967 Indonesia, Malaysia, the Philippines, Singapore, and Thailand established the **Association of Southeast Asian Nations (ASEAN)** to combat this political threat.

By the 1800s, European colonialism reached Australia, New Zealand, and the Pacific islands. The British settled Australia and New Zealand and attempted to reproduce their Western culture, taking little account of the indigenous people and keeping their distance from Asia. The local impact was nearly complete Westernization (Figure 6.4).

Historically, culture, economic relations, and geographic situation kept Australia and New Zealand relatively isolated from their Southeast Asian neighbors. In addition to current global connections with Europe and the United States, Australia and New Zealand are increasingly interacting with the countries of Southeast Asia. The Pacific islands attracted little attention or sympathy, apart from exploitative intrusions, from other parts of the region.

Situated at the southern extent of the global sphere, Antarctica comprises a unique, geographically distinctive site near Southeast Asia and the South Pacific. Although Antarctica has never had permanent settlements, many countries claim segments of it and fund scientific research on it. Their findings contribute to global research in such areas as ozone depletion and oceanic fishing resources. We ask, is Antartica a true geographic region?

Regional Economic Contrasts

Southeast Asia and the South Pacific is a region containing some of the world's wealthiest countries and others that are materially impoverished (Table 6.1). Australia and New Zealand are the most Western countries in the region in terms

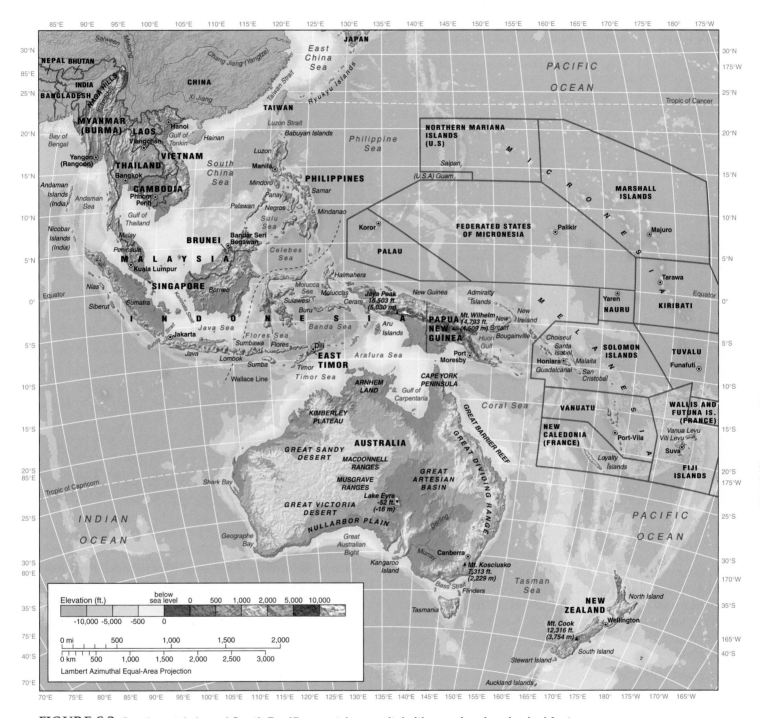

FIGURE 6.2 Southeast Asia and South Pacific countries, capital cities, and major physical features.

of incomes and lifestyles. Per capita wealth is highest in Australia, Singapore, and New Zealand. Thailand, Indonesia, Brunei, Malaysia, and the Philippines are in the middle group. Vietnam, Cambodia, Laos, Myanmar (Burma), Papua New Guinea, and many of the Pacific islands remain in material poverty, and some of the islands depend almost totally on external aid.

Cultural History and Colonialism

Prehistoric migrants came to Southeast Asia and the South Pacific from the north, filtering southward through the large river valleys and diffusing out across the island archipelagos to Australia, New Zealand, and other nearby islands. Darker-skinned indigenous peoples migrated to and concentrated in

FIGURE 6.3 Southeast Asia: Buddhist heritage. (a) Pagoda temples, Lake Inle, Myanmar. (b) The golden Phra Sri Ratana Chedi, part of Bangkok's Grand Palace Complex. **Photos:** (a) © age fotostock/Superstock; (b) Thomas Hutchinson.

(a)

(b)

isolated mountain or outback locations. The varied precolonial and colonial histories combined to provide the great diversity present in the contemporary human geography of the region. The current political boundaries of Southeast Asia and the South Pacific are almost entirely the legacy of colonial territories.

Khmer, Burmese, Thai, and Vietnamese Empires

Mon and **Khmer** people occupied present-day Cambodia from the north, and between the A.D. 800s and 1200s this was the center of the Khmer Empire. Later the **Vietnamese, Lao,** and **Burmese** arrived in the territories that now form their national centers. Indian traders brought Hindu and Buddhist

FIGURE 6.4 Southeast Asia and South Pacific: contrasts in cultures. (a) "Long-necked" Karen tribe children, western Thailand. (b) Australian children wave flags at an international cricket match. **Photos:** (a) © Ian Coles; (b) © Vol. 8/Corbis.

(a)

(b)

Table 6.1 Southeast Asia and South Pacific. Data by subregion, country, area, population, income (GNI PPP—Gross National Income Purchasing Power Parity), urbanization, Human Development Index (HDI), Human Poverty Index (HPI).

| Subregions | Land Area (km²) | Population (millions) | | GNI PPP 2003 | | % Urban | Index 2002 | |
		Mid 2005 Total	Est. 2025 Total	Total (US$ billions)	Per Capita* (US$)	2005	HDI (rank of 177)*	HPI (% population)*
Southeast Asia	4,483,656	556	694	2,095	6,838	38	102	23
Australia and New Zealand	7,984,350	25	29	648	24,705	89	11	
South Pacific Islands	527,909	8	13	21	4,125	42	108	29
Totals or averages*	12,995,915	589	736	2,764	11,889	56.1	73.7	25.9

Source: World Population Data Sheet, Population Reference Bureau (2005), World Development Indicators, World Bank (2005), Human Development Report, and United Nations (2004).

religions across the ocean from South Asia. The imprint of Indian traditions and architecture reflected in temples such as **Angkor Wat,** Cambodia, led to the area that includes Cambodia, Laos, and Vietnam being called "Indochina."

The Red River valley in the north became the center of a Vietnamese kingdom by the first century B.C. The area was later conquered and reconquered by various Chinese dynasties and took on Chinese methods of government and administration. The Vietnamese, however, always fought back and retained their separate identity. In the A.D. 1400s the Vietnamese not only fought off the Chinese but also extended their lands southward across older coastal kingdoms to take over some of the Khmer territories at the mouth of the Mekong River.

Myanmar became Buddhist as Sri Lankan monks disseminated Buddhism. The central Pagan kingdom situated around present-day Mandalay in a fertile, mountain-rimmed plain in northern Myanmar fought powerful tribes, including the **Shan** and **Karen,** who lived in the surrounding hills and today continue to resist the Myanmar government's attempts to draw them into the country's mainstream. The Mongols ended the Pagan Empire. After they left, Myanmar remained divided until the 1700s, when efforts to promote Burmese expansion led to conflicts with the British in India.

At the time of the Mongol advance, a group of Thai-speaking people moved from western to southern China and then into the western part of the Indochina peninsula, forming a unified political grouping by the 1300s with its center at Ayutthaya. After conflicts with Myanmar and Cambodia, the people established their main territory that was never again occupied by other powers and became modern Thailand.

Cultural Intersection

External influence brought frequent shifts in power to the island groups that now form Malaysia, Indonesia, and the Philippines. Phases of migration in the Indonesian islands brought together more than a hundred different ethnic groups and languages. Traders and invaders from India and China came to the islands prior to the establishment of a rice-based kingdom on Java in the A.D. 500s. From the 600s to 800s, ruling dynasties controlled the Malay Peninsula and much of present

Indonesia, erecting Buddhist monuments on Java. After around 1000, Buddhism added Hindu elements in some areas, as on Bali (Indonesia), but in most of western Malaysia and Indonesia, both Hinduism and Buddhism gave way to Islam. From mixed Malay, Indonesian, and Chinese influences, a Filipino culture emerged about the 400s.

After the arrival of Muslim groups during the 1200s, the local sultanates established at Malacca (now in Malaysia) and Brunei controlled the islands' trade. Malaysia, Indonesia, and Brunei remain predominantly Muslim in religion, with Indonesia being the world's largest Islamic state.

Australia, New Zealand, and the Pacific Islands

The indigenous people of Australia, New Zealand, and the Pacific Ocean islands included ethnic and culture groups whose origins are still debated. The populations of the islands are distributed over vast ocean distances and contain mixtures of Melanesian, Polynesian, Micronesian, and mainland Asian heritage.

There were between 200,000 and 500,000 indigenous **Aborigines** present in Australia in the late 1700s when the Europeans arrived (Figure 6.5). The Aborigines were

FIGURE 6.5 Australia: Aborigines. Aborigines after shopping at a store in Nangalala, Northern Territory. Tensions arise between traditional ways and Western lifestyles. **Photo:** © Penny Tweedie/Corbis.

nomadic hunters and gatherers living in communities or clans spread across the continent and speaking 200 different languages. Rock paintings, based on their **animistic** (nature worship) religious beliefs and social organization, are part of their legacy.

The **Maoris** of New Zealand came from the wider South Pacific around the A.D. 800s, with some of the final waves of people migrating from Tahiti in 1350. They replaced an earlier dark-skinned people, the Moriori, most of whom they drove out.

The inhabitants of the South Pacific oceanic islands are distributed over vast ocean distances. They are grouped in three broad geographic categories—the **Melanesian** ("black islands," so named by Westerners because of the presence of dark-skinned people), **Micronesian** ("small islands"), and **Polynesian** ("many islands") people. The Polynesian groups generally have lighter skin tones than the Melanesians. These groups are also distributed into the eastern islands of Indonesia, where they still form the population majority.

The Colonists and Independence

Philippines

The Portuguese were the first colonial force to establish trading centers in the Philippines. They were forced out in the late 1500s by the Spanish, who built their capital at Manila on Luzon (Figure 6.6). Spanish control attracted Roman Catholic priests and their monastic orders, who assumed control over vast lands and resources. Rapid conversions to this new faith united many disparate groups of Filipinos in what became the only largely Christian country in Southeast Asia. Spain retained its Philippine colonies until the end of the 1800s, when the United States placed them under its jurisdiction after winning a war with Spain over Cuba (the Spanish–American War). Local opposition to colonial rule increased during the 1930s and under Japanese occupation in World War II. The Philippine Republic was established in 1946; the United States retained its military bases until the 1990s, and then reactivated some after the attacks on September 11, 2001.

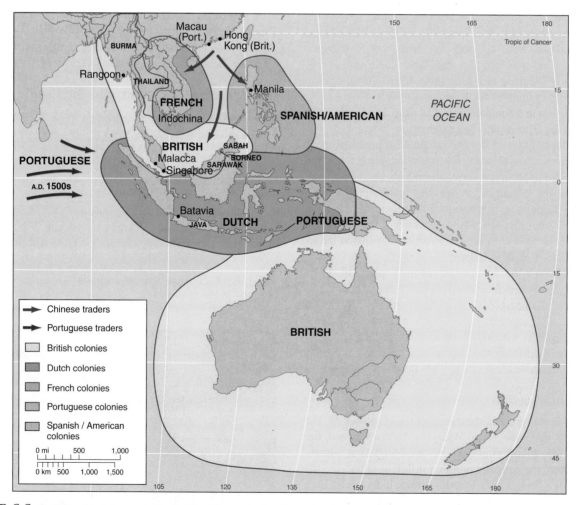

FIGURE 6.6 Southeast Asia and South Pacific: colonization. Americans, British, Dutch, French, and Portuguese assumed control of areas easily accessed by sea.

Indonesia

After 200 years of intermittent fighting among Spain, Portugal, the Netherlands, and Britain, Java and the surrounding islands became the Dutch East Indies colony in 1799. The Dutch made Batavia (modern Jakarta), the trading headquarters of their **Dutch East India Company** from the late 1500s, the capital of their colony. Local farmers were organized to produce coffee, rubber, and other new crops for export, but neglect of food production resulted in famines. Nationalist movements that began in the early 1900s were further motivated by harsh Japanese treatment during their occupation of World War II. Although the Dutch resisted granting independence after the war, it came in 1950 when the new country of Indonesia was formed.

Malaysia and Singapore

The Portuguese annexed Malacca on the Malay Peninsula in 1511 to control the spice trade but lost important regional control when the Dutch took over the Malay Peninsula in the 1600s. For over 100 years, the Dutch Batavia–Malacca link gave it a spice trade monopoly, contested by the (British) East India Company from 1601, which shared facilities at Bantam (west Java) until the Dutch took them over in 1682. In the late 1700s the British founded a port on Penang Island, a rival trading center linked to their sales of Indian opium. In 1804 the British and Dutch agreed to take control of Malaya and the East Indies, respectively.

The British built a new port at Singapore (1819) and controlled the Malay States by diplomacy. The sultanates on the northern tip of Borneo (Sabah) and the northeastern coastlands (Brunei and Sarawak) became British protectorates administered from Singapore. Tamil Indian and Chinese laborers were brought into sparsely populated areas to work, respectively, in the Malayan rubber plantations and tin mines. In World War II the Japanese took Malaya; but the British reestablished their rule afterward, fighting against Communist guerrillas until independence was granted to Malaya in 1957. Singapore, Sabah, and Sarawak joined the Federation of Malaysia on their 1963 independence. Brunei decided to stay out of that federation, and Singapore withdrew in 1965 after problems of maintaining its largely Chinese identity among the Malay majority in the rest of Malaysia.

Myanmar (Burma)

The British colonized Myanmar and made it a province of British India. The British developed a rice-growing industry in the Irrawaddy River delta and built the port of Yangon (Rangoon) to export the rice. Over a million Indians moved into this southern part of Myanmar, becoming the leaders of the country's commercial community. Attempts were made to unify the newly developed Lower Burma with the main Burmese population around Mandalay and the hill tribes. After riots against the British in the 1930s, the main independence group sided with the Japanese in World War II but changed sides toward the end and gained independence in 1948.

Indochina

The French occupied Indochina (Cambodia, Laos, and Vietnam) during the 1800s and early 1900s. They built roads and railroads and encouraged manufacturing. During World War II, Nazi Germany occupied France and persuaded the French colonial government to allow the Japanese forces to pass through Indochina. After the war, the French tried to reestablish control of the area, but Communist groups forced them to leave the northern parts of Vietnam in 1954. The divided country, with a Communist north and free-market south, was subject to further warfare in which the United States supported South Vietnam. In 1975 North Vietnam was victorious and reunified the country. Laos went through much strife along its border during the Vietnam War, slowing economic development. Cambodia became independent in 1953 but suffered 30 years of civil war and invasions by the Vietnamese.

Thailand (Siam)

From 1856 to 1939 and 1945 to 1948, Thailand was known as Siam. The term **Thai** came into use for the first time in the early 1900s. This country never became a colony, maintaining its independence through a strong monarchy. After the 1932 abolition of absolute monarchy, a rigorous bureaucracy ruled, despite many political changes (19 military coups, 53 governments, and 16 constitutions from 1932 to 1999). The influence of the bureaucracy is reflected in the identical school buildings in all 76 provinces.

Thailand's rulers established commercial treaties with Britain in the 1800s and encouraged modernization. In the 1890s, following a short war when France tried to make it part of Indochina, Thailand yielded land to France (now in Cambodia) and some of the Malay Peninsula to Britain. In World War II, Thailand at first sided with Japan, allowing the Japanese armies access to invade Burma and Malaya. Thailand regained territories it claimed in Malaya and Cambodia. Its new government in 1944 wisely supported the victorious allies but had to return the disputed territories at the end of the war. The fighting in Vietnam in the 1960s and 1970s made it seem that Thailand would be the next "domino" to fall to the worldwide Communist advance. However, Thailand stemmed that advance at its borders, attracting considerable financial aid from the United States to help it resist.

Australia and New Zealand

Terra Australis, the "Southland," was the last major inhabited continent unexplored by the Europeans. The southeast trade winds blew early European traders in the East Indies away from Australia. The Dutch discovered the continent in the 1600s, when their improved ships and occupation of the East Indies led them to explore southward. British ships visited the region, but most early assessments were of dismal lands with little economic or settlement potential for Europeans.

More promising reports concerning the potential of Australia came from the surveys of Captain James Cook and others in the 1770s. The British claimed Australia, first calling it New

South Wales. The British used Australia as a penal settlement for the first few decades, which served to replace the British Empire's loss of America to independence. Settlement increased in the early 1800s after initial problems of food supply led to improvements in government administration and the issuance of free land grants that encouraged sheep farming.

Most of the convicts placed in Britain's Australia-based penal system came from the most impoverished neighborhoods of British cities. Although tensions existed between the ex-convicts and other settlers, some former convicts did well in business or government in Australia. A gold-mining boom in the 1850s drew speculators and new settlers to Australia. By the time the last convict ship arrived in 1868, voluntary migrants outnumbered the convicts and their descendants by 10 to 1.

During the 1800s, British groups established new colonies around the Australian coasts, each with its own main port city and a competitive pride that generated rivalries among the colonies (Figure 6.7). The federal idea finally came to fruition in 1901, when the Commonwealth of Australia within the British Empire was created out of the five colonies that then became states.

The Europeans took little account of the Aborigines, and many died from disease and oppression. The Tasmanian Aborigines were wiped out by 1876, and numbers of Aborigines on the mainland were reduced to fewer than 75,000 by 1933. In the 1900s, attempts to integrate Aborigines into Australian life partly succeeded in health and education terms, and they now number 300,000. Of the current numbers, only about 10,000 follow traditional ways of life, and the rest live on reservations or in cities. Many are materially poor, and despite making up only 1.5 percent of Australians, they comprise as much as 30 percent of prison inmates, often being subject to more rigorous judgment for small crimes than people of European heritage.

Colonial settlement came later to New Zealand than it did in Australia. Maoris resisted British missionaries and whalers in the 1700s and early 1800s. It was not until after 1840 that the British government encouraged the immigration of farming settlers. As it took sovereignty in New Zealand, it agreed to respect Maori land ownership. After 1860 a short gold rush brought speculators. Technological advances in refrigerated shipping made it possible to export fresh meat to Europe from 1882, and more sheep farmers became established. In 1907 New Zealand gained dominion status within the British Commonwealth with a large degree of autonomy. Although subject to some segregation practices, the Maoris are more integrated in New Zealand life than the Aborigines are in Australia.

Pacific Islands

The United Kingdom was the main colonizer of the Pacific Islands, including Fiji, Kiribati (Gilbert Islands), Tonga, Tuvalu (Ellice Islands), some of the Solomon Islands, southeastern New Guinea, and parts of modern Vanuatu, which was shared with the French. Guam and the Marianas were Spanish colonies until taken over by the United States as protectorates just after 1900.

The French colonized New Caledonia and the islands around Tahiti, which, like some Caribbean islands, remain part of France. Political decisions continue to be made by the French government in Paris, as emphasized by recent nuclear tests in French Polynesia. Like French colonies in the Caribbean, the French South Pacific islands have access to European Union markets.

The Germans were active in the 1880s, when they took the Marshall Islands, together with Nauru, Western Samoa, northeastern New Guinea, and some of the Solomon Islands. All were lost to the United States, Britain, Australia, or New Zealand in World War I.

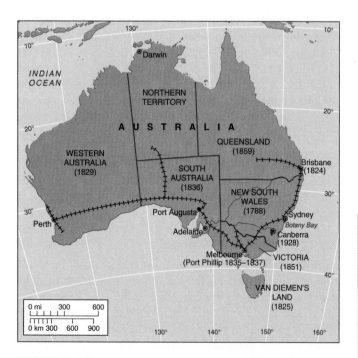

FIGURE 6.7 Australia: colonization patterns. Beginning in the late 1700s and continuing through the 1800s to dominion status in 1901, Australia developed political institutions around the separate colonies (later states). The dates in parentheses indicate when the colonies were established. The straight-line boundaries in many cases were drawn across the interior desert.

Summary

The continental and island countries of Southeast Asia, together with Australia, New Zealand, and hundreds of Pacific islands, comprise a varied region that had separate historic experiences but is becoming increasingly integrated into global affairs.

Many peoples of contemporary Southeast Asia trace their regional origin to groups migrating many centuries ago from more interior regions of the Asian continent. Traders from China and South Asia brought Hinduism and Islam to the region. European colonists claimed much of this territory in the 1800s, but small numbers of Europeans settled permanently, and the present countries achieved independence in the 1960s and 1970s.

Australia and New Zealand garner political and economic attention among some of the world's poorest and smallest Pacific island countries. Following colonization by mainly western European countries in the late 1800s, many of the islands of the South Pacific remain poor with few products to sell in global markets. Indigenous groups in Australia (Aborigines) and New Zealand (Maoris) fared badly under European settlement.

Questions to Think About

6A.1   Summarize the variety of external influences that contributed to the present human geography of this region.

6A.2   What have been the consequences, good and bad, for Australia and New Zealand of being at a distance from the materially wealthier Northern Hemisphere countries?

6A.3   How did differing colonial experiences in Southeast Asia, Australia, and New Zealand create different contemporary human geographies?

Key Terms

domino theory	Karen
containment	Aborigine
Association of Southeast Asian Nations (ASEAN)	animism
Mon	Maori
Khmer	Melanesian people
Vietnamese	Micronesian people
Lao	Polynesian people
Burmese	Dutch East India Company
Angkor Wat	Thai
Shan	*Terra Australis*

Natural Environments

The natural environments of Southeast Asia and South Pacific range from expansive landmass areas to thousands of small islands, from equatorial to midlatitude (to polar in Antarctica) climatic environments, and from volcanic islands and coral reefs that are forming now to some of the world's oldest landscapes. The long-term geographic isolation of some lands in this region gave them unique flora and fauna.

Equatorial, Arid, and Oceanic Climates

Equatorial Southeast Asia

Malaysia, Indonesia, and the southern parts of the Philippines have hot, rainy weather all year in the equatorial climatic environment (Figure 6.8) that covers up to 10 degrees of latitude on either side of the equator. Surface air is heated by the nearly 90° angle of the sun; the air mass converges and rises, and then condenses in cooler levels of the atmosphere. Massive clouds form, and gravity pulls the condensed moisture back to the surface in the form of **convective rain showers.** The islands and narrow peninsular environments are surrounded by water, which directly influences the climate and vegetation present. During the day, land areas near the coast heat up more quickly than the offshore waters. The warm air rises, and cooler air moves from the water to the land, in the form of a **sea breeze,** to replace the rising air. Tall clouds build up in the afternoon as humid air blows in from the ocean, and intense thunderstorm rains occur in the late afternoon and evening. Some places have over 300 days a year with thundershowers. Nights are mostly clear as air drains back from the cooling land to the ocean in the form of a **land breeze.**

The northern Philippines, Indochina's mainland, Thailand, Myanmar, and northern Australia experience a monsoon tropical climatic environment in which summer rains are brought by winds from the oceans. Winters are cooler and drier, dominated by winds blowing outward from Asia's interior.

The western Pacific is subject to frequent **typhoons** (hurricanes). Southeast Asia is at the western end of the oceanic and atmospheric circulations that cause the El Niño fluctuations off western Latin America. When Peru is dry, this region has plentiful rains and vice versa. During the major 1997–1998 El Niño, an area stretching from the South Pacific islands into Indonesia and northern Australia suffered intense drought and forest fires while Peru had unusual deluges of rain.

Tropical Ocean Climates

The islands of the South Pacific Ocean are nearly all in the tropical belt. Poleward of 10 degrees north and south, the trade winds blowing from the northeast (Northern Hemisphere) or southeast (Southern Hemisphere) are nearly constant factors. The windward east-facing sides of mountainous islands experience **orographic** enhanced precipitation. Low-lying **coral atolls** (fringing reef) lack hills to cause uplift and rain and are often arid with small and uncertain rains.

The tropical oceans have high water temperatures throughout the year, supplying moisture and heat to the air above, which fuel intense tropical storms and typhoons. Tropical disturbances and typhoons occur in the belt between about 10 and 25 degrees north or south of the equator (although they are more frequent north of the equator). The western Pacific is very susceptible to the development of tropical cyclones: an average of 32 such storms are generated each year.

Australia and New Zealand

Australia's climates are dominated by its arid continental interior. One-third of Australia is arid, and another one-third is semiarid. Water shortages are permanent characteristics of much of the continent, and even the areas that are normally well watered

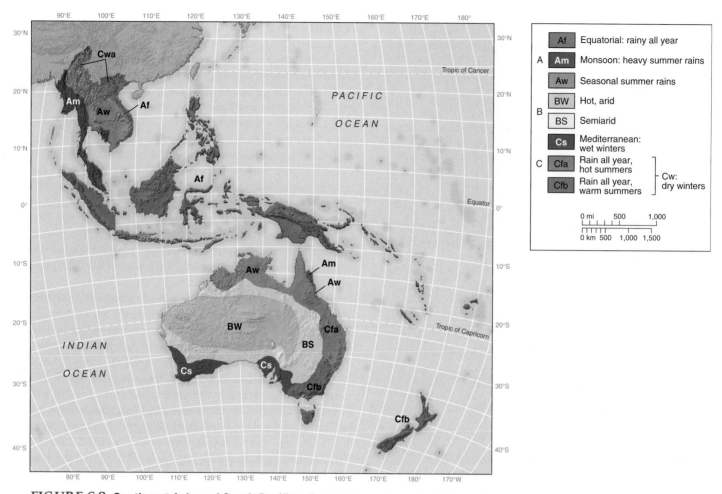

FIGURE 6.8 **Southeast Asia and South Pacific: climates.** From the equator to the south pole. Compare the significance of continents and oceans in the range of climates.

suffer from lengthy droughts. The aridity of the interior may be broken by storms and flash floods, occasionally filling some of the dried lakes such as Lake Eyre.

The coastal areas of Australia tend to have more regular rainfall. Winter rains of the Mediterranean climatic environment, brought by the **midlatitude cyclones** of the southern oceans, are characteristic of the southwestern corner of the continent and the Adelaide area. **Monsoon climatic environment** or seasonal summer rains and tropical cyclones occur along Australia's tropical northern coasts. In March 2005, tropical cyclone Ingrid damaged resorts as it brushed along the northern coast of Australia. The southeastern coastlands, where most Australians live, have a warm midlatitude climate with rains all year, most falling in summer.

New Zealand's climate is humid and similar to the British climate that many of its settlers left behind. It has fewer temperature extremes than the British Isles because there is no continental interior nearby to influence the range of average annual temperatures. The mountains on New Zealand's South Island create a rain shadow to their east, necessitating the irrigation of some of the farmland.

Continents and Islands

The surface features of Southeast Asia and New Zealand (see Figure 6.2) include mountain ranges and island chains that formed in the recent geologic past. Australia is an ancient continent. The Pacific islands are mainly the product of recent volcanic activity.

Plate Movements, Mountain Ranges, and Volcanic Activity

The region's dramatic relief and high mountains, frequent earthquakes, and volcanic eruptions result from the clashing of three major and one minor tectonic plates (Figure 6.9). **Convergent plate boundaries,** where plates slowly collide, form much of the western and all of the southern and eastern periphery of Southeast Asia. The boundary between the Indian and Eurasian plates turns southward after following the west–east line of the Himalaya Mountains. Where the Indian plate pushed into Asia from the west, it formed parallel north–south ranges in Myanmar (Burma). Folded mountains form Thailand's western border with Myanmar and extend into the Malay Peninsula.

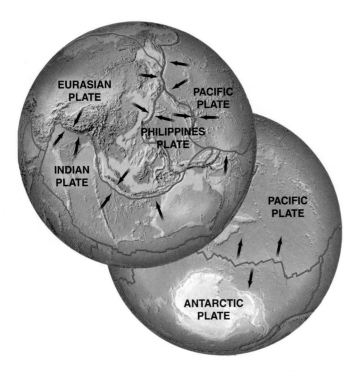

FIGURE 6.9 Southeast Asia and South Pacific: plate convergence. The convergence of tectonic plates produces landforms, earthquakes, and volcanic eruptions throughout the region.

The plate boundary swings eastward south of the line of Indonesian islands including Sumatra and Java but keeps north of Australia. The Indian Ocean floor (Indian Plate) is **subducted** beneath the Eurasian Plate and Indonesian islands, contributing to the eruption of volcanic material that forms numerous islands north and east of this plate boundary.

On the eastern margins of Southeast Asia, further groups of islands, from eastern Indonesia to the Philippines, formed where the Pacific plate was subducted beneath the eastern margin of the Eurasian and Philippine plates. The volcanic islands of Indonesia and the Philippines provide active landscapes with eruptions every decade or so. Between the convergence zones along the eastern and southern boundaries of Southeast Asia is a triangular area where the seas are shallow, as in the Sunda Shelf between Malaysia, Vietnam, and Borneo. The Indochina Peninsula and eastern Thailand (Khorat Plateau) are hilly areas formed from the erosion of ancient rocks brought to the surface by uplift.

The two collision zones of convergence meet in eastern Indonesia and form an active zone between the Indian and Pacific plates running through New Guinea and islands to the east. A **transform plate margin,** where plates are sliding horizontally along one another, marks the eastern boundary of the Indian plate and then angles southwestward through New Zealand. The unusual shape of Sulawesi (Indonesia) marks a front of uplift formed by pressures from the east.

New Zealand is the product of plate margin activity. Its two main islands are part of a fragment of the southern continent that broke away from the main mass of **Gondwanaland** (Figure 6.10) around 100 million years ago. The uplift of the land keeps pace with the glaciers and rivers that wear it down. In the Ice Age, glaciers carved deep valleys in New Zealand's rising Southern Alps, while rivers flowing from these mountains cut into the sloping eastern plateaus. The rising sea level as the climate warmed at the end of the Ice Age drowned the mouths of the deep glaciated valleys to form fjords (Figure 6.11).

Many volcanic islands also occur farther out in the Pacific Ocean, away from plate margins. These islands form above areas of crustal heating and may have a volcanic core

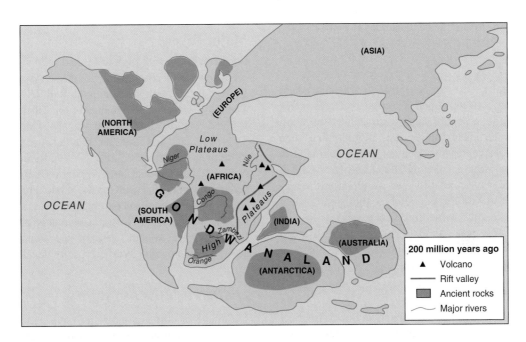

FIGURE 6.10 **Gondwanaland.** Earth's landforms 200 million years ago.

FIGURE 6.11 New Zealand: South Island coast. The area was carved by glaciers and then drowned by the rising sea level at the end of the last glacial phase. **Photo:** © Vol. 37/Corbis.

FIGURE 6.12 South Pacific: coral island. A typical South Pacific island, with a hilly volcanic core, an outer coral reef, and a shallow lagoon between the reef and mainland. The village is situated where a channel leads from a breach in the reef front. **Photo:** © Patrick Ward/Corbis.

surrounded by coral reefs (fringing reef), a volcanic island with a wide lagoon between it and the coral reef or **barrier reef** (Figure 6.12), or a central lagoon surrounded by coral reef but no volcanic island (atoll). It is thought that such coral reef formations represent a sequence that begins with active eruptions building a volcanic island and ends with it sinking beneath the waves under its own weight after the volcanic eruptions cease. Coral reefs colonize the islands when tiny animals secrete limy structures (limestone) at or just below sea level. The coral animals live near the tropical ocean surface, where it is warm enough and where algae on which they feed get access to light and oxygen.

Ancient Continent

Australia was once joined to Africa, Antarctica, South America, and the peninsula of India in the continent of Gondwanaland formed of ancient ore-laden rocks (see Figure 6.10), but now

lies on the eastern portion of the Indian plate at a distance from all plate margins. The relief features of the western half of the continent are low plateaus and plains on the ancient shield rocks. Mountain ranges formed over 600 million years ago were worn down to form landscapes of little relief today, and the lack of vegetative cover in the arid area exposes ancient mountain rock structures.

The Great Dividing Range along the eastern edge of Australia is formed of rocks that were deposited from the erosion of the ancient continent and were then uplifted and broken into blocks by faulting. Apart from the steep edges of these eastern uplands, Australia has fewer major relief contrasts than other large landmass areas. The world's largest continuous chain of coral forms the Great Barrier Reef off the northeastern coast of Australia.

Major Rivers

Long rivers carrying high water and sediment flows in summer are a feature of continental Southeast Asia. They include the Irrawaddy, Salween, Mekong, and Red Rivers (see Figure 6.2). The Salween, which originates in the Tibetan Plateau, and the Irrawaddy flow through Burma. The Mekong rises in the Tibetan Plateau and flows through the countries of Indochina to its delta in southern Vietnam. The Red River lowlands form the northern Vietnam heartland. Western Thailand is drained by a series of rivers that join in the Chao Phraya lowlands near Bangkok, forming the economic and cultural heart of the country. The continent of Australia is mainly dry, but the Murray-Darling River system west of the Great Dividing Range supports farming and hydroelectricity development.

Distinctive Ecosystems

The range of climatic and relief environments and degrees of long-term isolation from other land areas as the Gondwanaland continent split apart produced a variety of natural vegetation and animal types in the region. The main contrast is between the continental Asian species and the unique groups of species inhabiting Australia, New Zealand, and the islands of eastern Indonesia.

The tropical rain forest areas of equatorial Indonesia and Malaysia contain diverse species of trees and other vegetation. The monsoon forests, from Burma to Indochina and the Philippines, have somewhat fewer species; extensive areas are dominated by teak forest.

Separated by plate movements from the other southern continents some 35 million years ago, Australia's mammals were at an early stage of evolution that did not include a long womb-based gestation of babies. The **marsupials,** such as kangaroos, koalas, wallabies, and opossums, raise their young in pouches and compose about half the native animals. They are rare in other parts of the world. Regional bird life is particularly varied and colorful. Australian vegetation is dominated by species of eucalyptus (Figure 6.13) and acacia, and there are unique desert species in the dwarf mallee community. **Mallee** is formed of drought-resistant eucalyptus shrubs that grow into almost impenetrable thickets of many close-spaced stems rising to 8 or 9 m (25–30 ft.) high.

Some of the distinctive Australasian plant and animal species also occur in the eastern islands that are part of Indonesia today, separated by the **Wallace Line** from Asian species (see Figure 6.2). Although this line was drawn by botanist Alfred Russell Wallace in the mid-1800s, it was over 100 years before the origin of the separation was made clear. The line marks the edge of plate tectonic action some 10–15 million years ago that forced the eastern islands, fronted by Sulawesi, against the western islands and brought Australasian plants and animals with them.

During European colonization, species from this region were taken to adorn European gardens, and external species were introduced. Domestic animals from Europe found they had few local predators. For example, wild rabbits introduced in 1859 spread across Australia, destroying large areas of grassland. Efforts to control the rabbits continue.

FIGURE 6.13 **Australia: eucalyptus forest in humid Victoria state.** Photo: © Vol. 8/Corbis.

New Zealand had a unique rain forest–based flora and fauna before European settlement, but the immigrants' domestic animals and crops replaced many native plants and animals. Virtually all the original forest was soon cut but was later reforested by quicker-growing Douglas fir trees and pines for commercial uses. Introduced reindeer damaged trees and shrubs and are now contained.

Some of the larger Pacific islands are forested with species closer to those of Indonesia and East Asia than of Australia. Palms are particularly numerous. Few islands have many animals, although bird species are diverse. The surrounding waters contain a wealth of tropical fish varieties, but each is in relatively small numbers, and they are too easily overfished.

Natural Resources

Southeast Asia

The mineral resources of Southeast Asia include tin, iron, gold, and precious stones. Regional tin deposits occur where rivers eroded tin-bearing veins in the rocks and deposited this heavy mineral in concentrated **alluvial layers** lower down the valley or offshore. Deposits of oil and natural gas occur in Indonesia and in many recently explored offshore locations, such as between Indonesia and Australia.

The recent history of the Mekong River illustrates some of the regional challenges of resource utilization. The Mekong flows through or along the borders of six countries—China, Myanmar, Thailand, Laos, Cambodia, and Vietnam. It is one of the region's longest rivers, and its development potential includes hydroelectricity, agricultural irrigation, and transportation. Improved flood control would also aid economic development of the lands along the Mekong's length.

Plans for integrated, multipurpose development of the Mekong waters in the 1940s were abandoned during the subsequent wars in the area. Suspicions among the countries prevented further planning until an initial meeting of the four lower-basin countries in 1994, when an outline agreement was signed to investigate sharing the waters and protecting them from environmental damage. China and Myanmar did not initially sign the agreement; Cambodia and Vietnam are in dispute over transportation on the river; Laos interrupts the flow by dams that generate electricity, a major source of foreign revenue; China built the large Manwan dam on the river's upper waters; and Thailand is criticized for taking too much water from the river to irrigate its arid northeast. There are concerns over the potential for nutrient-rich sediment losses downstream from the dams and disruption of the freshwater fisheries in Tonle Sap, Cambodia. In 1995 Cambodia, Laos, Myanmar, Thailand, Vietnam, and China's Yunnan province signed an agreement for the future integrated development of the Greater Mekong Subregion based on a plan submitted by the Asian Development Bank. This adds formality to previous intentions but does not immediately overcome the difficulties of securing cooperation among former enemies and potential economic competitors.

Australian Resources

The ancient rocks of Western Australia—like the shield rocks of similar geologic age in Africa, northern Canada, and Siberia—contain many large deposits of iron ore and other metallic ores such as nickel, gold, platinum, uranium, and copper. The rocks that form the Great Dividing Range in eastern Australia are of more recent origin and contain coal, silver, lead, zinc, and copper ores.

Between the shield rocks and the Great Dividing Range, the lowlands are drained by the Murray-Darling River system in the south and form the Great Artesian Basin in the north where the worlds most significant opal deposits are located. Rain falling on the eastern mountains soaks into the rocks and drains westward and downward, accumulating in the sedimentary rocks of the basin (Figure 6.14). Because the rocks in the mountains remain filled with water, the pressure on water in the rocks beneath the lowlands is so great that wells drilled into the rocks cause water to flow out on the surface without pumping. The northern part of the lowlands was named for the **artesian wells.**

Pacific Islands

Some of the Pacific islands along the plate collision zone have mineral resources. The copper deposits on Bougainville, an offshore island that is part of Papua New Guinea, contain one of the world's largest copper reserves, which had been intensively mined until terrorist activities halted production. New Caledonia is the world's third largest producer of nickel ore. Most of the larger islands have a covering of rain forest, but some of the drier and flatter islands have only sparse vegetation.

Environmental Problems

Natural Hazards

Many people in this region live close to plate boundaries and have to contend with earthquakes and volcanic eruptions. The 1995 eruption of Mount Ruapehu on the North Island of New Zealand closed airports and highways and caused worries over the prospect of a larger eruption. Volcanic and earthquake activity continues along the plate boundaries, with 1990s eruptions in the Philippines and previous events in Sumatra and Java. Threats from typhoons, floods, and droughts also affect various parts of the region.

The tropical environments of Southeast Asia harbor many diseases. Human modification of vegetative cover, improvements in sanitation, better diets, and the wider availability of primary medical care continue to improve regional health conditions, although malaria, cholera, typhoid, and rabies remain serious problems. HIV/AIDS is an increasing threat in Southeast Asia. HIV/AIDS began primarily as an urban issue in the region in cities such as Bangkok but has now diffused widely throughout Southeast Asia.

For the future, the low-lying parts of islands, especially in the South Pacific but also in Southeast Asia and the Great Barrier Reef, face environmental disaster if global warming leads to a rising ocean level. The highest parts of some of the coral atoll islands are only a few meters above that level, and many have a concentration of settlement on low-lying coasts. Much of the region's coral, which both protects the coast and is a major tourist attraction, is under threat from rising sea level, predators, and pollution.

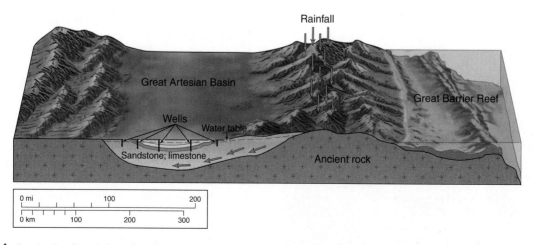

FIGURE 6.14 Australia: Great Artesian Basin and the Great Barrier Reef. The artesian basin of northeastern Australia is created by the geologic conditions that produce a major groundwater source as rain falls on the Great Dividing Range, seeps into the rocks, and flows downward to replenish the water in the rocks beneath the basin. The water in the deepest rocks flows out under pressure from the water moving down from the hills. The Great Barrier Reef is one of the world's largest developments of coral reef and forms a feature just off the Queensland coast.

THE INDIAN OCEAN TSUNAMI OF DECEMBER 2004

Geographic knowledge and geographic tools were utilized immediately to assist in recovery efforts following the devastating Indian Ocean tsunami in December 2004. The analysis and computer manipulation of remotely sensed images (satellite images and aerial photos from planes and helicopters) using geospatial tools like geographic information systems (GIS) began within hours of the impact of the tsunami's huge waves. The computer manipulation of satellite and aerial images of coastal zones (including the analysis of "before and after" imagery) made it possible to assess losses to the human landscape (houses, stores, hotels, roads, bridges, and port facilities). They also highlighted changes to the physical environment such as significant alteration of coastal morphology, loss of mangrove vegetation and the creation of new inlets and islands and the elimination of others, brought on by the force of the water (Box Figure 1).

An understanding of the human geography of Southeast Asia and other regions affected by the tsunami was and continues to be essential in the coordinated efforts of aid agencies, the international community, and regional governments. The ongoing recovery efforts require an in-depth geographic appreciation of regional and local culture (social relations, burial practices), geopolitics (insurgencies and regional governmental relations), and transportation and communications networks, among numerous other areas of importance.

Geographers and geographic tools continue to be used to assist with recovery and rebuilding, and to learn from the disaster in order to be better prepared in the future.

Box Figure 1 Indian Ocean Tsunami, Koh Raya, Thailand. People flee one of a devastating series of tsunami waves crashing on shore in Koh Raya (near Phuket) along Thailand's west coast. Photo: © John Russell/AFP/Getty Images.

Tsunami

The peninsular and island nature of the majority of the countries of Southeast Asia, situated where the Indian and Pacific Oceans meet, creates expansive coastal territory in the region. Extensive plate convergence zones exist along regional coastal stretches and underneath the region's waters (see Figure 6.9). Plate convergence beneath the ocean, coupled with the expansive regional coasts, makes Southeast Asia vulnerable to **tsunami** (Japanese for "harbor wave"). Tsunami are waves usually generated by water displacement from **megathrust** (substantial uplift) seafloor earthquakes. Although more tsunami are generated in the Pacific Ocean than any other of the Earth's oceans, a 9.0 magnitude earthquake on December 26, 2004, occurring under the Indian Ocean, produced a series of devastating tsunami. The December 26 earthquake (and its aftershocks) resulted from the subduction of the Indian plate under the Eurasian plate, causing uplift of the Eurasian plate and the displacement of a massive volume of water. The tsunami resulting from the water displacement caused extensive loss of life and property along the coasts of Indonesia and Thailand in this region, as well as in several other countries surrounding the Indian Ocean basin. (See the Geography at Work box above.)

Pollution, Erosion, and Mining Excavations

Human use of the natural environment in areas of high population density and modern industrialization significantly adds to the impacts of natural hazards. Expanding agriculture, together with cutting of forests for export lumber, resulted in extensive deforestation in many parts of the region. Rapid loss of forest contributes to soil erosion, watershed flooding, and destructive offshore sedimentation of fisheries and coral reef areas as silt is carried out to coastal waters. Warfare in Vietnam in the 1960s and 1970s defoliated forests and left chemical residues and barren hillsides that needed decades to recover.

Pollution of air, water, and soils is a serious problem in Southeast Asia as concentrations of people and industrial activities increase, more vehicles are used, and farming is made more productive by the use of fertilizers, pesticides, and mechanization. In the late 1990s, forest fires on Borneo and

247

Sumatra—lit to clear forest for planting oil palms, coupled with an El Niño–enhanced drought—covered much of Malaysia and Indonesia with dense smog and even contributed to the crash of an Indonesian commercial airliner.

Illegal and corrupt logging industries in several regional countries exacerbate existing environmental conditions. In Sarawak loggers finance local political parties that will allocate areas for tree felling when in power. International pressure for better logging practices in the region has influenced some structural changes in regional practices. In 2001 the Malaysian government agreed to take part in certifying that its timber exports come from well-managed forests. In 2002, Migros, a large Swiss supermarket chain, committed itself to buying palm oil from ecologically sound sources in this region, as opposed to the newer plantations that devastate the rain forest.

Australians are increasingly aware of the environmental damage caused to their fragile landscapes by farmers and miners. Farmers in Western Australia felled trees that previously helped to balance the water table and to slow the force of the wind near the ground. The resulting increase of wind erosion and rise in the water table led to surface salinization and made large areas unusable for farming. Replanting and mixing lighter soils with clays from mining waste helps to combat the dual menace. Overgrazing and overirrigation in the lowlands between the shield area and the Great Dividing Range have had similar negative results.

Mining extraction scarred many parts of Australia, causing environmentalists to join their concerns with land claims by Aborigines to delay and impose new conditions on applications for mining licenses. British nuclear tests carried out in the interior in the 1950s left scars on the land and on the Aborigines living in those areas. The United Kingdom still pays compensation for this damage.

When such problems came to public consciousness, the Australian federal government declared the 1990s a "Land Care" decade. Millions of new trees were planted in the early 1990s to make up for the halving of the limited Australian forest area following colonial settlement. It is anticipated that the capital being invested in attempts to restore soil quality may have positive results locally and in technologies that will be extended to poorer countries in arid parts of the world.

In New Zealand, the main forms of environmental degradation include soil erosion and the changes brought about in the fauna and flora by introducing new species. New Zealand takes an increasingly strong stance against potential polluters, and its desire for world peace makes it willing to endanger its relationship with the United States by keeping nuclear-powered naval ships out of its waters.

The common perception of the South Pacific islands as a pleasant and untouched part of the world is spoiled by the dumping of oil and other materials from commercial ocean vessels. In addition, several of the islands are uninhabitable after being mined for copper, nickel, or phosphate, or after nuclear testing. Nauru was left as an uncultivable skeleton following the extraction of its phosphate deposits.

Test Your Understanding
6B

Summary

Oceanic influences dominate the climatic environments of Southeast Asia and the South Pacific. These environments range from forested equatorial rainy areas through trade-wind climates to the stormy seas surrounding Antarctica. Interior Australia remains a desert.

Clashing tectonic plates produce mountains, volcanic islands, and earthquakes from Southeast Asia through Papua New Guinea to New Zealand. They are linked to productive soils and the deposition of mineral resources such as copper, gold, and nickel. The shield areas of Australia contain mineral resources such as iron, uranium, and gold. Environmental problems stem from attempts to establish European farming and tropical plantation crops in these new lands. Soil degradation and mining damage are among the main problems faced.

Questions to Think About

6B.1   What impact does the convergence of plates have on the region's relief patterns and on the regional human geography?

6B.2   Are natural environmental conditions present in Australia a causal factor in Australia's relatively small population? Explain.

Key Terms

convective rain showers	transform plate margin
sea breeze	Gondwanaland
land breeze	barrier reef
typhoons	marsupial
orographic	mallee
coral atolls	Wallace Line
midlatitude cyclone	alluvial layers
monsoon climatic environments	artesian wells
convergent plate boundaries	tsunami
subducted	megathrust

Globalization and Local Change

There is a wide range in per capita income figures for the countries of Southeast Asia and the South Pacific (Figure 6.15). The countries at the top of the diagram in Figure 6.15 have the highest incomes and the highest consumer goods ownership (Figure 6.16) and are the most involved in the global economic system.

Significant global connections began in the 1600s for Southeast Asia as European merchants and colonists established plantations and mines utilizing a combination of local labor and workers brought in from China and India. Economic links created between some parts of Southeast Asia and other world regions during the colonial era laid the foundation for

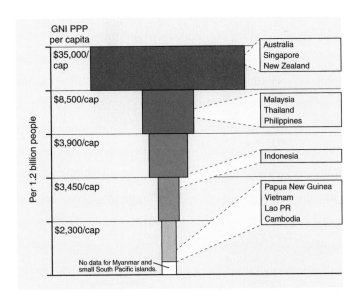

GNI PPP
per capita

$35,000/cap — Australia / Singapore / New Zealand

$8,500/cap — Malaysia / Thailand / Philippines

$3,900/cap — Indonesia

$3,450/cap — Papua New Guinea / Vietnam / Lao PR / Cambodia

$2,300/cap

Per 1.2 billion people

No data for Myanmar and small South Pacific islands.

FIGURE 6.15 Southeast Asia and South Pacific: country average incomes compared. The countries are listed in the order of their GNI PPP per capita. **Source:** Data (for 2000) *World Development Indicators,* World Bank, 2002.

many current global connections present in places such as Singapore. In Australia and New Zealand there were fewer pre-European inhabitants. Both Australia and New Zealand were settled by Europeans who composed the majority of the population and who maintained economic and cultural contacts with Europe. Global connections created in the islands of the South Pacific were fewer and did not have as lasting an impact on contemporary patterns.

The economic growth of the Southeast Asian countries in the late 1900s was at first based on exporting raw materials and later manufactures to American and European markets. By the 1990s it was also bolstered by intraregional trade. After 1970 much of the Southeast Asian, Australian, and New Zealand trade was oriented toward East Asia and the United States. The economies of Malaysia and Singapore were particularly linked to the U.S. and Japanese markets. In Australia and New Zealand, globalization resulted in less emphasis on a closed self-sufficiency with limited external market dependence and a more open attitude toward neighboring people, markets, and trade. Although many Pacific islands now have better connections to the rest of the world, few are so closely involved in the global economy apart from their involvement in the global tourism industry.

As a response to the late 1990s Asian economic crisis, the Asian countries agreed on steps to avoid future crises. Other proposals include an Asian Monetary Fund and a common currency unit (like the euro), both based on an Asia-wide group of countries that could act like the European Union to resolve historical tensions through economic links. As yet, internal Asian relationships remain difficult. Most ASEAN countries see fellow members as competitors rather than partners.

Although many economic and political events bring the countries of this region closer to each other, others threaten to reignite older conflicts and initiate new ones through cultural clashes, often within country borders. Indonesia, the country with the most people, faces threats from disparate island communities and intensified Islamist pressures after the tragic events of September 11, 2001. Other countries in the region fear that radical and terrorist movements will affect them.

Population Dynamics: Movement, Settlement, and Growth

Dramatic contrasts in the regional population distribution are evident in Figure 6.17. Population on the mainland portion of Southeast Asia is clustered in lowland river valleys and along coastal zones. Some of the islands of Southeast Asia exhibit very high population densities. The island of Java, in the country of Indonesia, has some of the most densely populated land in the region, as do many of the Philippine islands. Rugged terrain and steep slopes complicate settlement on numerous regional islands, resulting in even higher population densities. People in Australia are primarily concentrated in urban areas around the country's periphery along the coasts, while Australia's vast interior space has some of the world's lowest population densities. New Zealand's North Island has higher population concentrations and a higher percentage of people than its South Island. Only the largest Pacific islands show up on the map at this scale, but densities are often high on small land areas.

The South Pacific islands have small total populations, with the largest numbers on islands near Australia and New Zealand. In 2005 the group had a total population of some 8 million people, 5.9 million of whom lived in Papua New Guinea and 800,000 on Fiji.

Natural Increase

In 1965 the combined population of the countries of Southeast Asia was 275 million. It doubled to nearly 600 million in 2005 and is expected to reach almost 750 million by 2025. From 1965 to 1980, most countries had rapid population growth rates averaging over 2 percent per year, with Thailand and the Philippines close to 3 percent. The exceptions were Cambodia and Laos, with growth rates that reflected the ravages of war at around 1 percent or less. By the early 2000s, annual natural population increase rates in most countries were falling. Thailand, Indonesia, Singapore, and Vietnam were down to around 1 percent, while Laos, Malaysia, and the Philippines remained just above 2 percent.

The slowing population growth rates reflected declines in total fertility rates. All countries were over 6 percent in 1965, but by 2005 only Laos, Cambodia, and East Timor were over 4 percent, and the other countries were 3 percent or less. Thailand's family planning policies reduced fertility in that country to under 2 percent. Birth rate declines contributed to regional progress in

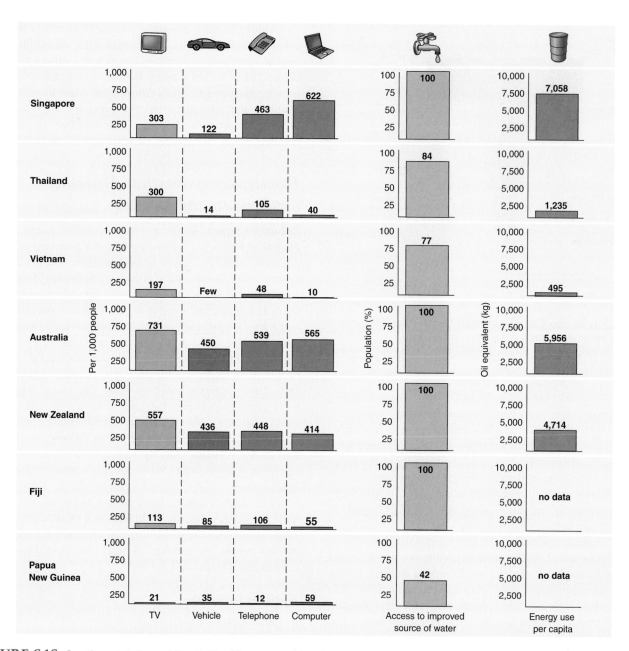

FIGURE 6.16 Southeast Asia and South Pacific: ownership of consumer goods, access to clean water, and energy usage.
Note the wide range of affluence among these countries (no water or energy data for Fiji, no energy data for Papua New Guinea). **Source:** Data (for 2002) *World Development Indicators,* World Bank, 2004.

demographic transition (Figure 6.18). The age–sex diagram for Indonesia (Figure 6.19) demonstrates the impacts of reduced numbers of births. The Vietnamese population profile was affected by wartime losses (those over 10 years old in 1975, over 25 in 2002), but the uncertain economic future after the end of hostilities in 1975 caused population numbers to grow slowly.

Until the 1960s Thailand encouraged births, and its population increased at more than 3 percent per year. A World Bank mission in 1958 recommended a program of birth control, and this became national policy in 1970. The Community Based Family Planning Service, begun in 1974, is a private, nonprofit organization that covers both urban and rural areas. This

organization worked both to educate the Thai people about the mechanics of family planning and birth control as well as to address the cultural constraints that favored large families. Buddhists monks contributed to the family planning efforts through their participation in a traveling road show, in which they used religious texts and blessed contraceptive pills. Bank loans were an incentive for farmers to take part. Thailand's rates of natural increase fell steadily from 1974 through 2001.

Indonesia also has family planning policies to reduce family size but placed huge resources into redistributing its population. Sixty percent of Indonesia's population of 222 million in 2005 lived on 7 percent of the land in the four largely metropolitan

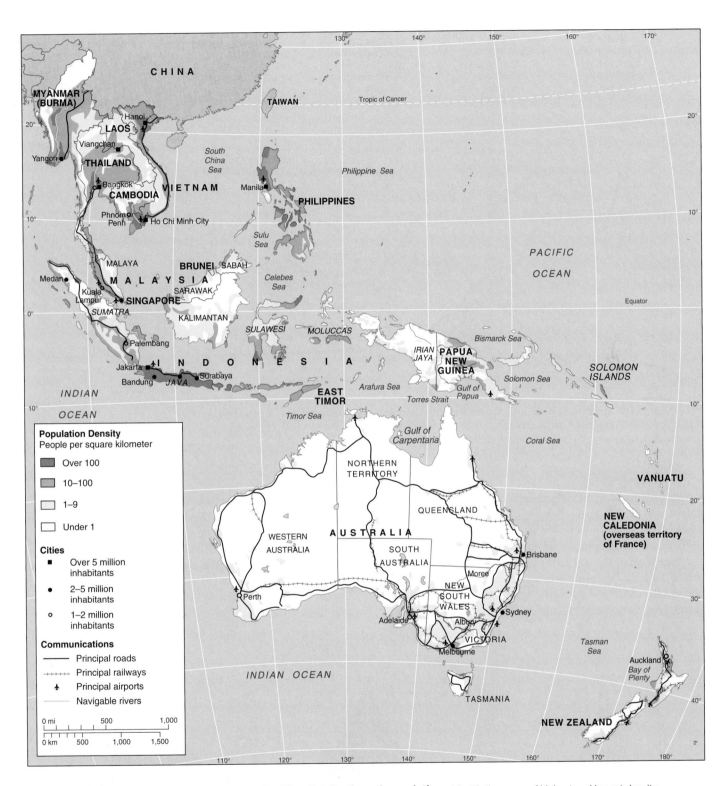

FIGURE 6.17 **Southeast Asia and South Pacific: distribution of population.** Identify the areas of highest and lowest density.

islands of "Inner Indonesia" (Java, Madura, Bali, and Lombok). Rural population densities on these four islands are often above 1,000 per km² (approximately 2,600 per mi.²), whereas they are mainly under 1 per km² (approximately 2.6 per mi.²) on the "outer" islands such as Kalimantan (the Indonesian part of Borneo), Sumatra, and the Moluccas—although none of these is "empty."

Australia's 2005 population of 20.4 million was double that of 1950. As with other materially wealthy countries, Australia's natural rate of growth is slow, with total fertility rates of less than 2 percent. Immigration results in annual population growth rates of over 1 percent, compared to less than 0.5 percent in European countries. Both Australia and New Zealand

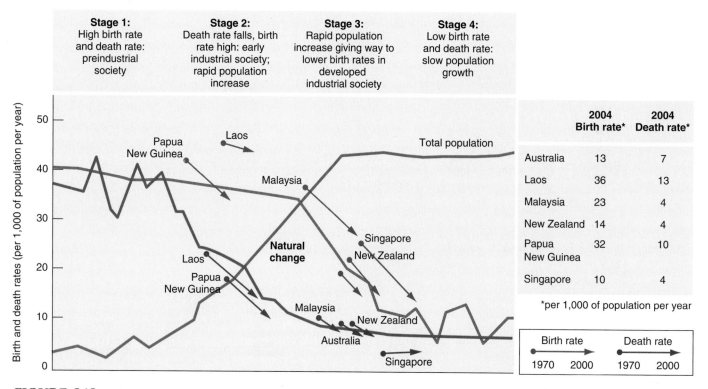

FIGURE 6.18 Southeast Asia and South Pacific: demographic transition. Compare the Southeast Asia range with that of the South Pacific countries. Papua New Guinea is typical of many of the islands. The Australian and New Zealand patterns resemble those of other rich countries. **Source:** 2004 update: *Population Data Sheet,* Population Reference Bureau.

are well advanced in the demographic transition process. Australia's age–sex diagram (Figure 6.20) reflects a slowing in births after a baby boom period from the mid-1950s to around 1970, as well as an aging population.

New Zealand's total fertility rate remains slightly above that of Australia, but its immigrants keep the annual population rate of increase at just over 1 percent. After gaining dominion status, New Zealand had to survive some years of labor force losses from its loyal support of Britain in two world wars. Today many younger New Zealanders in particular migrate to Australia for jobs in the professions.

Population growth continues on many of the islands due to high total fertility rates. The age–sex diagram for Papua New Guinea resembles those for poorer countries elsewhere and implies a large proportion of young people in the total population (Figure 6.21).

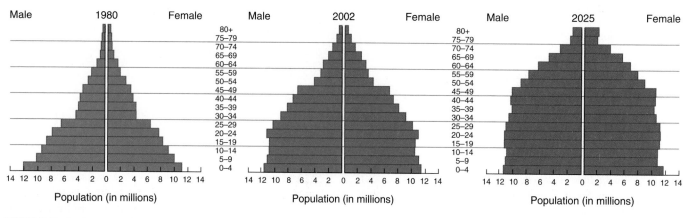

FIGURE 6.19 Indonesia: age–sex diagram. What factors cause Indonesia to have a different pattern than Australia? See Figure 6.20. **Source:** U.S. Census Bureau, International Data Bank.

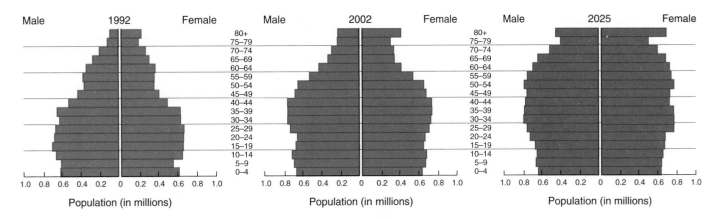

FIGURE 6.20 **Australia: age–sex diagram.** **Source:** U.S. Census Bureau, International Data Bank.

Many islands are at the point of overpopulation, where the land and its resources cannot support its people on a subsistence basis. Economic development has not occurred to support larger numbers of people. Although fishing and the cultivation of vegetables and fruit bring good harvests and a healthful diet, after World War II many islanders switched to imported canned meats and such delicacies as turkey tails, which are high in fat and contribute to a high incidence of connected obesity, diabetes, and heart disease. This has been called the "New World Syndrome." Only on Fiji, Guam, New Caledonia, and Palau is life expectancy close to 70 years; for most islands it is around 55 years. Genetic factors are also thought to influence the trend to obesity.

Migration and Resettlement in Indonesia

Indonesia's **transmigration** program began in 1950 and resettled more than 6.5 million people on the less populated islands. The selection process did not adequately consider age, skill level, ethnicity, or geographically appropriate farming systems (among others) in determining the destination and occupation

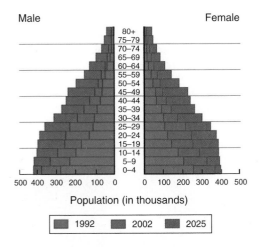

FIGURE 6.21 **Papua New Guinea: age–sex diagram.**
Source: U.S. Census Bureau, International Data Bank.

of those relocated. Despite this policy's democratic rationale, it was regarded by many as an insidious form of political manipulation and social engineering. People introduced to outlying islands were seen as government supporters and as importers of economic and social norms from Java. Nearly half of the migrants failed to raise their standard of living. From 1997, there were outbreaks of tribal groups murdering settlers as the settlers cut and burned forest habitats to extend palm oil plantings. Areas of rain forest were felled by illegal logging or mining, causing further social upheavals. Many migrants tried to move back to their home areas, but they no longer had homes or land and formed groups of politically unsettled peoples. The Indonesian government ended the program in March 2001.

The Indonesian experience asks whether imbalances of population distribution can be redressed on a large scale by officially sponsored migration. Few good tracts of agricultural land with lower densities of population remain in the islands (or elsewhere in the world), so there is little chance of accommodating major population movements in this way. Indonesia needs either to reduce population growth or to increase economic productivity, or both.

The Impact of AIDS

Population change in Southeast Asia is increasingly affected by HIV/AIDS. Thailand and Myanmar became the centers of the expansion of HIV/AIDS in Southeast Asia in the 1990s, despite government denials and optimistic statements about the disease being "self-contained" among drug users, prostitutes and their clients, and (though not mentioned) homosexuals. Asian women are less likely to have extramarital sex than women on other continents, but it is known that drug injectors have sex with prostitutes. There are large sex markets that may involve men who also have sex with their wives. So the disease is not contained, and 10 percent of adult Thais were infected before the government introduced a condom use program that cut new infections. In 2000 nearly 800,000 Thais had HIV/AIDS, and around 400,000 had died. In Myanmar 500,000 were infected, and nearly 200,000 had died.

Population Interaction in Australia and New Zealand

In the early 1900s wild claims were made that "empty" Australia could accommodate over 200 million people, although Griffith Taylor, a British geographer in the 1920s, doubted it could sustain more than 20 million. After decades of immigration, Australia's population in 2005 just exceeded the smaller figure.

Until the mid-1900s, most Australian immigrants came from the British Isles. In the early days of the Dominion of Australia, an informal **white Australia policy** acted against the pressures of Asians in particular. To increase its population and maintain its links with the United Kingdom, Australia encouraged immigrants from Britain by paying their passage until after World War II. As that source became insufficient, more immigrants came from other parts of Europe. By the 1990s only 12 percent of immigrants came from the United Kingdom, while 43 percent were from Asia. The legacy of immigration policies that limited admittance to whites (Europeans), a policy that ceased in 1972, still affects some Australian attitudes toward new immigrant communities and efforts to focus trade in Asia. Some Australians are challenged by the realization that countries they once looked down on now have economies as sophisticated as they perceive their own to be. Australia is becoming a multiracial, multiethnic, and multicultural country—an environment where tolerance for the rights of Aborigines and greater acceptance seem to be increasing.

Although high unemployment in the 1990s led to a halving of immigrants allowed into Australia, Australian businesspeople argue for greater numbers in the light of less-than-replacement levels of births and predictions of falling population totals and skill levels. Australia also aims to give 12,000 refugees a year the right to permanent residence; but in 2001 its insistence on offshore preentry interviews earned criticism from human rights groups such as Amnesty International, as the Australian navy intercepted boat people and transferred them to the isolation of detention camps on Nauru.

New Zealand's population continues to be more British in origin and allegiance than that of Australia. New Zealand receives fewer immigrants, although it does not discourage nonwhites in the way Australia did until 1972. The indigenous people, the Maoris, take a more significant part in New Zealand life than the Aborigines do in Australia. They total around half a million people, mostly living in North Island. Auckland has over 100,000 Maoris, many of whom have professional jobs.

Effects of Rapid Urbanization

Urbanization is increasing in all the countries of Southeast Asia (Table 6.2), although in 2005 Cambodia and Laos had less than 20 percent of their populations living in urban areas and Vietnam was only slightly higher at 26 percent. Southeast Asia's smallest countries, Brunei and Singapore, had very high urban percentages with 74 and 100 percent respectively. Most countries had a concentration of people, jobs,

Table 6.2 Southeast Asia and South Pacific. urban centers with over 1 million inhabitants. Numbers in millions of people.

City	Population	
	2003	2015 est.
Adelaide, Australia	1.1	1.2
Auckland, New Zealand	1.1	1.3
Bandung, Indonesia	3.8	5.3
Bangkok, Thailand	6.5	7.5
Brisbane, Australia	1.7	2.0
Haiphong, Vietnam	1.8	2.3
Hanoi, Vietnam	4.0	5.3
Ho Chi Minh City, Vietnam	4.9	6.3
Jakarta, Indonesia	12.3	17.5
Kuala Lampur, Malaysia	1.4	1.6
Manila, Philippines	10.0	14.0
Medan, Indonesia	2.0	2.7
Melbourne, Australia	3.6	4.0
Palembang, Indonesia	1.6	2.2
Phnom Penh, Cambodia	1.2	1.5
Singapore	4.3	4.7
Surabaya, Indonesia	2.6	3.5
Sydney, Australia	4.3	4.8
Yangon, Myanmar	3.9	5.3

Source: United Nations Urban Agglomerations 2003 (2003).

infrastructure, health care, and education opportunities in the largest **primate city.** The region's best example of urban primacy is in Bangkok (6.5 million), the capital of Thailand, which has 20 times the population of Thailand's next largest city, Chiangmai (Figure 6.22). Jakarta, Kuala Lumpur, and Manila also dominate in many aspects of the human geography of their respective countries.

Access to health care, education, business services, transportation, and global connections is greatest in the subregion's largest urban areas. For example, individuals and business

FIGURE 6.22 Southeast Asia city: Bangkok. The sprawling city skyline is dominated by modern high-rise buildings and punctuated with ancient temple compounds known as *wats*. The wat provides a sanctuary for Buddhist worship and relief from the busy city streets. **Photo:** © David Zurick.

firms in Bangkok have much better access to such services than those in Thailand's poverty-stricken northeast. Inside cities, governments subsidize urban transportation and provide tax incentives for industrialists to locate factories close to the main workforce. Most countries encourage manufacturing to a greater extent than agriculture—a further factor in urban growth at the expense of rural areas. The negative side of growing cities—congestion and pollution—appears to have little effect on their expansion. Bangkok's traffic jams are considered to be among the world's worst. Figure 6.23a shows the features of a typical Southeast Asian city.

Rapid growth of the subregion's urban systems outpaces governmental efforts to provide urban infrastructure. Government actions are often ineffective in controlling city expansion. For example, the uncontrolled growth of Jakarta, Indonesia, in the 1970s caused the Indonesian authorities to issue housing permits only to those who could prove employment. Such attempts to control urbanization actually augmented the areal coverage and densities of Jakarta's shantytowns.

Intraregional and interregional trade have fueled the growth of numerous city–regions in Southeast Asia and Australia, and even New Zealand. These city–regions are centers of international business and are the global or regional headquarters for numerous multinational corporations (see Figure 2.15). Among global city–regions, Singapore (see Point–Counterpoint: Singapore on page 256) leads the region's cities in global business connections (on a par, in this sense, with Chicago and Los Angeles in the United States). Sydney (Australia) is the region's second most connected city, while numerous other cities, such as Jakarta (Indonesia) and Kuala Lumpur (Malaysia), are increasing their connectivity.

Australia is one of the world's most urban countries, with 91 percent of its people living in urban environments in 2005 and almost two-thirds of the total population living in Australia's five largest cities. This reflects the historic primacy of these cities within their states and the growing significance of service industry employment and output. Australia's capital city, Canberra, has one-third of a million people. The major Australian cities retain their established port functions, although Melbourne grew further in importance by specializing in container trade. They have downtowns that combine state government and financial sectors with major cultural facilities such as the Opera House in Sydney (Figure 6.24). Industrial suburbs vie with residential areas for space.

Smaller towns are often linked to narrower functions, reflected in their urban landscapes. Thus Newcastle, New South Wales, is a coal and steel town, while Broken Hill is an inland mining center. Many small mining towns had short lives while the mineral resources or markets lasted and then became ghost towns. Towns along the Queensland coast have tourist functions, while inland towns such as Mildura are farm market centers.

In 2005, 86 percent of New Zealanders lived in towns and cities. New Zealand's main cities on both islands began as ports that acted as "hinges," giving access to the interior and back to the main overseas markets of the colonial homeland. Processing industries concentrated in the seaports, which became market, commercial, financial, and government centers. Auckland, with just over 1 million people in 2003, is the center of an urbanized area that stretches along the peninsula to its north (Figure 6.25). Wellington, with one-third of Auckland's

FIGURE 6.24 Australia: the center of Sydney. Sydney's harbor and Opera House are located on the near promontory. **Photo:** © Vol. 8/Corbis.

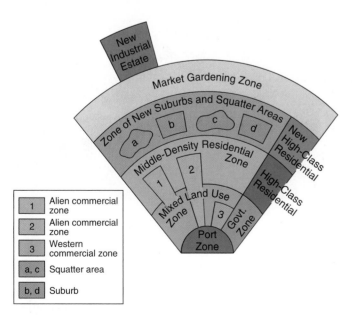

1	Alien commercial zone
2	Alien commercial zone
3	Western commercial zone
a, c	Squatter area
b, d	Suburb

FIGURE 6.23 Southeast Asian cities. Generalized plan of Southeast Asian cities in the 1970s and 1980s. Today extreme diversity exists in the urban centers of the subregion.

SINGAPORE

The Republic of Singapore is the most globally connected city in Southeast Asia (Box Figure 1). The population is well educated and mobile. The business operations and corporate ownership structures are intricately connected to major banks, business service firms, and stock exchanges around the world. But what is Singapore? Is it a city? A country? An island? Singapore is all of these things. It is a tiny urban country situated on a small island (which is surrounded by several smaller islets). Over 4 million people live on Singapore's 620 km² (239 mi.²) of land area off the southern end of the Malay Peninsula. Despite being the smallest country in Southeast Asia and without natural resources, it is by far the most materially wealthy in terms of GNI per capita. This is reflected in the comparisons of consumer goods ownership in southeast Asia (see Figure 6.16).

Hong Kong and Singapore have been entrepôts since colonial times, when the British recognized the value of their site and situation, and today they are the world's two busiest ports (each has more than double the traffic of any other world port). Singapore imports raw materials and unfinished goods from other countries in the region, adds value to them through processing, refining, and manufacturing, then sells them for a profit in various global markets.

Chinese people began migrating to Singapore during the colonial era, and today people of Chinese heritage comprise the country's largest ethnic group (Singapore's largest ethnic groups are Chinese—76 percent; Malay—15 percent; and South Asian—6 percent). Independence from the United Kingdom came to Singapore and Malaysia together as a federation in 1963. Disagreements between the Malay majority in Malaysia and Chinese majority in Singapore led to Singapore's declaration of independence from the federation in 1965.

Today Singapore is a multicultural, multinational, global city–state with international business services to rival major cities such as Hong Kong, China, Frankfurt, Germany, and Chicago in the United States. Religious groups, including Buddhists, Muslims, Hindus, and Christians, with smaller numbers of Jews and Sikhs, add to the global culture. The official languages are Malay, English

Box Figure 1 Singapore. (a) Sri Mariamman Hindu Temple with figures of gods on an external wall. (b) Waterfront at night. **Photos:** © Vol. III/Corbis.

(a)

(b)

population, is a major port and the center of government. Christchurch and Dunedin are the main towns of South Island. New Zealand towns reflect similar histories to those in Australia, including the most recent growth of manufacturing and service industries. All have port facilities.

Urbanization under Communism

The Communist countries experimented with distinctive urbanization policies aimed at reversing the uncontrollable urban growth and resettling people in rural areas to produce more food. The cities of Cambodia, such as Phnom Penh, were

(Continued)

(common language), Chinese (Mandarin), and Tamil (from southern India and Sri Lanka). Various Chinese dialects and other Indian languages (Punjabi, Hindi, Bengali, Tegelu, and Malayalam) are also spoken in specific communities. Occasionally tensions in race relations come to a head. In early 2002 the government was accused of ignoring the Malay Muslim minority group that makes up 14 percent of the population by cooperating with the United States in the Afghanistan war and arresting 13 Malay Muslims on suspicion of planning a bombing campaign. The Malay group is less schooled and earns less than the majority Chinese or many Indians, another minority group. Yet the Singapore government makes special provisions for them, such as Muslim schools.

The strong control exercised by the Singapore government redirected its economy from port-centered activities toward high-tech industry, including information technology and automation. In the 1990s Singapore firms manufactured 40 percent of world computer disk drives and included some of the leading makers of computer sound cards. As a transportation and communications hub, Singapore became the regional headquarters of many multinational corporations and one of the top 10 global city–regions. By late 2001, however, Singapore's employment was affected by cutbacks in U.S. and Japanese technology and financial services companies and multinational corporations. It is vulnerable to recession in the global economic system.

Singapore offered relocation facilities to Hong Kong businesses that feared the Chinese takeover in 1997. A few took up the offer. In the late 1990s, falling economic growth rates in the rest of East Asia and involvement in projects such as building in Shanghai slowed growth in Singapore. Its major thrust in the new century is as a regional business center, promoting investments. Other centers in Indonesia, China (Shanghai), and Vietnam are partially exports of the Singapore business environment.

Singapore's global connections are hindered in part by local pressures. Singapore's existence and prosperity are still resented by neighboring Malaysia. Both countries compete intensely for international business and investment. Malaysia opened a new port, PTP, at the southern tip of its peninsula, just opposite Singapore Island. The new Malaysian port immediately resulted in the loss of some of Singapore's major port customers. Singapore depends on Malaysia for nearly half of its fresh water.

The government of Singapore is involved in nearly all aspects of business and social life in the small country. Singapore is often referred to as "Disneyesque." It is one of the cleanest, safest, and most orderly urban areas in the world. International attention is periodically focused on the city–state when seemingly minor crimes are punished with what Westerners perceive as very harsh sentences. The government attempts to recruit well-educated businesspeople from around the world to live in Singapore while severely restricting the immigration, rights, and length of stay of less skilled laborers and domestic servants. The Singapore government at times has restricted the sale of chewing gum and has imposed and enforced fines on people for not flushing public toilets. Some of the most intense international scrutiny was focused on Singapore's government when it developed a formal plan to encourage reproduction among its most intelligent, educated, and financially successful citizens while discouraging the same from those the government perceived to be less "gifted."

The government of Singapore argues to its critics that the country has almost no crime, is a model of urban cleanliness, safety, and efficiency, and has extremely high living standards, a great quality of life, very high socioeconomic indicators, and high per capita material wealth.

POINT	COUNTERPOINT
Tight government control provides structure and order for the citizens of the country.	Citizens are not free to behave as individuals.
Government control produces extremely low crime rates; Singapore is one of the world's safest urban environments.	Singapore's punishments are extreme and inhumane— they are much too harsh and do not fit the crimes.
Singapore is one of the world's cleanest urban environments.	Singapore is sterile and contrived, it is more like a Disney theme park than a real urban environment or country.
There are no slums.	There are materially impoverished peoples in Singapore who live spatially segregated and restricted lives.
There is very low unemployment.	Those perceived to have lower job skills serve as street sweepers, "public toilet police," or household domestics and are often forced to live segregated and restricted lives.
Citizens of Singapore enjoy high living standards, great socioeconomic indicators, and a high quality of life.	High quality of life comes with a high cost of restrictions on civil liberties.

systematically depopulated in the late 1970s under Pol Pot. In early 2002 large areas of squatter settlements along the river in Phnom Penh suffered fires that displaced nearly 20,000 people. Those people were moved to fields outside the city and housed in tents, while the river land was designated for new buildings.

At its end, the Vietnam War focused on the American defense of Saigon. During the war, the frightened and war-torn people from surrounding rural areas moved into Saigon and other coastal cities such as Da Nang. Saigon grew from 2.4 million people in 1964 to 4.5 million in 1975.

FIGURE 6.25 New Zealand: urban landscape. Auckland harbor, North Island, is the main commercial center of New Zealand. Its waterfront has extensive yacht docking facilities and oil terminals close to the high-rise city center offices. **Photo:** © Vol. 121/Corbis.

When the Communist government took over Saigon after the 1975 reunification of North and South Vietnam, they renamed it Ho Chi Minh City and implemented a **deurbanization** policy to reduce the "unhealthy" overcrowding that led to insufficient water, housing, and power. Vietnamese families were directed back to their villages. Economic zones were set up in rural areas. However, ethnic Chinese businesspeople, who mainly lived in the city, were so persecuted that nearly 75,000 left the country between 1977 and 1982. Many fled by boat and were known as "boat people." Eventually the Vietnamese government realized that the initial target for reducing Ho Chi Minh City's population was too ambitious, and they raised the allowed total to 1.75 million. In the late 1980s the policy changed again, with controlled but slower urban growth alongside a welcome for foreign capital investment in new industries. The 2003 population of 4.9 million was forecast to reach 6.3 million by 2015.

Geographic Identity

Ethnic Variety

Following centuries of migration over land and sea, many of which were instigated by Chinese expansion and the European colonial powers, the people of Southeast Asia comprise very diverse groups, including the Burmese, Thais, Cambodians, Vietnamese, Malays, Indonesians, and Filipinos. Chinese communities are significant in all countries, making up 7 percent of the total population. Overseas Indians form important groups in Myanmar, Malaysia, and parts of Indonesia. Groups of indigenous people survive in the hilly and mountainous regions of the mainland, particularly in the uplands of Myanmar (Shan and Karen) and on the islands of Indonesia such as Kalimantan (Indonesian Borneo).

Significant ethnic, religious, and linguistic diversity exists in Southeast Asia. Animistic religions are prevalent among indigenous tribes. Islam dominates Indonesia, Malaysia, and Brunei. Buddhism is the chief religion of the majority of the mainland countries. The Philippines are mainly Roman Catholic. Hinduism is locally important, as in southern Myanmar (Bengalis), Malaysia (Tamils), and a remnant of the religion exists in Bali and Lombok (Indonesia). Even where a country's population is mainly of one religion, there are groups with other allegiances. For example, many of the Chinese businesspeople in Indonesia, Malaysia, and Singapore are Christians.

In Thailand, ethnic characteristics give identity to different regions, exemplifying the diversity of Southeast Asia. The people of the mountainous northern and western areas of Thailand are mainly Buddhists with historic links to those in Myanmar. Minority upland groups, such as the Karen, migrated from Myanmar, Laos, and China over the centuries. In Thailand's northeast, people share languages, artistic traditions, and Buddhism with the Lao people to the east. Centuries of relative isolation and autonomy shaped local identities. In the central Chao Phraya River basin, the cultural heartland of Thailand, people speak standard Thai. Their Buddhism is more akin to that of the Khmer people in Cambodia—the more conservative form that is similar to that in Sri Lanka (see Chapter 7). A small proportion of Hindu Brahmins living in the central region act as directors of royal and official ceremonies and use their astrological expertise to prepare the annual calendar. The rapid economic growth of this region adds to the mix by drawing in people from the poorer northeast. Along the hilly southern coastal area and extending eastward along the Cambodian border, people of Chinese descent settled in the 1800s to work on sugarcane plantations and now work in timber mills and run small stores. The southwest Pak Tai region has people linked closely to those in central Thailand, but Malay-speaking Muslims live in the extreme south.

Indonesia's many islands are home to varied ethnic mixes. In western Indonesia, strongly Hindu wet-rice growers on Java and Bali have a sophisticated culture of social and agricultural traditions that evolved over centuries. Islamic coastal groups include Malays on Sumatra and the Makasarese on southern Sulawesi. Kalimantan has many indigenous tribal groups, dominated by the Dayaks, often still bound to a shifting-cultivation economy.

Human Rights

Human rights abuses in Southeast Asia largely exist along ethnic lines. Such diversity often leads to tensions and conflict. Malaysia's 1969 race riots led to new policies to correct the economic "imbalance" where ethnic Chinese owned nearly all the businesses; by 1990 non-Chinese people owned 30 percent of Malaysian shares. However, the policy was relaxed when Chinese investments that could have been made in Malaysia began to be placed in other countries. Such policies to restrict Chinese economic influence, however, tended to turn racial differences into rigid, institutionalized social barriers. In Myanmar, the east remains a battlefield between the hill country people (Shan, Karen, and others) and the central government that tries to crush them, while the southern center around the capital, Yangon, is less troubled (Figure 6.26).

People living in the countries of Indochina suffered greatly from many conflicts in the latter half of the 1900s. When the Communist **Khmer Rouge** ruled Cambodia from 1975 to 1979, an estimated 1.7 million people died from starvation, disease, or execution. Many of the killers continued to live prominently in the country without being charged with ethnic crimes until King Sihanouk signed a law in 2001 to create a tribunal to put the remaining Khmer Rouge leaders on trial. The genocidal eradication of hill tribes in Cambodia and Vietnam was a further example of ethnic hatred being linked to political crusades. The Vietnamese hatred of the Chinese is exposed in occasional border skirmishes, in the expulsion of many Chinese businesspeople from southern Vietnam since 1975, and in competing claims for the Spratly Islands. Central Vietnam has a number of disadvantaged ethnic minority groups, many of whom became Protestant Christians before 1975. Some fled to Cambodia, but although a 2001 United Nations repatriation program expected to send them back to Vietnam, the program was ended in 2002 because the UN was not convinced it would be in the people's interests.

Subregions

Three subregions arise from the interactions of natural environments and cultural and economic histories of the countries:

1. *Southeast Asia:* continental Vietnam, Laos, Cambodia, Thailand, and Myanmar (Burma), and island-based Malaysia, Indonesia, East Timor, and the Philippines.
2. *Australia and New Zealand.*
3. *South Pacific islands.*

Southeast Asia

Southeast Asia (Figure 6.27) includes part of the Asian continental mainland and a series of islands between continental Asia and Australia. The subregion's insular and peninsular situation provides the majority of Southeast Asian countries with extensive stretches of coastline. The combination of spectacular coastal scenery with the subregion's rich cultural heritage and diversity formed the basis of growing tourist industries in the 1990s. The determination of several Southeast Asian countries to globally expand their industrial and corporate connections is challenged by ethnic and political instability. The operations of terrorist networks in the Philippines and Indonesia, coupled with internal conflict in Myanmar, Malaysia, Indonesia, and Mindanao (the southernmost island of the Philippines), divert resources from development assistance and economic goals. In this subregion, the **cultural fault lines** are local features and divide many communities.

FIGURE 6.26 Southeast Asia: a Burmese street in Yangon. Photo: © Ian Coles.

Along with friction among communities on the land are increasing numbers of pirate attacks at sea. Although incidents extend west to Bangladesh and India, the favorite pirate zone is in Indonesian waters, with over 100 recorded attacks in 2000. In particular, the Malacca Strait between Sumatra, Malaysia, and Singapore is used by around 200 ships each day and accounts for 40 percent of reported attacks. Ships slow down to pass through this busy and narrow waterway, making them vulnerable to attack. Although only one pirate attack occurred in the Malacca Strait in 1998,

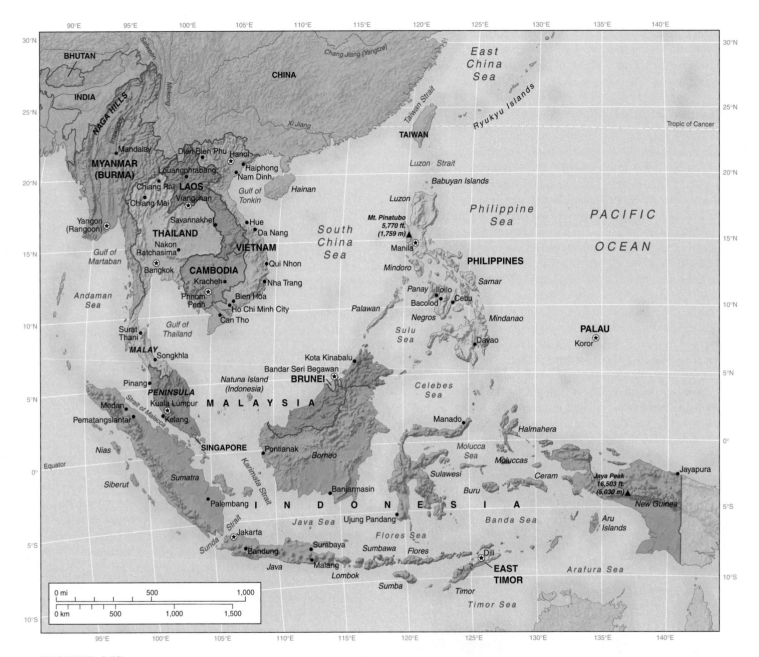

FIGURE 6.27 Southeastern Asia: countries and major cities. How important are ocean links in this region? In 1999, East Timor gained independence from Indonesia after 25 years of military occupation. It had been a Portuguese colony. The weakened Indonesian government is also faced with civil disturbances in other islands.

75 incidents occurred in 2000 after international controls slipped, linked to Indonesia's deteriorating political situation. New international piracy laws are being prepared, but their implementation requires more policing.

Countries

The countries of Southeast Asia are Brunei, Cambodia, East Timor, Indonesia, Laos, Malaysia, Myanmar (Burma), the Philippines, Singapore, Thailand, and Vietnam. In 2005 the populations of Southeast Asian countries ranged from fewer than 1 million people in Brunei and East Timor to 222 million

people in Indonesia (Table 6.3). Indonesia is by far the largest in overall area, covering three times the area of the next largest, Myanmar, and having an east–west extent equal to the distance across the United States.

The Mainland Countries

The continent-based countries of Myanmar, Thailand, Cambodia, Laos, and Vietnam each evolved from ancient kingdoms established primarily by migrants from the north (see Figure 5.7). Regional cultural diversity increased through long-term trade relations with Hindus and Muslims from India and farther west. Such trade interaction delivered religious traditions,

Table 6.3 Southeast Asia and South Pacific. Data by subregion, country, area, population, income (Gross National Income Purchasing Power Parity), urbanization, Human Development Index (HDI), Human Poverty Index (HPI).

Country	Capital City	Land Area (km²)	Population (millions)		GNI PPP 2003		% Urban	Index 2002		Ethnic Group %
			Mid 2005 Total	2025 Est. Total	Total (US$ billion)	Per Capita (US$)	2005	HDI (rank of 177)	HPI (% population)	
Southeast Asia										
Cambodia, Kingdom of	Phnom Penh	181,040	13.3	18.9	28	2,060	15	130	42.6	Khmer 90%, Vietnamese 5%
Laos, People's Democratic	Vientiane	236,800	5.9	8.7	10	1,730	19	135	40.3	Lao 50%, Thai 14%, Meo and Yao 13%
Myanmar, Union of (Burma)	Yangon (Rangoon)	676,580	50.5	59.0	No data	No data	29	132	25.4	Burman 68%, Shan 9%, Karen 7%, other tribes
Thailand, Kingdom of	Bangkok	513,120	65.0	70.2	462	7,450	31	76	13.1	Thai 75%, Chinese 14%, Malay 3%
Vietnam, Socialist Republic of	Hanoi	331,690	83.3	103.2	202	2,490	26	112	20.0	Vietnamese 90%, Chinese 2%, tribal groups
East Timor (Timor-Leste)	Dili	9,486	0.9	1.9	No data	No data	8	158	No data	
Indonesia, Republic of	Jakarta	1,904,570	221.9	275.4	689	3,210	42	111	17.8	Javanese 45%, Sundanese 15%, Madurese, Malays
Malaysia, Federation of	Kuala Lumpur	329,750	26.1	36.1	222	8,940	62	59	No data	Malay 59%, Chinese 26%, Indian 7%
Philippines, Republic of	Manila	300,000	84.8	115.7	379	4,640	48	83	15.0	Malay peoples 95%, Chinese 2%
Singapore, Republic of	Singapore City	620	4.3	5.1	103	24,180	100	25	6.3	Chinese 76%, Malay 15%, Indian 7%
Australia and New Zealand										
Australia, Commonwealth of	Canberra	7,713,360	20.4	24.2	563	28,290	91	3	12.9	White 95%, Asian 4%, Aborigine 1%
New Zealand	Wellington	270,990	4.1	4.7	85	21,120	86	18	No data	White 73%, Maori 12%, Asian 5%, Pacific Islander 4%
South Pacific Islands										
Fiji, Republic of	Suva	18,270	0.8	0.9	5	5,410	46	81	21.3	Fijian 50%, South Asian 45%
Kiribati, Republic of	Tarawa	730	0.1	0.1	No data	No data	43	No data	No data	Micronesian with Tuvaluan minority
Marshall Islands, Republic of	Majuro	179	0.1	0.1	No data	No data	68	No data	No data	Micronesian, German, Japanese, U.S., Filipino
Micronesia, Federated States of	Palikir	702	0.1	0.1	No data	No data	22	No data	No data	Chuukese 41%, Pohnpeian 26%, other groups
Nauru, Republic of	Yaren	20	0.01	0.02	No data	No data	100	No data	No data	Naurean 58%, other Pacific 26%, Chinese 8%

(continued on next page)

Table 6.3 *(continued)*

Country	Capital City	Land Area (km²)	Population (millions)		GNI PPP 2003		% Urban	Index 2002		Ethnic Group %
			Mid 2005 Total	2025 Est. Total	Total (US$ billion)	Per Capita (US$)	2005	HDI (rank of 177)	HPI (% population)	
South Pacific Islands *(continued)*										
Palau, Republic of	Koror	458	0.02	0.02	No data	No data	70	No data	No data	Malay/Melanesian/ Polynesian, Filipino, Chinese
Papua New Guinea, Independent State of	Port Moresby	462,840	5.9	10.6	12	2,240	13	133	37.0	Melanesian 96%
Solomon Islands	Honiara	28,900	0.5	0.7	1	1,630	16	124	No data	Melanesian 93%, Polynesian 4%
Tonga, Kingdom of	Nukualota	750	0.1	0.1	1	6,890	33	No data	No data	Tongan 98%
Tuvalu	Funafuti	30	0.01	0.01	No data	No data	47	No data	No data	Polynesian 96%
Vanuatu, Republic of	Port-Vila	12,190	0.2	0.3	1	2,880	21	129	No data	Ni-Vanuatu (Melanesian) 94%, white 4%
Western Samoa, Independent State of	Apia	2,840	0.2	0.2	1	5,700	22	75	No data	Samoan (Polynesian) 93%, Euronesian mixtures and others 7%

Source: World Population Data Sheet, Population Reference Bureau (2005), World Development Indicators, World Bank (2005), Human Development Report, United Nations (2004), and Encarta Microsoft (2005).

symbols, and architecture to coastal areas of Southeast Asia. In the late 1800s, the colonization of Burma by the British and of Indochina (Cambodia, Laos, Vietnam) by the French (see Figure 6.6) introduced further external demands.

After Japanese military occupation in World War II, Burma (1948) and later Indochina (1954) became independent. Burma was mainly governed by military rulers who changed the country's name to Myanmar in 1989. Vietnam, Laos, and Cambodia divided up Indochina and became Communist countries from the 1950s. The United States spent $150 billion and deployed nearly 600,000 troops in an effort to prevent the diffusion of Communism to South Vietnam. After several years and much loss of life of U.S. and Vietnamese troops and Vietnamese civilians, the United States ended its campaign of support in southern Vietnam; the northern and southern areas became a single Communist country.

The military and Communist governments kept Burma, Cambodia, Laos, and Vietnam aloof from increasing world trade, while internal conflicts slowed economic development. In Burma the military government fought persistent hill tribes wanting independence. In Cambodia the Communist government murdered over a million people in the late 1970s under Pol Pot. Vietnam and Laos were involved in warfare for nearly 30 years, from the 1950s against France and the United States and then with each other to around 1980. Reconstruction since that time has been slow.

In the late 1970s, support from the United States enabled Thailand to resist the Communist advance. With its constitutional monarchy and democratic government, Thailand adopted an open attitude to the expanding world economy, encouraging foreign investment, export manufacturing, and a growing tourist industry. Thailand became one of the economic leaders in Southeast Asia until the 1997 economic crash. Just as the 1997 crisis broke, Thailand adopted a new, far-reaching constitution focused on people's rights and challenging corruption and entrenched power at local and central government levels. However, the new measures instituted by the Thai government are contested by rival groups.

Island Countries

The island countries of Southeast Asia (see Figure 6.27) are Malaysia, Singapore, Brunei, Indonesia, East Timor, and the Philippines. Although Malaysia's two largest land components are continentally situated on the Malay Peninsula and on the island of Borneo, the country's more insular orientation dictates geographic analysis with the other island countries of Southeast Asia. Since becoming independent, most of the island countries of Southeast Asia have had histories of strong leadership, often alternating with periods of instability.

Indonesia is the subregion's most populous country, where 222 million people live in an **archipelago,** or island group, of more than 17,000 islands. Although some 6,000 of Indonesia's islands are inhabited, the country's population is concentrated on just four of the larger western islands, producing extremely high population densities. The combination of island fragmentation and the existence of more than 300 ethnolinguistic groups caused the Indonesian government to create a national motto of "diversity in unity" in an attempt to create a sense of national identity. Indonesia descended into violence and uncertainty in the late 1990s after long periods under the virtual dictatorships of Presidents Sukarno (1949–1966), who emphasized political control of the various islands in the federation but achieved little economic growth, and Suharto (1966–1997), who switched his government's emphasis to economic development. Both periods ended with economic crises and popular uprisings against corruption.

In the Suharto years, annual per capita income rose from US $70 in 1965 to $1,142 in 1996; more social services were made available than in many other Southeast Asian countries; and the number of people living below the poverty level decreased. In the mid-1990s the lavish lifestyles of Indonesia's political elite, centered on the extended Suharto family, drained the country's economy. The middle classes, who had supported Suharto, found upward social mobility capped by corrupt controllers. Parallel to these social changes, foreign debt doubled between 1990 and 1996 as both large and small companies engaged foreign capital. Some large companies owned Indonesian banks and lent to themselves, but when the 1997 Asian economic crisis grew, they could not repay those loans and had to seek foreign financing when worsening exchange rates multiplied the overseas debt. The politicized internal banking system collapsed. The financial crisis brought an end to Suharto's authority.

The overturn of the Suharto government in Indonesia was accompanied by social problems, including aggression against the Chinese business community in Indonesian cities, the resettled people in Kalimantan, and the Christian Chinese in Molucca. Several years of hesitant government followed. After Megawati Sukarnoputri became president in 2001, economic conditions began to improve, and the International Monetary Fund was able to reschedule the debts.

Problems for Indonesia continue, with ethnic strife threatening to lead the country—and possibly others in the subregion—into political chaos. Militant and pirate groups flourish on its islands and in neighboring countries. Several Indonesian areas, such as Aceh in western Sumatra and the Moluccas and Irian Jaya in eastern Indonesia, seek independence, threatening to break up the country. In 1999, in the northern Moluccas, a dispute over a gold mine turned into fighting between Christians and Muslims in which many died before the Indonesian president imposed a state of civil emergency. Thousands of Muslims and Christians now camp out, too scared to return to homes vulnerable to attack and burning, while the military stand aside or are accused of siding with the Muslims. By 2002, groups of militant Muslims from western Indonesia went to support Muslims in the eastern islands. Even on Irian Jaya, where Christian Papuans form a large majority, actions by

groups of Muslims increased after the September 11, 2001, attacks in the United States. Laskar Jihad and Laskar Kristus are two paramilitary groups that take sides in these disputes, with the latter linked to the Moluccan Sovereignty Front independence group.

In 1999, with United Nations help, the 800,000 people of East Timor began to work toward independence from Indonesia, which was achieved in 2002 after 24 years of military occupation. When Portugal left its colony in the eastern part of Timor Island in 1975, it was taken over by Indonesian military, who resisted allowing independence until the UN action. In 2001 the East Timor and Australian governments agreed to share the offshore oil and natural gas resources in the Timor Gap between the two countries. These resources should provide East Timor with plentiful funds for its development and defense.

In the Philippines, the colonial influences of Spain (1565–1898) and the United States (1898–1946) make it the most Westernized country in the subregion. The long-term dictatorship of President Ferdinand Marcos ended in 1986, and since then, a fragile democracy that involves military influences has struggled to maintain order. The economy advanced more slowly than others in the subregion. At present, government forces, assisted by U.S. advisers, seek to restrict an armed Muslim group, the Abu Sayaff, to Basilan Island off southern Mindanao.

Malaysia has a small population relative to its rich natural resources (see Personal View: Malaysia on page 264). It has a long-term, strongly led government that seeks democratic mandates under the supreme rule of the sultan, although opposition parties are tightly controlled. The government established policies to promote a competitive economy, enabling the country to experience the most rapid economic growth in the subregion after Singapore.

Singapore, small and urbanized, is the most materially wealthy of the Southeast Asian countries. Although officially a multiparty democracy, the socially controlling government of Singapore primarily represents a single political party (see Point–Counterpoint: Singapore on page 256).

Brunei is a small country on the northern coast of the island of Borneo. Its oil and natural gas riches and population of only 330,000 give it a GNI per capita of around $10,000. Although its geographic position appears to be off the main global routeways, it is central to growing Southeast and East Asia. A British protectorate until independence in 1984, Brunei refused to join the Malaysian Federation that surrounds it. However, the United Kingdom and Singapore made defense agreements with Brunei, which joined ASEAN immediately after independence.

Brunei's rule by a Muslim sultanate began in the A.D. 1300s. In 1990 Brunei introduced the concept of a Malay Muslim monarchy to promote Islamic values. Although two-thirds of the population are Muslims, the one-fifth who are Chinese felt alienated by this movement, as did the indigenous groups that inhabit the tropical rain forest interior and expatriates who

MALAYSIA

Malaysia became noticeably more prosperous during and since the mid-1980s. More people moved into the towns, especially the capital, Kuala Lumpur, because of the jobs in government offices, businesses, and banks, the greater range of shopping available in supermarkets and chain stores, and the increasing leisure opportunities from golf courses and swimming pools to badminton courts—the favorite sport. Kuala Lumpur experienced a huge expansion of high-rise apartment and office construction in the 1990s (Box Figure 1), including investments by corporations moving out of Hong Kong. Housing is more expensive in Kuala Lumpur than in other parts of Malaysia, so many people live in suburbs outside the city and commute by motorcycle, car, bus, train, and the light rail system. In the second half of the 1990s, Malaysia began the planning and construction of a new administrative center, Putrajaya, 40 km (25 mi.) south of Kuala Lumpur. In 2005, residents, businesses, and some government services occupied the new city. Malaysia has a democratic government and is a member of the Commonwealth (of former British colonies). Although select governmental administrative services moved to Putrajaya, the Malay Parliament convenes in Kuala Lumpur, which remains the Malay capital.

Most Malaysian houses have stone floors to help cool the equatorial air. Urban homes have refrigerators, TVs, and video players. TV programs include many from Australia and the United States, with a few British programs and local programs for Malay, Chinese, or Indian groups. There are radio stations for each of the main languages—Malay, Chinese, Tamil, and English. Most cooking is by gas, bought in cylinders delivered to the house. Water is metered, and most houses have showers rather than bathtubs. Life in the west coast cities and on the Malay peninsula in general contrasts with life in the rural villages of East Malaysia, where people live in houses of wood with coconut-leaf or tin roofs instead of brick and concrete.

Family ties remain strong, although family size decreased from about three or four children before 1980 to around two in the 1990s. Parents usually live with the oldest child, who looks after them when they retire. Government policy encourages high birth rates to generate a large domestic market for manufactured goods. Large families are promoted in an attempt to slow the fertility decline linked to increasing affluence and use of contraceptives.

Schools are mostly subsidized by the government, although some are still run by churches; the Chinese and Indian communities run independent schools. Even at the subsidized schools, parents pay fees and buy uniforms and books. Poor families often cannot afford to send their children to school. For most children, schooling lasts from 7 to 17 years of age, with the possibility of another two years for university entrance exams. The school day lasts from 8 A.M. to 12:30 or 1 P.M., with a break during the hot part of the day. School related recreation takes place after 4 P.M.

Most Malaysians use Western medicine, although some use herbal remedies. Doctor visits and hospital treatment have to be paid for. Poor people get free medical care but may have to pay for prescriptions.

Malaysia is a multiethnic society, with many people of mixed parentage, such as an Indian father and Thai mother. Malays, who are mainly Muslims, retain control of parliament and often regard themselves as first-class citizens and others as second-class citizens. The Chinese came to work in the tin mines during British rule, when Indians were also brought in for the rubber plantations.

Box Figure 1 Kuala Lumpur, Malaysia. The city center, with the historic Sultan Abdul Samad building and gardens, is backed by modern high-rise office blocks. **Photo:** © David Ball/The Stock Market/Corbis.

Chinese are now mainly middle class with commercial interests, while Hindu Indians include poor laborers, owners of small businesses, and professionals such as lawyers or doctors. Fearing that the Chinese and Indian groups would become too powerful, the Malaysian government passed laws that forced Chinese businesspeople to have Malay partners holding 50 percent control and enforced strict racial quotas in higher education, government jobs, and business ownership. Relationships among the different groups are not so tense as they were following government actions after race riots in the late 1960s, and are helped by the increasing well-being of Malaysians.

After 1986, economic growth exposed labor shortages. More women became involved in the labor force for high-tech manufacturing, and their wages contribute to rising family incomes. The Malaysian labor force grew faster than the total population as foreign workers flowed in to take increasing numbers of jobs. Immigrant workers from Indonesia, Bangladesh, and the Philippines extended the multiethnic nature of Malaysian society.

The orientation of Malaysian society changed with its economic focus. British influences remained strong until 1970, but since then, Japanese and U.S. firms established businesses in Malaysia that are seen as forming the foundation of recent economic growth. In addition to the manufactured products of Japanese and other international firms, palm oil, rubber, tin, and iron ore remain important exports, and oil comes from offshore wells in the South China Sea.

In 1997 Malaysians became aware of environmental quality loss as forest fires in Indonesian and Malaysian areas being cleared for oil palm plantings created intense smog over the country. A public health crisis followed in which schools closed. This event, repeated in early 1998, focused attention on lax environmental policies that encouraged timber cutting and mining at the expense of silting estuaries, disappearing coral reefs, and depleted fisheries.

have chosen to live in Borneo. The 1990 move followed the crushing of prodemocracy movements from 1962 and a state of emergency, after which the sultan ruled by decree.

The oil and natural gas wealth that produces Brunei's huge income was developed largely by the Shell Corporation since 1929 and was a target of Japanese occupiers in World War II. The sultan and his family are the main beneficiaries. Sultan Hassanal Bolkiah is one of the world's wealthiest men and built himself a palace costing $450 million in the capital, Bandar Seri Begawan. Inhabitants of the country do not pay any taxes, government jobs are plentiful, and the provision of health care and education is free and generous. A taxi driver serving a recent visitor had four children at university in Britain, two supported by the Brunei government and two sponsored by Shell.

After September 11, 2001, many Muslims in the subregion were arrested on suspicion of terrorist activities such as the planned (but foiled) attacks on U.S. military and commercial interests in Singapore. Jemaah Islamiah is a regional terrorist organization that seeks to establish an Islamist country, Dauliah Islamiah, to include Indonesia, Malaysia, and the Philippines. Ethnic conflicts in Indonesia following the 1997 Asian financial and social crisis radicalized some Muslims, and links were reported to al-Qaeda training camps on Sulawesi Island. The Indonesian government does little to control such developments because it fears a backlash from its Muslim majority if it acts strongly against suspected terrorists or Islamist organizations. Singapore, Malaysia, and the Philippines take stronger actions, although a lack of information undermines the confirmation of many accusations.

Traffic in People

The business of trapping people into laboring jobs or prostitution in other countries is a fast-growing scourge of this subregion, which supplies around half of the women and children trafficked every year across the world. Vietnamese are taken to Cambodia for the sex industry and to Taiwan and China for marriage. Cambodians shipped to Thailand, Taiwan, and Singapore become beggars, sex slaves, or domestic laborers.

People wishing to migrate to escape poverty or trying to get away from the traffickers may be caught by people smugglers who charge large sums (estimated to total over US $7 billion per year) to take them to a new country. Smuggled people often end up in the same occupations as those who are kidnapped. Such illegal smuggling and trafficking creates political tensions, as when Malaysia deports Indonesians or Australia intercepts boatloads of asylum seekers.

Some 10 percent of Filipinos, especially women, work abroad to support their families and educate their children, resulting in US $6 billion in payments in 2000. In the 1970s the Philippines was second only to Japan in wealth in this part of the world and employed many Chinese refugees escaping through Hong Kong. Today, as a result of transitioning challenges from the Marcos years, the Philippines is one of the subregion's poorer countries and the situation is reversed: Filipinos are common domestic help, or *amahs,* in Hong Kong, as well as

in Japan, and Saudi Arabia. Although the women may have college degrees and speak fluent English, the jobs they perform often mean virtual slavery.

Economic Development

The countries of Southeast Asia range from the oil-rich Brunei and the prosperous trading center of Singapore, to the economically growing Malaysia, Thailand, Indonesia, and the Philippines, to the military dictatorship of Myanmar and Communist regimes of Cambodia, Laos, and Vietnam with their widespread poverty.

Improvements in health, education, and income from 1960 to 2000 gave these countries increasing levels of gender equality, particularly in adult literacy. In the 1990s Malaysia made the reduction of poverty a major policy objective; by its own definition, poverty fell to 21 percent by 1985 and 14 percent by the late 1990s.

ASEAN and APEC

The Association of Southeast Asian Nations, or ASEAN group, was initially a defensive alliance rather than a trading one, but the peace it established formed a basis for economic growth. From the late 1960s, the ASEAN countries experienced a pattern of rapid industrialization stimulated by foreign investment and internal government encouragement. In addition to the five original members of ASEAN, Brunei joined in 1984, and Cambodia, Laos, Myanmar, and Vietnam all joined in the 1990s. ASEAN seeks to accelerate regional economic growth, social progress, and cultural development and to promote regional peace and stability. Although the organization achieved economic successes in recent years, cultural differences in religion, language, systems of governance, and justice, among others, challenge the cohesion and productivity of ASEAN. Further complicating the effectiveness of ASEAN is the extreme economic competition that exists between its member countries for foreign capital and foreign investment.

The **Asia–Pacific Economic Cooperation (APEC),** which is a much more geographically inclusive economic alliance, may overshadow any gains made by ASEAN. The APEC organization refers to its "member economies," not member countries. In 2005 APEC had 21 member economies including many countries in East Asia, Southeast Asia, Australia and New Zealand, and western Latin America, as well as Canada and the United States. Differing perceptions of what constitutes fair economic treatment of individual countries' agricultural products and what is the appropriate political interaction with Taiwan are among APEC's current challenges. The APEC organization seeks to link the growing economies around the Pacific Ocean margins and to provide an alternative, and formidable, major world trade grouping to those situated on either side the North Atlantic (Figure 6.28).

To be more competitive within APEC and against the European Union, the ASEAN membership is currently working on some common economic goals with a few countries outside Southeast Asia. Economic cooperation with the People's

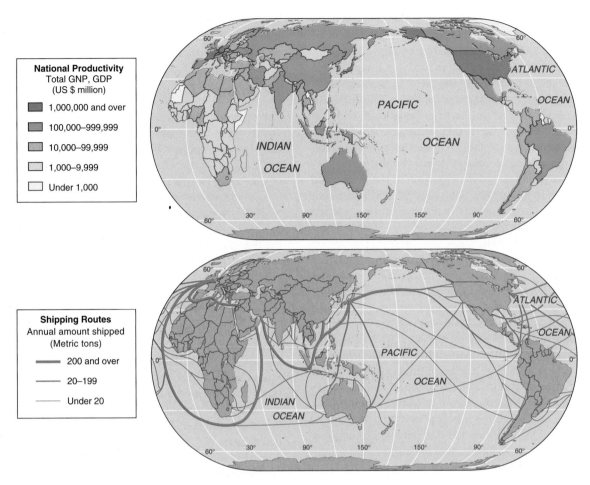

FIGURE 6.28 **The countries of Southeast Asia and South Pacific in a global context.** Compare their productivity and distance from main shipping routes with the positions of other wealthier countries.

Republic of China (ASEAN Plus One), and jointly with The People's Republic of China, Japan, and The Republic of Korea (ASEAN Plus Three) may strengthen the organization's global position, although all individual countries involved remain economically competitive against one another.

Economic Crisis: Late 1990s

Rapid economic growth in noncommunist ASEAN countries came to a halt in the late 1990s, when overlending by banks in the subregion caused a major financial crisis. Thailand, followed by the Philippines, Malaysia, and Indonesia, had to devalue their currencies. The problems were caused by high levels of foreign borrowing, government budget deficits, major bank loans to glutted property markets, and slower-than-expected economic growth. Huge Japanese investments in the late 1980s and growth in exports from Southeast Asian countries were linked to a high yen value against the U.S. dollar; by the late 1990s, however, the value of the yen fell, forcing the Japanese to slow investment and technical transfer to Southeast Asia. Southeast Asia's problem affected East Asia: too much economic growth, too soon, on too much unsecured credit.

Thailand's economy suffered the most. The financial crisis led to a change of government, which imposed austerity measures and distanced itself from the past with a new constitution. Indonesia and Malaysia, ruled by long-term authoritarian governments under President Suharto and Prime Minister Mahathir Mohamad, respectively, found it more difficult to admit fundamental mistakes in financial regulation. In Indonesia, Suharto was forced to resign, leaving the country's economic and social life in chaos.

By the early 2000s, new challenges came from China's growth, together with slowdowns in the American and Japanese markets for Southeast Asian goods. China provided both increased competition and increased markets. The redirection of foreign direct investment to China in the 1990s slowed Southeast Asia's recovery from the 1997 crisis as Chinese textiles (Vietnam, Indonesia) and higher-value products (Thailand, Malaysia) competed with those of the subregion. Southeast Asian countries may meet these challenges by placing greater emphasis on open government to attract investors, lowering trade barriers to open up markets, and upgrading domestic workers' skills.

Economic Changes in Thailand

Thailand's recent experiences are typical of other countries in the subregion. The 1997 financial crisis in Thailand left many unfinished, empty commercial and residential towers in Bangkok's skyline. After years of thrift and investment in new manufacturing industries, the returns dwindled, and restrictions on foreign investment loosened in the mid-1990s. Better potential returns from property investment made the prices of land and buildings rise as borrowing became more competitive and cheaper, expanding an asset "bubble." The bubble burst when a loss of confidence led to withdrawals of funds from banks, putting pressure on the baht (Thailand's currency), which was allowed to float and moved downward. The cost of foreign debt doubled and banks and businesses collapsed, leaving the building shells as worthless assets.

Five years later, in 2002, the debts still slowed the country's economy, leaving half the industrial capacity idle. The Thai Petrochemical Industries Corporation (TPI) had the largest debts of almost US $3 billion, had not paid interest since 1997, and was slow to impose restructuring and management changes. TPI and other Thai corporations claimed that they would be solvent if the value of the baht had not fallen in 1997, when the government removed the link to the U.S. dollar. In the wider Thai economy, tourism suffered from both the reductions of Asian visitors from 1997, the events of September 11, 2001, and the Indian Ocean tsunami of December, 2004.

Despite these difficulties, Thailand's major export industries of oil products and vehicles grew; the country quadrupled exports from its refineries between 1996 and 2000, and welcomed fresh investments from BMW, GM, and Toyota. Worries over China's growth and resultant losses of export markets were calmed by knowledge that exports to China trebled since 1993 and numbers of Chinese tourists to Thailand quadrupled. Thailand possesses comparative advantages that include its membership in ASEAN with a market of over 500 million people, a strong local car parts industry, and tax incentives. Thailand has higher labor costs than Vietnam and more expensive textile exports than Indonesia, but the food-processing industry, including tuna canning, is growing for the local middle-class market.

Thailand's other growing industries are the illegal sex and drug trades that thrive through boom and bust but generate a frightening public health crisis as HIV/AIDS and drug-related problems spread. *Ya baa,* the methamphetamine "crazy pill," was first produced alongside heroin in the early 1990s, when Thailand was merely a staging post for movements of heroin. Thais prefer *ya baa,* and countries such as the United States are less concerned about it than about opium. Popularized first by manual workers who take it to remain alert on long shifts, it is now used widely as a recreational drug in colleges and schools. Children as young as 6 become addicts, and hospitals treat more *ya baa* addicts than alcoholics. The worst symptoms after years of addiction are paranoia and hallucinations, producing uncontrolled violence. Cures take several months with no guarantee a relapse will not occur. The number of drug offenders in Thai prisons rose from just over 10,000 in 1992 to 100,000 in 2001, swamping prisons and the courts, with over 80,000 cases per year. Most of the drugs are manufactured in southern Myanmar, but when Thai military close the northern border, traffickers use roundabout routes through Laos and across the Gulf of Thailand.

Farming Changes in Southeast Asia

Most of the people in Southeast Asia still live in rural areas and gain their living from farming, although agricultural employment and agriculture's proportion of each country's income declined with the growth of manufacturing and service industries. As many districts and families changed from local subsistence to a cash-based economy, they linked to and became dependent on growing for urban and overseas markets. Such market demands led to new crops and new strains of traditional crops being grown and changing patterns of land ownership. The highest productivity occurs in the major river valleys and island coastal areas (Figure 6.29).

Wet-rice cultivation, or **padi,** remains the chief form of agriculture in Southeast Asia. A reliable water supply comes from rain in Thailand and the Philippines, but irrigation is used in much of Indonesia. From the mid-1960s, the **Green Revolution** was a term applied to the rapid increase in crop yields as a result of a combination of new, high-yielding varieties of rice with the use of fertilizer, pesticides, mechanization, and irrigation to raise production (Figure 6.30). A major part of the revolution was the **new rice technology** based on agricultural research at Los Banas in the Philippines. It doubled padi yields from 1969 to 1989 when population was increasing faster than food production, and this enabled Indonesia and the Philippines in particular to become almost self-sufficient in grains. The wealthier owners of larger farms in suitable environments with sufficient labor and good management benefited most. They had the capital to invest to obtain higher yields and multiple annual crops. Many poorer farmers could not meet such costs.

The commercialization of farm production changed the use of land and labor. Former communal land sold to a minority of rich landowners forced poorer people to rent land or work as landless laborers. Landlessness increased further with population growth and the expansion of urban areas across former farmland. For example, the amount of rice land per person in Java halved from 1940 to 1980. Communal rights in the rice harvest—vital in supporting many poor families—were replaced by the sale of the standing crop to middlemen, while mechanization displaced wage labor. In Malaysia, for example, the introduction of combine harvesters halved the numbers of workers involved in the rice harvest. As the traditional farming practices of supportive and informal labor exchange during planting and harvest gave way to wages and mechanization, social impacts multiplied. The growing numbers of landless people in the subregion provided support for Communist and Islamist groups, as in some of the islands of the Philippines, and were often associated with ethnic disturbances in Indonesia.

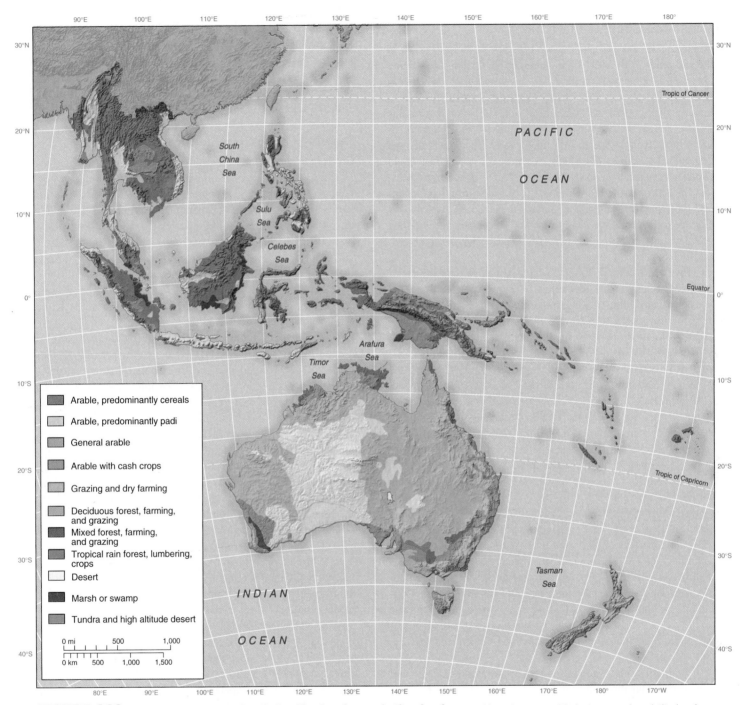

FIGURE 6.29 Southeast Asia and South Pacific: farming and other land uses. How do geographic features such as latitude, elevation, and proximity to water impact the spatial distribution of land use in the region?

While rice growing is the dominant use of farmland in all Southeast Asian countries, many also produce commercial tree and fruit crops, such as rubber, palm oil, and pineapples, in **plantations,** where single species are grown on large estates for commercial export purposes. The good management of tree crop plantations can produce an almost continuous harvest and year-round employment. The plantations were often established under colonial regimes. Although plantations were once owned and run by foreign corporations, Southeast Asian governments nationalized much of the production. Malaysia still produces a large proportion of the world's rubber crop, but a shortage of cheap labor led to falling production as southern Thailand and Indonesia increased their outputs.

Seeking an alternative product, Malaysia became the world's largest producer of palm oil. The crop is less labor-intensive, while the Chinese demand for cooking oil keeps world prices high. Oil palms now constitute the world's biggest vegetable oil crop, despite containing more "unhealthful" fats

FIGURE 6.30 Southeast Asia: farming. (a) Wet-rice cultivation in Bali. (b) The transition to commercial agriculture has introduced the widespread use of chemicals, including pesticides being applied by this young farmer in Thailand. **Photos:** (a) © Vol. III/Corbis; (b) © David Zurick.

(a)

(b)

than midlatitude vegetable oils. Malaysia contributed 58 percent and Indonesia 28 percent of the world total in the late 1990s. Malaysian policies, however, are diverting investment away from agricultural production and into manufacturing.

Indonesian plantation crops, including coffee, spices (cloves, nutmeg, cinnamon), tea, rubber, and palm oil products, constitute 13 percent of exports. In the Philippines, coconut products, sugarcane, pineapples, bananas, and coffee make up 16 percent of exports. In Thailand, plantation crops form 10 percent of exports. Myanmar remains (followed by Afghanistan—see Chapter 7) the world's leading producer of poppies for opium and heroin production. It supplies illegal drug traffic through Thailand and other routes.

In the 1990s Vietnam rose from a marginal coffee producer to rival Brazil and become the world's largest exporter of the cheaper robusta coffee, grown in the central highlands. Farmers from the crowded lowlands migrated into the highlands as world coffee prices rose in the mid-1990s, but Brazil and other coffee producers also boosted production, leading to an oversupply crisis. Vietnam is cutting robusta production but expanding higher-quality arabica coffee trees in the north of the country, as well as encouraging farmers on poorer land to change crops to rubber, peppers, cotton, and hybrid corn.

Southeast Asian Forest Products

Hardwoods from the tropical rain forest that covers much of this subregion are of great economic significance to Southeast Asian countries. Timber products from Malaysia and Indonesia, more plentiful than tropical timber exports from Africa

and Latin America combined, dominate world production. In the 1990s prices for tropical timber rose by 50 percent when prices for other commodities fell or remained the same, and this stimulated further cutting. Wood industries became more important in Sumatra and Kalimantan (Indonesian Borneo) as Indonesia increased the income from exploiting its forests. Much of Myanmar remains forest-covered. It has around 250 commercially useful tree species, of which teak is a major export. The Myanmar army forces slaves to fell tropical trees destined for world markets.

The limits of cutting without replanting are being reached throughout the subregion. Thailand banned logging in 1989 as its teak and yang forests diminished. In the Philippines, the forest cover was reduced from 60 percent of the country's area in 1950 to 10 percent in the 1990s. Fears of losing its major resource led Indonesia to curtail clear-cutting of trees and implement reforestation programs. Malaysia claims to be the most responsible country in the subregion for "sustainable logging"; yet it exports more tropical logs and sawn timber than any other Southeast Asian country, and annual cutting reduces the forest cover by 1–2 percent. Malaysia has not yet adopted international standards of forest management, with loggers disregarding the interests of indigenous forest people in Sarawak, who are losing their livelihood from tree resources, soils, and fish. Malaysia indicated in 2001 that it intended to adopt more stringent standards to placate its Western customers. Western objections to unmanaged cutting on environmental grounds are resented by governments in Southeast Asia, which cite their own controls over illegal logging and the historic European and American forest removal.

Mining in Southeast Asia

Mining adds to the primary commodity exports from Southeast Asia (see Figure 5.12). Indonesia is a major oil producer, and much of the world's tin production comes from the subregion. Oil from Sumatra and Kalimantan accounts for 20 percent of Indonesian exports, down from 80 percent in the 1980s because of a fall in new oil discoveries and the development of other income sources in Indonesia's manufacturing base. Growing energy demands, with increasing industrialization and without new oil discoveries, will soon convert Indonesia into an oil importer.

Malaysia was for long the world's largest tin producer and still exports some, together with oil and natural gas from Sabah and Sarawak on Borneo. Bangka and Billiton (Indonesian islands) and the Malay Peninsula of Thailand are other long-term tin producers. Indonesia's bauxite is mined in the Riau islands and southwest Kalimantan and is processed in the Kualatajuy smelter near Sumatran coal mines. Thailand is one of the world's largest exporters of gems and jewelry, with rubies and sapphires mined along the east coast of the peninsula.

Market-Led Industrialization

While the transition to commercial farming was gradual in the ASEAN countries, moves into manufacturing were more rapid. By the time the countries of Southeast Asia became independent, most had factory-based industries. At first, import-substitution industries supplied local markets and were protected by tariffs and quotas, but by the 1970s, the small domestic markets restricted further growth.

The combination of small markets, limits of local skills, and restrictions imposed by government licensing offices made the products expensive. In the Philippines, where the government's licensing role led to corruption, the term **crony capitalism** described close linkages among bureaucrats and entrepreneurs.

In Indonesia, the government owns large-scale processing facilities for agriculture and mining products. Medium- and small-scale factories, often Chinese-owned and based mainly in western Java, dominate consumer goods industries (furniture, household equipment, textiles, and printed matter). The Indonesian textile industry is divided into government- or multinational corporation–owned spinning mills using imported raw materials. Weaving is carried out by smaller firms concentrated in Bandung. Batik, the Indonesian hand-printed textile, is produced in a cottage industry that is being taken over by larger-scale factories.

Multinational corporations invested in new factories, employed cheap labor to produce goods for export, and used labor-intensive manufacturing processes. Malaysia and Thailand gained most from this process, although Indonesia and the Philippines also experienced related economic growth. In the 1980s countries such as Thailand and Malaysia moved from labor-intensive to high-tech industries. However, multinational corporations from the United States, Japan, and South Korea were slow in diffusing innovations to these growing countries and preferred to use them as a cheap source of labor and to open access in their domestic markets.

Much of the export-oriented industrialization relies on imported capital, skills, and management, causing it to be labeled **ersatz capitalism.** That suggests it is an inferior substitute for development generated by local entrepreneurs based on local savings and investment. The education gap between these countries and the wealthier countries, however, makes it difficult for the Southeast Asian countries to generate and develop their own innovations. Economic growth based on urban manufacturing emphasizes the growing contrasts between moderate-to-high-income urban industrialized areas and low-income rural areas.

Expanding Tourism in Southeast Asia

Some countries in Southeast Asia combine their cultural heritage, enhanced accessibility by airline, and increased visibility via the Internet to promote and develop growing tourist industries. Visitors to the primary tourist destination countries—Thailand, Malaysia, Singapore, Indonesia, and the Philippines—increased from 7 million people in 1983 to 20 million in 1990 and 37 million in 2002. Nearly 13 million people visited Thailand in 2002, just over 13 million in Malaysia, and 7 million in Singapore. The highest numbers reflect the political stability of the countries as well as their attractions, which include modern cosmopolitan cities, precolonial historic preservation, and spectacular beaches and coral reefs. Upland resorts and an increase in **ecotourism** based on natural phenomena (landforms, animals, and plants) and historic tourism spread visitors into places once closed by political actions, as in Myanmar and Vietnam. The industry brought in foreign currencies; fewer local people ventured abroad for vacations, and those who did went mainly to Hong Kong and Japan. By contrast, the United States had over 42 million visiting tourists in 2002 (despite September 11th), while nearly 60 million Americans vacationed abroad. Temporary visitor declines resulting from the devastating Indian Ocean tsunami on December 26, 2004, were expected to rebound rapidly with little overall economic impact.

Communist Economic Stagnation and Change

Economic growth in Cambodia and Laos remains slow. For most people, only bare survival is possible. Vietnam has larger areas of fertile soil, more mineral resources, and a larger population than the other two countries, but its national income grew more slowly than its population, causing per capita incomes to fall in the 1980s.

Much of Vietnam's poverty stemmed from the effects of the fighting on its territory that lasted from the early 1950s until 1975 and disrupted all aspects of the economy. Vietnam continued to spend heavily on military excursions into Cambodia (1978–1989) and on a border war with China. Internal policies produced little economic growth. The five-year plans for the late 1970s and early 1980s directed a shift into heavy industry and out of agriculture, but the centralized bureaucracy was unwieldy and unresponsive to demands, and it failed to gain the support and further sacrifices of the war-weary population.

The Soviet Union contributed approximately $1.5 billion per year in aid in the 1980s to Vietnam's economy but ceased assistance after 1989. The economic blockade by the United

States and the ASEAN countries prevented Vietnam and its neighbors from receiving overseas aid until 1991.

After 1985, however, Vietnam's leaders allowed crops to be sold directly in private markets instead of through the communes and cooperatives selling to government ministries that set prices. The Vietnamese currency was devalued and foreign investment encouraged. By the late 1980s, Vietnam exported rice again, becoming the world's third largest exporter in 1994—although bad weather in some years caused it to be displaced by India. Vietnam established economic leadership over the two smaller countries of Laos and Cambodia to the west. The shift to greater involvement in the world economy was easier in southern Vietnam—a capitalist country until 1975—than in the northern sector, which had been under Communist control longer. In the mid-1990s, corporations from Japan and South Korea in particular invested in Vietnam. In 1995 the United States "normalized" relations with the country when Vietnam became the first Communist member of ASEAN.

Myanmar's Economy

Myanmar's political isolation results from its lack of democracy. It keeps an unrealistic official exchange rate, 20 times higher than the black-market rate, making its products too expensive to sell on world markets but also making imports cheap. Such subsidies will be removed as the country seeks to join the global economic system. In the 1990s Myanmar brought in Chinese contractors to build bridges, railroads, airports, and hotels, and entered joint agreements with Japanese and European corporations to develop the economy.

Myanmar creates major problems for its neighbors, especially Thailand. The so-called "Golden Triangle," where Myanmar, Thailand, China, and Laos meet, is a major source of opium and amphetamine drug production. It is estimated that up to 800 million pills passed across the border into Thailand during 2001. The Wa people, a minority group that used to fight the Myanmar army, switched to selling drugs and fought another rebel group, the Shan people. Myanmar's leaders allow the drug trade, but the Thai authorities resist it along with the rising flow of refugees from Myanmar. The Thai resistance is not made easy by the bribing of their officials and the supplies of electricity it sells to eastern Myanmar, some of which is used to produce the pills.

Test Your Understanding
6C

Summary

Southeast Asia comprises countries at different levels of political and economic development. Groups of people with different cultures mix within each country, often leading to conflicts. Singapore is a wealthy but small country. Thailand, Malaysia, Indonesia, and the Philippines

are emerging economies. Laos, Cambodia, and Vietnam remain extremely poor following years of warfare. Myanmar's exit from military dictator–imposed isolation is slow. All Southeast Asian countries are developing the manufacture of goods for export and are attracting foreign investments. They mostly have governments that take an involved part in industrial and trading policy.

Questions to Think About

6C.1   Which are the ASEAN countries? Why was the organization set up, and what is its likely future?

6C.2   Assessing their locations and past and present economies, compare the future prospects of Hong Kong (see Chapter 5) and Singapore.

6C.3   What common issues do North Korea (see Chapter 5), Vietnam, Laos, Cambodia, and Myanmar face?

6C.4   What were some of the contributing factors in the Southeast Asian economic crisis of the late 1990s?

Key Terms

transmigration	padi
white Australia policy	Green Revolution
primate city	new rice technology
deurbanization	plantation
Khmer Rouge	crony capitalism
cultural fault lines	ersatz capitalism
archipelago	ecotourism
Asia–Pacific Economic Cooperation (APEC)	

Australia and New Zealand

Australia and New Zealand (Figure 6.31) are inhabited almost entirely by European people, who transferred their politics, cultures, and economies to the opposite side of the world (see Table 6.3). From the 1970s, the ties that bound these countries to the United Kingdom in particular were gradually severed and their trade interests reoriented to Asia and the wider Pacific. Australia became a focus of world attention when it hosted the 2000 Olympic Games. Many in more affluent Western countries regard both Australia and New Zealand as pleasant and exciting places to visit or relocate. Their low population densities and high levels of income make them desirable places for Asian migrants and for refugees seeking to escape political, ethnic, or economic challenges.

Australia has a diverse and affluent economy. Although it had only 3.5 percent of the region's 2005 population, it produced 20 percent of its GNI and has a GNI PPP per capita that is slightly higher than that of Singapore. New Zealand is a smaller country but also has a high GNI PPP per capita; with less than 1 percent of the region's population, it produces 3 percent of its output. The two countries face similar problems, including great

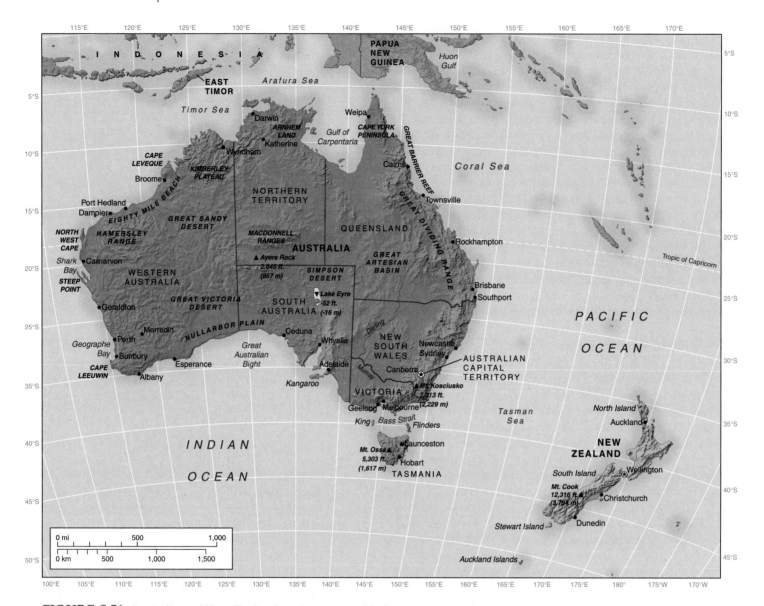

FIGURE 6.31 Australia and New Zealand: major geographic features. The political divisions, physical features, and major cities.

distances from their traditional markets, a small local market, and the transition from historic colonial ties with Europe to a closer involvement with Asian and Pacific Ocean countries. Australia is often referred to as "Asia's farm and quarry."

Countries

Australia

Australia is the only country that is also a continent. Its size produces internal contrasts of physical and human geography.

Political Regions: The States

Separate colonies were established around developing port cities, giving access to inland areas and connecting them back to the ruling country. The early colonies determined later geographic differences by setting their own patterns of inland expansion, choosing different railroad gauges, and concentrating major developments in their chief cities. In 1901 the Commonwealth of Australia came into being as a federation of the states that emerged from the separate colonies. Sydney and Melbourne competed for the role of national capital, but the question was resolved by building a specially planned city in a new Australian Capital Territory centered on Canberra, almost midway between them. The federal government moved there in 1928.

Southeast Australia

Most Australians live in the temperate, humid southeastern coastal region from near Brisbane in the north to Adelaide in the south. This area includes three of the five largest cities (Sydney, Melbourne, and Adelaide), as well as the federal capital, Canberra.

For most of its length, the southeastern coastal region is bound inland by the hills and plateaus of the Great Dividing Range, which formed a barrier to inland movement in the early years of colonization. Three states and their dominant cities occupy this area. Sydney, the largest city, lies at the heart of

New South Wales. There are several industrial cities near the New South Wales coast, and Australia's sheep farming is concentrated on the western side of Australia's only significant series of mountains, the **Great Dividing Range.** Victoria is focused on Melbourne but is a smaller state, and its inland farming area is concerned mainly with more intensive irrigated farming for grapes, fruit, and grain crops. South Australia has its main city, Adelaide, in a farmed area at the southern edge of the state. Much of the rest is desert.

Northern Australia

From Brisbane northward, the more tropical climate and the offshore Great Barrier Reef—the world's largest continuous coral reef formation—form an area much favored by tourists. Tropical Queensland is the second largest state in area with its main city, Brisbane, in the southeast corner. Queensland contains the majority of Australia's cattle ranching, has major mining operations, and is developing tourist attractions.

Northern Territory is the least populated of Australia's primary political divisions. In the far north, the tropical climate brings monsoon rains along the coasts. Most of the few remaining Aboriginal people who live in traditional ways are in this territory, but it has fewer inhabitants than Australia's Capital Territory.

Interior and Western Australia

Inland of the Great Dividing Range, the Murray-Darling River lowlands have a subhumid climate that supports a major farming region producing cattle, sheep, grain, and fruit in western Queensland, New South Wales, and northern Victoria. Westward of this sparsely populated farming region, the virtually empty **Great Australian Desert** covers almost all of the rest of the continent. In the southwestern corner, rains come in winter in a Mediterranean climatic regime and provide a basis for agricultural settlements centered on Perth.

Western Australia is the largest state in area but is mostly desert, and its main settlements around Perth are separated by over 1,600 km (1,000 mi.) of uninhabited land from the large cities of southeastern Australia. Perth has farming and mines in its immediate hinterland and large iron ore mines farther north. Parts of South Australia and Northern Territory are also included in the vast Australian interior.

New Zealand

Internal rivalries in New Zealand are less marked than those within Australia. North Island is distinguished by its central upland area that is affected by volcanic activity. It is home to nearly all the Maoris, most of the people of European origin, the largest city, Auckland, and the country's capital, Wellington. South Island has fewer people. South Island's high western mountains, the Southern Alps, are occasionally rocked by earthquakes, and its eastern plains provide good farming land.

Economic Development

European market demands, including the impact of the European Union and subsequent reduction of trade, determined products and economic structures in Australia and New Zealand until the 1970s, when they became increasingly supplanted by growing Asian markets. Along with their attempts to develop wider trading links with Asian countries in ASEAN and APEC, Australia and New Zealand are economically connected to other countries inside the South Pacific region. Australia's attempts to form closer economic ties to New Zealand in the 1980s, however, slowed in the 1990s, possibly because the attempts to enter Asian markets were seen as more important or because Australians saw New Zealand as benefiting the most from such economic connections. The **South Pacific Forum,** established in 1971, links 13 countries: Australia, New Zealand, Papua New Guinea, the Solomon Islands, the Cook Islands, Fiji, Kiribati, Nauru, Niue, Tonga, Tuvalu, Vanuatu, and Western Samoa. Its aim is to develop regional political cooperation. This agency has had some success in confronting regional problems for the islands in particular. Australia has championed a number of causes, but these often bring it into confrontations with Asian nations, such as when it opposed the exploitative Malaysian logging of Papua New Guinea.

Changes in Australia

Australia's production of wool for British factories from the late 1800s led to diversification into fruit, lamb, beef, and dairy products that followed the same routes to markets in Britain and Europe. During the 1900s, mining products grew in significance. The high cost of transporting low-value goods from Europe to Australia enabled the imposition of tariffs to protect the growth of **import-substitution manufacturing** (government-imposed structure fostering the domestic consumption of domestically produced goods) in the new country, with foundries, textile mills, breweries, and other food- and drink-processing enterprises. Other industries, such as steel and automobile manufacturing and assembling, were later set up under similar or increased protection levels.

By the mid-1900s Australia had a two-tier economy. The first tier depended on exports from the natural resource base of farms and mines to provide income that paid for imports. The second tier consisted of protected manufacturing and government-owned enterprises such as airlines, railroads, and docks that were often inefficient but had a role in spreading the wealth from the export industries' income over a wider group of people.

In the 1970s and 1980s, the protectionist policies collapsed as exports of agricultural and mineral commodities obtained lower prices on world markets and the expansion of the European Union cost Australia its markets in Britain and Europe. Australia could no longer campaign for free markets for its exports while keeping high tariffs on its imports as it shifted its export markets to Asian countries and especially Japan. One implication of the lower tariffs is that many Australian manufacturers will not survive. Of the five automakers that sheltered behind protective tariffs, Nissan closed its Australian operation.

In the late 1980s and through the 1990s, Australia's economy boomed by supplying the growing Asian markets with coal, iron ore, and other minerals and some grain and livestock products. Between 1970 and the late 1990s, farm exports fell from 44 to 20 percent, while mine exports rose from 27 to 42 percent of total exports. By the mid-1990s Japan bought more goods from Australia than from Europe and the United States combined. A mid-1990s World Bank study that took account of "natural capital" in terms of land value, water, timber, gold, and other minerals, in addition to human resources, ranked Australia as the top world country, just ahead of Canada. Clearly, mineral resources played a major part in this designation. In the early 2000s Australia's GNI was coming less from agriculture and manufacturing and increasingly from service industries.

Problems of Trade Dependence

Australia's economy is increasingly tied to Asian markets. As it accepts more imports of manufactured goods from Asian countries, it needs to export more to them. Australia wishes to add value to its exports by processing its agricultural and mining products in Australia, but Japan and other Asian countries retain import restrictions on processed materials. Certain specialty products created their own growing markets abroad, but Australia is a relatively small market for Asian goods.

The countries of ASEAN have a regional free-trade agreement among themselves but are reluctant to admit Australia and are more interested in forging agreements with China and Japan. While Asian governments and businesses work closely together, and environmental and civil rights issues assume less importance, Australians remain attached to Western values that sometimes upset Asian countries and place political barriers in the way of regional trade.

Australia lobbies in APEC, which is dedicated to trade liberalization. The increasing links between the United States and Latin American countries, through the North American Free Trade Agreement (see Chapter 11), APEC, and other relationships, threaten Australian access to markets in the United States.

Australia's Dominant Mining

Mining provides nearly half of the value of goods exported by Australia. It is the world's largest exporter of coal, amounting to three-fourths of its record production of 200 million tons in the 1990s, and the world's largest supplier of high-quality (and low-polluting) coking coal. Plans for expansion could increase this level of production if world demand for coal remains high and the production of iron and steel increases.

New energy resources include natural gas fields along Australia's west coast, tied to the construction of liquefaction plants that export the gas to Japan. Such projects attract investment by oil companies but compete with producers in nearby Malaysia, Brunei, and Indonesia. Their viability depends on world oil and gas prices.

Australia produces a wide variety of other minerals, but expanding production often faces the complexities of obtaining new mining licenses, particularly in areas that are environmentally sensitive or might have Aboriginal sites. For example, Australia produces 10 percent of the Western world's output of uranium and has 30 percent of the Western world's reserves. Australian uranium production is restricted to three mines at present, and there are companies that wish to expand production and opposing groups who wish to curtail it further or see

usage restrictions placed on buyers of uranium from Australia. Other minerals of importance to Australia include iron ore, of which it is the fourth largest world producer, bauxite (first), nickel (fourth), and gold (third). In the 1990s Australia also opened the world's largest diamond mine.

Australia's Farm Output

Two percent of Australia's agricultural land is cultivated, and the remainder is devoted to livestock grazing (see Figure 6.29). Australia's farmers export 80 percent of their production, selling more in Asia following protectionism in Europe and the United States. Wheat, oilseeds, beef, veal, and wine are the main exports (Figure 6.32). Occasional interruptions of market access to Japan and the United States and currency devaluations in countries such as Brazil and Argentina are among the global challenges Australia faces in maintaining and increasing a stable market for its agricultural products.

Australia's Underdeveloped Northlands

Only 28 percent of Australia's exports are produced in the northern half of Australia and only 6 percent of Australia's population (slightly more than 1 million people) lives there. The northern part of Australia is far less developed than the country's southern half in part because of arid and tropical monsoon climates and

FIGURE 6.32 **Australia: farming.** (a) Herding sheep. (b) Harvesting grapes. **Photos:** © Vol. 8/Corbis.

(a)

(b)

distance from other population centers. The region still lacks good transportation infrastructure, including a rail link south to join up with the country's system at Alice Springs.

Tourism

Numbers of foreign visitors increased from around 1 million in the early 1980s to 5 million in 2002. Over half of these visitors came from Asia, attracted by the beaches, golf courses, and the theme parks along Queensland's **Gold Coast** (Figure 6.33). It is clear that the industry's prosperity depends on growing affluence in Asia. Many visitors come from Japan and Taiwan partly to purchase goods at lower prices than are available at home.

Tourists from North America and Europe come not so much for mass tourist facilities as for the outdoor experience, including viewing wildlife and rock formations. Visits to such features form the basis of ecotourism for more affluent tourists who have the time and money to take the long journey to Australia.

New Zealand

New Zealand's economy is based on natural resources like that of Australia, as well as exports of farm and forest products to Europe. New Zealand's main economic products entering world markets are wool, lamb, and dairy products. Over half of New Zealand is in pasture, and over one-third of its exports are from livestock (Figure 6.34). In New Zealand, sheep outnumber people by 12 to 1. Its income from meat products fell in the 1990s, leading to major reductions in herds of beef cattle and sheep, but dairy product demand and deer farming expanded.

In the development of farming, pastureland replaced much forest. The New Zealand government replanted large areas of forest with Radiata pines and Douglas firs, softwoods that thrive in New Zealand's climate and environment, from the early 1900s. This **afforestation** policy provides an increasing harvest that finds markets in Asia as those countries cut their natural forests but do little replanting. **Sustainable forestry** is now as profitable as farming in New Zealand, and the state-

FIGURE 6.34 New Zealand: farming. Sheep show in Rotorua with 20 breeds of sheep, each having advantages and many originating in Scotland (U.K.). **Photo:** © David C. Johnson.

owned forests have nearly all been privatized, albeit with strong conservation laws. Processing industries use the capital accrued from foreign sales of timber.

The contributions of agriculture and manufacturing to New Zealand's economy fell as that of services rose to account for around two-thirds of GNI. In the 1980s the New Zealand government instituted economic reforms to revive a stagnant economy and produce a trading surplus together with lower unemployment and inflation. Tariffs and restrictive port practices were removed and government spending reduced as a proportion of GNI. Lower labor, transportation, and utility costs brought increases in manufacturing productivity. New Zealand's GNI per capita in 2000 fell behind that of Australia and West European countries, although its high quality of life continued.

New Zealand now sells more goods to Japan than it sells to Australia, the United States, and the United Kingdom combined. Most of its imports still come from Australia and the United States. Like Australia, New Zealand is a member of the South Pacific Forum, and ties with the Pacific islands continue that originated when New Zealand was the United Nations agent in those island countries before their independence.

New Zealand expanded its tourist industry through its outdoors attractions in both North and South Island. Greater flight coverage and decreasing airfares compensated for the great distances from the main sources of wealthier tourists. In 2002 New Zealand had 2 million tourist visitors, double its 1990 number. The unprecedented success of the motion picture trilogy *The Lord of the Rings* (and to a lesser degree the film *Whale Rider*) brought significant global attention to New Zealand. *The Lord of the Rings* films, which set global box office records, showcased the spectacular scenic beauty of New Zealand. Some aspects of New Zealand's tourism grew in direct response to the success of the film trilogy. Also stemming in part from the successes of *Whale Rider* and *The Lord of the Rings* is positive growth in New Zealand's film industry, through which a number of internationally viewed television series were already filmed (Figure 6.35).

FIGURE 6.33 Australia: tourism. The Gold Coast area of Queensland attracts many visitors from Asia. The Great Barrier Reef lies offshore. **Photo:** © Jack Fields/Corbis.

FIGURE 6.35 *Lord of the Rings* **Premier Party, Wellington, New Zealand.** The producers and stars of *The Lord of the Rings: The Return of the King* pose for photos at the film's World Premier in Wellington, New Zealand. The film trilogy contributed to growth in New Zealand's motion picture and tourism industries. **Photo:** © Getty Images.

Test Your Understanding
6D

Summary

Although Australia has vast areas of desert and unoccupied land, the majority of its 19 million people are concentrated in higher-density urban areas, giving Australia one of the highest urban percentages of any country in the world. Australia's economy ranks it with the world's wealthier countries and is based on exports of a wide range of minerals and farm products, together with growing manufacturing and tourism industries. Its five largest cities hold nearly 60 percent of the total population.

New Zealand is another materially wealthy country. It consists of two main islands. Its 3.9 million people export wool, lamb meat, timber, dairy products, and fruit. New Zealand is currently experiencing growth in its tourism and films industries.

Questions to Think About

6D.1 Why does such a high proportion of the Australian population live in the major coastal cities of the southeastern quadrant of the country?

6D.2 How have cultural relations changed between Aboriginal Australians and European Australians from the colonial era to the present?

6D.3 What changes were involved in Australia's economic migration from European to Asian markets?

6D.4 How does New Zealand use its natural environment to increase the country's global connections?

6D.5 What are some of the environmental impacts faced in Australia and New Zealand?

Key Terms

Great Dividing Range
Great Australian Desert
South Pacific Forum
import-substitution manufacturing

Gold Coast
afforestation
sustainable forestry

South Pacific Islands

The South Pacific islands stretch from the eastern part of New Guinea—the independent country of Papua New Guinea with nearly 6 million people—to tiny islands with a few hundred people thousands of kilometers to the east (Figure 6.36). The smaller islands are grouped in independent countries or colonies. Their small sizes and markets place them at a disadvantage in disagreements and negotiations with Asian countries or the United States over timber, mineral, agricultural, or fisheries exploitation, and the standards of living of their people suffer accordingly (see Table 6.3). Occasionally local tension garners global media attention, as in the case of the eruption of violence on the French Polynesian island of Tahiti over the French nuclear testing in mid-1995. Although these islands were perceived by early travelers as a "Garden of Eden," day-to-day life was not as idyllic as they wrote (Figure 6.37).

The South Pacific Forum, formed in 1971 with Australia and New Zealand as its major members, has its headquarters in Suva, Fiji, and is concerned with the need for new rules to manage the resources in this region. The growing problems faced by the small islands brought closer cooperation on an economic and political agenda that includes achieving independence of remaining colonies, increased investment, and trying to stop French nuclear testing.

Island Countries

Most of the South Pacific islands gained their independence in the 1970s. Western Samoa achieved it in 1962 and the Marshall Islands in 1991. The Micronesian group of islands remains a United Nations trust territory. New Caledonia and French Polynesia are part of France for government purposes.

Although independence appeared desirable to many of the islands, economic difficulties, internal tensions, and dependence on continuing economic aid and protection keep them linked economically to the United States, France, the United Kingdom, Australia, or New Zealand. Kiribati, Tonga, Tuvalu, Western Samoa, and the Solomon Islands remain among the world's poorest countries, with few products in world demand.

Modern country groupings of the islands cut across ethnic divisions:

- The southwestern group approximately coincides with former Melanesia and includes Papua New Guinea, the

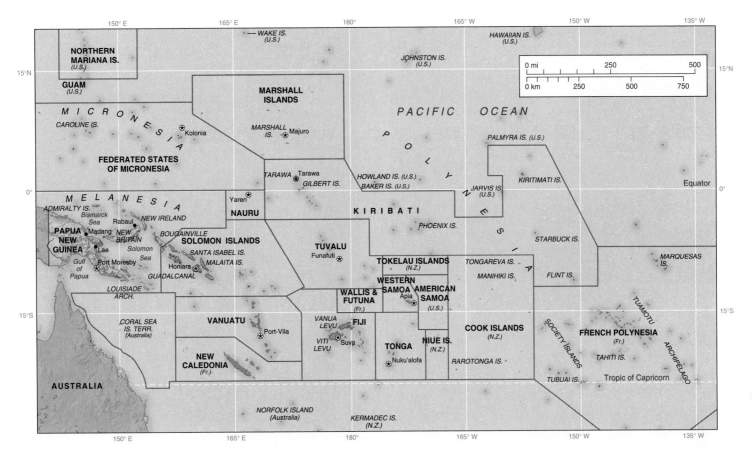

FIGURE 6.36 South Pacific islands: major geographic features. The country groups and main cities. These islands are spread across huge distances.

Solomon Islands, Vanuatu, Fiji, and New Caledonia. These are the largest and most populated islands and are the nearest to Southeast Asian markets and Australia–New Zealand links. Papua New Guinea includes the eastern part of the largest of the islands; the western part of this island, Irian Jaya, is in Indonesia.

- The northwestern group, roughly coinciding with former Micronesia, includes Guam, the Northern Mariana and Marshall Island groups, Nauru, Palau, and the Federated States of Micronesia.

- The largest of the three groups in terms of ocean area but the poorest in economic development contains the islands of the central South Pacific such as Kiribati, Tuvalu, Wallis and Futuna, Tonga, Tokelau, Western and American Samoa, Niue, the Cook Islands, Pitcairn, and French Polynesia.

The islands vary considerably in their well-being, as the consumer goods ownership in Fiji and Papua New Guinea shows (see Figure 6.16). The French colonies of French Polynesia and New Caledonia are heavily subsidized by France and have above-average incomes. Others, such as the Marshall Islands, Palau, Micronesia, Guam, and the Northern Marianas, have U.S. support.

Small Towns

Few South Pacific islands are large enough to have major cities. Most island towns began as colonial ports and continued as political capitals in the postcolonial independent countries (Figure 6.38). Most towns in the South Pacific islands have populations of a few thousand people and contain a mixture of facilities dating from the colonial era, some processing industries, port facilities, and shantytown areas. Many islands link by air or sea to the cities of New Zealand and Australia or to the Hawaiian Islands for greater access to world transportation routes. Suva on Fiji and Port Moresby in Papua New Guinea, with their international airports, are the main exceptions.

Economic Development

Island communities had complex economies prior to colonization, often linked across wide ocean expanses. Exchanges took place among the better-watered islands, where vegetables and fruit were grown, and the arid islands that relied on fishing. Unique maps of direction and distance were constructed, but storms and crop failures brought death and destruction to many.

(a)

(b)

FIGURE 6.37 **South Pacific.** (a) An idyllic view of Ralarea, French Polynesia, which is close to French nuclear testing sites. (b) Traditional dress (plus sneakers) in Papua New Guinea. **Photos:** (a) © Vol. 112/Corbis; (b) © Mission Aviation Fellowship.

FIGURE 6.38 **South Pacific: urban landscape.** In Levuka, Fiji, where colonial touches remain in the landscape, a British-style church tower and wooden store buildings line Beach Street. **Photo:** © Robert Holmes/Corbis.

Commercialization of interisland trade occurred in the early postcolonial years. Some of the islands entered the world economy with sales of coconut palm products such as **copra**— the dried white meat that lines the inside of the coconut shell and provides an oil used in soaps and candles. Manufactured goods from Europe and the United States and expensive diesel oil for power were imported. Many islands needed subsidies from prior colonial powers and remained materially poor. Some islands, including Tuvalu, relied on external aid for nearly 100 percent of their income.

Many South Pacific islands faced economic ruin in the 1990s. Some see external aid as a compensation for past colonial exploitation; others see it as demeaning and sapping self-reliance. With a low economic base, the islands still suffer from increased exploitation rates by foreigners of their few natural resources. Some islanders see tourism as an income prospect, while others view it as a further form of economic colonialism.

Farm, Forest, and Mine Products

Apart from Papua New Guinea (copper and gold), New Caledonia (nickel), and Nauru (phosphate), few islands have natural resources except for a warm climate, nutrient-rich volcanic or coral limestone soils, and the surrounding ocean (Figure 6.39). The small areas available for agriculture allowed few islands to diversify into commercial crops beyond coconuts and copra. Fiji produces sugar, and Tonga produces bananas and vanilla in addition to coconuts.

World attention was drawn to the cutting of tropical hardwoods on Papua New Guinea and the Solomon Islands by a statement at the South Pacific Forum in Brisbane in the mid-1990s that the islands were being "ripped off." The imposition of stricter environmental laws in Malaysia and Indonesia turned their logging companies to the islands that had fewer regulations. Cutting is now at levels that could remove all the timber, and little replacement planting is taking place. Local landowners receive US $2.70 for timber that fetches $350 on world markets.

Similar problems face the local fishers as fishing fleets from Japan, the United States, South Korea, and Taiwan take fish from this region to make up nearly 40 percent of the world catch. Foreign fishers pay access fees to the fishing grounds around some islands, but it is impossible for the small islands with limited resources to patrol the whole area.

Mining occurs on a few islands but often brings problems. Papua New Guinea suffered an income reduction when a separatist group on Bougainville forced the closure of copper mining that made up 30 percent of Papua New Guinea's exports. After protesting the copper mining and its environmental impacts since 1969, the people of Bougainville declared the island's independence in 1990. Papua New Guinea used military force to prevent secession, but the Australian-owned mining operation in a separatist-controlled area remains closed. The Papua New Guinea government is trying to replace this income by encouraging the development of gold mines in other parts of its territory.

For a few years in the 1970s, Nauru, only 21 km^2 (8 mi.^{2}) in size, had one of the highest per capita incomes in the region. From the mid-1900s, strip mining for phosphate rock (used in fertilizers) left a barren, jagged surface over much of the island as mining companies cleared the tropical vegetation that was then unable to reestablish itself.

Opting for independence in 1968, the island bought out the phosphate company. New buyers came from Japan and South Korea, injecting income that funded schools and free medical treatment facilities; few Naurians needed to work. The wealthy times attracted Australian managers, Chinese shopkeepers, and miners from other Pacific islands, who now make up one-third of the 12,000 population. Air Nauru has bought five Boeing 737s.

Surplus revenue was invested in Pacific Rim properties, but in the 1990s the value of these assets dropped from over US $1 billion to $130 million. By the 1990s the depleted phosphate resources and falling world prices also reduced income. In the 1990s and early 2000s, the Nauru government made some money by suing former colonial governments for mining-related degradation, allowing offshore banking that encouraged the deposits of unrecorded Russian mafia accounts (estimated at US $70 billion in 1998 alone), and leasing land to Australia for a detention camp for asylum seekers. Such sources of income are temporary, and the once-affluent Naurians suffer perceived indignities including rationed electricity, grounding of the last Air Nauru plane, and cancellation of plans to hold the International Weightlifting Federation world championships on the island.

Tourism

Tourism is one of the few ways open to the people of the subregion to increase their incomes. However, the islands have different levels of access to the rest of the world and of interest shown by tourist companies. Tourism is important on some islands that are on international air routes, have particular attractions, or have well-organized tourist facilities, such as Fiji, Guam, the Marshall and Northern Mariana Islands, Toga, Vanuatu, and Western Samoa. It has little or no impact on those that lack such facilities, such as Kiribati, Micronesia, Nauru, New Caledonia, Palau, the Solomon Islands, or Tuvalu. Tourism can, however, destroy or merely exploit the last remnants of traditional culture and often pays low wages to local people. Yet it may be a better industry to

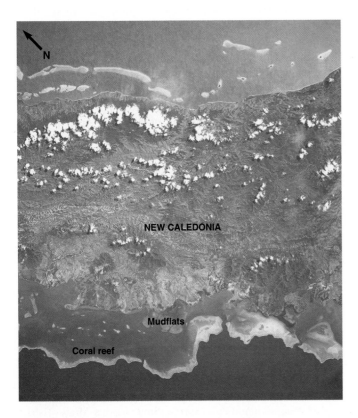

FIGURE 6.39 South Pacific: island landscape. A space shuttle photo of New Caledonia shows evidence of environmental degradation. Several fires are burning what little remains of the forest cover to provide cattle pasture. Much of the land has been degraded by open-pit mining for nickel. Offshore, the coral reef is being killed as mud from eroded land fills the lagoon to produce mudflats off the southern coast. **Photo:** NASA.

work in than the low-wage factories found in other poorer countries. Tourism in this region remains dependent on the retention of an attractive environment and political stability and so is likely to encourage good environmental practices. Where tourism has become important, as in Fiji, it has forced a greater openness to foreign visitors' requirements, creating fresh opportunities for local businesses.

Antarctica: A Region?

Antarctica occupies 10 percent of Earth's land surface but is not settled on a permanent basis (Figure 6.40a). It remains a frozen, empty continent apart from access limited to scientists and occasional small numbers of tourists. Is it a major world region because it is so distinctive? Or does its lack of people and country jurisdictions exclude it from such status? Can it be ignored in a world regions course? Its inclusion as part of another world region is problematic because it is so different and far removed from the other southern continents.

Antarctica's Global Status

International expeditions to Antarctica from the late 1800s concentrated on the personal physical feat of crossing the frozen continent on foot. They gave way to international claims on sections of the continent during the mid-1900s (Figure 6.40b). Some of the claiming countries established scientific surveys of importance to climate, oceanographic, and glaciological studies.

In 1961, 39 countries signed the **Antarctic Treaty** as a basis for nonmilitary scientific cooperation, environmental safeguards, and international control. In 1991 the subsequent components of the Antarctic Treaty banned commercial mining activities and introduced protection regulations. Some countries did not sign this agreement because they did not have an interest in the continent or did not agree with the restrictions imposed.

Antarctica's natural resources are not exploited, and the international agreements stipulate that they should not be. Only the future will tell how long this protective situation will last or whether Antarctica will turn into an open-pit mine for the rest of the world. The main interest in Antarctica in the 1990s concerned its important role as a laboratory for monitoring global climate change and the study of the development of the ozone hole above it, the extent of ice on the continent and the surrounding oceans, and the bounty of the ocean ecosystems. Research carried out in Antarctica finds atmospheric pollutants that indicate a decline in world environmental quality. In their small but increasing manner, however, research stations also degrade the local environment by pouring untreated sewage into the ocean and dumping oil drums on sites that seabirds use for nesting.

Antarctica and the Southern Oceans

Without people, Antarctica's significance is in its natural environment. The divergent plate margin between the Indian and Antarctic plates separates Australia and Antarctica. The Antarctic continent formed the ancient rock core of Gondwanaland. After the combined continent broke apart, Antarctica remained at the south pole. As Earth's atmosphere cooled some 20 million years ago, ice accumulated on Antarctica, burying the mountain ranges that continue the line of the Andes from South America (see Figure 6.40a). A large part of the continent sank under the weight of ice and is now below sea level. If the ice melted, Antarctica would form several landmasses.

The frozen continent forms its own climate with extreme cold and a heating deficit (see Figure 2.21) throughout the year that built up an ice cover over the last 15 million years. Winters have almost total darkness for several months; summers are all daylight. In winter, the ice-covered area of the oceans around Antarctica doubles in size as the sea surface freezes; in summer, glacier ice calves off icebergs into the surrounding ocean. The waters are very cold in all seasons. Perhaps Antarctica's most significant boundary is between 50 and 60 degrees south, where cold and warm ocean water and atmosphere converge. The contrast in the atmosphere above creates a sharp boundary between Antarctic air and warmer midlatitude air, generating a frontal zone and a succession of midlatitude cyclonic storms with high winds. The southern oceans surrounding Antarctica have little land to interrupt these strong airflows and cyclonic storms. This zone of confrontation of contrasting air types effectively cuts off Antarctica from warming influences.

Despite the barrier to climatic elements entering Antarctica's atmosphere, small quantities of chlorine gases penetrated from lower latitudes to produce the **ozone hole** above the continent at the end of the Antarctic winter in October. This "hole" is a thinning of the protective ozone layer high in Earth's atmosphere that filters out harmful ultraviolet radiation from the sun. The hole enlarged in the 1980s and 1990s as the chlorine gases reacted with the ozone but appeared to stabilize around 2000.

During the late 1990s and early 2000s, some of Antarctica's ice shelves, where ice flowed off the continents to cover the ocean surface, melted and broke up—seen by many as a sign of global warming. But a report in 2001 showed that temperatures in the heart of Antarctica are falling rather than rising.

Antarctica's Resources

Antarctica is not a country and so does not have an economy of its own. Although there is no authority to regulate resource usage, international agreements prevent the known deposits of coal and other minerals from being exploited.

While the natural resources in its rocks are off-limits to exploitation, Antarctica's surrounding oceans draw fishers from all over the world. The oceans surrounding Antarctica are among the world's most profuse life zones. Antarctica's living organisms are dominated by a huge variety of seabirds, including penguins that rely on the rich ocean life of plankton, fish, seals, and whales (Figure 6.40c). The Antarctic oceans are an

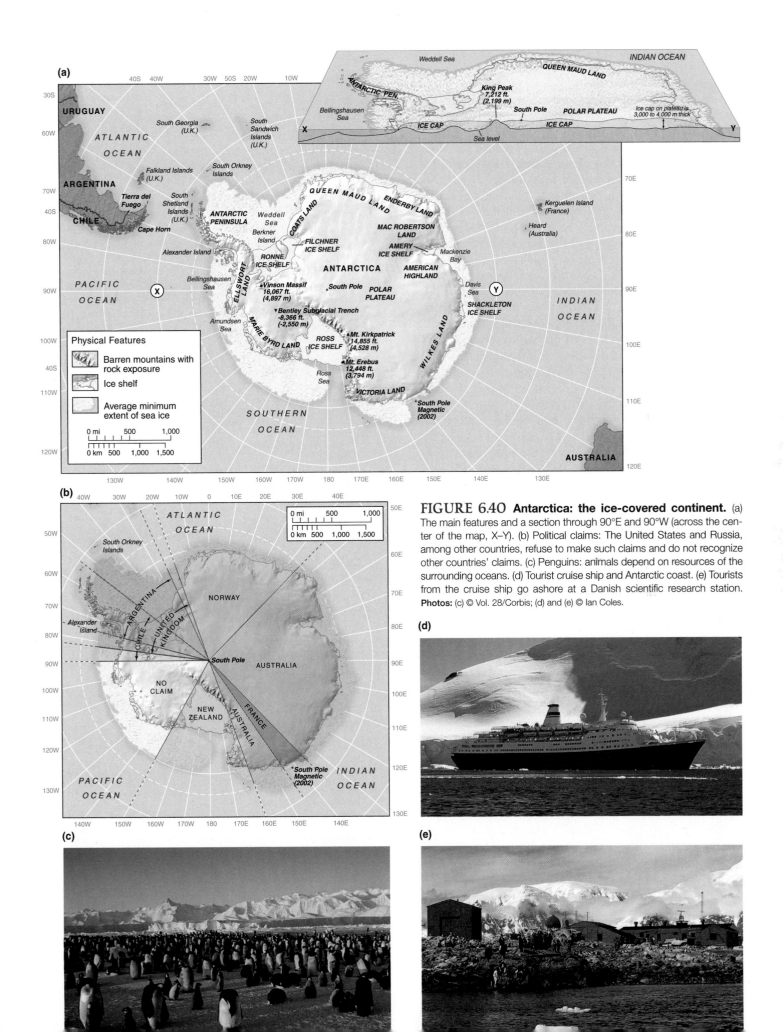

(a)

URUGUAY

ATLANTIC OCEAN

ARGENTINA

South Georgia (U.K.)

South Sandwich Islands (U.K.)

Falkland Islands (U.K.)

South Orkney Islands

CHILE

Tierra del Fuego

Cape Horn

South Shetland Islands (U.K.)

ANTARCTIC PENINSULA

Weddell Sea

Berkner Island

Alexander Island

COATS LAND

QUEEN MAUD LAND

ENDERBY LAND

FILCHNER ICE SHELF

MAC ROBERTSON LAND

AMERY ICE SHELF

Mackenzie Bay

Kerguelen Island (France)

Heard (Australia)

RONNE ICE SHELF

ANTARCTICA

AMERICAN HIGHLAND

Davis Sea

PACIFIC OCEAN

Bellingshausen Sea

ELLSWORTH LAND

▴Vinson Massif 16,067 ft. (4,897 m)

✛South Pole

POLAR PLATEAU

SHACKLETON ICE SHELF

INDIAN OCEAN

Amundsen Sea

MARIE BYRD LAND

▾Bentley Subglacial Trench -8,366 ft. (-2,550 m)

ROSS ICE SHELF

▴Mt. Kirkpatrick 14,855 ft. (4,528 m)

WILKES LAND

▴Mt. Erebus 12,448 ft. (3,794 m)

Ross Sea

VICTORIA LAND

✛South Pole Magnetic (2002)

SOUTHERN OCEAN

AUSTRALIA

Physical Features

- Barren mountains with rock exposure
- Ice shelf
- Average minimum extent of sea ice

0 mi 500 1,000
0 km 500 1,000 1,500

Inset section:

Weddell Sea

QUEEN MAUD LAND

INDIAN OCEAN

ANTARCTIC PEN.

King Peak 7,212 ft. (2,199 m)

South Pole

POLAR PLATEAU

Ice cap on plateau is 3,000 to 4,000 m thick

Bellingshausen Sea

ICE CAP

Sea level

ICE CAP

X — Y

(b)

ATLANTIC OCEAN

South Orkney Islands

Alexander Island

ARGENTINA

CHILE

UNITED KINGDOM

NORWAY

AUSTRALIA

South Pole

NO CLAIM

NEW ZEALAND

FRANCE

✛South Pole Magnetic (2002)

PACIFIC OCEAN

INDIAN OCEAN

0 mi 500 1,000
0 km 500 1,000 1,500

FIGURE 6.40 Antarctica: the ice-covered continent. (a) The main features and a section through 90°E and 90°W (across the center of the map, X–Y). (b) Political claims: The United States and Russia, among other countries, refuse to make such claims and do not recognize other countries' claims. (c) Penguins: animals depend on resources of the surrounding oceans. (d) Tourist cruise ship and Antarctic coast. (e) Tourists from the cruise ship go ashore at a Danish scientific research station. **Photos:** (c) © Vol. 28/Corbis; (d) and (e) © Ian Coles.

(d)

(c)

(e)

important basis for wider ocean food chains and are being studied to gain an understanding of the sustainable levels of fishing, sealing, and whaling.

In 1982, as commercial fishers from around the world increased their exploitation of the marine resources, it was agreed internationally to regulate such fishing. Fish stocks, such as cod, together with some groups of whales, were declining. Some fishing fleets, however, claimed that they fish outside the Antarctic convergence zone that forms the northern boundary of the partially protected area or professed ignorance of the quotas that were established. Additionally, it is almost impossible to monitor fishing in this extensive area of ocean that has few ships passing through and is not the responsibility of a particular country. Related problems that cause worries over the future of the total ocean ecology include, for example, the long-line hooks used to catch Patagonian toothfish in the mid-1990s, which also snared albatross and petrel birds.

Tourism

Tourists, usually based on cruise ships (Figure 6.40d), visit Antarctica to view the scenery, wildlife, and some research stations (Figure 6.40e). By 2005, up to 40 ships carrying from 100–600 tourists each were visiting the Antarctic Peninsula annually. Equipped with Zodiac landing craft, passengers could land for short periods. Self-imposed controls limit the number of tourists ashore at one time to 100. At present, the Antarctic Treaty system does not have a code regulating the tourism industry.

Test Your Understanding
6E

Summary

The South Pacific islands include large numbers of small islands grouped in mainly independent but poor countries. Papua New Guinea is the largest, having over half the total population of all the islands. Many South Pacific islands have no commercial products; others produce coconuts and copra; some export a particular mineral; and some have economies maintained by French colonial support or U.S. grants. Antarctica is a continent without a permanent population. Its resources are not exploited, and its main role is as a scientific laboratory for weather and marine ecosystem study.

Questions to Think About

6E.1   What attractions do the countries of this subregion offer to tourists? What are the advantages and disadvantages of the tourism industry for the inhabitants?

6E.2   Should Antarctica be included as a subregion of a world region, be a world region on its own, or be ignored by regional geographers?

Key Terms

copra   Antarctic Treaty   ozone hole

Chapter 7

South Asia

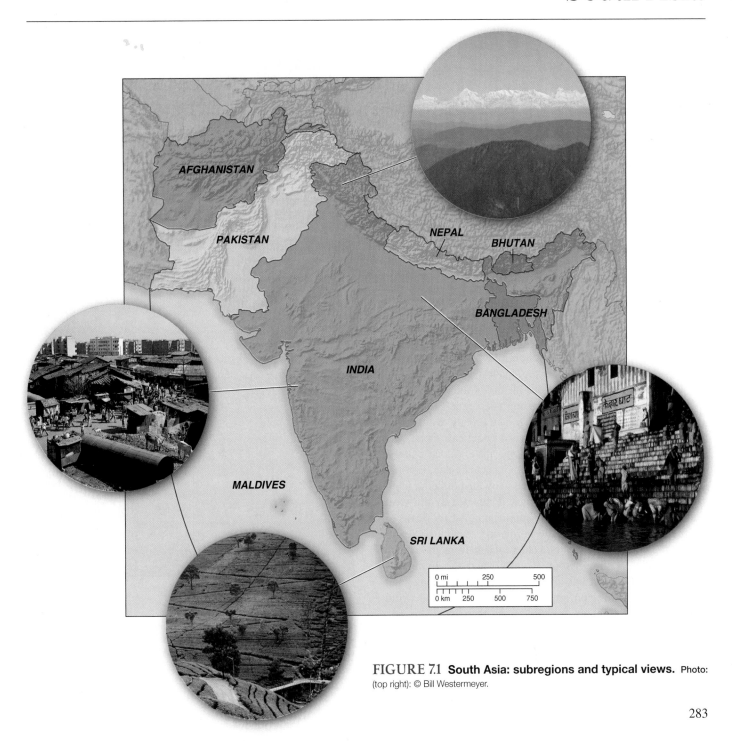

FIGURE 7.1 South Asia: subregions and typical views. Photo: (top right): © Bill Westermeyer.

Themes in This Chapter

South Asia has a rich history that spawned major religions and material wealth for many. Following colonial control and partition, its countries developed their own approaches to independence. During the Cold War they were marginalized, but since 1990 they have been through major internal conflicts and increasing world involvement. The major themes are:

A region rich in history and culture.

Diversity as the outcome of invasions, colonization, and the thrust toward independence.

A natural environment of monsoon rains and aridity, of the world's highest mountains and wide plains.

Geographic contrasts and modern conflicts in the context of global trends.

Three subregions (Figure 7.1):

- India: soon to become the world's most populous country.
- Pakistan and Bangladesh: contrasting Muslim countries.
- Mountain and Island Rim: small countries isolated in the mountains or located on ocean trade routes.

Point–Counterpoint: Afghanistan, Pakistan, Kashmir, and India

Personal View: Kerala: An Anomaly in India

Geography at Work: Battling Infectious Diseases

(a)

Past and Present

South Asia consists of the land from the Himalayan peaks to the Indian Ocean—the part of Asia often referred to as the Indian subcontinent. The region is home to 1.4 billion people, a total that may rise to nearly 2 billion by 2025, more than any other world region. And yet it is the smallest world region.

A varied and often tumultuous history of interactions with the natural environment and invading peoples left the region with a common culture interacting increasingly with global influences (Figure 7.2) but with uneven political, social, and economic geographies. Historically, the material and cultural wealth of the region vied with that of China and, later, Europe.

Physically, the region is hemmed in by the northern mountain wall (Figure 7.3). Internally, it is dominated by the great plains of the Indus, Ganges, and Brahmaputra rivers. Waves of people entered first through the northern mountain passes to access the rich plains. Later invaders came from the ocean, including the Muslims (from A.D.1100) and the Europeans (from A.D.1500). Each left its marks on the land, people, and culture; today this region interacts with the rest of the world from a base of complex traditions and profound ideas and values.

Today, India's media industry illustrates how people with such ancient cultures respond to modern globalization. By the 1990s, satellite delivery of TV programs enabled the globalizing

(b)

FIGURE 7.2 South Asia: contrasting cultures. (a) An Internet café is close to a Hindu temple in Bangalore. (b) Posters for movies, Mumbai (Bombay). The Indian movie industry is the world's most prolific, and most production is in this city (Bollywood). **Photos:** (a) © David Wells/Corbis. (b) © Catherine Karnow/Corbis.

284

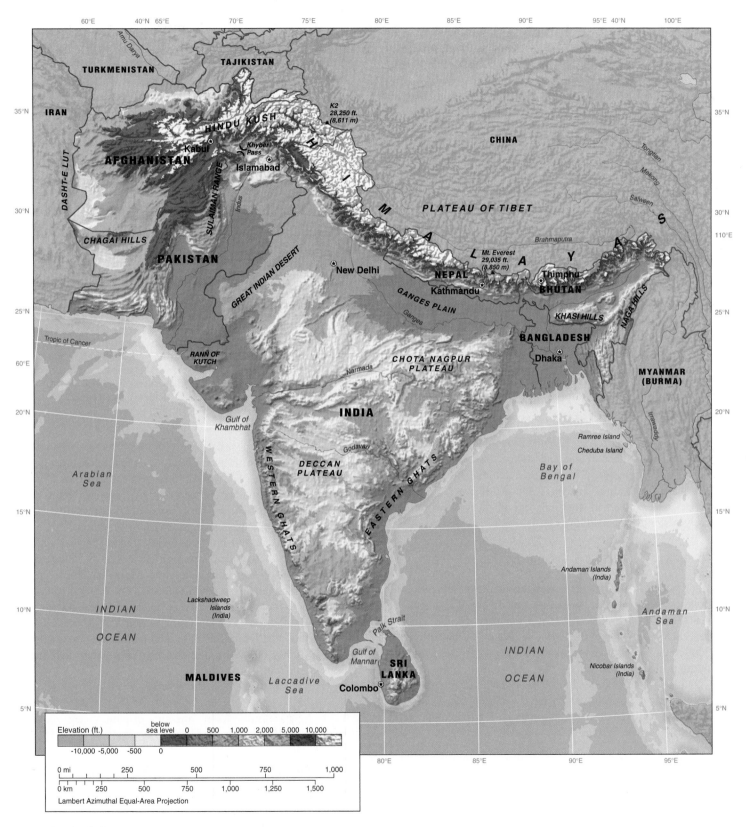

FIGURE 7.3 South Asia: physical features, countries, and capitals. The map shows how the Himalayas form the northern boundary and are separated from the plateaus of southern India by the Indus, Ganges, and Brahmaputra rivers.

media moguls to replace the single-channel Indian government programs and provide a diet of U.S. and Western films, soap operas, and sports, mainly in languages spoken by fewer than half the total population. In response, new Indian TV stations, such as ZeeTV, brought a mainly Indian emphasis and now provide programs in 15 languages. The Indian movie industry, which already produces more films than the U.S. industry, diversified into TV soap operas with Indian scenes. The new range of TV channels are consumerist, focusing on Indian interest groups and especially the large number of young adults. Soap operas deal with issues of local concern such as violence against females. One result of the wider media coverage is to encourage a pan-Indian awareness within an extremely diverse country and an increased confidence in being Indian in a West-dominated world.

In this chapter, the essential understanding of South Asian cultural and political history is followed by a study of the region's diverse natural environments that define some possibilities and limitations for human activities. An overview of the recent global context is followed by the subregional studies that focus on what is geographically distinctive in the modern countries, including their political status, ethnic characteristics, and economic development.

Diverse Cultures

The history of South Asia is marked by attempts to unify the people of the subcontinent and protect them from external invaders. A strong social culture emerged that survived many political and economic changes. Cultural differences, including gender issues, emerged from religious allegiances that became enshrined in political attitudes.

Precolonial Cultures

Little is known about the earliest events in the evolution of South Asian societies. The dark-skinned Dravidian people of southern India (Figure 7.4) are thought to be the modern descendants of some of the oldest inhabitants of the region. By 3000 B.C. the Dravidians most likely established irrigation farming in the Indus Valley. They left ruins of cities (Figure 7.5) and irrigation works, but their written heritage was only a small number of seals, which have not been translated. The Harappan civilization rivaled those in the Tigris-Euphrates and Nile valleys (see Chapter 1).

Hindus and Caste

By 1500 B.C., but possibly many years earlier, lighter-skinned and taller Indo-Aryan people invaded through the northwestern passes and drove Dravidians southward into the Indian peninsula. As a result of Aryan influences, Hinduism crystallized in the Indus River valley around 1200 B.C. The Aryans brought together traditional myths and gods to form Hinduism as an inclusive nationalistic religion, based on the Sanskrit Vedas (1500–1000 B.C.), which rank among the world's greatest religious literature.

FIGURE 7.4 South Asia: Dravidian peoples. Indian people engaged in a workshop industry in Bangalore, southern India. The dark skins often originated from the Dravidians and other possibly pre-Dravidians such as the Veddas from Sri Lanka. Dravidians occupied India before the invasions of lighter-skinned Aryan groups. **Photo:** © Michael Bradshaw.

Hinduism can most accurately be described as "the religion of the people." This makes it an ethnic religion that people are born into. Today nearly all of the world's 800 million Hindus live in India. Hinduism is not a centrally organized religion with a single sacred text but has a series of "Great Traditions" as set out by religious experts and "Little Traditions" about local gods, beliefs, and practices. Hinduism recognizes millions of gods, although local groups often see one as dominant, and incorporates beliefs of reincarnation. More than 80 percent of the Indian population is classified as Hindu, although many, particularly in rural areas, do not recognize that Hinduism is the title given to their local devotions: it is their way of life. Hinduism is not totally confined to India. Missions took it to Southeast Asia 2,000 years ago, and the modern Hare Krishna groups recruit worldwide.

FIGURE 7.5 South Asia: ancient origins. Archaeological remains of the ancient city of Modenjodaro. **Photo:** © Munir Khan/Harappa.com.

Although caste is a Hindu concept, it may have permeated all South Asian society before the arrival of the Aryans and was adopted by later invading groups. The **caste order** is similar to ethnic or class divisions elsewhere. Aryans imposed social groupings on the conquered: the priests (*Brahmins*—the top group, who decided the membership of other groups); warriors and rulers (*Kshatriyas*); and commoner merchants and artisans (*Vaishayas*). Those outside these divisions, mostly non-Aryans, were the lowest caste of menials and servants (*Shudras*) and outcasts, who were believed to be by nature "disgusting, polluting, and unworthy" and were labeled "untouchables." Both the menials and the outcasts live separately and use separate wells. In addition, the tribal peoples are also outside the caste order—groups who were perceived as the "primitive" forest dwellers or descendants of the original population. They have similar physical features to Hindu Indians and may include many who were rejected or were not absorbed by Hindu groups.

The religious base of the major caste divisions (*Varna*) was paralleled by the common interests of groups of people (*Jatis,* or subcastes) who worked in similar craft industries of roughly equal status. These economic and social divisions are thought to have evolved from tribal kinship groups into specialist feudal roles. Membership became hereditary, based on intermarriage within the group; birth determined status in society. In the typical South Asian village of the past, landowners, tillers, carpenters, potters, barbers, priests, and other groups worked within local hierarchies for the common (or landowner's) good.

Buddhism and Jainism

Buddhism and Jainism were founded in the Ganges River valley in the 500s B.C. in reaction to aspects of Hinduism. Buddha began life as a Hindu. A major early attraction of his teachings was the rejection of the millions of Hindu gods. Buddha also questioned the rigid caste system and believed in the possibility of greater social mobility. When the Mauryan Empire expanded to incorporate most of the Indian subcontinent from 321 to 181 B.C., its most celebrated emperor, Asoka, adopted Buddhism and propagated the *dharma* policy, a code of conduct focusing on social responsibility, human dignity, and socioreligious harmony. After Asoka's death, weak successors allowed the empire to fall apart. Brahman Hinduism took over, envisaging a country as a divine creation rather than a Buddhist concept of a contract between ruler and people to control the chaotic and demoralizing effects of mortal possessions. The renewed dominance of Hinduism forced Buddhists to move out of the Indian mainland, converting much of Sri Lanka and spreading their faith eastward and northward into Southeast and East Asia.

Jainism followed a nonviolent code that was later taken up by Mahatma Gandhi in his early-1900s campaign for India's independence from the British. Forbidden to farm (and kill worms and insects), Jains became an exclusive group. Many, especially in the Mumbai (Bombay) area, are wealthy traders. They make up a small proportion of the Indian population today, but include over 3 million people.

Many Invasions

Invasions kept historical Indian societies in a state of flux, especially in the northern plains, where a mosaic of distinctive peoples, landscapes, and anarchic governments resulted. For example, the short-lived invasion by the Greeks under Alexander the Great in 326 B.C., at the end of his conquests in western and central Asia, left behind military and legal influences. The Gupta Empire from A.D. 319 to 950 brought peace, economic growth, and a new flowering of art, music, and literature.

Muslims, Mughals, and Sikhs

Over the next 600 years, invading Mongol groups, Muslims, and Turks were unable to establish lasting or total rule. After Muslims invaded the north in the A.D. 1100s, they pushed southward from the Indus and Ganges Plains. However, it was not until the 1500s that the **Mughal (Mogul) dynasty** extended Muslim beliefs to the rest of the region. The Mughals were Persian Turks led by Babar, a descendant of the Mongols (see Chapters 4 and 5). Under the Mughals, India experienced developments in such fields as architecture; the Taj Mahal and many examples of Mughal buildings built in the 1600s (Figure 7.6) survive throughout the region.

Sikhism emerged in the early 1500s as one of several responses by some Hindus in the Punjab to the Mughal dynasties. It later resisted absorption into mainline Hinduism. Sikhism is marked by a strict code of conduct and many temples that have kitchens providing food for all. The Golden Temple at Amritsar is the holiest temple. Constructed by Guru Arjan Singh in the late 1500s, it contains the Sikh scriptures that he compiled. Although few in number, many Sikhs became wealthy from their bumper crops in Punjab and continue to lobby for a separate state or even an independent country (Sikh Khalistan, the "land of the pure").

FIGURE 7.6 South Asia: historic buildings. The entrance to the mausoleum of Akbar, the greatest of the Mughal emperors, near Agra. **Photo:** © Bill Westermeyer.

Mountain Isolation and Island Openness

Places along the mountainous northern margin of the region—the present countries of Afghanistan, Nepal, and Bhutan—mixed isolation with occasional intrusions of wider influences. Tribes in the Himalayas established their own ways of living, but Nepal and Bhutan were occasionally invaded from Tibet.

The area that is now Afghanistan became a mountain refuge for many ethnic groups. The Pashtuns formed the majority in the southern parts, but Tajiks and Kyrgyz people dominated the north (Figure 7.7). In medieval times, Herat near the present Iranian border was home to a fusion of Buddhist and Persian culture that left buildings and statues. Until the 1980s, the rich cultural heritage from these experiences was on display in the Kabul National Museum: Islamic art, Roman bronzes, Egyptian glass, Chinese lacquerware, Indian ivories, and Buddhist collections. However, the Soviet military, looting and export through Pakistan, and vandalism under the Taliban government from 1996 destroyed this collection.

Ceylon's (modern Sri Lanka) closeness to the Indian peninsula and its openness to ocean-borne trade and conquest attracted competing groups. Indo-Aryan peoples with Buddhist beliefs from northern India developed a civilization based on irrigated rice agriculture in the north center of the island that lasted from around 100 B.C. to the A.D. 1200s. During the later part of this period, Tamils—Dravidian peoples from southern India—established kingdoms in northern Sri Lanka based around Jaffna. As Tamils expanded their influence, the majority Sinhalese peoples drifted southward, abandoning the irrigation systems and relying on rain-fed agriculture. They grew spices such as cinnamon. Arab traders settled in the country and controlled the overseas spice trade.

The growing Arab (Muslim) control of medieval Indian Ocean trade routes included the Maldive Islands, which stretch southward into the Indian Ocean. Chinese and Arab merchants set up trading posts there to exchange their goods for local fish, coconuts, and shells. The islands became solidly Muslim by the A.D. 1100s and remain so.

Colonial Impacts

Trading Expansion

The great wealth of this region, built on internal gem and metal deposits and also on external trading links to Africa, Arab, and southeast Asian lands, became a goal for European adventurers in the phase of mercantile expansion from the mid-1400s. Searching for an alternative route to get Indian products and riches to Europe without traveling through the increasingly hostile Muslim countries of southwestern Asia, the Europeans arrived by sea and found few defenses. In 1498 the Portuguese explorer Vasco da Gama reached India after sailing around southernmost Africa. During the 1500s, the Portuguese and Dutch established trading stations, such as Goa, often based on previous ports (Figure 7.8). During the 1600s, the Dutch forced the **(British) East India Company** out of the East Indies (see Chapter 6), and the British switched to India, ousting their Portuguese and Dutch rivals from most trading centers. The British East India Company took the Portuguese port of Bombay ("good bay," now Mumbai), built a new port at Madras (Chennai), and began trading with the most populous area of Bengal around modern Calcutta (Kolkata), which became the company's main center.

In the early 1700s, political chaos caused by internal factions in the Mughal Empire splintered South Asia into small and large kingdoms, often ruled by Muslim or Hindu leaders who styled themselves as maharajas or princes. Foreigners took control of the increasing overseas trade in cotton, cloth, rice, and opium. The French contested British control but were defeated. Over the next 100 years, the British East India Company, backed by the British Indian army that employed Indians as foot soldiers (*sepoys*), increased its hold on South Asia, taking a particularly strong position in Bengal and the peninsula. Many of the company directors and employees became extremely rich, buying huge homes and honors back in England. British ports expanded to handle the cargoes from the new trade. The company's area of influence was extended southward into Ceylon in 1798 and westward into Punjab in the mid-1800s.

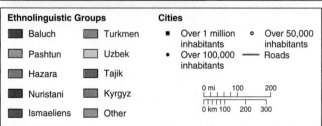

FIGURE 7.7 Afghanistan. The complexity of ethnic groups and their links across the political borders is a major factor in the difficulties of governing this country. The small number of border crossings in this mountainous country link some of the groups.

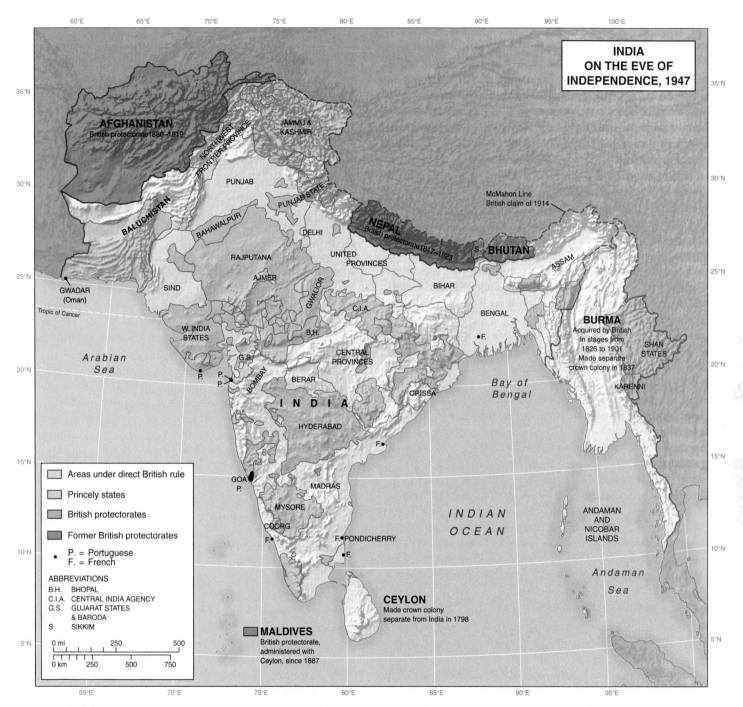

FIGURE 7.8 South Asia: the British Indian Empire just before independence in 1947. The princely states had working arrangements with Britain but were not ruled as part of the Indian Empire. At independence, they agreed to be part of the new countries of India and Pakistan and subject to their governments. Kashmir is still disputed between India and Pakistan.

British Indian Empire

Following several uncoordinated revolts, a major mutiny of its Indian sepoy troops in 1857, which many Indians see as "The First Independence War," led the British government to take full political control of South Asia. It abolished the British East India Company and established the **British Indian Empire,** often referred to as the "British Raj" (*raj* means rule or government). It included almost all the subcontinent. However, 40 percent, with almost one-fourth of the total population, remained under "independent" princely family governments, who ordered matters in harmony with British policies. The 600 princely states ranged from Hyderabad, with 14 million people in 1947, to Jathiamar, with only 200.

The British saw their role as "civilizing" India through Western education, new technology, public works, and a new system of law. However, benefits went both ways. British interests redirected India's farms to produce raw materials for British industries, while the Indian textile industry was suppressed in favor of British cotton goods. The region entered the expanding global economy of the later 1800s, when the export of Indian cotton compensated for negative British trade balances. Furthermore, the British Indian army became a tool of attempted imperial expansion into Afghanistan and Burma in the 1880s and into Tibet in 1903–1904. Burma became part of the British Indian Empire in 1885, but Britain established Afghanistan as a separate country.

From the 1850s, Nepal had a close relationship with Britain, and many Nepalese served in the British Indian army. Following British attacks in the 1800s, Nepal remained independent under pro-British rulers, who hired out Gurkha soldiers to the British. Bhutan was annexed by Britain in 1826 and given autonomy in 1907.

British imperial rule had massive effects on the geography of the peninsula. Its mercantile economy focused on primary products to be exported to Europe, defined the resources that could be developed, determined what was produced, altered patterns of land control, selected areas for development and cities for growth, and controlled external trade. British engineers irrigated land in the Indus and Upper Ganges river basins to produce cotton for export. They built railroads from the main ports to move troops and exports. Former communal land was reallocated to larger and smaller landowners, forming interest groups that were expected, in return, to support the colonial administration. The cities of Calcutta, Bombay, and Madras grew faster than others after the British East India Company designated them as focuses where lines of overseas trade and internal communications met. The English language became the unifying language among the traders, lawyers, and elite families.

Ceylon and the Maldives

As British colonies, Ceylon and the Maldives were also incorporated into the world economic system that focused on the demands of the colonizing country and other wealthier markets rather than on local needs. In Ceylon, the Portuguese and Dutch had been the early traders and colonists. The Portuguese left a strong legacy of Roman Catholic missions and of Portuguese as a trading language. In 1795 the British took Ceylon, which remained a crown colony until its 1948 independence. British companies set up plantations for growing tea, rubber, and coconuts, instead of the traditional spices. More Indian Tamil labor was brought in to work on the plantations, increasing local ethnic tensions. Their better education in Christian schools in the north enabled the Tamils to take most of the colonial clerical jobs. A Western-educated elite of local plantation owners attempted to emulate the social life of the colonials. The emphasis on export crops, however, was accompanied by the neglect of traditional agriculture, leading to a decline in the rice crop and a need to import half the island's food needs.

Afghanistan

After a century of attempted dominance over the Northwest Frontier area of India, the British created the country of Afghanistan in the late 1800s. The northern border defined at that time established a limit, or buffer, to Russian expansion, but the new country brought together groups fragmented by language, creed, geography, and historic cultural traditions. King Abdur Rahman (1880–1901), backed by British money and weapons, tried to create the basis of a centralized Afghanistan. His and later attempts to build and unify the new country were often ruthless, resented by the intensely independent groups, and generally unsuccessful.

Paths to Independence

The Origins of India and Pakistan

The extension of British rule during the late 1800s created an awareness of an emerging national identity across the subcontinent. In 1885 a combination of cultural background and personal interest led a mainly Hindu elite to form the **Indian National Congress Party.** It had a secular and multicultural basis with the aim of an inclusive India, in which Hindus were a majority but coexisted with Muslims, Sikhs, Jains, Buddhists, Christians, Jews, and others. Leaders of the Muslim population, the largest minority, were skeptical of the Congress Party motives. They formed the **Muslim League** in 1906 with the aim of establishing an organization devoted to promoting Muslim interests in British India.

During World War I (1914–1918) the Indian army fought alongside Europeans in a European war, leaving many in India to suffer from disrupted trade, injuries, and deaths. After the war, India became less vital to Britain's economic interests of trade, investment, and political empire. In the 1920s Mahatma Gandhi welded together a coalition of interests seeking independence, and his policy of nonviolent resistance brought the country to a standstill on several occasions. A sharp population increase began after 1920, when the Indian Empire population numbered around 250 million people, but the British government did nothing to slow the rise at a time of frequent famines. A more outspoken group, the Rashtriya Swayamsevak Sangh (RSS, National Volunteers Association), founded in 1925, wanted national unity on the basis of Hindu supremacy.

By 1940 Britain and the National Congress Party favored a single country in South Asia with a political structure based on a weak central government and strong provinces. Muslim fears that they would be an underprivileged, even persecuted, minority in a single country escalated. The Muslim League insisted on a separate country formed of provinces in which there was a majority of Muslims. However, the two large provinces of Punjab and Bengal had Muslim majorities of only some 56 percent and large numbers of Hindus and (in Punjab) Sikhs. The British government proposed dividing the provinces of Bengal and Punjab between India and Pakistan, thus placing the great majority of territory and people in a new India and leaving the Muslims with two separate areas to the west and east of India.

At the end of World War II, pressures for a rapid handover of government resulted in a grudging Muslim acceptance of the smaller East and West Pakistan.

The partition of British India into the two countries of India and Pakistan took place in 1947. Many Muslims and Hindus caught in the wrong country fled across the borders to avoid ethnic bloodshed. Some 12 million people were displaced, but over a million died in clashes. Kashmir remains a major border issue between India and Pakistan (see Point–Counterpoint: Afghanistan, Pakistan, Kashmir, and India on page 292).

Ceylon Becomes Sri Lanka

The majority Buddhists in Ceylon pressed for independence early in the 1900s. Ethnic differences and resentments surfaced before independence in 1948 as Buddhist revivalism among the Sinhalese alerted the Tamil population to the need to preserve their separate identity, which had been protected in colonial times. Although Ceylon was not partitioned at independence, tensions were increased by the dominance of Sinhalese nationalists in the new government and their change of the country's name to Sri Lanka.

Natural Environment

The natural environments of South Asia have major impacts on the human geography. Long dry seasons, uncertain rainfall, and large arid areas make water basic and crucial in sustaining the livelihoods of the large population of the region. The mountains, plains, and hilly areas influence the distributions of settlement and economic activities. Furthermore, such a large population places stresses on the natural environments.

Monsoon Climates

The monsoon climatic environment of South Asia (Am and Aw in Figure 7.9) brings heavy summer downpours over much of the Indian subcontinent but little rain at other times of the year.

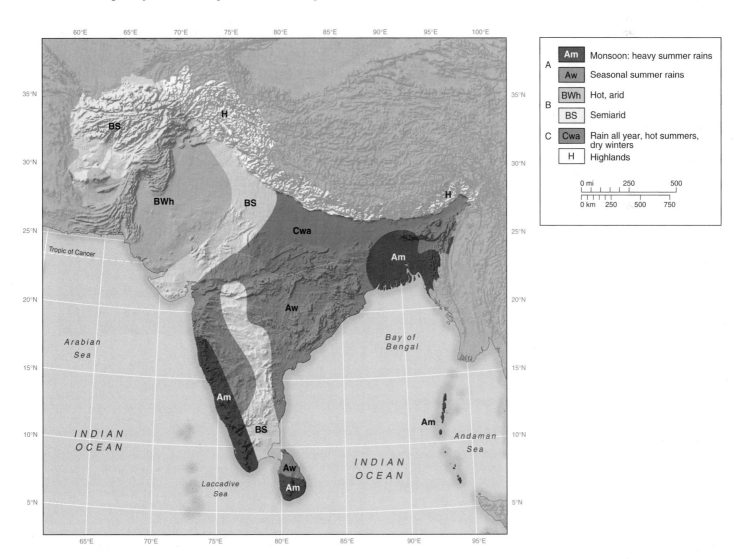

FIGURE 7.9 South Asia: climates. Climatic map, showing the extent of full monsoon conditions, lesser summer rains, arid areas, and highland climates.

POINT
COUNTER
POINT

AFGHANISTAN, PAKISTAN, KASHMIR, AND INDIA

The events of September 11, 2001, and the subsequent attacks on Afghanistan brought to light many links of agreement and conflict across sets of countries. One of these was the long-lasting dispute between India and Pakistan over the Kashmir Province at the northern end of their common border. By early 2002, India and Pakistan were drawing military forces to the border and war looked possible. As in its relationships with the former Taliban government in Afghanistan, however, the "War on Terrorism" caused the Pakistan government to hold back and even repudiate groups of Islamic radicals it supported and trained for many years. The situation along the Indian–Pakistani border remained tense—as it had since the partition adopted at independence in 1947.

This is a situation where geographic principles may assist in deeper understanding. First, it is important to know where Kashmir is and how it relates to the surrounding countries (Box Figure 1). Second, the nature of the countries, including those surrounding India and Pakistan, and of their often conflicting central beliefs and political aims is very significant. Third, the influence of external pressures, such as that from the United Nations or from major powers such as the United States, Russia, and China, is pertinent. After September 11, 2001, and its aftermath, it is increasingly difficult for any country to operate without reference to others—be it Afghanistan, Pakistan, India, or the United States. Like many others throughout the world, the relationships among the countries in this region are complex and debated. It is unlikely that there will be a rapid solution.

Some history of the Kashmir situation is basic to present events. In the mid-1800s, the British sold the rulership of the Muslim state of Kashmir to a Hindu maharajah for around US $50 million at today's prices. In 1947 at independence, Kashmir opted for being part of India. Although the maharajah tried to avoid the need to join India or Pakistan, an invasion by Pakistani tribesmen forced his hand, and he chose India in return for military help. The Indian prime minister, Jawaharlal Nehru, in his eagerness to get popular ratification for the accession, brought Pakistan's 1947 invasion to the notice of the United Nations and urged Pakistan forces to withdraw until a referendum allowed Kashmiris to choose between India and Pakistan. After the 1948 cease-fire, Pakistan held onto the one-third it conquered, much of it close to the city of Rawalpindi and later the new capital, Islamabad. At that stage, Kashmir's main Muslim leader, Sheikh Muhammad Abdullah, who became prime minister of Kashmir in 1948, preferred India's secular socialism to the Muslim ideology of Pakistan. The Pakistanis, however, never withdrew, and Nehru never put the matter to a vote.

This is often where the debates begin. Did the maharajah persecute Muslim subjects? Did the tribesmen attack spontaneously for their defense, or were they pushed by Pakistan's new government? Did the British support India's new government in acquiring Kashmir?

At first the maharajah, who handed over control of only defense, foreign affairs, and communications to India, leaving Kashmir with its own prime minister until 1965, encouraged local politics. In 1953 suspicions (disputed) that he plotted with Pakistan and the United States earned Sheikh Abdullah 11 years' imprisonment.

India progressively integrated Kashmir, moving many of its Hindu population to Delhi and setting them up with craft industry employment making rugs, woolen clothes, and decorated plates and selling them in tourist store outlets. In 1964 India extended to Kashmir the right of the federal government to dismiss state gov-

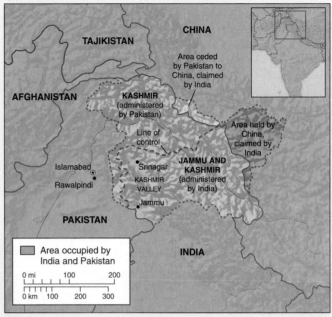

Box Figure 1 Kashmir. (a) Map showing the complex geography. The line of control separates Indian and Pakistani areas of present occupation. There are also border disputes with China. (b) A mountain road in Kashmir. **Photo:** (b) © Alasdair Drysdale.

ernments. Each decision, however, seemed to thwart the popular will of Kashmiris. India and Pakistan fought over Kashmir in 1965–1966, when India accused Pakistan of backing insurgents.

War broke out again in 1971, when Bangladesh (East Pakistan) became independent, backed by India. Kashmir was another source of fighting in that war, after which the 1972 Simla Agreement divided Kashmir into the Pakistan state of Kashmir and the Indian state of Jammu and Kashmir. This agreement established today's line of control.

(Continued)

Any goodwill generated by the first genuinely free Jammu–Kashmir elections in 1977 that brought back Sheikh Abdullah collapsed with the ousting of his son from power in 1984. Antigovernment Muslim groups contested the 1987 state elections, but Delhi annulled some of their successes. From 1989, separatist violence that India alleges is backed by Pakistan was matched by Indian military repression.

By the late 1980s, Pakistan looked friendlier to Kashmiris than India despite having a very different culture based on Islamic rules and military dictatorships. Indians believe their democratic constitution can provide liberation for Kashmiris. In fact, most Kashmiris see themselves as prisoners of either country and are weary of continuous fighting. They prefer independence.

It is difficult to see who speaks for Kashmiris. The state government based in Srinagar is very unpopular. Separatist politicians based around the All-Party Hurriyat Conference have yet to test their popularity in elections but do not refute suspicions that they repeat Pakistan's policies. Militant groups formed as groups of local young men went to Pakistan for training and arming, and returned with other Pakistanis linked to Afghanistan's Taliban.

Some Kashmiris look to the United States and its allies as potential saviors that can push India into a settlement. But India and Pakistan are the real arbiters. India shifted from inaction to attempts to talk with Pakistan or separatists. The early 2000s Pakistani crackdown on terrorists causes Indian cynicism about Pakistan's promises when banned groups such as Lashkar-e-Taiba continue to act after merely changing their names. India, however, keeps a security force of 400,000 troops in Kashmir and hopes the crackdown will moderate the militancy. There is speculation that Pakistan will return support to more moderate Kashmiri groups such as Hizbul Mujahideen.

Any election or referendum faces major problems. If separatist groups, which deeply distrust India, participate, they have to declare themselves Indian citizens to qualify to vote. Without the separatists, however, Kashmiri Muslims will mock the result as worthless. The Indian government rejects proposals for foreign election observers or outside mediation as questioning its authority. This attitude continues a high-handed and superior approach based on the original partition decision, Indian democratic institutions, and resistance to terrorist attacks but avoids discussion of how India alienated the Kashmiri people.

Partition of Kashmir between India and Pakistan with a degree of autonomy for both parts of Kashmir might then occur. Pakistan conceded much after September 11, 2001, withdrawing support for Taliban in Afghanistan and shutting down a potential war against India after terrorists attacked the Indian parliament in Delhi. It remains to be seen whether Pakistan can accept a more democratic and humane version of the status quo Pakistan fought against for over 50 years. President Pervez Musharraf of Pakistan is a military dictator who gained international disapproval when he dismissed the previous democratic (but corrupt) government in 1997, ordered a nuclear bomb test in 1998, and oversaw training camps for Islamic extremists. Against this, he ended ties with the Afghanistan Taliban and banned Islamist groups likely to annoy India, showing that he is at least preparing the ground to act on his stated wishes to modernize Pakistan and bring in a democratic government again. Most Pakistanis see the winning of Kashmir and supporting the "freedom fighters" there as of much greater importance than backing the former Taliban government and its links to terrorist al-Qaeda in Afghanistan.

Both India and Pakistan took advantage of U.S. policies arising from the war on terrorism from late 2001, but the Kashmir question threatened to undermine the uneasy peace. The buildup of Indian troops along its border with Pakistan took Pakistani units from their patrols along the Afghanistan border, allowing Taliban and al-Qaeda elements to escape more freely. India used the situation to defuse Islamic militant terrorism in Kashmir and Delhi, getting U.S. President George W. Bush to freeze assets of Pakistan-based groups. The war on terrorism broadened to include cooling antagonisms between two nuclear powers who became allies. Already seeking a new relationship with India as a counterweight to China, the U.S. administration took the opportunities after September 11, 2001, to reshape relations with Pakistan as well. The complexity of having to deal with terrorist groups operating in Kashmir is matched by international support for Pakistan and longings inside Pakistan for a period of peace and development.

This account is written in the midst of the continuing tensions centered on Kashmir and Afghanistan. By 2005 the Indian and Pakistan leaders were talking to each other again. Add new information, gleaned from newspaper websites, to debate the following Point–Counterpoint issues.

POINT

Pakistan: Kashmir should be ours; India took it unfairly.

Kashmiris: We prefer independence to either secular Indian rule or Islamic Pakistani rule.

India: Pakistan trains terrorists to disrupt society in Kashmir.

India: Pakistan held onto one-third of Kashmir despite a United Nations ruling that a referendum should decide Kashmir's future.

Kashmiris: We feel that we live in a prison under Indian military guard and do not like the alternative of Pakistan military dictatorship control. If we take part in elections, India will declare we are their citizens, and we have little confidence that India will moderate its high-handed attitude.

India: After September 11, 2001, Pakistan sent terrorists into India and nearly precipitated war.

COUNTERPOINT

India: We assumed control of Kashmir after a request by the Hindu maharajah following a precipitate Pakistani invasion.

India and Pakistan: Kashmiri independence is not feasible or offered by either country.

Pakistan: Those who resist Indian domination within Kashmir are local tribespeople fighting for their rights of self-determination.

Pakistan: India delays a referendum for no good reason except it . knows it will lose. It even annulled the 1977 elections that would have given a good indication of support for Pakistan control of Kashmir.

India: The Kashmiris would prefer a democratic constitution and have already experienced good results of Indian provisions.
Pakistan: The years of Indian suppression makes the people feel that we are friendlier.

Pakistan: The government is against terrorism and did not send those who attacked the Indian parliament in Delhi. We demonstrated this by withdrawing support for the Afghan Taliban against the wishes of many Pakistanis.

In the winter season of the dry monsoon, the Himalayan Mountains cut off South Asia from Central Asia and its freezing air (see Chapter 4). South Asia remains warm and very dry as winds flow outward from high atmospheric pressure over the northwest. Only northern Sri Lanka and southeastern India receive rain at this season—from winds that blow out from the continent over the Bay of Bengal, become moist by evaporation from the ocean surface, and turn toward the land.

As the land warms in early summer, temperatures in South Asia become unbearably hot and air rises, lowering atmospheric pressure. When the wet monsoon breaks in June or July, the winds change direction, being drawn into the low atmospheric pressure area over the continent. Southwesterly winds from the Indian Ocean bring moisture, causing heavy rainfall on the western coastal mountains of the peninsula. Precipitation occurs as snow at high elevations on the Himalayas. Some of the world's largest annual rainfall totals, over 10,000 mm (400 in.) per year, are recorded in the Assam hills of northeastern India. The lift forced by humid air flowing up and over mountains (the orographic effect) adds to the amount of condensation and precipitation. In the peninsula, most of the monsoon rains fall on the Western Ghats mountains, leaving the eastern lands in a rain shadow, which receive lower and more irregular summer rainfalls that vary sharply from year to year. Each year during the summer, a small number of tropical cyclones occur in the Bay of Bengal, bringing flooding and death to the Ganges–Brahmaputra Delta in Bangladesh.

The monsoon rains also miss most of the northwestern parts of South Asia, including Pakistan and Afghanistan, which remain dry throughout the year, forming one of the world's major arid regions, the **Thar** (or Great Indian) **Desert.** The main source of water in this part of the region is the rivers, such as the Indus and its tributaries that are fed in early summer by the melting snows of the Himalayas.

Surface Features

World's Highest Mountains and Deep Valleys

The Himalayan Mountains, forming the northern boundary of the region (see Figure 7.3), include Mount Everest (8,850 m; 29,035 ft.), the world's highest mountain, and nearly 50 other peaks that rise over 7,500 m (25,000 ft.). This mountain wall is the southern margin of the high, broad plateau of Tibet (China). The combination forms a wide barrier to both climatic influences and historic incursions of all but the most determined people.

The Himalayas resulted from the continental plate fragment that is peninsular India crashing into the Eurasian continental plate (Figure 7.10). The collision thickened the crustal rocks to produce the world's highest mountains. The continuing thrust of peninsular India beneath Asia, signaled by earthquakes, raises the mountains by 6 cm per year. As they rise, however, glaciers and rivers cut into them and wear them down. At present, an approximate balance exists between the rate of rise and the rate at which these mountains are being

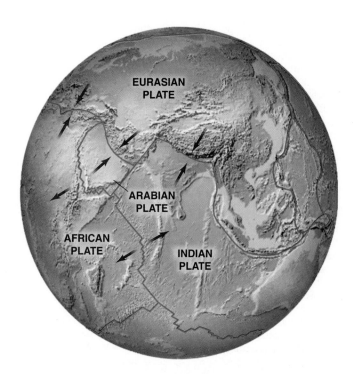

FIGURE 7.10 South Asia: plates and plate margins. The Indian plate crashed into the Eurasian plate, forming the Himalayan Mountains and Tibetan Plateau. The process continues, and these mountains rise at the same rate that erosion wears them down—so the height varies little.

eroded. Many of the plentiful minerals of South Asia (see Figure 5.12) were injected during the geologic activity along these ancient plate margins.

The headwater streams of the Indus and Ganges river systems breach the Himalayan wall in the far northwest. Deep valleys, carved by glaciers and rivers, provided lower routes, such as the Khyber Pass through the Hindu Kush range, that give access to Afghanistan and Central Asia and formed historic points of entry for Aryan and Central Asian groups. The passes remain a focus of tensions between India and Pakistan, which vie for control of the Kashmir region. At the northeastern end of the Himalayas, the ranges are lower as they turn southward into Myanmar, and they include breaks eroded by the Brahmaputra river system that afford Indian access to and from China, although there are border disputes.

Peninsular Hills and Plateaus

Peninsular India is a tilted continental plate fragment of very ancient rocks. The highest points along the western coast—the Western Ghats—rise to just over 2,500 m (8,000 ft.). Much of the peninsula consists of plateaus and hills sloping eastward to the Bay of Bengal—a gradient followed by many rivers. Along the east coast, a broken line of hills, the Eastern Ghats, rises in places to 1,500 m (5,000 ft.). Coastal plains of varied widths encircle the peninsula, at their widest where rivers such as the Krishna and Godavari create deltas on the eastern coast.

As the continental plate fragment forming the peninsula of South Asia broke away from the other southern continents over 150 million years ago (see Figure 6.10), layers of volcanic lava poured out through cracks in Earth's crust, covering large areas of older land. The lava flows form the Deccan Plateau in the northwestern peninsula.

Major River Basins

Between the Himalayas and the peninsular plateaus, three major river systems—the Indus, Ganges, and Brahmaputra—cross a wide lowland zone formed by their deposits. The melting Himalayan snows and the monsoon rains combine in powerful flood flows in these rivers. The strength of flow can be gauged from the fact that the Ganges and Brahmaputra sweep debris 3,000 km (nearly 2,000 mi.) out to sea in the Bay of Bengal.

Rock particles and fragments worn from the Himalayas and carried by these rivers built deposits of alluvium up to 3,000 m (10,000 ft.) thick beneath the plains. The surface of the deposits is generally flat but marked by small-scale relief of a few or tens of meters where the rivers cut new channels in older material. Along the northern margins of this plain, large **alluvial fans** of gravel mark the junction of the steep mountain courses with the lowlands. The sudden lowering of gradient as they enter the lowlands causes the rivers to drop the coarse gravel and sand they carry along their steeper mountain valleys. In the drier areas of the northwest are sections of **badlands topography,** where occasional rainfall runoff cuts dense networks of steep-sided gullies into unvegetated alluvial deposits. At their mouths the Ganges and Brahmaputra join to form a huge delta of low-lying, floodable land that occupies most of Bangladesh. These plains are now densely populated and intensively farmed, with the river water being used for irrigation. The Ganges River has great religious significance for Hindus (Figure 7.11).

FIGURE 7.11 South Asia: the sacred Ganges River. The river at Varanasi draws Hindu pilgrims to bathe in its waters. The river is regarded as a goddess. It is supplied from Himalayan glacier meltwater, is prone to flooding, and pollution is an increasing problem. **Photo:** © Alasdair Drysdale.

Forests and Soils

Much of South Asia was originally forested, including the peninsula, northern river plains, and Himalayan foothills. After centuries of clearance by expanding populations for fuel and to increase the cultivated area, the teak forests of southern India and the forests on the Himalayan slopes are almost all that remain. These are being reduced further.

The most fertile soils occur beneath the former forests, in areas subject to annual flooding, or around the lava plateaus. Long use for farming, however, reduced the soil quality. The drier area soils have been degraded by soil erosion, waterlogging, and salinization.

Natural Resources

The ancient rocks of the South Asian peninsula contain precious stones and mineral ores including iron and uranium, while the newer rocks on top contain one of the world's largest coal reserves. Deposits of oil and natural gas occur in the thick sediments beneath the major valley areas and just offshore.

For most people in this region, the critical natural resource is water. In the many thousands of villages across South Asia, access to water is a constant focus of efforts to manage the environment. Early in history, the monsoon rains made it possible to support large populations at subsistence level. The Ganges and Brahmaputra brought economic life to the lowlands south of the Himalayas, especially after irrigation and water storage techniques were developed. In the west, the Indus tributaries brought water to its arid lower valley, where one of the world's earliest civilizations developed. The waters flowing northward from Sri Lanka's central hills also fostered a civilization based on irrigation.

Water was instrumental in the Green Revolution. In the states of Punjab and Haryana, and parts of Uttar Pradesh, production rose and prosperity increased by irrigating more land, applying more fertilizer, and using more pesticides. The chemicals washed from fields into rivers. Irrigated cropland across South Asia increased from 18 percent (132 million hectares; 326 million acres) in 1960 to 32 percent (174 million hectares; 430 million acres) in 1980.

Irrigation in South Asia is of many types. Tanks, excavated hundreds of years ago by local communities, are most common in the peninsula. Large dams and canal systems, irrigating over 4,000 hectares (10,000 acres) each, are supplemented by medium-sized dams that have been built in planned, coordinated projects over the last 100 years. Over a thousand large or medium-sized dams were constructed after 1947. Most of the remaining sites involve grandiose construction plans for the upper Ganges and Indus in the Himalayas, often linked to large hydroelectricity projects that have political implications for India–Pakistan relations.

Natural Environmental Problems

People living in South Asia face a number of environmental hazards. Some are an integral part of the dynamic natural environment, but others are the outcomes of human clearance of land, population increase, and the context of the global economic system.

The dynamic natural environment is brought about partly by the clashing plates that produced the Himalayas and continue to set off earthquakes, such as those at Latur, Maharashtra, in 1993 and Bhuj, Gujarat, in January 2001. Both caused loss of life and destroyed poorly constructed housing. More frequent earthquakes along the base of the Himalayas are less devastating, although very large ones occurred in 1905, 1934, and 2005.

Flooding is a major problem in the lower Ganges and Brahmaputra valleys and especially over their combined delta. Peak floods in 1998 and 2004 affected over 30 million people. It would be impossible, however, to either move so many people to unfloodable areas or construct effective flood prevention measures, but improved flood preparedness is needed for the vulnerable poor population. Increased flooding levels result from deforestation in the upper reaches of these rivers and their tributaries in India, Nepal, and Tibet, but there is no regional cooperation to reduce this effect.

In other parts of South Asia, drought is the main problem, and the availability of water is of crucial significance. Water shortages raise major social issues in both urban and rural areas of India. Although the urban middle classes can install storage tanks against times of shortage, the slum dwellers wait in line with buckets. While members of the village upper castes use good wells, the untouchables have to find their water elsewhere. The combination of flooding and drought with so many people and strained intergovernmental relations makes a regional water plan unlikely, but may result in armed conflicts.

In the future, the Maldive Islands and much of the Ganges–Brahmaputra Delta, which are only a few feet above sea level, face drowning by a rising ocean level as global warming proceeds. The high costs of raising dikes and building sea defenses make such measures unlikely.

At the end of 2004, southeastern India and Sri Lanka were struck by the tsunami generated by a massive earthquake off northern Sumatra (Indonesia). Damage and loss of life were great among fishers and those in the tourist industry occupying coastal sites. After this disaster, the United Nations intended to establish an early warning system for such events.

Human-Induced Environmental Problems

In India, Hinduism, Jainism, and Buddhism proclaim the importance of care for the environment. High-caste Hindus avoid killing animals for food, and Jains abhor any killing. Mahatma Gandhi's philosophy was built around a lifestyle that had a low impact on nature and an acceptance that humans have a place within nature instead of over it. Despite such views, the huge rise in India's population and the necessary growth in its economy leave landscapes of exploitation, degraded resources, and pollution. The union (federal) and state governments are concerned first with providing subsistence and jobs for their people.

Air Pollution. Making the herbicides and pesticides used in the Green Revolution resulted in toxic concentrations of chemical factories. In 1984, at the Union Carbide pesticide plant in Bhopal, India, water leaked into a methyl isocyanate storage tank, triggering a chemical reaction and a cloud of toxic gases. It killed 3,000 people immediately and another 12,000 subsequently in the adjacent slum areas (Figure 7.12). These official figures are thought to have been around one-third of the true numbers. The seriously injured numbered at least 50,000, with long-term effects on eyes, lungs, and immune systems. Over half a million people qualified for compensation, although payments were slow. Over 20 years after the event, the site, now owned by the state of Madhya Pradesh, remains highly contaminated. Both the state and Dow Chemical (the U.S. parent company of Union Carbide) deny responsibility for

FIGURE 7.12 South Asia: pollution sources. The Union Carbide chemical plant and nearby poor housing in Bhopal, India. **Photo:** © Jagdish Agarwal/The Image Works.

cleaning up the site. This was a well-publicized, but not isolated, instance of the multinational corporation siting "dirty" manufacturing facilities in India.

Increasing damage to the Taj Mahal because of air pollution led to the closure of metal workshops in the surrounding area to reduce the sulfur gases that damage building stones. However, it is only possible to take such expensive action on a local scale. The agglomerations of industrial development north of Mumbai and Delhi produce massive air pollution that destroys farming livelihoods and makes winter smog worse.

Water Pollution. Nearly 40 percent of the Indian population lives in the middle and lower Ganges River basin. The land is intensively cultivated, and there are many factories and urban areas. The faithful believe Mother Ganga will preserve them from harm, but health dangers from various sources increase. For example, heavily polluted effluent from a concentration of leather tanning works enters the Ganges River close to some of the main Hindu ritual bathing sites. In recent years, an upsurge of trash dumping in the Ganges, combined with funeral pyres along its banks, further worsened the water quality. Washing of cattle adds to the water pollutants swept into rivers. The 1980s Ganga Action Plan aimed to clean up the river, but, although a few projects were implemented, lack of funding, continuing industrial malpractice, illegal dumping of wastes into the river, and illiteracy brought the program few successes.

The environmental problems associated with water-based irrigation and hydropower projects occur in both the source and using areas. Upstream water storage requires building dams and flooding. Even where dams are constructed in sparsely inhabited mountains, they eventually fill with sediment or may be destroyed by earthquakes. In 1993 local activists, supported by foreign environmentalists, influenced the World Bank to withdraw its funding for a dam on the Narmada River in the power-hungry states of Madhya Pradesh and Gujarat. In 2000, however, work resumed on some of the major sections of this project as the National Hydro Power Corporation became involved. Ninety thousand people being displaced received improved compensation.

Downstream, the extraction of water for irrigation alters river flow patterns. For example, the Ganges River floods are reduced, but dry season flow is much lower. In areas using irrigation, poor management of water sometimes waters fields where the terrain or subsurface conditions are unsuitable. Soils become waterlogged or saline. Wells already provide nearly half the irrigation water used and remain the main source of irrigation growth. Tube wells drilled by modern rigs multiplied from 200,000 in 1960 to over 4 million by the 2000s. However, the sinking of deep wells lowered water tables, cutting out smaller farmers who could not afford deep wells. Water is lifted by diesel or electric pumps, in contrast to the human or animal power used in older wells. Wells have environmental advantages over canal-fed irrigation in that they use local water supplies, keep the water table low, and encourage downward water movement in the soil to avoid salinization. They are increasingly combined with canal irrigation so that local groundwater sources can be replenished.

Test Your Understanding
7A

Summary

South Asia has a huge and growing population and large numbers of extremely poor people. Its cultural diversity provides an identity but also fuels internal conflicts. South Asia occupies the Indian subcontinent, hemmed in by the Himalayan Mountains. Passes through the mountains provided routes for invaders. Incoming Muslims confronted the Hindu and Buddhist religions that originated in South Asia. The British Indian Empire was the world's most intense expression of modern colonialism. The region's economy was oriented to the needs of Britain until independence was achieved in 1947.

The physical environment of South Asia is marked by monsoon and arid climatic environments, the world's highest mountain system in the Himalayas, the wide plains of the Indus, Ganges, and Brahmaputra river basins, and the hills and plateaus of the peninsula. The dynamic physical environment produces earthquakes, summer river floods, and tropical cyclones. Human economic development makes use of the mineral and water resources and results in increasing modifications of the natural environment, including irrigation systems and industrial pollution.

Questions to Think About

7A.1 Which major cultural elements contributed to the human diversity of South Asia?

7A.2 Assess the impacts of the British Raj on the region.

7A.3 How would you illustrate the statement that South Asia's natural environments are dynamic and full of contrasts?

Key Terms

Hinduism	British Indian Empire
caste order	Indian National Congress Party
Buddhism	Muslim League
Jainism	Thar Desert
Mughal (Mogul) dynasty	alluvial fan
Sikhism	badlands topography
British East India Company	

Global and Local Changes

During the Cold War, the newly independent countries of South Asia (Figure 7.13 and Table 7.1), with India in the lead, wished to remain unaligned. However, events and rivalries drew them into dependence on other countries while the region remained a backwater in world affairs. The external and internal rivalries raised defense expenditures and affected economic progress.

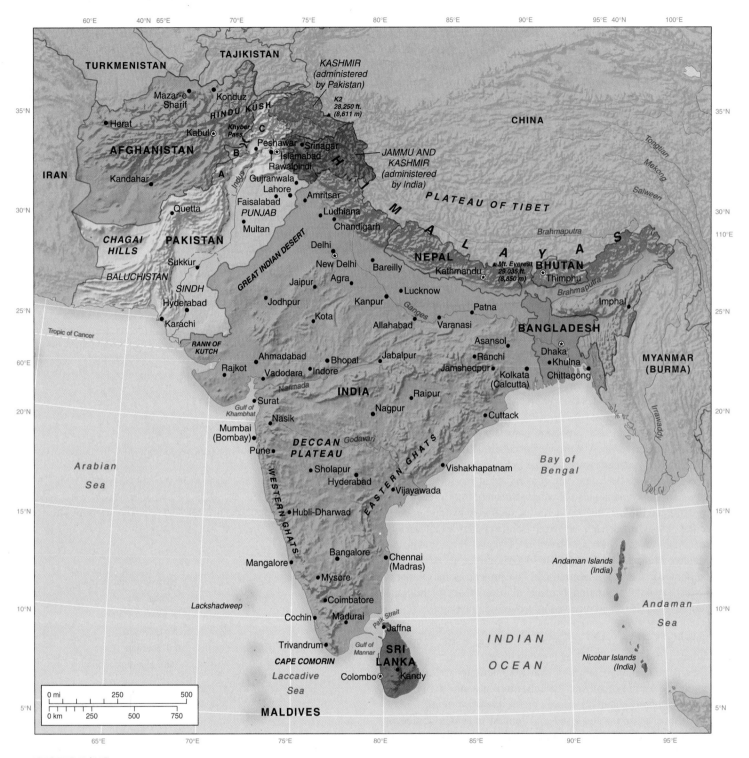

FIGURE 7.13 South Asia: major geographic features, countries, and cities. The Indian subregion dominates the region by its size and centrality; Bangladesh and Pakistan are two Muslim countries, created as one at independence but since broken apart. Afghanistan, Nepal, and Bhutan form the Northern Mountain Rim, while the Maldives and Sri Lanka are the Southern Island Rim. This is a region of huge, growing cities. In Pakistan A = Waziristan; B = Tribal Areas; C = Northwest Frontier.

The end of the Cold War changed the external political and economic linkages as South Asian countries were drawn more fully into the global economic system. From 1950, the populations of countries in South Asia increased rapidly, placing increased pressure on the limited resources.

Politics of Independence

Cold War Alignments

When the newly independent South Asian countries tried to keep out of the Cold War, they discovered that the United States and the Soviet Union would not accept them as neutrals. India tried to coordinate a "third force" of Asian and African countries, but a nonmilitary pact was clearly unworkable.

At first, disputes between India and Pakistan over Kashmir and rights to the waters of the upper Indus and Ganges rivers loomed larger than the Cold War. Moreover, India also had continuing conflicts over its northern borders with China (Tibet). For example, the northeastern state of Arunachal Pradesh was ceded in 1913 by Tibet to India, an arrangement that modern China refuses to accept.

Pakistan gained support from the United States and worked with the Chinese. Ceding border dispute lands to China led to building the Karakoram Peace Highway from Pakistan to western China and transfer of Chinese nuclear technology. India balanced these influences by linking with the Soviet Union, especially after the Soviet–Chinese split in the early 1960s. During the India–Pakistan War of 1965, the United States and the United Kingdom imposed an arms embargo on both countries, but the combatants received arms from the Soviets and Chinese, respectively. India abandoned its previous antinuclear policy as international rivalries increased.

When the Soviet Union invaded Afghanistan in 1979 to support the Marxist government against Islamic and local warlord rebels, the United States and Saudi Arabia supported groups opposing the Soviet forces. Around 3 million refugees fled over the border into Pakistan, living mainly in camps close to the Afghanistan border. During the 1980s, the Pakistani Jemiat-e-Ulemi-i-Islam religious party taught and trained Afghan refugee children, who later flocked to support the Taliban, a hardline, fundamentalist resistance group that included students (*taliban* is Arabic for "students").

A smaller conflict with global implications remains over the island people of Diego Garcia. Diego Garcia was detached from the Maldives in 1965 as a separate colony and was leased to the United States to function as a long-range bomber base. The former inhabitants of Diego Garcia were moved to Mauritius, but lobby to get their homes back.

Internal Conflicts or Democracy after the Cold War

When the Soviet Union broke up and the Soviet military left Afghanistan in 1989, India was left without a major client, but the United States stopped aid to Pakistan because of the latter's nuclear weapons. India and Pakistan continue to express rivalry at a military level (although cricket tournaments between the two were reinstated in 2004 as a hopeful sign of better relations).

Pakistan encouraged the Taliban group in Afghanistan throughout the 1990s by recruiting supporters and supplying armaments. This was partly from Sunni Muslim religious sympathizers, partly to ensure a friendly country to the west so that it could challenge India over Kashmir, and partly in hopes that an oil pipeline could be built from Central Asia across Afghanistan and Pakistan. The association with the hardline Islamic Taliban group, however, fed back into Pakistan politics with demands for stronger Islamic influence, often backed by terrorist activity.

The events of September 11, 2001, placed Pakistan in the forefront of international actions against Afghanistan's Taliban rulers who harbored bin Laden's al-Qaeda organization. Despite internal demonstrations in support of the Taliban, the Pakistan government led by President Musharraf joined the international antiterrorist collaboration, a step that led to a rise in its international status, removal of sanctions, and renewal of foreign aid. The United States returned to Pakistan after 2001 to oust the Taliban regime and al-Qaeda terrorists from Afghanistan, but this inflamed the India–Pakistan conflict over Kashmir. India was not interested in advice from a United States that became involved again with Pakistan. Much of the remote Pakistan territory in the mountainous Northwest Frontier province bordering Afghanistan, where Pakistan military control is incomplete, remains a haven for Afghan–Pakistani terrorists.

Other more local rivalries and conflicts occurred. In 1992 militant Hindus pulled down a Muslim mosque at Ayodhya, a place that is thought by some to be the major historic center of Hinduism and the birthplace of the god Rama. In 2002 Hindu–Muslim tensions erupted in violence in Gujarat as Hindu mobs burned Muslim areas after 58 Hindu pilgrims were killed in a train fire. In Sri Lanka the northern minority group of Tamil people continued a civil war as "Liberation Tigers" against the dominant Sinhalese until a number of ceasefires slowed the momentum of this conflict from 2002 (Figure 7.14). The Nepal government is threatened by Marxist control of rural areas.

In the early 2000s new concerns emerged in the region. Whereas India had been the dominant country that influenced policies in other countries, especially the smaller ones, threats to Indian political stability came from those surrounding countries. The events in Afghanistan and the Kashmir situation continue to be perceived as a major threat to India. In Nepal, after eight years of insurgency, Maoist guerrillas control the rural areas and are likely to take over the government in Kathmandu. Similar Maoist groups, known as "Naxalites," operate within India; although they talk to state governments (as in Andhra Pradesh), they refuse to disarm. Bangladesh is becoming involved with the spread of Islamic extremism linked to attacks in northeastern India. Neighboring Myanmar is believed to harbor insurgents as well as exporting drugs and HIV/AIDS. Sri Lanka, the scene of civil war between Tamils and the main Sinhalese peoples, is quieter at present, but India fears that its own Tamils may seek independence. A plan to dredge the Sethusamudram Ship Canal through the shallow Palk Strait between Sri Lanka and India is causing fresh tensions following India's decision to go ahead despite Sri Lankan opposition. This canal would save over a day's sailing for ships between east and west Indian ports, but would harm local fishers, Colombo's transshipment trade, and local environments.

South Asia may be turning into a "shatterbelt" of conflict and disruption linked to the troubled areas of southwestern Asia. The intraregional conflicts hold back the type of cooperation and regional trading agreements that are common in other world regions. The South Asian Association for Regional

Table 7.1 South Asia. Data by country, area, population, income (GNI PPP—Gross National Income Purchasing Power Parity), urbanization, Human Development Index (HDI), Human Poverty Index (HPI).

Country	Capital City	Land Area (km²)	Population (millions) Mid 2005 Total	Population (millions) Est. 2025 Total	GNI PPP 2003 Total (US$ billions)	GNI PPP 2003 Per Capita (US$)	% Urban 2005*	Index 2002 HDI (rank of 177)*	Index 2002 HPI (% population)*
Afghanistan, Islamic State of	Kabul	652,090	29.9	50.3	No data	No data	22	No data	No data
Bangladesh, People's Republic of	Dhaka	144,000	144.3	190.0	258	1,870	23	138	42.2
Bhutan, Kingdom of	Thimphu	47,000	1.0	1.4	No data	No data	21	134	No data
India, Republic of	New Delhi	3,287,590	1,103.6	1,363.0	3,068	2,880	28	127	31.4
Maldives, Republic of	Malé	600	0.3	0.4	No data	No data	27	84	11.4
Nepal, Kingdom of	Kathmandu	140,800	25.4	36.1	35	1,420	14	140	41.2
Pakistan, Islamic Republic of	Islamabad	796,100	162.4	228.8	306	2,060	34	142	41.9
Sri Lanka, Democratic Socialist Republic of	Colombo	65,610	19.7	22.2	72	3,730	30	96	18.2
Totals or averages*		5,133,790	1,487	1,892	3,739	2,392	25	123.0	31.1

Source: World Population Data Sheet, Population Reference Bureau (2005), World Development Indicators, World Bank (2005), Human Development Report, United Nations (2004), and Encarta Microsoft (2005).

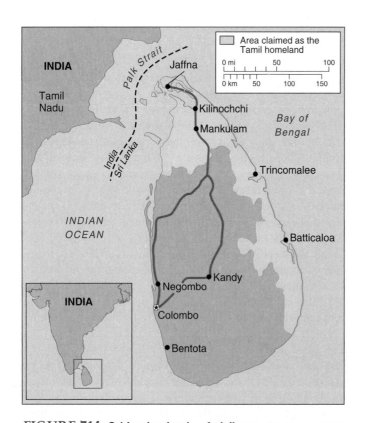

FIGURE 7.14 Sri Lanka: basis of civil war. The Tamil population lives mainly in the north and northeast of the island. **Source:** *The Economist*, December 5, 1998. Copyright © The Economist Newspaper Group, Inc. Reprinted with permission. Further reproduction prohibited. www.economist.com

Cooperation (SAARC), set up in 1985, generated little economic interaction apart from some academic studies of the potential.

Population Pressures

Ethnic and Social Contrasts

South Asia is a region of diverse peoples and cultures with its many languages and distinctive religious orientations, including Hinduism, Sikhism, Islam, Buddhism, Jainism, and Christianity. Global religions (Islam, Christianity, Buddhism) clash with ethnic religions (Hinduism, Sikhism, Jainism). Central and state governments find that applying their rules locally often clashes with the interests of feudal-type warlords and landowners. All countries have sizable ethnic and social groups that provide political opposition or antagonistic groups that promote internal conflict.

A Small Space for So Many People

As a relatively small world region that is home to the world's largest population, South Asia's population numbers and distribution reflect interactions between global and local influences (Figure 7.15). Apart from a few almost uninhabited areas, this region has continuously high densities of population. The greatest concentrations of people are in the lowlands of the Indus, Ganges, and Brahmaputra rivers, around the southern coasts of India, and in southeast Sri Lanka. The

Country	Ethnic Groups %	Languages O=Official N=National	Religion %
Afghanistan, Islamic State of	Pashtun 38%, Tajik 25%, Hazara 19%	Afghan-Persian 50%, Pashto 35%, Turkic languages 11%	Sunni Muslim 84%, Shiite Muslim 15%
Bangladesh, People's Republic of	Bengali 98%	Bangla (O), Urdu, English	Muslim 88%, Hindu 10%
Bhutan, Kingdom of	Bhhutia 50%, Nepalese 35%, Sharchops 10%	Dzongkha (O), Tibetan/Nepali dialects	Lamaistic Buddhist 75%, Hindu 25%
India, Republic of	Indo-Aryan 72%, Dravidian 25%	Hindi (O) 40%, 14 official regional, English	Hindu 82%, Muslim 12%, Christian 2%, Sikh 2%
Maldives, Republic of	Sinhalese, Dravidian, Arab, African	Divehi (Sinhala, O), English	Sunni Muslim dominant
Nepal, Kingdom of	Nawar, Indian, Tibetan, Gurungi, Sherpa	Nepali (O), 20 others, English	Hindu 90%, Buddhist 5%, Muslim 3%
Pakistan, Islamic Republic of	Punjabi, Sindhi, Pashtun, Muhajir (Indian origin)	Urdu (O), English (O), Punjabi, Sindhi	Sunni Muslim 77%, Shiite Muslim 20%
Sri Lanka, Democratic Socialist Republic of	Sinhalese 74%, Tamil 18%, Moor (Arab) 7%	Sinhalese 74%, Tamil 18%, English	Buddhist 69%, Hindu 15%, Christian 8%

areas of very low population density include the desert areas of Pakistan and the western area of India near the southern Pakistan border, the mountainous areas of western and northern Pakistan, Afghanistan, the swampy Rann of Kutch that is mostly within India, and the Himalayan northernmost India, Nepal, and Bhutan. These areas hold little prospect for future population expansion. Such areas of very low population form breaks along the region's political borders and provide a clear definition of South Asia. The rest of the region has moderate densities of population, which often reflect relatively high densities for the resources that support rural employment. Over most of the region, improved water supplies for irrigation scarcely keep up with rising numbers of people, and so these areas of moderate densities offer few prospects for supporting more people.

Lots of People and Rapid Increases

South Asia's total population of around 1.5 billion in the early 2000s is predicted to rise to nearly 2 billion by 2025, compared to less than 1.7 billion in East Asia. This represents a major growth in numbers of people when the region has few current or prospective means of supporting them at improved levels of life quality. India's population of 1.1 billion in 2005 is the world's second largest after China's, and its rate of increase is greater. By 1961 India's population was 440 million, and Indian government estimates of future population growth caused it to institute policies to slow population increase. Estimates for 2025 raise India's population to 1.363 billion. Growth continues because of the huge number of people in the child-bearing ages (Figure 7.16a, b) and increased life expectancy. In the demographic transition process (Figure 7.17), India remains in the phase of population increase.

The rapid increases of population in South Asia are caused by birth rates that remain higher than death rates, although both are falling. Birth rates are highest among the materially poorer groups, especially in rural areas, leading to significant changes in the social balance and political pressures from hitherto disadvantaged groups.

Bangladesh had a slightly higher population than Pakistan until 1980. By 2005, however, Pakistan's population exceeded that of Bangladesh (162 million compared to 144 million), and estimates for 2025 suggest a further increase in the difference, with Pakistan having 230 million and Bangladesh under 200 million people. The greater rate of population increase in Pakistan is the result of a continuing high total fertility rate of nearly 6, compared to 3.3 in Bangladesh. Both have life expectancies in the upper fifties.

Given its smaller area, Bangladesh has greater problems feeding its growing population, an issue that stimulated its government to institute effective family planning programs in the 1970s with significant reductions in fertility. By the late 1990s, half of married women used contraception, up from 8 percent in 1975. Despite a continuing subordinate role for women, new policies encourage their social and economic development.

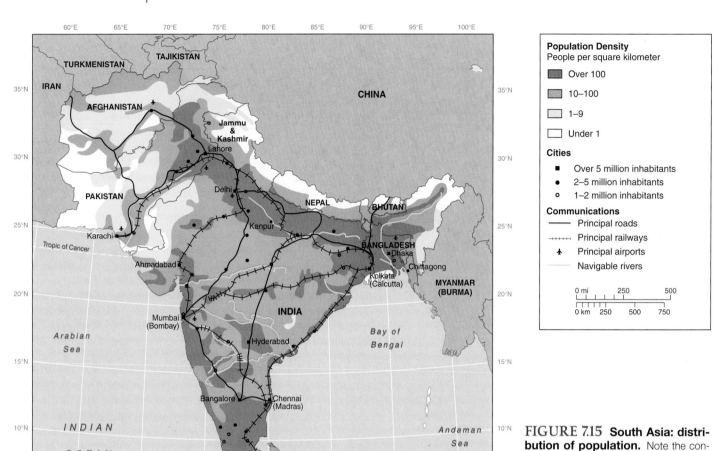

FIGURE 7.15 South Asia: distribution of population. Note the contrasts within South Asia, linking the areas of high population density to river valleys and of low population density to arid areas and mountains. **Source:** Data from *New Oxford School Atlas,* Oxford University Press, U.K.

Pakistan's need of an educated workforce is undermined by the combination of influences from Islamic religion, feudal landlords, and poor government that kept Pakistan's school population low until the late 1990s. Its enrollment rates in primary and secondary education were only 30 percent, compared to 70 percent in India and over 80 percent in Indonesia and China. In the mid-1990s, aid donors' demands led to new national policies for education, health, and population. Government funding shifted from elite university education toward universal primary education.

Of the countries on the margins of this world region, Bhutan, the Maldives, and Afghanistan populations grow most rapidly at around 3 percent per year. Sri Lanka's population increases at only 1.2 percent. In most countries, the total fertility rates are declining but remain over 6 in Afghanistan and over 5 in the Maldives and Bhutan. All the countries are at a stage when the population increases rapidly and there are many young people. For example, Nepal has a strong family planning program, widely disseminated, but only 29 percent of women and 2 percent of men use birth control methods; fertility rates decline slowly. This is linked to Nepalese female adult literacy rates, which were 3.7 percent in 1971 but rose to only 26 percent by 2002.

Birth Control, Gender Inequalities, and HIV/AIDS in India

In India, efforts to spread the use of birth control methods gained acceptance slowly among a largely illiterate people. Male sterilization, an early "solution," was widely perceived as economic suicide for families. The threat of its forced imposition was a major political issue in the 1977 elections that ousted Indira Gandhi's government until 1980. The greater effectiveness of family planning in the 1980s and 1990s built on rising levels of literacy, increased urbanization, and improvements in the status of women. Hindu women of higher castes use family planning more than those among the poorer castes and less educated Muslims. The new urban middle class provides signs of the Indian population stabilizing around fertility rates of 2–3 children, but this applies to a minority of the population. More women adopt family planning, including sterilization, once they have given birth to a boy, and this is the most successful aspect of the program. Many families in rural areas, however, continue to want large families to provide extra manual labor and care of parents in later life. Moreover, many of the 120 million Muslims living in India disapprove of birth control.

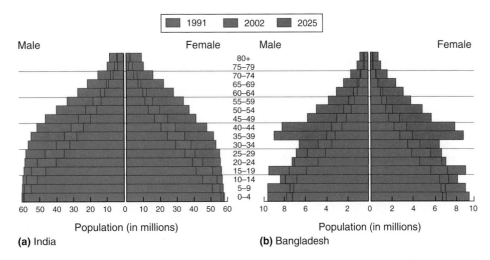

FIGURE 7.16 India and Bangladesh age–sex diagrams. (a) India. (b) Bangladesh. The large numbers of young people will produce continuing rapid population growth when they reach childbearing age for many years to come—even if fertility declines. **Source:** U.S. Census Bureau, International Data Bank.

The Indian age–sex diagram records fewer females than males. The 2001 Indian census showed that India has a sex ratio of 107 males per 100 females, slightly lower than in 1991 but higher than that of its neighbors. Within India, sharp differences exist across the states, from 95 males per 100 females in Kerala to 116 in Haryana. Furthermore, in the under-7 age population, the sex ratio increased from 105.8 in 1991 to 107.8 in 2001, with the sharpest rises in the more prosperous states. It has been suggested that there are millions of missing females in India because of female infanticide, excused by some as saving the girl from a lifetime of suffering. In 1997 the *Times of India* estimated that mothers or village midwives kill 16 million girl babies a year. In 1994, after discovering that most aborted fetuses were female, the Indian government outlawed tests that could determine the sex of a fetus. Laws now prohibit doctors from revealing the sex of an unborn baby, although they are flouted among wealthier families. The poor cannot afford surgical abortions and carry out infanticides. In part of Rajasthan state, there are only 550 women to every 1,000 men, and newborn girls are often buried without question.

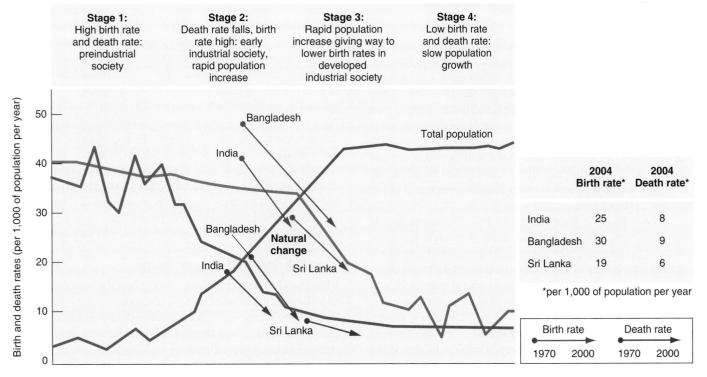

FIGURE 7.17 South Asia: birth rates and death rates related to demographic transition. All countries have a wide gap between falling birth rates and death rates. Sri Lanka is farthest ahead in reducing this gap. **Source:** 2004 update: *2004 Population Data Sheet,* Population Reference Bureau.

From the 1990s, HIV/AIDS emerged as a major health and development threat. An estimated 1 percent of India's adult population (at least 10 million but probably many more) is affected by the disease—the highest number for a country after South Africa. Worst affected are the southern Indian states (apart from Kerala) and Manipur in the northeast next to the Myanmar boundary. Together these states account for 80 percent of the reported cases of HIV/AIDS in the country. The cities of Mumbai, Chennai, and Delhi also have high prevalence of the disease. However, rates in some of the northern Indian states that do not have checks may be just as high.

Alarmed by the predicted effects of an unchecked epidemic on its population and economic development, the government launched an HIV/AIDS prevention campaign. Its mission is to educate the Indian people on the means of transmission of the virus and offer support for prevention measures. To date, most action comes from private organizations and funds from abroad, although the Indian government is releasing funds. Many of the reactions typify Indian social situations. While vulnerable groups, such as prostitutes and truck drivers, are particularly prone to infection, the low status of women in families assists transmission. Those infected by husbands are cast out by their in-laws. Many HIV-positive men lose their jobs. Much of the data is based on surveys of vulnerable groups, and conclusions are drawn that the disease is concentrated and contained. However, wider data are needed to investigate whether HIV/AIDS is spreading to the general population. Another Indian phenomenon is the Cipla drug company, which produces antiretroviral drugs and sells them abroad cheaply. However, the Indian government hesitates to accept offers from the company.

Increasing Urban Populations

Indian cities continue to hold under 30 percent of the country's growing population. However, the total population of urban areas is projected to rise from 250 million people in the mid-1990s to over 600 million by 2025. Urban areas provide most of South Asia's economic opportunities. New jobs in factories and offices, better access to education and health services, and most opportunities in the informal sector of the economy occur in urban—especially large urban—areas. From 1980 to 2003, the number of Indian cities with over a million people jumped from 12 to 37. The largest cities are becoming immense. By 2015 Mumbai, Delhi, and Kolkata (Calcutta) will be 3 of the world's 10 largest cities (Table 7.2).

Like India, both Bangladesh and Pakistan remain largely rural countries, with 34 percent of the population living in the towns of Pakistan in 2005 and only 23 percent in Bangladesh. However, both figures understate the true extent of urbanization. The major cities struggle to cope with the arrivals of rural people and the shantytowns and social tensions that the influx creates. Although separated from the political power centers farther north, Karachi is the largest city in Pakistan and its major port and commercial center. Violence in the 1990s among rival ethnic Muslim groups was made worse by the glaring contrasts between rich and poor, the large numbers of unemployed teenagers, and the availability of guns following

Table 7.2 Major City Populations in South Asia

City	Population 2003 (millions)	Population Estimate 2015 (millions)
Mumbai (Bombay), India	17.4	22.6
Delhi, India	14.1	20.9
Kolkata (Calcutta), India	13.8	16.8
Chennai (Madras), India	6.7	8.1
Hyderabad, India	5.9	7.5
Bangalore, India	6	8.4
Ahmadabad, India	5	6.6
Karachi, Pakistan	11.1	16.2
Lahore, Pakistan	6	8.7
Faisalabad, Pakistan	2.4	3.5
Dhaka, Bangladesh	11.6	17.9
Chittagong, Bangladesh	3.8	6.2
Kabul, Afghanistan	3	5.4

Source: United Nations Urban Agglomerations 2003 (2003).

years of warfare in Afghanistan. Other major Pakistani cities are inland, where Lahore is a major center of Muslim culture and Faisalabad is the cotton industry center. Islamabad, the new capital, is close to the Kashmir border, while Peshawar is the commercial center of the far north at the eastern end of the Khyber Pass into Afghanistan.

Bangladesh has fewer large cities. Dhaka is the capital, Chittagong is the main port, and Khulna is a growing industrial center. The pressure of people on scarce farming land and lack of farm jobs for adults has forced migration into cities, where jobs demand education qualifications that most migrants do not have. From the 1980s, the shortage of jobs in Bangladesh resulted in many laborers seeking employment outside the country, such as in the Persian Gulf countries and Malaysia.

In the mountainous countries and islands, the main concentrations of people are in lower areas that are accessible to major routeways. Capital cities Kabul (Afghanistan), Colombo (Sri Lanka), and Kathmandu (Nepal) are the largest in the subregion and contain extensive shantytowns. In Afghanistan, the valleys around the capital, Kabul, the northern plains bordering on Turkmenistan, and southwestern valleys connected to Iran have the highest concentrations. The Himalayan states have most people near the Indian border. Sri Lanka's population is focused on the rainier southwestern coastlands around Colombo.

South Asian Urban Landscapes

The urban landscapes of South Asia reflect the waves of cultural influences that washed across the region (Figure 7.18). Many ordinary buildings were constructed of materials such as wood that do not have a long life, but most towns have more permanent religious, royal, or military buildings that have lasted for centuries. Hindu temples are often very ornate (see Figure 7.2a). Muslim mosques and tomb gardens (see Figure 7.6) also provide distinctive landscape elements.

(a)

(b)

FIGURE 7.18 South Asia: Indian urban scenes. (a) Slum shantytown conditions in Mumbai with dwellings made of sacking, plastic and metal sheeting, and boards. Compare these dwellings with those in the distance. (b) Karol Bagh, a shopping area of New Delhi, showing Western influences.
Photos: (a) © B.P. Wolff/Magnum Photos, Inc. (b) © Parvindar Sethi.

Older precolonial sections of towns left a heritage of high housing densities with poor transportation access along winding alleys. Shops and artisan workshops encroach on walkways. Few buildings are more than two stories, and often there is no clear distinction of residential, commercial, and industrial functions. People of similar caste or trade live and work together in localities.

Sections added to towns during the British rule from the 1700s to the mid-1900s included a central market, administrative offices, and frequently a clock tower. British-built churches and civic buildings were European in style. New residential areas were built on European plans. Wide streets and open places separated business and residential functions. Britain built and developed major ports such as Kolkata, Mumbai, Chennai, Karachi, and Colombo initially as a means of establishing trading security. Where the railroads from ports to inland centers passed through a precolonial town, the town grew in administrative and trading significance; where one was bypassed, it often declined. Some smaller cities, such as Bhopal, have no planned infrastructure, and, even in major cities, funds for maintaining infrastructure are scarce. Shantytowns are common.

During the British Raj, hill station towns provided escapes for British families from the worst of the summer heat and humidity and the diseases that spread in that season. Simla, due north of Delhi, is typical of these hill stations and is still accessed by a mountain railroad line built at great initial expense. The hill stations often resemble small British towns in their architecture. They are now tourist centers and have English-language boarding schools.

After independence and partition, cross-border migrants between India and Pakistan moved into large cities and were accommodated in refugee camps that themselves became cities. For example, Ulhasanagar near Mumbai now has 600,000 people. Karachi contains large numbers of those who moved out of India into Pakistan.

Modern cities have major industrial areas, including steelworks and aluminum plants, chemical factories, textile mills and garment-assembling factories, vehicle makers, and electrical goods production. Concentrations of factories are most common around Mumbai in the Indian state of Maharashtra and nearby Gujarat. Up to the mid-1990s, the wealthy lived in the city centers and the poorer people toward the outskirts, but many of the wealthy have moved to new suburbs.

South Asians Abroad

Around 15 million ethnic Indians live abroad, with the largest groups in Nepal, South Africa, Malaysia, Sri Lanka, the Persian Gulf states, the United Kingdom, Canada, and the United States. Under the British Empire, Indians provided a source of cheap labor and were moved to work on plantations and railroad construction in the Caribbean, Eastern and Southern Africa, Burma, Malaysia, and Pacific islands such as Fiji. Many stayed as shopkeepers and businesspeople.

More recently, flows of workers are to the oil-producing countries of the Persian Gulf. Professionals move to the United Kingdom and United States for better-paid opportunities, creating a significant "brain drain." Many of the Indian "diaspora" come from Gujarat and Andhra Pradesh. They increasingly work in high-tech centers such as Bangalore and Hyderabad before moving abroad.

Many Pakistanis, Bangladeshis, and Sri Lankans also live and work abroad, and there was a major period of migration to the United Kingdom from the 1960s when former colonial people were allowed a British passport. More recently, workers from these countries moved to jobs in the Persian Gulf countries and Southeast Asia. Their remittances to families in South Asia form important income contributions.

Economic Challenges

Self-sufficiency and Isolation

The countries of South Asia remain among the world's poorest (Figure 7.19), as is shown by their low possession of consumer goods, access to clean water, and energy usage (Figure 7.20). At independence in 1947, India and Pakistan, with major influences on the other new countries of the region after their independence, moved away from the mercantile colonial involvement in the global economy that they believed had held them back. They preferred to attempt self-sufficiency and isolated themselves from world markets by protective tariffs, export taxes, and an emphasis on internal investment. For 40 years, economic growth was too modest to establish greater material wealth. The 3.5 percent GDP growth per year on average from the 1950s to the 1990s, which Indian leaders saw as a success, did not match the East Asian countries' achievement of twice that rate over the same period.

Complex bureaucratic rules and a lack of basic education, health care facilities, transportation, or power supplies left South Asian countries on the margins of the world economic and political systems. In the bipolar Cold War world, even the largest country, India, did not hold a major strategic position by virtue of its military power in comparison with the countries of East Asia, Southeast Asia, or Arab Southwest Asia (see Chapter 8).

Since 1990, the loss of Soviet (India) and Chinese (Pakistan) support and trading forced South Asian countries to enter the global economy. India doubled its rate of economic growth in the 1990s, although it is taking time to reorient its institutions and bureaucratic outlook. All countries became more open to foreign investment and internal private enterprise but attracted less than China. Several Indian cities, including Mumbai and Delhi, have increasing global corporate services such as accounting, advertising, banking, and law, while Bangalore and Hyderabad have global connections through their high-tech industries.

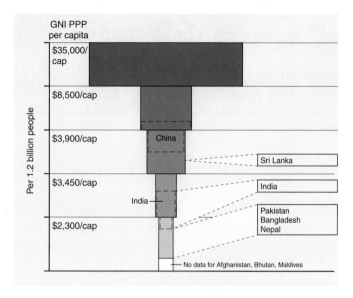

FIGURE 7.19 South Asia: country average incomes compared. The countries are listed in the order of their GNI PPP per capita. **Source:** Data (for 2000) from *World Development Indicators,* World Bank, 2002.

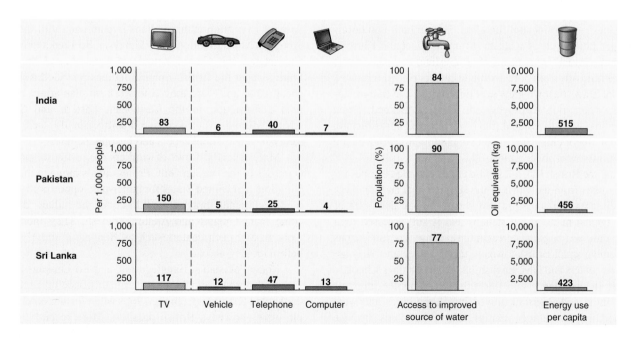

FIGURE 7.20 South Asia: ownership of consumer goods, access to clean water, and energy usage. How do these figures compare with other Asian countries? **Source:** Data (for 2002) from *World Development Indicators,* World Bank, 2004.

Subregions

A new diversity within South Asia emerged in separate countries following the end of the British Raj, leading to the recognition of three subregions:

- *The Republic of India,* the world's second largest country in total population, has a dominant role within the region.

- *Bangladesh and Pakistan* are two Muslim countries resulting from the breakup of the original Pakistan in 1971.

- *The Mountain and Island Rim,* comprising Afghanistan, Nepal, and Bhutan in the mountainous north and Sri Lanka and the Maldives in the island south, includes a group of smaller countries on the margins of South Asia.

India

In 2005 the Republic of India's population of 1,104 million was second only to China's, but its total GNI PPP in 2003 placed it fourth in the world (after the U.S.A., China, and Japan). Forecasts expect India's population to have a growing working-age group (in contrast to most other countries) and its economic growth to continue at a high rate well into the 2000s, possibly eclipsing other major emerging countries such as Brazil or Russia (ninth and tenth in 2003). Approaching 60 years of independence in 2007, India remains the world's largest democracy.

The combination of population size and productive potential continues to make India a regionally dominant presence. India's increasing significance is a product of its history; and although it has limited resources to underwrite change, India has an expansive vision of its future. It has a mixture of high-tech and basic manufacturing, military resources of modest technology, and links to other developing countries.

India became an independent country in 1947 as a republic with a federal union government; many of its institutions have a Western flavor. The upper house of the union government is formed of state-nominated representatives; the lower house, or people's assembly, is directly elected from India-wide constituencies. The former princely states and British Indian provinces were admitted as states within India's federation (Figure 7.21). There are 29 states and 6 union territories, each

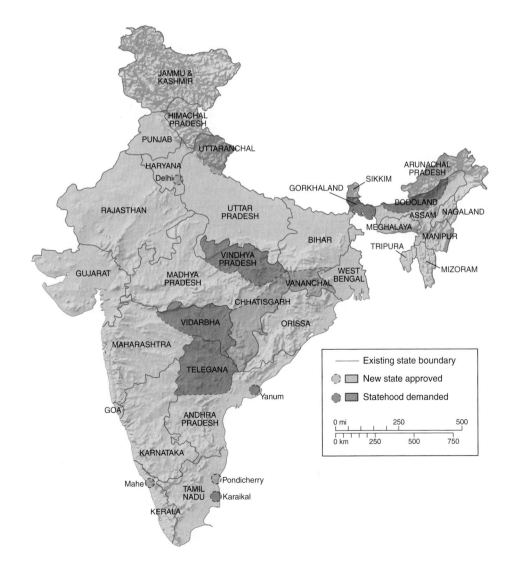

FIGURE 7.21 India: new states. In 1956 Madras Province was divided into Tamil Nadu and Andhra Pradesh. In 1960 Bombay Province became the states of Gujarat and Maharashtra. In 1998 the union government approved the formation of three new states—Utteranchal (Uttarakhand), Vananchal (Jharkhand), and Chhatisgarh—to be carved out of Uttar Pradesh, Bihar, and Madhya Pradesh, respectively. The status of a state of the union was also conferred on Delhi and Pondicherry. **Source:** Based on information published in *Frontline,* July 31, 1998.

with its own elected assemblies. The state governors are appointed centrally to represent the union. State boundaries are largely linguistic, reflecting the huge range of languages spoken in the country. The states' roles in administration, law and order, education and welfare, and economic development are supported by local taxes and grants from the central government. Each state is divided into village-based administrative units. The number of states increased after independence, mainly as the result of the original states not coping with the interests of ethnic groups.

Every adult has a vote. Women have a stipulated minimum number of seats in local government. In both federal and state governments, proportions of seats and roles are reserved for the scheduled castes ("untouchables") and the scheduled tribes. Before independence they had no place in mainstream Hindu society and were greatly exploited, but now they have protected status. Today both groups are significant political forces, although many social taboos still operate against them, especially in rural areas.

As in other federal countries, relations between the central government and the states are complex. The union government has responsibilities for defense, foreign affairs, and currency. It raises most of the taxes and makes grants to the states. In the 1990s states increased their functions, particularly in the delivery of social programs including health and education. Some states took initiatives in foreign relations and defense and strove to attract foreign capital, but others did not, intensifying the uneven nature of Indian economic development.

States vary in their approaches to government. For example, Bihar, the poorest state, retains remnants of its feudal legacy with land ownership by elites, political and administrative corruption, and inefficiency; private armies control local areas. Bihar finds it difficult to attract foreign investment. On the other hand, nearby West Bengal instituted land reforms that led to considerable economic and social gains based on smaller farm units and higher levels of popular participation. Some states are governed by Hindu nationalists, others by secular socialists.

Self-Sufficiency or Global Involvement?
Self-Sufficiency and "Slow" Development

At independence, India's leaders wanted to establish a democracy among a multireligious and multilingual people and return to precolonial "glory" and a new international respect. The government's intention was a benevolent role (*Maa Baap* = "mother and father"). First the prime minister, Jawaharlal Nehru, transferred land from the aristocracy to the peasants in a move to rid India of the rural feudalism that was encouraged in the colonial era: princes and rulers had taken 20 percent of the national income while millions of poor people without rights starved. Second, the government planned to make India an almost self-sufficient industrial power.

Although the private economic sector remained important, contributing 75 percent of total output, foreign competition was excluded by high import duties. The "license raj," "permit raj,"

or "quota raj" needed many bureaucratic permissions before any economic activity could be undertaken. The civil service expanded to administer and regulate all aspects of the economy.

There were limited internal funds for investment in manufacturing enterprises and infrastructure support. Power shortages led to regular outages of two to three hours per day. The government encouraged firms to employ as many people as possible to the point of overemployment. Many firms could not update the technology applied in textile, vehicle, and other manufacturing. Education and health facilities were not geared to supporting economic growth in agriculture, industry, or services. Moreover, the system dampened entrepreneurial enthusiasm and was open to bribes and other forms of corruption. Government investments in large, often unprofitable, and virtually unmanageable manufacturing and infrastructure projects (such as large dams) absorbed much public money. These measures resulted in overcapacity, uneconomic units, corruption, and a fall in industrial productivity, while the low priority given to food production led to shortages and famines.

New Approaches

Nehru's daughter, Indira Gandhi, became prime minister in 1966 and reversed the neglect of agriculture by encouraging Green Revolution techniques supported by better distribution facilities. The famines of the 1960s ended. Later, in the 1980s, Indira Gandhi and her son and successor, Rajiv, began to open India to world economic forces. However, the internal subsidies and the huge costs of repeated elections in a country with so many people took India into deep indebtedness. Furthermore, the 1980s changes allowed in only foreign assembly industries and contributed little to government or personal income.

Crisis and Reform

In 1990 the Gulf War caused oil price increases and the cessation of remittances from Indian workers in the Persian Gulf countries. India faced a balance of payments crisis. In 1991 reformist parties took over the government, reducing government food and health care subsidies. India's middle classes, multinational corporations, overseas Indian investors, and others who gained immediately, welcomed the reformed program.

The need to ease the balance of payments crisis involved liberalizing foreign trade and internally deregulating markets. Measures to achieve this included devaluing India's currency, the rupee (making exports cheaper and more competitive but raising the prices of imports such as petroleum), reducing loans to small businesses and grants to scheduled classes and scheduled tribes, and increasing casual employment. The Indian government lost its command of the internal economy and politics, and the poor lost government support. However, annual economic growth increased from the 4 percent that could not get people out of poverty to around 8 percent (still less than China).

Economic growth came slowly, but by the early 2000s, more foreign capital was being invested in revitalizing industry and infrastructure, and new entrepreneurs emerged from all kinds of business sizes and social classes.

The elite upper caste groups who had controlled the government ministries now saw privatization and market links as the way to become wealthy and maintain their influence. Foreign links, often stemming from higher education in the United States or United Kingdom, linked them to global cultural values. Those who developed high-tech and computer software skills often built on links with defense research and development in the Bangalore and Hyderabad areas. Some moved to the United States and set up companies there.

In the 1990s politics became increasingly caste-based and regional. The Brahmin-dominated Congress Party gave way to local groups in the states, and some countrywide political groupings began to have greater roles. The RSS, which had been banned by Nehru after being linked to Mahatma Gandhi's assassination in January 1948, formed a political wing, the Bharatiya Janata Party (BJP), in the 1980s. In 1998 the BJP became the new government with a strong Hindu nationalist mandate. It detonated five nuclear weapon devices, causing horror around the world and drawing sanctions from major funding nations. However, the nuclear weapons proved popular with many Indian voters as a gesture of identity and confidence in a changing world. In 2004 the Congress Party regained control of the union government. India's problems of poverty, unemployment, weak demand, and low levels of investment continue to be major challenges. The state of Bihar—the poorest with 83 million people—remains badly governed, caste-ridden, and subject to Naxalite terrorism.

People: Religion, Language, and Ethnicity

India's population is predominantly Hindu: it is the religion of over 80 percent of the population (see Figure 2.17). India is also a country of sizable minorities, with over 11 percent (120 million) Muslims (making it the fourth largest Muslim population in the world) and just over 2 percent (up to 20 million) each Christians and Sikhs. The proportions of Buddhists and Jains are under 1 percent.

Religion, with diverse expressions, remains important in Indian life. Popular Hindu festivals and ritual bathing in rivers at holy cities draw many thousands. Religious leaders (from young Anna who hugs devotees and has global enterprises to many respected older gurus) and the Art of Living Foundation of Sri Sri Ravi Shavkar are popular. The rise of political Hinduism through *Hindutva* ("Hindu view") appears to threaten other religions.

Within India Muslims remain vulnerable, while Christian groups in "Nagaland" in northeastern India wish to form their own independent country. In 1980 extremist Sikhs in Indian Punjab occupied their Golden Temple in Amritsar and used it as a base for killing rival Sikhs until they were ousted in 1984 by a major military offensive. In revenge, they murdered Prime Minister Indira Gandhi, after which Sikh areas of Delhi were burned by mobs.

The diversity in languages is even greater. **Hindi** is spoken by only 30 percent of the population, mainly in the north. It is merely one of 15 official languages recognized by Indian states and the one most used in movies and TV. Of the official languages, the Indo-Aryan languages are prevalent in the north and those of Dravidian origin in the south. In all, over 1,600 languages and dialects are spoken in India. English remains the common language used by the legal system, is spoken by the educated elite—often being used in modern Indian literature and university courses—and is a common form of communication in commerce and national politics. Speakers of minority languages are at an increasing disadvantage as globalization affects more of India.

Caste remains a complex and unequal set of practices of great significance to Indian people. In a modernizing era, many Indians see it as providing places for all in society and a diversity that politicians have to manage. However, although rapid urbanization blurs some of the caste divides and suggests a "forward movement of the backward," caste remains an important consideration. In mid-2001 in Alipur, Uttar Pradesh, a young couple was hanged by their parents because they insisted on going ahead with an intercaste relationship that their families tried to ban; several of the family members faced murder charges.

The question of "untouchability" strikes at the heart of the Indian social system. The word itself is avoided as insensitive but is still used in anger or merely to make others recognize the existence of the extremely poor and rejected. Mahatma Gandhi preferred *harijan* (meaning "children of God"), but this was seen as paternalistic and impractical. Today *dalit* (meaning "ground down" or "oppressed") is more commonly used. Officially, people in this group are the scheduled castes.

Dalits have long questioned their social position, with rebellions since the A.D. 600s. After Indian independence in 1947, special consideration for the newly designated scheduled castes gave them access to government employment and rural land ownership. The Indian Congress Party assisted them with special favors, such as reserving college and civil service places for them. The few that obtained reserved places or jobs often used their new status to further political ambitions for the group. Jagjivan Ram, a dalit, became India's Deputy Prime Minister in 1979.

Today caste discrimination is most obvious in rural areas, where one's status is known to all. Dalits are still barred from using the same wells as upper caste Hindus, dalit homes are still burned, and caste-based political control remains in many areas. In Bihar, war broke out when the former landowning families, confronted by dalit-based guerrilla groups, set up their own private armies.

Urban Contrasts of Wealth and Poverty

Major contrasts among the lives of Indian people arise from their urban or rural locations. In the cities, clogged street traffic, high-rise office blocks and prestige apartments, international hotels, and TV ads for consumer goods portray a society moving into affluence.

The materially very wealthy include around 1 million people (only 1 percent of the total urban population) who consume at the level of Western countries, buying imported cars, appliances, clothing, cosmetics, and food. They are visible politically. Those with "old money" acquired it from land ownership or established industries. After independence, wealth taxes and ceilings on land holdings reduced the prominence of this group by policies aimed at creating a more egalitarian society. Under British rule, men of the upper castes became prestigious university professors; their children often apply their knowledge and personal links in the corporate world and the software industry. Wealthy traders also emerged from *jati* clans, such as jewelers and money lenders. A newly wealthy group is linked to the high-tech industries such as the software industry in Bangalore and Hyderabad. It also includes profiteers from the "black" economy, to which influential public office, unlicensed business, trade in drugs, smuggling, or extortion contribute. Liberalization in the 1990s and a 1997 amnesty on undeclared wealth made it possible for materially wealthy families to invest in entrepreneurial projects or multinational corporations, or to spend their money on luxury holidays, interior design, and imported goods.

The very wealthy live in guarded colonies, bus their children to private schools, and send them abroad for higher education. They use air-conditioned cars and have backup water and power supplies. They pay for the best hospital care (including surgery in foreign facilities) and travel first class by rail or air. They are cosmopolitan, even global, in education, livelihood, and social activities. They increasingly break with traditional Indian culture, and many move away from the social conscience enshrined in the Congress Party policies at independence. Today's wealthy young go to Western-style nightclubs and drive sports cars. They take leading roles in the media and literature.

The "patriarchal joint family," in which the family head, several sons and their wives, unmarried daughters, and elderly relations resided together, was typical of landowning families and family businesses. The whole was taxed as a unit. Smaller nuclear families are becoming the norm, with, at most, the addition of an elderly parent or unmarried daughter.

The importance of marriage is reflected in huge wedding celebrations, for which materially wealthy families save and invest. The bride takes a dowry, often in the form of jewelry, money, and consumer goods with her into the marriage, but in the wealthiest families she is more likely to retain the right to it.

The urban middle classes have jobs in factories, offices, social work, schools, police, or telephone exchanges. Those in government posts saw their incomes erode in the 1990s reforms. Job security outside government is poor and the competition fierce. However, the new technology and call centers provide jobs paid at rates above other local employment. A growing young, fairly affluent set is expanding the middle class. Estimates of middle-class numbers suggest as many as 300 million Indians are included, but this is the widest definition.

Households dependent on government jobs, such as the railways or police, occupy subsidized housing in colonies of two- to three-bedroom houses or apartments. Upwardly mobile young people rent their own apartments. At the other end of the middle-class housing spectrum, many live in overcrowded tenements. Families buy some consumer durables (TV, radio,

tape recorder), but refrigerators are luxuries. They buy food from local markets, but convenience foods (in the Indian context such as packaged flour, instant coffee, or baked loaves) are uncommon, and women expect to prepare meals based on grains and vegetables. If there are spare funds, they are used to educate the sons in private English-language schools.

Many middle-class families belong to the "forward" castes and resent the reservations for the "backward" castes in civil service jobs. One response is to join political movements—particularly the *Hindutva*—that reaffirms the worth of national culture but is fertile ground for right-wing Hindu chauvinism. Many vote for the BJP party.

Major consumers of media products in film, TV, and popular press, these families are part of an emerging mass culture drawing on both global and pan-Indian values. While younger members try to emulate those who benefit from globalization and liberalization in the economy, others look back to a "Golden Age" of Indian culture before the Mughal emperors or the British Raj and try to restore Indian "purity." They regarded the testing of the "Hindu nuclear bomb" as a national achievement.

Many girls in middle-class households marry in their teens and men in their twenties, nearly all by arrangement. Dowries transfer money to the man's family, often for use as his sisters' dowries. Families with daughters start saving at birth but still require loans. The practice of dowry is on the rise in all sectors of Indian society, causing a further devaluation of girls and women. Wives may be badly treated if their dowries are not up to expectations, and they may be murdered by being burned alive so the husband can remarry for a better dowry. It is also a wifely duty to have sons.

The largest group forms the marginalized urban poor, struggling for existence in shantytown slums (*bustees*) or on the sidewalks. Many slum dwellers are rural migrants with no title to housing or rights to establish businesses. They work in the unorganized, unprotected informal economy where children and elderly women are an integral part of the workforce. Children work for their living in tea stalls, selling newspapers, scavenging, begging, or in industrial workshops. Few can afford to obtain an education, and there are few schools in the slums. Crime is the only way most young people can exist. The very poor suspected of even minor offenses are imprisoned for long periods, unable to afford bail.

Some slums have brick or block dwellings with windows, doors, and corrugated roofs; others are shacks made of any available material, often at risk from fire. The poorest ragpicker settlements adjoin and merge into refuse heaps. Where slums are recognized by municipalities, they have piped water, legal electricity connections, and shared sanitation. Many are not recognized officially. There is no street lighting, cleaning, or repairs. Municipal policy is generally to control and raid, rather than protect, the slums. Water is supplied by tankers and public standpipes, and there are few toilets apart from open spaces and the gutters. Yet homes are kept clean. Disease (diarrhea, tuberculosis) is endemic or occurs in rampant outbreaks of cholera, hepatitis, and typhoid. HIV/AIDS is a major scourge, as are drug dealing and drinking illicit alcohol.

However, some slums have entrepreneurs. For example, in Dharavi, near Mumbai airport, over 100,000 people produce $500 million worth of goods a year from small businesses such as bakeries, metal workshops, recycling, tanneries, and potteries. Families who profited and moved out continued the business connections.

Caste remains the most significant social division among the very poor, and bustees are often dominated by one group. Most slum dwellers are dalits, but some are Brahmins—remaining aloof in their poverty. Bustee inhabitants of all castes have voting rights and are organized by politicians as vote banks in exchange for small favors and gifts. During the 1990s, however, government grants to help poor people were reduced or ended; many in-town slums were cleared for new development, but alternative accommodations were not provided.

Women in bustees face frequent attacks from the high proportion of men, most of whom leave their wives behind in villages. Some women work as domestic laborers in middle-class homes, where they are housed comfortably and protected but have little personal freedom. Many others enter prostitution, often bonded and with little chance of escape.

People with physical and mental disabilities—who were previously looked after in their villages but are now marginalized beggars in the cities, rejected by relatives—form another group of destitute poor. Some are hidden in huge mental institutions where they are controlled but not treated.

Mainly Materially Poor Rural Population

India's population remains largely rural. Between 1960 and 2004, the population in rural India decreased from 82 to 72 percent of the country's total, but their share of India's total GDP went from 50 to under 30 percent. Rural areas still have large proportions of the semisubsistence and low-paid farming lifestyles that are linked to long-term poverty and lag behind urban areas in education and health care provision. By 2025 it is projected that India's rural population will be down to 55 percent of the total, but this will comprise a greater number of people than at present (762 million instead of 630 million) because of the overall population increase.

Ways of life in rural India altered substantially during the late 1900s. Even the remotest village became part of the global economy, often selling cotton crops to European mills. However, there is still a huge diversity among states, across areas of different climate and topography, and under different social structures, indigenous cultures, and local market conditions. Most villages remain extremely poor. Although Mahatma Gandhi envisaged an India of "village republics" based on cooperative peasant holdings, this happened only in a few places such as West Bengal.

The *zamindari* system, under which the British Raj gave "aristocracy" land grants to elites who practiced a feudal way of life tying cultivators and craftsmen to their village, ended after independence. Land reforms put a ceiling on the amount of land one family could own, although they were not universally applied. Today communities of landowning peasants are the norm. Few old landowning elites remain.

Rural poverty involves indebtedness. Laborers have little chance of escaping a lifetime of poverty, and debts are inherited by children. Cash incomes are small, and often unexpected obligations, such as sickness, the death of plowing oxen, or a demand for a dowry on the marriage of a daughter, deplete savings. A manual laborer pays up to four times his annual wages for a daughter's dowry. Money lenders help with loans, but inability to repay leads to mortgaged and often lost land, while members of the family may be taken into bondage.

Countermovements exist to combat poverty. The most encouraging are the women-run cooperatives and small rotating loans projects linked to literacy campaigns (see the "Grameen Bank" section on page 320). From the 1970s, India's union (federal) government provided nonagricultural income opportunities for landless peasants in its five-year plans. Many projects were successful, but scandals affected others. In a Bihar fodder scam, a group of landowners and bureaucrats sold cattle several times among themselves, taking advantage of grants for veterinary care, subsidized fodder, and long-distance transport. In 2005 the Indian government moved to extend a guarantee of employment to all rural areas, with jobs mainly in public works projects at minimum wage levels.

Rural education is poor in amount and quality, apart from the notable exception of Kerala State, where all children are in school. The lowest educational involvements are in the "backward" castes. Without literacy, people cannot pursue their rights or take advantage of helpful projects.

Not all Indian farmers are poor. The Green Revolution, especially in Punjab, Haryana, and parts of Uttar Pradesh, led to the emergence of wealthy farmers who enlarged their holdings and invested in commercial production. They employ labor from the poorer states and campaign for even larger government subsidies for seeds, pesticides, fertilizers, and irrigation water. Agribusiness is replacing some semisubsistence farming, bringing higher living standards, mainly for farm owners and merchants. The wealthiest farmers provide many members in the union government's people's assembly.

Economic Development

In 2003 India's total GNI PPP was fourth in the world. However, a 2002 human poverty index value of 31.4 percent shows that many Indians remain very poor, tied to traditional low-productivity farming and work in urban factories.

Parts of India's economy are expanding. In the early 2000s India's broadening outlook toward global trade included increased private investment to and from the United States, increased involvement in Europe, and trading and investment agreements with Japan and China. The Indian government scrapped many export restrictions, expecting to double its sales abroad by 2007 and favoring agricultural products and computer hardware. It also abolished price controls on the telecommunications and oil refining industries, apart from the subsidies for cooking gas and oil.

India remains deficient in transportation infrastructure. Its railroad system has a major role, but needs updating to compete with the gradual improvement of roads and the efficient new private airlines.

The economic profile is moving from a farm base to factories and offices. Between 1965 and 2002, farming decreased its contribution to GDP from 44 to 23 percent without causing famines. Industry rose from 22 to 29 percent but then declined a little as services rose from 34 to 51 percent.

Food Production and the Green Revolution

India produces enough food to support the basic needs of its population, although prices and methods of distribution do not guarantee that everyone is fed adequately (Figure 7.22). The vagaries of the monsoon, however, still determine the harvest: good years enable India to export food; bad years require imports. Important regional variations exist in land use and farm productivity.

In 1966 high-yielding varieties of wheat and rice, developed at research institutes in Mexico and the Philippines, were introduced as part of Indian agricultural policy and set off the Green Revolution. Indian production of food grains almost quadrupled, from 50 million tons in the 1950s to over 190 million tons today. In 2002 India became the world's second largest exporter of rice in good monsoon years (after Thailand) and second largest producer of wheat (after China).

Benefits from the Green Revolution were geographically and socially uneven. In Punjab and Haryana, wheat and irrigated rice yields increased most. The larger land holdings and wealth of the Sikh and Jat farmers enabled them to purchase the expensive seeds, fertilizers, pesticides, and tractors. The local infrastructure of available irrigation water, 90 percent rural electrification, 70 percent all-weather roads, good bus and truck transportation, many medium-sized towns, and regulated grain markets provided a beneficial basis for the farmers' investments. Per capita incomes increased, but so did family size, reducing the economic benefit of the increased grain harvest.

FIGURE 7.22 India: rural scene. Farmers in north India plains share a new tractor for fieldwork. **Photo:** © David Zurick.

Elsewhere in India, the outcomes of the Green Revolution were less positive. Results were negligible in areas where irrigation water was not available, farm units were small, farmers had low incomes, and infrastructure access to markets at home and abroad was poor. Where the yields did not increase enough to match the increased costs, debts put farmers out of business. Furthermore, the switch to intensive grain production from peas and beans increased levels of malnutrition among local populations.

Even in Punjab and Haryana, diminishing returns from grain caused farmers to consider more profitable crops for export, causing India's agricultural exports to quadruple in value. The new crops included durum wheat (used in pasta), vegetables, fruit, and flowers. Agribusiness linked these crops to local processing facilities for making pasta, french fries, tomato puree, and wines. Nearly three-fourths of the exports go to the oil-rich, but arid, countries around the Persian Gulf.

India is the world's third largest cotton producer, after the United States and China, but has low yields despite having the largest acreage. In early 2002 the Indian government approved the use of genetically modified "BT cotton" seed that is resistant to the boll worms that destroyed 15 percent of annual yield. It is expected that the new seeds will double India's cotton production and reduce pesticide usage. The decision was taken against environmentalist objections based on "transgenic pollution." Further government measures encourage the wider role of modern agribusiness in grains, oils, cashew nuts, and butter.

Traditional Farming Areas of India

India has some long-established specialist areas for commercial crops:

- Tea continues to be an important plantation crop in the Assam hills of northeastern India and the ranges of Tamil Nadu and Kerala in southern India. Multinational corporations control marketing, processing, packaging, and distribution of the product in the world's wealthier countries, but pay low prices to producers, resulting in low local wages.

- The coastal deltas around the peninsula were drained for rice growing and cotton production. The most successful farming areas in southern India rival output from Punjab and Haryana in the northwest. They also have electrification, good roads, and local markets.

- Jute is a commercial fiber crop on the western margins of the Ganges–Brahmaputra Delta. Its growth there was developed when independence separated the Kolkata factories from the Bangladesh jute-growing areas.

Farming in other parts of India is less intensive and commercial (Figure 7.23). The Ganges River plains east of Delhi retain traditional farming methods for semisubsistence on mainly small farms. Large areas of the hilly peninsula have inadequate irrigation and grow coarse grains such as millet and sorghum that withstand dry conditions.

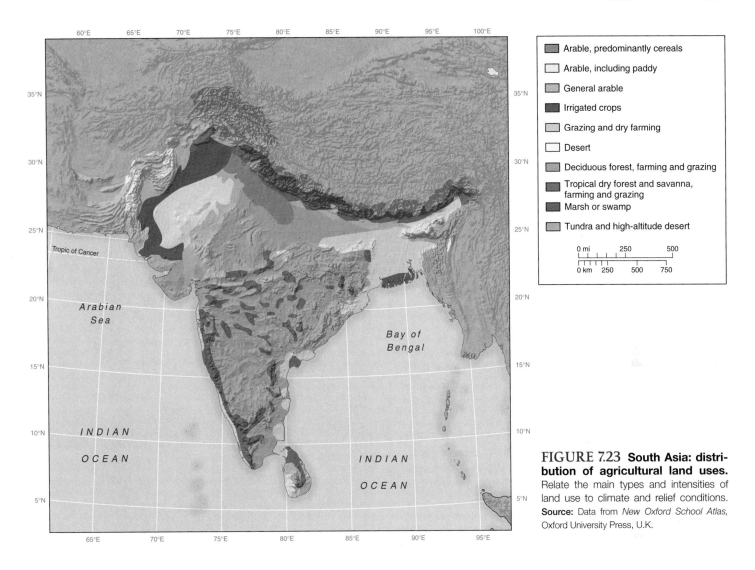

FIGURE 7.23 South Asia: distribution of agricultural land uses. Relate the main types and intensities of land use to climate and relief conditions. **Source:** Data from *New Oxford School Atlas*, Oxford University Press, U.K.

Legend:
- Arable, predominantly cereals
- Arable, including paddy
- General arable
- Irrigated crops
- Grazing and dry farming
- Desert
- Deciduous forest, farming and grazing
- Tropical dry forest and savanna, farming and grazing
- Marsh or swamp
- Tundra and high-altitude desert

Large numbers of livestock are a major feature of India. The 200 million cattle are the greatest number for any country in the world. Many are used as working animals because most religious and high-caste Hindus do not eat any meat, and all Hindus and Sikhs avoid beef. India, however, is a meat exporter. Water buffaloes work in the wetland environments of rice-growing areas. The laws against killing cattle because of their sacred status allow many to wander freely through villages and suburban areas. Cows are kept for milk and making *ghee* (liquid butter), both of which are an increasingly important part of the "white revolution" in Indian diets. Dung from cattle is an important fuel and a flooring plaster.

Mining

The most mineral-rich part of India is the Chota Nagpur Plateau in the northeastern part of the peninsula, where plentiful iron ore and coal deposits formed the basis of local steelmaking. India's rocks also contain other metal ores such as copper, gold, and manganese. By 2000 India raised its coal mining output to around 200 million tons each year, making it one of the world's leading producers. India also has the world's fifth largest reserves of bauxite (aluminum ore) and has the advantage that the metal can be extracted from its ore at lower temperatures than are required in many other major deposits. The thick sediments of the Ganges lowlands contain deposits of natural gas and oil, and by the early 2000s India produced 75 percent of its oil needs.

Until the 1700s India was the world's only source of diamonds. Some of the world's most celebrated diamonds, such as the Koh-i-noor, were obtained from the Deccan's Golconda mines. India also produces a wide variety of semiprecious stones. Considered one of the wonders of the world, the Taj Mahal, built by Mughal emperor Shah Jahan in the 1600s as a memorial to his favorite wife Mumtaz Mahal, contained mosaics of precious and semiprecious gems inlaid in the white marble on the interior walls.

Fishing and Forestry

India is one of the top 10 fishing countries in the world, and fish are important in the Indian diet. Many coastal areas depend on income and food from the fishing industry. Most of the industry still uses traditional methods of catching, with mainly unskilled labor and small boats.

Most of India's forest resources were cut centuries ago, and many villages now find it difficult to produce sufficient firewood for their needs. Forests still grow on the southern slopes of the Himalayas, from which softwood is cut, and on some of the hilly areas of the peninsula where hardwoods such as teak are exploited.

Growing Manufacturing

India has a large, growing manufacturing sector ranging from small-scale craft industries to large-scale steelworks and modern high-tech facilities. The distribution of manufacturing in India reflects historic craft specialties, raw material locations (mines, farms), the British legacy of transportation hubs, ports, and administrative centers, varied encouragements offered by the union and state governments since independence, and entrepreneurial skills.

Small-Scale Manufacturing

Small-scale industries in India employ 140 million people and produce 35 percent of all manufactured goods (Figure 7.24a). The union government and some state governments often favor small local businesses over large investments by multinational corporations. Two specialties in Uttar Pradesh illustrate the range and conditions of such industries:

- The Kanpur area has many small tanneries that invest little in technology, training, health and safety, or environmental treatment of the toxic effluent. They pay low wages, employ children, and rely on unskilled labor for their "quality handmade" goods. Workers go barefoot without protective clothing in an unfiltered atmosphere of toxic fumes and liquids to make safety boots for global markets.

- At Mirzapur, carpet-making began in the A.D. 1500s, when the Mughal emperor brought in weavers from Persia. Most modern factories were set up by British owners, but closed following labor union militancy over the application of labor laws. Only basic processes such as dyeing wool and packing carpets are still carried on in factories. Around 10,000 independent weavers work on 3,500 looms in surrounding villages. They rely totally on this income but face increasing competition in world markets.

Large-Scale Manufacturing

As part of its self-sufficiency policies after independence, the Indian government developed a large-scale, heavy industrial sector of steel, chemical, and aluminum works. The main steelworks were built near the large iron and coal deposits in Vananchal, Chhatisgarh, and Orissa.

Aluminum works are distributed more widely across India close to bauxite deposits and hydroelectricity sources. Since the Indian economy was opened to foreign investment in 1991, the aluminum industry received major investments from multinational corporations. Many large chemical plants were established in the 1960s in conjunction with Western corporations such as the Union Carbide Corporation and ICI. These have

(a)

(b)

FIGURE 7.24 India: aspects of industry. (a) Craft industries. A merchant displaying copper and bronze "mini-temples" in a New Delhi suburban market. (b) Modern technology. The Infosys Technology campus, Electronics City, Bangalore. Infosys employs 14,000 staff and is India's top software exporter with clients that include Bank of America and Citigroup. **Photos:** (a) © Parvindar Sethi. (b) © Reuters/Corbis.

been joined by an increasing number of Indian firms such as Reliance Industries Ltd., one of the 10 most profitable chemical companies in the world.

Car and vehicle assembly became major parts of the self-sufficient Indian manufacturing diversity. Up to the late 1980s, the state-owned factories producing Ambassador cars (based on a 1960s British model in the foreground of Figure 7.2b) and Tata trucks (at Pune) dominated production. After the 1990s reforms, the output and range of manufacturers increased, with new auto and truck factories linked to major multinationals such as General Motors, Ford, Chrysler, Peugeot, Volvo, Hyundai, Mitsubishi, Daewoo, and Volkswagen. The Indian market leader, Maruti, has links to Japanese Suzuki and produces the most popular car. Tata Motors lost money for many years until it cut its workforce and is now expanding in South Korea, China, Russia, and South Africa. Indian metal-forging companies make car parts, and the Indian pharmaceutical industry is challenging world markets.

Textile Manufacturing

Textiles and garments comprise over one-fourth of India's exports, mostly to the United States and European Union countries. Collaboration between Indian manufacturers and foreign customers brought maker and market closer together. Although cotton remains the dominant textile fiber grown and manufactured in India, coarser textile fibers including *coir* (from coconuts) and jute are cheaper than cotton and are blended with cotton and synthetic fibers for use in denim cloth.

Engineering textiles used in road construction and in preventing soil erosion on slopes are other jute and coir products. The traditional process of manufacturing coir from coconuts polluted large areas of coastal lagoons during the *retting* (soaking) phase. It lasted 10 months and left quantities of waste products that could not be disposed of. New concrete tanks and machinery for retting cut the time and improve the quality of both the fiber and the local environment.

High-Tech Industries

The total output of India's modern high-tech industry increased from US $2 billion in 1995 to US $12 billion in 2001. In the Bangalore area in Karnataka state, southern India, the location for many large public companies established after independence, the state government attracted multinational corporations such as 3M, AT&T, Digital, Ericsson, Hewlett-Packard, IBM, Motorola, and Texas Instruments. The main new industries are electronics, computer engineering, software and services, telecommunications, aeronautics, and machine tools. Bangalore became India's first "electronic city" (Figure 7.24b), and parallel developments are taking place at Mysore and Dharwad, and in Hyderabad in the state of Andhra Pradesh.

India's English-speaking, partly college-educated workforce is the basis of a low-cost service industry in "back office" work such as call centers (Figure 7.25) and other information-technology-enabled services. For example, GE Capital Services

FIGURE 7.25 India: call center. A call center in Bangalore. This service industry is a major part of the Indian development of IT outsourcing. Jobs in this area are set to rise to over 1 million and sales to $14 billion by 2007. **Photo:** © Sherwin Crasto/Reuters/Corbis.

opened its Indian call center in the mid-1990s and employs more than 5,000 people collecting from delinquent credit card users. British Airways runs frequent flier programs and handles computer message errors in India. American Express has a large office in Delhi. Savings of up to 50 percent on costs may grow as India's high telecommunications charges fall. Independent call centers have had more fluctuating fortunes, as has the market for U.S. medical transcription, although this area is growing fast. Companies are developing a range of activities, including data entry and conversion, rule-set processing (Can an airline passenger upgrade to business class?), problem solving (Should an insurance claim be paid?), direct customer interaction, and expert knowledge services using specialists linked to databases.

India's high-tech industries follow the pattern experienced in mass mechanization in the 1900s. The IT industries began with computers mechanizing office work. As administrative work components spread from accounting to law, employee records, advertising and other sophisticated and specialist areas, aspects were outsourced. The process became global in extent, with countries such as India, as well as Ireland, the Caribbean, Russia, and China, taking back-office and later more complex and higher-value business process outsourcing (BPO). In Bangalore and Hyderabad, thousands of Indian software engineers took over many basic tasks in software design and now move into the most sophisticated areas.

Mumbai: Center of Manufacturing and Movies

The main geographic concentration of manufacturing growth in India is around Mumbai, formerly Bombay, in both its own state of Maharashtra and adjacent Gujarat. Maharashtra state accounts for over one-fourth of India's manufacturing output and is home to over half of India's top 100 companies. Mumbai handles one-fourth of India's trade in its port and 70 percent of its stock exchange transactions. It is the headquarters of nearly all of India's commercial banks. As well as having factories making a wide range of products from textiles to pharmaceuticals, Mumbai is the center of India's very active movie industry (often called "Bollywood"). Over half of the 500 to 600 titles produced annually are filmed in this area. Alongside the growth of new industries in the Mumbai area, however, are many old textile mills with aging and inefficient machinery.

Gujarat, to the north of Mumbai, is the second most industrialized state. The coastal zone between Ahmadabad and Mumbai is marketed as the "Golden Corridor," and over 70 percent of its recent investment is in the southern sector close to Mumbai. Ahmadabad was once the main textile manufacturing center in Asia, but textiles now compose only 10 percent of its manufacturing output, equal to engineering. New investments from the 1990s almost equaled those in Maharashtra, with many in petroleum, petrochemicals, and pharmaceuticals. The people of Gujarat have a reputation for business acumen, and many Indian businesspeople in other countries come from this state.

Northeastern India: The Kolkata Focus

The other main manufacturing region in India is in the northeast. Kolkata remains the Indian headquarters of jute manufacturing, which is entering a revival after years of decline and labor union strangulation. At partition in 1947, Kolkata's mills were separated from the main production areas of jute in the Bangladesh section of the Ganges Delta. Although Indian farmers in West Bengal were encouraged to grow more jute, India still buys large quantities, especially of higher qualities, from Bangladesh.

The Jamshedpur iron and steel area, 250 km (150 mi.) west of Kolkata, is an enclave within a rural and mining region of great poverty. India's first iron and steelworks opened at Jamshedpur in 1914, built by the textile entrepreneurial Tata family, whose interests later expanded into engineering, truck manufacture, hotels, software, and an airline. Output increased from 10 million tons in 1980 to over 20 million tons today. The Jamshedpur factory operates well-maintained old blast furnaces together with the most modern steelmaking and steel-rolling mills and produces steel at a cost undercut only by the best South Korean mills. Much is sold to China.

Service Sector Growth

The Indian service sector is expanding rapidly. Government employment remains a major source of jobs for the well-educated middle class, although the economic reforms will reduce this source of jobs. India is well supplied with some groups of professionals, from lawyers to accountants and economists.

The education and health sectors both need further development to deliver a good basic education to the whole population and to improve health care and public health levels. The scope clearly exists for the expansion of employment in these areas, but it needs the political will to shift investment into both physical infrastructure and human resources. The facts that 300,000 children die of diarrhea each year and infant mortality remains high highlight the problem of health in India. Public health services do not work well. For example, Delhi generates nearly 4,000 tons of trash each day but clears only 2,500 tons, leaving the rest on the streets. Of 1,800 tons of sewage, only two-thirds are collected.

Education is good in parts of India. Kerala claims 90 percent literacy in contrast to the overall Indian figure of half that level. Education policy in India needs to achieve a better balance between the basic levels of education that are still lacking and the oversupply of university graduates. The elementary, secondary, and higher education sectors divide the education budget into three almost equal parts, and higher education is subsidized for all who qualify.

India's tourist industry attracted 2.4 million international visitors in 2002, up from 1 million in 1980. With its riches of historical buildings, mountains, and beaches, together with its exciting culture, India is a magnet to many from wealthier countries. Visitor numbers could increase with improved infrastructure and public health provisions.

Reducing Poverty

When India became independent in 1947, its leaders blamed colonialism for the poverty of its people. Over the next 50 years, famines ceased, literacy rates doubled, life expectancy rose from 33 to 59 years, infant mortality fell (from 165 per 1,000 live births in 1960 to 70 in 2001), poverty was reduced, and overall income growth was substantial.

Economic growth over this period explains half of the reduction in poverty based on income. Other factors affecting income poverty are less clear. For example, the state of Kerala (see Personal View: Kerala, an Anomaly in India on page 317) is a paradox with one of the lowest incomes per capita and slow economic growth but some of the best education and health care statistics and a human poverty index of 15 percent, compared to 50 percent in another poor state, Bihar. Haryana has the fastest economic growth, but discrimination against women is such that the ratio of females to males in the state is 86 to 100.

By the early 2000s India's overall human poverty index was still high at 31.4 percent; 60 million children under age 4 were undernourished, 50 percent of the population was illiterate, women remained at a disadvantage in society, and rural poverty affected about 40 percent of the population. Progress is slow: between 1990–92 and 2000–02, the hungry in India decreased from 25 to 21 percent, but the numbers increased from 216 to 221 million.

KERALA: AN ANOMALY IN INDIA

Thirteen-year old Ambili Nair waits excitedly at the international airport in Cochin, the largest city in the coastal state of Kerala in southwestern India. The new airport, which was built to cater to the increasing international air traffic to the state, is an island in a sea of coconut groves and rice paddies. The entire family is there to welcome Ambili's father, Mohan, who has been a guest worker in the Middle Eastern emirate of Abu Dhabi for over a year.

Ambili's home state is known for its lush natural landscape and its high level of social development. In fact, in many of the indicators of development, Kerala is similar to developed parts of the world. It boasts long life expectancies for both women (74 years) and men (68 years), which are well above the national average. In a country where only 65 percent of the population is literate, the state has an overall literacy level of over 90 percent and only a 6 percent gap between male and female literacy. Unlike in the rest of India, child labor is rare in Kerala. It is much more common to see children in uniforms make their way to government and private schools on weekday mornings than to see them work in small businesses or stay home to help with household duties.

Fertility is also considered an indicator of development. Kerala's fertility rate is not only low as is the case in many developed parts of the world, but has been below the replacement level of 2.1 for almost two decades. This means that most couples here have fewer than two children. Ambili's parents are fairly typical in that they planned to have only two children and are happy that they have a girl and a boy. Low fertility is in part due to greater use of contraceptives by the educated populace in the state, an exceptionally low infant mortality rate, and fairly good access to health care.

Kerala's women and girls also fare exceptionally well when compared with their counterparts in the rest of the country. Of India's 29 states, Kerala is the only one with more females than males. According to the 2001 Census of India, the country had 107 males per 100 females, while Kerala had 94.5 males per 100 females. The high social position that women are said to enjoy in

Kerala is often credited to the state's long tradition of matriliny among more than half its Hindu population. The Nair subcaste to which Ambili's family belongs is among the numerically strong matrilineal groups in the state. In such communities, property and the family line are passed on from mother to daughter, not through the male heirs as is usually the case in India. In consequence, girls and women are highly valued in matrilineal societies.

Social scientists have speculated on the reasons for Kerala's special position within India. Some credit progressive native rulers who made efforts to provide education to their subjects. Others conjecture that the state's coastal location, its position as a major spice trading center for centuries, and its long history of international contact have led to an openness to different and progressive points of view. Followers of three major world religions live peaceably in the region. Hindus are in the majority at 57 percent of the population, while Muslims and Christians form significant minorities at 23 percent and 19 percent respectively. Syrian Christians from Kerala believe that their ancestors were converted by St. Thomas the Apostle in the first century A.D. The Christian population here also includes those who were converted by Dutch, Portuguese, and English colonizers. Islam was brought to the state by Arab traders who used the monsoon winds to sail to and from India's southwestern coast. Although social contact between different religious groups was limited historically, the prevailing attitude was one of open-mindedness. Kerala was one of a few places in the world where Jews were not discriminated against. Cochin even had a thriving Jewish population until the mid-twentieth century. This attitude of religious tolerance prevails to this day.

Additionally, an elected Communist government that has been in power for much of the state's existence is seen as the driving force behind the social, political, and economic reforms that set the stage for human and social development. The success of the Kerala government in providing universal education and primary health care to its residents is often pointed to as a cause for Kerala's remarkable advancements.

However, Kerala is not without its problems. Limited job opportunities for even its educated men and women have caused many people to migrate to other areas in India and even the world in search of work. Ambili's father is among the several thousand skilled and unskilled migrant workers from Kerala who work in the oil-rich countries of the Middle East. Remittances from guest workers in the Gulf countries are critical to the state's economy, and even exceed total government expenditures. The money is important for families as well. The money Mohan sends every month has allowed the family to finally repair and renovate their house. It also pays for luxuries such as a washing machine, microwave oven, and color television they could not afford when he was working for a local company.

Economic activities that have traditionally been important in Kerala include agriculture, particularly the cultivation of commercial crops such as tea, coffee, rubber, and spices on plantations (locally called estates), and the production of rice, cashews, and coconuts. Fisheries are important to the economy, and much of the seafood caught is exported. Recently the natural beauty of the region, its tropical climate, and its rich cultures have attracted tourists from within India and abroad. Still, Kerala's most important money earner and export is its educated and skilled people.

Ambili wishes her father did not have to leave the family for long periods. She says that she would like to get a college degree and become a science teacher someday. Her parents support her educational and career goals. Even her grandmother, who says that she hopes to live to see Ambili married and have her own children, agrees that it would be best for her granddaughter to concentrate on gaining a good education first.

Box Figure 1 Nair girls decorating the entrance to their home for the Onam festival in Kerala, India. **Photo** © Lindsay Hebberd/Corbis.

317

Test Your Understanding
7B

Summary

India has the world's second largest population and dominates South Asia in land area, resources, and economic activity. India remains a mostly rural country, although the major cities such as Mumbai, Kolkata, and Delhi are growing rapidly. Most of the people are Hindus and are still affected by the caste system.

India's economy is characterized by low productivity in agriculture and manufacturing that produces goods mainly for local markets, but the northwestern agricultural area and the west coast industrial area around Mumbai are developing rapidly as global involvements increase.

Questions to Think About

7B.1   What were the advantages and disadvantages of the Green Revolution in India?

7B.2   How do the poorer and richer areas of India contrast with each other? What factors are responsible for the differences?

7B.3   How are India's relations with the rest of South Asia changing?

Key Term

Hindi

Bangladesh and Pakistan

Bangladesh and Pakistan share much of the same cultural heritage as the Republic of India and were part of the British Indian Empire until 1947. Their central areas occupy lowlands along the Indus River valley and at the mouth of the Ganges and Brahmaputra rivers (see Figure 7.3). They were partitioned from India at independence on the basis of their Muslim majorities and were part of the single country of Pakistan until 1971. East Pakistan became convinced that the Punjabi military leaders controlling Pakistan were taking East Pakistan's resources to benefit West Pakistan. They revolted and, with India's military help, established Bangladesh as a separate country.

Pakistan is a country of arid lowlands and high mountains. Bangladesh is mostly low-lying, apart from the small hilly eastern region inland of Chittagong, and is well watered by rain and rivers. Water is scarce in Pakistan, but flooding is a major problem in Bangladesh.

Countries

The differences between Bangladesh and Pakistan outweigh the influence of their common Islamic faith. Bangladesh is a secular republic, while the republic of Pakistan is an Islamic country.

After the 1971 civil war, Bangladesh had to build a new country, integrating Muslims expelled from Myanmar and Bengalis returning from India. Huge quantities of aid came from wealthier countries. Its internal politics involved military takeovers, coups, and assassinations. In the 1990s a settled democracy brought major changes.

Although Pakistan had an established base of agricultural and manufacturing products at independence, its people are little better off than the Bangladeshis. After the loss of East Pakistan in 1971, Pakistan's rulers worked out an Islamic religious base for its governmental, industrial, financial, and educational institutions. The 1973 constitution included a basis of Islamic law. In October 1999 General Pervez Musharraf seized power and dismissed the democratically elected Pakistan parliament. Although the Pakistan Supreme Court backed this move, it insisted that an election be held within three years; the general was reelected in 2002, but in 2004 agreed to stand down as head of the military so that he could remain in office until 2007.

In the early 2000s opposition to the Punjabi origins of many military and government personnel came from other parts of Pakistan. Immediately north of the capital, Islamabad, the Northwest Frontier Province seethes with discontent over the outcome of the Afghanistan war. In 2002 elections the MMA, an alliance of Islamist parties, won control of the province and is introducing policies reminiscent of the Taliban zealotry. In Baluchistan Province along the Afghan border in the south (see Figure 7.13), the MMA now shares a coalition government, while local Bugti tribesmen disrupt gas pipelines to signal their dissatisfaction with the Islamabad government support for the local mullahs rather than the tribal leaders. The Tribal Areas and Waziristan in west-central Pakistan are notorious for harboring Taliban members from Afghanistan and are scarcely under Pakistani military control. Then there is Kashmir, which also highlights Pakistan's internal differences and problems with its neighbors.

People: Ethnic Character

Bangladesh is the country of the Bengalis. **Bengali** (Bangla) is the language of 98 percent of the people, who originated from an Indo-Aryan stock that mixed with local ethnic groups. The **Biharis** form a smaller group of non-Bengali Muslims who speak Urdu and migrated to East Pakistan after independence in 1947.

As the site of the main historic entry routes of different groups of people to the Indian subcontinent, Pakistan has greater ethnic diversity than Bangladesh. **Urdu** is the official language of the country, a form of Hindi with Arabic script used by Muslim people before independence. The 1947 **Muhajir** immigrants from Hindu India to Pakistan speak Urdu. **Punjabi** and **Sindhi** are also widely spoken in Pakistan. Many Pashtun groups live in the hills bordering Afghanistan. Rivalries among groups and the elitism of Urdu-speaking feudal landowners and military officers continue to dominate social and political life in Pakistan.

Economic Development

Bangladesh and Pakistan are among the world's poorest countries. The human poverty index is over 40 percent for both countries. Pakistan's GNI PPP total is greater than that of Bangladesh, but both have adverse trade balances and depend on aid donations.

British governments built irrigation canals and railroads so that the Indus valley could be developed as a source of cotton for British industry. However, both they and the Pakistan government from 1947 to 1971 largely ignored the East Pakistan (Bangladesh) economy. From the 1970s, the Bangladesh government tried to develop its highly populated area in the face of few local resources and major environmental hazards.

In Pakistan, from 1980 to the early 2000s, the literacy rate doubled, although it was still only 45 percent; electrification of rural households rose from 16 to 61 percent; and terrestrial television reached 75 percent of city dwellers and 50 percent of the rural population. In late 2001, as a result of its support for the U.S. antiterrorist coalition, sanctions against Pakistan (imposed after its 1998 nuclear tests) were dropped, and it was given increased access to European markets as a reward for opposing terrorists and the drug trade.

Prominence of Agriculture

Over half (75 percent of females) of the Bangladeshi labor force is engaged in farming, producing 23 percent of GDP in 2002, down from 50 percent in 1980. Bangladesh grows over half of the jute that enters world trade (Figure 7.26), as well as rice, tea, and sugarcane. As in India, livestock are kept mainly as beasts of burden. Expansion of the farming area in Bangladesh occurred on new delta islands formed by deposition at the mouth of the Ganges River and reclaimed by building costly dikes. These areas are below sea level and vulnerable to ocean surges during tropical cyclones.

FIGURE 7.26 South Asia: jute harvest near Tangail, Bangladesh. The fiber is stripped from the plant stalks using the plentiful water of the Ganges River delta area. What is jute used for? **Photo:** © James P. Blair/National Geographic/Getty.

In Pakistan in 2002, agriculture employed a similar percentage of the labor force, and its products also accounted for 23 percent of GDP. Cotton is the chief commercial crop, supported by government credits to farmers producing the raw material of the main manufacturing sector. Heavy rains and poor crops in some years have caused farmers to question whether they should grow different crops. The battle between virus-resistant strains of cotton and new viruses continues. Other land-intensive crops, such as wheat, rice, and sugarcane, are grown as alternatives to cotton. Some farmers are beginning to look to more profitable labor-intensive crops, such as vegetables and fruit.

Much of Pakistan's best farmland is owned by wealthy people, who treat it as a source of political power by controlling local voting rather than as a basis for greater productivity. Such political support gains the patronage of many jobs in the government bureaucracy as well as the ability to veto any measures that might change their dominant place in Pakistani society.

Mining and Forestry

Both Bangladesh and Pakistan have reserves of oil and natural gas, although neither exploited them fully until the 1990s because they did not want multinational corporations to carry out discovery and production. Pakistan's important chemical industry is built on deposits of gypsum, rock salt, and soda ash.

Increasing Manufacturing and Service Industries

In Bangladesh, the slow expansion of agricultural production at 2 to 3 percent per year cannot support the faster population growth, and manufacturing growth needs to rise above the current 7 percent increase per year. The jute industry requires restructuring, but the Bangladesh government faces the opposition of trade unions in trying to privatize the industry. However, Bangladesh's garment industry expanded in the 1990s, becoming a major exporter. Bangladeshi apparel manufacturers work closely with their retailing customers in Europe, the United States, and East Asia. Enterprise zones at Chittagong and Dhaka (Dacca) are further attempts to expand industrial income and employment. Bangladeshi apparel workers are paid 9 to 20 U.S. cents per hour (20 to 30 cents in Pakistan and India and over $8 in the United States).

Pakistan has a longer experience of manufacturing, which grew to 16 percent of GDP in 2002. In exports, primary products decreased from 48 to 10 percent of the total between 1975 and the early 2000s, while manufactured goods rose from 52 to 89 percent. Textile manufacture, especially of cotton goods, dominates industries that include food processing, chemicals, and car assembly. The vested interests of tractor manufacturers within Pakistan prevented the introduction of cheaper tractors made in Poland and Belarus.

Both countries lack many basic provisions that might encourage more manufacturing industry to locate there. Power supplies are subject to shortages, although Pakistan hopes to double its power provision with the aid of private investment. In the 1990s it completed major hydroelectricity projects in the Himalayas, while a new thermal power station is being built with foreign capital and expertise at Hub near Karachi.

Service industries employ a growing proportion of the labor force in Pakistan (36 percent of women and 18 percent of men) and are expanding in Bangladesh (30 percent of men, 12 percent of women). Health, education, and financial services are growth areas. Neither country attracts many tourists.

Grameen Bank

Bangladesh developed the concept of *microcredit*—making small loans to individuals and groups to enable them to start businesses. The idea began in 1974, when Professor Muhammad Yunus, an economist at Chittagong University, lent US $25 of his own money to 42 basket weavers and found they could then pull themselves out of extreme poverty. In 1983 he founded the Grameen Bank. Controversially, women were allowed equal access to the loans, and by 2004, 95 percent of the 3.5 million borrowers were women. Also confounding early critics, 99 percent of the loans are repaid on time. The Grameen Bank offers 6,000 scholarships each year to encourage children of borrowers to stay in school.

Loans are available for income-generating activities such as simple processing (rice husking or lime making), manufacturing (pots, weaving, garment sewing), storage, marketing, and transportation. Average household incomes of Grameen Bank members are 25 to 50 percent higher than their neighbors'. Landless people gain the greatest benefits, and the shift from agricultural wage labor to self-employment in petty trading brings a feeling of social improvement. A related result has been better employment conditions and wages for agricultural laborers as their numbers dwindle. One recent development is the Village Phone Project, in which 100,000 women use cell phones to provide telephone services in 90 percent of Bangladeshi villages.

Similar methods are now applied in 58 countries, and the World Bank acknowledges that millions of individuals have worked their way out of deep poverty using the loans, although the total effect on each country's economy is small.

Economic Restructuring

In the early 1990s Bangladesh adopted structural adjustment measures imposed from outside. It reduced trade restrictions and encouraged external financing of garment factories, doubling its export of goods. During the 1996 elections, political strikes of half the 1 million poorly paid garment workers led to economic reforms that returned people to work and the economy to growth. Privatization continues, but severe infrastructure bottlenecks (roads, power stations, communications) slow the economy's expansion.

At the same time in Pakistan, the growth in output and productivity in manufacturing increased jobs, even in rural areas, at a rate that equaled the rate of new entrants to the labor force.

The new government elected in 1997 gave signs that it would help to make its producers more competitive internationally. It also wished to tackle problems that kept so many of its people poor and socially deprived. The detonation of nuclear weapon devices in 1998 was a response to the Indian demonstration of military power. But it was a costly event, financially and politically, because sanctions lost markets up to September 11, 2001.

Poverty and Social Services

In Bangladesh, the contributors to poverty are mainly those of overpopulation (very high densities with a low resource and infrastructure base). Less than half the population has access to sanitation or legitimate electricity supplies and only 14 percent to garbage disposal.

In Pakistan, the inequitable distribution of wealth still leaves large numbers very poor, although the average per capita income is higher than that of Bangladesh. Pakistan lags behind other low-income countries with low levels of literacy among women, high infant mortality (85 per 1,000 live births, compared to the average of 60 in all low-income countries), and continuing high total fertility.

Mountain and Island Rim

The countries of the northern mountains—Afghanistan, Nepal, and Bhutan—have environments that isolate them from global connections and foster internal strife through tribal rivalries. Events in these mountain countries are influenced not only by what happens in the rest of South Asia but also by interactions with neighboring countries in Central Asia, Iran, Russia, and China. The island countries of Sri Lanka and the Maldives have easier ocean connections with the global economic system but experience other problems that restrict their fuller development.

People: Ethnic-Based Conflicts

The Mountain and Island Rim countries of South Asia have local fault lines between dominant ethnic groups and minorities. Muslims dominate in the west (Afghanistan and Maldives), Hindus in Nepal, and Buddhists in the east (Bhutan and Sri Lanka).

Afghanistan is 99 percent Muslim, of which 84 percent are Sunni and 15 percent Shiite. Fifty percent speak the main language, an Afghan form of Persian. Pashtuns dominate the south, while groups of Central Asian peoples, such as the Tajiks and Uzbeks, inhabit the north (see Figure 7.7).

Nepal is the only country in the subregion where nearly 90 percent of the people are Hindus. It also includes small proportions of Buddhists who have had a major impact on the local character of Hinduism. Like Afghanistan, however, Nepal has ethnic diversity arising from peoples moving into the mountainous country from all directions—Tibetans, Indians, and many local hill tribes.

Geography at Work

BATTLING INFECTIOUS DISEASES

Mohammed Ali's interest in the geography of epidemiology was stimulated by his diverse academic and professional backgrounds. Graduating in economics from Rajshahi University in Bangladesh in 1978, he started his professional life as a research analyst for the government of Bangladesh. The research involved identifying barriers to agricultural production in Bangladesh. Here is how Mohammed describes his work:

"In late 1984, I joined the International Center for Diarrheal Disease Research, Bangladesh (ICDDR,B) as a computing professional. I became involved in health and population data management and analysis, and worked with various research groups coming from different disciplines and from various parts of the world. However, there were problems for a full geographic analysis until 1994, when we were able to use the geographic information systems (GIS) introduced by a Belgian scientist. We created a household-level GIS for the Matlab field study area in Bangladesh. By analyzing the health and population data at different geographical/ecological scales, the GIS offered new ways of studying the health problems. And that was the beginning of my research career in geography and health.

"The region of Matlab experiences a higher rate of cholera than other parts of Bangladesh. However, my understanding was that the entire Matlab area is not equally exposed to the disease risk. People's environments and their health behaviors vary in space. Following completion of a master's program in geographic information sciences, I was encouraged to undertake this issue in my Ph.D. research. I researched the spatiotemporal patterns of cholera in rural Bangladesh and identified environmental risk factors for the disease. I also conducted geographic research on other diseases such as acute lower respiratory infection (ALRI) and dengue fever.

"Presently, I am working in the International Vaccine Institute (IVI) based in Seoul, Korea. The mission of the institute is to introduce new and improved vaccines in developing countries at an affordable cost. We evaluate the vaccines through multidisciplinary research programs. After joining the IVI in early 2001, I introduced GIS in vaccine evaluation studies for our research sites in China, Vietnam, Pakistan, and India. The congestion of buildings and narrow lanes in Pakistan and India created problems that we resolved using remote sensing technology. The household GIS database we created was used to make geographic area studies of vaccine and control agents. Subsequently, we used the GIS data to map geographic variation in vaccine coverage and to find out the ecological factors that influenced them.

"My newest spatial epidemiology studies investigate the effectiveness of cholera vaccines. Being aware of the limitations of the analytic methods favored by the regulatory agencies, my new approach demonstrates that the use of geographic variations in vaccine coverage can help evaluate the indirect protective effects of the vaccine. This new method is applicable to trials of a wide range of vaccines before they are introduced into public health practice. Our findings suggest that if the indirect effect of vaccines is not evaluated, then the potential of vaccine in public health programs may be underestimated. Geographic

Box Figure 1 Dr. Mohammed Ali working with colleagues in Guanxi Province, China.

analysis of vaccine efficacy and the ecological factors leading to spatial variation of efficacy need to be known for better interpretation of a trial outcome.

"Neighborhood-level studies are sometimes essential to identify important public health problems and to generate hypotheses about their potential causes. This is because people do not live in isolation; they live in groups and neighborhoods that may influence their lifestyle, health, and health-seeking behavior. Also, some issues do not make sense at the individual scale. For instance, a household owning a good sanitation system can be exposed to bad sanitation from neighbors. Several diseases such as tuberculosis, dysentery, typhoid, and malaria have been studied within our ecological study programs.

"I continue to use GIS, satellite remote sensing, and spatial modeling techniques in research that is grounded in geographic theories of human–environment interaction. Recently we received a grant from NOAA to investigate the emergence and fluctuation of cholera in Vietnam (two study areas along the central coast at Hue and Nha Trang) and Matlab, Bangladesh. We will model the emergence and fluctuations of cholera in the study areas by integrating spatial data sets including satellite imagery (sea surface height derived from TOPEX Poseidon, sea surface temperature derived from AVHRR, chlorophyll concentration derived from SeaWiFS, rainfall derived from TRMM, land-use land cover derived from Landsat TM and MSS) with climatic variables (monthly temperature, monthly rainfall) and sociodemographic data (population distribution and socioeconomic status). Associations between these variables and cholera incidence in Vietnam and Bangladesh can be used to predict future epidemics in other parts of the world."

Bhutan is a Lamaistic Buddhist country like its neighbor, Tibet, but 35 percent of its population are Hindus from Nepal and India. The main group of people, the Bhote, drove out Indian peoples in the A.D. 800s and built impressive fortified monasteries in the 1500s.

Sri Lanka has a Buddhist majority, which is the dominant Sinhalese-speaking group. Nearly 70 percent of Sri Lankans are Buddhists, 15 percent Hindus (mainly Tamil), and 8 percent each Muslims and Christians. The Muslims and Tamils are the main business groups, and since independence, both have been subject to repression by the dominant Sinhalese.

In the Maldives, Arab traders established Islam from the 1100s. The islands have a diversity of Muslim ethnic groups from southern India, Sri Lanka, Arabia, and Africa.

Countries and Political Issues

In 2005 Afghanistan, Nepal, and Sri Lanka each had 20–30 million people. Bhutan and the Maldives had tiny populations of fewer than 2 million people between them. The mountain countries are among the world's poorest, whereas the Maldives and Sri Lanka have better living conditions.

Afghanistan

From its independence in 1920, internal feuding and external influences held back attempts to modernize Afghanistan. Tensions with Pakistan emerged over the northwest frontier lands that Pakistan swiftly annexed after its own independence in 1947. Afghanistan then linked more closely to the Soviet Union until 1979, when the latter invaded and occupied it. During the period of Soviet occupation, local warlords, or mujahideen (mujahedin), supplied with weapons by the United States through Pakistan, fought the Soviet armies and established their own military units. After the Soviet Union withdrew in 1992, the local mujahideen and their armies fought each other for control of the country. Most government systems broke down. Education became irregular, while public health and health care were weakened.

In the late 1990s the Taliban group conquered virtually all Afghanistan, although opposed fiercely by northern tribes who resisted the imposition of hard-line Sunni Islamic tenets. From the west, Iran supported Shia groups, including Hizbollah, but they failed to overthrow the Taliban. The rest of the country lived under the repressive Taliban order, which expelled most international aid agencies. Iran and Uzbekistan closed their borders with Afghanistan. However, Pakistan supported the regime.

The association of Afghanistan's Taliban rulers with Osama bin Laden's terrorist network began during the Soviet occupation in the 1980s, forming a group of "Arab Afghan" exiles from all over the world who shared the militant Wahhabist Islamic beliefs (the Sunni Muslim group that dominates Saudi Arabia; see Chapter 8). The 1992 withdrawal of the Soviet forces added credibility to the Taliban, which claimed the victory. U.S. attacks on Taliban strongholds and al-Qaeda positions in late 2001 built on internal opposition to the Taliban,

deposing the Taliban government and establishing a new government of national unity and the first "democratic" elections in 2004 and 2005 (though marked by local coercion).

Afghanistan faces a difficult and uncertain future in the light of its history of factional strife, occasional oppressive rulers, a devastated economy, and difficult relationships with surrounding countries. Iran and Pakistan continued to support (or allowed continued resistance by) groups around Herat in the west and the former Taliban groups around Kandahar in the center. Tajiks and Uzbeks expelled Pashtuns from the north into Pakistan in a phase of ethnic cleansing. The Afghan International Office for Migration encourages the return of qualified Afghans who fled the Russian- or Taliban-dominated country but has difficulties in attracting professionals who settled comfortably in the United States, Europe, or Australia.

Political tensions continue. The Taliban remains active in the south and southeast; although unable to take back the towns, its guerrilla fighters, armed by drug money, control the opium routes and areas between major centers. In the west, Ismail Khan dominates the Herat region with Shiite, Iran-style morality patrols. The local government is paid for by customs dues at the Iranian border. In the north, local warlords control events and trade across the borders with Central Asian countries, and oversee the renewed search for oil. Money from the sale of opium continues to fund local groups and undermines overall political stability.

Nepal and Bhutan

Nepal emerged from isolation and became a democracy in the 1980s, but alternative political parties were not allowed until 1990. Left-wing (Maoist) guerrillas caused major disruption in occupying rural areas and attacking remote army and police barracks. In 2001 Nepal's monarchy was almost all murdered in a family argument by the heir to the throne, who suspended moves toward democracy. In 2005 the embattled king suspended all constitutional freedoms, but had little hope of peace through talks with the Maoists.

Bhutan remains a tiny buffer state between India and China. It has few cross-border contacts, preferring to maintain its own culture and to limit Western tourists and influences.

Sri Lanka

After independence, Sri Lanka's government by the dominant **Sinhalese** people instituted reforms that were disliked by the Tamil minority, leading to a civil war. From a position where Sinhalese and Tamils lived side-by-side in Colombo, the capital, there were increased separations of people, over 60,000 deaths in the fighting and urban terrorism, and the emigration of many Tamils to India, Europe, and Canada. The Tamil Tigers (Liberation Tigers of Tamil Eelam, where *eelam* means "homeland") became a very aggressive terrorist group fighting for what they saw as their minority rights and building links with foreign terrorist organizations. Around 50 million Tamils live in India, which at first gave training support to the Tamil Tigers. However, India banned the organization from 1992 after it assassinated the former Indian Prime Minister Rajiv Gandhi in 1991. In early 2002 there was a cease-fire between the Tamil Tigers and the Sri Lankan government.

Economic Development

Amid the physical and political isolation of the mountain countries and continuing internal conflicts, agriculture remains important in all the countries. Most people still depend on subsistence farming and occasional wage labor. The highland environments provide hydroelectric potential, but construction of dams is expensive and may cause environmental destruction.

Shattered Afghanistan

The rebuilding of Afghanistan's economy is a major project. Afghanistan is destitute and a threat to its neighbors as long as it remains in chaos. Repairs to roads, power lines, schools, hospitals, functioning police and judicial systems, an acceptable currency, and a central bank are needed alongside housing, factories, and offices to replace the piles of rubble. Millions of land mines need clearing. In 25 years of fighting, most doctors and teachers left the country.

Agriculture Bases

Sri Lanka (tea, rubber, and coconuts) and the Maldives (coconuts) have commercial plantations that sell cash crops to world markets (Figure 7.27). These provide 35 percent of Sri Lanka's exports, down from over 90 percent in 1950 as a result of diversification in the economy since independence. Sri Lanka's adoption of new rice varieties in the late 1960s converted the major food deficit—which arose out of the colonial focus on export crops—into a virtual self-sufficiency in rice today. In Nepal, 82 percent of workers are in agriculture, mainly on the Terai Plain along the Indian border.

FIGURE 7.27 Sri Lanka: tea plantation. How are the tea bushes arranged on the hills? **Photo:** © Alasdair Drysdale.

Noncommercial agriculture is based around the nomadic keeping of sheep in Afghanistan and the growing of rice and other cereals elsewhere. In a politically motivated act in 1998, the Taliban destroyed the fertile central farming region of Afghanistan, the Shardi Plain near Kabul. They cut fruit and nut trees, burned villages and wheat fields, blew up irrigation channels, and left crops to wither. Around 200,000 people from the region took refuge in Kabul or fled northward.

Despite international pressures, opium poppies remain an important crop in Afghanistan, which in the 1990s supplied 70 percent of the world's heroin as the country's other farm products dwindled through war and Turkey, Iran, and Pakistan enforced strict drug control laws. Wheat was replaced by poppy fields, leaving city dwellers hungry. Afghanistan's isolation, disrupted government, lack of alternative commercial products, and warlord control of the drug trade make it difficult to stop the production of opium and heroin and their entry to illegal drug traffic routeways. Although Afghanistan was threatened with the withdrawal of aid because of its increasing production of these drugs, aid payments continued for fear of angering the warlords and as part of projects to persuade peasants to stop growing poppies. In 2000 the Taliban government stopped the growing of poppies as ungodly, and by the following year, few poppy fields remained. After September 11, 2001, however, the Taliban stopped cooperating with the United Nations, and planting took place on a large scale. When the Taliban government was deposed in late 2001, planting continued with the farmers becoming indebted to the opium traders, who take their land and even their women.

Mines, Forests, and Fisheries

Apart from natural gas in northern Afghanistan (which was formerly exported by pipeline to Russia), these countries are poor in mineral resources. Several produce gemstones. Sri Lanka is a major source of graphite.

The forests in Nepal and Sri Lanka are cut and used locally, mostly for firewood, leading to deforestation and erosion as populations have increased. From 1980 to 2001, Afghanistan's woodland area was reduced by five-sixths, being cut for firewood or by Pakistan-based loggers. The resultant barren hillsides succumbed to rapid soil erosion and gave rise to rapid runoff and flooding. Fishing is important to the economy of the Maldives, where tuna accounts for 60 percent of exports.

Small-Scale Manufacturing

Sri Lanka has a more diversified manufacturing base than the other countries as the result of government policies since independence, including the production of steel, textiles, tires, and electrical equipment. By the 1990s, however, expansion placed pressure on existing transportation and telecommunications facilities, while power shortages reduced output and exports of garments and other textiles. Much is in small-scale factories or **cottage industries** based on the dispersal of manufacturing processes in homes. Local agricultural processing together with the making of carpets and footwear are the most common industries.

Nepal (Mahakali project) and Bhutan (Chukha) encouraged external investment in dams for hydroelectricity and other uses. This resulted in additional local industry—making garments in Nepal and steel alloys in Bhutan.

In Afghanistan, Kabul's industrial district had 200 factories in 1992, but they were reduced to 40 under the Taliban, of which only 6 still operated in late 2001. The 1992 fighting destroyed most of the Jangalok steelworks outside Kabul, and the machinery was taken to Pakistan. Even the army factory that made cannons, guns, and ammunition is largely abandoned. One factory sends its employees to gather discarded plastic shoes and melts them to produce new cheap shoes.

Tourism

The mountain and coastal environments of these countries suggest tourist potential. Before the recent internal strife, Nepal attracted half a million tourists in 2000 (Figure 7.28) but few

visited in subsequent years. Sri Lanka was trying to attract visitors again after its civil war but some of its coastal areas were devastated by the 2004 tsunami.

In the Maldives, tourism increased from a few thousand visitors in 1980 to nearly half a million by the early 2000s. The limits of expansion were being reached, signaled by coastal pollution, groundwater contamination, and the need to bring in labor to cope with the increasing demand. The tourist industry employed over 25 percent of the working population, up from only 4 percent of the total labor force in 1985. Bhutan keeps down the numbers of tourists entering to reduce the impact of foreign cultures on its people and its natural environment.

FIGURE 7.28 South Asia: tourism in Nepal. The peak of Annapurna in the Himalayan range rises behind the hotel in Kaski village. How would a trekking vacation in Nepal differ from one in the Rockies or Appalachians? **Photo:** © John Burbank/The Image Works.

Test Your Understanding
7C

Summary

Pakistan and Bangladesh were a single country from partition in 1947 until 1971. They remain largely agricultural, but are developing more diversified economies. Pakistan has a long-established cotton textile industry.

Afghanistan, Nepal, and Bhutan are mountainous countries on the northern margins of South Asia. They have limited connections to the world economy. Wars and political tensions have destroyed Afghanistan, and Nepal is disrupted by Maoist guerrillas.

Sri Lanka and the Maldives produce plantation crops and fish for sale on world markets, and the Maldives focus on tourism. Sri Lanka's attempts to diversify its economy have been hampered by civil war.

Questions to Think About

7C.1 To what extent were Pakistan and Bangladesh incompatible as parts of a single country?

7C.2 Write a short Point–Counterpoint that focuses on the Sri Lankan civil war.

7C.3 Assess the political and economic needs of Afghanistan.

Key Terms

Bengali   Muhajirs   Sinhalese
Biharis   Punjabi   cottage industry
Urdu   Sindhi

Chapter 8

Northern Africa and Southwestern Asia

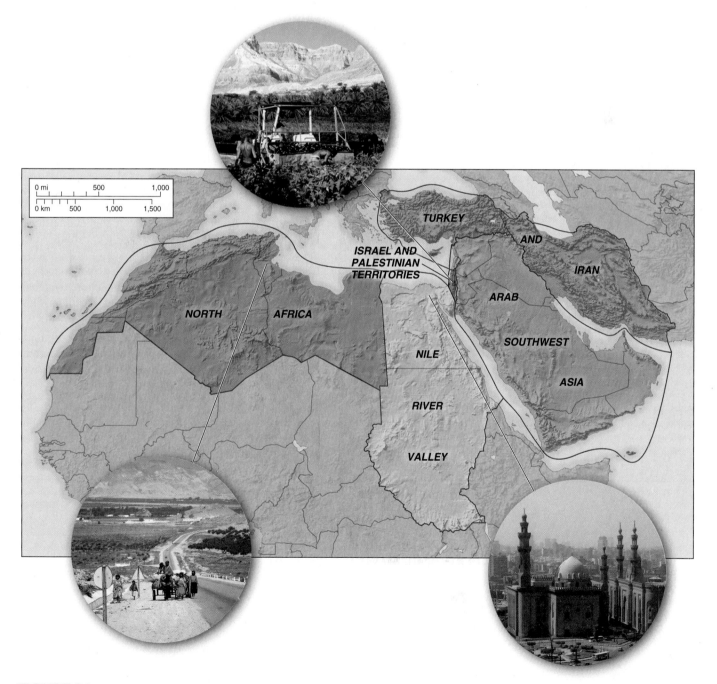

FIGURE 8.1 Northern Africa and Southwestern Asia: subregions and aspects of the geography.

Themes in This Chapter

Northern Africa and Southwestern Asia is the hearth of early civilizations such as Egypt and Mesopotamia and also the three Western religions of Judaism, Christianity, and Islam. Today, Western influences have resulted in conflict and are unwelcome by some segments of society. Arid natural environments have kept population numbers low throughout history. High population growth rates in recent decades have put increasing stress on the natural environment, especially on freshwater supplies. Oil has created great wealth in some of the region's countries. The themes are:

The region's cultural influences, particularly the Islamic religion and Arabic language.

Arab and Israeli politics.

The world's main oil-producing region.

Arid climates, water shortages, major rivers, high mountains, and dangers of pollution.

A significant role in the global economy.

Five subregions (Figure 8.1):

- North Africa: between the Sahara and the Mediterranean Sea.
- Nile River Valley: dependence on water flows.
- Arab Southwest Asia: desert heartland of Islam, oil production, and country rivalries.
- Israel and the Palestinian Territories: small enclave of polarized claims for land and religious and political control.
- Iran and Turkey: contrasting large countries in arid and mountain environments.

Personal View: Oman

Geography at Work: Operation Iraqi Freedom

Point–Counterpoint: Israelis versus Palestinians

In the News

The countries of Northern Africa fronting the Mediterranean Sea and those of Southwestern Asia occupy nearly 12 percent of the world's land and have 6 percent of the total population (Table 8.1).

Although the smallest world region, it is one of the most influential parts of the world, often at the center of historic events. Today's strategic economic interests make it one of the most politically contentious and diverse world regions (Figure 8.2).

This region incorporates Africa north of the Sahara and what Westerners commonly call the "Middle East." The "Middle East" is an ethnocentric, specifically a Eurocentric, term that describes lands by their location relative to Europe. The term originated from an early 1900s French view of the "Near East" incorporating modern Turkey, Syria, Lebanon, and Palestine, and the British concept of the "Middle East" that included Egypt, Arabia, and the Persian Gulf area. The "Far East" (Pakistan and farther east) indicated greater distance from Europe. Rather than the Eurocentric term of "Middle East," the more global and culturally sensitive term "Southwestern Asia" is used in this book. In this chapter, the boundaries of Northern Africa and Southwestern Asia are primarily coastlines with a line drawn through the Sahara in the south. Mountain ranges mark the northern and eastern limits of Turkey and Iran. Sudan is included in the region because of its strong links with Egypt along the Nile River valley (Figure 8.3).

The region is often in the news: the wars in Iraq, al-Qaeda, Islamic fundamentalism, the Israeli–Arab conflict, the Islamic Republic of Iran, the struggles of the Kurdish people, and the politics of oil and water resources. Each issue ignites local conflicts and has wider impacts. Many of the issues reflect the dominance of the Islamic religion in the region and attitudes its adherents take in facing the challenge of Western world dominance.

Geographic conditions and forces are extremely important in Northern Africa and Southwestern Asia, which is geographically at one of the world's great human and physical junctions. It provides the connecting tissues between three continents.

- The region was the center of the earliest urban civilizations. The urban civilizations of Mesopotamia (approximately modern Iraq) and Egypt developed in major river valleys here.
- Judaism, Christianity, and Islam all originated in this region before spreading worldwide, and the region remains

Table 8.1. Northern Africa and Southwestern Asia. Data by subregion, area, population, income (GNI PPP—Gross National Income Purchasing Power Parity), urbanization, Human Development Index (HDI), Human Poverty Index (HPI).

	Land Area (km²)	Population (millions)		GNI PPP 2003		% Urban	Index 2002	
		Mid 2005 Total	Est. 2025 Total	Total (US$ billions)	Per Capita (US$)	2005*	HDI (rank of 177)*	HPI (% population)*
North Africa	4,751,440	78.5	99.6	376	5,577	63.8	79.3	99.3
Nile River valley	3,507,260	112.5	164.5	328	2,910	37.0	114.2	162.4
Arab Southwest Asia	3,726,330	110.9	178.8	566	5,577	75.4	113.2	178.3
Israel, West Bank, Gaza	21,060	6.8	9.3	128	19,200	65.0	7.1	9.3
Iran, Turkey	2,427,450	138.7	173.6	950	6,940	63.0	142.4	179.2
Totals or averages*	14,433,540	447.4	625.8	197,445	7,925	60.8	456.2	628.5

Source: World Population Data Sheet, Population Reference Bureau (2005), World Development Indicators, World Bank (2005), Human Development Report, and United Nations (2004).

FIGURE 8.2 Northern Africa and Southwestern Asia: diverse landscapes and peoples. (a) McDonald's in central Cairo, Egypt, is one example of global connections. (b) The Great Sphinx among the pyramids of Giza, Egypt, illustrate in the landscape the long and rich history of the region. (c) Weekly animal market at Sinaw, Oman. Although some camels are sold, most of the animals traded are goats. Oman is an oil-producing country, but many Omanis still depend on agriculture and pastoralism for their livelihoods. (d) Village of Bilad Sayt in the Hajar Mountains of Oman. Although all of Oman's cities and towns are connected by good roads, this and many other villages can still only be reached by dirt roads. Note that because good farmland is scarce here the village is built on a bluff. The major crop here is dates. **Photos:** (a) Joseph P. Dymond; (b) © Vol. 97/Corbis; (c) & (d) Alasdair Drysdale.

important to Judaism and Christianity while likewise serving as the center of the Islamic world.

- European control in the early 1920s resulted in the imposition of new political boundaries and the creation of new countries that reshaped identities and relationships.

- Muslim reactions against Western influences often center on the 1948 establishment of the country of Israel as a homeland for Jews in the midst of Arab/Muslim countries that opposed its presence from the beginning.

- From 1950, oil wealth and then resurgent Islamism in some countries brought further global significance. Material wealth generated by oil revenues, however, is unevenly distributed both between and within countries.

- The region has a largely arid natural environment, making fresh water as politically significant as oil. Nevertheless, despite extensive deserts, more productive mountain and river plain environments are significant as well.

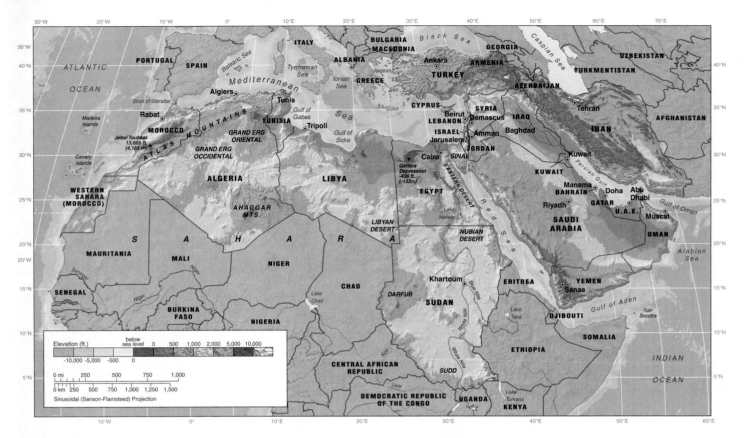

FIGURE 8.3 The physical features of Northern Africa and Southwestern Asia. As well as being arid, the region is mountainous, especially along the northern margins of the Atlas, Turkish, and Iranian ranges. Locate the two main river basins.

Cultural and Political History within a Wider World

This region was a hearth for early technical developments, three monotheistic religions (Judaism, Christianity, and Islam, related to one another through worshipping the God of Abraham), further technical and artistic developments in the medieval Muslim empires, and a focus of geopolitical strategies by major world powers from the 1800s. Located between other world regions, this region has influenced and has been influenced by neighboring regions. Religion and language are as fundamental as politics and economics in this world region.

First Civilizations

The Tigris–Euphrates River Valley of Mesopotamia (modern Iraq) and the Nile River Valley formed two of the world's early cultural hearths (see Chapter 1). Early human achievements diffused to the surrounding continents. Even in ancient times of slow and limited transportation, this region acted as a hub with constant movements of people to and from Northern Africa, Europe, and China. The internal empires of the Assyrians, Babylonians, and Egyptians gave way to external control from Persia, Greece, Rome, and then the Byzantines before the A.D. 600s expansion of the Muslim Arabs.

Religions

Religion and related traditions contribute greatly to peoples' identities in Northern Africa and Southwestern Asia. The early animist religions had many gods linked to natural phenomena. Some emperors were treated as gods. After approximately 1000 B.C., dominance by religions with many gods gave way to religions based on a single god: *monotheism.* Judaism was the first monotheistic religion. Christianity grew out of Judaism, and then Islam emerged and is dominant today, but Judaism and Christianity still exist in the region today. These three monotheistic religions share common religious figures and religious texts. All three diffused from their hearths in Southwestern Asia to Europe, Africa, and Asia (Figure 8.4).

Judaism

Judaism is a religion whose adherents worship Yahweh, seen as the only God, creator, and lawgiver. It began in the area now known as Israel and the Palestinian territories around 2000 B.C., where Abraham and his descendants settled after moving from Mesopotamia. Jewish beliefs focus on the historic role of family based on the line from Abraham, persecution beginning with slavery in Egypt, redemption as Moses led the people out of Egypt, and their occupation of the promised land. Yahweh intervened to support and punish through times of trouble and deportation, promising a messiah to save Jews from domination by

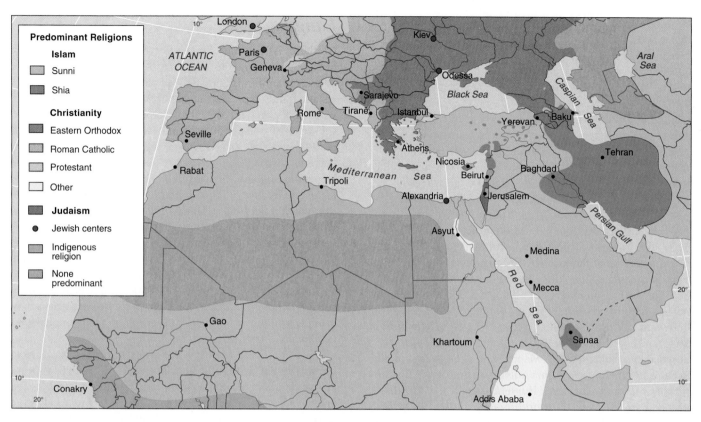

FIGURE 8.4 Major religions in Northern Africa and Southwestern Asia. Beginning in Mecca and Medina, Muslim Arabs conquered the area shown in green after Muhammad emerged as a new religious leader around A.D. 600. Islam also spread over trade routes. Christianity became the religion of the Roman Empire, spreading through Europe. Jews were dispersed at the fall of Jerusalem in A.D. 70, and the modern country of Israel was not formed until 1948.

others. At the time of Jesus ("Christ" is a Greek translation of the Hebrew "messiah"), many Jews expected a military leader messiah to end the Roman occupation. In A.D. 70, the Roman army destroyed Jerusalem and dispersed Jews through Southwestern Asia, Northern Africa, and Europe to form a major diaspora.

Christianity

Christianity stemmed from new interpretations of the beliefs and teachings of Judaism in the early years A.D., beginning with the life, teachings, death, and claimed resurrection of Jesus of Nazareth. Christians began as Jews who believed that Jesus of Nazareth was the messiah. Christianity gave a fresh impetus to monotheism against the many gods of the Greeks, Romans, and Arabs. Christian churches spread across Southwestern Asia and into Africa and Europe. In the late A.D. 300s, the Roman Emperor Constantine declared it to be the empire's official religion. Controversies over how much Jesus was man and how much God resulted in divisions within Christianity. A later division occurred between the eastern (Orthodox) and western (Catholic) groups of churches in Europe. From A.D. 395 to 1453, the eastern church was centered in Constantinople (modern Istanbul, Turkey). The largest Christian sect in this region today is the Coptic Church, with a pope who resides in Alexandria, Egypt.

Islam

Muhammad founded Islam in Arabia, notably in Mecca and Medina, during the early A.D. 600s. **Islam** means "submission to the will of Allah (God)" and the followers of the religion, **Muslims,** are "those who submit to Allah." Muhammad adopted many Jewish and Christian beliefs such as monotheism. Seen as the prophet of Allah, Muhammad is believed to succeed earlier prophets such as Moses, David, and Jesus. The **Qu'ran** (holy book) is believed by Muslims to be the word of Allah revealed to Muhammed.

After Muhammad's death in 632, Arabs spread Islam rapidly westward to Northern Africa and Spain as well as eastward into central Asia, uniting the Arab peoples and creating a series of empires, converting Persians, Turks, and people in India through conquest and trade. It led to a Muslim "Golden Age" of artistic and scientific achievements alongside further military expansion from the 800s to the 1100s. New forms of art and architecture developed. Muslim mosques remain dominant features of town landscapes, often doubling as centers of religious and secular activities (Figure 8.5). One of the oldest is the Al-Aqsa Mosque in Jerusalem, begun in 692. Islamic visual art, often characterized by abstract patterns that avoid depicting human forms, ranges from pottery and metalwork to paintings and textiles. Calligraphy became the most important form of art because of the religious significance of the Qu'ran. Poetry and music were generally more important than the visual arts but less significant in the landscape.

Early divisions within Islam continue to be significant today. The **Sunni Muslims,** or Sunnis, are the majority Muslim group and base their way of life on the Qu'ran, supplemented by traditions. Political power was at first given to a succession of leaders, or *caliphs,* descended from the historic Muslim leaders. Many Sunni Muslims are moderates in maintaining traditional Islamic practices. Exceptions include the Wahhabi sect that dominates Saudi Arabia and the Taliban faction that ruled Afghanistan from 1996 to 2001.

A minority of Muslims also follow the Qu'ran but dispute the caliph leadership succession accepted by Sunnis. They look to descendants of Ali, the fourth caliph of Islam and a cousin and son-in-law of Muhammad. His son, Hussein, was defeated and killed by the Sunni caliph of Damascus in 680. Supporters of Ali see Ali as the only *imam*—authoritative interpreter of the Qu'ran—and are known as "partisans of Ali," "Shi'at 'Ali, **Shia Muslims,** or Shiites. They look to the return of the twelfth *imam,* Ali's descendant, Muhammad al-Mahdi, who disappeared in 874. For most of their history, Shia Muslims were known for commemorating and lamenting Hussein's death by flagellation and withdrawal from the world as they awaited al-Mahdi's return. Politically, they were quiet, regarding political control as evil and looking to clerics, supported by alms, for leadership. Today Shiites comprise large proportions of the population in a few countries, including 90 percent of the Iranian population and 60 percent in Iraq. From 1979 Shiites governed Iran, tried to establish Iranian isolationism from Western ways, and supported extremist groups such as Hamas (Palestine) and Hezbollah (Iran, Syria).

FIGURE 8.5 Islam and architecture. Mosques are a common feature of urban landscapes in Muslim countries. Older (1300s) and more recent (1800s) mosques tower over Cairo, Egypt. **Photo:** © Joseph P. Dymond.

Languages

Northern Africa and Southwestern Asia are linguistically diverse with numerous languages spoken among four major language families (Figure 8.6; see also Figure 2.16). The Afro-Asiatic family represents most language speakers of the region with such languages as Arabic, Berber, and Hebrew. **Arabic** is spoken by just under 50 percent of the people in the region. It is also the preferred language in the Qu'ran and Muslim prayers. "Arabs," who were once defined as living in the Arabian Peninsula, now include all those using Arabic as their first language. **Berber** is found primarily in Morocco and Algeria, especially in the Atlas Mountains and the Sahara Desert where it best resisted the spread of Arabic. **Hebrew** is the official language of Israel. The language of the Hebrew Bible and religious services is an archaic form, superseded in everyday usage by modern Israeli Hebrew. The movement devoted to creating a Jewish state from the 1800s, known as *Zionism,* revived the use of Hebrew, modernizing it for secular usage to provide a common religious and secular element among Jews from many countries.

Indo-European languages are also significant in the region. **Farsi (Persian)** is the official language of Iran. **Kurdish** speakers straddle Turkey, Syria, Iraq, and Iran. Their numbers are unknown because the language is at the heart of demands for a separate nation–state: non-Kurds talk in terms of 10 million people, while the Kurds claim their numbers to be over 20 million.

Turkish is a member of the Altaic family (see Figure 2.16), which spread westward from Central Asia. Turkish speakers in Turkey use the Roman script, reflecting Europe's influence on Turkey. The Azeri (or Azerbaijanis) in Iran, who are also Turkish speakers, use the Arabic script. A large number of the world's Turkish speakers live in Central Asia (see Chapter 4).

Political Organization and Control: Empires and Colonies

Persian Dynasties

Invading Mongols ruled Persia until the 1400s, when the Shia-dominated Safavids gradually reclaimed Persia and established a dynasty that lasted until 1722. The Shiite *ulama* (cleric scholars) asserted an early doctrine that only they were true interpreters of the Qu'ran and therefore leaders of Islamic communities. By 1722 they took political control from the Safavids but had no military force to resist the Afghans, who rose against the governor of Kandahar and then took much of Persia. Chaotic governing followed.

Ottoman Turks

In the 1500s the Turkish Ottoman Empire under Sulayman the Magnificent conquered the Mediterranean coastlands as far as Morocco and also took much of Balkan Europe (see Chapter 3). The early vigor of the Ottomans subsided into maintaining their position and even attacking other Muslims as they established

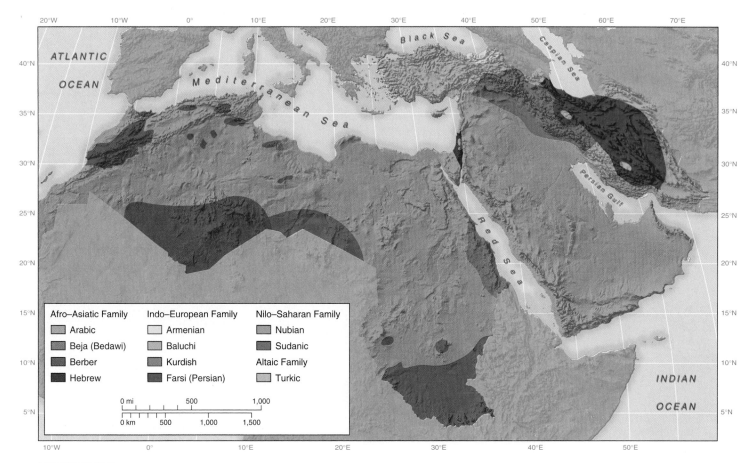

FIGURE 8.6 **Language families and languages of Northern Africa and Southwestern Asia.**

control of Arab lands, but the overseas colonial expansion of European countries eroded their empire. The Ottomans did little after the 1700s to encourage modernization or involvement in the expanding global economic system.

European Colonies and Protectorates

From the 1800s, the North African coastlands became colonies or protectorates of France (Tunisia, Morocco, Algeria), Italy (Libya), and Britain (Egypt, Sudan, and much of the southern Arabian Peninsula and Gulf coasts after the 1869 opening of the Suez Canal). Russia took land in the north. Some of the more isolated tribes in Arabia became largely independent under their own kings or sultans. During World War I, the British and French built on Arab nationalism to expel the Turks from the areas that became Palestine, Syria, Lebanon, Jordan, Iraq, and the Arabian Peninsula.

Arab Southwest Asian countries gained independence in different ways from colonial overlords. The area remained under the disintegrating Ottoman Empire until World War I, after which the League of Nations made Syria and Lebanon French protectorates. The rest of the area was placed under British protection. The French encouraged republican governments. British protection guaranteed the survival of the kingdoms they established, including Jordan (still a monarchy today), Iraq (a monarchy until 1958), and the small emirates along the Persian Gulf under traditional local rulers, or *emirs*. In the 1970s seven emirs united as the United Arab Emirates (UAE). Animosities against former Turkish and European colonial powers that were generated by colonial decisions such as the establishment of the country of Israel remain significant factors in understanding internal tensions.

Natural Environments

Alongside culture and politics, the natural environments of Northern Africa and Southwestern Asia are basic to the region's geographic distinctiveness and internal diversity. Hot, dry plains contrast with snow-capped mountain ranges and well-watered river valleys (Figure 8.7).

Dry Climates and Desert Vegetation

Arid climatic environments, in which evaporation rates are greater than precipitation, dominate virtually the whole region (Figure 8.8). The region boasts the world's highest recorded

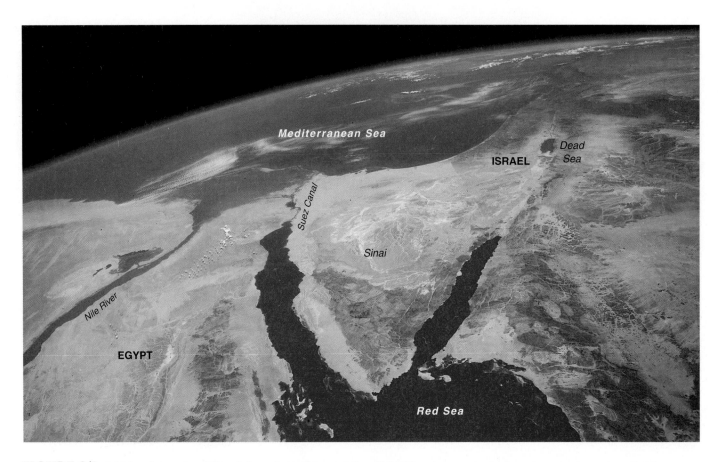

FIGURE 8.7 Arid lands: parts of Egypt, Israel, Jordan, and Saudi Arabia. A space shuttle view taken over the northern Red Sea across desert lands to the Mediterranean Sea. On the left, the delta mouth of the Nile River and the Suez Canal; in the center, the Sinai Peninsula; on the right, the Gulf of Aqaba leading northward to the Dead Sea. **Photo:** NASA.

shade temperature (58°C, or 136°F, at Al'Aziziyah, Libya), but the lack of cloud cover makes nights cool or cold, and freezes are possible in winter.

Most places have some rain, but it falls irregularly and with increasing uncertainty as aridity increases. The rainy winters of the coasts of North Africa and the eastern Mediterranean lands are caused by the seasonal southward shift of midlatitude frontal weather systems. The mountains of Turkey and Iran force air to rise, in turn causing rain and snow. This winter snowfall provides meltwater that feeds the Tigris–Euphrates River system in spring. In southern Sudan, summer rains come from the northward movement of the equatorial rainy belt.

The arid climatic environments support only drought-resistant desert plants that form a partial ground cover. Large areas of desert have little vegetation and are gravel-strewn, rocky, or sand-covered: sand seas cover one-fourth of the Sahara, which mostly has a rocky or gravel-covered surface. The vegetation cover thickens and becomes denser on uplands, where higher precipitation totals and lower rates of evaporation occur. In the uplands of North Africa, Lebanon, western

Syria, Turkey, and Iran, grassland and woodland vegetation increase with altitude as temperatures and evaporation levels fall and make the low-to-moderate precipitation more effective.

Soils are poor and undeveloped through most of the region; the best occur in the rainy coastal areas and along the valley floors of rivers, where annual floods deposit fertile alluvium. Soil erosion following plowing for grain and cultivation on steep slopes is a major problem in most of the region.

Climate change left its mark on the region and affected the history of human settlement. The present aridity is one stage in a sequence of fluctuations between more and less arid conditions. Pollen analysis, animal distributions, dried lakes, cave paintings, and written records suggest that the climate was more humid and cattle herders more widespread across northern African and Arabia a few thousand years ago. The stocks of underground water that accumulated in humid phases provide much of the water being used today, but current precipitation does not replenish them. The present aridity began some 5,000 years ago, forcing many people into the watered valleys of the Nile and Tigris–Euphrates Rivers, thus

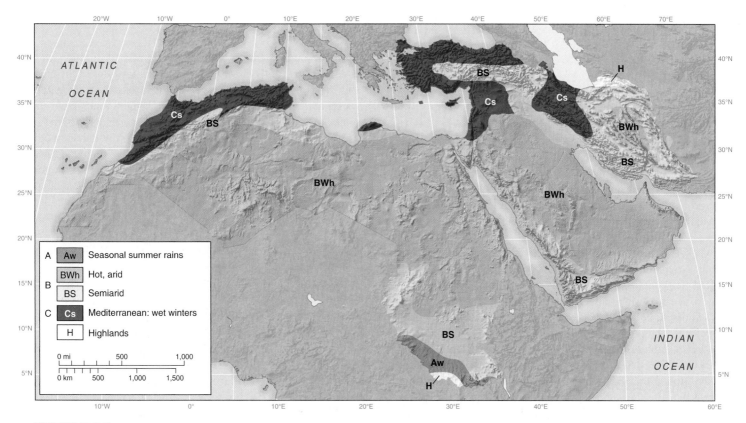

FIGURE 8.8 Northern Africa and Southwestern Asia: climate regions.

A	Aw	Seasonal summer rains
B	BWh	Hot, arid
	BS	Semiarid
C	Cs	Mediterranean: wet winters
	H	Highlands

concentrating populations. The desert margins continue to fluctuate. They retreat in wetter years and advance in drier years—or as human actions remove vegetation and lower groundwater levels.

Clashing Plates

Contrasting landscapes from high mountain ranges and extensive plateaus to wide river valleys (see Figure 8.3) stand out where an arid climate supports little vegetation. It is more common to see bare rock than grassed or wooded slopes. Such landform contrasts are related to geological activity along tectonic plate boundaries (Figure 8.9) as well as to climatic conditions.

The African and Eurasian continents and the Arabian Peninsula are the tectonic plates of this region. Such rigid sections of Earth's crust underlie the extensive plateaus of Northern Africa and Arabia. The ancient rocks are often covered by flat layers of sedimentary rocks, such as the limestones and sandstones forming the plateaus on either side of the Nile River Valley.

Collisions of the African and Arabian plates with the Eurasian plate formed the highest mountains: the Atlas in North Africa and the Taurus, Zagros, and Elburz ranges in Turkey and Iran. The Atlas Mountains rise to nearly 4,000 m (12,500 ft.) and the Elburz to over 5,500 m (17,000 ft.). Both are parts of

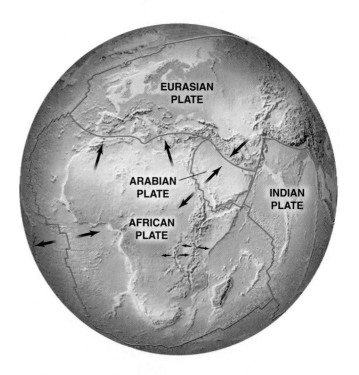

FIGURE 8.9 The physical features of Northern Africa and Southwestern Asia. The plate margins of this region coincide with mountain ranges and seas. The rift valleys of Eastern Africa (see Figure 9.11) are thought to be a divergent margin.

the young folded mountain ranges that extend from Northern Africa and Southern and Alpine Europe, through Southwestern Asia, to the Himalayas in South Asia. Rocks forming in the seas that filled the basins between the plates were caught up as the collision closed the seas and raised them as parts of the mountains. Volcanic eruptions along the geologically active belt injected mineralized fluids that solidified as veins in the rocks; earthquakes continue today. The compression folded and lifted up these rocks. Large quantities of eroded rock were deposited in downwarped sections of the crust, where oil and natural gas could be trapped. The Persian Gulf area has the world's largest concentration of these fuels, whereas the Mediterranean Sea and the Tigris–Euphrates River Valley occupy the site of a closing ocean.

The Red Sea is considered an opening ocean. Steep slopes and mountain ranges also occur along the lands bordering the Red Sea. The tectonic plates here are pulling apart to form a new ocean. Rising molten rock beneath the surface first raised the crustal rocks into a dome that cracked to form a rift valley. Volcanic activity occurred along its margins as the molten rock erupted at the surface. Then the sides pulled away from the center and the Red Sea opened. At its northern end, the rift splits in two around the Sinai Peninsula (see Figure 8.7). The eastern branch forms the valley into which the Jordan River flows from the uplands around the Sea of Galilee toward the Dead Sea some 400 m (1,312 ft.) below sea level.

Major River Valleys

Rivers are unusual in the arid climatic environment of Northern Africa and Southwestern Asia. Most flow from water sources in the surrounding mountains or from rainy areas outside the region. For example, snowmelt on the high mountain ranges of Turkey and Iran feeds the Tigris–Euphrates River system, supplying water to arid parts of Syria and Iraq. The spring floods gouge sediment from the upland areas, depositing the coarser materials in giant fans where they leave the mountains and depositing the finer sand and mud across the Syrian and Iraqi lowlands.

The Nile River system has two major headwater branches that begin in the rainy equatorial area around Lake Victoria (White Nile) and the seasonal rainy area of the Ethiopian Highlands (Blue Nile). Two major Nile River branches join near Khartoum (Figure 8.10). The White Nile River has tributaries that are fed by rains throughout the year and supply a fairly constant flow of water. On reaching southern Sudan, the White Nile River flows through the Sudd Swamp, in which it loses nearly half its water by evaporation. The Blue Nile River flow is more important because it is less subject to evaporation and produces the annual September Nile River flood in Sudan and Egypt.

A narrow zone just a few kilometers wide in a valley carved by the river into the surrounding plateaus allows cultivation and settlement as the river flows northward across Sudan and Egypt. Away from the Nile River on either side and northward of central Sudan is arid land that has little prospect of development apart from the river waters.

Natural Resources: Water

Water resources have been highly significant through this largely arid region's history. The major civilizations of the Nile and Tigris–Euphrates River Valleys were based on irrigation water, and many local developments brought water from its source to areas of use (see Personal View: Oman, page 336).

The renewable supplies that come from rain and snowfall feeding rivers are meager over most of this region. Other sources include underground reservoirs that accumulated over millennia (virtually nonrenewable) and desalination plants that make seawater usable (expensive). The demands are mainly from irrigation farming, the oil industry, and growing urban centers.

Political issues arise when the precipitation falls in one country and feeds rivers flowing through another. The Euphrates River rises in the Turkish mountains and flows through Syria and Iraq, but Turkey's planned water resource projects assume that it will use most of the water and that only modest amounts will be released downstream. Turkey, Syria, and Iraq compete for the waters of the Tigris and Euphrates. In Israel, the Jordan River system supplies many highly productive farms, while the use of groundwater stored beneath the West Bank hills is a further source of political controversy.

Although Egypt once divided the Nile River waters with its neighbor Sudan, it is now having to take account of increasing use in Ethiopia and the other headwater countries, including Uganda, Kenya, and Tanzania (see Chapter 9). By 1988 the expectation that the Aswan High Dam (completed in 1970) would ensure Egypt's water needs for 50 years was proved wrong, and discharges below Aswan had to be reduced to maintain flow through the year.

Egypt and Sudan seek to enlarge the scope of their agreement on the use of Nile River waters by involving Ethiopia—the source of 80 percent of the long-term Nile River flow—and other upper-basin countries, such as Tanzania, that wish to extract water from Lake Victoria. The outcome will cause tensions and reduce the water available to both Egypt and Sudan. Sudan seeks to develop water projects in its southern region, where easing the flow of the White Nile through the swamps would reduce evaporation.

The competition for use of Nile River waters demonstrates that political and social factors are as important as environmental provision in the economic development of natural resources. Action on such projects is delayed by political upheavals in Ethiopia, Sudan, and the upper basin. For many years, these upheavals halted development in those countries, allowing Egypt to have maximum use of Nile River water in the 1970s and 1980s. At present, international

FIGURE 8.10 The Nile River Valley: rainfall and river flow compared. The relatively small areas of irrigated land are shown. How do the climate graphs relate to the river flow at the places shown? The rainfall at Addis Ababa joins the Blue Nile at Khartoum. The White Nile flow is low but consistent.

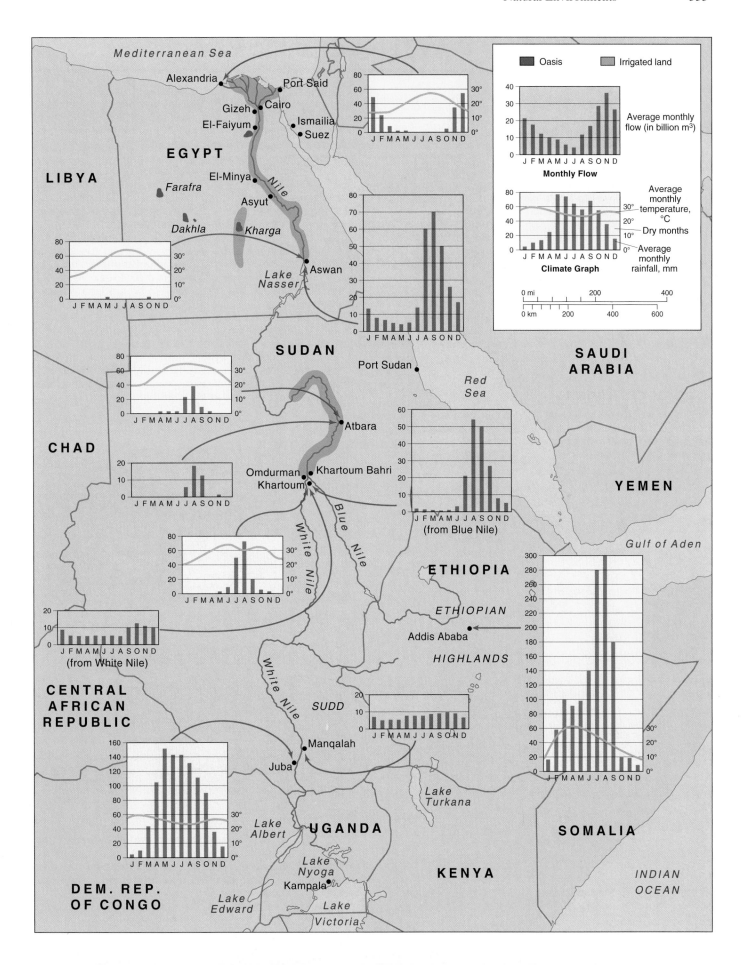

Oasis | Irrigated land

Monthly Flow
Average monthly flow (in billion m³)

Climate Graph
Average monthly temperature, °C
Dry months
Average monthly rainfall, mm

0 mi 200 400
0 km 200 400 600

Mediterranean Sea

Alexandria
Port Said
Gizeh Cairo
El-Faiyum Ismailia
Suez

EGYPT

LIBYA

Farafra

El-Minya

Asyut

Dakhla *Kharga*

Nile

Lake Nasser Aswan

SUDAN

Port Sudan

Red Sea

SAUDI ARABIA

CHAD

Atbara

Omdurman Khartoum Bahri
Khartoum

White Nile
Blue Nile

(from Blue Nile)

YEMEN

ETHIOPIA

ETHIOPIAN

Addis Ababa

HIGHLANDS

Gulf of Aden

(from White Nile)

CENTRAL AFRICAN REPUBLIC

White Nile

SUDD

Manqalah

Juba

Lake Turkana

SOMALIA

DEM. REP. OF CONGO

Lake Albert

UGANDA

Lake Nyoga
Kampala

Lake Edward

Lake Victoria

KENYA

INDIAN OCEAN

OMAN

Oman is a country at the entrance to the Persian Gulf (Box Figure 1). It produces oil and has a history of trading across the Indian Ocean, still retaining links in eastern Africa, where Zanzibar Island was owned by the sultans of Muscat, and in the island of Gwadar off the Pakistan coast. Although the 2005 population of Oman was only 2.7 million people, it is a country of contrasts, and its peoples live with both increasing global connections and continuing local voices. Traditional ways meet the modern in Oman.

Oman is a mainly hot and arid country, most of which hardly sees a cloud, let alone rain, from one year to another. After a rare rain, the normally dry valleys fill with water that may spread out over the flatter coastal areas to depths of 30 cm (1 ft.) for a short while. Lower land separates mountains in the north facing the Gulf of Oman and in the Dhofar region in the south. It is a 10 hour drive from Muscat to Salalah across desert landscapes or around 90 minutes by air. Dhofar is the exception to the unrelenting arid climate, with its August monsoon (*khareef*) that brings rains, cool air, and mists to some of Earth's hottest places.

Traditional Oman

The traditional past of Oman is reflected in the agriculture of areas near the coast, where narrow aqueducts (*falajs*), constructed centuries ago by Persians, bring water from springs in the hills, often by underground routes. The water is used to grow dates of many varieties and colors, and fruit from limes to papaya and coconuts in the far south (Dhofar). Fishing for tuna and kingfish, among others, is also part of the traditional economy. Omani food combines rice, meat, and fish in spicy dishes. The brown *halwa* dessert is made from spices, sugar, and nuts.

Until modern changes arrived, settlements were small villages, and port towns were homes to merchants and traders as well as farmers and fishers. The towns have souks (traditional markets) where foods, spices, and craft goods are sold. Standing above some villages and ports are forts, often constructed by Portuguese traders (see the banknote in Box Figure 2). Although Arabic is the national language today, many Omanis speak Swahili (former eastern African creole language of trade) or Baluchi (language of western Pakistan). The landlord of the expatriate university teacher mentioned later has Swahili as his first language (and mother and brothers who live in Zanzibar), English as his second, and Arabic as his third.

In the traditional clothing, men wear the white "dishdash" robe, sandals, and an embroidered cap (*kuma*) or turban (*masa*), or sometimes both. The women wear long dresses and head-scarves. They love bright colors but wear a black outer robe (*abaya*) in public. It is not customary for all women to cover their faces; this is not demanded by Islam but is a traditional extra

(a)

(b)

Box Figure 2 **Views of Oman provided by banknotes.**
(a) Omani banknote with the front information in Arabic and the head of Sultan Qaboos bin Said, ruler of Oman since 1970, date palms, and a *falaj* water channel. (b) Omani banknote, reverse side, in English. View of Nizwa with its souk, mosque, and fort.

Box Figure 1 **Map of Oman.**

(a)

(b)

Box Figure 3 Oman and its people. (a) Coastal mountains southeast of Muscat. (b) Thuwara spring above Nakhl, a vegetated valley where the stream flows from time to time (wadi). The valley is popular with Omanis and expatriates. **Photos:** © Gerald Selous.

practice for some Omanis and in countries such as Saudi Arabia. The Ibadi Muslim sect that most Omanis follow allows more freedom for women.

Modern Oman

Modernized Oman emerged after 1970, when the present sultan, Qaboos bin Said, deposed his father with the help of the British. He used the oil and gas wealth developed by Shell, British Petroleum, and Petroleum Development of Oman, the local oil company, in a paternalistic development of the economy that is referred to as a "renaissance" leap forward. In particular, the oil income funded new hospitals, including the Sultan Qaboos University teaching hospital, which is rated highly by the World Health Organization. The university dates from 1986 and has schools of agriculture, arts, commerce, education, engineering, medicine, and science. The growth of higher education in Oman also attracted private institutions such as the Caledonian College of Engineering linked to a Glasgow, Scotland, university. New elementary and secondary schools continue to be built in which male and female enrollments are equal; boys and girls are taught separately. The high levels of achievement by many girls mean that they have to obtain better grades than boys to enter the most prestigious higher medical education—which also has strictly equal male and female enrollments and where the sexes are taught together.

Modern technology is used wherever possible. Cell phones are common with a system of pay as you go, called *hayyak,* with no deposit or monthly bills. Computers are commonly available, and e-mail is much used. Modern shopping includes French-owned supermarket chains (Carrefours, Prisunic) and American fast food chains (McDonald's, Pizza Hut, Starbucks, Subway, and others).

Cultural tourism is a recent trend aimed at wealthy visitors from the United States, Japan, and Europe that takes advantage of good airline linkages, the sunny climate, the quaint souks, and the historic forts.

Oman has 75 percent of its population claiming a local origin, higher than most Gulf countries. Only 25 percent are foreigners, including low-wage workers from South Asia and the Philippines employed in factories, shops, gasoline stations, hotels, and gardens. They may live in outside rooms attached to larger homes or in group accommodations. One villa with four apartments housed 141 young Sri Lankan women for a year who were bused each day to work in a clothing factory. Many overseas workers come from Kerala state in southern India and are known by their language as "Malayalees." There is also some abject poverty, including female beggars, although begging is prohibited by the government because in a Muslim country, women in need should ask for *zakat* from the mosque. Some poor people live in suburbs isolated from the rest of the towns.

Newer suburbs have large white villas with walled gardens and ornate gates. A strip of blue-and-white tiles along the tops of walls is common; roofs are flat; and some windows may have stained glass. These are the homes of the wealthier Omanis, many of whom take bank loans and rent the property to expatriate engineers, teachers, geologists, diplomats, military advisers, or airline pilots to help pay off their loans.

The Student and the Expatriate University Teacher

Ahmed is a medical student, studying in what is regarded as the "cream" of Omani professional courses. He is one of seven children, and going to university in Muscat was his first time away from home. He began the medical training directly out of high school. It took some adjustment and several weekend trips home—a three-hour drive taken in relatives' cars or taxis. Public minibuses called *biaza* are cheap and stop anywhere along the main routes.

In his first year, Ahmed lived in a hostel for men. He enjoyed sports and a variety of clubs in mixed groups on campus. At Sultan Qaboos University, the men enter lecture rooms from the front and the women from the back. The men have walkways at ground level, the women on the level above. All students wear national dress, except when engaged in sports. The women at university may not wear face coverings for security and identity reasons.

The facilities for study are as good as anywhere in the world, with modern laboratories and huge libraries. Everything is free. There are no tuition fees, room and board costs, or charges for books. Ahmed now lives off the university campus and receives 120 rials (US $300) a month toward his expenses.

Fred is a U.K. national who teaches English as part of the medicine faculty team of eight with colleagues from India, Pakistan, the United States, and Canada. His team is part of a group of 150 teachers of English who provide courses for almost all the students and use 70 percent of all classroom space.

Fred lives in a rent-free villa. It is about 10 km (6.2 mi.) outside Muscat, giving him a daily 10- to-15-minute drive. He leaves his auto about a mile from his office and walks for exercise—although the early morning effort means he has to dry out in the air conditioning. He has to wear light clothing and a hat to protect himself from the strong sun. He works from 8 A.M. to 4 P.M., with an hour for lunch, from Saturday to Wednesday and has the local Thursday and Friday Muslim "weekend."

Living in Oman has many pleasant features for Fred. The traditional Omani natural and cultural environments are overlaid with the trappings of the West and with welcome local features such as cheap fuel and no taxes (Box Figure 3). There are reminders of the United Kingdom in the BP gasoline stations, British Airways flights, the embassy, and the British Council (which funds cultural links to the United Kingdom). The Protestant church thrives. The supermarkets contain more selections than those in the United Kingdom. When in the United Kingdom, the HSBC bank gives Fred an immediate readout of his Omani account in rials (1 rial = UK £ 1.80).

He also has many opportunities not found in the United Kingdom. Cinemas show not only the latest Western films but also Indian and some Arabic films. There are camping opportunities ("wadi bashing") and other outdoor activities in the cooler months (between November and March), visits to the old forts, diving, snorkeling, rock climbing, and eating out in a wide range of cheap cosmopolitan restaurants.

Omani Women: The Traditional and Modern

From the 1970s, the sultan of Oman encouraged women to be more involved in society. Today Omani women both maintain the traditional culture and, particularly the younger ones, look for jobs appropriate to highly educated people. In 1975 hardly any schools existed, and girls did not attend them. Since then, all girls and boys go to school.

Although many hard-working and intelligent women emerge with college educations, they find it difficult to get jobs in what remains a man's world. Many women become teachers or doctors, and others are now allowed to be waitresses in restaurants or nurses in hospitals, overcoming previous taboos against such work outside the home. Omani women are particularly successful in businesses and banking. There is a specific place for them on government advisory councils, where they are the ones concerned with social and family affairs. As with other aspects of this thrust to involve women, including allowing them to drive cars, the Omani approach is different from that found in Saudi Arabia and many of the Gulf countries.

Women continue traditions through their roles of raising children and providing hospitality, and in dressing and adorning themselves for special occasions. In the past, most Omanis were brought up in large families of nine or more children that kept the mothers in the home for much of their lives. Although some women became physically weak and ill and were advised by their physicians not to become pregnant, they continued to have children if their husband and extended family expected it. The extended family and wider kinship group forms the Omani "tribe," and most Omanis continue to marry within that circle. Newly married couples will often live in the same house as the man's parents, and this cements family relationships. Because young people marry within a known group, there is no arrangement of marriages. This system is considered to work well, but some educated families now question the wisdom of closely related marriages if specific physical or mental defects continue through interbreeding. Moreover, many modern families, encouraged by the government, have a maximum of three children, spaced out over a woman's childbearing years.

Women's clothing is traditional, colorful, and elegant with different emphases in specific parts of Oman. Traditional jewelry is based on older, chunky silver craftwork linked to Oman's silver-mine products, and each of many rings worn has a meaning. Today gold is more common. Henna, a copper-colored extract from leaves, is applied to feet and hands in intricate patterns for religious feast and wedding days during a preceding "henna day" time of preparation. Perfumes based on oils, as well as Western commercial products, are most important: many shops in the souks sell the oils in glass bottles. Their application plays an important part in social life, as does the burning of frankincense (*bakhour*) from trees grown around Salalah. All houses have a burner that is lit in the late afternoon before visitors come.

New visitors are welcomed with a special ceremony that brings together many aspects of Omani traditions. On entry, women and men go to separate rooms. The lady of the house brings into the women's room a large round tray on which are placed oranges, dates, and other fruits, a coffee jug and cups, and a water bowl (augmented today with a box of paper tissues), and all the guests and family gather round. They are offered fruit and newly baked bread, rinsing their hands. Coffee is poured into very small cups, which are refilled until a person shakes the cup to indicate he or she does not want any more. Then the incense burner is brought in, and each lady stands and places it under her outstretched skirt until the smoke rises upward and out through her neck and into her hair. All the company take part in this one by one. Finally, a tray of perfume bottles is passed, and guests apply a mixture before they are "free to leave" (the party is over). This complex welcome makes newcomers feel like royalty and may lead on to a continuing series of visits with others in the neighborhood. Later visits to the same household will be less formal.

The Sultan of Oman's aim is to modernize his country without losing the values inherent in the traditional ways of life. Although the country remains exotic in many ways, the people can take advantage of educational opportunities and health care. Omanis remain conscious of their antecedents and see modern life as an extension of the old. Each holiday and religious festival involves returning to their village and meeting the extended family.

funding is not available for major water projects in the upper basin to meet the rising needs there. Such needs, however, will eventually increase water extraction before the Nile River reaches Sudan. At the same time, population growth in the lower basin countries adds pressures for making more water available there.

In Saudi Arabia, most water comes from underground sources. The growth of the oil industry and urbanization in the arid areas placed greater demands on water resources, leading to increased groundwater extraction. Coastal settlements have built costly **desalination plants** to provide fresh water. A large proportion of the land is without fresh water, however, and has little potential for economic development.

Environmental Problems

The rapid population growth and attempts to raise economic output have focused attention on whether the region is becoming "overdeveloped." This region has little space to expand agriculture, relies on fluctuating oil revenues with few other potential income sources, requires foreign labor, and builds new cities in sensitive arid environments.

The limits of land and water may bring greater problems than using up the oil. Arid environments not only pose problems of water supply but also are particularly fragile in response to human activities. In this region, the availability and management of water resources are becoming political issues as the population increases and higher living standards raise the demand for water.

Irrigation farming in arid areas requires good management so that the high rates of evaporation do not draw so much salt to the surface soil that crop productivity is reduced or ended. This process is known as **salinization.** Under careful management, maintaining good drainage allows water to flush the salts downward. Adding too much water waterlogs the soil and concentrates salts at the surface. In ancient irrigation schemes, 60 percent of the land in the Tigris–Euphrates River lowlands became unusable because of poor management. Modern usage has led to loss of farmland for similar reasons.

The oil industry in the Persian Gulf area pollutes the atmosphere and waters of areas around oil wells. Leakages pollute the waters near ocean terminals. Particles in the atmosphere cause fogs that create or worsen respiratory problems. Even before the Gulf War in 1991, the Persian Gulf had lost most of its plant and animal life as a result of pollution since the first oil production began in the early 1900s. At the end of the Gulf War, the retreating Iraqis set oil wells on fire, adding carbon gases and particles to the atmosphere, but the particles fell on desert sands. They also released a huge oil slick, damaging the plant and animal life, but it had less impact than it would have if the Persian Gulf had not already been heavily polluted.

Global Environmental Politics

When measuring CO_2 emissions from fuel combustion per person, the top emitters in the world in 1997 were Qatar, Bahrain, United Arab Emirates, and Kuwait, followed by the United States in 7th place and Saudi Arabia in 10th place. Though the oil-producing countries of Northern Africa and Southwestern Asia are contributing significantly to global CO_2 emissions, none of the countries are on the Kyoto Protocol's list of Annex I countries, the key countries necessary for the implementation of the protocol. Nevertheless, countries of the region have shown their support for the Kyoto Protocol. Morocco was first country of the region to ratify the protocol (47th country overall to ratify), followed by Jordan (103rd), Israel (121st), and Yemen (125th). After Russia ratified the protocol in November 2004, thus satisfying the minimal requirements for the protocol to go into effect, Qatar, Egypt, Oman, United Arab Emirates, and Saudi Arabia ratified the Kyoto Protocol soon afterward in January 2005.

Test Your Understanding
8A

Summary

Northern Africa and Southwestern Asia have historical significance as the hearth of human civilization. Great empires arose in the region, but from the 1800s, Europeans exerted their control over the region. Judaism and Christianity also emerged in the region. However, the Arab language and Islamic religion are important to most countries today, although Turkey and Iran have their own languages. Israel stands apart from the rest in both language and religion.

An arid environment is common in these countries, influencing where people live, and making fresh water a precious commodity and a source of conflict. The region's oil, which is burned as fuel, contributes to global warming. Nevertheless, many of the region's countries support the Kyoto Protocol.

Questions to Think About

8A.1 What are the similarities and differences of the region's religions?

8A.2 What are the similarities and differences of the region's languages?

8A.3 What empires have controlled the region and what have been their influences?

8A.4 What are the sources of fresh water, and where is water particularly scarce?

8A.5 Which countries support the Kyoto Protocol and which do not? Is there a pattern to support and opposition?

Key Terms

Judaism	Sunni Muslims	Farsi (Persian)
Christianity	Shia Muslims	Kurdish
Islam	Arabic	Turkish
Muslims	Berber	desalination plant
Qu'ran	Hebrew	salinization

Global Significance

Following a medieval history of trading connections across Asia, eastern Africa, and the Indian Ocean, Europeans monopolized the region's extension of trade and political dominance from the 1800s. The discovery of oil and the founding of new countries after World War I brought the region into world prominence again. Its significance affected World War II strategies and grew with a huge postwar rise in world demand for oil. Wars between the Arab countries and Israel, and the Cold War rivalry between the United States and the Soviet Union, added to that significance. The 1973 war between Israel and the surrounding countries triggered massive increases in the price of oil, and the oil-producing countries in this region built huge financial surpluses in the 1970s. Later declines in oil prices brought problems of indebtedness, but many of the region's countries remain wealthier than other formerly poor countries in the world (Figure 8.11). Ownership of consumer goods, access to good water, and energy usage (Figure 8.12) reflect a range of material wealth across the region. Istanbul, Turkey, has the greatest claim to being a global city–region, but Tel Aviv–Jaffa (Israel), Cairo (Egypt), Riyadh (Saudi Arabia), and some of the Gulf cities such as Abu Dhabi have growing business services with global connections.

Pan-Arabism

In the mid- and late 1900s, postindependence movements, generally opposed to Western economic colonialism, sought to unite Arab peoples on the basis of (mostly secular) nationalism

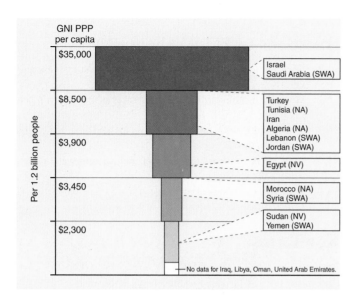

FIGURE 8.11 **Northern Africa and Southwestern Asia: country incomes compared.** The countries are listed in the order of their GNI PPP per capita. **Source:** Data (for 2000) *World Development Indicators,* World Bank, 2002.

into a single country with increased world influence. None succeeded in uniting the Arab world, though their efforts are noteworthy.

The **Arab League** was created in 1945 to encourage the united opposition of Arab countries to the establishment of Israel. Its seven founding members were the only independent Arab countries at the time, but its membership increased to 21 as more gained independence. The members of the Arab League eventually included the **Palestine Liberation Organization (PLO),** a political organization providing an umbrella for many smaller groups that demand a country for Palestinians, the people living in the lands used to create the state of Israel. The Arab League linked Arab countries in the western Mediterranean with those in the eastern Mediterranean and southwestern Asia. After Egypt's 1979 accord with Israel, the Arab League expelled Egypt and transferred the Arab League headquarters from Cairo to Tunis. From 1958 to 1961, Egypt and Syria joined as the United Arab Republic under the leadership of Colonel Jamal Nasser with the intent of persuading other countries to commit their futures to a single **Pan-Arab country.** It did not attract others and soon broke up.

Disunity, military defeats, and tensions over the Israel–Palestine issue and among countries with different resource bases weakened the Arab League and its ability to foster unity among Arab countries in the 1980s and 1990s. A major blow to the Arab League came in the Gulf War of 1990–1991, when one Arab country (Iraq) invaded another (Kuwait) and was defeated by a coalition of other Arab countries backed by the United States and other Western countries. Many Arabs felt betrayed by both Saddam Hussein's invasion of another Arab

country and their own need to rely on outside help. It was clear that the interests of individual countries remained more significant than an overriding Pan-Arabism, feelings that were repeated in later Arab economic summits.

Islamism

In the 1970s and 1980s, as individual countries preferred their independence to a Pan-Arab identity, pressure increased to base political actions on religious affiliations. The influence of extremist Islamist groups grew in opposition to the global political and economic order.

Political Islamism: Organization of the Islamic Conference

As the moves toward the unification of Arab countries died away, links through the Islamic religion began to replace them. In 1970 foreign ministers of Muslim countries set up the **Organization of the Islamic Conference (OIC),** which now has 45 members, including such countries as Pakistan, Indonesia, and even Nigeria. However, it is more successful in advancing individual member countries' interests than in defining and pursuing a common agenda to rival Western-dominated globalization. One of the OIC's most important affiliates is the Islamic Development Bank, which is dedicated to economic development among OIC members.

Although closer economic integration was encouraged from the late 1970s, many Muslim countries carry out most of their trade with the materially wealthier countries in Europe, the United States, and Japan. For example, the December 1997 meeting of the OIC in Tehran (Iran) was unusually well attended by all members but focused on pragmatic world trading issues rather than on Islamic religious policies. It was chaired by that country's new moderate president, who spoke against plans for Muslims to unite against Israel and the West, urging the conference to avoid rejection of the West and calling for a "dialogue of civilizations."

Attempts at Politicizing Islamism

In the 1970s Islamic political groups came to the fore, basing their ideology on the Qu'ran, interpretations of *jihad* (applied widely by Muslims to many differing causes) as holy war, and references to past Islamic triumphs. They call for the reestablishment of an Islamic country, rejecting the traditional view that Islam relegated political combat to a secondary concern. They hate the fragmentation of Islamic lands into separate countries, in which the religious establishment is subservient to those educated in Westernized ways. They liken the situation to the ignorance and barbarism among Arabs before the early 600s, when Mohammad brought the worship of Allah and new purpose to Arabs.

The main modern flowering of political Islam occurred after the 1967 and 1973 Israeli–Arab wars and the OPEC price increases that brought huge surpluses of dollars to the region.

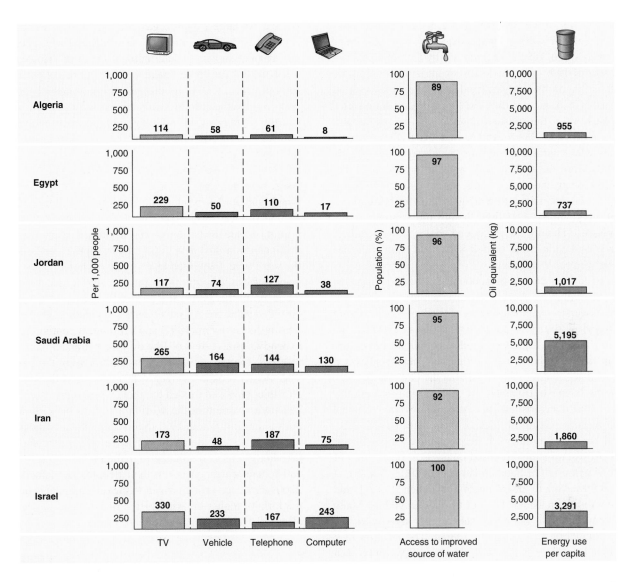

FIGURE 8.12 **Northern Africa and Southwestern Asia: ownership of consumer goods, access to good water, and energy usage.** Compare these levels with those of other countries in Africa. **Source:** Data (for 2002) *World Development Indicators,* World Bank, 2004.

Its main success was the 1979 Islamic revolution in Iran under Khomeini, which managed to gain support from radical young people in the cities; the more conservative, God-fearing middle classes that valued a strict religious and social order; and professionals who had become wealthy from oil income but had no political voice. Islamic law (*sharia*) became official in Iran in 1983, with over 100 offenses carrying the death penalty.

Although the success in Iran stimulated Muslims in other countries to consider similar action, most Muslim countries, including Libya, Syria, and many small Gulf countries, stifled such ideas and imprisoned religious activists. Through the 1980s, Iran and Saudi Arabia struggled for dominance. As Iran tried to export revolution, the Saudi princes wished to maintain the system by which they had grown rich. When Saddam Hussein of Iraq declared war on Iran in 1981, he had support from the Saudis, other Gulf countries, and

Western countries. Iran responded with terrorist acts, such as the taking of Western hostages in Lebanon and disrupting pilgrimages to Mecca. In 1989, when an armistice was reached, Khomeini declared a *fatwa* (legal ruling) against a British citizen, Salman Rushdie, the author of *Satanic Verses,* a novel that criticized Islam. This action assumed that Islamic law was valid in other countries.

The 1979 invasion of Afghanistan by the Soviet Army generated another jihad financed by the Persian Gulf and Western countries. This unified Islamists around the world, taking over from the Palestinian cause for some years. International brigades came from Egypt, Algeria, the Arabian Peninsula countries, Pakistan, and Southeast Asia and were trained in guerrilla warfare as part of the Islamist armed struggle. In 1989 the Soviet Army withdrew from Afghanistan, causing Islamists to declare a great victory and gain immense self-confidence.

The 1989 fall of the Berlin Wall that signified the beginning of the end for world Communism gave Islamists hope that their way might fill the vacuum left by Communism, but the Gulf War of 1990–1991 and subsequent failures to impose their military solutions in Algeria and Egypt were major blows for Islamists. In particular, the Gulf War demonstrated how Saddam Hussein of Iraq miscalculated possible Arab support in threats to invade Saudi Arabia and expel the "infidel" Americans who guarded the soil on which so many Islamic holy places were built. That war and the cessation of the Afghanistan jihad ended the Saudi Arabian–led consensus, and a radical Islamic fringe turned against the Arab kingdoms and their international networks, losing the support of the middle classes. Having defeated one major power, radical Islamists turned against the United States and Arab countries with rulers who repressed them.

Islamist failures caused most countries that had supported guerrillas to withdraw from extremist actions. In Iran, for example, moderate candidates were elected president during the 1990s and in 2001 (but not in 2005). Turkey and Pakistan turned from Islamist leaders to moderate military or secular leaders. In Jordan, radicals lost their parliamentary seats when Jordan began negotiating with Israel after the Gulf War.

Extreme Islamists frustrated by political impasse continued and expanded terrorist actions. Attention reverted to Israel and Arab resentments at the lack of movement by the U.S.-supported Israeli government to accommodate the wishes of Palestinians. Attention also focused on the U.S. occupation of Iraq following the fall of Saddam Hussein in 2003. But terrorism, despite resorting to the events of September 11, 2001, and suicide bombings in Israel, has yet to yield anything. Many Muslim countries held memorial services for the victims of 9/11, including Iran in its capital, Tehran. The election of a conservative hard-liner as Iran's president in 2005 can be seen as a recent victory for Islamists but the victory resulted more from voter dissatisfaction with corruption allowed by reformers, and reformers not having redistributed the wealth to poorer segments of society as they had promised. Indeed, the new conservative president redistributed wealth to the poor when he was mayor of Tehran.

Oil Resources

The huge oil and gas production from Saudi Arabia and the Gulf states amounted to 22 percent of world oil production from the 1960s to the early 2000s. Despite such extraction and the addition of new producers in Latin America, Asia, and Africa, the proportion of total known world reserves located around the Persian Gulf increased (Figure 8.13a, b and Figure 8.14). Continued discoveries increased reserves from 61 (1965) to 67.8 (2003) percent of world total reserves despite increased usage (from 215 billion to 560 billion barrels). These figures do not include Iran, Libya, Egypt, or Algeria, which are also major oil and natural gas producers in this region. Algeria, for example, has the fifth largest natural gas reserves in the world and is the second largest natural gas exporter. Libya was politically shunned by most countries in the 1980s and 1990s because its government supported international terrorists. However, after 2004 the Libyan government became more cooperative, and foreign investment in its oil industry subsequently increased. New discoveries of oil fields in the Caspian Sea area (see Chapter 4) may rival the Gulf in output, and Iran and Turkey may try to attract oil pipeline traffic across their territories from this new source.

Not all the countries of Northern Africa and Southwestern Asia, however, are major oil producers. Morocco, Turkey, Israel, and Jordan produce no oil. Tunisia, Sudan, Syria, and Yemen produce and export modest amounts. Countries that import oil face the burden of purchasing oil, whatever its price, and many went into debt during the high prices in the 1970s. At the same time, they often received an economic boost from the wages of workers who filled the major oil producers' need for labor. From the 1970s, the oil-rich countries of the Gulf also provided aid to other Arab countries and causes. However, paying for the Gulf War (1991) and subsequent continuing military expenditures at a time of low oil prices drained their financial reserves, forcing cutbacks in expenditures by the mid-1990s. The poorer countries received fewer donations from the wealthier ones and missed the money sent home from their citizens who worked in the oil-rich countries but returned home during the Gulf War in 1991. Such cutbacks caused their inhabitants to question the quality of leadership, particularly where hereditary princes continued to maintain extravagant lifestyles and remain unresponsive to their people. The late 1990s and 2000s, however, brought relief to the oil-producing countries in the form of higher oil prices. High oil revenues, however, may also stall economic and political reform.

Organization of Petroleum Exporting Countries

From the early 1900s, international oil companies kept oil prices low to consumers in the world's wealthiest countries by paying little to the producing countries. In 1960 the producers around the Persian Gulf, together with Venezuela, formed the **Organization of Petroleum Exporting Countries (OPEC).** Today 76 percent of OPEC oil reserves still are in Arab countries. OPEC's main purpose is to work as a **cartel**—an organization that coordinates the interests of producing countries by regulating oil prices.

The successful oil embargo during the Arab–Israeli War in 1973 established OPEC for a few years as the controlling factor in world oil distribution. Prices rose fourfold. Revenues of the oil-producing countries boomed in the mid- and late 1970s. Countries such as Saudi Arabia, Kuwait, United Arab Emirates, and Iraq gained over 90 percent of their export earnings from oil sales. They invested their money locally in new roads, hospitals, government buildings, airports, and military hardware, as well as in projects and banks outside the region.

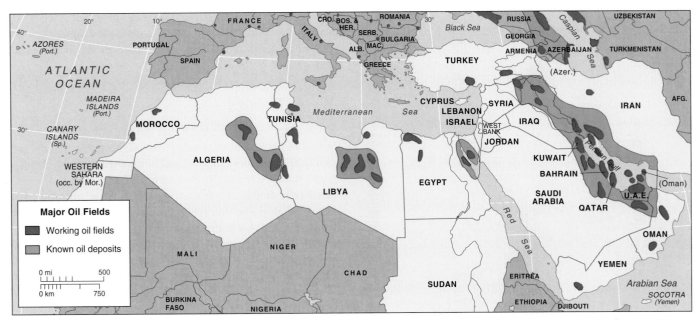

(a)

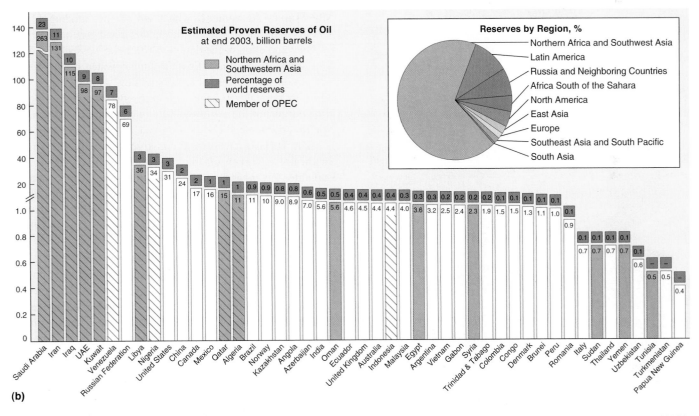

(b)

FIGURE 8.13 **Oil resources in Northern Africa and Southwest Asia.** (a) Map of major oil fields. (b) World oil resources at the end of 2003, showing the continuing significance of this region. The Persian Gulf share, 67.8 percent of world reserves, is a little greater than 10 years ago, and reserves in this region are being used more slowly. Note which countries in the region have major oil reserves and which have little or no oil. **Source:** (a) *The Economist* (b) British Petroleum.

The world's wealthier countries that used most of the oil were affected first by higher oil prices and then economic recession. It was soon clear, however, that the high oil prices and recession also adversely affected the world's poorer countries. Western oil companies opened new oil fields outside the OPEC area, including those in the North Sea (Europe) that had previously been too expensive to develop. Beginning in the 1990s, Russia exported increasing quantities

interests lay in a moderately high, but stable, oil price. War, other crises, and increased demand from growing economies like China pushed oil prices to record highs in the early 2000s.

Water Politics

Much attention is given to oil, but the issue of water is crucial in this arid region of the world. Like oil, however, water is not evenly distributed. Eighty percent of the region's fresh water is found in the Nile and Tigris–Euphrates River basins. The Jordan River, though much smaller, is crucial to countries that depend on it. However, each of these rivers flows through more than one country (see Figures 8.3 and 8.10) and requires cooperation between governments, creating and sometimes exacerbating political conflict where agreements cannot be reached. Conflict may increase as fresh water scarcity is becoming more of an issue as populations grow rapidly. Between 1975 and 2001, the amount of fresh water available to each individual dropped by more than half.

In 1959 the Egyptian and Sudanese governments signed the **Nile Waters Agreement,** which led to the building of the Aswan High Dam to store sufficient water and generate electricity. On completion of the dam in 1970, Lake Nasser behind the dam stored three times Egypt's annual water usage; Sudan receives only 13 percent of the annual flow. Along with disputes they may continue to have with each other, Egypt and Sudan also now contest the use of the water with countries in the upper Nile River watershed such as Ethiopia, Uganda, and Tanzania.

The Turkish government constructed a series of dams in Southeastern Turkey on the upper reaches of the Tigris and Euphrates Rivers as a part of a greater plan to irrigate lands to increase agricultural production (Figure 8.15). Because these irrigation projects divert water from Iraq and Syria downstream, tensions have increased between Turkey and these two countries.

To provide fresh water, the wealthier countries of the Arabian peninsula built desalinization plants to turn seawater into drinking water, but such efforts are extremely expensive. Around 60 percent of the world's desalinization capacity is in the Arabian peninsula, with 30 percent of the world's total in Saudi Arabia alone.

Israel, Gaza, and the West Bank occupy a small area of land between the Mediterranean coast and the Jordan River valley. The terrain includes a coastal plain and mostly hilly land with lower areas around Lake Tiberias (Sea of Galilee) and along the Jordan River to the Dead Sea. Although Israel receives winter rain, the total rainfall is low, and summer drought brings water shortages. Great efforts made it possible to supply water to dry areas and to manage the environment efficiently. Despite such careful management, internal groundwater sources are now fully used, and Israel relies on external sources for 25 percent of its water. Brackish (slightly saline) water is used for some farming but can damage the soils if too much is used over many seasons. Urban water supply relies increasingly on coastal desalination plants.

FIGURE 8.14 Oil: Iraq. Shaiba oil refinery, 20 kilometers south of Basra. **Photo:** © AFP/Getty Images.

of oil and natural gas as its main source of currency. A world oil glut brought very low market prices and financial problems to the Arab producers. By the late 1990s, the OPEC oil producers and major industrial countries saw that their

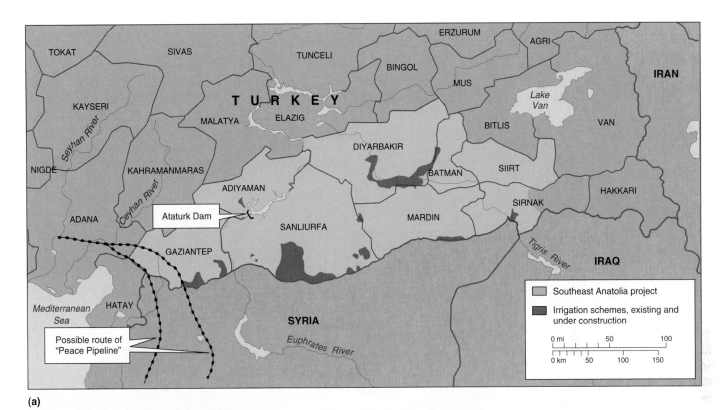

(a)

(b)

FIGURE 8.15 Turkey: southeast Anatolian water project. (a) Storage of water behind huge dams, such as the Ataturk Dam, makes it possible to irrigate large new areas of land and provide hydroelectricity for local industry. Such uses of water, however, deprive Syria and Iraq of water they have been using from the Tigris and Euphrates Rivers. The "Peace Pipeline" was part of a project to supply water to Israel and its neighbors but was not implemented. (b) The Ataturk Dam on the Upper Euphrates River as it neared completion in 1992. **Source:** (a) Southeast Anatolian Water Project from *The Economist,* December 14, 1991. Copyright © The Economist Newspaper Group, Inc. Reprinted with permission. Further reproduction prohibited. www.economist.com. **Photo:** © Ed Kashi.

The water shortage enters politics. Israel retains jurisdiction over the West Bank and the Golan Heights both for defense and access to water. Peace negotiations with Jordan included agreement on the use of the Jordan River basin waters, which have become central in the negotiations over the future of the West Bank.

Agriculture

The arid lands of Northern Africa and Southwestern Asia leave little opportunity for farming, requiring most countries to import food. Most of the arable land extends from the shores of the Mediterranean (Figure 8.16) and Red Seas and the Persian

FIGURE 8.16 Northern Africa and Southwestern Asia: rural landscapes. Bedouin women taking goods to market near Foudouk al Aouerab, Tunisia. The cultivated valley behind them and the bare hillsides above are typical of North Africa and much of the wider region. **Photo:** © Kess van der Berg/Photo Researchers, Inc.

Gulf. The Nile and Tigris–Euphrates Rivers provide irrigation. Interior deserts have given rise to nomadic herding. Some agriculturally productive areas are on the fringes of the region such as in Turkey (Figure 8.17). Indeed, the building of dams in southeastern Turkey has enabled Turkey to increase its production of cotton, soybeans, grains, fruits, and vegetables.

Many of the problems in agricultural production stem from the type of economy established by colonial countries. The colonial system involved land appropriation for European settlers and local elites, who used irrigation water for intensive commercial farming tied to markets in Europe. Consequently, agriculture is geared in many areas to export crops such as citrus fruits, olive oil, and cotton. Iran is still known for its export of nuts, especially pistachios. In recent decades, governments have encouraged greater cultivation of domestically needed crops such as cereal grains, vegetables, and dairy and poultry products. Productivity has increased with the introduction of mechanization, the application of fertilizers, and double cropping, which allows some areas

to harvest two crops within a year. As noted in the "Environmental Problems" section, improper irrigation techniques in places leads to continual problems of salinization.

Population Distribution and Dynamics

The population of Northern African and Southwestern Asia tends to be concentrated near the Mediterranean and Red Seas, the Persian Gulf, and in the fertile valleys of the Nile and the Tigris–Euphrates Rivers (Figure 8.18). Moderate densities are also found in the rest of Turkey and western Iran. Much of the region is almost empty of people.

The populations of all the region's countries continue to grow rapidly, with the population increasing by more than one-third over the next 20 years (Table 8.2). High birth rates combined with falling death rates result in the rapid population growth (Figure 8.19). As the population pyramid in Figure 8.20a indicates, the large groups of young people will produce

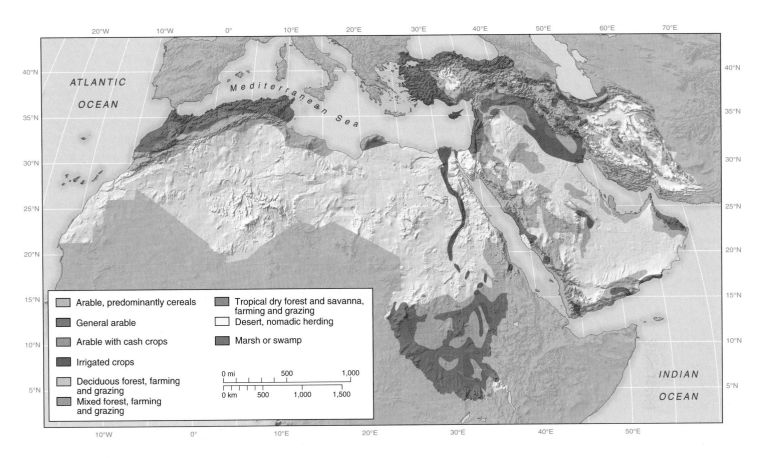

FIGURE 8.17 Northern Africa and Southwestern Asia: agricultural land use. Note the extent of desert and other dry-farming uses. Data from *New Oxford School Atlas,* Oxford University Press, U.K.

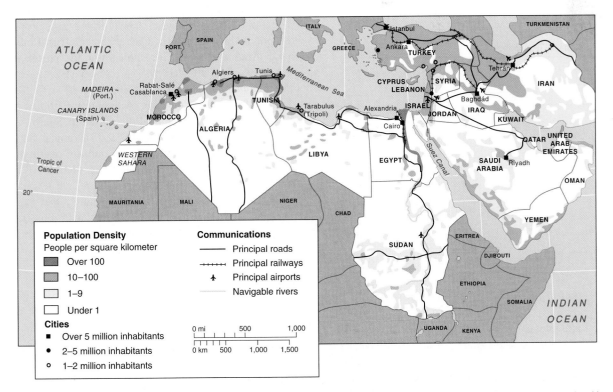

FIGURE 8.18 Northern Africa and Southwestern Asia: population distribution. Note where the highest and lowest densities of population occur in each subregion. **Source:** Data from *New Oxford School Atlas.* Oxford University Press, U.K.

Table 8.2 North Africa and Southwestern Asia: Major Cities and Their Populations

	Population	
North Africa	2003 (in millions)	2015 (estimated in millions)
Algiers, Algeria	3.1	4.2
Casablanca, Morocco	3.6	4.6
Tripoli, Libya	2.0	2.5
Tunis, Tunisia	2.0	2.4
Rabat, Morocco	1.8	2.3
Nile River Valley		
Cairo, Egypt	10.8	13.1
Alexandria, Egypt	3.7	4.5
Khartoum, Sudan	4.3	5.6
Arab Southwest Asia		
Baghdad, Iraq	5.6	7.4
Amman, Jordan	1.2	1.5
Beirut, Lebanon	1.8	2.2
Riyadh, Saudi Arabia	5.1	7.2
Damascus, Syria	2.2	2.8
Sana'a, Yemen	1.5	2.7
Basra, Iraq	1.1	1.4
Mosul, Iraq	1.2	1.6
Jeddah (Jidda), Saudi Arabia	3.6	4.9
Mecca, Saudi Arabia	1.4	2.0
Aleppo (Halab), Syria	2.4	3.1
Kuwait City, Kuwait	1.2	1.4
Israel and Palestinian Territories		
Tel Aviv–Jaffa	2.9	3.5
Iran and Turkey		
Tehran, Iran	7.2	8.5
Meshed (Mashhad), Iran	2.1	2.5
Isfahan (Esfahan), Iran	1.5	1.9
Tabriz, Iran	1.3	1.7
Istanbul, Turkey	9.4	11.3
Ankara, Turkey	3.4	4.2
Izmir, Turkey	2.4	3.0
Adana, Turkey	1.2	1.5
Bursa, Turkey	1.3	1.8
Karaj, Iran	1.2	1.6
Shiraz, Iran	1.2	1.5

Source: U.N. "Urban Agglomerations (2003)" (2003).

continuing high rates of population increase as they reach childbearing age. Oman, Saudi Arabia, and Yemen have some of the world's highest total fertility rates with over 6.5 children per woman. The Palestinians living in Gaza have total fertility rates around 6 and an annual natural increase of 3.7 percent. In most of the smaller and richer Gulf countries—Bahrain,

Kuwait, Qatar, and UAE—2001 fertility was under 4. When the Iranian religious leaders gained power in 1979, they encouraged more births but later reversed the policy. They now find that the preponderance of younger adults is a major source of opposition, both politically and culturally.

In recent years, many of the region's countries have implemented policies to slow birth rates. For example, Tunisia set a minimum age for marriage and instituted a successful family planning program soon after it obtained independence. Morocco and Algeria implemented family planning and maternal and child services. Egypt's support for family planning resulted in a drop of that country's total fertility rate from over 5 to 3.5 and to under 3 in cities between the 1970s and 2001.

Variations in immigration make Israel's population growth irregular. The annual growth rate fell from 2.8 percent in the period 1965 to 1980 to 1.8 percent in the 1980s but rose again in the mid-1990s to 3.3 percent as a result of immigration. In the early 1990s, over 1 million Russian Jews moved to Israel. In 2001 Israeli natural increase was 1.6 percent. The 2001 resident Israeli population of 6.4 million included 18 percent non-Jews, most of whom were Arabs. The fluctuating number of immigrants and the higher proportion of older people means that Israel's population structure differs from that of the region's Arab countries (Figure 8.20b).

Immigration, nevertheless, plays a significant role in many countries. The booming oil business requires more labor than many of the oil-producing countries possess. Much needed imported labor comes from other parts of the region (such as Jordan, Lebanon, and Yemen) and from South and East Asia. Foreign labor made up nearly 30 percent of the total labor force in the mid-1980s. Immigrant labor formed large proportions of the population in Kuwait, Qatar, United Arab Emirates, and Saudi Arabia. Figure 8.20c shows how the presence of male foreign workers affected the age and sex structure of Kuwait. The reconstruction of Kuwait after the 1991 Gulf War also depended on immigrant labor. Although the origins of foreign labor shifted from other Arab countries to South Asia in the 1990s, the total remains high. In 1995, estimates put the number of migrant workers in the Gulf countries at around 1.6 million, over 90 percent coming from India, Pakistan, Bangladesh, and Egypt.

Most of the region's countries have life expectancies around 70 years. Israel has the highest with almost 80 years. Life expectancy in Yemen remains lower at 60 years. In Iraq it fell below 60 after years of war and deprivation.

Urban Patterns

Urban Population Growth

The natural environment and recent nature of economic growth in a globalizing world led to a dominance of urban living environments in the region. Today most countries in the region are very highly urbanized (Figure 8.21). The high level reflects the direct shift from nomadism to urbanism in countries that lack farmland, changes brought by the oil industry in numerous countries, and

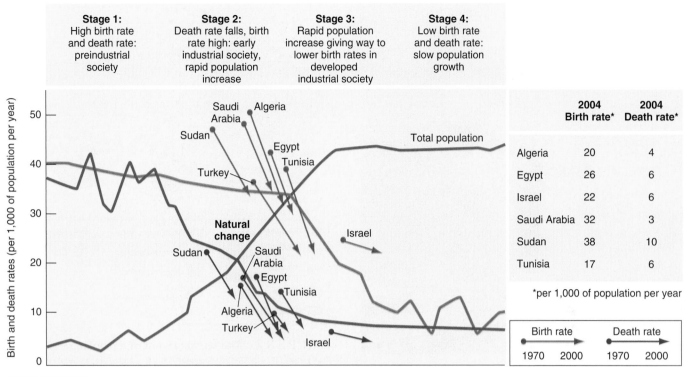

FIGURE 8.19 **Northern Africa and Southwestern Asia: demographic transition.** Birth and death rates in selected countries. Compare the experiences of these countries with those in other parts of Africa. **Source:** *2004 Population Data Sheet,* Population Reference Bureau.

	2004 Birth rate*	2004 Death rate*
Algeria	20	4
Egypt	26	6
Israel	22	6
Saudi Arabia	32	3
Sudan	38	10
Tunisia	17	6

*per 1,000 of population per year

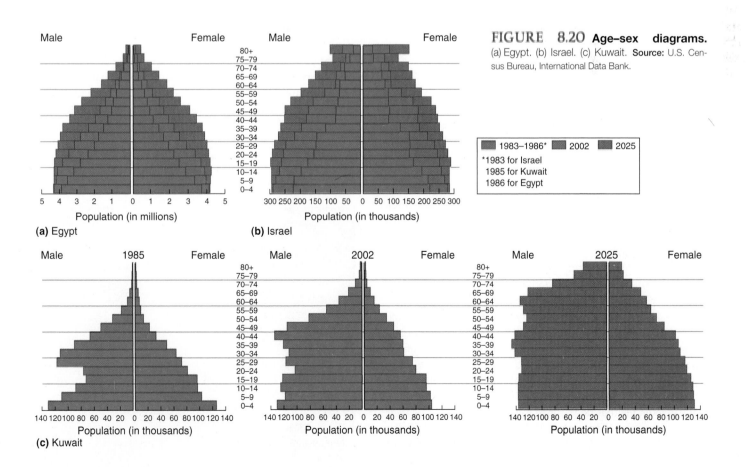

FIGURE 8.20 **Age–sex diagrams.** (a) Egypt. (b) Israel. (c) Kuwait. **Source:** U.S. Census Bureau, International Data Bank.

1983–1986* 2002 2025
*1983 for Israel
1985 for Kuwait
1986 for Egypt

(a) Egypt

(b) Israel

(c) Kuwait

FIGURE 8.21 **Arab Southwest Asia: Dubai.** New buildings along Sheikh Zayed Road. **Photo:** © Alasdair Drysdale.

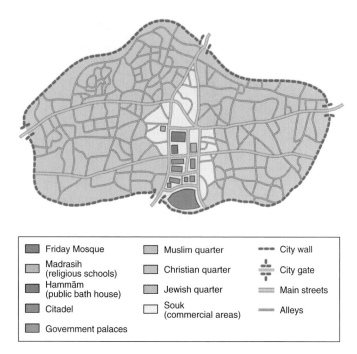

Friday Mosque	Muslim quarter	City wall
Madrasih (religious schools)	Christian quarter	City gate
Hammām (public bath house)	Jewish quarter	Main streets
Citadel	Souk (commercial areas)	Alleys
Government palaces		

FIGURE 8.22 **The typical layout of a medina.** **Source:** Graph from *Iranian Cities* by Masoud Kheirabadi, figure 1, page 7. Reprinted by permission of the author.

the development of urbanized manufacturing and government employment. For example, between 1950 and 2001, Kuwait and Qatar went from 50 percent to over 90 percent urban, while Saudi Arabia went from 10 to 83 percent urban. The rate of urban expansion means that many cities are dominated by new buildings. Amman, Jordan, lost most of its historic small-town character as it grew to be a capital city with over a million people; land speculation, combined with lack of planning, replaced the old town with wide avenues and featureless blocks. In Kuwait City after the destruction of the 1990–1991 Gulf War, only the ruler's palace remained of the old town.

As the populations grew and crowded into cities, the expanding demand for housing by poorer people was often more than governments could meet. Shantytowns are features of cities across the subregion, known as *bidonvilles* in Casablanca, Morocco, and as *gourbivilles* in Tunis.

The Fertile Crescent, which runs along the Tigris–Euphrates Rivers, along the eastern shore of the Mediterranean, and up the Nile River Valley, is the location of some of the world's oldest and most historically significant cities (Table 8.2). These older cities, which were central to agricultural and trading economies, survive as enclaves within newly expanded cities of the early 2000s. Some are much changed by the clearing of crowded buildings for new highways. High densities of homes, commercial premises, and public buildings inside city walls marked the traditional small towns of the region and their central **medinas** (Figure 8.22). Medinas, named after the sacred Muslim city in Saudi Arabia, are historic sectors of

cities, valued for their distinctive structure and social fabric. Their labyrinthine alleys, *souks* (commercial areas) (Figure 8.23), and artisan shops relate them to the past and attract tourists. Within these sectors existed a rigid pattern of land use; prestigious craft workers, such as religious artisans, had premises close to a central mosque, castle, and square; those in lower occupations, such as leather workers, existed on the edges. The residential population tended to concentrate in

FIGURE 8.23 **Urban patterns.** Street scene in the souks of Medina, Marrakech (Marrakesh), Morocco, North Africa. **Photo:** © Lee Frost. Robert Harding World Imagery/Getty Images.

"quarters," usually based on religion so that separate Muslim, Jewish, and Christian quarters were common. The World Heritage list of historic cities contains many medinas across the whole region, from Fez and Marrakech in the west to the tall, brown mudhouses of Yemen in Southwest Asia. Today city walls remain only in cities where tourist interests are economically important, such as Fez and Marrakech (Morocco).

As traditional trading and craft industries gave way to manufacturing and service industries, urban landscapes changed with the construction of prestigious government offices in capital cities, high-rise apartments, hotels, and offices just outside the old centers. Beyond are extensive suburban spreads of housing, factories, and shopping facilities. Many older town areas were abandoned to poorer people when merchants and businesspeople moved to more spacious homes and commercial premises in the new suburbs. The factories and supermarkets changed patterns of working and social movements. Segregation by wealth and social group is common within the cities, especially in oil-rich Arab Southwest Asia, where Arabs often live apart from immigrant workers.

Historically, Fez, Cairo, and Constantinople (today Istanbul) were three important global cities and ranked as the third, fourth, and fifth largest cities in the world as recently as the A.D. 1200s. Though having fallen in global significance, Istanbul (Turkey) and Cairo (Egypt) are the two largest cities in the region today and have many global city–region characteristics (Figure 8.24). Abu Dhabi, Dubai (both in UAE), and Riyadh are considerably smaller but have many global connections. On the other hand, the isolation of Iran from much of the global economy leaves Tehran with few global city features, though it is the third largest city in the region.

Human Rights

General Concerns

Gender inequalities are a continuing issue in Muslim countries. Male attitudes and the law work against equal opportunities for women. As with other aspects of Muslim culture, local interpretations and practices vary. Considerable debate exists in Muslim countries over what the Qu'ran says on these matters. Saudi Arabia boycotted the 1995 fourth World Conference on Women in Beijing on the grounds it was anti-Islamic. In Iran, it is a criminal offense for a woman not to wear a headscarf. And yet Muslim-dominated countries such as Turkey, Pakistan, Bangladesh, and Indonesia have elected women leaders in recent years. While Saudi Arabia excludes women from all public activity, Tunisia promoted women's rights after its independence in 1957.

FIGURE 8.24 Nile River valley: Cairo, Egypt. Cairo, on the Nile River, is a modernizing city—the political and economic center of Egypt. **Photo:** © Vol. 145/Corbis.

Literacy and education are a measure of human rights and women's empowerment. In general, a great discrepancy exists in the region between female and male literacy. For example, only 25 percent of the women in Yemen are literate, while 67 percent of the men are. However, literacy rates are not uniform throughout the region. In Qatar the literacy rates are 83 percent and 80 percent, respectively. Overall literacy rates have improved dramatically for women in the region over the last 30 years (Figure 8.25).

Basic human rights for all citizens are a concern in some countries. For example, although no countries have abolished the death penalty, countries like Turkey use it only in exceptional circumstances, and countries like Algeria and Tunisia have not imposed the death penalty in more than 10 years, though it is still legal. However, all the other countries retain the death penalty. According to Amnesty International, Iran ranks number two in the world for number of its citizens that it executes, behind China and just ahead of the United States and Vietnam.

A wider variety of human rights abuses result from specific conflicts within countries. The ones that have created the greatest number of deaths and refugees are centered on the conflicts between Israelis and Palestinians (see the section on Israel and the Palestinian territories on page 367), the Civil War in Sudan, and repression involving the Kurds, who are found in Turkey, Syria, Iraq, and Iran.

Sudan

Not long after Sudan became an independent country in 1960, internal conflict ensued within the country, causing a series of numerous civil wars well into the 2000s. Much of the conflict has been between the Arab Muslim north and the black Christian and animistic south. The warfare created famine and misery in this drought-prone country. As many as 2 million people died; many others were displaced and left homeless. Arab Muslims have controlled the government in recent times. However, years of war weakened the country's economy and made it difficult for the government to assert effective control over the country. The Sudanese government relies on undisciplined troops and paramilitary groups to fight those who oppose it. Consequently, human rights abuses during campaigns have been widespread. Northern soldiers taking southern captives as slaves is just one example. Many Christian organizations in the United States have raised funds to buy the freedom of southern Sudanese Christians. This is a controversial practice because it may actually encourage more slave taking.

In the early 2000s armed conflict erupted between the government and non-Arabic tribes of western Sudan. To put down the rebellion, the government armed local Arabic tribes who were steadily moving into the western province known as Darfur (meaning "land of the Fur people"), a source of anger for the Fur people and a partial cause of the rebellion. These armed individuals earned the name *Janjaweed,* an Arabic word meaning "devil on horseback with a gun." The name reflects the fact that these nonprofessional troops, though acting on behalf of the government, terrorize the local population, burning villages and raping women in what the international community labels as genocide. The international community condemned the Sudanese government, and some international peacekeepers were sent to the region. However, the Sudanese government denies its support of the Janjaweed despite deputizing many Janjaweed as police with the assignment to restore order in the western region. This policy encourages human rights violations.

Kurds

Many Kurds have long desired to have their own nation–state called Kurdistan and have struggled and been persecuted for trying to establish their own state. In 1984 the Kurds of southeastern Turkey launched a military campaign to free the Kurdish areas of Turkey. The Turkish military responded with force, leading to many human rights abuses. After the capture of a Kurdish leader who was then sentenced to life in prison, the Kurdish fighters changed their military struggle to a political one. In the early 2000s the Turkish government, desiring membership in the European Union, succumbed to pressure from the European Union to grant more rights to its citizens. Kurds now hold political office and are allowed to speak Kurdish in schools.

While the situation of the Kurds in Turkey is much better today, life for Kurds in Iraq is far less certain. In the early years of Saddam Hussein's reign, Kurds enjoyed a number of rights. However, after a number of Iraqi Kurds joined with the Iranians in the Iran–Iraq war in the 1980s, Saddam Hussein persecuted Kurds, even dropping lethal mustard gas on them. After the Gulf War in 1991, the United States and its allies set up a no-fly zone in northern Iraq to prevent the Iraqi airforce from bombing Kurdish areas. The act allowed the Kurds of northern Iraq to effectively set up their own governmental structures. Following the ouster of Saddam Hussein in 2003, the United States proposed that Kurds join in a new federally organized Iraq.

Kurds in Iran have suffered less persecution in recent years and also have not engaged in armed conflict. The last armed struggle led by Iranian Kurds occurred right after the Iranian

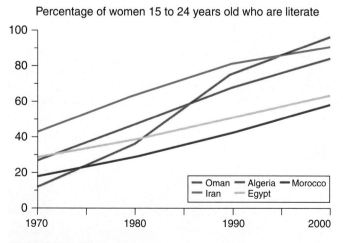

Percentage of women 15 to 24 years old who are literate

	Oman	Algeria	Morocco
	Iran	Egypt	

FIGURE 8.25 Literacy rates among young women in selected countries, 1970–2000. Source: Graph from Literacy Statistics from UNESCO, www.uis.unesco.org. Reprinted with permission of UNESCO.

Revolution in 1979. Though Iran's Kurds have engaged in fewer armed struggles than Kurds elsewhere, they succeeded in creating the only modern Kurdish nation–state, the State Republic of Kurdistan with the capital in Mahabad. This nation–state was short-lived: it was founded and destroyed in 1946.

Kurdish rights have been increasingly violated in Syria since 1963 when the Baathist Party took power. The Baathist Party is similar to the party by the same name in neighboring Iraq that governed under the leadership of Saddam Hussein. The Syrian Arabization program deprives Kurds of Syrian citizenship, ownership of their land, and since 1992, the right of children to use their Kurdish names when registering for school. False imprisonment, torture, and mass arrests of Kurds are common in Syria.

Test Your Understanding
8B

Summary

The countries of Northern Africa and Southwestern Asia have a strategic geographic position between Europe and the rest of Africa and Asia. Pan-Arabism and Islamism drive much of the region's politics, in conjunction with vast oil resources that led to the creation of the Organization for Petroleum Exporting Countries (OPEC).

Scarce water resources exacerbate conflict, limit agriculture, and heavily influence the distribution of populations. Indeed, the extensive arid areas have encouraged high levels of urbanization.

Women's rights have increased in many countries but remain restricted in others. Peoples in Sudan and the Kurds in Iraq, Iran, Syria, and Turkey struggle to have their human rights recognized.

Questions to Think About

8B.1   What is the Arab League and what does it seek?

8B.2   In which countries have Islamic political groups focused their efforts?

8B.3   To what extent does water limit agriculture and economic development in the region? What water agreements exist?

8B.4   How are population growth rates for the region best characterized?

8B.5   What are the typical features of medinas?

8B.6   Which groups have experienced greater recognition of human rights, and which have experienced an erosion of their human rights?

Key Terms

Arab League	Organization of Petroleum
Palestine Liberation	Exporting Countries
Organization (PLO)	(OPEC)
Pan-Arab country	cartel
Organization of the Islamic	Nile Waters Agreement
Conference (OIC)	medina

Subregions

The region is divided into five subregions (see Figure 8.1 and Tables 8.1 and 8.3):

- The countries of *North Africa* face the Mediterranean Sea and southern Europe west of Egypt and have desert interiors.

- The *Nile River Valley* links Egypt and Sudan.

- The region's heartland of *Arab Southwest Asia* extends from Syria in the northwest to Oman in the east. The western section consisting of Syria, Lebanon, Jordan, plus Palestine, and Israel, sometimes including Turkey, is referred to as *Al-Mashrig* ("Levant" in English).

- *Israel and the Palestinian Territories* form a continuing focus of tensions within the region.

- *Iran and Turkey,* two of the largest countries in population and area, form the northern and eastern borders and have mostly mountainous terrains. They differ from the rest and each other in language, with Farsi and Turkish, respectively, instead of Arabic.

North Africa

The four countries of North Africa are Algeria, Libya, Morocco (with Western Sahara), and Tunisia (Figure 8.26), each with differing population sizes (Table 8.3). Over 80 percent of Algeria's and Libya's territories are desert, but Morocco and Tunisia do not extend so far into the arid Saharan environment. The northern parts of Morocco, Algeria, and Tunisia are dominated by the Atlas Mountains (Figure 8.27), and that area is known as the Maghreb. It includes high ranges (Mount Toubkal in Morocco is 4,165 m, or 13,665 ft., in altitude), broken by internal plateaus and river valleys. The harsh, largely arid, and often mountainous natural environments of the North African countries restrict agriculture and most human settlement to a small percentage of the territory along the northern coasts and in the immediate mountainous hinterland. Problems of water supply affect all these countries.

Although located at a considerable distance from the heart of the Muslim/Arabic world in Southwest Asia, this region shares an adherence to Islam as the almost exclusive religion, and Arabic is even more dominant as the official language. In 1979 the link with other Muslim Arab countries was strengthened by the move of the Arab League headquarters from Cairo (Egypt) to Tunis.

North Africa faces Europe across the Mediterranean Sea. Many of its past political ties and present economic links are northward to Spain, France, and Italy. Algeria, Morocco, and Tunisia retain particularly close ties with France and have strong links to markets in Europe for selling products, buying goods, and sending emigrant labor. Libya sells much of its oil and gas to Italy. The European Union's 1986 extension to include Spain, Portugal, and Greece—which compete with North Africa for similar agricultural produce markets—placed stresses on these relationships.

Table 8.3 Data by subregion, country, area, population, income (GNI PPP—Gross National Income Purchasing Power Parity), urbanization, Human Development Index (HDI), Human Poverty Index (HPI).

	Capital City	Land Area (km²)	Population (millions)		GNI PPP 2003		% Urban	Index 2002	
			Mid 2005 Total	Est. 2025 Total	Total (US$ billion)	Per Capita (US$)	2005	HDI (rank of 177)	HPI (% population)
North Africa									
Algeria, Democratic and Popular Republic of	Algiers	2,381,740	32.3	40.5	189	5,940	49	108	21.9
Socialist People's Libyan Arab Jamahirya	Tripoli	1,759,540	5.6	8.3			86	58	15.3
Morocco, Kingdom of	Rabat	446,550	30.6	39.2	119	3,950	57	125	34.5
Tunisia, Republic of	Tunis	163,610	10.0	11.6	68	6,840	63	92	19.2
Nile Valley									
Egypt, Arab Republic of	Cairo	1,001,450	73.4	103.2	266	3,940	43	120	30.9
Sudan, Republic of	Khartoum	2,505,810	39.1	61.3	63	1,880	31	139	31.6
Arab Southwest Asia									
Bahrain, State of	Manama	680	0.7	1.0	11	16,190	87	40	No data
Bahrain, State of	Baghdad	438,320	25.9	41.7	No data	No data	68	No data	No data
Jordan, Hashemite Kingdom of	Amman	89,210	5.6	8.1	23	4,290	79	90	7.2
Kuwait, State of	Kuwait City	17,820	2.5	4.6	42	17,870	100	44	No data
Lebanese Republic	Beirut	10,400	4.5	5.7	22	4,840	87	80	9.5
Oman, Sultanate of	Muscat	212,460	2.7	4.0	32	13,000	76	74	31.5
Qatar, State of	Doha	11,000	1.7	1.0	No data	No data	92	47	No data
Saudi Arabia, Kingdom of	Riyadh	2,149,690	25.1	40.1	281	12,850	86	77	15.8
Syrian Arab Republic	Damascus	185,180	18.0	27.6	60	3,430	50	106	13.7
United Arab Emirates	Abu Dhabi	83,600	4.2	5.4	79	21,040	78	49	No data
Yemen, Republic of	Sanaa	527,970	20.0	39.6	16	820	26	149	40.3
Israel, West Bank, Gaza	Jerusalem	21,060	6.8	9.3	128	19,200	92	22	No data
Iran	Tehran	1,648,000	67.4	84.7	477	7,190	67	101	16.4
Turkey	Ankara	779,450	71.3	88.9	473	6,690	59	88	12.0

Source: World Population Data Sheet, Population Reference Bureau (2005), World Development Indicators, World Bank (2005), Human Development Report, United Nations (2004), and Encarta Microsoft (2005).

	Ethnic Group %	Languages O-Official N-National	Religion %
Algeria, Democratic and Popular Republic of	Arab 83%, Berber 16%	Arabic (O), Berber dialects, French	Sunni Muslim 99%
Socialist People's Libyan Arab Jamahirya	Arab–Berber 97%	Arabic (O), Italian, English	Sunni Muslim 97%
Morocco, Kingdom of	Arab–Berber 99%	Arabic (O), Derija and Berber dialects, French	Muslim 99%
Tunisia, Republic of	Arab–Berber 98%	Arabic (O), French, Berber	Muslim 98%
Egypt, Arab Republic of	Egyptian–Bedouin–Berber 99%	Arabic (O), English, French	Sunni Muslim 94%, Coptic Christian 6%
Sudan, Republic of	Black African 49%, Arab 39%, Nubian 8%	Arabic (O), Nubian, Dinka, others, English	Sunni Muslim 70%, local 25%, Christian 5%
Bahrain, State of	Bahraini 63%, Asian 13%, other Arab 10%, Iranian 8%	Arabic (O), English, Persian, Urdu	Shiite Muslim 60%, Sunni Muslim 25%, Christian 9%,
Bahrain, State of	Arab 75–80%, Kurd 15–20%, Assyrian, Turkmen	Arabic, Kurdish, Assyrian, Armenian	Shiite Muslim 60–65%, Sunni Muslim 32–37%, Christian 3%
Jordan, Hashemite Kingdom of	Arab 98%, Circassian, Armenian	Arabic (O), English	Sunni Muslim 95%, Christian 5%
Kuwait, State of	Kuwaiti 45%, other Arab 35%, Indian, Pakistani 9%, Iranian 4%	Arabic (O), English	Sunni Muslim 45%, Shiite Muslim 30%, other Muslim 10%, Christian 10%
Lebanese Republic	Arab 93%, Armenian 5%	Arabic (O), French, Armenian, English	Muslim (Shiite, Sunni, Alawite, Druze, Ismailite) 70%, Christian 30%
Oman, Sultanate of	Omani Arab 75%, Indian, Pakistani 21%	Arabic (O), English, Urdu	Muslim 98%
Qatar, State of	Arab 40%, Pakistani 18%, Indian 18%, Iranian 10%	Arabic (O), English, other	Ibadi Muslim 40%, Hindu 28%, Christian 15%, Sunni Muslim 14%
Saudi Arabia, Kingdom of	Arab 82%, Yemeni 13%	Arabic (O), English	Sunni Muslim 85%, Shiite Muslim 15%
Syrian Arab Republic	Arab 90%, Kurd, Armenian, Turkmen and others 10%	Arabic (O), Kurdish, French	Sunni Muslim 73%, Alawite, other Muslims 13%, Christian 10%, Druze 3%
United Arab Emirates	UAE Arab 19%, other Arab 23%, South Asians 50%, other expatriates 8%	Arabic (O), Persian, English, Hindi, Urdu	Sunni Muslim 80%, Shiite Muslim 16%
Yemen, Republic of	Mainly Arab, with African–Arab and South Asian	Arabic (O)	Muslims dominant; Shafii (Sunni), Zaydi (Shiite)
Israel, West Bank, Gaza	Jewish 82% (born in Israel 62%; white 26%, African 7%, Asian 5%), non-Jew (mainly Arab) 18%	Hebrew (O), Arabic, English	Jewish 82%, Muslim (mostly Sunni) 14%, Christian 2%
Iran	Persian 60%, Azerbaijani, Turkic 25%, Kurd 7%	Persian (Farsi), Turkic, Kurdish	Shiite Muslim 95%, Sunni Muslim 4%
Turkey	Turkish 80%, Kurd 17%	Turkish (O), Kurdish, Arabic	Muslim (mostly Sunni) 99%

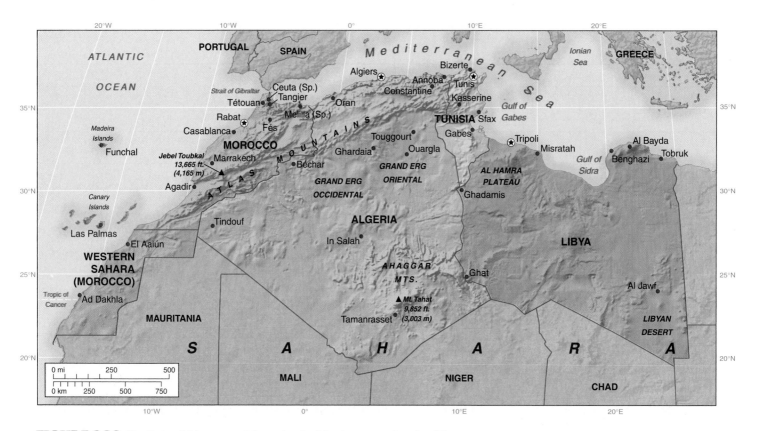

FIGURE 8.26 Northern Africa: countries, physical features, and main cities. "Ergs" are major sanddune areas.

Countries

Political History

In the 800s B.C. the coast was settled by traders from the eastern Mediterranean, who established the kingdom of Carthage. This was conquered by Rome, then overrun by Vandals in the A.D. 400s, but retaken by the Christian Byzantines. Arabs invaded in the A.D. 600s, converted the local Berber tribes to Islam, and imposed new ways of life. The Muslim (Moor) invasion of Spain followed in A.D. 711 (see Chapter 3), producing the architectural riches of Córdoba, Seville, and Grenada as reflections of the wealth and civilization they brought, including toleration of local Christians and Jews. By 1500 internal divisions and weaker control allowed Spanish Christians to expel the Muslims, forcing over a million to move back to North Africa.

From the 1200s, pirate groups controlled most of the subregion's port cities. The coast (apart from Morocco) became known as the Barbary Coast, making the western Mediterranean dangerous for merchant shipping and contributing to the decline of Mediterranean ports in Southern Europe. The pirates continued their activities even after the Ottoman Empire took control of this area in the 1500s. During the 1800s European countries and the United States stopped the pirate attacks, and France annexed territory in Algeria and later in Tunisia and Morocco.

In Algeria, Tunisia, and Morocco, French settlers and local people established commercial farms that grew citrus fruits and produced wine for European markets. The settlers took political and administrative control of the colonies. After World War II, nationalist groups fought for and obtained independence—in 1956 in Morocco and Tunisia, and in 1962 in Algeria. Libya remained poor, a mainly desert area of little economic or political outside interest until Italy occupied it in 1911. After World War II, Italy was replaced by a British–French protectorate until Libyan independence was achieved in 1952.

Recent Changes

Independent Algeria has been ruled by democratically elected governments with socialist policies based on central planning. The first-round election success of the fundamentalist Islamic Salvation Front (FIS) in the 1992 elections led to an army takeover of the government and civil war with the dispossessed Islamic militants. Terrorist activities and army repression through the 1990s devastated Algeria's economy and people with over 100,000 deaths and the army's destruction of the FIS as a political party. In early 2000 an amnesty for armed terrorists was mainly successful, reducing monthly deaths from 1,000 to 200. After September 11, 2001, Algeria regained some international credibility for its own war on terrorism, although its terrorists never expressed enmity to the United States and fed on the disregard of social needs and human rights by the

FIGURE 8.27 Northern Africa: the Atlas Mountain environment. A Berber village in the Atlas Mountains, with desert vegetation (cacti, thorn bushes) in the foreground, olive trees and arable land around the village, and trees planted on the hillside to prevent flash floods by slowing the passage of rainwater into the gullies. **Photo:** © Wolfgang Kaehler/Corbis.

military government. In 2001 anti-Islamic Berber youths in eastern Algeria served as another source of disquiet in the country.

Morocco and Tunisia view the self-destruction in Algeria with fear and take steps to avoid involvement. Morocco has political stability under its moderate but strongly involved king, Muhammed VI. His predecessor, King Hassan II, introduced gradual political reforms in the 1990s and gained international Muslim credibility through his mediating role in Arab issues and the construction of a massive new mosque. The former Spanish Sahara, an arid land known as Western Sahara, was annexed by Morocco in 1976, but the Saharawi people and their independence movement, the Polisario Front, seek control over their territory. By 2002 the United Nations had proposed a plan that seemed to accept Moroccan control with limited local autonomy and a referendum in Western Sahara that would include all residents such as the recent settlers from Morocco. While UN resolve weakens, so does that of Polisario's supporter, the Algerian government, which was bought off by French aid. A

further step was taken when Morocco awarded licenses to two international oil corporations for exploring areas offshore of Western Sahara.

Tunisia had 30 years of one-party rule when Islamic extremists were repressed but women's rights were established. Women in Tunisia have a better position than those in other Arab countries, resulting in lower adult illiteracy (40 percent in 1999) and a prohibition on polygamy.

Libya remains under the strong direction of Colonel Muammar al Qadhafi, who seized power in 1969 and runs the country as a mixed military, socialist, and Islamic republic. In the 1970s and 1980s he supported anti-West terrorism, including the destruction of a PanAm airliner over Scotland. The U.S. bombing of Tripoli and UN sanctions from 1992 reduced such activities. More recently, Qadhafi moved toward a role as mediator and leader of new political linkages in the rest of Africa (see Chapter 9). After September 11, 2001, he signaled an intention to be part of the allied antiterrorism campaign and offered Libyan assistance to the United States, but he also took the opportunity to post website lists of so-called anti-Libyan agitators who have little or no involvement in terrorism. In 2004 he renounced terrorist activities and began working more closely with Western governments.

People

Ethnic Variety

Arab and Berber people with Muslim faiths constitute over 95 percent of the populations of these countries (see Figures 8.6 and Table 8.3). Speakers of local dialects and traditional groups of nomadic pastoralists now make up a small proportion of the population. Colonial legacies include many French speakers among educated middle classes in the Mahgreb countries, while Italian has a similar role in Libya.

Employment and Migration

The current rapid population growth creates problems for the region's education systems and employment prospects. Only in Libya is there a shortage of labor and need for immigrants. The other countries have more entrants to the labor force than jobs. Algeria and Morocco have more than 20 percent unemployment. And yet, despite intensive education programs to provide primary school literacy and to redress the male–female differences in opportunity, shortages of skilled labor continue.

From the 1970s, many North Africans migrated to France and other European countries, where low-paying, mainly manual, jobs were available; more recently there have been fewer vacancies, and resistance to immigration from North Africa has increased. Today around 1.5 million North African workers send home wages from European countries, especially France and Spain. When workers return from Europe, they often set up a store or business and buy a good house. In the 1980s migrant workers from Morocco and Tunisia also worked in Libya and in the Persian Gulf. Each country has a growing university-educated group, but these people find few well-paid professional employment outlets; that, combined with the oppressive political regimes, spurs many to emigrate to Europe and the United States, creating a significant brain drain.

Economic Development

Algeria, Morocco, and Tunisia are rising middle-income countries, but their economic growth faltered from the 1990s. Algeria's civil strife halved its per capita income, while Morocco's and Tunisia's rates of increase slowed. Libya does not report income, but its oil reserves and a relatively small total population probably put it into the upper-middle-income group of countries. In spite of UN sanctions, which may soon ease, Libya's oil and gas are sold to Europe, and the income is used to subsidize housing, basic foods, oil products, education, and health care. The subsidies make up for low wages.

Morocco's Primary Products

Only in Morocco is as much as half the population still dependent on agriculture, although it constitutes only 15 percent of GDP. Morocco's farmers use irrigation water to produce citrus fruits, vegetables such as tomatoes and potatoes, and cut flowers for European markets. Morocco's economy also rests on other primary products. Its exports of phosphate (for fertilizer) are the most important. Its fishing industry has grown to the point where it provides 15 percent of Moroccan exports. At present, Morocco is contesting the management of fishing grounds off western Africa with the European Union and particularly Spain. The main fish caught are squid (for export to Japan) and tuna. Morocco continues to export cork from the bark of oak trees in the northern area of the country. Algeria and Tunisia also export phosphate rock, but it is a less important component of their overall exports.

Expanding Manufacturing and Services

Manufacturing is expanding in all countries but is overshadowed by oil income in Algeria and Libya. Morocco's manufacturing sector is the least developed but includes substantial craft industries. All countries have basic food processing, construction materials, and textile industries located in and around major cities. Except in Morocco, oil-refining and petrochemical industries are important. Algeria also has light industries, including the manufacture of electrical components; Libya makes steel and aluminum; and Tunisia has a small steel industry.

North African countries have growing government bureaucracies and service occupations in education and health care. In Morocco and Tunisia, tourism is a major source of income, based on their sunshine, coastal locations, historic and cultural attractions, and stable political environments. Most tourists come from northern Europe, with its cool and rainy climate and high personal incomes. Tunisia had over 5 million visitors in 2002, up from 1.6 million in 1980 and 3 million in 1990, while Morocco had over 4 million. Terrorist activity in Algeria after 1992 inhibited tourist expansion there, but there were 1 million visitors in 2002. Libya does not encourage tourism.

Trade with Europe

Much trade of the North African countries, both imports and exports, is with Europe. Algeria and Libya supply one-fourth of the European Union's oil and natural gas, while Morocco sells over half its phosphate exports to the EU. Morocco, Algeria, and Tunisia sell agricultural products and textiles to EU markets, but 1992 EU rules made these exports more difficult. This is partly because the addition of Portugal and Spain to the EU led to surpluses of products that are also the main outputs of North Africa. Such setbacks came at an unfortunate time when remittances from workers in Europe were unlikely to increase and compensate for the lost export earnings.

Economic Policy Changes

Governments in the North African countries placed economic growth high on their agendas. The setting of five-year plans, however, became wastefully bureaucratic and inflexible at a time of rapid change. Only the oil-exporting countries receive adequate foreign currency to pay for major developmental projects. Algeria, however, borrowed so much on the strength of its oil and natural gas sales that its debts rose to one-fifth of its GDP. Despite the smaller role of Tunisia's oil income, drought and locust attacks caused a financial crisis when its revenues decreased as a result of lower world oil prices. The collapse of world phosphate prices and the costs of conflict over Western Sahara upset Morocco's overambitious planning in the late 1970s, leading to a massive increase of its debts.

In the 1980s the International Monetary Fund required Tunisia, Algeria, and Morocco to carry out measures that would have cut internal budget deficits by reducing the public sector of the economy and encouraging trade in export products. The resultant unpopular measures of structural adjustment included cutting civil service employment and removing subsidies on basic foods (flour, oil, and sugar). They increased poverty and sparked riots. In some places, literacy and access to health care fell.

In the 1990s North African countries began to privatize large sections of their economies. Tunisia is furthest ahead and even has its own stock exchange. Morocco followed with unparalleled sales of state holdings.

Nile River Valley

The Nile River provides a life-giving water supply to Egypt and Sudan (Figure 8.28) that has sustained a human presence in the dry eastern Sahara since the early days of farming and civilization (see Figure 8.10). The histories of the two modern countries were linked over many centuries by the Nile River and political control. Although the Nile River waters provide a basis for living in an arid climatic environment, the resource is finite. Pressures from increasing populations make it difficult for Egypt and Sudan to continue to modernize and diversify their economies. Both are economically poor countries (see Table 8.3) with major balance-of-payments problems. Neither produces sufficient export income to pay for the imported needs of its people. Both rely on aid from the world's wealthier countries and the major Arab oil producers.

Countries

Egypt and Sudan share some commonalities, but their geographic positions, political environments, and products provide contrasts. Egypt is the largest Arab country in population (74 million in 2005) and one of the foremost in international relations. It retains control over the major global choke point of the Suez Canal, which has been widened to accommodate increasingly large shipping vessels. It also controls the Sinai Peninsula, which is the (little used) land route from Africa into Southwestern Asia. By contrast, Sudan has twice the area of Egypt but only half the

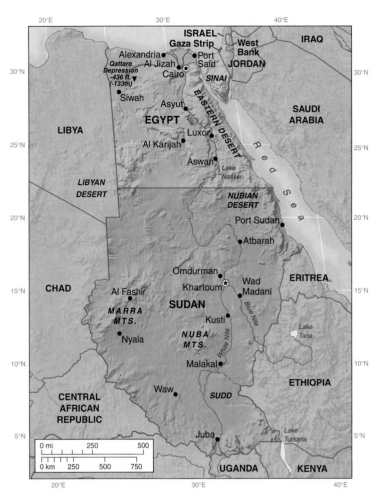

FIGURE 8.28 **Nile River valley: countries, physical features, and main cities.**

population, and has an isolated position relative to global trading. Part of the southern Saharan rim, it has conditions somewhat similar to those of its neighbors Chad and Ethiopia and resembles other African countries in many ways.

Political History: From Empires to Colonies and Independence

The earliest Egyptian empire dominated the lower Nile River valley from around 3200 B.C. The centers at Memphis in the north and Thebes in the south of modern Egypt expanded influence over the Nubian kingdom in the area of modern Sudan. The complex imperial history ended when Alexander the Great conquered Egypt in 332 B.C., and the Romans and Byzantines imposed control up to the A.D. 500s. During the later part of this period, the Christian Coptic church spread through the area and southward into Ethiopia.

In the A.D. 600s Muslim Arabs conquered Egypt, but they did not fully control the area of modern Sudan until the 1500s. Later, Mongols and Turks invaded Egypt, and it became part of the Ottoman Empire in the 1500s. Egypt's position at a crossroads of international trade brought some prosperity to its markets. In the 1800s the United Kingdom and France vied for control of the country. The British encouraged farmers to grow cotton for British textile mills. The opening of the Suez Canal in 1869 enhanced Egypt's strategic position and French involvements, but the British then occupied Egypt until the latter gained full independence in 1952.

After becoming a Muslim country in the 1500s, Sudan extended its territory southward, although many of the tribes in the southern area retained their traditional beliefs and some became Christians. In the 1800s Egyptian forces attacked Sudan, but it held out until the British joined in an 1899 invasion. After this, Sudan came under the joint control of the United Kingdom and Egypt. British engineers constructed irrigation works in the Gezira area south of the junction between the White Nile and Blue Nile Rivers.

Recent Events

The year 1952 marked a turning point in Egypt's modern history. After centuries of Ottoman and then British domination, Egypt exercised its full independence and became a socialist state with new priorities focused on its own internal needs..

President Gamel Abdul Nasser hoped Egypt would lead the Arab world and for a few years combined Egypt and Syria in a single country. By getting the Soviet Union to help build the Aswan High Dam on the Nile River, he also exploited Cold War politics. Under Nasser's leadership, Egypt developed more rapidly than the rest of Africa in the 1960s and 1970s. Nasser died in 1970, and by the late 1970s, Egypt's foreign policy shifted by recognizing Israel. Then generous aid came from the United States, but Egypt was shunned by the rest of the Arab world.

Sudan regained its independence in 1956, despite Egypt's claims to keep control. The tensions resulting from the period of Egyptian occupation of Sudan and fears of future Egyptian expansion remain in the minds of many Sudanese and affect negotiations over the use of Nile River water. Internally, the Arabic Muslim peoples have controlled the government in a series of military regimes that have tried to impose Arabic and Islamic culture and traditions upon all the peoples of the country, including the south, which is primarily black African and either Christian or animist. Consequently, a civil war erupted from 1972 to 1982. A ruling military junta composed of a mixture of the military and Islamists took control in 1989. Sporadic violence continued but began to die down in 1999 when Sudan's government moved toward multiparty politics internally and better international relations. Peace talks in 2002 and 2003 led to the signing of peace accords and a cease-fire; but episodes of violence have not completely ceased, with the latest violence erupting in the western Sudan (see the earlier "Human Rights" section).

People

Ethnic Diversity

With an ancient history, Egyptians constitute a discrete subgroup within the larger Arab population spread throughout the region. Although most are Muslims (see Figures 8.4 and Table 8.3), Coptic Christians are a sizable minority of Egyptians, primarily in Alexandria and Cairo. They are a very old Christian group, originally converted by Saint Mark, author of one of the four gospels of the New Testament. Egyptians speak Arabic but have a distinct dialect known as Egyptian Arabic, although this is becoming widely understood through Egyptian broadcasting.

In contrast to the Egyptians, Sudan is very diverse, with over 600 ethnic groups and 400 languages spoken (see Figure 8.6). Islam and Arabic diffused from Egypt over the centuries and assimilated many peoples (such as the Beja). Arab Muslims and Arabized Muslims tend to be in the north. The peoples in the south are related to the peoples of Africa South of the Sahara (see Chapter 9) and practice traditional animistic religions or Christianity. The Dinka and the Nuer are among the larger of the southern, non-Muslim, non-Arabic speaking groups. Transitioning from north to south, many of the black African ethnic groups have partially or fully assimilated into Muslim–Arabic society. Nubians illustrate this point as they are a Nilo–Saharan people who converted to Islam centuries ago and adopted Arabic, though many still speak their traditional Nubian language.

Economic Development

Egypt was the one country in Northern Africa that doubled its GDP in the 1990s. Sudan is poor and still has a tenuous hold on external economic links because of its narrow economic base, internal conflict, and isolation from major transportation routes.

Economic Diversification in Egypt

Following the 1959 Nile Waters Agreement between Egypt and Sudan, the High Aswan Dam was constructed on Lake Nasser in 1970. The Egyptian government hoped that the electricity generated from the dam would fuel economic development through manufacturing. However, the Egyptian government's plans were not fully realized. The manufacturing base of iron, steel, and chemicals production, together with the assembly of cars, food processing, and bus, textile, tire, and television manufacturing, is indeed more diverse than in most other African countries, but the factories are aging. The largely nationalized industries need modernization and considerable investment. Egypt's armament production received a boost in the Iraq–Iran War (1980–1988) but was less busy afterward.

Egypt's agreement to a cease-fire and closer relations with Israel earned it massive aid from the United States, amounting to $10 billion per year, or 10 percent of Egyptian GDP, at its height. This made it second to Israel in U.S. aid. Furthermore, the Gulf countries paid for Egyptian military involvement in the 1991 Gulf War.

The wages sent home from 2 million Egyptian workers in the Persian Gulf equaled agriculture and manufacturing as a source of income for a few years in the 1980s. Over a million workers with their families returned to Egypt during the Gulf War, but few returned to the Gulf after the war. Within Egypt, the Nile River valley upstream (to the south) of Cairo is designated "Upper Egypt." Incomes in Upper Egypt are half those of the rest of the country, and this area ranks poorly in infant mortality, adult literacy, health services, schools, and unemployment. Many Egyptians who worked abroad came from Upper Egypt, and those who returned at the time of the Gulf War came back to a penniless area. Much of the civil unrest in Egypt and nearly all the terrorist killings occur there.

Tourism based around Egypt's historic treasures, such as the pyramids and temples along the Nile River, formed another growth area until Islamic fundamentalists began to attack foreigners in the 1990s, causing a major slump in 1993. By 2000 over 5 million tourists visited Egypt, up from 1.25 million in 1980; but the September 11, 2001, events reduced numbers at first, although overall 2002 still had nearly 5 million visitors. The Suez Canal revenues, static for a long time at just below $2 billion per year, also fell after September 11 as world trade slowed for a while, and the low level of Egypt's oil exports does not make up for these losses of income.

Sudan's Economic Plight

In contrast to Egypt, Sudan never had much support from the United States or the Soviet Union, and the little it received from other Arab sponsors fell away after the 1991 Gulf War. Wages sent home by its small number of workers overseas also declined in the 1990s. Economic disaster was enhanced by the attempts to integrate the southern part of Sudan into an Islamic state by force. Years of civil war destroyed production in the south and diverted northern resources to military expenditure. The internal policies, civil war, and failure to implement required reforms interrupted external funding from the World Bank.

Until the 1990s, Sudan's main products were from farming. Cotton provided 50 percent of exports in better times, and the southern part of the country produced livestock products for export. Although Sudan is normally able to feed its population, civil war, drought, and the pressures of refugees from Ethiopia combined to cause famine from the 1990s. Sudan's manufacturing sector is at an early stage of development. Textile factories, paper mills, and an oil refinery, all based around Khartoum, contrast with small-scale production for local needs in the rest of the country and together make up a small part of the country's total product.

Discoveries of oil in Sudan in the late 1990s enabled the country to supply its home market and export enough to offset one-fourth of its import costs after a pipeline to the coast was opened in 1999. The oil deposits occur in the middle White Nile River basin in southern Sudan. Chinese, Malaysian, Canadian, and Swedish oil companies share the development. Although oil may be part of the solution to Sudan's problems, it also brings new problems. There is evidence that the Sudanese government used the oil income to fund armaments purchases, killed or removed inhabitants from the oil-producing area and oil pipeline zone, and used the cleared area as a base for attacking other parts of the south.

Test Your Understanding
8C

Summary

North Africa has close ties with the former colonial powers in Europe, particularly France. It continues to supply Europe with energy but faces economic difficulties as the European Union expands to include countries that market similar agricultural products.

The Nile River Valley countries of Egypt and Sudan still rely heavily on the river's water, but the supply cannot be increased to match the population growth. Economic diversification in Egypt scarcely keeps income above population growth rates. Civil war slows development in Sudan.

Questions to Think About

8C.1 Which five geographic characteristics do you think make each of the North African countries distinctive?

8C.2 Why should Egyptians be pessimistic about their future water supply?

Arab Southwest Asia

Arab Southwest Asia is the heart of the Arab and Islamic worlds. It comprises the Arabian Peninsula and the Fertile Crescent that includes the Tigris–Euphrates River basin and the Lebanon coast (Figure 8.29). The world centers of Islam that are Muslim pilgrimage sites are located in the Arabian Peninsula at Mecca, where Muhammad was born in A.D. 570, and at Medina, which became his power base after he was expelled from Mecca. Despite having few people, the countries of this subregion play a major part in world affairs because of their oil wealth and involvement in the Arab–Israeli peace process.

Countries

The Arab Southwest Asia countries are differentiated by forms of government, emphases within Islam, and natural resources of oil and water. The differences in natural resources and economic management produce a wide range of economic status—from countries that remain poor to those that rival the world's wealthiest (see Table 8.3).

Arab Southwest Asia is rich in history and culture. Some of the earliest known towns are here. Ancient Mesopotamia, situated along the Tigris–Euphrates Rivers, has long been considered the "cradle of human civilization." The subregion was also the location of early Hebrew and Persian states. In medieval times, the expansive and prosperous Muslim empires were centered in the subregion, first under the Umayyad dynasty with its capital in Damascus and then under the Abassid dynasty with its capital in Baghdad. Coupled with its role as the hearth of the three

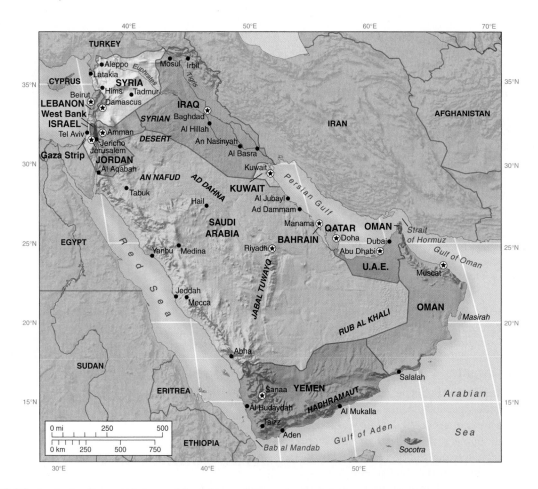

FIGURE 8.29 Arab Southwest Asia and Israel: countries, physical features, and cities. The Fertile Crescent stretches from the Mediterranean coastlands through Syria and Iraq to the Persian Gulf. Israel declared Jerusalem as its capital in 1950. However, most countries of the world do not completely recognize this declaration and have their embassies in Tel Aviv.

Western religions of Judaism, Christianity, and Islam, this subregion has played a major role in the development of both Western and wider world civilizations.

Governments

The governments of the countries in this subregion mostly move slowly, if at all, toward democracy. Many of the oil-rich countries remain dominated by ruling families or dictators: Saudi Arabia is the only country named after a family, the Sauds, who still rule with the title of king and support a huge range of related princes. Other Gulf countries have sheikhs, emirs, and sultans with similar roles, and Jordan has a king. Syria has a military dictatorship. Until the fall of Saddam Hussein, Iraq had a military dictatorship but is now moving toward democracy (see Geography at Work: Operation Iraqi Freedom on page 363). In Lebanon a greater degree of democracy now exists after a destructive civil war in the 1980s and 1990s between the Muslim and Christian groups spurred on by Islamic extremists, although Syria still exercises a major influence.

Syria's dictatorship under Hafez Assad lasted for 30 years. The minority Shiite Alawin people took control of the army and the socialist Baath Party. Supported by the Soviet Union during the Cold War, Syria fostered extremist groups that took part in the Lebanese civil war and actions against Israel and European countries. After the Soviet link ended in 1991, Syria's inward-looking economy suddenly sought to enter the global economic system. Privatized businesses expanded their activities. However, the poor banking system and lack of political freedom attracted little foreign investment, and economic growth stood still as the population increased. Following Hafez Assad's death in 2000, his son Bashar slowly relaxed the government's tight grip on society and the economy.

The tensions in the subregion led to conflicts among the countries. Iraq and Iran fought a war through the 1980s. The Gulf War in 1991 resulted from Iraq's invasion of Kuwait the previous year. Internal conflicts sprang up in Oman from 1970 to 1975 and in Yemen in the 1980s. The latter resulted in the merger of North Yemen with the Yemen People's Democratic Republic in 1990. The 1994 renewed outbreak of violence between the two parts of Yemen highlighted remaining differences; victory went to the northern Islamists over the southern socialists. After siding with Saddam Hussein of Iraq in the Gulf War, Yemen lost foreign aid and saw its workers deported from Saudi Arabia. When the USS Cole was bombed in the Yemeni

OPERATION IRAQI FREEDOM

Ask the soldiers in the 450th Movement Control Battalion (MCB) about their experience in Operation Iraqi Freedom (OIF), and they'll tell you with a smile, "It's all about geography." Their commander, Lieutenant Colonel (LTC) Mark Corson (Box Figure 1), told and showed them that geography mattered in everything they did. When not active as a U.S. Army Reserve officer, Mark is a geography professor at Northwest Missouri State University in Maryville, Missouri.

Mark notes that military transportation has three parts: terminal operators, mode operators, and movement controllers. Terminal operators work the airports, seaports, and trailer transfer points to stage cargo for movement. Mode operators drive trucks, fly helicopters, and operate railroads and watercraft. Movement controllers are the brains of the operation, telling the mode operators what to haul, where, and to whom (they also control the roadways to prevent traffic jams). The 450th MCB served as the movement controllers in southern Iraq during the opening phases of the war. They later controlled transportation operations in Kuwait, where they oversaw the major airports, seaport, and staging bases.

Understanding the geography of Iraq helped soldiers of the 450th MCB. Their knowledge of physical geography prepared them for the weather, climate, and terrain of the desert. Their well-designed uniforms and equipment allowed them to brave searing heat, sandstorms, and desert sand so they could accomplish their mission. Understanding the terrain, the location of key cities, and the nature of the connecting roads allowed them to site their logistics bases and pick the best transportation routes.

Knowledge of human geography was also valuable to Mark and his soldiers. Understanding Iraq's three major cultural/religious groups (Shia Muslims, Sunni Muslims, and the Kurds) ensured that Mark and his soldiers respected these peoples' traditions and did not make enemies of them. Their knowledge of the political geography and history of the region was helpful in negotiating sensitive diplomatic agreements to allow their trucks to cross the Kuwait–Iraq border.

Geography provides powerful technical tools such as cartography, remote sensing, geographic information systems, and the global positioning system. Mark and his soldiers used these geographic tools extensively. Mark's Highway Regulation Teams conducted route surveys using GPS to get exact coordinates of facilities, bridges, intersections, and key terrain. His operations section then created a simplified map that they distributed widely. Anyone who needed to go anywhere in their area referred to that map so he or she could travel safely and without getting lost. Mark and his soldiers used a sophisticated navigation and communications system called the Military Tracking System (MTS) in their vehicles and in their headquarters. MTS has a digital map display linked to GPS that shows where the user and other MTS users are on the map. Mark's soldiers used the satellite text-messaging capability to communicate over long distances. In many cases the soldiers saved lives by text-messaging medical evacuation requests that could not be passed by radio. Mark's unit also maintained in-transit visibility of convoys and cargo using the Joint Distribution Logistics Model (JDLM), which is a type of geographic information system (GIS).

From controlling traffic on the main highway, to sending trucks with the right supplies to the right people at the right time, to transporting cargo on the Iraqi Republican Railroad, to surviving

Box Figure 1 LTC Mark Corson (right) and 1LT BJ Vincent of the 450th Movement Control Battalion at Saddam International Airport (now Baghdad International Airport) shortly after its capture by the U.S. Third Infantry Division in April 2003. **Photo:** Courtesy of Mark Corson.

sandstorms, searing heat, and the threat of enemy action, geography influenced everything. Understanding the geographical aspects of Iraq and the war enabled Mark's soldiers to accomplish their missions better, faster, and more safely. Geographic knowledge and techniques are powerful tools. Ask Mark's soldiers and they will tell you, "It's all about geography."

harbor of Aden in late 2000, Yemen was suspected of housing al-Qaeda terrorists, especially as the government scarcely controls a number of rural hideouts. In 2002 the Yemeni government avowed support for the U.S. war against terrorism.

People: Ethnicity

The people of this subregion are mainly Arabs and Muslims, but important variations exist (see Table 8.3). The Kurds, who are in Iraq, Syria, Turkey, and Iran, are Muslims but are not Arabs. Lebanon contains a very diverse set of Christian as well as Muslim groups. Among the Christians are Maronites, Greek Catholics, Greek Orthodox, Roman Catholics, Jacobites, Armenian Orthodox, Nestorians, and Protestants. Both Sunni and Shia Muslims along with subgroups such as Ismailis, Alawites (found in larger numbers in neighboring Syria), and Druzes live in this country. Jews also live in Lebanon. Yemen contains groups of African origin. The Gulf countries have important groups of Asian people, mainly foreign workers with temporary residence but some who are allowed to stay longer, while Oman has many links from its former trade with eastern Africa (see Chapter 9).

Economic Development

Arab Southwest Asia has major economic differences between countries with high oil revenues and those that produce little or no oil. The different levels of income are reflected in the ownership of consumer goods (see Figure 8.12) and total income. The Saudi Arabian economy produces about four times the total income of any other country in the subregion (GNI PPP $281 billion in 2003), while Jordan, Lebanon, and Yemen have GNI PPP of $25 billion and under per year.

The richer countries bordering the Persian Gulf sit on huge oil reserves (see Figure 8.13), and exploiting them brought high incomes to the population in the 1970s and early 1980s. Then as oil prices fell until the late 1990s, incomes fell. However, oil prices rose again in the early 2000s. These countries have small total populations and rely on immigrant labor. Wealth is not distributed widely or evenly among their populations, either through increased wages or social services such as education and health care. The menial, low-wage jobs and much of the commerce, especially in retailing, are left to Indians and other Asian immigrants.

The small, oil-poor countries of Lebanon, Jordan, and especially Yemen are much less developed. They have more available labor and some capacity for agriculture. Iraq and Syria are the exceptions to this rich–poor duality: until the mid-1980s they both had oil revenues, significant water resources and agriculture, and large labor forces. The Iran–Iraq War, Gulf War and subsequent UN sanctions, and the U.S. invasion in 2003, however, almost destroyed Iraq's economy, though the United States and its allies are attempting to rebuild the economy.

Impacts of Fluctuating Income on Oil-Producing Countries

Before the oil boom beginning in the 1900s, the very small population totals in the lands of Arab Southwest Asia had economies based on low-intensity farming wherever water was available, pastoralism where water and vegetation were scarce, and extensive empty desert areas. A few crops, such as dates and citrus fruits, were exported from the best-watered places in southern Iraq and along the Mediterranean coasts.

The oil income changed countries' economies by adding wealth but creating dependence on oil. OPEC's actions in the 1970s resulted in huge increases in the price of oil that brought the newly rich oil-producing countries of the Gulf to a strategic place in the global political and economic systems. In 1965 Saudi Arabia's GDP was $2 billion at today's price equivalents, but it rose to $156 billion in 1980, when oil prices reached their highest point, fell to $105 billion in 1990, and then rose to over $200 billion in the early 2000s as oil prices recovered again partly because of OPEC pressures. Kuwait's GDP rose from $2.1 billion in 1965 to $28 billion before the Gulf War in 1990.

Internally, the income from oil enriched the ruling elites and was used partly to generate industrialization, intensify agricultural output, provide more and better roads, airports, health services, and education, and increase living standards. Up to one-third of the income in some countries went to purchase military hardware. The oil income made it possible to shift economies toward sustainable development based on diversification. A **diversified economy** is one in which manufactured goods are more important than primary products and a variety of manufactured products is joined by a growing service sector (see Chapter 2).

As world oil prices fell in the 1980s, many oil-producing countries borrowed heavily to maintain the levels of internal investment that had been possible in the 1970s. For example, falling oil prices cost Saudi Arabia $2.5 billion per year for every $1 drop in the price per barrel. As crude oil and oil products still make up between 80 and 95 percent of exports from many oil-producing countries that have not greatly diversified, the price declines devastated internal programs of change, although military expenditures fell less. Record prices for oil in the early 2000s are filling government coffers again. It is unclear if governments will spend lavishly again or remain fiscally cautious to prepare for another downturn in oil prices.

Kuwait, a country with a very small population, had less need or opportunity to invest its huge oil wealth internally, and by the late 1980s, its overseas investments generated more income than oil sales. Such income was the major source of Kuwaiti rebuilding funds for several years after the destruction in the Gulf War of 1991. Iraq's oil production was also reduced by military action during the Gulf War. Following the war, with continuing international suspicions over the Iraqi production of nuclear, biological, and chemical weapons, heightened by the refusal of Saddam Hussein's regime to allow UN inspections, Iraq was subjected to sanctions that prevented exports and imports apart from food and medical supplies. After 1994, Iraq could have sold some oil on world markets but did not choose to do so until 1998, and in 2002 it threatened to withhold its quota in protest against Israeli oppression of the Palestinians. U.S. President George W. Bush named Iraq as one of the "Axis of Evil" rogue countries because it refused to allow UN inspections. On the grounds that Iraq allegedly possessed weapons of mass destruction, President Bush sent the U.S. military to Iraq to overthrow Saddam Hussein. First under American administration and then with American help, the new Iraqi government is trying to resume oil exports, but militants within the country routinely sabotage oil facilities and pipelines.

Meanwhile, by the late 1990s, Kuwait's government largely restored the oil production that was wiped out in the 1990–1991 Iraq invasion and occupation. The 742 oil wells torched by the retreating Iraqi armies were repaired and output restored to pre-1990 levels. The costs of reconstruction and funding the Gulf War coalition, however, reduced the country's overseas assets from $100 billion to $35 billion. As the oil output increased, the employment and social welfare conditions of its people improved. The dominance of the ruling Al-Sabah family in both political and economic areas was reduced by an outspoken legislature. They demanded the privatization of Kuwait Airways and the government oil company. Other Gulf countries also faced major changes. During the 1980s and 1990s, some smaller states, such as Bahrain, Qatar, and Oman, began to experience falling oil production, while the UAE had a dramatic rise in output. Saudi Arabia took over the position of leading oil producer from Iraq and Kuwait.

Though oil prices have risen and fallen in cycles, the world demand for oil is dramatically increasing, especially from rapidly industrializing countries like China that have an increasing hunger for oil. Thus oil prices may never fall again as low as they have in the past. If this is true, the future will bring more income to the countries of this subregion.

Diversification in Oil-Producing Countries

The oil-exporting countries spend a large proportion of their income on food imports, a need that grows as the population increases. Although major investments were made in home-based irrigated agriculture, costs are very high and production is uneconomic compared to world prices.

Manufacturing development began with the building of oil-refining capacity to replace the direct export of crude oil to European refineries. Petrochemicals and other secondary activities based on the materials produced from oil take advantage of the value added by processing the raw materials. Bahrain and UAE used the local energy source to operate aluminum smelters; Oman built a copper smelter, while Qatar and Saudi Arabia produce steel. Several countries also manufacture construction materials, including cement, to provide the materials for upgrading roads and building new housing and factories. Countries developed food and consumer goods industries. Services also became important in many of these countries, where banking and government employment are increasing rapidly.

From the 1980s, the Gulf countries invested some of their oil revenues in higher education facilities, and the many graduates are now employed in banking, education, health care, government, and new industries, including the media and information technology. In 2002 Dubai's Internet City had 200 firms employing 4,000 workers and linked similar centers from Europe to Bangalore, India. The adjacent Media City, Festival City, conference center, expanding major airport and booming Emirates airline, luxury housing, and high-quality tourist facilities all signal Dubai's investment in the future.

In Saudi Arabia, new industrial towns are being developed at Al-Jubayl on the Gulf coast and at Yanbu on the Red Sea. Al-Jubayl is halfway toward its planned population of around 300,000 and has major industrial zones with airport, port, and highway linkages. It will cover an area the size of a large U.S. city such as Atlanta. From the late 1990s, Saudi Arabia began to allow foreign investment, including the construction of a giant desalination plant by Japan's Sumitomo Corporation, the building of 3,000 schools by a U.S. consortium, and natural gas exploration linked to power, desalination, and petrochemical developments by international oil corporations. Yet the Saudi Arabian government still controls two-thirds of the internal economy, and its refusal to open internal markets denies it a place in the World Trade Organization.

In 1981 the Gulf oil countries, except for excluded Iraq, formed the **Gulf Cooperation Council** in the context of the Iran–Iraqi War. It focused on common economic and political interests. Saudi Arabia dominates the organization, which was effective in bringing together other countries to resist the 1990 Iraqi invasion of Kuwait.

Tourist and Pilgrimage Sites

The hot, sunny climate and status as a major stopover for international passenger airlines from Europe to Asia, as well as the cultural attractions of the subregion, formed the basis for a growing tourist industry in some countries that contain religious centers or have a welcoming political stability. For example, UAE had over 5.5 million tourists in 2002 following international marketing of their sunny climate and shopping opportunities. In the 2000s, new islands for tourist resorts were built off the coast of Dubai (UAE) (Figure 8.30).

Although it refused entry to tourists, Saudi Arabia had 7.5 million foreign visitors in 2002. Some were businesspeople, but most were pilgrims destined for the Islamic sites in Mecca and Medina that every Muslim wishes to visit at least once in his or her lifetime (Figure 8.31).

Countries with Little or No Oil

Syria has small oil deposits in its eastern area that are its major export at present. Failure to find new sources will lead to the end of oil income in the early 2000s. Syria also has lands watered by the Euphrates River on which its farmers grow cotton, cereal grains, and fruit. It is not self-sufficient in food, however. After centuries of soil erosion on its northern hills, the slopes are being reseeded to expand local sheep farming. In the 1990s Syrian markets were opened to foreign products, and private enterprise and exports were encouraged. Syria's tourist industry grew in the 1990s with the greater openness to foreign visitors, rising from 0.5 million in 1990 to 1.6 million in 2002. The interest is particularly in religious and historic sites, and most tourists come from Europe.

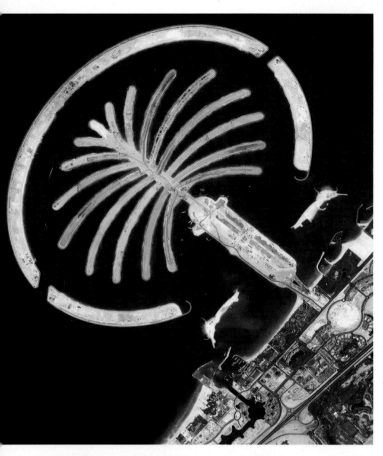

FIGURE 8.30 Dubai. Palm-shaped artificial islands in the Persian Gulf built from 130 million cubic yards of rock and sand create 75 miles of shoreline and new houses, hotels, and a marine park. **Photo:** *Courtesy Space Imaging Middle East.*

FIGURE 8.31 Arab Southwest Asia: Mecca, Saudi Arabia. Al-Haram mosque, the main pilgrimage site for Muslims who seek to visit it at least once in their lifetimes. **Photo:** © Nebeel Turner/Stone/Getty Images.

Jordan, Lebanon, and Yemen continue to have small economies with limited prospects of expansion. When Israel appropriated the West Bank area in 1967, Jordan lost half its agricultural land and much of its water and labor force. Jordan's high debts led to austerity measures and social unrest. Exports of phosphate, its main source of income, are affected by fluctuating world prices. Attempts to reestablish manufacturing industries around Amman when Israel took the West Bank were not successful. Aid income from oil countries supported Jordan as a "frontline state" adjoining Israel. After the 1991 Gulf War, Arab countries withdrew their aid, overseas workers returned home, and a flood of refugees arrived from Iraq. As in Syria, Jordan's tourist industry expanded in the 1990s, from 0.5 million to 1.6 million visitors in 2002, with special interest in historic sites such as the rock-hewn city of Petra.

In 1986 Yemen experienced a political crisis that led to uniting the two parts of the country in 1990. The discovery of oil in the north gave hopes of an improving future, but renewed internal conflict in 1994 showed that major difficulties still stand in the way of concerted efforts to improve the economy. Yemen remains the poorest country in the subregion.

Before the civil war that began in 1975 tore it apart, Lebanon's mountainous landscapes and banking system had earned it the title "Switzerland of the Middle East." Most of its income was generated from banking and financial corporations based in its capital, Beirut, the commercial center for much of the region and an early global city. This status was lost when many Beirut financiers fled abroad during the civil war, although some are beginning to return. By the early 2000s, Beirut was being rebuilt (Figure 8.32), and much of the financial capital that went abroad during the civil war was returning. Farming production was being restored in the coastal plain (fruits, vegetables, tobacco) and the Bekaa Valley (grains).

Prospects for Regional Interchanges

The economies of all Arab Southwest Asian countries are linked closely to the world's wealthier countries and their demands for oil and other products. Although an Arab Common Market was established in 1965, only Iraq, Syria, Jordan, and Egypt are full members. Thus the countries of Arab Southwest Asia have little trade with one another. They remain vulnerable to price instability in world oil markets, and some face European Union protectionism against agricultural imports. In the 1990s, increases in trade with East Asia offset oil exports and stimulated hopes that markets there might develop further, reducing dependence on Europe and the United States. Egypt, Israel, Jordan, and Syria are at present considering new forms of economic cooperation with each other.

(a)

(b)

FIGURE 8.32 Arab Southwest Asia: Beirut, Lebanon.
(a) damaged building along the Green Line that separated predominantly Christian East Beirut from mostly Muslim West Beirut during Lebanon's civil war. Although Beirut has largely been rebuilt, occasional reminders of the war are seen in the landscape. (b) Section of the waterfront in West Beirut, Lebanon. Most of the buildings are apartments, although some embassies are in this neighborhood. **Photos:** © Alasdair Drysdale.

Israel and the Palestinian Territories

Israel is a major anomaly in Southwestern Asia. It is a unique example of a country created by the United Nations for a particular ethnic group, despite opposition from those living in and around it (Figure 8.33). Subsequent territorial disputes, terrorism, and the creation of refugee groups have major geographic impacts locally and implications throughout the world. The Palestinian territories of the West Bank and Gaza are Israeli-occupied territories following conquests in the 1967 war, although the United Nations ruled that they should be returned to Syria, Egypt, and Jordan. This subregion is marked by a dual society of different opportunities for Jew and non-Jew.

In 1998 Israel celebrated 50 years of very hard work, growing prosperity, democratic government, and the settling of more than one-third of the world's 15 million Jews in a separate homeland. Still surrounded by antagonistic countries and subsidized heavily by the United States, Israel has a strong economy and a powerful military, including a range of nuclear weapons. None of the Arab countries has the same firepower.

Countries

Origins of Israel as a Modern Country

For some Jews, Israel reestablishes a religious national way of life after centuries of dispersal; to other Jews, it is a distinctive secular ethnic community. To Palestinians, Israeli Arabs, and increasing numbers of Muslims worldwide, the establishment and defense of Israel and the need to free Palestinians focuses their need to assert an Islamic presence.

The idea of a separate country for Jews arose out of Zionism, a movement that began in Europe and Russia in the 1800s, which called for the creation of a separate Jewish nation–state. Increasing anti-Semitism, including the imprisonment and murder of Jews, led wealthy Jews to finance initial waves of settlers, numbering 60,000 from 1880 to 1914, in Palestine. The settlers taught the modern Hebrew language in schools and established socialist institutions such as labor organizations and farming in communal **kibbutzim.** In the kibbutzim, land is communally owned and decisions are made collectively. Settlers aimed to prepare for mass migration into a new, independent country as a safe haven from a persecuting world.

During World War I, the United Kingdom, France, Italy, and Russia made secret agreements to divide the Ottoman Empire after the defeat of Germany and its allies. Of these, the Sykes–Picot Agreement recommended internationalizing Palestine and giving much of Turkey to Russia, the Persian Gulf area to the United Kingdom, and Syria–Lebanon to France. The United Kingdom negotiated separately with Arab leaders, promising them a separate country on lands won from the Turkish Ottoman Empire.

However, the 1923 Paris Conference settlement incorporated the 1917 British Balfour Declaration that favored "the establishment in Palestine of a National Home for the Jewish people." It proposed that the British carve an area for this purpose from within its protectorate of Palestine. Arabs agreed to the principle on the basis of their obtaining a new Arab country out of the old Ottoman Empire lands that Turkey lost in World War I. However, many Arabs expressed disappointment when the future Saudi Arabia was the sole politically independent Arab country and the rest of the former Ottoman Empire was divided into separate protectorates under European ruling countries.

The issue was still unresolved in the 1930s, when many Jews fled Nazi Germany and Europe. Before 1940, 350,000 Jews moved to Palestine despite prohibition under the British mandate (Figure 8.33a). The largely impoverished Jewish immigrants were supported from Germany and donations by Jews in the United States. After World War II and the genocide of the Nazi Holocaust, a million more joined them. Violence among Jews, Arabs, and British forces increased. Pressure mounted for an independent Jewish state while Arabs resisted turning over lands to such a state. The United Kingdom could no longer control its protectorate and handed jurisdiction to the United Nations, leaving in 1948 even though Arabs had rejected a 1947 United Nations Partition Plan (Figure 8.33b) that was proposed to separate Jews and Arabs.

After the partition plan was rejected, Arab protests led to war when in 1948 the Jewish settlers declared the independence of their new state, which they called Israel. Arab forces attacked the Jewish settlers, but Israeli forces not only repulsed the attackers but also expanded the country's original territory by the end of the war in 1949 (Figure 8.33c). During the conflict, some 600,000 Palestinian Arab refugees were forced out or fled for their lives and moved mainly into Jordan and Lebanon. By 2000 some 3.6 million Palestinian Arabs were living in Jordan (1.5 million), West Bank and Gaza (1.4 million), and Lebanon and Syria (600,000 together).

Further Wars and Negotiations

Arab countries refused to accept the existence of independent Israel and fought it unsuccessfully again in 1956, 1967, and 1973. In 1956 Israel assisted France and the United Kingdom in attacking Egypt, and in 1967 the Israelis attacked first amid rising tensions. During the 1967 war, Israel extended its territory southward into the Gaza Strip and Egypt's Sinai Peninsula, eastward across the West Bank area of Jordan, and northward

(a) Palestine was a U.K. protectorate after World War I. Jewish immigrants made up 11% of the population by 1922, 29% by 1936, and 32% by 1946. Palestinian Arabs objected to the influx and rioted in 1920, 1921, 1929, and 1935–1939 in the face of land purchases and exclusive labor policy by Jews.

(b) The 1947 U.N. Partition Plan envisaged the orange-shaded area as Israel, but was rejected by Arabs. It was a U.N. initiative after the U.K. withdrew, finalized in May 1948, when Israel declared itself to be a country.

(c) Palestinian Arabs, supported by Egypt, Jordan, and Syria, invaded the newly declared country of Israel, but were repulsed as Israel extended its land area to that outlined in red. Two-thirds of the Palestinians became refugees. Further wars took place in 1956, 1967, and 1973.

■ Israel in 1967

■ Controlled by Israel after Six Day War

(d) After the 1967 war, Israel extended its frontiers to include the West Bank (from Jordan), the Sinai Peninsula (from Egypt), and the Golan Heights (from Syria). In 1973 a Syrian and Egyptian attack failed, leading to a desire for a diplomatic settlement: the 1978 Egypt–Israeli Camp David Peace Accord in which Israel returned Sinai.

FIGURE 8.33 **History of modern Israel: a timeline of changing maps.** Source: © The Economist Newspaper Group, Inc. Reprinted with permission. Further reproduction prohibited. www.economist.com

to the Golan Heights of Syria (Figure 8.33d). Despite UN resolutions, Israel refused to give up the occupied lands, apart from Sinai (Figure 8.33e), for security reasons. Israel wanted surrounding countries to accept the existence of Israel and renounce military activity against it. Syria would not acknowledge Israel's existence, so Israel held onto the Golan Heights. Unsuccessful in open warfare, Arab groups, both secular and religious, turned to terrorist methods in the 1970s.

In 1994, following the Oslo Accord between Israel and the Palestine Liberation Organization (PLO), a Palestinian Authority was established and given limited jurisdiction and autonomy in Gaza and the West Bank. Israeli settlements in these areas (Figure 8.33f) were maintained, however, and their populations expanded. After 1996 changes in the Israeli government, progress toward the development of a Palestinian country halted, as Israel resisted further devolvement of power and transfers of land to Palestinians. Terrorist activity by a Palestinian group known as Hamas increased. The Israeli military entered Palestinian territories to punish people for atrocities, such as suicide bombings, committed against Israelis.

After decades of conflict over the position and status of a Jewish state, many outside the situation see an almost insoluble problem (see the Point–Counterpoint: Israelis versus Palestinians box on page 370). Palestinians and other Arabs continue to wish for Israel's destruction, causing Israel to focus on defending itself. Revenge for one set of killings sets off more revenge, leading to economic devastation and cultural warfare that could become widespread.

Continuing Stresses

Many Israeli citizens found it difficult to celebrate wholeheartedly their country's 50th anniversary in 1998 because they were still the target of terrorist acts that many Palestinians regard as steps toward freedom. Furthermore, strong disagreements exist among Israeli Jews over issues that intertwine in practice to generate a pluralistic but polarized society:

- Should the territories gained in 1967 be traded for peaceful relations or kept as part of Israel?

- How should relationships among nationalism, government, land, and religion be resolved in conflicts between Jewish religious and secular interests?

- How should Israel manage the move from a rural focus that made the desert bloom to a high-tech, urbanized country?

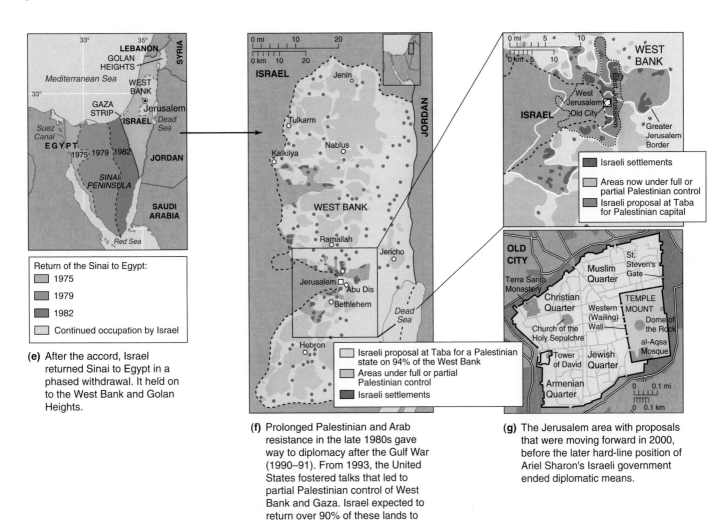

(e) After the accord, Israel returned Sinai to Egypt in a phased withdrawal. It held on to the West Bank and Golan Heights.

(f) Prolonged Palestinian and Arab resistance in the late 1980s gave way to diplomacy after the Gulf War (1990–91). From 1993, the United States fostered talks that led to partial Palestinian control of West Bank and Gaza. Israel expected to return over 90% of these lands to form a Palestinian country.

(g) The Jerusalem area with proposals that were moving forward in 2000, before the later hard-line position of Ariel Sharon's Israeli government ended diplomatic means.

ISRAELIS VERSUS PALESTINIANS

[The following is an attempt to stimulate informed discussion of a sensitive issue. The views expressed in the table at the end are those of imagined disputants, not the authors of this text. The most extreme views are held by hard-liners on either side, but that includes only about one in five of the people in Israel and the Palestinian Territories.]

Perhaps the most significant world conflict today is between the Jews living in Israel and the Arab Palestinians living in the occupied Palestinian Territories of the West Bank and Gaza. Although the United States is at the forefront of trying to reconcile the two sides, the Arab countries see it as the main supporter of Israel. The conflict's origins go back in history, before Israel was established as a new country in 1948. The continuing resistance of Palestinians to the heavily armed Israelis suggests an irreconcilable conflict, at least one that will not be resolved until an independent homeland is created for the Palestinian Arabs.

The land in question is a small part of what is known as the "Fertile Crescent" that cradled and linked the early civilizations of Mesopotamia and Lower (northern) Egypt. The hilly coastal lands provided a home for the ancestors of the Israeli nation. When they escaped from slavery in Egypt, they took the land from the Canaanites to establish their own country. Conflicts with neighbors continued, including with the coastal Philistines and the major inland empires of Babylon, Assyria, and Persia. Always fiercely independent, many Jews were taken into slavery each time their country was occupied. They returned and fought to establish their independence, but Roman armies later subdued the country. Eventually, the Romans tired of Israeli plots and rebellions, sacked Jerusalem, and dispersed most of the people in A.D. 70 to create the Jewish diaspora.

For the next 2,000 years, Jewish communities established themselves in most European, North African, and even Asian countries. Small in number and exclusive in outlook, they often attracted suspicion, jealousy, and opposition, becoming subject to oppression and segregation in ghettos. The lands in Southwestern Asia became known as Palestine and were occupied by people from surrounding areas, mainly Arabs, together with the few Jewish families who had not been evicted. In the A.D. 600s, Arabs quickly converted to Islamic beliefs and Muslim armies spread Islam. Jerusalem became the third most holy site for Muslim pilgrimages after Mecca and Medina. Challenges by European armies that were called "crusades" had only temporary effects but resulted in deep hatred between Muslims and Christians.

From medieval times, the Turkish Ottoman Empire governed these lands without any great emphasis on economic development. Siding with Germany in World War I, Turkey lost most of southwestern Asia to groups of Arab tribes who had fought alongside the British and French to free the lands. After the war, protectorates were given to France and the United Kingdom, which established the present Arab countries. The new countries gained independence after World War II.

Palestine was one of the territories under British mandate. Although the authorities resisted too many Jews coming to Palestine, numbers had been building since the late 1800s, when the Zionist movement emphasized the need for a Jewish homeland after oppression in many countries of eastern Europe and consequent migrations to the United States. Jewish voices became more powerful. The Holocaust of World War II, when Germany and its allies murdered millions of Jews, led to further pressure for establishing a new country of Israel on Palestinian land. However, Arabs rejected the 1947 United Nations Partition Plan. In 1948, when Jews took matters into their own hands and declared Israel's independence, Egypt, Syria, Jordan, Palestine, and Iraq declared war, but Israel succeeded in resisting their attacks. Millions of Arab Palestinians were displaced in temporary camps in Jordan and southern Lebanon.

Since 1948, in further wars between Israel and Arab countries, the Israeli military power usually triumphed and took over more Arab lands. "Terrorist" activities by Palestinians—seen by them as attempts by an oppressed people to establish their own country—were ruthlessly countered by Israeli military action (see Box Figure 1). In the 1970s and 1990s there were attempts, led by U.S. presidents, to reconcile differences. After the 1967 war, Israel returned the Sinai Peninsula to Egypt but retained control of the Israeli-occupied Palestinian territories of the West Bank and Gaza. The Jordanian and Egyptian governments gave up their rights to the West Bank and Gaza lands (which were previously parts of their territories) as a basis for creating a new Palestinian country by negotiation.

The following views are those that might be expressed by an Israeli Jew and Arab. Is either totally right? Should the Arabs recognize Israel and live peaceably with it, or should the Jews give back the land they govern in Israel? Should the Arabs stop attacking Jews, or should the Israeli army stop entering the Palestinian territories? Does a knowledge of geography and the distributions of people and resources help in searching for answers to these questions? You may wish to add other opposing views to this list. Is there any prospect of a way ahead that does not involve bloodshed?

A Hard-Line Jewish View	A Hard-Line Arab (Palestinian) View
We were here first and have a longer history of occupying this land; it is ours, as claimed by Abraham and Joshua.	Our ancestors were here from the time of Abraham, whom we also recognize as a father of our people.
Our religion started first in this area. Our holy sites include Hebron (where Abraham is buried) and Jerusalem, both in, or partly in, the West Bank.	Jerusalem and Hebron are sacred to Muslims.
The United Nations agreed to the partition of Palestine and recognizes Israel with its Jewish majority.	The decision resulted from a combination of weak Arab support and strong U.S. and British pressure. We were ignored, although we made up most of the population here in 1948. Israel has taken large areas of land from us that were not part of the UN plan.

(a)

(b)

Box Figure 1 Jerusalem: aspects of conflict. (a) Dome of the Rock and the Western Wall ("Wailing Wall"). The Western Wall (sometimes referred to as the "Wailing Wall") is the holiest site for Jews because it is all that remains of the Herod's Temple complex destroyed in A.D. 70 by the Romans. Note the gold dome of the Dome of the Rock, the holiest Islamic site in Jerusalem, just 150 meters away in the Temple Mount (Harem al-Sharif) complex. (b) Israeli soldiers relaxing in the Old City. **Photo:** (a) © Mike Camille and (b) © Michael Bradshaw.

We have returned some territories we took in the 1967 war but hold onto the rest as a matter of national security and survival.

Arabs deny our rights to exist and want to wipe us off the map; we are exercising our right to defend ourselves. When joined, the surrounding Arab countries outnumber us, so we have to ensure strong security.

We regard them all as possible terrorists who blow up our restaurants and nightclubs, kill our athletes, assassinate our leaders, and drive suicide bombs into our neighborhoods.

Jerusalem is our real capital city and is central to the Jewish faith. Some Jews want to remove the mosque on the holy mount.

No strong Palestinian nationality was expressed here before 1948. This area of land was merely a British protectorate carved out of the former Ottoman Empire, and most people knew of the intention to create a land for Jews. The present Palestinians are Arabs who should have been taken in by existing Arab countries. They have invented Palestinian nationalism as part of a plot to eliminate Israel.

They are poor workers and earn only low wages. We go out of our way to employ them, but it would be better to employ only Jews.

The Palestinian Arabs have not repaid all the help we have given to them, raising their well-being above that of other Arabs in this region.

After the 1967 war, the United Nations ordered Israel to hand back the occupied territories, but it has not done so more than 35 years later.

We were forced out of our land, we lived in crowded camps without amenities, and we now live in poverty. The Jews close checkpoints with no notice and interrupt our lives. We cannot argue with them because they are supported by the United States, and we now hate that country as well.

They refuse to recognize our presence and nationality, suppressing our language, religion, and culture. Most of us want to live peaceful lives, but they treat us all as spies and criminals, abusing our human rights.

Jerusalem is sacred to us, and any moves to destroy the mosque would be a declaration of all-out war that would unite Muslims.

We want our own lands and independence from Israel.

They are well fed and materially wealthy. If we want to study for the qualifications that would earn us better jobs, we cannot do so in our country and have to go elsewhere. A doctor friend of mine, who works in a Jerusalem hospital, had to go to Greece to qualify.

We can do nothing that is legal to improve our situation, so it is not surprising that some of us take to the gun and bomb.

- How can the original melting pot required by national cohesiveness in the face of early enemies accommodate the divisive ethnic interests of Russian, Moroccan, Western, and Asian Jews as well as Israeli Arabs?

A further cause of tension is that although Israel proclaimed Jerusalem to be its capital in 1950 and built the Gnesset (parliament) in Jerusalem suburbs, the United States and most other countries do not recognize Jerusalem as the Israeli capital and maintain embassies in Tel Aviv. Jerusalem presents a major problem for Israeli–Palestinian negotiations. After Israel took East Jerusalem in the 1967 war, it confiscated land in the occupied area and made Palestinians second-class citizens by denying them property rights. The democratically elected Israeli government reflects the complex grouping of interests, conservative and liberal, religious and secular, as well as those of varied ethnic origins. Many parties receive small percentages of the election votes, forcing alliances in the formation of governments and giving small parties scope to assume greater power when voting is close.

One of the consequences of prolonged conflict between Israelis and Palestinians has been increased geographical separation resulting from the attempt to provide security. Because Israelis are in the position of power, they have dictated the course of separation. To protect their settlements in the West Bank they have been constructing a wall that runs through the territory (Figure 8.34a). While the wall prevents Palestinians from easily entering Israeli settlements, the land for the wall comes from Palestinians who are then displaced. Moreover, whereas the wall is constructed to give easy access by the Israelis to the rest of Israel, many Palestinians find themselves

cut off from their communities and workplaces. The wall has resulted in much international criticism and has even been rejected by Israel's highest court, causing the Israeli government to modify its placement in many locations; but the government is forging ahead with its construction in the hopes that it will provide security for Israelis. At the same time, the Israeli government decided to remove Jewish settlers from the Gaza strip in 2005 (Figure 8.34b). Many Israelis supported the move in the hopes that it would foster peace. Many thought it too costly to provide security for 9,000 settlers occupying more than 25 percent of the Gaza strip, overall inhabited by more than a million Palestinians mostly hostile to the settlers.

People

Ethnic Differences

Israel's population is composed of Jews, Israeli Arabs (which include Druze and Bedouin), and Palestinian Arabs (see Table 8.3). The Arabs and some of the Jews trace their family existence in the country to before Israel's independence. The Sephardic Jews from southern Europe form the majority, but the Ashkenazi Jews of Central and Eastern European origin play important roles in politics. Other Jews are of Asiatic or African origin or arrived from Russia in the 1990s.

Israeli Arabs, which include Druze and Bedouin, accept the Israeli state and rarely conflict with Israeli Jews. On the other hand, Palestinian Arabs conflict greatly with Israelis because they reject the state of Israel and would prefer to have their own Palestinian state. Most often simply called Palestinians, they go by the name of the land they occupied before the Israeli state was carved

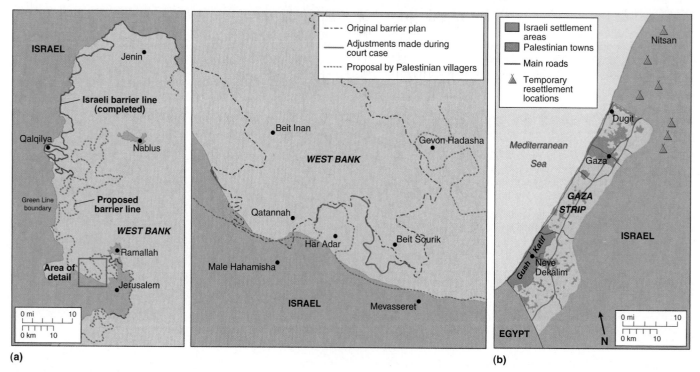

(a)   **(b)**

FIGURE 8.34 **Israeli barrier in the West Bank.** (a) Israeli Barrier in the West Bank. (b) Jewish settlements abandoned in Gaza in 2005. **Sources:** (a) Mideast Barrier changes from *The New York Times*, July 1, 2004. Copyright © 2004 The New York Times Co. Reprinted by permission. (b) *The Economist*, August 20, 2005, p. 36.

out of it in 1949. Interestingly, it was common to also call Jews Palestinians before Israel was created. Palestinian Arabs compose 90 percent of the population in the West Bank and Gaza and resent not having full governmental powers. As soon as a terrorist or external threat occurs, Palestinians are confined to their living areas. As Israeli government policy, Jewish settlements were established with Israeli government funding in both the West Bank and Gaza (see Figure 8.33f) and populated by hard-line Jews committed to retaining these areas.

A major source of the tensions between Arabs and Jews is the continued building of Jewish settlements in the Palestinian areas. The intrusive Jewish settlements split the Palestinian territories and cost the Israeli government heavily to build them and maintain their security. Palestinians are mainly restricted to low-wage jobs and denied access to higher-earning professions; Palestinians who wish to become doctors, for example, have to train abroad.

Rural versus Urban Emphases

The Israeli population density is higher than that of other countries in the region because of the country's small size, special ethnic characteristics, and urban–industrial economy. Most people live along the coastal plain, but political considerations support settlements close to frontiers. The West Bank is also densely populated, mainly by Palestinians but with Jewish settlements.

Kibbutzim (singular: kibbutz)—that is, democratically run communes—provided much of the initial thrust of Israeli settlement in rural areas. Socialist Zionists created the kibbutz idea and established the first kibbutzim in 1909. Kibbutzim provided a spiritual, agricultural, and social basis for the economy and in the formative days of Israel as a country tied families to the land (Figure 8.35). More recently, development of the rural areas has been through cooperative (*moshav*) farms and smallholdings (*moshawa*). Many Israelis, however, moved into the growing towns, and less than 5 percent of the Israeli labor force now works on kibbutzim or *moshav* farms.

Israel changed its goals as increasing numbers of immigrants came from urban areas in other parts of the world and sophisticated manufacturing and commercial functions developed. By 2005, 92 percent of the Israeli population lived in towns. Tel Aviv–Jaffa, with nearly 3 million people, is the largest urban area, with some features of global cities, while Jerusalem and Haifa are other large cities. Jerusalem has immense significance for Jews, Muslims, and Christians. Relations in and around the city remain tense over such issues as the holy sites and Israeli prevention of Arab movements at times of emergency.

Economic Development

Israel's economy places its per capita income in the same league as countries of southern Europe. Ownership of consumer goods is high, good water is available, and energy use is high (see Figure 8.12). In the mid-1980s the steady growth and

FIGURE 8.35 Israel: communal farming. Harvesting eggplant on Kibbutz Ein Gedi, located between the desert hills and the Dead Sea. Bananas are in the background. **Photo:** © George Holz/The Image Works.

diversification of the economy stalled in crisis. Israel's large 1985 budget deficit and rapidly depleting currency reserves required U.S. assistance to reduce foreign debts to manageable proportions. Israel remains a highly centralized and heavily taxed country—an outlook supported by workers and managers fearing job losses, an influential group of politicians, and a large bureaucracy.

Diversified, High-Tech Economy

Israel's economy is diversified. The agricultural sector, using intensive reclamation and irrigation farming methods, produces fruits, vegetables (Figure 8.36), and flowers for export to Europe but now constitutes only 5 percent of total GDP.

Israel possesses a well-educated workforce and access to foreign aid and investments. Providing a sophisticated defense capability prepared a generation of engineers for work in Israel's high-technology industries. Although the influx of Russian Jews in the early 1990s generated worries about rising unemployment, they brought professional skills and caused a new boom in house building.

Manufactures make up 44 percent of exports and include diamonds, machinery, military equipment, and chemicals. In the 1990s Israel became a major center and leader of high-technology development in manufacturing areas such as telecommunications, electronic printing, diagnostic imaging systems for medicine, and data communications. Such products now account for 50 percent of industrial output, compared to 15 percent in 1990. Most of the manufacturing industries are based around the coastal cities such as Tel Aviv and Haifa. Industrial estates along the borders with Gaza and the West Bank employ cheap Arab labor.

FIGURE 8.36 Israel: agriculture. Sprinkler irrigation on a farm in the Negev Desert of southern Israel. Assess the importance of the presence or absence of water for landscapes in this region. **Photo:** © Paul Souders/Corbis.

Services

Israel gains over half of its GDP from the services sector. The quality of education and health-care provision places it on a par with Western Europe and ahead of most Arab countries. The financial sector grows increasingly as Israel's economy opens to privatization and foreign investment.

Tourism is a major industry that attracted over 2.4 million visitors in 2000 but conflict and war reduced the number of tourists to 862,000 in 2002. Many visitors combined visits to, for example, Jerusalem, Petra (Jordan), and the pyramids (Egypt), but the 2002 invasions of Palestinian areas caused both Jordan and Egypt to close their borders with Israel and suspend diplomatic relations, temporarily ending such combined itineraries.

Diversifying Trade Links

Israel is now actively pursuing free-trade policies and has opened its borders to products from neighboring Jordan and countries farther away in Asia and Latin America. In a peaceful time, Israel could have a leading economic role in the eastern Mediterranean. Conversely, conflict necessitates more arms spending, interrupts trade, and generates long-term hatreds.

At present, half of Israel's trade is with the European Union, and the proportion is rising. Its attempts to join the EU as a trading partner—with access to the free-trade area, rather than as a political partner—move forward more slowly than Israel would like. Israel is particularly keen for its agricultural and telecommunications products to gain equal entry alongside those of European countries. It argues that joining the European free-trade group of countries would compensate for opening its home markets to Jordanian produce and the future return of the Golan Heights to Syria.

Poverty in Gaza and the West Bank

While Israel has a growing economy that places it ahead of its neighbors in development and lifestyles, the Palestinian areas of Gaza and the West Bank continue to have poorer conditions for human development. Palestinians accuse the Israelis of paying unequal attention to the needs of Palestinians in these territories compared to Israelis—a form of apartheid. Israelis accuse Palestinians of harboring terrorists. Though billions of dollars in international aid was given to Palestinians in the late 1990s, the intense and very destructive conflict between Palestinian terrorists and the Israeli army in the 2000s has destroyed much of the infrastructure for Palestinians in both Gaza and the West Bank, in turn ruining the economy and leading to a social crisis. In Gaza, for example, 50 percent of the workforce is unemployed, and 40 percent of its population lives in refugee camps. Many families depend on emergency food supplies.

Turkey and Iran

Turkey and Iran occupy the northern and eastern margins of this region (Figure 8.37). The two countries are influential in Southwestern Asia and the wider world, sharing economic

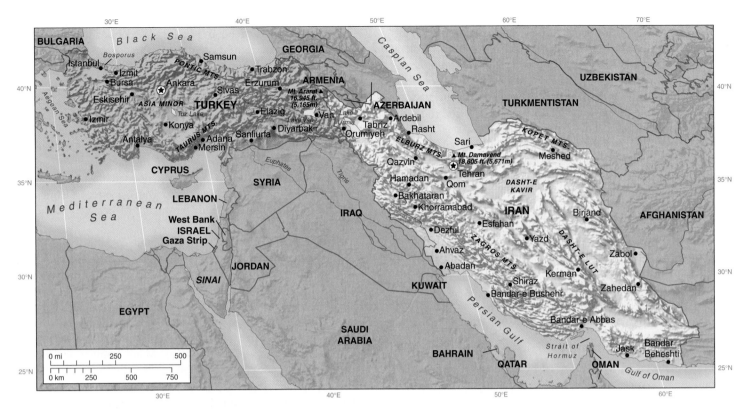

FIGURE 8.37 Turkey and Iran: countries, physical features, and cities.

leadership of the region with Saudi Arabia, Egypt, and Israel. The populations of Iran and Turkey together comprise nearly one-third of the total population of the Northern Africa and Southwestern Asia region (see Table 8.3).

Turkey and Iran have crucial strategic positions between the southern boundary of Russia and Neighboring Countries (see Chapter 4) and the Persian Gulf oil fields. Their control of three major choke points on sea routes—the Bosporus and Dardanelles in Turkey linking the Black and Aegean Seas and the Hormuz Strait at the entrance to the Persian Gulf—adds to their strategic significance. The increasingly democratic nature of their governments, after years of military and religious dictatorships, contrasts with those of other countries of Southwestern Asia and gives them greater potential for success within the world economic system, although Iran is still the subject of U.S. sanctions. In 2004 the EU deemed it acceptable to open talks with Turkey on its bid to join the EU.

Iran and Turkey share other characteristics. They are largely mountainous countries along the plate collision margin between Arabia and Asia. This makes them subject to frequent, devastating earthquakes. The mountains also cause the uplift and cooling of humid air, attracting precipitation, much of which falls as winter snow. Meltwater in the spring and early summer feeds rivers flowing southward to the Persian Gulf and northward into the Black and Caspian Seas.

Both countries have Kurdish minorities and must deal with the fight for Kurdish independence, a problem they share with Iraq and Syria. All these countries resist demands for a separate country of Kurdistan.

Countries

Although Iran and Turkey have some similarities such as their shared belief in Islam, they are long-term rivals with very different religious emphases, types of government, basic resources, and approaches to the world economy. Iran is occupied mostly by Persians, a people who speak an Indo-European language, Persian (Farsi). Most people in Turkey are Turks and their language is in the Altaic family (see Figure 8.6 and Table 8.3). Thus, linguistically and more broadly culturally, Iranians and Turks are culturally distinct from the Arabic peoples who dominate this world region. Separate histories of empire building add to their distinctiveness. Iran further stands out in the region as a country where Shiite Islam dominates.

In the late 1990s Turkey diversified its economy in the context of its secular and democratic government and slowing population growth. In 1979 Iran moved from a military dictatorship to an Islamic constitution, but its suspected support for terrorist activities led to U.S. sanctions. Iran spent major resources on the 1980–1988 war with Iraq and has to contend with a faster growing population than Turkey.

Historic Rivalries

Although Iran and Turkey are Muslim countries, both are proud of their historic and pre-Islamic traditions. Modern Iran occupies part of a wider Persian Empire, which existed in many different forms under various dynasties from 648 B.C. until A.D. 1935, when the name "Iran" was adopted. Beginning in the later 1800s, rivalry between Russia and the United Kingdom raised Iran's significance in world geopolitics. British discoveries of oil in Iran and its extraction since 1901 began the Southwestern Asia oil era. During World War II, Britain and the Soviet Union occupied parts of Iran to guard its oil fields against German attack.

The land that is now interior Turkey had its own civilization, developed under the Hittites around 1600 B.C. Later, Persians, Greeks, and Romans took it over. It was Christianized as part of the eastern Roman Empire of Byzantium. The Muslim expansions during the A.D. 600s reduced Byzantine influence to the peninsula that is modern Turkey. In 1071 the Seljuk Turks defeated the Byzantines and superimposed Islam on Christian traditions. In 1281 the Ottoman Turks established a new empire, which grew very large in the 1500s and 1600s. At its height, the Ottoman Empire included southeastern Europe (see Chapter 3) and most of the Arab world. At the end of the 1600s, the Ottoman Empire began to decline. An alliance with losing Germany and Austria-Hungary in World War I brought about the empire's final destruction and establishment of modern Turkey after the war.

Modern Developments

In the early 1900s Iran was ruled by a shah who had elevated himself from a military career to royal status. He and his successor son kept the country largely under military control but with a semblance of freedom to adopt Western education and develop a wider range of economic activities. Iran's oil wealth and control of the Strait of Hormuz (Figure 8.38) resulted in involvements with the global economy but in less economic diversification at home. Though the economy of Iran grew under the shah, he was a repressive leader and resentment toward him grew. In 1979 nationalist religious leaders, led by the Ayatollah Khomeini, seized political power; the shah fled, and Iran shifted to Islamic religious leadership and isolationism. Because the U.S. government supported the shah and refused to turn him over to stand trial, anti-American feelings grew in Iran, culminating with the taking of hostages at the American embassy in Tehran.

The new constitution of the Islamic republic in Iran depends on a Shiite interpretation of Islamic government. Clergy are expected to establish a just social system and implement Islamic laws. In practice, this system raises tensions between the president and the members of the Islamic Assembly (Majlis), all of whom depend on a popular vote, on the one hand, and the religious priorities of the controlling Council of Leadership, on the other. Many Iranians see the latter as obstructionist over internal reforms and external relations, preventing the elected president and Majlis from enabling Iran to take a full place in the modern world. In the

FIGURE 8.38 Iran and the Persian Gulf. Space shuttle view of the Strait of Hormuz with the Zagros Mountains of Iran to the left (north) and the United Arab Emirates and Oman on the south side of the strait. All marine traffic into and out of the Gulf must pass through the strait, which has great strategic significance. **Photo:** NASA.

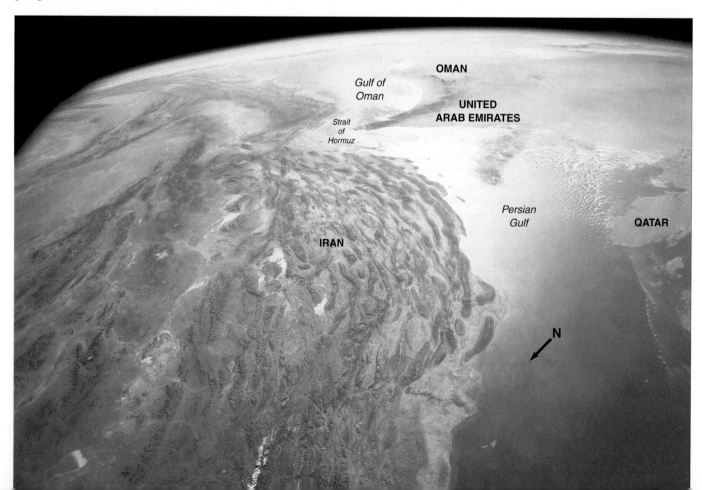

Cold War, Iran avoided close links with the West, however, and developed agreements with the Soviet Union, especially after 1979.

Turkey's political system is very different from that of Iran. After the disastrous defeat of the Ottoman Empire in World War I, Turkey became a nationalist and secular republic, putting the country before religion in questions of government. From the 1920s until 1936, the new leader, Mustafa Kemal Ataturk, ruled with a single-party government. Turkey gradually modernized and became increasingly involved in the global economy, although most economic developments came after World War II, in which it remained neutral. After the war, pressures from the Soviet Union caused Turkey to become a member of NATO (see Chapter 3). The country received U.S. support as it provided locations for U.S. air force facilities. The control of the Bosporus made it possible to monitor Soviet ship movements from Turkey's southern flank of the Black Sea. U.S. bases in Turkey were close to the Soviet heartland.

Following the breakup of the Soviet Union in 1991, rivalry between Turkey and Iran continued as both sought links with Muslim countries in the former Soviet republics of the Caucasus Mountains and Central Asia. Turkey claims historic and language links with these new countries, but Iran is developing more practical links with the landlocked countries in trade and transportation (see Chapter 4).

In the late 1990s both Iran and Turkey faced tensions between liberalizing Western influences and pressures from Islamic-oriented political groups. Young Iranians in particular dislike the strict Islamic rules imposed by a fundamentalist few on a moderate majority. The election of a more liberal cleric as president in 1997 and again in 2000 resulted in conflicts between his edicts and those of the conservative Council of Guardians. In contrast, Turkey faces continuing challenges from Islamic political groups to its long-term secular state principles. When a militant Islamic group was elected to government in the late 1990s, the military took over and installed a secular group in power.

People: Ethnic Variations

Though Iran and Turkey are dominated by Persians and Turks, respectively, both are ethnically diverse (see Table 8.3). Of the minority groups, the Kurds are the largest. Stretching from southeastern Turkey across northeastern Syria, northern Iraq, and western Iran, they are a minority group with sizable population numbers. Governments feel threatened by them, especially the Turkish government (see the earlier "Human Rights" section).

Economic Development

The two countries have different types of economic development based largely on oil income (Iran) or water resources (Turkey). With 11 percent of world oil reserves, Iran experienced rapid oil income growth from the 1960s, reaching 40 percent of the value of national production in the oil boom of the 1970s. Much of the additional wealth was at first spent on encouraging urban–industrial, rather than agricultural, expansion and importing foreign goods. The takeover of the country by religious leaders in 1979, U.S. sanctions, and the war with Iraq in the 1980s led to isolation from the world. The low oil prices of the 1980s and 1990s worsened Iran's financial position. In the 1990s, however, Iran began to look outward and restructure its economy and world links. The country sought links with major multinational oil corporations for marketing and technical updating, but Iran's constitution forbids foreigners from owning mineral concessions. Although Iran possesses other mineral resources, few have been explored or extracted. In the early 2000s the Iranian government invested in nuclear power plants but was condemned by the United States for such activities because nuclear technology employed to generate electricity also can be easily used to develop nuclear weapons.

Turkey has little oil but invested heavily in the development of its water resources for more agricultural output and hydroelectricity. Industrial expansion increased from 1950 and especially in the 1970s and 1980s. The 1990s were a period of political and economic confusion, partly due to the internal war against the Kurds that absorbed large amounts of money and manpower.

While Iranian national income fluctuated after 1980 with world oil prices, Turkey's real income rose steadily to exceed that of Iran until the late 1990s—partly as a consequence of lower oil prices and the sanctions affecting Iran. By 2005, rising oil prices made Iran equal to Turkey in total GNI PPP; they lead the region (see Table 8.3).

Manufacturing Differences

Iran was the first major oil exporter in the Persian Gulf, but internal economic development was slow until the 1960s. Craft industries, such as carpet making and weaving, remain significant. Oil revenues, which now provide 90 percent of exports, were used to fund public-sector manufacturing ventures such as oil refineries and petrochemical plants on the Gulf coast, the iron and steel works at Isfahan, and machine building at Arak. In the 1960s and 1970s new roads were built to link the capital, Tehran, with the Gulf coast ports.

The private sector, often supported by public investment, developed car assembly, textiles, leather goods, plastics (Figure 8.39), and other light industries around Tehran and other major cities. Small industrial estates were established in most medium-sized and larger towns. In the early 2000s Iran launched its own car, to be produced mainly by Iran Khodro, one of the world's top 20 carmakers. The motor industry makes up 20 percent of Iran's manufacturing output but stalled behind protectionist trade barriers, relying on old designs.

The first Turkish manufacturing developed out of mining for chromite and copper in the mountain ranges. In the 1990s gold was added to mineral exports. Import-substitution industries, such as car assembly, textiles, and steel, were also established.

FIGURE 8.39 Iran: manufacturing. The Irana plastics factory, in which the machinery is basic and aging. **Photo:** © Don Smetzer.

As an oil importer, Turkey's manufacturing economy grew more steadily than that of Iran in the later 1900s and is more diversified. The problems of high oil prices in the 1970s caused Turkey to reform its economy, and industrial output increased in the 1980s and 1990s at a time of low oil prices. Between 1980 and 2000, real GDP per capita rose by 50 percent while Iran's stayed the same. Turkey diversified into light industrial products, which now make up some 80 percent of exports. In the 1980–1988 Iraq–Iran War, Turkey supplied both countries with food and manufactured goods in exchange for oil.

Services and Tourism

Turkey has a more developed services sector than Iran. Government employment is very important in the economy. International tourism grew and annually brings in over a billion dollars. In 2002 Turkey attracted 13 million visitors (up from 1 million in 1980 and under 5 million in 1990). The development of tourist resorts along its sunny coasts and the availabil-

ity of historic, often religious, sites made Turkey a major venue for Europeans. The rising income from tourism partly compensated for the falling level of wages sent home by 1.5 million Turkish workers in European and Arab countries. In the late 1980s they contributed nearly half of Turkey's foreign exchange income. Turkey is the only country in Northern Africa and Southwestern Asia (outside of Israel) with such a diversified economy.

Test Your Understanding
8D

Summary

Arab Southwest Asia is the heartland of the Arab world but has internal divisions based on whether oil is produced. Few countries have sufficient water for agriculture. Israel faces a future of continuing conflict with its Palestinian population and its neighbors and, like them, has problems of finding resources to handle an increasing population. Israel has a Western economy emphasizing human resources and technologic skills.

Turkey and Iran are large but very different countries from historic times to the present cultural and political regimes. Turkey's water resources are the basis for much of the country's development, and it has a more diversified economy than most countries in the region. Iran's oil income and reserves provide continuing potential.

Questions to Think About

8D.1 How do the economies of the oil producers and non-oil-producing countries in this region compare with each other in terms of income fluctuations and levels of diversification?

8D.2 What has been the impact of wars and civil strife on economic development in the countries of this region?

8D.3 How do Israel's geography and development contrast with those of the other countries of this region?

Key Terms

diversified economy

Gulf Cooperation Council

kibbutzim

Chapter 9

Africa South of the Sahara

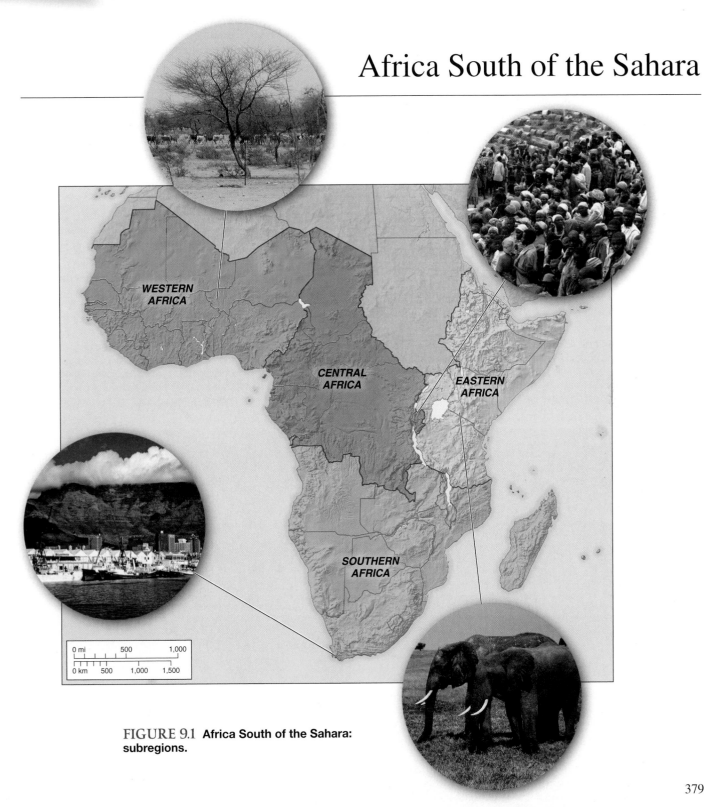

FIGURE 9.1 **Africa South of the Sahara: subregions.**

WESTERN AFRICA

CENTRAL AFRICA

EASTERN AFRICA

SOUTHERN AFRICA

| 0 mi | | 500 | | 1,000 |
| 0 km | 500 | 1,000 | 1,500 | |

Themes in This Chapter

The people of Africa South of the Sahara face so many challenges that it is not easy to focus on the great achievements and resources of the region. Political, economic, health, and environmental problems cause African people to be among the poorest and most deprived in the world today. But this continent is where human beings began and where many human cultural contributions developed. Today much life and enthusiasm—personally, musically, in sports, and in public affairs—extend from the region to places where Africans visit or live in other parts of the world. In this chapter we examine

The region's past and present achievements, resources, and challenges.

Ethnic diversity, colonial regimes, and prospects at independence.

Natural environments of tropical climates, changing climates, ancient rocks, plateaus and rifts, tropical forests, grasslands, and deserts.

Human interactions with the natural environment.

Global and local changes: the current situation across the region—political, population, and economic trends.

Four subregions: distinctive geographies within the region.

- Central Africa: challenges at the heart of the continent.
- Western Africa: contrasting natural environments, history of empires, crop and mineral exports.
- Eastern Africa: Indian Ocean and Arab historic orientations; modern conflicts.
- Southern Africa: most potential, but feeling the main effects of HIV/AIDS.

Personal View: A Brighter Future in Ghana

Point–Counterpoint: Africa at the Crossroads

Personal View: Rwanda

Geography at Work: Social Justice in South Africa

A New Dawn?

A Heritage of Resources and History

Africa South of the Sahara covers most of the world's second largest continent (Figure 9.1). This region was the cradle of the human species. Skeletons of the oldest known *Homo sapiens* are found only in Africa, and African peoples contain much greater genetic variety than the rest of the world's population. People in other continents are probably all descended from a group of *Homo sapiens* that moved out of Africa after the initial phases of human genetic diversification (see Figure 1.9). Modern Africans number over 700 million people.

FIGURE 9.2 Ancient African achievements. (a) The ruins of Great Zimbabwe record the existence of a major trading center from around A.D. 1100. It was based on a sophisticated political organization and economy before European exploration and colonization. Each layer of rocks in the walls is approximately 7 to 10 cm (3–4 in.) high. Great Zimbabwe became a symbol to both sides in the 1960s struggle for majority rule in Southern Rhodesia (now Zimbabwe). For Africans, it was a symbol of African historic achievement; for white Rhodesians, it symbolized the unthinkable triumph of black Africans. (b) San (Bushmen) rock paintings that are found in caves in Southern Africa and are up to 6,000 years old. **Photo:** (a) © Michael Bradshaw; (b) © Vol. 145/Corbis.

(a)

(b)

The history of African peoples before the coming of Arabs and Europeans was rich and sophisticated. It included organized kingdoms and empires with trading connections across the continent (Figure 9.2). However, the extension of global links with Arab, Asian, and European people in the last 1,500 years was often to the disadvantage of Africans. The arrogance of the external powers brought undeserved low expectations of African abilities. African people, whose ancestors were forcibly moved as slaves, now form significant populations in other parts of the world. In the Americas, particularly Brazil and the Caribbean but also the United States,

they contributed lasting cultural features despite local discrimination. Folklore and folk tales persisted from the rich African story-telling traditions. Musical influences from spirituals and blues to reggae are basic to much modern Western music, while African art underlies many modern abstract paintings and sculptures.

Africa South of the Sahara has a great diversity and beauty of natural landscapes from plateaus to rift valleys and mountains (Figure 9.3). Its climates range from tropical arid to equatorial rainy; its ecosystems include tropical rain forest, savanna grass-lands, and desert—a stock of biodiversity. The rocks underlying

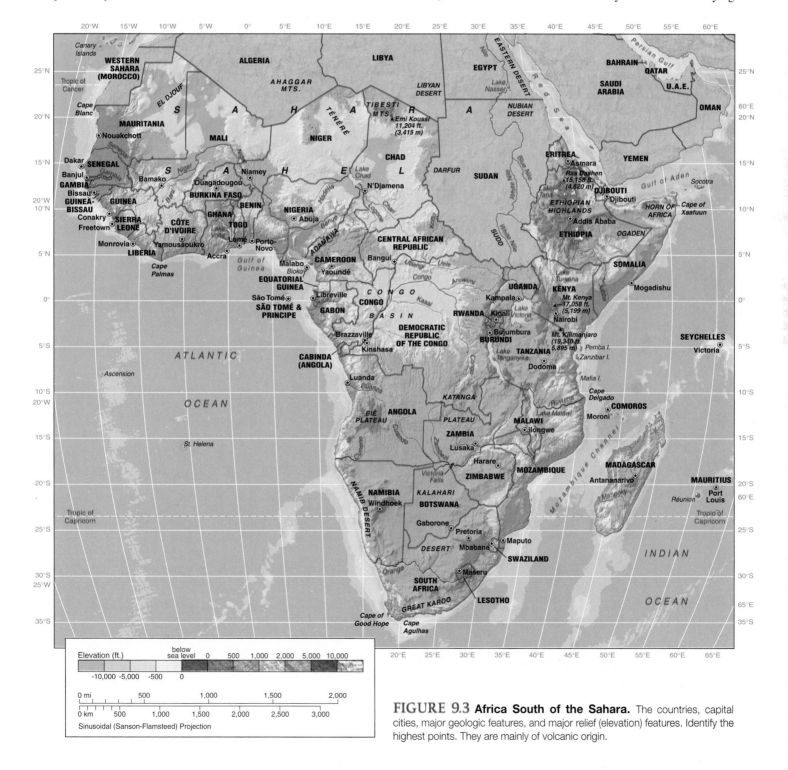

FIGURE 9.3 Africa South of the Sahara. The countries, capital cities, major geologic features, and major relief (elevation) features. Identify the highest points. They are mainly of volcanic origin.

these landscapes are mostly ancient and contain some of the world's largest deposits of bauxite (for aluminum), cobalt, copper, gold, and diamonds as well as newly valuable platinum and coltan (tantalum, used in mobile phones). Increasing amounts of oil are being found around the coasts.

Clear Boundaries

This world region has clear, mainly coastal, boundaries. The northern boundary is through the almost empty Sahara. In the colonial late 1800s and early 1900s, trade routes and connections moved away from the land crossing of the Sahara that connected northernmost Africa with the rest of the continent and toward the ocean routes linked with the expanding European global economy. The Sahara remains a significant boundary, although some countries could be placed in different regions. Thus Mauritania, but not Sudan, is included in Africa South of

the Sahara. The former has long-term links with Western Africa, while the latter is part of the lower Nile River valley with Egypt.

The Challenges of the Present

Materially, Africa South of the Sahara is the world's poorest region, where half the population lives on less than one U.S. dollar a day. Most of its countries are not gaining in wealth, and some even lost ground in the later 1900s. In 1955 Africa South of the Sahara accounted for 3.1 percent of world trade; by 2000 its proportion had fallen to 1.2 percent. Despite continuing, though modest, support by former colonial countries such as France and the United Kingdom, this region's problems were mostly ignored by the West until the G8 countries faced the obvious needs of the region in the early 2000s. In 2000, the U.S. Congress passed the Africa Growth and Opportunity Act, which reduced import duties on African imports, especially in textiles and clothing.

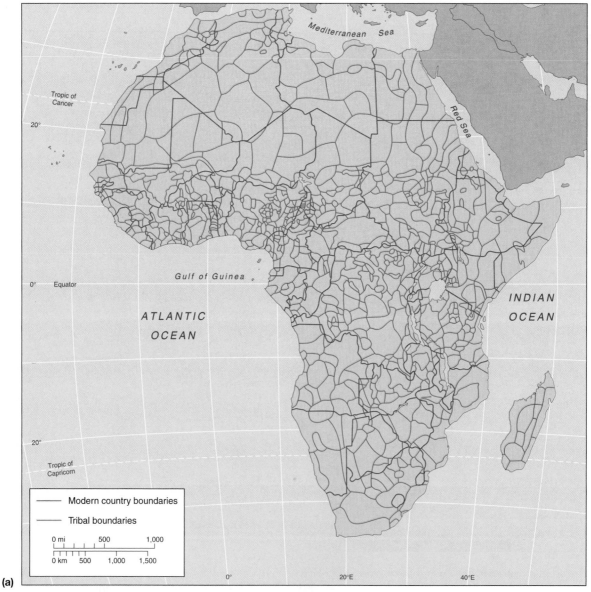

FIGURE 9.4 Africa South of the Sahara: cultural features. (a) The relationship of colonial-imposed boundaries of the modern countries to ethnic group areas **Source:** (a) From *World Regional Geography: A Question of Place,* by Paul W. English and James A. Miller, 3rd ed., 1989, © Paul W. English and James A. Miller.

Major internal problems include the civil strife and poor government that hold back many countries politically, economically, and socially. In 2001 there were reports of a slave trade operating to force children to work on cocoa plantations in Western Africa and to export them to Europe and the Arab countries. Always a problem, tropical diseases, such as malaria and river blindness, together with the new scourge of HIV/AIDS, infect many people across the continent. Other environmental problems include the further expansion of the Sahara as the climate becomes more arid under future global warming.

African Cultures

The present challenges have to be seen in the context of the region's past. Africa was the stage for the generation of the first human cultures. Africans mixed with Arabs, Asians, and Europeans, all of whom now call Africa South of the Sahara home. The geographic variety today resulted from the interactions among traditional indigenous cultures, varied natural environments, the expansion of Arab Muslim cultures, the European colonial cultures, and the Westernizing cultural invasions of the late 1900s.

African Ethnic Diversity and Shared Cultures

The members of ethnic groups forming the basic social and political units of the indigenous people share kinship and territorial links (Figure 9.4) and frequently a language and cultural

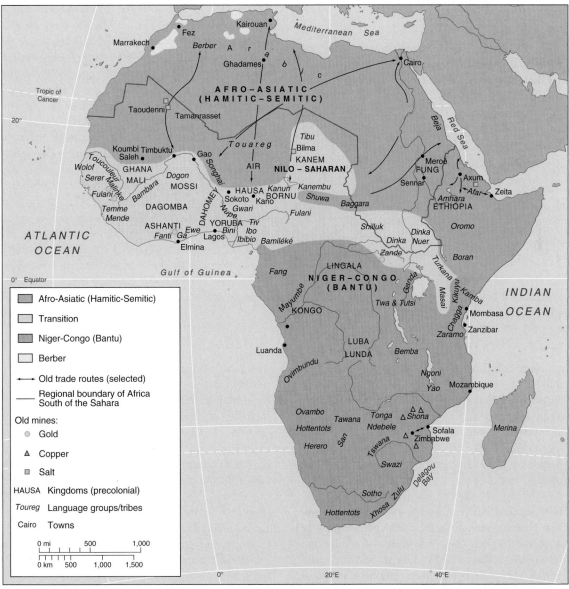

(b)

FIGURE 9.4 continued (b) Ethnic groups, languages, and precolonial kingdoms. In the zone between the area dominated by the Afro-Asiatic groups of northern Africa and the area dominated by the Niger-Congo peoples to the south, a diversity of languages exists. The advance of the Afro-Asiatic groups into regions south of the Sahara in medieval times helped to encourage southward movements of Bantu groups. By the late 1800s, they were replacing longer-established groups in the far south. **Source:** (b) From *The Changing Geography of Africa*, by A. T. Grove, 2nd ed. Oxford University Press, 1993.

institutions. European colonists called these groups **tribes,** but it is difficult to define "tribe" with precision. Sometimes distinctive appearance marks a tribe, such as the tall Masai herders of the grassy plains or the small pygmy groups of the forest. However, tribal identities often override physical and cultural differences; separate identities may occur within groups having similar physical characteristics; and, in some places, individuals may move from one group to another.

Languages reflect the diversity of groups. More than 1,000 languages are spoken in Africa South of the Sahara (Figure 9.4b). The Niger–Congo group is the largest and includes the languages of Western Africa and those of the Bantu groups in the south. The Afro–Asiatic group is important along the southern Saharan margins. Smaller groupings include the Nilo–Saharan languages in the north and the Khoisan languages in the far south.

Shared features of indigenous African groups are all being challenged today:

- Traditional African beliefs have strong relationships between humans, nature, and spiritual forces. Gods and spirits influencing human success or failure inhabit rivers, rock outcrops, and tree groves. Such beliefs are collectively known as animism.

- All humans are seen as part of a continuing chain of life, with reverence for ancestral spirits and the family unit. Large families are a blessing and childlessness a tragedy. The wider family supports each member through difficult times and educates the young in traditions.

- Artistic expressions in sculpture, music, dance, and storytelling are central to African cultures.

- Wisdom and strong leadership are respected. Traditionally, the chief is a feared, trusted spiritual leader, acting for the common welfare.

- Most Africans still get their living from the land through cultivating, herding, or hunting. Land is traditionally in communal, not individual, ownership—seen as an inheritance from the past and responsibility to the future.

External practices influence many of the indigenous cultures, especially in language and religion. Common trading languages became significant: older **creole languages,** combining several languages, include Swahili (combining African, Indian, and Arabic elements) in Eastern Africa. English, French, and Portuguese are still used in former colonies and, along with Swahili, are common in the TV and newspaper media.

The great diversity of religious allegiance includes many versions of traditional animistic religions, together with strong influences from Islam, especially in the north and east parts of Africa South of the Sahara, and Christianity—Catholic and Protestant—elsewhere. Many faiths coexist without strife between their members. Some syncretism (mixing of religious forms) occurs among Muslim, Catholic, and traditional expressions of faith. Conflicts often result where political interests play on deep religious differences.

African Groups and Empires

For thousands of years, Africa was home to a great variety of peoples, mixing and migrating with their cultures. Eastern Africa was a major crossroads of African peoples. Cattle herders from the Nile River valley, such as the Masai, lived in tension with farming Bantu peoples, such as the Kikuyu of Kenya who migrated from Western and Central Africa. In Southern Africa, the San and Khoikhoi hunters and collectors with low population densities were gradually displaced by the agricultural and cattle-herding Bantu tribes from the north, who reached the present South Africa in the 1700s.

Indigenous African groups, particularly in Western Africa (see Figure 9.4b), established empires based on the wealth created by trade in salt, gold, ivory, and slaves. The Western African empires of Ghana (A.D. 700–1240), Mali (1050–1500), and Songhai (1350–1600) had widespread influence. When the Malian Muslim emperor, Mansa Musa, visited Cairo, Egypt, in 1324, on his pilgrimage from western Africa to Mecca, he had 500 porters, each bearing a golden staff. As an example of the influence of these empires, Timbuktu in modern Mali lies near the northernmost bend of the Niger River on the southern margin of the Sahara. It grew as a market for local crops and cattle. By the A.D. 700s, its salt trade linked mines in the Sahara with markets down the Niger River. Local gold fields made its merchants very wealthy. Universities with libraries containing large numbers of imported books were established at Timbuktu and Djenne before any existed in Europe. Scholars from Greece, Egypt, and Arabia were employed as teachers.

In the south, Great Zimbabwe (see Figure 9.2a) was the center of a widespread trading empire. By the early 1800s, the Bantu peoples in southernmost Africa—the Swazis, Zulus, Xhosas, and Sothos—fought each other for territory and built strongly defended kingdoms.

Muslims in the North and East

The Arab Muslim expansion from the A.D. 600s affected the northern and eastern parts of the region. The Muslims brought their Islamic religion but also accommodated local practices, including polygamy and the use of the African drum. The introduction of camels enabled Arab and African traders to exchange goods along a few well-marked routes across the Sahara. It reached a major level of activity from 900 until the 1800s. In the west, the routes went south from Fez and Marrakesh (modern Morocco) to the middle Niger River valley (Figure 9.5); in the center, routes connected Tripoli (modern Libya) and the Lake Chad area; in the east, the Nile River route connected Egypt, Sudan, and Ethiopia. Later Islamic expansion set off holy wars as Fulani warriors and zealots grabbed grazing lands. Although Muslims tried and failed to dislodge the Ethiopian monarchy on many occasions, they devastated much of the area. In 1523 they established coastal settlements that are now the countries of Eritrea, Djibouti, and Somalia.

FIGURE 9.5 Africa South of the Sahara. A Moroccan sign that tells camel caravans it will take them 52 days to reach Timbuktu across the Sahara. **Photo:** © Vol. 145/Corbis.

Eastern Africa's first orientation to the outside world was toward the Arab countries bordering the Red Sea and the Indian Ocean. Arabs established trading centers along the eastern coast of Africa, often on islands such as Zanzibar and Pemba. They constructed impressive palaces. From the A.D. 700s until the 1800s, ivory and gold were traded and around 5 million slaves were exported to Arabia, Persia, and even China. An African–Arabian Islamic culture developed with Indian and Persian influences. Communication through the Swahili creole language became common.

Colonial Regimes

European influence in Africa grew from the mid-1400s as improved ship technology made it possible to reach India and avoid the Muslim countries of southwestern Asia. Europeans built forts at coastal trading posts in Western Africa, where ships could be loaded with the local slaves, gold, ivory, and palm products in exchange for alcohol, guns, and sugar. Sectors of the Western African coast were called the Ivory Coast, Gold Coast, and Slave Coast for their chief products. Portuguese ships took gold from Eastern Africa to pay for the silks and spices of Asia.

Slave Trade

The first European colonists were the Portuguese, who established relations with the local Kongo ruler in northern Angola in 1483. They reached the Mozambique coast in 1498 and ousted many Arab traders. From the 1600s, the new American colonies producing sugar and tobacco and later cotton for Europe needed labor. European ships transported African slaves to the Americas. Slaves entering the triangular Atlantic trade (Figure 9.6) were brought to the coast along the inland supply routes pioneered by African empires and Arab traders. The Portuguese slave trade devastated both Angola and Mozambique; nearly 4 million slaves were taken to Brazil from Angola alone. Western and Central Africa contributed millions of slaves to the Americas.

The slave trade enriched European shipowners and merchants, financing the growth of ports in Europe, which then dominated world trade. Extracting such "wealth" caused the underdevelopment of Western Africa. The payments for slaves to the slave-trading aristocracies inside Africa enabled them to buy guns and luxuries from Europe, dominate their African neighbors, and resist European colonizing incursions.

Europeans justified enslaving peoples of different skin color and culture through assumptions of racial superiority. Eventually, somewhere between 6 million and 30 million (with 10 million as a widely accepted figure) African slaves were transported across the Atlantic Ocean to the United States, the Caribbean, and Latin America. The slave trade slowed after Britain abolished slave shipments in 1808, but some countries did not abolish slavery until 1880. Humanitarian efforts in Europe and the United States during the 1800s resulted in the return of African families freed from slavery to people the new country of Liberia, the port of Freetown (in modern Sierra Leone) and Libreville (in modern Gabon). However, the returnees found it difficult to integrate with the Africans who had not been enslaved, creating long-term social tensions.

Explorers and Colonies

As the slave trade declined from the early 1800s, all of Africa South of the Sahara, except Ethiopia and Liberia, became colonies of European countries. Explorers such as Henry Stanley (United States) and David Livingstone (United Kingdom) penetrated the inland regions along major rivers, meeting many of the African empire rulers and passing on knowledge from their travels. While they did so, the Industrial Revolution developed in Western Europe, resulting in demands for raw materials such as tropical tree crops and minerals. During the later 1800s and early 1900s, French and British companies established plantations to produce cocoa and palm oil in the coastal forests, while mining companies explored the interiors. Roads and railroads connected interior mines and plantations with ports to facilitate exports. Although European countries professed an intent to "civilize the dark continent" through missionaries and administrators, they were also motivated by commercial interests and competition for overseas possessions.

In the late 1800s, the opening of the Suez Canal caused the United Kingdom to expand the port of Aden on the Arabian Peninsula to refuel its ships on the way to and from India. It established British Somaliland on the African coast. Italy colonized Eritrea and the Indian Ocean coast of present Somalia in the late 1800s.

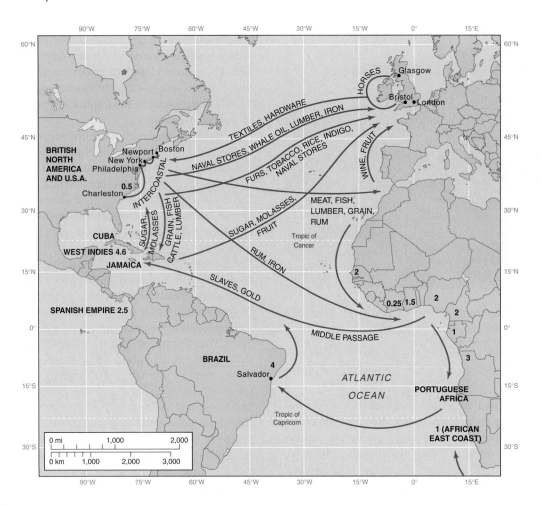

FIGURE 9.6 Atlantic Ocean: the slave trade. The 1700s Atlantic economy, trading colonial commodities with home countries and bringing slaves from Africa to the Americas. Numbers in millions from Africa are estimates of slave movements to sectors of the Americas. The British ended most of the Atlantic slave trade in the early 1800s. The country boundaries are those of today. **Source:** Data from Hugh Thomas, *The Slave Trade,* Simon & Schuster, 1997.

In 1884–1885, the European countries competing for world power met at the Berlin Congress and divided the African continent into French, British, German, Portuguese, Belgian, Italian, and Spanish spheres of influence (see Figure 1.12). There was no consultation with Africans. Occupying and controlling the new colonies, however, was often difficult and met strong resistance. Colonies became inland extensions of coastal bases, and boundaries between the colonial territories divided tribes and African territories. For example, the Kanem–Bornu sultanate, founded around A.D. 1100, was divided among Nigeria, Cameroon, and Niger/Chad. Germany lost its colonies (modern Togo, Tanzania, and Cameroon) to the British and French after World War I.

Colonists Who Settled: South Africa

In a few places, Europeans stayed, settled, and made lasting impacts on the local political and economic life. The evolution of the Republic of South Africa is particularly significant. From the mid-1600s, Dutch settlers built the port of Cape

Town (Figure 9.7) to support ships trading to and from the Dutch East Indies (now Indonesia). Farmers came to supply the port and town with crops and meat. The settlers forced the indigenous Khoikhoi, a people they called Hottentots, northward and in places enslaved them.

After Britain bought the Cape Colony from the Netherlands in 1814, it introduced new colonists, who demanded the use of English, the end of slavery, and the protection of the Khoikhoi. Many Dutch settlers moved to the more isolated interior to preserve their culture. They undertook their "Great Trek" to the Orange and Vaal River valleys, where they established the Orange Free State and Transvaal. The Boers (Dutch for "farmers") retained their Dutch-based language (Afrikaans) and were later called Afrikaners. Their new settlements displaced the Ndebele people, who moved north of the Limpopo River, and the Zulus, who moved southward into Natal, creating tensions with the local tribes.

The Boers declared a South African (Afrikaner) Republic in their new lands, but the discovery of diamonds and gold there in the 1860s caused British entrepreneurs such as Cecil

FIGURE 9.7 South Africa: Cape Town. The port, established by Dutch settlers 400 years ago, remains busy. The fish dock is a growing specialization. The city with its high-rise offices and hotels is backed by Table Mountain. **Photo:** © Alasdair Drysdale.

Rhodes to intrude. The occupation of South West Africa (modern Namibia) by the Germans in 1884 led Britain to annex Bechuanaland (modern Botswana) to block German–Boer links. The British also extended protectorates to Basutoland (modern Lesotho) and Swaziland after Boer attacks. In 1899 the Boers declared war on the British, who had instigated local conflicts. Although the British soon took the major centers, a costly and inconclusive guerrilla war followed.

After the South African (Boer) War, the four colonies—Cape, Natal, Orange Free State, and Transvaal—joined in 1910 to form the Union of South Africa as a self-governing dominion within the British Empire. However, the rights of black Africans were never part of the arrangement, and black struggles for self-determination took longer than in other African countries.

Other Colonists Who Settled

In the late 1800s, the United Kingdom developed the swath of land between Angola and Mozambique. Primarily aimed at developing mineral wealth, the British colonization sometimes coincided with cries for help from indigenous peoples who perceived the threat of Portuguese colonial repression to be greater than that of British administration and loss of land to farming settlers. For example, tribes in Nyasaland (modern Malawi) in 1891 and Northern Rhodesia (now Zambia) in 1889 requested British protection, whereas the Ndebele in western Southern Rhodesia (modern Zimbabwe) signed 1888 contracts with Cecil Rhodes to allow his British mining company to exploit the minerals on their land. European settlers came to take up the good farming land, particularly in Southern Rhodesia.

Portugal discouraged settlement by Europeans in its colonies until the early 1900s, when there was a push toward greater economic exploitation of mineral resources and plantation crops led by Portuguese technology, finances, and administrators. Little attempt was made to provide schooling for the African population, and any dissension was brutally repressed. As European settlement increased, African opposition gained strength, with rival groups fighting each other in both Mozambique and Angola. In 1975, a revolution in Portugal led to the end of a dictatorship that had held onto the colonies. The new government gave independence to Angola and Mozambique. Most people of Portuguese origin left in the late 1970s, when their scorched-earth policy destroyed much cropland and machinery.

When European colonists arrived in Eastern Africa in the late 1800s, they met less organized resistance to settlement than in Western or Southern Africa. Britain got involved initially to end the slave trade based on Zanzibar Island under the sultans of Oman and Muscat (Figure 9.8), who also controlled the Indian

FIGURE 9.8 Eastern Africa: Zanzibar. Zanzibar was a center of Arab trade along the coasts of Eastern Africa and across the Indian Ocean before Europeans arrived. The view from the top of the Arab fort takes in part of the old city. Now that Zanzibar is part of Tanzania, many in the city campaign for its independence to be renewed. **Photo:** © Volkmar Kurt Wentzel/National Geographic Image Collection.

Ocean trade in cloves and palm oil. In 1886 Britain annexed Kenya and Uganda, and in 1902 built a railroad from Mombasa to Lake Victoria in Uganda. It brought in laborers from India and redirected trade toward European countries. British settlers farmed the fertile Kenya highland area and lands around Lake Victoria. In Uganda, local cotton was processed for export. A smaller number of German colonists settled German East Africa for coffee and tea production, but after World War I it became a British protectorate as Tanganyika. The British did little to develop Tanganyika and did not encourage white settlement there.

Colonial Government

Some colonial powers, such as France and Britain, provided basic education and health care. They also gave limited experience to Africans in local government, especially where traditional political institutions were in place and the chiefs or kings amenable. Elsewhere military strength imposed control. The new colleges, hospitals, and other social infrastructure built by the colonizers in the centers they established educated professionals in European ways. The French policy of assimilating Africans into the French way of life linked their colonies more closely to French institutions. Although colonial powers claimed to legitimize their rule by bringing peace to ethnic rivalries, they often took sides that deepened the rivalries. Local ethnic groups were labeled as "dependable" or "lazy,"

encouraging regional differences of investment, forcible recruitment of some groups as laborers, and widespread resentment. However, Portugal, Belgium, and Germany (before 1918) provided little infrastructure outside the mining and plantation areas and limited the extent of education prospects.

African colonies played small parts in the world economy by exporting raw materials to manufacturers in colonizing countries. Africans worked on farms and in mines, often in brutal conditions. Where local labor was insufficient, migrants were used—often in new forms of semislavery. The emphasis on commercial crops for exports caused a decline in subsistence farming of traditional food crops, increasing food shortages. The mercantile colonial system ruled out manufacturing, apart from local needs, in case it should compete with the products of the colonizing countries. Preferences for European products and ways of living brought cultural as well as economic dependence.

Natural Environments and Resources

The dominant tropical natural environments in Africa South of the Sahara include climatic contrasts from arid to all-year rain, ecosystems from desert to rain forest, some of the world's longest

FIGURE 9.9 Africa South of the Sahara: climate and vegetation regions. (a) Tropical climates dominate the region.

A
Af	Equatorial: rainy all year
Aw	Seasonal summer rains

B
BWh	Hot, arid
BS	Semiarid

C
Cs	Mediterranean: wet winters
Cfa	Rain all year, hot summers

H	Highlands

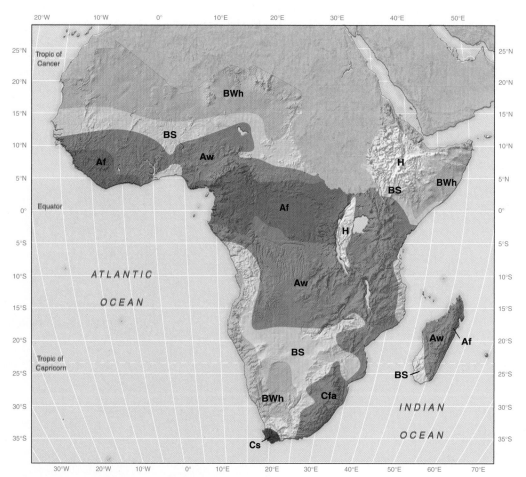

(a)

rivers, and some of its grandest scenery. The range of big animals, from herbivores such as elephants and gazelles to carnivores such as lions, attracts tourists from around the world. Water and mineral resources are among the world's greatest.

Mainly Tropical Climates

The tropical climatic environments range from arid through tropical seasonal to equatorial (Figure 9.9). A satellite view (Figure 9.10) shows the dense thundercloud clusters over the equator, the cloudless arid areas, and the midlatitude weather systems with frontal cloud patterns affecting the southernmost parts.

The equatorial climatic environment (Af) dominates the basin of the Congo River in Central Africa. Temperatures remain high all year, and rain comes in all months. The large rainfall totals and huge basin area within this environment cause the Congo River to carry to the oceans the second greatest volume of water in the world (after the Amazon River in a similar climatic environment in South America). The high plateaus in Eastern Africa, however, restrict the extent of the equatorial rains and often suffer droughts outside of rainy periods in April and November.

In Western Africa, the coastal areas have a monsoonlike climatic environment (Aw) with contrasting wet and dry seasons caused by alternating airflow directions. Moist southerly air from the equatorial Atlantic Ocean brings heavy summer rains that penetrate inland. In winter, dry northerly winds, known as *harmattan,* blow from the Sahara and restrict rains to a coast strip.

Most of Africa South of the Sahara has tropical seasonal climatic environments (Aw, BS) in which temperatures remain high throughout the year and rains come in the season of the overhead sun. Rainfall totals range from low to moderate, and all areas are subject to variations from year to year. The dry seasons increase in length and the rains become more variable closer to the arid areas.

The tropical arid climatic environments of the Sahara, Kalahari, and Namib Deserts (BWh) have little rain in any season. The little that falls is rapidly returned to the atmosphere by high rates of evaporation.

Winters become cooler toward the southern tip of Africa. Winter cold in southwestern Africa comes from a combination of heat loss through clear skies in the dry climatic environment and the cooling of air above the cold Benguela ocean current along the Atlantic coast.

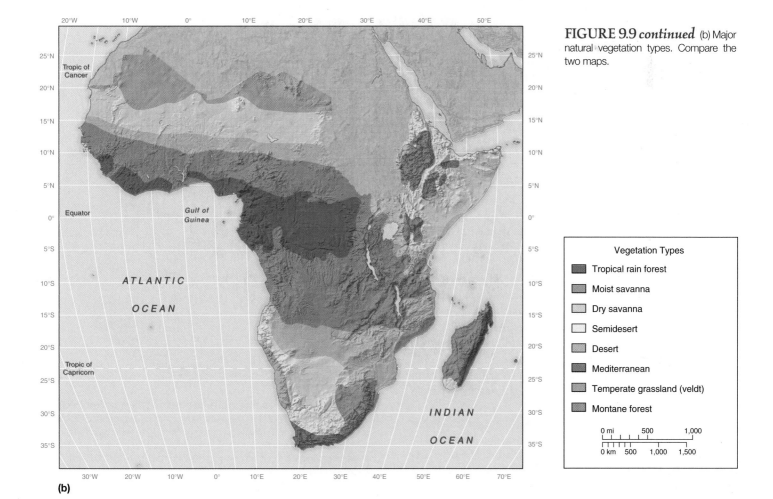

FIGURE 9.9 *continued* (b) Major natural vegetation types. Compare the two maps.

Vegetation Types
- Tropical rain forest
- Moist savanna
- Dry savanna
- Semidesert
- Desert
- Mediterranean
- Temperate grassland (veldt)
- Montane forest

(b)

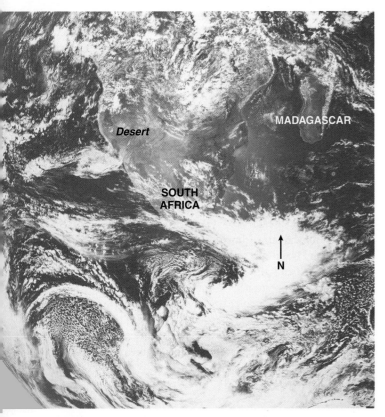

FIGURE 9.10 Southern Africa: weather patterns. A space shuttle photo shows the cloudy (rainy) zones along the equator in the north. Moist air rises near the equator, condensing into clouds and spreading northward and southward at higher levels in the atmosphere. The air then descends at around 30 degrees of latitude, drying out and preventing cloud formation over the desert. To the south of the continent, the swirls of midlatitude weather systems and frontal clouds over the ocean move from west to east. **Photo:** NASA.

Changing Climates

Over periods of a few decades, the southern boundary of the Sahara shifts north and south by up to 160 km (100 mi.), forming the **Sahel,** a zone of very low rainfall. In periods of drier years, the grasses covering old sand dunes die and expose the sands to wind erosion. In wetter periods, the grasses extend their coverage.

Over longer periods, the changes are greater. Five thousand years ago, the Sahara was more humid than today, as shown by the dry lakes and riverbeds around Lake Chad and the remains of settled human communities. Around 3000 B.C., increasing drought forced people southward toward the Western African savannas and forests and eastward into the Nile River valley. The margins of the Namib and Kalahari Deserts in Southern Africa experienced similar changes, but fewer people lived there.

Today uncertainty over the climatic future fuels debates about whether the drying of the continent is a human-induced or natural phenomenon. Although people can do little about the natural changes, it might be possible to modify their own activities

and so reduce the impacts of change. For example, tree-planting programs can help where overgrazing or the removal of vegetation for cultivation caused the expansion of desert conditions.

In the future, this region is likely to face the worst effects of global warming, especially in the drought-stricken areas on the arid margins. The world region that adds least to greenhouse gases may thus be the greatest sufferer of the consequences of an enhanced greenhouse effect.

Ancient Rocks, Rifts, and Volcanoes

Ancient rocks, hundreds and thousands of million years old, dominate Africa. As recently as 200 million years ago, Africa was at the center of a huge worldwide continent (see Figure 6.9). The world continent broke apart as divergent plate margins opened up ocean basins; its fragments moved to the positions of the present continents (Figure 9.11). The rift valleys that cross Eastern Africa from north to south are extensions of the divergent plate margin activity along the line of the Red Sea. **Rift valleys** lie along zones where Earth's crust arched, pulled apart, and broke, collapsing the central part. The rift valleys contain deep, elongated lakes and have volcanic activity along their margins. Geologists identify the combination of the East African rift valleys, their occurrence across Ethiopia, the Red Sea, and the Dead Sea–Jordan River valley line, as being part of a new ocean that is beginning to form.

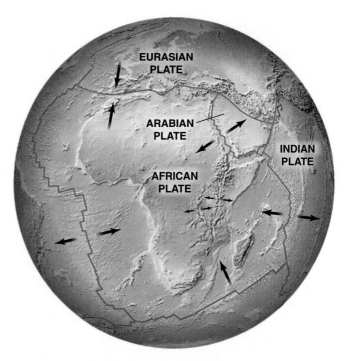

FIGURE 9.11 Africa and plate tectonics boundaries (red). The arrows show whether the plates are moving apart (divergent) to create new ocean floor crust, or clashing (convergent) to form mountain ranges. Both boundaries are linked to volcanic and earthquake activity. The dashed line along the East African rift valley is thought to be a divergent plate margin that will open farther.

The highest mountains in Africa include Mount Kilimanjaro (5,895 m; 19,340 ft.) and Mount Kenya (5,199 m; 12,058 ft.) in Eastern Africa, which are extinct volcanoes close to the rift valley. The eruption of Mount Niriganga that engulfed the town of Goma, Democratic Republic of Congo, in January 2002 was part of this pattern. The Ethiopian Highlands are built partly of lava flows, also linked to the rift valley volcanic activity. On the other side of the continent, the Cameroon Highlands on the Nigeria–Cameroon border include active volcanoes that are distant from plate boundaries and rift valleys, and result from the local melting of deeply buried rocks and their surface eruption.

Plateau Landscapes and Major Rivers

Extensive plateau surfaces on ancient rocks dominate African landscapes (see Figure 9.3). **Plateaus** are elevated areas with relatively flat tops. The African plateaus were formed by long-term erosion that lowered landscapes (planation) in seasonally wet tropical climates. Across whole landscapes, consistent high temperatures and the availability of water in deep soils encourage chemical reactions that break down rocks. During the dry season, the grasses and small shrubs die and the soil surface dries. At the start of the wet season, the combination of bare, loose soil and water flow carries fine sand and clay across the landscape and into the rivers. This process lowers the surface and leaves individual hills with bare rocky

sides, or **inselbergs,** as prominent landforms. Steep rocky slopes also characterize the steps from one plateau level to another.

The major rivers such as the Niger, Nile, Congo, Zambezi, and Orange flow across the plateaus, enabling boat transportation. However, seasonal variations of flow, or waterfalls and rapids where the rivers descend from one plateau level to the next or from a plateau level to the coast, interrupt navigation.

Long-term climatic changes affected the processes producing the extensive plateaus. The most recent glacial phases produced shifts in the climate zones in Africa but left glacial landforms only on the highest mountains. The ice caps on top of Mounts Kilimanjaro and Kenya are now melting rapidly. The desert landscapes probably formed under seasonal rainfall conditions during periods of greater humidity. Increasing aridity desiccated the soil and dried the rivers. The wind then blew away the finest soil particles and concentrated the sandy fractions into large dune seas.

Forests, Savannas, and Deserts

The natural vegetation of Africa South of the Sahara (see Figure 9.9b) follows closely the patterns established by the climatic regime. Equatorial climatic areas are covered by dense tropical rain forest containing a huge variety of tree and other plant species and of birds and insect species (Figure 9.12a). The tropical seasonal climate areas between the forests and deserts are

(a)

(b)

(c)

FIGURE 9.12 Africa South of the Sahara: contrasting ecosystems. (a) Congo. Small village by a river in the heart of tropical rain forest—an ecosystem based on plentiful rainfall. (b) Savanna woodland of the tropical summer rains zone with Mount Kilimanjaro in the background. (c) Rocky desert with vegetation supported by subsurface water. **Photos:** (a) © Torleif Svensson/The Stock Market/Corbis; (b) © Vol. 35/Corbis; (c) © Vol. 145/Corbis.

characterized by savanna grasslands with varying amounts of tree cover (Figure 9.12b). The savanna grasslands are noted for their large herbivore animals such as elephants, giraffes, zebras, and a variety of deerlike forms, such as antelopes and springboks, together with their predators, the lions, leopards, and wild dogs. The long-term presence of humans in these regions suggests that they used fire to expand the savanna grasslands and so increased the animals that were their meat sources. Arid areas are deserts with sparse vegetation growth that support only a few small animals (Figure 9.12c).

The majority of soils in Africa, especially in the tropics, are poor in nutrients because of rapid chemical weathering and removal of the nutrients as water flows through the soils and washes out the soluble matter and finest particles. Some soils are workable for agriculture, however, if care is taken to cope with the high clay and iron contents that are liable to becoming cemented if allowed to dry out. More fertile areas occur around some of the volcanoes, where soils form from the breakdown of rocks rich in plant-supporting nutrients.

Resources

Africa has a wealth of natural resources. Many African countries depend for their foreign income on mining and exporting the minerals contained in the ancient rocks. Overall, however, Africa's potential mineral resources are underused, and the economic benefits from their exploitation seldom return to Africans. Most profits from mining and processing the ores go to companies in the materially wealthier countries. Some large deposits, such as the iron ores of Equatorial Guinea, remain unused at present because of civil wars or the lack of internal transportation.

Until the 1990s, many African countries possessed few sources of fossil fuels. Known deposits of oil and natural gas were limited to areas of coastal subsidence and sediment accumulation such as the Niger River Delta. In the 1990s, evolving offshore oil exploration technology identified major oil fields along the Atlantic coast of Africa. Gabon and Angola became important producers based on multinational oil corporation investments in areas judged to be safe from internal conflicts. In the 1980s, South Africa developed coal mining when international sanctions prevented imports, but coal deposits in other countries remain remote from transportation.

The tropical climates make it possible to grow a variety of crops, including yams, rice, cassava, corn, millet, and sorghum that are the basis of life for most Africans, who also sell them in the growing urban markets. Commercial crops for export to midlatitude countries include bananas, cocoa, coffee, tea, palm oil, rubber, cotton, tropical fruits, and peanuts. More research and farmer education are focused on the export crops than on the subsistence crops.

Forest resources are extensive in the equatorial countries, and plentiful fish resources occur in the major rivers and in the areas of cold ocean currents off northwestern and southwestern Africa. Both forests and fisheries are experiencing increased rates of exploitation and depletion through demand from the world's wealthier countries.

Africa also possesses the natural resources for tourism. These include sunshine, coastal beaches, and large numbers of big animals that can be viewed in the protected national parks of countries such as Kenya, Tanzania, Zimbabwe, Botswana, Namibia, and the Republic of South Africa. Tourist interests also center on scenic wonders such as Victoria Falls on the Zambia–Zimbabwe border and historic sites such as the former slave markets of Eastern and Western Africa. Many opportunities for tourist development, however, await political stability and the development of better accommodation and transportation facilities. Moreover, tourism brings mainly low-wage employment to local Africans.

Environmental Problems

Drought and Desertification

Water resources range from plenty in the Congo River basin to scarcity in many areas where there is seasonal rainfall, frequent droughts, or continuous aridity. Shortages of water lower crop productivity and hydroelectric project efficiency. The Sahel zone suffered increasing droughts from the 1970s, leading to overgrazing and removal of woody plants for firewood—a case of desertification. It caused many livestock deaths in Western Africa and Chad and local famine conditions.

Soil Quality Losses

Many tropical soils lack nutrients or are difficult to work with simple tools. The decline in soil quality and workability following agriculture is one of the most widespread problems in Africa. For example, removal of the forest cover in the Ethiopian Highlands led to rapid erosion of the soils, forcing people to move elsewhere and impose further pressures on land resources in already crowded areas. Soil exposure in the seasonal rainfall areas may lead to a cementing of the soil into hard laterite that resists plowing. In semiarid areas, the overgrazing of slow-growing vegetation and the removal of woody plants for firewood expose soils to erosion by water and wind.

Threatened Wildlife

The tropical African flora and fauna are threatened by expanding logging, farming, and poaching. The level of the problems varies from place to place. For example, the killing of elephants for their ivory tusks in Kenya caused a fall in the elephant population and international action to ban the ivory trade in the late 1900s. Meanwhile, in Southern Africa, better conservation policies resulted in an expansion of the elephant population until it exceeded the carrying capacity of the land and forced governments to institute annual culls. The ivory from the culls fills many warehouses but cannot be sold because of the world ban.

Killer Tropical Diseases

Despite massive efforts and advances in health care, tropical diseases remain among the main environmental problems of Africa for people and food production. Malaria, river blindness (see Geography at Work: Mapmakers and GIS Analysis, page 8), and other human diseases favored by the tropical climates remain endemic, while cholera flares up from time to time. Some areas of savanna grassland have very low populations because of the prevalence of sleeping sickness affecting humans and heavy tsetse fly infestations infecting cattle; people escaping from that threat move into river valleys where river blindness attacks. Diseases such as measles and tuberculosis are returning in the wake of HIV/AIDS, which weakens immune systems.

Great strides in immunology and antibiotics allow recovery today from many previous killer diseases. However, new deadly viruses, such as Ebola, which affected the Democratic Republic of Congo in 1995, keep appearing and cause immediate panics. Other major diseases are still killers on a larger scale than the Ebola virus. Sleeping sickness, for example, kills 200,000 people a year in the Democratic Republic of Congo. Although a cure for sleeping sickness is known, few people are treated for it. The high cost of continuing treatments for diseases such as sleeping sickness and malaria makes them too expensive for most Africans. The continual risk of infection from waterborne diseases causes death rates to remain higher than those in other parts of the world. In many parts of Africa, a lack of sufficient food causes malnutrition and makes people less resistant to disease.

Test Your Understanding 9A

Summary

Africa South of the Sahara has a rich history and a wealth of natural resources, but it is one of the world's poorest regions and faces major problems of engaging with the global economic system. The legacies of slavery, colonialism, and ethnic divisions are negative influences on modern political, economic, and social conditions.

The African natural environment is dominantly tropical in climate and vegetation, from rainy equatorial climate with tropical rain forest, through areas of seasonal rain and savanna grassland, to arid areas of desert vegetation. The ancient rocks contain mineral resources, while the scenery and animals attract many tourists. Droughts, poor soil quality, and diseases are negative factors, and this region is the world's worst for the spread of HIV/AIDS.

Questions to Think About

9A.1 How can a region with such a rich history and wealth of resources end up as the poorest in the world?

9A.2 How do the histories of Asian and African people compare in their relationships with European colonizers?

9A.3 How are climate, natural vegetation, and landforms linked in Africa South of the Sahara?

9A.4 What would be the Western countries' response if they experienced the African levels of HIV/AIDS?

Key Terms

tribes
creole languages
Sahel
rift valley
plateau
inselberg

Global and Local Changes

From the 1950s, countries in Africa South of the Sahara gained political independence in the context of the Cold War. As the Cold War waned from the 1980s, new patterns of activity centered around growing African confidence and attempts to solve their own problems. Democratic intentions spread to most countries. However, many local wars and the expansion of HIV/AIDS held back economic and personal independence.

Politics of Independence

Agitation for independence from European colonial powers began in the 1920s. The Italian invasion of Ethiopia in 1936, World War II isolation, and the example of independence gained by India and Pakistan in 1947 increased the demands by nationalist movements. The independence of new African countries began with Ghana in 1957 and ended with Namibia in 1990.

Cold War and Subsequent Pressures

On gaining political independence, most countries of Africa South of the Sahara began with a limited experience of government and restricted sources of income. In many countries, the methods of the authoritarian colonial governments and bureaucracies led to new dictatorships, single-party governments, and military rule. From the 1990s, changes from single-party to multiparty constitutions and open elections made it conceivable that entrenched leaders might be removed by ballots instead of bullets. However, many established leaders found it difficult to hand over power and resisted implementing the new constitutions. People are discouraged if new governments do not produce economic growth, tribal rivalries surface in open conflict, and countries return to authoritarian governments.

During the Cold War, the U.S. and Soviet rival superpowers propped up bad governments or encouraged rebel groups. They sold weapons to help their African clients maintain or achieve power. For example, Ethiopia became a battlefield as Soviet forces supported its Communist government against rebels from Somalia or those fighting for Eritrean independence.

In 1980, the other countries in Southern Africa formed the Southern Africa Development Coordination Conference (SADCC) to oppose South African apartheid. In 1992 the name was changed to **Southern Africa Development Conference (SADC)**, and the focus moved toward reducing economic dependence on South Africa. After the abandonment of apartheid, SADC welcomed the 1994 membership of South Africa to strengthen mutual linkages in trade and to help control traffic in illegal drugs and arms.

When the Cold War ceased in the early 1990s, the African countries lost strategic interest. Heavy debts made it difficult for them to attract new investment or aid. By the early 2000s, African countries attracted only 3 percent of the foreign investment flowing to developing countries, while countries in Latin America attracted 20 percent and those in East Asia 60 percent.

Population Pressures

Growing Numbers

In the period from 1980 to 2005, the total population of Africa South of the Sahara rose from 380 million to over 700 million at rates of increase that equalled or exceeded rates of economic growth. The annual rate of population increase remained around 3 percent. Death rates in most countries plummeted as modern medicine was applied, but birth rates in many countries fell slowly if at all (Figure 9.13). However, agricultural production growth was only 2 percent, and per capita food production dropped by 12 percent between 1960 and 2000—a contrast to the Green Revolution increases in Asia. Population pyramids (Figure 9.14) reflect the high fertility rates and short life expectancies. High proportions of younger age groups dominate the population structures even in the more developed South Africa, while differences of lifestyle affect life expectancy (higher for whites) and total fertility (lower for white women).

Reducing population growth is one major key to Africa's future. However, efforts to encourage smaller families have moderate success because of cultural, economic, and often political pressures. Large families are seen as a sign of male virility, and women have little say in the numbers of births. In fact, women often prefer to have many children, since they remain a central element in social security for old age, while barrenness and childless marriages are unacceptable. Sadly, the main curbs on population growth and life expectancy are the HIV/AIDS pandemic, warfare, ethnic cleansing, and famines.

Flows of refugees within and among countries are increasingly significant. Examples include the movements across the borders of Democratic Republic of Congo (DROC), Uganda, Rwanda, and Burundi and also those among Ethiopia, Sudan, Somalia, and Uganda. Refugees are often made part of military activities, including guerrilla groups, adding to disruption in countries without means of resettlement apart from special camps.

Population Distribution

The population distribution within the region (Figure 9.15) links cultural history, environmental challenges, and external influences. Central Africa has the lowest density of population: on average 21 per km² in 2005 (50 per m²). It will almost double by 2025, although economic output is unlikely to increase. At present the small countries of Burundi and Rwanda have some of the highest population densities among African countries. By contrast, the Saharan north of Chad and the tropical forests of other countries have large areas with fewer than one person per square kilometer. Only coastal Cameroon, the mining areas of southern DROC, and the immediate vicinity of towns and cities have more than 10 people per square kilometer.

Western Africa is the most populous subregion: the 264 million people there in 2005 compared to 150 million in 1980 and may reach over 400 million by 2025. Densities exceed 100 people per km² along most of the coast from

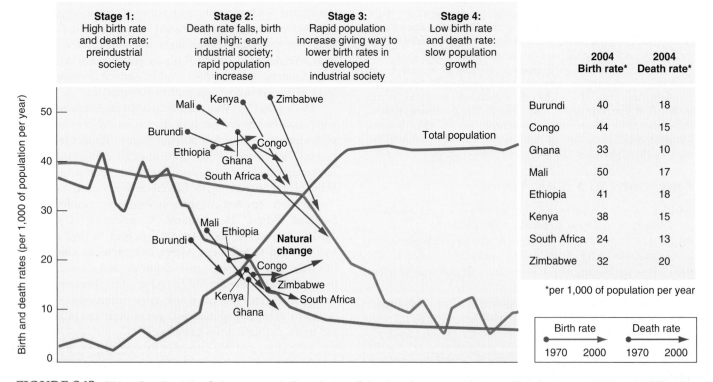

FIGURE 9.13 **Africa South of the Sahara: population change linked to demographic transition between 1970 and 2000 with 2004 update.** Although the death rates fell comparably to the experience of core countries in the 1800s, many birth rates remain very high. Some death rates increased, especially in countries subject to HIV/AIDS. **Source of 2004 update:** 2004 Population Data Sheet, Population Reference Bureau.

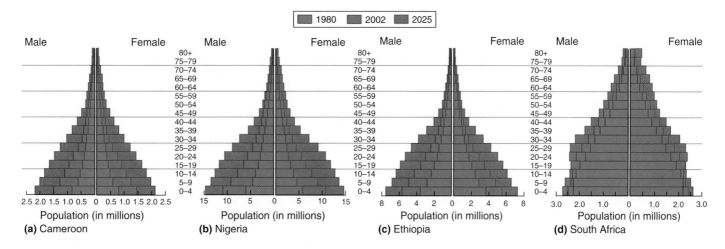

FIGURE 9.14 Africa South of the Sahara: age–sex diagrams. These are typical of the materially poor developing countries with a high proportion in younger age groups and fewer older people. How is South Africa different? Compare them with those for European and East Asian countries. (a) Cameroon, Central Africa; (b) Nigeria, Western Africa; (c) Ethiopia, Eastern Africa; (d) South Africa, Southern Africa. **Source:** U.S. Census Bureau: International Data Bank.

Nigeria to Côte d'Ivoire. The northern parts on the Saharan margins have densities of fewer than one person per km². Between the coasts and interior desert, areas of moderate population, including the highest rural densities in interior Nigeria, have better transportation links and more commercial farming.

Most people in Eastern Africa live in the better-watered upland areas of Ethiopia, Kenya, and Uganda, including the lands bordering Lake Victoria and the major routes linking inland areas of commercial activity to ports. Few inhabit the semiarid areas between the Ethiopian uplands and Indian Ocean coast, where nomadic herding supports the people.

In Southern Africa, the population is very sparse over a large area including the desert-bordered southwestern coasts. The main populated area is in South Africa between the mining and industrial area around Johannesburg and the southern coasts from Cape Town to Durban. Other well-populated areas are along the northeastern coast of Mozambique and the railroad lines to inland mining centers in Zimbabwe and Zambia.

Urban and Rural Shifts

Two-thirds of the populations of most countries in Africa South of the Sahara remain rural, linked to the dominance of subsistence farming in the economy. However, the growth of urban centers (Table 9.1) reflects the rising influence of global connections. Such places contain more prospects for waged employment and better educational and health facilities than rural home villages. Urban populations also increase as people migrate into towns for perceived safety in a time of civil war. Official figures of urbanization are lower than the fact. The

Table 9.1 Major Urban Centers in Africa South of the Sahara

City, Country	2003 Population (millions)	2015 Projection (millions)
Kinshasa, DROC	5.3	8.7
Lubumbashi, DROC	1	1.7
Yaoundé, Cameroon	1.6	2.2
Douala, Cameroon	1.9	2.5
Brazzaville, Congo	1.1	1.6
N'Djamena, Chad	1	
Lagos, Nigeria	10.1	17
Abidjan, Côte d'Ivoire	3.3	4.4
Dakar, Senegal	2.2	3.1
Accra, Ghana	1.8	2.6
Ouagadougou, Burkina Faso	1	
Bamako, Mali	1.3	2.2
Addis Ababa, Ethiopia	2.7	4.1
Nairobi, Kenya	2.6	4
Kampala, Uganda	1.2	2
Cape Town, South Africa	3	3.2
Johannesburg, South Africa	3.1	3.7
Durban, South Africa	2.6	2.7
Pretoria, South Africa	1.2	1.4
Maputo, Mozambique	1.2	1.9
Luanda, Angola	2.6	4.3
Harare, Zimbabwe	1.5	1.8
Lusaka, Zambia	1.4	1.8
Antananarivo, Madagascar	1.6	2.6

Source: U.N. Urban Agglomerations 2003 (2003).

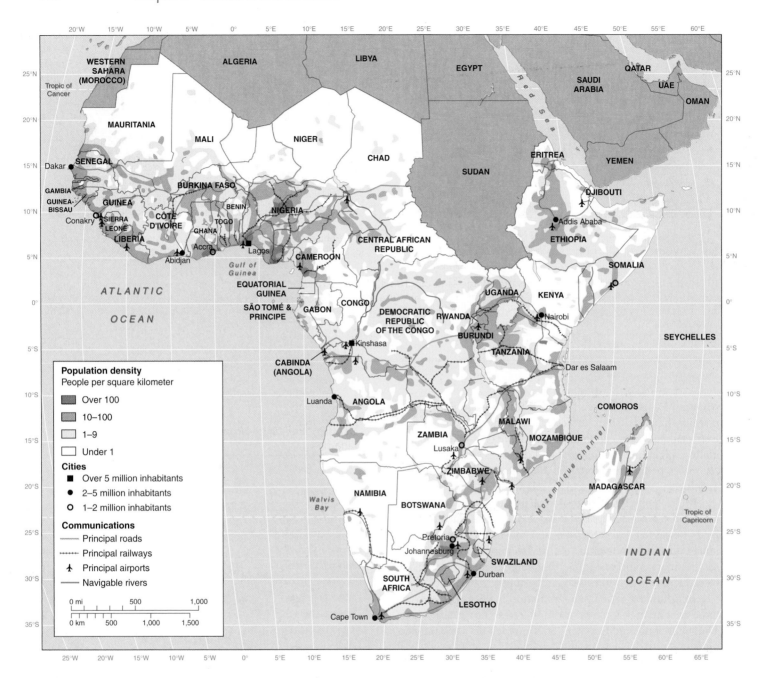

FIGURE 9.15 Africa South of the Sahara: population distribution. What is the relationship between the population densities and the natural environments (see Figures 9.3 and 9.9)? **Source:** Data from *New Oxford School Atlas,* Oxford University Press, U.K.

informal urban economy becomes the only means of livelihood for many people who are cut off from their home villages and subsistence food production.

African cities have distinctive geographies. A few, such as Timbuktu and Ibadan, have central areas that date back to the precolonial cross-Sahara trade and empires. Many of the oldest African urban landscapes occur in Western Africa. They include the trading centers of Timbuktu, Sokoto, and Kano at the southern end of trade routes across the Sahara. These

Islamic cities typically have central markets, mosques, citadels, and public baths, and are still dominated by craft workshops rather than modern industry. Ibadan, the center of the Yoruba culture in southwestern Nigeria, preserves a pre-European urban landscape. The central palace and nearby market are at the focus of streets radiating toward other towns. Fortifications were added later to resist Muslim attacks. The modern populations of these older towns remain much more ethnically unified than those in many of the newer urban areas.

The colonial era was a time of port building and interior control centers that left often grandiose buildings of European design. Accra in Ghana illustrates some common features (Figure 9.16). The largest African cities are often capital cities, but few can claim the status of a global city–region. Brazzaville (Congo) and Kinshasa (DROC), on opposite banks of the Congo River, illustrate different forms of colonial urbanization. Brazzaville became the chief city of French Equatorial Africa, where French diplomats and educated African elites built large houses on quiet avenues. Its population grew after 1960 independence, from around 100,000 to over 1 million today. Kinshasa reflects grandiose

Belgian purposes—continued in the Mobutu era—with its wide boulevards and postindependence construction of prestigious tall office blocks. It grew from around 100,000 people at independence to over 5 million people, with huge sprawling shantytowns. Independence brought desires for new capital cities apart from the colonial centers.

Shantytowns are slum areas that occur in all African cities. Shantytowns are unplanned, constructed of any materials that come to hand—from packing cases to cement blocks and corrugated iron—and basic in their services. Some condemn families to a hopeless future of poverty. Their inhabitants are often involved in the informal economy. In many cases, however, people move into them on arriving from a rural area but eventually find better accommodations. Governments may supply utilities and build schools, hospitals, and roads to integrate shantytowns with the rest of a large urban area, usually after a considerable period in which they become established. Shantytowns were encouraged under apartheid in South Africa (Figure 9.17), and a long process of improving housing conditions there is under way.

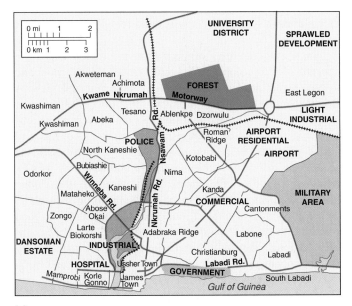

(a)

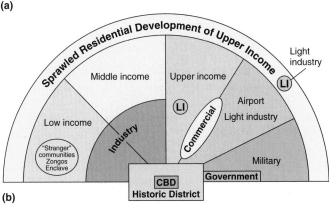

(b)

FIGURE 9.16 African urban landscapes: Accra, Ghana.
(a) The port city of Accra, Ghana, expanded from a village for colonial trade, with port facilities along the lower river, flanked by government offices to the east and poor housing to the west. Account for locations of military, airport, and university districts. (b) A generalized pattern based on Accra that is repeated (with variations) in other Western African port cities.
Source: (a) Data from S. Aryeetey-Attoh, *Geography of Sub-Saharan Africa*, Fig. 7.4, pp. 192–93. Prentice-Hall, 1997.

FIGURE 9.17 South Africa: Soweto Township, near Johannesburg, in 2002. The contrasts within Soweto can be seen between the squatter shacks in the foreground and the more permanent housing in the rear. The sprawling township became a symbol of the struggle by South African blacks against the racist apartheid rule. Now it is an important tourist destination. Little change appears to have occurred in the 10 years since the end of apartheid. However, postapartheid restrictions on labor migration and the high AIDS/HIV infection rates among their workers have prompted mining companies to replace hostels with family homes in some areas. **Photo:** © AP/Wide World Photos.

In the absence of the wealth common in Western countries, many Africans create ways of living that enable them to survive and even enjoy life. Items that might be regarded as waste become children's toys; bicycles or walking are common modes of travel, while cheap rides on crowded vans and buses enable wider circulation; and loads from water to many goods and possessions are often carried on people's (especially women's) heads or shoulders (Figure 9.18).

Human Rights and Women's Roles

With its many dictatorships, local wars, and a prevalence of military expenditure over health and education, Africa South of the Sahara has generally poor human rights environments. In many societies, women have a low priority and in some areas female mutilation is common. It is estimated that 100 million women are affected, mainly in the northern countries such as Nigeria (25% of women) and Mali (90%). Complications from the cutting include bleeding that spreads HIV/AIDS, painful intercourse, and childbirth difficulties. Some countries have banned the practice.

Women in Africa are expected to bring up children, draw water, raise crops, and cook meals. Few avoid such a life, and hardly any attain high political office. Only 13 percent of African members of parliament are women. Before the colonial period, ethnic groups such as the Kongo people in Central Africa had matriarchal inheritance, but the colonial powers ended many female institutions and reduced their rights, even the right to complain through ecstatic *vimbuza*. However, although women are limited to modest roles in society, wives of dictators often make their presence felt. The wives of Nigerian dictators in the 1980s and early 1990s built their own personal fortunes. In Rwanda, the wife of the dictator from 1994 is suspected of links to the groups who carried out the genocide. In Gabon and Zambia, estranged wives returned to embarrass their husbands as pop stars or in court cases. The wife of the leader of Liberia from 2003 claims that she, not her husband, is in charge.

Better Education

After independence, all countries increased educational achievement. By the 1990s, countries such as Botswana, Cameroon, Kenya, South Africa, Zambia, and Zimbabwe had virtually total enrollment, male and female, in elementary schools, up from 50 percent in 1965. Burundi, Chad, and Mauritania made major strides by increasing primary education from under 20 percent in 1965 to between 70 and 80 percent in the early 2000s. However, Burkina Faso, Ethiopia, Guinea, Mali, and Niger in the northern Muslim belt still have only half of children in elementary school, and female education lags behind male.

Increasing numbers of people in the region also have opportunities to become fully literate in secondary school and to earn higher academic qualifications (see Personal View: A Brighter Future in Ghana, page 399). It is often disappointing to many who gain higher qualifications that there are few jobs available in their home countries. African doctors, lawyers, and airline pilots are increasing in numbers, but many enter the brain drain and find their employment abroad in the world's wealthier countries. Emigration from Africa to the United States more than doubled from the 1990s, disproportionately in the professional, managerial, and technical occupations, and continues to grow. Many decide to live in the wealthier countries, although they may benefit home African countries by sending money to their families. Some of the skilled and experienced personnel return. Meanwhile, African countries pay expatriates from wealthier countries high salaries to carry out professional jobs.

HIV/AIDS Pandemic

The World Health Organization lists AIDS as the third main cause of global deaths. By destroying the immune system, HIV/AIDS makes its victims more vulnerable to diseases such as tuberculosis, pneumonia, toxoplasmosis, fungus infections, and cancers. It is transmitted in blood and other body fluids, including through semen during sexual activity, unhygienic injection needles, and transfusions of infected blood. It is passed from infected mothers to babies at birth or in breast milk. The causes of the high levels of HIV/AIDS in Africa include poverty, the breakdown of traditional family support systems, the apartheid policy in South Africa that brought miners into male-only camps serviced by prostitutes, continuing promiscuity at a time when traditional polygamy gives way to the taking of sexual partners outside monogamous marriages, and mistaken government policies. HIV/AIDS spreads quickly in cultures that value male sexual prowess.

FIGURE 9.18 **Women in Nigeria.** Govari women carrying firewood across a busy road in Gwagwalada village in Nigeria's middle belt. The increased 2004 usage of firewood followed increases in kerosene prices. **Photo:** © George Esiri/Reuters/Corbis.

A BRIGHTER FUTURE IN GHANA

When meeting Yaa Boadi, one is readily drawn in by her natural, high-wattage smile—an appealing west African trait. But it is not merely because she is Ghanaian that this young woman smiles so freely; it is also because of how she feels about her life—one in which she has made remarkable strides from her rural, impoverished upbringing.

At the age of 26, Boadi's most vivid childhood memories are of beginning each day by carrying pails of water on her head. Sometimes she made three grueling trips to the village well before school.

Boadi's village, Nkawkaw, was on a red-earthed mountain in the eastern region of Ghana. It is inaccessible to motor vehicles, so getting there involves a walk of 30 minutes or more along a footpath from the main highway linking the regional capital of Kumasi with the national capital, Accra.

During the dry season, it could be so cold in the morning that Boadi warmed herself beside an outdoor fire before setting off for water. Then it could become so hot and dusty during the day that her grandmother would rub cocoa butter into her black hair as a moisturizer.

Water collection also became more arduous during the dry season. As the level in the well dropped, villagers would jostle in line. When it was her turn at the well, Boadi tried to lower her pail gently to avoid stirring up the muddy bottom.

Inevitably, the well dried up before the spring rains returned, forcing villagers to travel far to a deeper source or to buy water from a government truck in a distant town.

On a Saturday morning recently near the central market in Accra, Boadi wore jeans and a royal blue, tie-dyed shirt with white lace embroidering the collar and short sleeves. She inherited this striking shirt from her father, a man she hardly knew.

While she was still a toddler, her parents left home: her father took a teaching position in Cameroon. Boadi and a brother were raised by their grandparents. Lots of cousins lived in the compound, as well as Boadi's uncles and aunts. Everyone slept on floor mats in two rooms, except her grandfather, who had his own room and a mattress. Boadi says it was a typical, traditional village upbringing.

In a region where many girls never attend elementary school, very few attend college, and fewer still ever achieve professional status, a doting uncle encouraged her to achieve. She could become an engineer one day, he told her. Then she could return home to build wells, so that other girls would no longer have to haul water.

Box Figure 1

A Ghanaian woman similar to Yaa Boadi.

Photo: © Reuters/Corbis.

Boadi made it to the university in Kumasi, the regional capital, one of only four women to enroll for civil engineering alongside 36 men and the only one in her class to stick with it as others switched to less rigorous majors. In time, Boadi found she could hold her own with male classmates from more privileged families who had been groomed in prestigious secondary schools.

On graduating four years ago, she became the first professional woman hired by an Accra engineering company, and she has worked on design teams for World Bank–funded national highway projects. She enjoys her work but also detects some bias when the men get all the field assignments while she remains office-bound with design work.

She has also become active in a support group for women engineers in Ghana and has traveled to South Africa to speak to an association of women engineers. Boadi was among seven Ford Foundation fellows selected from almost 600 applicants in Ghana. She plans to enroll in a master's degree program at the University of Southampton (United Kingdom), focusing on engineering for rural development. Within a few years, she hopes to form an NGO dedicated to attracting funding for infrastructure projects in neglected regions of Ghana. Perhaps, she says, one day she'll return home to build wells—just as her uncle foretold.

From "Getting the best out of Africa" by Todd Shapera as appeared in *Financial Times (London) Weekend*, 27–28 April 2002. Reprinted by permission of Todd Shapera.

Although reduced (for a time) in Europe and North America in the 1990s by expensive triple-drug therapy centered on monitoring clinics, the disease diffused rapidly through Africa from the 1960s. Africa South of the Sahara has the world's highest and most rapidly increasing concentration of this disease (Figure 9.19). It affects up to 40 percent of the population aged 15 to 49 years. In 2003 there were over 30 million African adults and children with HIV/AIDS, making up 66 percent of the world total. Nearly 4 million new infections and 2.5 million deaths from HIV/AIDS occurred in this region. Life expectancies in the worst-affected countries of Botswana and Zimbabwe, which rose to 60 years in 1990, fell to 40 years in the following decade. Other African countries experienced smaller but growing impacts unless special measures were adopted (see the Point–Counterpoint: HIV/AIDS box, page 406). Only in Uganda did infections fall in the early 2000s.

As well as the demographic impacts, HIV/AIDS has many geographic social implications. Half the miners in South Africa are HIV carriers, and millions of orphans, often carriers themselves, constitute a growing need for help in the region. By 2010, orphans with HIV/AIDS will rise from 1 to 2 million in Nigeria and from under 4 million to nearly 6 million in Eastern Africa. Throughout Africa South of the Sahara, the lives of young, often skilled, workers and their families are being shattered. Military personnel, migrant miners, and their wives and prostitutes have the highest proportions of infection. The lack of medical understanding and panic at not having the funds or expertise to do anything about it generate false taboos and myths, such as the one that men can cure themselves by having sex with a virgin girl, a basis for many child rapes. Although HIV/AIDS mainly occurs in urban areas, it also threatens rural communities in Southern Africa, where one-fourth of farm workers have died.

Possible measures for controlling the spread of AIDS are complex. One approach is to provide cheap or free condoms and needles because HIV is spread by sexual activity and drug addicts reusing needles. These treatments work more effectively than moralistic "better behavior" policies. In Senegal, religious groups supported a program that included sex education in schools, the "social marketing" (at low prices) of condoms and needles, and a focus on at-risk groups (prostitutes and young men in the army). HIV infections remained below 2 percent, compared to nearly 10 percent in nearby Côte d'Ivoire. Successful actions result from prompt responses, open discussion of sensitive issues, and repeated targeting of vulnerable groups. Even where the cost of the drugs is reduced to affordable levels, African countries still spend more on armaments than on the clinics and laboratories needed to administer and monitor the use of the drugs. HIV/AIDS requires global action that is as vital as controlling terrorism, the armament and drug trades, and slavery. However, current U.S. laws prohibit programs that fund condom distribution or abortions, cutting off a major funding source of help.

Economic Challenges

The countries of Africa South of the Sahara remain among the world's poorest (Figure 9.20). Many of their problems are economic, related to tiny internal markets because many countries have small total populations and all have few people who are wealthy enough to purchase consumer products.

Global Connections

Much of Africa South of the Sahara remains a plantation or quarry providing cheap raw materials to the materially wealthy Western world—often a hangover from trade established in colonial times. Some African countries and cities are more integrated in the global economy as their ports and airports link commercial farms and mines to markets in wealthier countries. Cities such as Nairobi (Kenya), Johannesburg and Cape Town (South Africa), and Lagos (Nigeria) exhibit some signs of becoming global city–regions as they provide headquarters for the subregional centers of commerce, multinational corporations, and NGOs.

However, many African countries are part of the global economic system as dependent debtors and recipients of aid. In their postindependence focus on self-sufficiency to develop the well-being of their people, they often spent their small resources on what they perceived to be important in raising their country's identity through lavish new capital cities and military purchases. However, in some countries, poorly managed projects, together with the corrupt siphoning of funds into personal bank accounts abroad, led international lenders and bankers to impose conditions that slowed economic opportunities.

In the 1980s the World Bank and the International Monetary Fund—the two major lending institutions for developing countries—established new guidelines for grants and loans. This was a response to the low rates of success achieved by previous loans and the large proportions that were absorbed by high exchange rates, the cost of internal government bureaucracies, and corrupt mismanagement of funds. To date, these structural adjustment policies have had little success in the countries of Africa South of the Sahara, although at times Ghana, Tanzania, Burkino Faso, Nigeria, and Zimbabwe came close to following their precepts. The policies relied too much on exporting commercial crops instead of growing food for local use. The reduction in government employees reduced health care and education provision. At present, countries in Africa South of the Sahara seem to lose out whether they decide to follow structural adjustment policies or not. If they do not adopt them, they lose access to funds from the World Bank, International Monetary Fund, and aid agencies. If they adopt the stringent policies, they often alienate their people.

Raw Materials, Multinationals, and the World Trade Organization

At the time of independence in the 1950s and 1960s, Western economists claimed that countries would grow economically by following the world's wealthier countries in moving from the primary sector into manufacturing and services. Few African countries achieved this progression.

Multinational mining companies and makers of coffee, tea, and chocolate products continue to buy African raw materials cheaply. For example, multinational aluminum manufacturers helped arrange funding for the Volta River project in Ghana that generated hydroelectricity to refine bauxite, the ore of aluminum, as cheaply as possible. However, such reliance on producing raw materials and low levels of processing keeps local incomes low.

Countries reliant on primary products are kept poor by low prices and restrictive practices in the wealthier countries over farm products. Thus Nigerian and Ghanaian coffee producers receive around 50 U.S. cents per pound. Each cappuccino served in U.S. coffee chains takes about one ounce (4.5c), but

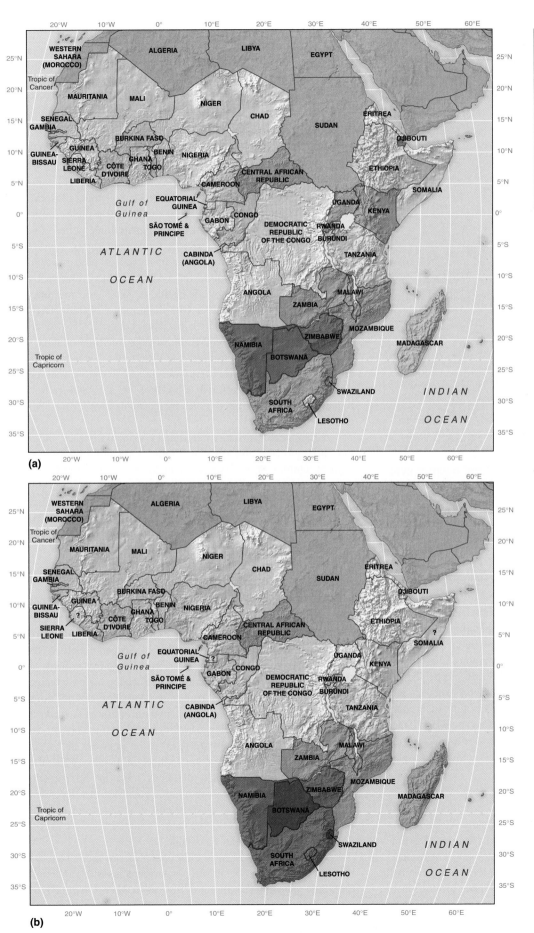

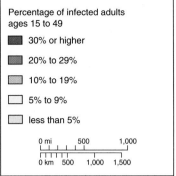

Percentage of infected adults
ages 15 to 49

- 30% or higher
- 20% to 29%
- 10% to 19%
- 5% to 9%
- less than 5%

0 mi   500   1,000
0 km   500   1,000   1,500

FIGURE 9.19 Africa South of the Sahara: HIV/AIDS. The percentage of adults (age 15 to 49) infected with HIV/AIDS. Some countries are unlikely to report the full extent. What factors might account for the higher prevalence in Southern Africa? (a) Data for 1997. (b) Data for 2003. **Source:** Data from United Nations and Population Reference Bureau.

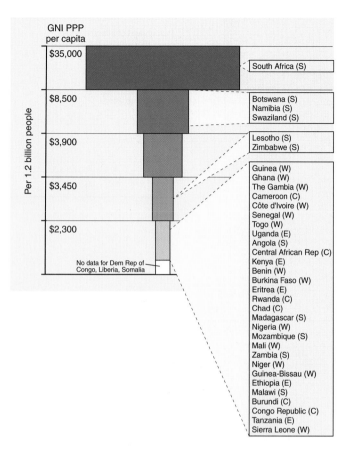

GNI PPP per capita

$35,000 — South Africa (S)

$8,500 — Botswana (S) / Namibia (S) / Swaziland (S)

$3,900 — Lesotho (S) / Zimbabwe (S)

$3,450 — Guinea (W) / Ghana (W) / The Gambia (W) / Cameroon (C) / Côte d'Ivoire (W) / Senegal (W)

$2,300 — Togo (W) / Uganda (E) / Angola (S) / Central African Rep (C) / Kenya (E) / Benin (W) / Burkina Faso (W) / Eritrea (E) / Rwanda (C) / Chad (C) / Madagascar (S) / Nigeria (W) / Mozambique (S) / Mali (W) / Zambia (S) / Niger (W) / Guinea-Bissau (W) / Ethiopia (E) / Malawi (S) / Burundi (C) / Congo Republic (C) / Tanzania (E) / Sierra Leone (W)

No data for Dem Rep of Congo, Liberia, Somalia

Per 1.2 billion people

FIGURE 9.20 Africa South of the Sahara: country average incomes compared. The countries are listed in the order of their GNI PPP per capita. (C)=Central Africa; (W)=Western Africa; (E)=Eastern Africa; (S)=Southern Africa. **Sources:** Data (for 2000) from *World Development Indicators,* World Bank, 2002.

customers pay around $2. Most of the markup goes to coffee traders, blenders, grinders, and retailers in the United States (and other wealthy countries). Purchasers of African raw materials often maintain low world prices by opening up new areas of production as growing markets absorb what established areas produce. Soaring raw material prices in the 1970s (minerals) and in the early 1980s (beverages) were short-lived but often enabled the producer countries to take out loans for economic development projects; when the prices fell, such countries faced debts they could not repay. In 2004 the World Trade Organization championed the cause of African products to gain wider access to world markets.

Tourism

Tourism—the world's largest industry—provides a major potential for earning foreign currency to many African countries through attracting visitors to view the unrivaled scenic grandeur and wildlife. The commitment of some governments to conservation in designated game and national parks resulted in Africa South of the Sahara having a higher proportion of

such land uses than any other continent. Unfortunately, the governments have little money to spend on maintaining the parks and their wildlife, and they cannot finance realistic management policies. Parts of Eastern and Southern Africa attract most international visitors.

Interregional Cooperation

In July 2001 the heads of African governments, meeting in Lusaka, Zambia, changed the "Organization of African Unity" to the "African Union." At the same conference, the Millennium Action Plan, proposed by President Thabo Mbeki of South Africa, had a twofold thrust. First, it restated the policies previously urged by Western countries and institutions: better government, more democracy, respect for human rights, market reforms, and recognition of the advantages of globalization. Second, the plan highlighted the need to reduce poverty by improving education and public health. It asked for continuing, more accountable aid together with the removal of trade barriers and agricultural subsidies in richer countries. However, subsequent actions suggest that the world's wealthier countries demand the first part but contribute little to the second. When the wealthiest (G8) countries met in 2002 and discussed African needs, they offered $1 billion of the $64 billion requested when the United States increased its own farm subsidies by $190 billion.

African countries also attempt to work with each other through regional trading groups (Figure 9.21) along subregional lines. Unlike the EU (see Chapter 3) or NAFTA (see Chapters 10 and 11), however, the African groups are loosely organized and often overlap. Many start with enthusiasm but then remain dormant or achieve little for want of political support from members, credibility among the wealthier countries, or difficulties in administrating their activities.

French Links

Arising from their history as French colonies, many countries of Central and Western Africa continue financial links with France (Figure 9.22). After independence, the **Communauté Financière Africaine (CFA)** franc was shared by 14 African countries. Its value was tied to the French franc and guaranteed by the Bank of France. This link held currencies at artificially high exchange rates but allowed French companies to retain dominant positions in local contracting. The high exchange rates, however, made it difficult for French-speaking African countries to export goods, used up their foreign exchange reserves on imports, and made them increasingly dependent on France. In January 1994 the French government devalued the CFA franc by 50 percent, causing initial hardship to people in the countries that depended on France, especially by raising the cost of imports. For its part, France wrote off the debts of the poorer countries, and the World Bank made increased grants available.

By the early 2000s, most CFA countries saw some benefits in lower inflation and better foreign currency reserves, although economic growth was still not as high as in other African coun-

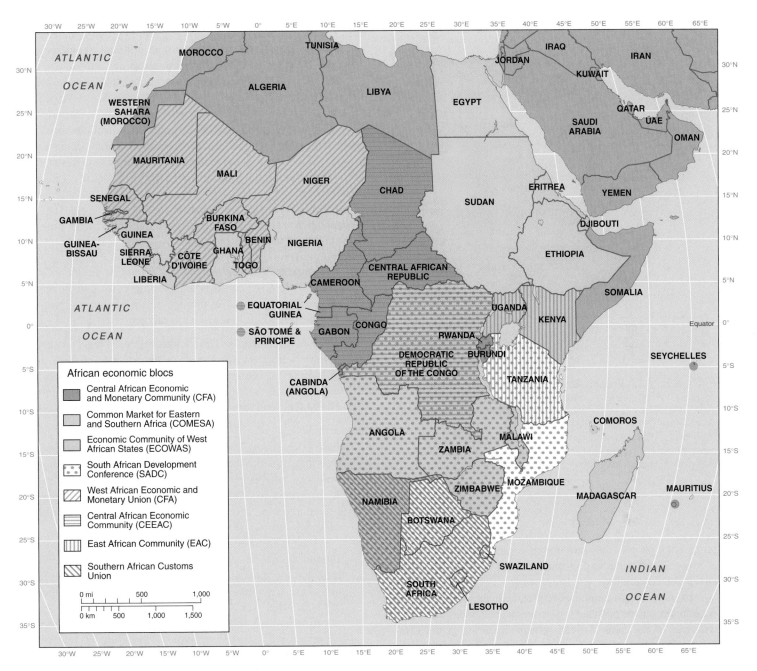

FIGURE 9.21 Africa: economic blocks. Few influence world markets. There are also overlaps between regional groupings that cause problems: for example, the ECOWAS considers monetary union, but the former French colonies in the French franc (CFA) zone are expecting to change to euros. **Source:** © *The Economist Newspaper Group, Inc.,* Reprinted with permission. Further reproduction prohibited. www.economist.com.

tries. The introduction of the euro currency in the European Union and the moves of the Economic Community of West African States (ECOWAS) toward monetary union in Western Africa resulted in confusion over the future of the CFA franc.

Local Emphasis

Globalization places an emphasis on export goods and foreign trade, but most people in the countries of Africa South of the Sahara rely on the local economy. Consumer goods remain

unusual or communal (Figure 9.23). Many rural Africans live with little reference to the global economy. However, it now reaches into the remotest villages through the use of motor vehicles or clothes made of synthetic fibers. Even small towns experience a growing mobility of people, increased levels of commercial exchange, and rising demands for the consumer goods seen advertised on global TV channels. Some villages mushroom into small service centers with rapidly built shops and market stalls where food and consumer goods are sold and

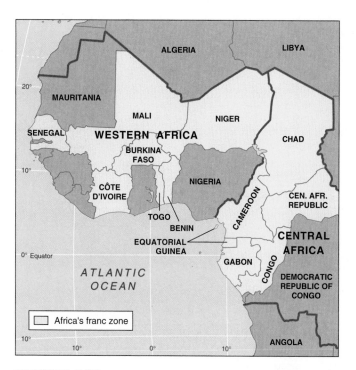

FIGURE 9.22 French franc (CFA) area of Africa. These countries, formerly French colonies, were supported by French loans, and their currency was linked to the French franc. The French government devalued the CFA franc by 50 percent in 1994, causing major changes in economic policies in the countries. The 2002 French adoption of the euro currency resulted in some confusion among CFA countries over their regional or European links.

buses bring people from surrounding rural areas. Low levels of access to good water and of energy usage also link to poorly developed economies.

In some areas a trend toward greater dealings in the global economy has been reversed. For example, in Zimbabwe, which was one of the main African growth countries in the early 1990s, President Mugabe encouraged the takeover of large white-owned commercial farms by war veterans. This changed the farming emphasis from tobacco and vegetable crops for export to the growing of corn for local consumption. This in turn cut off much of Zimbabwe's foreign exchange and reduced many former farmworkers to unemployed poverty.

Can Africa Claim the Twenty-First Century?

On the assumption that there is a wish to end poverty around the world, Africa South of the Sahara presents a challenge. Many of the world's worst development problems (see the discussion of development in Chapter 2) are concentrated there. Many countries are short of the basic human resources

and infrastructure needed to slow population growth, increase economic development, and encourage political democracy. This is a local problem with global implications.

In early 2005 world leaders proclaimed that they would focus on poverty reduction with special reference to Africa. The Millennium Development Goals (see Table 2.2), increased aid, debt forgiveness, and WTO trade liberalization should make it possible for poor African countries to emulate countries in Asia. However, to date, reductions in poverty have resulted from better domestic government in a few countries, rather than external aid.

External (France, the United Kingdom, the World Trade Organization, and the World Bank) critiques of Africa's problems highlight four areas of action:

1. *Improved governance and conflict resolution* are the most basic needs. Civil conflicts impose huge costs at home and in neighboring countries through deaths, maimings, property destruction, and refugee migrations.

2. *Investment in people.* The vicious circle of high fertility and mortality, low enrollment in education (especially of girls in some countries), high numbers of young people dependent on the working age group, little action against HIV/AIDS, and low savings lies behind much of Africa's slow or static development.

3. *Economic diversification* makes countries more competitive in world markets. The region's countries need new products and better terms of trade, new incentives, and wider access to markets in wealthier countries. The perceived risks of investing and doing business in Africa make job creation slow.

   Internal reforms needed include the reduction of corruption, improvement of infrastructure and financial services, and the provision of better access to the information economy. Countries do not have sufficient all-weather roads or other forms of internal transportation. Ports are poorly equipped and expensive. People lack clean water supplies and adequate sanitation. Spending is low on health and education. There are shortages of electricity and telecommunications.

4. *Reduced aid dependence and debt, and stronger intraregional partnerships.* Africa remains the world's most aid-dependent and indebted region. Programs of debt relief became more significant from the 1990s as aid donors channeled investment to countries, but still insisted on approved development policies to avoid corruption. The World Trade Organization tries to help African and other developing countries improve their access to markets for agricultural products and reduce farm subsidies in wealthier countries, but the United States and the European Union make small concessions and increase their own farm subsidies to maintain their own farming communities.

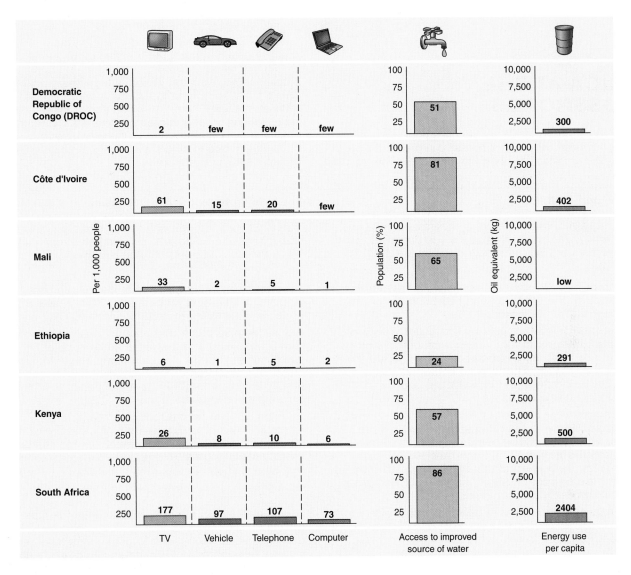

FIGURE 9.23 Africa South of the Sahara: consumer goods ownership. It is very low in African countries, along with access to good quality water and the use of energy, apart from South Africa: compare levels with those in other world regions. How does this diagram reflect Western values? Are these items the sole criteria for prosperity? **Source:** Data (for 2002) from *World Development Indicators,* World Bank, 2004.

To date, Africa South of the Sahara has been a loser in the global economy. The countries of this region produce only 1 percent of global GNI PPP. Most people have little access to the consumer goods that are the signs of material well-being or financial ability to develop entrepreneurial skills or engage with global connections. These issues are discussed in the Point–Counterpoint: Africa at the Crossroads on page 406. As in other developing countries, cell phones provide a communications breakthrough. Already there are twice as many cell phones as fixed landline phones. It is significant that cell phone companies are international MNCs, whereas landline phones are linked to bureaucratic government institutions that make it difficult to get one.

Subregions

Four subregions of Africa South of the Sahara have distinctive patterns of human geography (see Figure 9.1 and Table 9.2):

- *Central Africa:* the countries of the Congo River drainage basin and equatorial climatic environments.
- *Western Africa:* the countries in the western bulge of Africa with varied tropical environments ranging from rain forest on south-facing coasts to interior desert.
- *Eastern Africa:* the plateau and hilly countries of the eastern Horn of Africa.
- *Southern Africa:* the countries south of Congo and Tanzania, influenced strongly by the Republic of South Africa.

AFRICA AT THE CROSSROADS

Although Africa South of the Sahara has many positive geographic qualities, the recent history of almost all countries in this world region is discouraging. The jury is still out over whether African people and countries can overcome the present difficulties and make a better future in the light of what they perceive as important priorities. Whatever happens will affect the changing geography of this region and will have impacts on the rest of the world. Nobody living in our world today can ignore this major issue.

Assess different opinions about the likelihood of the outcome through the following discussion.

Africans can do it.	There is no hope for Africa's future.
There are signs that Africans are moving toward a better future. The peaceful transition in South Africa, the Uganda success with AIDS, fewer wars, and increasing democracy are examples of good trends.	Such signs are few and temporary. Most things get worse. Civil strife springs up in new places and most democracy is a facade. These are basic reasons for crippled African development.
Younger "born frees" (i.e. since the end of colonial rule) are better educated and more inclined to expect good leadership from the current leaders rather than blaming the past.	The present problems stem from the colonizers and their racism. Whatever Africans do to put them right is futile. (This is a common complaint of the older people born into colonial rule.)
African countries can produce goods in addition to the crops and minerals that others want to buy if trading terms are improved with the wealthier countries. Some countries are growing rapidly, some as a result of oil windfalls; a few others such as Mozambique, Rwanda, and Uganda have seen a decade of economic growth after decades of strife and poverty.	Few leaders place much importance on sustained economic growth, and their actions tend to reduce the ability of governments to develop education, health, and employment prospects. Few African leaders allow widespread involvement in government that reduces their powers of patronage and ability to make arbitrary decisions.
Smart businessfolk can do well, as in the oil companies on the west coast and the mining corporations. Those making cheap luxuries such as bottled drinks and soap powder prosper, and mobile telephones increased from almost none 10 years ago to 25 million today. The business climate could improve with more privatization in such areas as utilities (telephones, electricity) and with shorter periods of business registration.	Too many countries place difficulties in the way of foreign corporations wishing to do business in African countries, including the continuation of the bribe culture and lawlessness. The past has left too many examples of squandered opportunities.
The rise in the numbers of urban Africans is leading to wider political participation and, hopefully, to demands for better government. Better government is particularly important at this time, but has to be widely wanted and supported. Power is based in country governments. But improved political involvement has seldom been linked to greater prosperity.	There is still too much rule and control by a few "big men" who turn the law and finances to their own ends. There are many examples of this, with President Mugabe of Zimbabwe getting in the news recently. His once fairly prosperous country is now among the poorest in the world. Too many governments are predatory and few are competent. There are still few leaders who left after electoral defeat as compared with those overthrown by war or coup.
Aid agencies are now putting more research into funding activities. More philanthropic donors are needed, overseeing their giving in relation to criteria such as "saving the maximum number of lives at minimum cost" (Bill Gates).	In the past, too many aid agencies left behind more harm than good. Dependence on outside resources, and spending aid such as World Bank grants in profligate ways, do not lead to local entrepreneurial actions.
Land reform is occurring, although there is a need to ensure that people who can farm get title to farming land and government favorites do not take over productive land.	Most countries suffer from a lack of security in property rights. People with communal or tenant rights cannot use that land to underpin bank financing.
African countries are at last realizing that they have to deal with the HIV/AIDS problem, and there are encouraging signs of governments' intentions.	HIV/AIDS is out of control with rapid spread of devastating social problems, including a generation of orphaned and infected children. Even temporary palliatives are too expensive, and there is little infrastructure to monitor their use.
South Africa acts as an example of democracy, modernization, involvement of black talent in major corporations, free press, strong labor unions, independent judiciary, and large middle class.	South Africa does little to help other African countries improve their systems, partly because it is also struggling. Its recent dominance by a single political party (African National Congress) could result in greater corruption.
Africans are beginning to pay more attention to marrying the local and global trends.	The change from traditional approaches, ways of doing things, and expectations to the trappings of modernization has been too great.

Table 9.2 Africa South of the Sahara. Data by subregion, area, population, income (GNI PPP—Gross National Income Purchasing Power Parity), urbanization, Human Development Index (HDI), Human Poverty Index (HPI).

| | Land Area (km²) | Population (millions) | | GNI PPP 2003 | | % Urban | Index 2002 | |
		Mid 2005 Total	Est. 2025 Total	Total (US$ billions)	Per Capita (US$)	2005*	HDI (rank of 177)*	HPI (% population)*
Central Africa	5,419,340	113.5	189.7	105	2,466.7	38.6	150.2	42.4
Western Africa	6,120,380	263.9	403.1	285	1,345.0	35.4	160.5	46.8
Eastern Africa	3,644,040	188.7	299.4	147	1,181.7	32.7	156.0	40.3
Southern Africa	6,571,950	142.0	189.8	606	3,655.5	32.0	147.1	44.0
Totals or averages*	21,755,710	708.1	1,082.0	1,143	2,162.2	34.7	153.5	43.4

Source: World Population Data Sheet, Population Reference Bureau (2005), World Development Indicators, World Bank (2005), Human Development Report, and United Nations (2004).

Central Africa

Central Africa (Figure 9.24) is the least developed subregion of the least developed world region. The largest part of this subregion remains isolated from world trade networks. The natural environment, marked by equatorial rains, dense forest, and diseases, was sparsely populated before European intrusions and remains so. Belgium and France, the main colonizers, encouraged limited economic development through selected extractive industries. Since independence, governments struggle to cope. Self-serving dictatorships and widespread civil warfare have destroyed the prospects for improving people's lives.

The coastal countries (Table 9.3), including tiny Gabon and Equatorial Guinea, together with Cameroon and the Congo Republic, have some direct links to international markets. The Democratic Republic of Congo (DRC), Central African Republic, Chad, Burundi, and Rwanda have no coastal outlets and rely for transportation links on the Congo River waterway, which is interrupted by rapids from its mouth to its source.

Countries: Politics and Ethnicity

Internal political conflicts are linked to an ethnic mosaic that sits uneasily with the country boundaries established by the colonial powers (see Figure 9.4a). For example, the Fang people dominate Equatorial Guinea, a small country, but Fang people also live in neighboring Cameroon and Gabon. In Gabon, a pact among smaller groups keeps the Fang—the largest group—out of government.

Following independence, the Central African countries mostly became dictatorships. Through the Cold War period they aligned to the Soviet Union or the United States. This provided some stability, but they were often harshly administered. The superpowers encouraged rebel groups in countries aligned with their rival. After the end of the Cold War, political order broke down, economies collapsed, and the social order was disrupted. Attempts to establish groupings of cooperating countries, such

as the Central African Economic Community and the Central African Economic and Monetary Community (see Figure 9.21), made little progress because of the civil strife.

Rwanda and Burundi

In 1994 the genocide of nearly 1 million people brought these two small countries in the heart of Africa to the world's notice. Ten years later, Rwanda is quiet under the virtual dictatorship of Paul Kagame, who ended the massacres by force. However, the conflicts have spread to neighboring countries.

Rwanda's main rival tribes are the shorter, flat-nosed majority **Hutu** cultivators and the taller, thinner, sharp-nosed **Tutsi** cattle herders. They skirmished with each other before the colonial period. However, tribal priority was never an issue, intermarriage was common, and many Hutus socially upgraded to become Tutsis (see Personal View: Rwanda, page 410). The Belgian colonists deposed Hutu chiefs in favor of Tutsis and issued ethnic identity cards, leading to ethnic resentment. At independence, the Hutu majority won the elections and imposed a 9 percent quota on Tutsis (their proportion of the population) for salaried jobs. General Juvenal Habyarimana seized power in 1973 as one of "the majority people" and further discriminated against the Tutsis—many of whom left the country. In 1990 Kagame gathered these exiles and invaded from Uganda. Despite a 1993 peace accord, some members of the previous Hutu regime recruited and indoctrinated thousands of Hutu militiamen and armed them with machetes. When Habyarimana's plane was shot down on April 6, 1994, his most bigoted associates took control and set off the genocide through local meetings. The property of those killed was given to enthusiastic murderers.

Kagame's Rwanda Patriotic Front (RPF) formed of exiled Tutsis won the short war, killing many Hutus and chasing others into the Zaire (now DRC) rain forests. The RPF continues to govern Rwanda, using massive foreign aid to rebuild the school and health care systems. Hundreds of thousands of Tutsis returned, bringing cash and skills to replace the lost middle class. Many had been born abroad

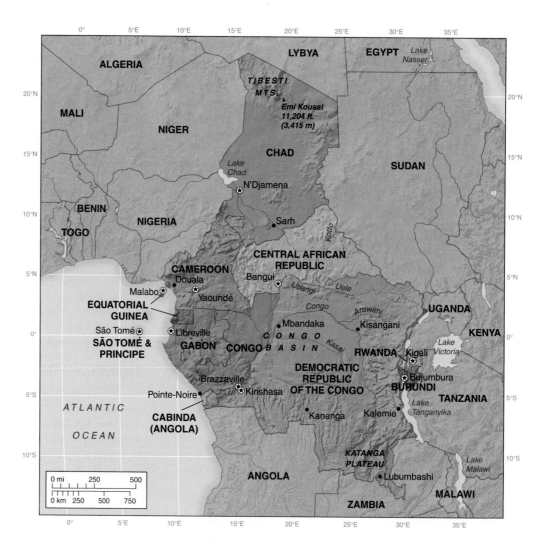

FIGURE 9.24 Central Africa: main features. The countries included in the subregion, the major cities, and rivers.

Table 9.3 Central Africa. Data by country, area, population, income (GNI PPP—Gross National Income Purchasing Power Parity), urbanization, Human Development Index (HDI), Human Poverty Index (HPI).

Country	Capital City	Land Area (km²)	Population (millions) Mid 2005 Total	Population (millions) Est. 2025 Total	GNI PPP 2003 Total (US$ billions)	GNI PPP 2003 Per Capita (US$)	% Urban 2005	Index 2002 HDI (rank of 177)	Index 2002 HPI (% population)
Burundi, Republic of	Bujumbura	28,000	7.8	13.9	4	620	9	173	45.8
Cameroon, Republic of	Yaoundé	475,440	16.4	22.4	32	1,980	48	141	36.9
Central African Republic	Bangui	622,980	4.2	5.5	4	1,080	41	169	47.7
Chad, Republic of	N'Djamena	1,284,000	9.7	17.0	9	1,100	24	167	49.6
Congo, Republic of the	Brazzaville	342,000	4.0	7.4	3	710	52	144	31.9
Congo, Democratic Rep of the (Zaire)	Kinshasa	2,344,860	60.8	108.0	34	640	30	168	42.9
Equatorial Guinea, Republic of	Malabo	28,050	0.5	0.8		9,110	45	109	32.7
Gabonese Republic	Libreville	267,670	1.4	1.8	8	5,700	81	122	No data
Rwanda, Republic of	Kigali	26,340	8.7	12.9	11	1,260	17	159	44.5

Source: World Population Data Sheet, Population Reference Bureau (2005), World Development Indicators, World Bank (2005), Human Development Report, United Nations (2004), and Encarta Microsoft (2005).

following large-scale emigration after 1959 and had little knowledge of the local language. The RPF's main objective is to maintain peace and involve all Rwandans. It tries to reeducate the murderers who survived. However, security is tight, with no press freedom or freedom of association; party loyalty is imposed; and Rwanda prisons are "life-threatening" (U.S. State Department).

In Burundi, similar divisions and tensions exist. In 1993, in the latest of many conflicts since independence, an attempt to hand power from a Tutsi president to a Hutu led to an army revolt inspired by Tutsi officers. The new president was assassinated and there was civil war, in which 300,000 were killed and another million fled the country, mainly into Tanzania. The 1994 events in Rwanda caused Tutsis to retain power until Nelson Mandela's 2000 mediation and message of reconciliation, but the process of handing over power remains fraught with problems.

Democratic Republic of Congo (DRC)

The combination of large size and larger population, centrality in the Congo River basin, and many exploited mineral resources might be expected to make the Democratic Republic of Congo (DRC) a focal country in this subregion. It is almost twice as large as the next country in area, Chad, and with 61 million people (2005) had three times the population of Cameroon, the second country in population. However, instead of forming a central driving force for development in the subregion in the late 1990s and early 2000s, DRC was largely taken over by private armies. Government of the whole country became impossible.

Soon after independence in 1960, the Democratic Republic of Congo army under Mobutu Sese Seko seized control from the Communist Patrice Lumumba. Supported by the United States, he remained in power, building up his family position and private bank accounts abroad, until a 1997 rebellion deposed him. Mobutu's robbery of national wealth resulted in a description of his government as "kleptocratic" ("klepto" infers stealing). Laurent-Désiré Kabila led the 1997 rebellion from his base in eastern DRC, using exiled Tutsi and local related groups with the backing of Rwanda. The demoralized DRC armies offered little opposition; Mobutu went into exile, and Kabila became president.

However, Kabila could not control DRC from the capital, Kinshasa. Other countries became involved, widening the impact of the internal conflict. Old Marxist ties led Zimbabwe and Namibia to send troops and equipment to support Kabila. Uganda and Rwanda backed the eastern (Tutsi) groups to attack the Kabila regime, who now used Hutu troops as the basis of his Congolese army. Angolan troops entered from the west, outwardly to support Kabila but mainly cutting off the supply routes to Angolan rebels. Once there, all the countries extracted mineral wealth, from diamonds to gold and copper, but by 2003 stresses at home caused them to withdraw.

After Kabila was assassinated in 2001, Joseph Kabila, his son, replaced him. Ethnic rivalries again erupted into local wars, such as the Lunda fighting the Luba in the south (Katanga). By 2004 ceasefires became more common between flare-ups in the fighting. As many as 2 million Congolese people had been killed, and many thousands of refugees were displaced. The DRC government is a coalition of rebels formed in 2003 in moves toward peace and democracy. However, many local

Country	Ethnic Groups Percentages	Languages O=Official N=National	Religion Percentages
Burundi, Republic of	Hutu (Bantu) 79%, Tutsi (Hamitic) 20%	Kirundi (O), French (O), Swahili	Roman Catholic 62%, local 32%, Protestant 5%, Muslim 1%
Cameroon, Republic of	200 groups: Fang, Barnileke, Fulani, Pahouin	English (O), French (O), 24 African	Christian 53%, local 25%, Muslim 22%
Central African Republic	Baya 34%, Banda 27%, Mandjia 21%, Sara 10%	French (O), Sango (N)	Local 60%, Protestant 18%, Catholic 17%, Muslim 5%
Chad, Republic of	Muslim groups N and C, non-Muslim in S	French (O), Arabic (O), Sara in south	Muslim 50%, Christian 33%, local 17%
Congo, Republic of the	Kongo 48%, Sangha 20%, Teke 17%, Mboshe 12%	French (O), Kikongo, Lingala, Teke	Christian 50%, local 48%, Muslim 2%
Congo, Democratic Rep of the (Zaire)	Over 200 African groups: Mongo, Luba, Kongo	French (O), Lingala, Kikongo, Tshiluba, Swahili (all north)	Roman Catholic 52%, Protestant 20%, Kimbanguist 10%, Muslim 2%, local 16%
Equatorial Guinea, Republic of	Fang 80%, Bubi 15%	Spanish (O), Fang, pidgin English	Roman Catholic 90%, local practices
Gabonese Republic	Fang 60%, Mpongwe 15%, M'bete 14%, Punu 12%	French (O), Fang, others	Christian 60%, local animist 39%, Muslim 1%
Rwanda, Republic of	Hutu 90%, Tutsi 9%, Twa 1%	Kinyarwanda (O), French (O), Kiswahili	Roman Catholic 48%, Protestant 9%, local 34%, Muslim 9%

RWANDA

Rwanda is one of the smallest and poorest countries in Africa, but it has often been in the world news over the last 30 years because of civil wars and horrific tales of hatred expressed in violence. Ntwari is a Rwandan who grew up in the country during this period. His experience of living in Rwanda reflects some of the events that affected people's lives and changed the human geography of the country in the later 1900s.

Rwanda is a hilly to mountainous country close to the equator (known as "the land of a thousand hills"), but it is landlocked and surrounded by other countries—Congo, Uganda, Tanzania, and Burundi (Box Figure 1a). The highest volcanic peak, Karisimbi (4,507 m; 14,787 ft.), is on the edge of the rift valley that contains Lake Kivu, and most of the country is over 1,220 m (4,000 ft.). This elevation modifies the equatorial climate, bringing temperatures down to just over 20°C (70°F) and resulting in a distinctive type of tropical rain forest vegetation.

Ntwari grew up in southern Rwanda, one of nine children on a typical family plot of land with the house surrounded by banana trees and segments of the land devoted to crops such as corn, beans, potatoes, tropical root crops, a little pasture for a few head of livestock, and land for growing the cash crop, coffee or cotton. From time to time, cousins might come and stay for a year or so, but any grandparents still alive maintained their own plots. Some friends lost their parents in the civil war, but the children continued to live on the family plot, with teenagers raising their younger brothers and sisters.

With a rapidly growing population, the family plots typical of Rwanda cut into almost all the preexisting forest. Even the Akagera National Park on the eastern border was largely taken over by farmed plots of land. Although the plots can provide both subsistence and cash-crop income, many on steeper slopes suffered soil erosion, reducing the harvest as soils lost their nutrients or flooding destroyed crops. Low world market prices for coffee reduced incomes and made families indebted. Rwanda became dependent on international aid to supply food and to fund rural development programs fighting soil erosion. Although the fertility rate remains high, total population growth slowed in the 1990s because of the civil war, a high incidence of HIV/AIDS, and the continuing impact of poor nutrition and tropical diseases such as malaria.

Ntwari went to the local first-level school up to age 15. There was no public transport, and so he walked the 5 km (3 mi.) to and from school each day with his friends. He was one of three from his

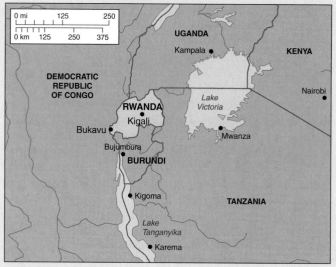

(a)

(b)

Box Figure 1 (a) Rwanda. This tiny country is located where Central and Eastern Africa meet, but has a mighty impact on its neighbors. **(b) Displaced Rwandans following the 1994 Civil War.** Mucaca Camp, 90 km (55 mi.) north of Kigali, the Rwandan capital, in 1998, showing newly built huts and some 45,000 people living there. People here included refugees returning from surrounding countries and those who came from Rwandan villages. The United Nations estimates there are currently over 600,000 displaced persons in Rwanda. **Photo:** © AP/Wide World Photos.

class of 45 to go on to the second-level school up to age 21. Again one of a small proportion who progressed further in their education, he then studied for a bachelor's degree in sociology and anthropology at the National University in Butare. Jobs were available for his fellow students in government and aid agencies, and for Ntwari in teaching. There was even a shortage of educated local personnel, requiring expatriates to be brought in.

Following a struggle for independence that involved a civil war and increasing tensions between Hutu and Tutsi peoples, Rwanda gained its independence from Belgium in 1959. Many of the former ruling Tutsi groups, including the king, went into exile in Europe and America. At first, people from the south, the Nduga, controlled the country's government through their political party, the Democratic Republican Movement. A mixture of Hutu and Tutsi peoples living together, they provided most university students and so gained most government jobs. In 1973 a military coup d'état was led by northerners, the Rukiga, mainly of Hutu peoples, who governed for 20 years through their political party, the Republican Movement for National Development. During this period, the president took control of the military and trained a militia force of northerners. At the same time, southerners, backed by the exiled Tutsis who had formed the Rwandan Patriotic Front (RPF), trained their own militias. While the northerners were in government, the mixture of southern peoples were increasingly labeled as "Tutsis," "friends of Tutsis," or simply "enemies of the state."

Changes in the 1980s and 1990s resulted in the greater social and political polarization of groups of people within Rwanda. Rural conditions still predominate over most of the country, with traditional ways of life continuing and a social life based around neighborhood parties. As Kigali, the capital, expanded its role after independence, the increasing numbers of educated young people in government jobs there established new social groups based on professional and business interests or on college links. Increasingly, too, social gatherings brought together peoples in groups from either the north or south of the country. This division reflected increasing political rivalries.

In 1990 matters came to a head as the RPF invaded Rwanda through Uganda in the north. They advanced southward, pushing back the Rwandan forces into the southern part of the country. This led to a massacre of southerners, who were suspected of working closely with the RPF. Ntwari, working for a church helping increasing numbers of fleeing refugees in Gitarama, was beaten up because he had taught in a school with a Tutsi head teacher (who was murdered) and had a Tutsi wife (many of whose family were killed).

Peacekeeping efforts prevailed to bring this part of the conflict to an end, but strife erupted again in 1994, when hopes for a democratic government were destroyed by the assassination of the newly elected Rwandan president. The country once more descended into civil war. Ntwari and his wife moved westward to a town near the Congo border, where they stayed for a month. Then the former (northerner-based) official Rwandan army and militias disintegrated under the RPF attacks and fled into Congo. Ntwari and his wife moved with them and entered a refugee camp at Bukavu. Almost all the people in the camp were Hutus who hated Tutsis. When Ntwari's wife was threatened with death, she fled and was cared for in hiding by a local family.

Many Tutsis were killed at this time. The remainder fled into the hills or into Tanzania. The intervention of the largely French "Operation Turquoise," backed by the United Nations, saved many Tutsis from the slaughter. They were brought together in camps within Rwanda.

Ntwari decided that there was no immediate future in Rwanda or a refugee camp outside the country. Although most of the non-militia Hutus from the camp returned to Rwanda (Box Figure 1b),

he found his way via a boat across Lake Tanganyika into Tanzania and then took a train and bus to Nairobi, Kenya. Having found that he could gain admission to Kenya through this route, he returned to bring out his wife. They persuaded the United Nations Refugee Commission that they would be in danger if they returned to Rwanda and were airlifted from Bukavu camp to Nairobi. Once there, they looked for an opportunity for Ntwari to pursue his studies in another part of the world.

The Hutu–Tutsi ethnic differences are blamed for the civil conflict, but it is clearly not just a tribal war. Colonial and postindependence events heightened previous rivalries. Before the colonization of this area by Europeans, the Tutsi tribe had the status of nobility, but people from other tribes could be transferred into the "Tutsi" group following marriage or effective military service, for example. The leadership group became broadly based in tribal terms and was identified as *imfura*—civilized people able to take leadership and speak in public.

The European influence was partly political, with a short-lived German occupation (1897–1918) and a longer Belgian protectorate (1918–1959), and partly religious. The Roman Catholic Church and Protestant missionaries had major influences, probably the greatest in any African country. In 1900 the white-robed and white-faced European Roman Catholic priests ("white fathers") entered the country. In 1943 the Rwandan king was baptized and declared his lands to be a Christian country. New social status was gained by those who became like Europeans in income or education: *umunzunga* indicates a person who is both educated and wealthy; *umusilimu* is one who is educated and not wealthy but still dresses like white people (such as a schoolteacher).

The various strands of social and cultural development led to friction between the better-educated and ethnically more varied southerners and those in the north who felt underprivileged. The military coup in 1973 and the civil war in the 1990s arose from such frictions and destroyed the fledgling economy of this small, poor country.

Ironically, despite attempts to throw off foreign influences, the country remains even more dependent on outsiders and particularly on the European countries and aid agencies that bring funding and expertise for water projects, house building, road and bridge construction, and forestry and advice for farmers. The Roman Catholic Church still runs most of the hospitals and clinics, although other churches are also involved. Any use of modern technology, from computers to telecommunications, occurs in the offices of United Nations and European agencies. All these organizations and projects are designed to work in partnership with local Rwandans and provide jobs for many of those educated to college level. But the civil war destroyed or set back many projects that had begun to improve people's lives.

Rwandans identify a need for improved education as a basis for overcoming the problems they face. This is in the widest sense. There is a need for people to understand the origins of frictions among groups of people. Who arrived first in the country—the Twa (pygmies), Hutu, or Tutsi? What were the grounds of claimed superiority? What did each group contribute? How can tolerance be reestablished? What does "democracy" mean and how can it be restored? What is the nature of development, and how can Rwandans take advantage of offered aid? What are the advantages of family planning in a country where population growth exceeds growth in economic provision? What is the truth about HIV/AIDS in a country with a high incidence and growing mortality rate? Unless the Rwandan people can come to terms with such issues openly, knowledgeably, and democratically, the future of the country will remain bleak.

Tutsi warlords and other militia groups resist invitations to join a government of national unity. Elections set for June 2005 were postponed, and funds designated for development have been spent on military hardware.

Other Central African Countries

Of the other Central African countries, Chad, Central African Republic, and Congo experienced 30 years of upheaval after their independence from France in 1960. In Chad there were wars between the Libyan-backed Muslim northerners and the southerners. In the Central African Republic, misrule and misspending by a flamboyant dictator were replaced by an elected civilian government with a new constitution in 1995, but military rebellions from 1996 again dislocated the economy. Congo had an authoritarian Marxist government from independence but also opted for more democratic status in the 1990s until civil war returned the former president in 1997. Long-term autocratic governments continue in Gabon and Equatorial Guinea.

People: Ethnicity

The peoples of Central Africa are almost all of black African ethnic groups who hold to their long-term allegiances. The positive roles of extended family and kinship connections work together with normally harmonious interreligious group relations in societies subject to major stresses. Most people belong to groups pushed southward into the forests when Muslims took over the northern parts of the region. The small numbers of the pygmy groups displaced or ethnically cleansed by the incursions continue to decline in numbers and now make up less than 2 percent of the population.

The proportion of Muslim population declines southward in the subregion. Muslim and Christian groups frequently work together, including high levels of syncretism, in which religions borrow from each other. For example, traditional practices are often adopted or allowed by Muslim and Roman Catholic groups.

Economic Development

The combination of physical isolation, repressive colonial history, continuing civil strife, and lack of foreign investment hinders the economic development of Central Africa. The subregion is almost ignored by the global economy. In real terms, incomes in all the countries except Equatorial Guinea and Gabon fell since 1980. Human poverty indexes are high. Despite improvements in education and health since the 1960s, levels of life quality in Central African countries hardly rose. Several countries depend on external aid.

Dominant Agriculture

Agriculture remains the main economic occupation of two-thirds of the populations of these countries. Most farming is for subsistence, using traditional methods to grow tropical root crops, cereals, fruits, and vegetables. However, only a small part of the surface is cultivated because of a combination of soils that are difficult to work with, plant and animal diseases, and poor transportation facilities. For example, only 3 percent of DRC is cultivated. Cattle farming in Central Africa is severely restricted by diseases, particularly those borne by the tsetse fly. Modern veterinary treatments for animals are expensive and seldom used. Some groups continue practices of gathering forest products or of shifting agriculture (Figure 9.25).

Commercial farming for export crops is poorly developed in Central Africa. Colonial powers introduced some plantation agriculture for tropical tree crops such as cacao (for chocolate), coffee, and rubber. Tree-based plantation crops replicate the form of the natural equatorial forest vegetation, but concentrations of a single species often fall victim to disease. Growing a single commercial crop (monoculture) may bring in a good income at a time of high world market prices, but farmers suffer when low prices combine with lower yields as soil quality declines. However, tree crop plantations have the advantage that they reduce the soil fertility over 30 to 40 years instead of the 5 to 10 years under field crops. DRC nationalization in the 1970s led to a decline of commercial farming within its borders. Farmers in the drier northern regions of Chad and the Central African Republic produce cotton for export, but poor transportation from these landlocked countries makes competing in world markets difficult.

Despite the dominance of farming occupations, most Central African countries find it difficult to feed their populations. Government policies that overvalued their currencies maintained low prices for food in urban areas but lowered farmer incomes and discouraged exports. The food imports made necessary by the consequent fall in local produce add to overseas debts.

Forestry, Fishing, and Mining

Timber products from the tropical rain forest, including mahogany and ebony wood (Figure 9.26), are of increasing importance to the exports of Cameroon and Congo, while Gabon produces a softwood that is used in plywood. Depletion of the tropical rain forest is an issue. Deep-sea fishing increases in the coastal countries such as Cameroon and Equatorial Guinea. Chad obtains fish from its lake.

Mining brings in most foreign exchange to the countries of Central Africa (Table 9.4). It could bring more if transport infrastructure were available or world markets paid higher prices. Before the civil disorders, southern DRC was one of the world's major copper-mining regions, and the country mined diamonds and cobalt and produced some oil. Gabon's relatively high per capita income comes from its (declining) oil and manganese production. Oil makes up 90 percent of Congo's export revenues. Cameroon, Chad, Sao Thomé and Principe, and Equatorial Guinea also have oil deposits that are now being exploited. Gabon possesses the world's largest unexploited iron ore deposit.

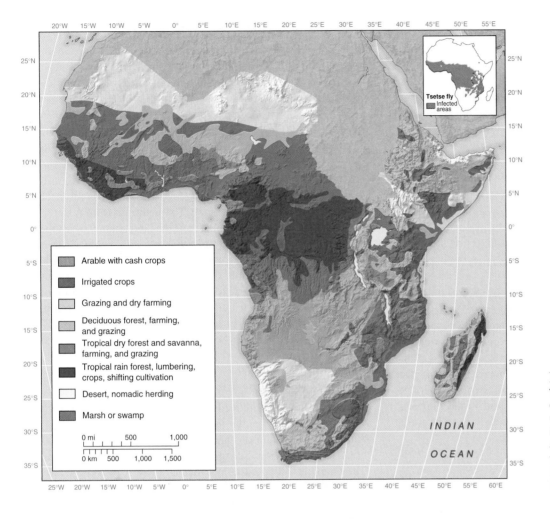

FIGURE 9.25 Africa South of the Sahara: land use. Note the distribution of intensive, commercial uses and of unusable land or low-intensity uses. The insert shows the area affected by tsetse flies, an area where cattle rearing is difficult. **Source:** Data from *New Oxford School Atlas.* Oxford University Press, U.K.

Legend:
- Arable with cash crops
- Irrigated crops
- Grazing and dry farming
- Deciduous forest, farming, and grazing
- Tropical dry forest and savanna, farming, and grazing
- Tropical rain forest, lumbering, crops, shifting cultivation
- Desert, nomadic herding
- Marsh or swamp

Insert: **Tsetse fly** Infected areas

FIGURE 9.26 Central Africa: timber industry. Logs from the Gabon rain forest come to Libreville harbor awaiting export. **Photo:** © Gallo Images/Corbis.

Table 9.4 Major African Mineral Products

Country	Mineral Products
Democratic Republic of Congo	Copper, cobalt, diamonds
Gabon	Oil, major unexploited iron ore
Ghana	Bauxite (aluminum ore), gold
Guinea	Bauxite, iron ore, gold and diamond reserves
Mauritania	Iron ore
Nigeria	Oil, tin
Niger	Uranium
Senegal	Phosphate
Togo	Phosphate
Sierra Leone	Diamonds
Angola	Diamonds, oil
South Africa	Gold, diamonds, iron ore, coal, platinum, chrome, manganese, vanadium
Namibia	Uranium, diamonds, zinc, lead, tungsten
Botswana	Diamonds
Zambia	Copper
Zimbabwe	Coal, chrome, gold, nickel, platinum

Oil started flowing from Chad's wells in 2003, exported via a pipeline through Cameroon. The income will help to pave roads and repair war damage, and an independent government-sponsored watchdog attempts to funnel the income to better uses after farmers squandered their payments for spoiled farmland as the pipeline was built. Along with Sao Thomé, Equatorial Guinea, and Angola, Chad is not an OPEC member and sells much of its oil to the United States.

Emerging Manufacturing

Manufacturing industries are little developed in Central African countries. Most factories are small and partially process local mine, forest, and farm products or make bulky products that do not stand transport (cement, bottled drinks). Many were destroyed by civil strife. Cotton textiles are made in Chad from the local crop.

The relative size of economies in this subregion is reflected by the low availability of energy (see Figure 9.23). The huge water resources of the Congo River basin are estimated to contain one-sixth of the world's potential hydroelectricity potential. This could be a source of power for manufacturing and other forms of economic development. However, although the Inga Dam downstream from Kinshasa harnesses some Congo River power, the electricity generated is sent to the mines in southern DROC and to countries farther south without intermediate allocations of electricity to local uses.

Transportation Is Vital

Prospects for economic development rest to a large extent on improving internal transportation links. Gabon was long the only country to invest some of its oil revenues in constructing all-weather roads and a railway into the interior. For example, the railways on either bank linking Kinshasa and Brazzaville with ocean ports are narrow gauge, limited in capacity, and unconnected to each other. The road from Kinshasa to the port of Matadi below the rapids, on which trucks once took five hours to cover the 350 km (220 mi.), is now in such poor repair that the trip takes five days. Elsewhere, the extensive, but impeded, navigable waterways of the Congo River system are the main means of inland transport.

Western Africa

The external orientation and environmental setting of Western Africa contrast with Central Africa. Its long coastline makes Western Africa (Figure 9.27) physically more open to the world than Central Africa, and this was the first part of Africa to receive European maritime contacts. The natural environments vary from the equatorial all-year rainy climate of the south-facing coasts to the margins of the Sahara in the north. A belt of tropical rain forest along the south-facing coasts merges northward through savanna grasslands, where seasonal rains support frequent trees, to dry savanna with little rain and semidesert. Human groups responded differently to these natural zones.

The precolonial African empires occupied the moister parts of the savanna grasslands. Colonial countries focused activity on coastal ports linked to plantation crops in the southern forests and mining wherever there were rich ore deposits.

Countries

In 2005 the 15 countries of Western Africa had populations that ranged from Nigeria, Africa's largest (132 million), to Guinea-Bissau and The Gambia, which had just over 1 million people each (Table 9.5). Nigeria ranges across all three natural climate/vegetation zones. The countries with a southern coast have strips of rain forest backed inland by moist savanna. The west-facing coastal countries become drier northward. The large land-locked countries of Mali, Burkina Faso, and Niger have only dry savanna and semidesert zones (Figure 9.28a). Attempts to bring together the countries of Western Africa for their mutual economic benefit centered in the Economic Community of Western African States (ECOWAS) and the West African Economic and Monetary Union, but these links grow slowly.

Changes Following Independence

Fourteen countries achieved independence around 1960, with Guinea-Bissau becoming independent of Portugal in 1974. Although many countries are small in area with small populations, attempts to join Senegal and Mali, and later Senegal and The Gambia, collapsed as politicians failed to work together. Except for Côte d'Ivoire, all countries until the early 1990s had long periods of control by dictators of military or socialist background. After that, most countries moved to multiparty democratic constitutions. Democracy worked most successfully in Côte d'Ivoire, Ghana, Mali, Senegal, and The Gambia. In others, the military retained real control, the move to democracy ended after military coups, or the elections were contested as having been unfairly conducted by the majority party. Even in Côte d'Ivoire, with its long history of political stability, internal north–south ethnic and immigrant worker quarrels damaged its tolerant and prosperous image. A 1999 military coup led to ethnic and antiforeign violence in the early 2000s. The combination of traditional intergroup rivalries, colonial favoritism, Cold War antagonism, and personal ambition lies behind the current conflicts.

In the 1990s, civil war in Liberia, Sierra Leone, and Guinea-Bissau disrupted their economies and social systems and threatened to spread into Guinea. In Liberia, the differences between local peoples and the slaves resettled from the United States in the 1800s were reflected in military rank. Officers came from the minority group descended from returning slaves and nonofficers from the majority indigenous groups. Sergeant Samuel Doe of the indigenous group led a coup in 1980 and became the new president. Further coups occurred into the 1990s, including that by President Charles Taylor, who supplied arms to Sierra Leone rebel groups in exchange for diamonds that he banked until he was deposed in 2003 (Figure

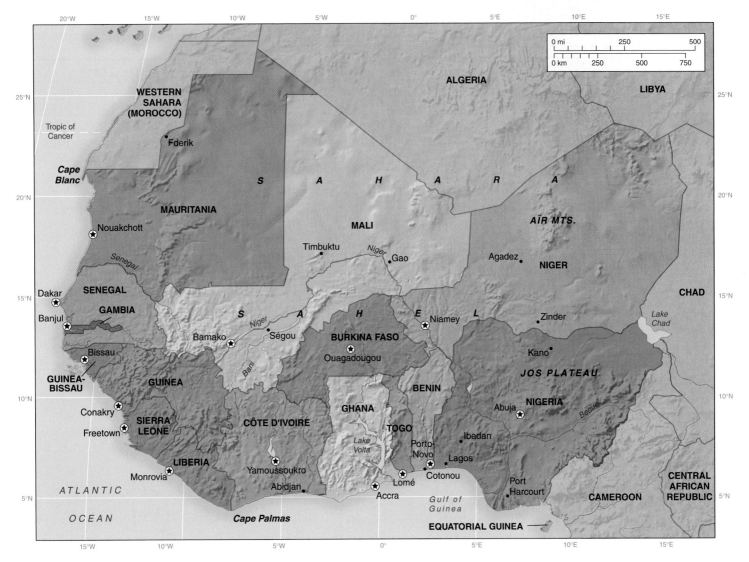

FIGURE 9.27 Western Africa: main features. The countries included in the subregion, the major cities, and rivers.

9.28b). The civil strife in Sierra Leone continued and spread into Guinea, where the resettling of refugees sparked fighting and encouraged local dissidents (Figure 9.28c).

Nigeria

The largest and most populous country in the region (Figure 9.28d) exemplifies many of the problems facing the countries of Western Africa. Nigeria experienced so much bad government from dictator generals in the 1980s and 1990s that it changed from an oil-rich, middle-income country to one of the world's poorest. A lot of the problems arose from the continuing north–south ethnic and religious conflict. The northern Muslim Hausa and Fulani people took control of the military and government, with the southern Yoruba being outvoted and the Igbos being widely disregarded after losing the 1960s civil war in which they had tried to assert their independence. Although the 1966–1979 military rule brought life to the

economy and united peoples with very different interests inside the country, the military rule from 1984 to 1999 was a disaster. The northern Muslims made Nigeria a member of the Organization of the Islamic Conference, annulled the 1993 presidential election, and repressed the peoples of the oil-producing Niger River delta region. This government deepened cultural divisions and mismanaged the economy through neglect and corruption. In 1998 it was estimated that three-fourths of official GDP was generated by the informal economy. Some US $12.4 billion of government funds was paid out without proper accounting, while the military dictator in the mid-1990s stole $3 billion.

After 1999, Nigeria tried to make a hasty transition from military to democratic government. The newly elected democratic government of Nigeria led by Olusegun Obasanjo faced huge problems over the legitimacy of the 1999 constitution that was quickly put together by the military before it ceded power.

Table 9.5 Western Africa. Data by country, area, population, income (GNI PPP—Gross National Income Purchasing Power Parity), urbanization, Human Development Index (HDI), Human Poverty Index (HPI).

Country	Capital City	Land Area (km²)	Population (millions)		GNI PPP 2003		% Urban	Index 2002	
			Mid 2005 Total	Est. 2025 Total	Total (US$ billions)	Per Capita (US$)	2005	HDI (rank of 177)	HPI (% population)
Benin, Republic of	Porto-Novo	112,620	8.4	14.3	7	1,110	40	161	45.7
Burkina Faso, People's Democratic Republic of	Ouagadougou	274,000	13.9	22.5	14	1,180	17	175	65.5
Côte d'Ivoire, Republic of	Yamoussoukro	322,460	18.2	25.1	23	1,390	46	163	45.0
Gambia, Republic of the	Banjul	11,300	1.6	2.6	3	1,820	26	155	45.8
Ghana, Republic of	Accra	238,540	22.0	32.8	45	2,190	44	131	26.0
Guinea, Republic of	Conakry	245,860	9.5	15.8	17	2,100	33	160	No data
Guinea-Bissau, Republic of	Bisau	36,120	1.6	2.9	1	660	32	172	48.0
Liberia, Republic of	Monrovia	97,750	3.3	5.8	No data	No data	45	No data	No data
Mali, Republic of	Bamako	1,240,190	13.5	24.0	11	980	30	174	58.9
Mauritania, Islamic Republic of	Nouakchott	1,025,520	3.1	5.0	5	2,010	40	152	48.3
Niger, Republic of	Niamey	1,267,000	14.0	26.4	10	820	21	176	61.4
Nigeria, Federal Republic of	Abuja	923,770	131.5	190.3	122	900	44	151	35.1
Senegal, Republic of	Dakar	196,720	11.7	17.3	17	1,660	43	157	44.1
Sierra Leone, Republic of	Freetown	71,740	5.5	8.7	3	530	37	177	No data
Togo, Republic of	Lomé	56,790	6.1	9.6	7	1,500	33	143	38.0

Source: World Population Data Sheet, Population Reference Bureau (2005), World Development Indicators, World Bank (2005), Human Development Report, United Nations (2004), and Encarta Microsoft (2005).

Problems included the fragile federal system, the military undermining of the judicial system, and the reduction of the police force (in case it competed with the military).

Some specific political challenges illustrate these tensions. First, in 1999 and 2000, several states in the Muslim north declared that Islamic religious (sharia) law should take precedence over the colonially inherited common-law order. This raised questions about the relationships between public institutions and religious traditions throughout Nigeria. Second, most of Nigeria's oil comes from the delta of the Niger River, inhabited by several small, poor ethnic groups. The oil revenues, however, go to the federal government, which takes the largest share and divides the rest among the 36 states on the basis of population and area. The oil-producing areas demand that more funds be returned to them, and local people cause pipeline disruptions. The federal government scarcely listens and replies with repressive measures such as the murder of the activist Ken Saro-Wiwa, who protested

oil company and government attitudes. Demands from the delta peoples increased when world oil prices rose sharply in 2004. Third, the 17 southern states complain that the federal government makes most important decisions without listening to the states. They want more local power over the police force, education, revenues from resource exploitation, and infrastructure provision. But the 19 northern states resist this proposal.

By 2005 Obasanjo had achieved more democracy and had appointed technocrats to key posts. The civil servants' benefits were increased and are now advertised on the Internet. There is more public bidding for government contracts.

People: Ethnicity

Ethnicity in Western Africa overlays tribal loyalties with religious attachments and experiences during the colonial period and since. The outcomes have strong influences on recent

Country	Ethnic Groups Percentages	Languages O=Official N=National	Religion Percentages
Benin, Republic of	42 African groups: Fon, Adju, Yoruba	French (O), Fon, Yoruba	Local 65%, Christian 20%, Muslim 15%
Burkina Faso, People's Democratic Republic of	Mossi 25%, Gourounsi, Senufo, Lobi, Bobo, Mande, Fulani	French (O), 90% Sudanic tribal languages	Muslim 50%, local 40%, Christian 10%
Côte d'Ivoire, Republic of	60 African groups: Akan, Kru, Mande, Senufo, Lebanese	French (O), Akan, Dioula, local dialects	Muslim 39%, local 35%, Christian 26%
Gambia, Republic of the	Mandinke 42%, Fulani 18%, Woluf 16%, Jola 10%	English (O), Mandinke, Woluf, Fulfulda	Muslim 90%, Christian 9%
Ghana, Republic of	Fanti, Ashanti, Ga-Adangbe, Ewe, Hausa	English (O), various African	Christian 43%, local 38%, Muslim 12%
Guinea, Republic of	Fulani 35%, Malinke 30%, Susu 20%	French (O), African languages	Muslim 85%, Christian 8%, local 7%
Guinea-Bissau, Republic of	Balante 27%, Fulani 23%, many others	Portuguese (O), Kriolu, French, local	Local 54%, Muslim 38%, Christian 8%
Liberia, Republic of	Bassa, Gio, Kpelle, Kru 95%, U.S. Liberians 5%	English (O) 20%, Mande, Kru-Bassa, others 80%	Local 70%, Muslim 20%, Christian 10%
Mali, Republic of	Mande 50%, Peul 17%, Voltaic 12%, Tuareg/Moor 10%, Songhal 6%	French (O), Bambara, others	Muslim 80%, local 18%
Mauritania, Islamic Republic of	Mixed Moor/black 40%, Moor 30%, Fulani, Wolof	Arabic (O), Wolof (N), French	Muslim 100%
Niger, Republic of	Hausa 56%, Djerma 22%, Fulani 8.5%, Tuareg 8%	French (O), 10 other official	Muslim 95%
Nigeria, Federal Republic of	Hausa, Fulani, Yoruba, Ibo total 71%	English (O), Hausa, Yoruba, Ibo, others	Muslim 50%, Christian 40%, local 10%
Senegal, Republic of	Wolof 44%, Fulani and Tutulor 24%, Serer 15%	French (O), Wolof, Serer, others	Muslim 92%, local 6%, Christian 2%
Sierra Leone, Republic of	Mende, Temne, Limba, Creoles, others	English (O), Krio, Mende, Temne, others	Local 60%, Muslim 30%, Christian 8%
Togo, Republic of	37 African groups: Ewe, Kabre, Gurma main	French (O), Ewe, Kabre, others	Local 50%, Christian 35%, Muslim 15%

political and economic events, as in Nigeria and Liberia. They were major factors in Senegal's frustrated attempts to link with its neighbors.

Economic Development

The newly independent Western African countries began with much greater hopes of an improving future than those in Central Africa or poorer countries in Asia. The combination of restricted economic bases, rapid population growth, drought hazards, political conflict and mismanagement, and fluctuating world market conditions dashed their hopes. After the 1970s, when the prices of oil, metallic minerals, and agricultural exports remained high, the 1980s fall in world prices brought economic disaster to many countries in Western Africa. For example, former Western European buyers of palm oil and peanuts as sources of vegetable oil

began to produce their own oil crops, causing a decline in demand for the tropical products. Moreover, crops are also affected by weather. From the 1970s, droughts affected peanut and cattle production in the northern parts of the sub-region.

From 1980, almost all countries in Western Africa experienced declining real income. They now have some of the highest human poverty indexes in Africa, mostly around 45 percent in 2002, with Burkina Faso, Mali, and Niger around 60 percent. Although involved in the global economy, the people in the countries of this subregion benefit little.

Western Africa supplies world markets with a wide range of products from farms, forests, mines, and ocean waters. Although some countries diversified their output, all primary products remain sensitive to world markets and exchange rates. Localities within the countries often depend on a single product.

FIGURE 9.28 Western Africa: contrasting geographies. (a) Herd of cattle in dry Sahel region of western Niger. (b) Nigerian peacekeeping force unveils weapons cache in Monrovia, Liberia. Nigeria has a raised profile in the region for supporting peace initiatives. (c) In Makemi, Sierra Leone, former child soldiers at Islamic Studies class. (d) Oshodi Market, Lagos, Nigeria. **Photos:** (a) © Bill Westermeyer; (b) © Robert Grossman/Corbis Sygma; (c) © Louise Gubb/Corbis SABA; (d) © James Marshall/Corbis.

Primacy of Agriculture

Agriculture remains the main source of employment (50–90 percent) and income (25–50 percent of GDP) for most countries in the subregion. For many, this means subsistence—based on growing crops in the south but increasingly on herding livestock toward the drier lands of the north. Nigeria, Ghana, and Côte d'Ivoire are world leaders in palm oil, cocoa, rubber, tropical fruits, rice, and coffee. Benin and Togo produce smaller amounts. Grain production in the drier countries of Burkina Faso, Mali, Niger, and Senegal depends on rain and a lack of pests. In 2004 output declined slightly despite good rains toward the south because of locust swarms and drought in the north. Senegal, The Gambia, Nigeria, and Mali

in the drier zone export peanuts, and Mali also exports cotton. The interior countries have few commercial farm products apart from their livestock, which provided meat for the coastal countries until the major droughts of the 1970s decimated the herds.

Commercial farming has experienced major shifts. Through the 1980s, Côte d'Ivoire diversified its established products such as cocoa, coffee, and palm oil (of which it is Africa's leading producer) into bananas, pineapples, cotton, and rubber. In the 1990s the government privatized the largest rubber plantations. For a few years growers also benefited from the CFA devaluation, with exported crops earning up to twice the previous price. Liberia's civil strife disrupted its output and trade in rubber produced on plantations established by the Firestone Rubber Company in 1926. Plantation owners moved their operations to countries such as Nigeria and Côte d'Ivoire. Until the 1990s, Ghana's agriculture sector grew slowly. The crop marketing boards established in colonial times set quotas and took a percentage of income for administration. By contrast, Nigerian farmers enjoyed greater incomes after the government abolished its crop boards and their charges. Nigeria invested more in crop research and farm management education and has a better rural infrastructure, enabling farmers to get their produce to market more easily.

In the 1970s, countries that neglected food production found themselves paying for expensive food imports. Attempting to produce more wheat, corn, and rice for local and national markets, they invested in large-scale water projects in the drier parts of northern Nigeria, Senegal, and The Gambia. However, poor management of human and natural resources resulted in lower-than-expected crop yields, higher costs, and disaffected farmers. The shift to commercial crop production using high-priced new strains of crops and heavy applications of fertilizer was unsustainable environmentally and financially. Subsequent development projects focused on involving local communities in sustainable development based on small-scale, appropriate technology that fed people as well as producing crops for sale in world markets. Funds were used to sink water wells worked by hand pumps and to provide rural infrastructure, advice, and credit. Projects combined commercial and subsistence crops and keeping animals, as well as better environmental practices.

Forestry, Fishing, and Mining

In many countries of Western Africa, the former coastal forests were largely cut and replaced by tree crop plantations. In Nigeria, attempts to maintain the remnant of woodland in national forest reserves face continual challenges by farmers and timber companies.

Fishing is of growing importance in the northern coastal countries, which are close to the Cold Canaries Current and the nutrients it brings to the surface. However, there is increasing competition from Spanish trawlers. Although Western African countries negotiated fishing licenses with the European Union and the revenue makes up one-third to one-half of their foreign exchange earnings, enforcing the agreements is difficult. EU boats—along with those of the Japanese, Russian, and Chinese—take more than their allocations, depleting the stocks. Lake and river fish are important foods in the interior countries such as Niger and Mali.

The main mineral developments (see Table 9.4) are the oil and natural gas fields of southern Nigeria (Figure 9.29) and smaller quantities in Ghana and Côte d'Ivoire. Oil was discovered in Nigeria in 1956, and production expanded to a maximum of 114.2 million tons at the height of world demand and high prices in 1979, making it a major world producer, before falling back to 73 million tons. Oil provides over 90 percent of Nigerian

FIGURE 9.29 Western Africa: oil and gas in Nigeria. Nembe Creek and the Agip flow station in the Niger delta. The canoes carry timber from the forest to port. The gas flares cause pollution and breathing difficulties for local people, who campaign for better environmental control and funding for more diverse economic development of this area. **Photo:** © Ed Kashi/Corbis.

exports, but much of the oil income was used for grandiose infrastructure projects funded by loans that turned into debts. Dependence on oil income slowed economic diversification. Despite the great hopes Nigeria placed on using its oil income to boost its economy, its living standards are now lower than before the oil boom of the 1970s.

In the 2000s, Ghana and Côte d'Ivoire wish to fuel thermal power stations by using the natural gas that was previously flared off from Nigerian oil wells. Plans to build a gas pipeline along the coast await agreement on whether it should be on land (Nigeria's preference for increased control) or offshore along the seabed (Ghana's preference to diminish terrorist actions).

In Ghana (formerly the colony of Gold Coast), production in the gold-mining industry increased from the 1990s following major new investments by South African corporations. Many countries have mineral resources and mining operations.

Manufacturing

The manufacturing sector produces only 5 to 15 percent of Western African countries' GDP. Import-substitution factories for food and drink products, construction materials, and other low-price or high-bulk goods supply local markets. The lowest proportions of manufacturing industry occur in the landlocked countries. Export crops and minerals are sometimes processed locally in oil-refining, textile, fertilizer, aluminum, rubber, or fishmeal factories. Nigeria and Côte d'Ivoire have car assembly factories.

The devaluation of the CFA franc encouraged some local manufacturing industries. Côte d'Ivoire now manufactures the highly colored printed cloth used in women's dresses across Western Africa (Figure 9.30)—after years of importing it from Europe.

Services

The service sector of the economy in Western African countries grows slowly, largely through the government employment that is a major factor in many countries. However, structural adjustment demanding the reduction of civil service bureaucracies causes pain to many countries because of the importance of family and political links into such employment.

Tourism, a service industry that is part of the development strategy of many poor countries, developed slowly in Western Africa, apart from servicing small groups of travelers crossing the Sahara. When African Americans began seeking their family roots, groups visited countries such as Ghana, Côte d'Ivoire, and The Gambia.

Some Western Africans finance small-scale enterprises by microlending. Small sums of money are lent to poor people to help them set up or expand a business. This type of program began in the 1970s in Bangladesh (see the section on the Grameen Bank in Chapter 7) and Latin America. For example, in Burkina Faso, a small restaurant in a marketplace near Ouagadougou is built simply of log and thatch and serves simple meals based on rice. The owner took out a loan to buy the rice wholesale and now employs seven people, pays her children's school fees, and owns a motorcycle. As elsewhere, microlending funds mainly businesses run by women, who are likely to use the additional income to feed and clothe children. Such lending offers independence and rewards enterprise.

FIGURE 9.30 Ghana. Girls wearing kente cloth, the signature fabric of the Ashanti people (Kumasi area). The cloth is handwoven in strips that are then sewn together. **Photo:** © Lisa Kahn Schnell.

Test Your Understanding
9B

Summary

Central Africa is isolated from the world economy by its physical environment and by the warfare that consumes the Democratic Republic of Congo, the largest country. Both physical and human resources are poorly developed. Most people depend on subsistence agriculture.

At independence, Western Africa was involved in the world economy through its tropical tree crops and mineral exports. Commodity trade established in colonial times continues with tropical crops and minerals but is increasingly disrupted by civil wars in the subregion and poor government.

Questions to Think About

9B.1   Summarize the differences between Central and Western Africa through five geographic characteristics (such as distribution of people, ethnic groups, climates and landforms, resources, economic activities, governments).

9B.2   Why is there a lack of engagement with the rest of the world in Central Africa?

9B.3   Why do many Western African countries have poorly developed manufacturing sectors?

9B.4   Why are tropical tree crops an uncertain source of income for Central and Western Africa?

Key Terms

Southern Africa Development Conference   Hutu   Tutsi
Communauté Financière Africaine

Eastern Africa

Eastern Africa's distinctiveness (Figure 9.31) comes partly from its history of contact with Arabian and Indian influences to the east. Its natural environment is marked by a plateau area in the south and the Ethiopian Highlands in the north. Although the region lies partly across the equator, the plateau reduces rainfall, and savanna grasslands are the dominant natural vegetation, trending toward desert in the north. On the western boundary, rivers flow to Lake Victoria and form the headwaters of the White Nile River. Today the subregion has a major tourist industry, attracting millions each year from wealthier countries to watch its large animals in game parks.

Countries

The countries of Eastern Africa form two groups (Table 9.6). In the north, Ethiopia is separated from the Red and Arabian Seas by Eritrea, Djibouti, and Somalia. The uplands of Ethiopia (see

Figure 9.3) force moisture to rise and cause seasonal (summer) rains that feed into the Blue Nile River and support agriculture. The surrounding lands are mainly arid.

To the south, the plateau area of the former British colonies of Uganda, Kenya, and Tanzania is broken in the west by major rift valleys and has volcanic peaks rising above.

Ethiopia and Neighboring Countries

Ethiopia has the distinction of being the only large African country that was not colonized by a European power. Although occupied by the Italian military from 1936 to 1941, it was soon returned to the Emperor Haille Selassie's rule by British forces early in World War II. Ethiopia remained an isolated feudal country, and Selassie was the last in a line of Ethiopian emperors dating back to the Coptic Church conversion in the A.D. 300s. He reigned from 1930 to 1974, when a military coup brought a Communist government to power.

After further coups, devastating drought, and movements of refugees in and out of the country, rebel groups replaced that regime in 1991. A new constitution in 1994 led to multiparty elections.

After World War II, the United Nations federated Eritrea with Ethiopia and joined British and Italian Somaliland as Somalia. Disappointed at being reduced to a province within Ethiopia, Eritreans fought Ethiopian and Soviet Union forces. After the Communist government in Ethiopia was deposed and the Soviet forces departed, Eritrea gained independence in 1993, making Ethiopia landlocked. Eritrea is Africa's newest country. Tensions between Ethiopia and Eritrea over the border location continue, and costly wars from 1998 into the early 2000s sapped both countries' resources and made it difficult to implement the new Ethiopian constitution.

The former British Somaliland declared its independence from Somalia in 1991 but is not recognized by other countries. In the early 2000s, chaos remained in Somalia. A United

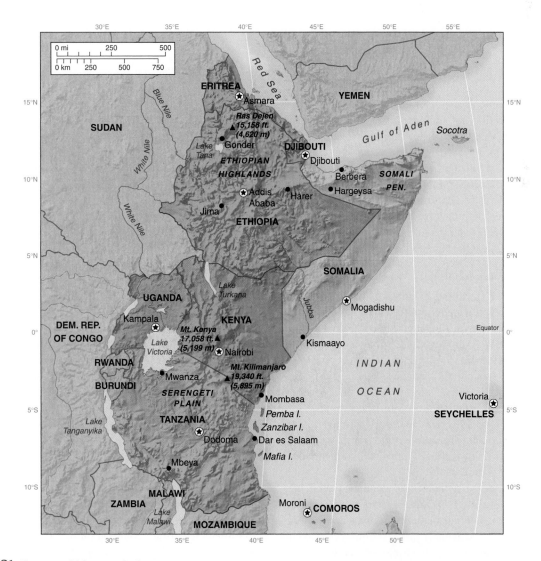

FIGURE 9.31 **Eastern Africa: main features.** The countries included in the subregion, the major cities, and rivers.

Table 9.6 Eastern Africa. Data by country, area, population, income (GNI PPP—Gross National Income Purchasing Power Parity), urbanization, Human Development Index (HDI), Human Poverty Index (HPI).

Country	Capital City	Land Area (km²)	Population (millions)		GNI PPP 2003		% Urban	Index 2002	
			Mid 2005 Total	Est. 2025 Total	Total (US$ billions)	Per Capita (US$)	2005	HDI (rank of 177)	HPI (% population)
Djibouti, Republic of	Djibouti	23,200	0.8	1.1	2	2,200	82	154	34.3
Eritrea, Republic of	Asmara	121,140	4.7	7.2	5	1,110	19	156	41.8
Ethiopia, Federal Democratic Republic of	Addis Ababa	1,100,760	77.4	118.4	49	710	15	170	55.5
Kenya, Republic of	Nairobi	580,370	33.8	49.4	33	1,020	36	148	37.5
Somalia	Mogadishu	637,600	8.6	14.9	No data	No data	33	No data	No data
Tanzania, United Republic of	Dodoma	945,090	36.5	52.6	22	610	32	162	36.0
Uganda, Republic of	Kampala	235,880	26.9	55.8	36	1,440	12	146	36.4

Source: World Population Data Sheet, Population Reference Bureau (2005), World Development Indicators, World Bank (2005), Human Development Report, United Nations (2004), and Encarta Microsoft (2005).

Nations peacekeeping force, largely made up of U.S. military, left in 1995 without restoring order. The southern part of Somalia remains ungovernable in a state of anarchy marked by random banditry and fighting among clans. The official government controls only part of the capital, Mogadishu. A loose alliance of warlords based in Baidou and supported by Ethiopia controls most of southern Somalia. Puntland, the northeastern corner of Somalia, remained peaceful throughout the 1990s, exporting livestock and frankincense. After September 11, 2001, however, activities by Islamist extremists gave the local political struggles an international emphasis, enabling Ethiopian intrusion against the supposed extremists.

Former British East Africa

Tanganyika, Uganda, Kenya, and Zanzibar gained their independence from the United Kingdom in the early 1960s following guerrilla warfare in Kenya. Tanganyika and Zanzibar united to form Tanzania in 1964, although a Zanzibar independence movement still exists. Zanzibar was one of the richest countries in Africa before it became part of Tanzania.

After independence in 1963, Kenyans elected their governments. President David Toriotich arap Moi ruled since 1978, but his last reelection in 1997 was marred by violence and fraud, and he retired in 2003. The successors to arap Moi have strong support from their own tribal groups and political allegiancies decide many government jobs.

In Tanzania, one-party socialist rule continued until the first multiparty elections in 1995. Zanzibar has its own president and legislature for internal matters. In Uganda, dictators Idi Amin (1971–1979) and Milton Obote (1980–1985) both governed by oppression, killing hundreds of thousands of people. The first postindependence multiparty elections were

held in 1996. The Tanzanian and Ugandan legislatures are elected by the whole adult population, with some seats reserved for women.

Uganda provides a modern African story of contrasting events. After decades of repressive dictatorships, the present government has notable achievements including a reduction of HIV/AIDS. However, people living along the Sudan border and in the western area abutting DROC are subject to intense violence. In the north, the Lord's Resistance Army cult led by Joseph Kony claim to bring back the Ten Commandments. However, supported by Sudan, they massacre village populations and abduct children who are then forced to make up their main fighting force. Almost 2 million people have been displaced. The Ugandan government has had some successes against the group, but fearful children still huddle in the main towns and special camps.

In early 2001 Kenya, Tanzania, and Uganda reestablished the East African Community that they abandoned in 1977. It is intended as a means of long-term economic integration, starting with a customs union. Fears have already arisen of Kenyan products replacing local ones in Tanzania and Uganda. Further movement toward political union, including an East African court of justice and legislative assembly, is a distant dream.

People: Ethnicity

The coastal people in the north are mainly Muslims, while those in Ethiopia and the southern countries are mainly Christian or traditional animist groups. Arabs, Arab–African mixes, and South Asians are common on the eastern coasts of Kenya and Tanzania, with Asians taking many of the business opportunities. In the 1970s, most Ugandan Asians were expelled by

Country	Ethnic Groups Percentages	Languages O=Official N=National	Religion Percentages
Djibouti, Republic of	Somali 60%, Ethiopian 35%	French (O), Arabic (O), Somali, Afar	Muslim 97%, Christian, others 3%
Eritrea, Republic of	Tigrinya 50%, Tigre-Kunama 40%, Afar 4%	Tigre, Afar, others	Muslim, Coptic Christian, other Christian, local
Ethiopia, Federal Democratic Republic of	Oromo 40%, Amhara, Tigrean 32%, Sidamo 9%, others	Amharic (O), Tigrinya, Arabic, English	Muslim 45%, Ethiopian Orthodox 40%, local 12%
Kenya, Republic of	Kikuyu 21%, Luhya 14%, Luo 12%, Kalenjin 11%, Kamba 11%	English (O), Swahili (O), Kikuyu, Luo, many others	Protestant 40%, Catholic 30%, local 22%, Muslim 6%
Somalia	Somali 85%	Somali (O), Arabic, Italian, English	Sunni Muslim 99%
Tanzania, United Republic of	Over 120 African culture groups	Swahill (O), English (O), local groups	Christian 45%, Muslim 35%, Hindu, local
Uganda, Republic of	Ganda 18%, Nyankole 10%, Kiga, Soga, Iteso, Langi, others	English (O), Luganda, Swahili, others	Roman Catholic 30%, Protestant 30%, local 17%, Muslim 7%

Idi Amin and moved to the United Kingdom or to other countries. Many returned since 1990, reclaimed property, and helped to revive Uganda's economy. Inland traditional groups include the cattle-herding Masai and cultivators such as the Kikuyu. Long-standing intergroup rivalries are modern political antagonisms. Languages include many of local significance, but those with wider use range from Arabic in the north and Amharic in Ethiopia, to Somali, Swahili, and English.

Economic Development

The countries in Eastern Africa have made progress in human development basics such as education and health care. Apart from Ethiopia (56%) and Eritrea (42%) other countries had 2002 HPIs in the mid-thirties. All countries face the disadvantages of inadequate infrastructure and the relicts of conflict (Figure 9.32).

Agriculture Is Central

In the general absence of mineral resources and the slow development of manufacturing and service industries, most countries in Eastern Africa depend on agriculture as an economic base. Commercial farming is important, and much foreign exchange comes from exports of farm products. Coffee makes up 90 percent of Ethiopian exports; coffee, tea, and tobacco are 97 percent of Uganda's exports; coffee, tea, sisal, cotton, cashews, and cloves are 85 percent of Tanzania's exports. Even in Kenya, which has greater economic diversity, coffee, tea, sisal, fruit, and vegetables provide around 50 percent of exports. In the 1990s, higher-priced tea, fruit, and vegetable exports replaced or augmented Kenya's coffee and sisal exports. The early 1980s boom in beverage crop prices followed a shortfall in world production, but production rose again and prices fell, remaining low in the 1990s.

Most people in Eastern Africa gain a livelihood by subsistence farming or nomadic herding. Long droughts since the 1970s and government actions led to major famines in Ethiopia. The central government not only fanned civil war but long prevented aid getting to famine-affected opposition areas. The famine of 1983 to 1985, when over half a million people died of starvation or diseases contracted at the feeding sites, attracted worldwide publicity. This famine arose after soil erosion in the formerly fertile highlands area of Ethiopia forced migration of people northward into the zone of drier conditions and civil war. Ethiopian farming continues to be poorly developed because of droughts, bureaucratic obstruction, little land-tenure security, and poor infrastructure (such as roads and telephones). Many parts of the country remain susceptible to famine.

The arid areas in the north of the subregion rely on exports of livestock. Before the current anarchy, Somalia's livestock made up 65 percent of its total exports, and much of the rest came from bananas grown in the south. Few of the arid countries balance their imports by exports, and most rely on external aid.

Kenya has the most diversified farming community. Following independence, the government redistributed as smallholdings the land from the European-farmed highlands around Nairobi. It encouraged Kenyans to grow commercial export crops, such as coffee and tea, and food crops for the home market. In the Machakos district east of Nairobi, a region of periodic drought that suffered famine and soil erosion in the colonial period, land-reform smallholders invested in high-value crops, earning money to support this by off-farm jobs.

(a)   (b)

FIGURE 9.32 Eastern Africa. (a) A road in Tanzania—straight but potholed and dusty in the dry season and a river in the wet season. (b) Ugandan farmer tends his crop as children play on an abandoned tank, a relic of recent oppression. **Photos:** © Mission Aviation Fellowship.

Former badlands were transformed into terraced hills and fenced fields. Good road links to Nairobi markets made it possible to sell fruit and vegetables there. By the early 2000s, the area had 1.4 million people, and output had risen 15 times with yields per hectare (2.47 acres) increasing up to tenfold.

The Tanzanian government applied socialist principles, regrouping the scattered farmers into communal farms and centralized villages. However, guaranteed low prices in food markets across the country made many farmers lose interest in increasing their productivity. Tanzanian exports are half those of Kenya. From the late 1980s, however, efforts to improve farm management included more use of fertilizers. Experimental corn crops raised yields from 2–3 kilograms per hectare in 1988 to 18–20 kilograms in 1994. Lack of storage, the high cost of fertilizer imports, and poor roads are problems in the way of further advances.

Variable Manufacturing Development

Manufacturing makes up between 5 and 13 percent of GDP in Eastern African countries. Although several countries gave emphasis to manufacturing in their early development plans, only Kenya achieved much. In addition to processing its crops, Kenya has oil refineries and medium-sized consumer goods factories. It is a growing center of production and distribution to the surrounding countries. The other countries do not have

the home demand for manufactured products, the international market links, or the foreign exchange needed to buy machinery and replacement parts. Small-scale enterprises are important, based on processing local crops, manufacturing food and drinks, and producing goods for local markets. In its Communist era, Ethiopia built metal and chemical industries, but such large-scale industries contrast with the more common small-scale craft manufacturing.

Services and Tourism

For most countries, the service economy is poorly developed apart from extensive government bureaucracies. Once again, Kenya is ahead of the others. Its relatively stable political environment attracted to Nairobi offices of the United Nations and the regional and continental headquarters of multinational corporations and nongovernmental organizations (Figure 9.33).

Kenya has also made the most progress in its tourist industry. In years when coffee and tea suffer poor prices, tourism is Kenya's main source of foreign exchange—making up 20 percent of export earnings in 1990 but only 10 percent in 1999.

Around 1 million people from the wealthier countries of Europe, the United States, and Japan visited Kenya in 2002, mainly to see the large animals in the national parks (Figure 9.34). The hotels in Nairobi have a twofold clientele of tourists and businesspeople. Other countries, especially Tanzania, also

FIGURE 9.33 Eastern Africa: Nairobi, Kenya. A combination of global links from British colonial buildings to modern high-rise offices, hotels, and apartments and some local influences in a city that began as a railroad terminus. **Photo:** © Yann Arthus-Bertrand/Corbis.

have tourist industries, but the egalitarian Tanzanians resisted building large luxury hotels, and their tourist industry remained small until the government encouraged a major expansion of park resorts in the mid-1990s. It had half a million visitors in 2002, up from 80,000 in 1980, and tourism receipts made up over 60 percent of export earnings.

Although many of the staff of the parks and hotels are poorly paid by the standards of materially wealthy countries, tourism brings an injection of foreign currency. Like other

FIGURE 9.34 Eastern Africa: Masai Mara National Park, Kenya. Tourists view elephants from the safety of their bus. Tourism based on such viewings is a major source of foreign income in Kenya and other countries in Africa, but its methods are often criticized by those who wish to preserve the animals. **Photo:** © Vol. 145/Corbis.

African enterprises, however, tourism is subject to fluctuations according to cycles of prosperity and recession in the wealthier countries.

Southern Africa

Southern Africa (Figure 9.35) lies at the southern margin of the tropical climatic environments. Seasonal tropical climates in the northern parts have summer rains that decrease southwestward toward the aridity of Namibia. The Republic of South Africa has midlatitude climates with winter rains at the Cape of Good Hope and summer rains along its southeastern coasts. This made the subregion attractive to European colonial settlement. A series of plateaus is drained by major rivers such as the Zambezi (Figure 9.36), Limpopo, and Orange-Vaal system and cut by the southernmost extension of the east African rift valleys.

Countries

Southern Africa includes countries that have some of the best records of economic progress in the continent based on mineral wealth, diversified agriculture, and manufacturing. Over a third of Africa's rail mileage is in this subregion.

Such potential, however, rests on a fragile immediate past and present. For most of the second half of the 1900s, civil wars devastated Angola and Mozambique. When the Republic of South Africa was isolated from the rest of the subregion and the world by its apartheid policy and resulting sanctions, the separated white and black communities diverged in well-being. Several countries remain among Africa's poorest (Table 9.7). The Republic of South Africa dominates the subregion. South Africa has one-third of Southern Africa's population but produces over three-fourths of its GNI PPP. Many families in the rest of this subregion derive considerable income from the wages sent home by their members who work in South Africa. Although perceived as an adversary through the apartheid years, countries such as Botswana, Malawi, Zambia, and Zimbabwe made public pronouncements against apartheid but maintained economic relations with South Africa. Lesotho and Swaziland, virtually encircled by South Africa, maintained political relations. To destabilize its potentially antagonistic neighbors, the South African military supported guerrilla groups in the civil wars in Angola and Mozambique. After its rejection of apartheid policies, South Africa had democratic elections of a national government and restored its part in the global economic system following decades of isolation and sanctions.

Republic of South Africa

After nearly 40 years as a self-governing dominion in the British Empire, the white Afrikaners who led the South African government imposed apartheid policy in 1948. In 1961 South Africa declared itself a totally independent republic without ties to the United Kingdom or the British Commonwealth. Apartheid laws and linked informal measures increased the

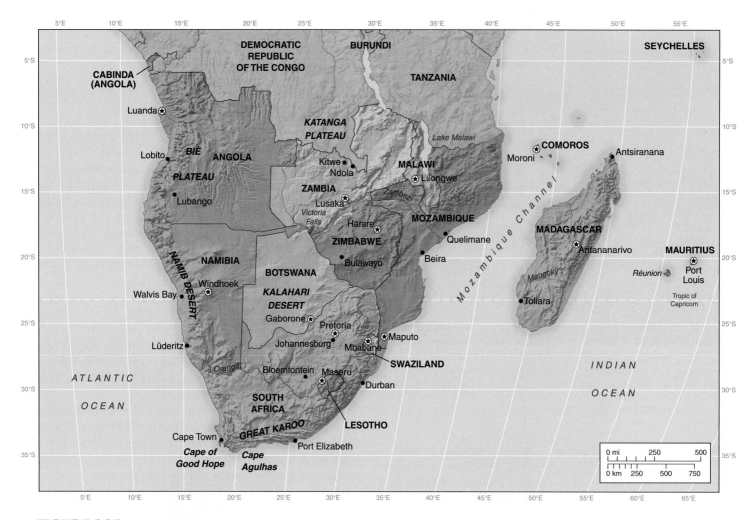

FIGURE 9.35 Southern Africa: main features. The countries included in the subregion, the major cities, and rivers.

separation of white, black, and mixed races. Black Africans and other "colored" (mostly Asian) people lost many human rights. Each had to carry an identity pass. At one level, "petit apartheid" legalized segregation equivalent to that in the American South up to the mid-1960s; at another, "grand apartheid" relocated black people to "independent" homelands. Their economic role was as labor for white people. Each group had a designated housing area, and each black person was assigned to a homeland area (see Geography at Work: Social Justice in South Africa, page 428).

South Africa's racial policies under apartheid generated hatred in the rest of Africa, and the country defended itself by intervening in the surrounding countries. When Germany lost Namibia after World War I, it became a protectorate of South Africa, but South Africa refused to give up the territory when the United Nations tried to cut this tie after apartheid became South African policy. For a time, the West tolerated South African apartheid, but then it largely isolated South Africa from the world political, economic, and sports competition systems.

The combination of internal resistance to apartheid by leaders of the African National Congress (ANC), such as Nelson Mandela, and external isolation and sanctions forced the ending of the apartheid policy in the early 1990s, signaled by the release of Mandela from long-term imprisonment. Despite many forecasts that South Africa would descend into anarchy, the transition orchestrated by Nelson Mandela and F. W. de Klerk (the South African president at the time) worked, and both received Nobel Peace prizes. Democratic elections in 1994 resulted in a "Government of National Unity" with Mandela as its first president.

South Africa struggles to change the geographic expressions of apartheid, including the townships, homeland policies, and separate racial schooling. The opposing communities of wealthy whites and poor blacks; ANC Xhosas and Inkhata Freedom Party Zulus; local Africans and those from other countries; and the parties that represent Afrikaners, other whites, and Coloureds barely tolerate each other. The Truth and Reconciliation Commission, however, dealt in a positive manner with many sensitive issues that whites and blacks held

(a)

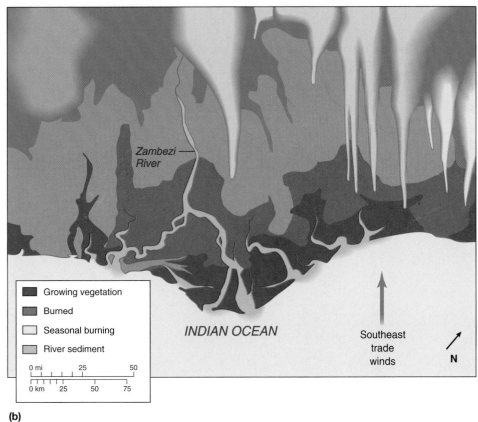

Zambezi
River

■ Growing vegetation
■ Burned
☐ Seasonal burning
☐ River sediment

0 mi   25   50
0 km   25   50   75

INDIAN OCEAN

Southeast
trade
winds

N

(b)

FIGURE 9.36 Space shuttle photo: the coast of Mozambique, Southern Africa. (a) The false-color reds are from the reflection of infrared radiation from growing vegetation, picking out wetlands near the coast. Red colors are absent from areas of seasonal burning. The smoke from burning vegetation shows the constancy of the trade winds blowing from the Indian Ocean. The map (b) provides details and scale. **Photo:** (a) NASA, Michael Helfert.

Geography at Work

SOCIAL JUSTICE IN SOUTH AFRICA

Many geographers work on projects to encourage social justice. One of the places where people's rights are being restored is in South Africa. The CD-ROM "Interactive World Issues of Place and Planet" that comes with this text includes a study that you will find valuable as an insight to changes taking place. Select Disc 1 and the study "South Africa: This Land Is My Land." You may find it interesting to work through Part 1, but the main section we wish to refer you to is Part 2. In this study Section 1 is a video of people returning to Doornkop, from which they had been forcibly removed years before; Section 3 is a video discussion by Brent McCusker, a geographer at the University of West Virginia, that refers to the importance of land rights and the process of restitution. You may also find it helpful to follow through the map exercise of Section 2.

Box Figure 1 Brent McCusker, who presents the CD-ROM item based on his research.

Table 9.7 Southern Africa: Data by country, area, population, income (GNI PPP—Gross National Income Purchasing Power Parity), urbanization, Human Development Index (HDI), Human Poverty Index (HPI).

Country	Capital City	Land Area (km²)	Population (millions) Mid 2005 Total	Est. 2025 Total	GNI PPP 2003 Total (US$ billions)	Per Capita (US$)	% Urban 2005	Index 2002 HDI (rank of 177)	HPI (% population)
Angola, Republic of	Luanda	1,246,700	15.4	25.9	26	1,890	33	166	No data
Botswana, Republic of	Gabarone	581,730	1.6	1.6	14	7,960	54	128	43.5
Lesotho, Kingdom of	Maseru	30,350	1.8	1.6	6	3,120	13	145	47.9
Madagascar, Republic of	Antananarivo	587,040	17.3	28.2	13	800	26	150	35.9
Malawi, Republic of	Lilongwe	118,480	12.3	23.8	7	600	14	165	46.8
Mozambique, Republic of	Maputo	801,590	19.4	27.6	20	1,070	32	171	49.8
Namibia, Republic of	Windhoek	824,290	2.0	2.1	13	6,620	33	126	37.7
Republic of South Africa	Pretoria	1,221,040	46.9	47.8	465	10,270	53	119	31.7
Swaziland, Kingdom of	Mbabene	17,380	1.1	1.0	5	4,850	25	137	No data
Zambia, Republic of	Lusaka	752,610	11.2	15.8	9	850	35	164	50.4
Zimbabwe, Republic of	Harare	390,760	13.0	14.4	28	2,180	34	147	52.0

Source: World Population Data Sheet, Population Reference Bureau (2005), World Development Indicators, World Bank (2005), Human Development Report, United Nations (2004), and Encarta Microsoft (2005).

against each other, although black Africans gave it greater credibility and support than many whites. White-owned businesses now employ black Africans at all levels. When Mandela retired in 1998, other ANC leaders continued the reconciliation policies as demands increased from black Africans for greater equality of access to facilities and jobs.

South Africa has extremes of wealth and poverty. Apartheid created a huge black underclass of voters, who will take decades to enjoy improved conditions. Some benefits, such as domestic water and electricity supplies, are becoming available to areas of poorer housing. Income growth enables an expanding black middle class to enjoy the fruits of affluence. However, jobs are not expanding at a time when 20 to 30 percent of adults have no formal employment. Many white managers and professionals moved out of the country for apparently better opportunities, living quality, and less crime in Australia, Canada, the United Kingdom, and the United States. Although many whites stayed for the sunny, mild climate and lower costs of living, the "white flight" of employers reduced other employment opportunities.

Other New Countries

In 1953 the white settlers in Zimbabwe first tried to establish a united and white-dominated Federation of Rhodesia and Nyasaland, in which they could dominate the former colonies of Nyasaland (now Malawi) and Northern and Southern Rhodesia (now Zambia and Zimbabwe). It was resisted by black Africans and dissolved in 1963, with Zambia and Malawi achieving independence in 1964. In 1965 the whites in Southern Rhodesia unilaterally declared independence, but Britain and other countries did not accept this move and imposed sanctions. Years of internal guerrilla warfare and international sanctions delayed Zimbabwe's full independence until 1980. In the early 2000s, Zimbabwe's leader, Robert Mugabe, made decisions that destroyed his country's prospects in economic and human development, leading to bloodshed, famine, and bankruptcy. He retains power, however, and has admirers across Africa for his defiance of former colonial powers.

Botswana and Lesotho gained independence in 1966 and Swaziland in 1968. Apart from Botswana, most countries had long-term single-party governments, with some moving slowly to multiparty systems. In the monarchies of Lesotho and Swaziland, hereditary kings are slowly divesting their autocratic powers to democratic systems. Lesotho's first multiparty constitution was adopted in 1993, retaining the king as figurehead. However, internal disruption in 1998 destroyed much of Maseru, the capital. In Swaziland, student and labor unrest in the 1990s added pressures for democratic reform.

In the 1990s Madagascar, Malawi, and Zambia moved from restrictive single-party governments toward multiparty democracies with varied degrees of openness. Madagascar added to its problems of underfunded health and education facilities by the erratic imposition of reform measures that are opposed by

Country	Ethnic Groups Percentages	Languages: O=Official N=National	Religion Percentages
Angola, Republic of	Ovimbundu 37%, Mbundu 25%, Bakongo 15%, others	Portuguese (O), Bantu languages	Roman Catholic 65%, Protestant 20%, local 10%
Botswana, Republic of	Tswana 75%, others	English (O), Setswana	Christian 50%, local 50%
Lesotho, Kingdom of	Basotho 79%, Nguni 20%	Sesotho (O), English (O), Zulu, Xhosa	Christian 90%, local 9%
Madagascar, Republic of	Merina 27%, Betsimisaraka 15%, Betsileo 12%, others	French (O), Malagasy (O)	Local 52%, Christian 41%, Muslim 7%
Malawi, Republic of	Chewa, Nyanja, others	English (O), Chichewa, others	Protestant 55%, Roman Catholic 20%, Muslim 20%
Mozambique, Republic of	Makua-Lomwe, Yao, Makonde, Chewa, others	Portuguese (O), Swahili, local	Local 55%, Christian 30%, Muslim 15%
Namibia, Republic of	Orambo 50%, other black 36%, white 6%, mixed 7%	English (O), local Afrikaans	Christian 90%
Republic of South Africa	Black African 76%, white 13%, mixed 9%, Indian 2%	12 official: Afrikaans, English, Zulu, Xhosa, others	Christian, Hindu, Muslim, local, Jewish
Swaziland, Kingdom of	African 97%, white 3%	English (O), Siswati (O)	Christian 60%, local 40%
Zambia, Republic of	African (over 70 ethnic groups) 98.7%, European 1.1%	English (O), Bemba, Luapula, Nyanja, others	Christian 72%, local 27%, Muslim–Hindu 1%
Zimbabwe, Republic of	Shona 71%, Ndebele 16%, others, European 2%	English (O), Shona, Ndebele	Syncretic Christian/local 50%, Christian 25%, local 24%

antigovernment strikes and demonstrations. Namibia was occupied by South African troops until 1990, when the United Nations finally made Namibia independent; it then moved toward democracy, although still dominated by a single party.

Former Portuguese Colonies

In Angola, the recognized government since independence in 1975 controlled only part of its country until 2002. The National Union for the Total Independence of Angola (UNITA) opposed the socialist government that was backed by the Soviet Union and Cuban military during the Cold War. Until 1990, South Africa backed UNITA, after which it funded itself by illegal diamond sales. Following a 1994 peace accord and elections, fighting resumed in 1998 until the death of UNITA'S leader in 2002 ended the fighting. For most Angolans, their economy is in disarray, and millions of landmines make farmers reluctant to return to their fields. In the late 1990s, the promise of more secure offshore oil income attracted external investment to Angola, but enthusiasm waned in the early 2000s as few productive wells emerged.

Mozambique also suffered civil war from before the Portuguese left in 1975 until the 1990s. This war devastated its government and economy. From being one of the world's poorest countries at the end of that period, it has experienced peace and economic growth for 10 years, although many long-term tensions exist, and some former rebel leaders have taken to smuggling and money laundering.

People: Ethnicity and Migration

The people of Southern Africa include a wider variety of major ethnic groups than the other African subregions, reflecting the last four centuries of history that witnessed major movements of people of all colors.

The level of European presence varies. In the former Portuguese colonies (modern Angola and Mozambique), few Europeans settled, and most of those left after independence in 1975. Today South Africa is the only country in Africa South of the Sahara with a sizable nonblack population: it is 75 percent black, 14 percent white, 3 percent Asian, and 8 percent of mixed race.

The island country of Madagascar has a mixture of African and Asian people and cultural influences. Peoples from the East Indies (Indonesia today) and Africa occupied Madagascar, but the Portuguese, French, and English all attempted to colonize the island. They found it difficult to overcome the forces of the powerful rulers in hilly and forested terrain. By the end of the 1800s, the French established a colony, but internal dissent gradually rose until independence was granted in 1960.

The migration history of South Africa has continuing human rights implications. Up to 1994, the country recognized only white migrants. Africans entered the country as temporary workers. The populations of neighboring countries were involved in arrangements ranging from formal mine contracts to unregulated movements. By the early 1970s, contract workers made up 80 percent of the South African mine workforce (now around 55 percent). South Africa later became a mass recipient of refugees from the 1980s wars in Mozambique and Angola.

In the 1990s South Africa became a major destination for African refugees, although there is no government support for asylum seekers. Undocumented (unauthorized) migration continues, and South Africa now includes 4–8 million people from other countries. Since 1994, the South African government has struggled to reverse the migration direction, deporting over 1 million undocumented migrants with methods criticized by human rights groups. A new Immigration Act (2002) grew out of public intolerance of foreigners and focused on control and exclusion rather than management and opportunity. Since 1994, while the rise in permits for visiting, tourism, and business from under 4 million to 10 million reflected improving international relations, levels of immigration and granting of permanent residence fell. Yet South Africa needs skilled workers in industry, education, and medicine.

Economic Development

In the early 2000s Southern Africa had 20 percent of the population of Africa South of the Sahara and produced 53 percent of its total GNI PPP. The Republic of South Africa itself produced 41 percent of the region's total. However, the subregional economic environment deteriorated because of factors such as the Zimbabwean collapse, the Zambian reliance on stalled copper mining redevelopment, and the spread or HIV/AIDS.

Agricultural Diversity

Subsistence farming remains the economic mainstay of over 85 percent of the population in the very poor countries. Even in countries where mining or tourism are important, subsistence farming occupies over 50 percent of the workforce. Corn (maize) is a common staple food, grown on thousands of small farms in Southern Africa.

Commercial farming is most significant in the more diversified economy of South Africa. Landscapes of commercial farming (Figure 9.37) exist alongside those of semisubsistence.

FIGURE 9.37 Southern Africa: Cape Province vineyard. Wine is one of South Africa's main export products. **Photo:** © Alasdair Drysdale.

South Africa made itself largely self-sufficient in food during the years of international sanctions. It produces a variety of temperate and subtropical grains, vegetables, and fruits, together with sugar, cotton, and livestock products.

Commercial crops provided nearly 40 percent of Zimbabwe's exports by value until the early 2000s. Although Zimbabwe's tobacco output fluctuated, commercial farmers diversified into vegetables and flowers for European markets. In Malawi, tobacco, tea, and coffee make up 70 percent of the country's exports. In Madagascar, coffee, sugar, vanilla, cloves, and cacao make up 70 percent of the exports.

Land ownership reform is a major issue in the parts of Southern Africa where white colonial settlers took land. Newly independent governments have to weigh the political advantages of returning that land to black Africans against the loss of income from white-run commercial farms. In Zimbabwe, most of the reduced numbers of commercial farms are efficient, employ local labor, and provide valuable export crops. When the government took over some white-owned commercial farmland for redistribution from the early 1980s, many independent African family farmers on smaller plots were more productive than large commercial farms because they used the land more intensively—albeit for subsistence rather than commercial export crops. In other cases, soil erosion caused the abandonment of the smaller landholdings, or new owners favored by political gifts of the land allowed it to revert to bushland. Farmers on smaller land areas find it difficult to obtain the finances necessary for commercial farming.

In the early 1990s and 2000s, drought affected the southern part of Zimbabwe where most of the communal areas and small farms are situated but had less impact on the commercial farms in the center and north. The commercial farms gained further advantages when marketing controls were dropped as part of structural adjustment and small farmers lost various forms of government assistance. However, these events emphasized the differences between the two types of landholding and created pressure from a growing population for political action to convert more large commercial farms to smallholdings. In the late 1990s and early 2000s, the Zimbabwean government allowed the appropriation of the white-run commercial farms, and groups of former rebel fighters ("war veterans") took over a number of them, expelling or killing the occupants and black farm workers and greatly reducing commercial production.

Mining Wealth

The extraction of mineral resources dominates the economies of Angola, Botswana, Namibia, South Africa, Zambia, and Zimbabwe (see Table 9.4). It involves them deeply in the global economic system. The possession of such resources, however, is not always a recipe for prosperity because cycles of economic boom and recession in wealthier countries cause fluctuating global markets to affect local employment and income. Mining also has negative environmental impacts, including air and water pollution and the desertification of local areas poisoned by fumes. Without stable political conditions,

adequate transportation facilities, and good management of the national economy, attracting multinational corporation investors with the necessary finances is difficult. Multinational corporations, however, do not always return wealth they gain to local communities, while their license payments and taxes go to central governments that may direct spending away from the mining areas.

In some cases, wealth from mineral extraction is returned to the indigenous peoples. In the 1990s, the Bafokeng people (just west of Pretoria, South Africa) changed from watching multinational mining companies extracting wealth from platinum mines on their land to obtaining an agreement with one company that ceded to them 22 percent of mining profits and a million shares. The massive income was invested in community facilities and basic necessities (water, sanitation, electricity, housing, and education).

Mining Wealth: Dominant South Africa

South Africa is the world's top producer of platinum and a major producer of gold, diamonds, iron ore, and several strategic metal ores (Figure 9.38). Mining products, often refined in South Africa, make up two-thirds of the country's exports. Platinum is of growing importance because of its use in aerospace construction and pollution control (catalytic converters) in modern vehicles. South Africa also has 40 percent of the world's gold reserves and 90 percent of the platinum reserves. It is the world's largest exporter of gold and platinum. Its

FIGURE 9.38 Southern Africa: mining. The Okiep Copper Mining Company at Nababiep, Northern Cape Province, South Africa. This is typical of many mines in the subregion: the piles of waste after the ore is separated from its containing rock, the pithead buildings, the smelter with its tall stack; location in a rural area with the surrounding vegetation killed by gases from the smelter. **Photo:** © Hubertus Kanus/Photo Researchers, Inc.

chrome, manganese, and vanadium are important in special steels. The diamond industry depends on the fashions that keep diamond jewelry in demand at high prices around the world. The South African firm of De Beers maintains a cartel of producers who restrict production to this end, although challenges are increasing from new producers in other parts of Africa and the wider world (Canada, Australia, and Russia).

South Africa's mining industry, however, faces growing challenges. In 1970 it mined 70 percent of the world's gold but only 27 percent in 1995. Over that period, Australia, Canada, Russia, and the United States increased their shares, and new mines opened in other African countries such as Mali and Ghana. As the world gold price fell, South Africa's costs of deep mining rose. Some major South African corporations now invest in gold mines in other parts of Africa, rather than in expanding their own mines. Employment in South Africa's mines is still a major source of income in the homes of migrant workers, mainly from Lesotho, Swaziland, Botswana, and Malawi. These workers are also at the greatest risk from HIV/AIDS, which affects around one-third of the mineworkers, adding US $10 to the cost of producing one ounce of gold.

Mining Wealth: Other Countries

Namibia (Figure 9.39) is the world's fourth largest exporter of nonfuel minerals and a major producer of uranium, together with diamonds, zinc, lead, and tungsten. It has the world's largest diamond-polishing firm (an aspect that does not interest De Beers). In Botswana, diamonds make up 80 percent of exports, and the country became wealthy by exploiting these resources from the late 1970s. In 2003 it produced the world's highest mined value of diamonds.

Zambia was formerly a major world producer of copper, sharing a rich ore field with southern Democratic Republic of Congo (DROC). Nationalization in Zambia led to a halved mineral output, and in the 1980s income was reduced by falling global copper prices. The resulting lack of investment to replace exhausted mines, combined with heavy debts, made the Zambian mines uneconomic in competition against Chilean mines that produce more copper with half the workers (see Chapter 10). The need to restructure its copper industry lies at the heart of the Zambian future. By the early 2000s, foreign corporations bought Zambian copper mines, expecting to redevelop them but with much smaller workforces.

Zimbabwe's mineral output, making up 40 percent of its exports, includes coal, gold, chrome, and nickel in a wide belt along a major igneous intrusion, the Great Dyke, between Bulawayo and Harare. After some years of standstill, output increased in the 1990s, when a new platinum mine was the largest foreign investment in Zimbabwe in 25 years. It was expected to make Zimbabwe the world's second largest producer of platinum after South Africa, but political problems inside Zimbabwe caused it to close in 2000.

Angola has oil fields along its northern coast, together with diamond mines inland. After the civil war destroyed its northern land-based oil installations, oil companies recognized the

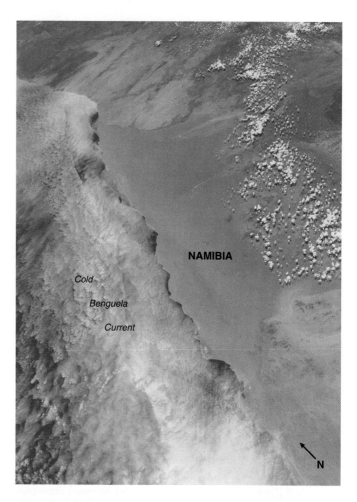

FIGURE 9.39 Southern Africa: Namibian Desert. The red sands of the desert dunes are a distinctive feature of this arid area, made drier by the offshore cold Benguela current that flows northward toward Angola in the distance. The clouds provide some rain to the higher inland areas, the cold ocean current is a rich fishing ground, and some of the world's richest diamond mines are found in the foreground of this space shuttle view. **Photo:** NASA.

deepwater (over 1,000 m, or 3,281 ft., deep) offshore resources as being of major world significance. As nine new oil fields opened, oil exports increased tenfold, making up 90 percent of Angola's export revenues and 20 percent of Africa's total oil exports. Angola was once the world's fourth largest diamond producer, but the UNITA occupied the interior diamond lands in the civil war; there is now potential for restoring legal production.

Manufacturing Contrasts: South Africa

South Africa dominates manufacturing in Southern Africa, together with the linked power utility and transportation systems. As soon as mining began a century ago, there was a need for engineering services and chemical supplies because the ores had to be refined near the mines to reduce the bulk for transportation. During World War II, sources of machinery and other manufactured goods in Europe were cut off, and South Africa diversified its manufacturing base into food products, textiles, clothing, armaments, and motor vehicle assembly in a process of import substitution. International sanctions resulting from its apartheid policy caused South Africa to increase its manufacturing base.

As South Africa raised its coal output from 50 million tons in 1970 to 200 million tons by the 1990s, it developed manufacturing technologies for using coal as a source of chemicals. The main coal-mining area is on the high veld east of Pretoria, where coal-burning plants generate 80 percent of South Africa's electricity. They cause acid rain downwind.

The great majority of South Africa's manufacturing is concentrated in the urban–industrial area around Johannesburg and around the ports of Cape Town and Durban. During the period of apartheid policy and the establishment of tribal homelands, factories were built just outside the homeland areas for access to cheap labor.

Six large South African corporations run almost totally by white South Africans continue to dominate mining and manufacturing. The current South African government has a commitment to enable black people to enter the white-controlled business world throughout the country. At present, most black businesses, apart from a major brewery and an insurance company, are small and engage in trade rather than manufacturing. Unemployment remains a great threat to political stability in the country as the business sector expands cautiously and the government spends money on training but not on creating many new jobs.

As South African markets open to foreign manufacturers, the large local business conglomerates and continuing political uncertainty deter some foreign investments. Multinational corporations, however, bought back companies they pulled out from in the 1980s. New investment was slow to start, but by 2000, corporations such as Levi Strauss, Nestlé, Coca Cola, Toyota, other auto manufacturers, and some South Korean electronics manufacturers established factories. Pharmaceuticals are a growing area for generic drugs that the government can buy cheaply, especially the anti-retroviral protection for HIV/AIDS sufferers. There is considerable capital within South Africa for investing in local and wider industrialization and infrastructure, but much goes abroad for investment, including to other African countries, China, and Vietnam.

Manufacturing Contrasts: Other Countries

Zimbabwe had the only other development of diversified manufacturing in Southern Africa. Industrialization began before the fight for independence, when cotton textile plants (Kadoma) and an iron and steel mill (Kwe Kwe) were established. During the period of independence declared unilaterally by the white minority, factories were built to produce goods that sanctions kept out. Manufacturing grew from 5 percent of GDP in 1965 to 26 percent in the 1990s. With the reopening of its South African trade and easier access to world trade, Zimbabwe needs to move from import-substitu-

tion industries to compete with other world manufacturers. The country possesses good power sources from its own coalfields at Hwange and hydroelectricity at the Kariba Dam. Ten years of drought in the 1990s lowered the Kariba lake level to the point where only a small fraction of the potential hydroelectricity could be generated, but better rains filled it in 2000. Political actions by the Mugabe government and one of the highest HIV/AIDS infection rates in the world (25 percent of the population aged 15 to 45) also weakened the economy, which is now static or declining.

Among the other countries, agricultural products are processed in Malawi and Swaziland, and minerals are refined in Namibia and Zambia. Portugal did not encourage manufacturing in Angola or Mozambique until the 1960s, but by independence, Angola had vehicle assembly and chemical factories, while Mozambique produced steel and textiles using hydroelectricity from Cabora Bassa Dam on the Zambezi River. After independence, skilled managers and technicians departed, some sabotaging the factories, and further decline occurred during the Angola and Mozambique civil wars. In the mid-1990s it was estimated that Mozambique industries operated at less than half their capacity, although restoration of some with South African assistance began later in the decade.

In the early 2000s, grounds for optimism in the economic future of this subregion and the significance of South Africa included the fact that infrastructure destroyed during civil wars in Angola and Mozambique was being reconstructed. In the latter, transmission lines take power from the Cabora Bassa Dam on the Zambezi River across 1,400 km (868 mi.) of land containing few people to reconnect South African users—and bring income to Mozambique. Port facilities at Maputo, Beira, and Nacala in Mozambique were expanded, together with the repair of railroad links into Zimbabwe and Malawi. In Angola, the line from Lobito into the southern mining area of Democratic Republic of Congo is being rebuilt. The development of major industrial zones along new roads between Johannesburg (Republic of South Africa) and Maputo (Mozambique) and between Windhoek (Namibia) and southern Angola testifies to increasing confidence by external investors and to the creation of regional marketing and development strategies.

Services

Service industries, from shops and banks to government jobs and tourist facilities, increase throughout Southern Africa, but especially in South Africa. Tourism is a growing feature of the economies in Botswana, South Africa, and Zimbabwe. There is a special interest in the national parks, where large animals are protected and able to live in something like natural conditions. In 2002 South Africa received 6.5 million tourists, Zimbabwe 2 million, and Botswana 1 million. In Zimbabwe, tourism was the fastest-growing sector of the economy before the political troubles of the early 2000s reduced it. Offshore island countries, such as the Seychelles and Mauritius (0.7 million visitors) in the Indian Ocean, are also important tourist venues.

Test Your Understanding
9C

Summary

Eastern Africa received cultural influences from Arabia and was later colonized by the United Kingdom, Germany, and Italy after the opening of the Suez Canal. Plateau landscapes are carved by rift valleys and topped by volcanic peaks. Arid conditions and periods of drought are common. Kenya has the most diversified economy based on agriculture, tourism, and service industries. Civil war and single-party governments hold back other countries.

Southern Africa contains poor countries such as Angola, Malawi, Madagascar, and Mozambique but also the wealthiest African country of South Africa. Botswana, Namibia, Zambia, and Zimbabwe have mineral wealth. This subregion has subtropical and midlatitude climates but also suffers from aridity and drought. It is where the future of Africa South of the Sahara may be worked out.

Questions to Think About

9C.1   Suggest five geographic characteristics that help to distinguish Eastern and Southern Africa (such as population, cultural history, economic products, government, or environment).

9C.2   Assess the impact of the precolonial and colonial histories on the present human geographies of the countries in Southern and Eastern Africa.

9C.3   Discuss whether increasing democracy is making a better life for the peoples of these two subregions.

Chapter 10

Latin America

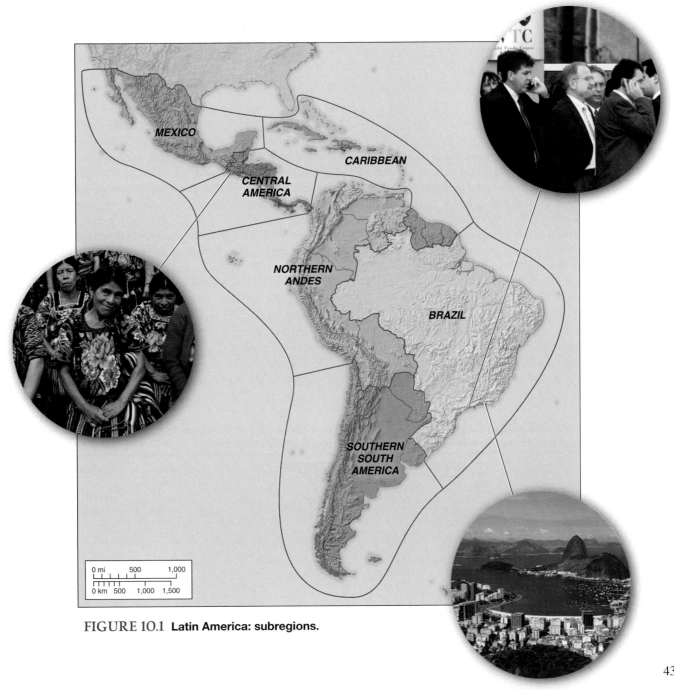

FIGURE 10.1 **Latin America: subregions.**

435

Themes in This Chapter

Dynamic cities, ancient ruins, tropical beaches, and volcanic eruptions are but a few of the dramatic aspects of the human and physical geography of Latin America. Latin-based Romance languages (primarily Spanish and Portuguese) dominate throughout the region and provide a degree of unity among the varied populations. Latin America today seeks greater global connections as it continues efforts to overcome its legacy from the colonial era, decades of postindependence dictatorships and military rule, and ongoing struggles for indigenous rights, land tenure, and socioeconomic class and gender equality. The major themes are:

Spanish and Portuguese colonial impacts on human geography.

Tropical, Southern Hemisphere, maritime, and mountain natural environments.

Six subregions (Figure 10.1):

- Mexico: regional power, U.S. neighbor, migrant communities, increasing global connections and local resistance.
- Central America: diversity amid a common past.
- The Caribbean Basin and Environs: island countries, national identities, and isolation.
- Northern Andes: mountain environments and the international drug trade.
- Brazil: burgeoning economic giant, national pride, and controversial resource use.
- Southern South America: temperate lands and a European culture.

Point–Counterpoint: Tropical Forests and Deforestation

Geography at Work: Geographical Research on Rivers in Eastern Mexico

Personal View: Bolivia

Geographic Contrasts

Latin America is a region of exciting contrasts in its human and physical geography. Countries range in size from population giants such as Brazil (184 million people in 2005) and Mexico (107 million) to the Caribbean Basin, where many countries have fewer than 100,000 people (Figure 10.2). Latin American countries range from high to low incomes, from dependence on a single economic product to a diverse and integrated economic base, and from involvement in to isolation from the global economic system. In the early 2000s several countries and subregions in Latin America were rising out of extreme poverty and into greater levels of global economic activity (Table 10.1), while others continued a long cycle of economic challenge. The total GNI of the region rose slowly during the 1980s and 1990s, while in many countries the income gap between the materially wealthy and materially poor worsened during the same period.

The second highest mountain range in the world (the Andes Mountains) contrasts with the huge, low-lying basins of the world's largest river system (the Amazon River). The Earth's largest tropical rain forest (the Amazon Tropical Rain Forest) is situated across the mountains from one of the world's driest deserts (the Atacoma Desert). The region's great range in latitude produces a variety of climate regimes, which are further altered dramatically by changes in elevation and proximity to mountain ranges. Some countries, such as Mexico, have both tropical beaches and snow-capped mountain peaks.

The contrasts present in Latin America also exist within each of the region's countries—inside cities and between the growing cities and hinterland rural areas. Mexico, Brazil, Argentina, Peru, and Chile contain large urban–industrial areas around their major cities, such as Mexico City, São Paulo, Rio de Janeiro, Buenos Aires, Lima–Callao, and Santiago–Valparaíso. The São Paulo and Mexico City metropolitan areas

Table 10.1 Latin America. Data by subregion, area, population, income (GNI PPP—Gross National Income Purchasing Power Parity), urbanization, Human Development Index (HDI), Human Poverty Index (HPI).

Subregion	Land Area (km²)	Population (millions)		GNI PPP 2003		% Urban	Index 2002	
		Mid 2005 Total	Est. 2025 Total	Total (US$ billions)	Per Capita* (US$)	2005*	HDI (rank of 177)*	HPI (% population)*
Mexico & Central America	2,477,700	147.4	187.7	1,084	5,509	56.1	89	14.0
The Caribbean Basin	689,910	35.2	42.9	106	7,728	55.7	75	13.2
Northern Andes	4,718,100	122.5	158.9	616	4,448	71.8	88	11.2
Brazil	8,511,970	184.2	228.9	1,219	7,450	81.0	65	11.4
Southern South America	4,108,000	64.3	77.9	628	8,363	80.8	53	6.1
Totals or averages*	20,505,680	548.5	678.6	3,652	6,700	69.1	74	11.2

Source: World Population Data Sheet, Population Reference Bureau (2005), World Development Indicators, World Bank (2005), Human Development Report, and United Nations (2004).

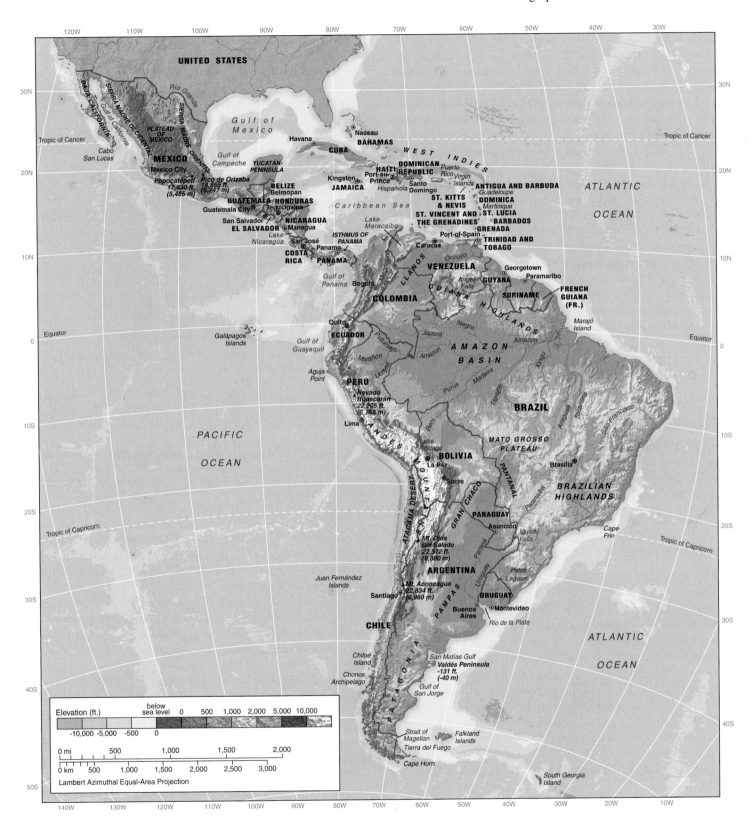

FIGURE 10.2 Latin America: physical features, countries, and capital cities. Mountain ranges, plateaus, islands, and river basins present contrasting conditions that influence human geography.

each contain nearly 18–19 million people and are two of the world's largest urban centers. Extreme contrasts between the materially wealthy and materially poor of the region are most dramatic within Latin America's cities.

International demand for selected agricultural products and mined resources from the region connects some rural areas to the global economic system while leaving others isolated from global connections. Brazil's coffee, soybean, and citrus fruit crops remain an essential component of its export economy. Argentina and Uruguay produce and export high-quality meat and wool from efficient commercial farms. Chilean farms exploit seasonal markets for fruit and vegetable products in Northern Hemisphere countries. Much of the exported Brazilian iron and gold, Chilean copper, and Venezuelan oil are extracted from rural areas. Although selected rural areas are enjoying global attention, widespread static or declining rural economic activity remains interspersed throughout Latin America. Heated debate, both within the region and in the outside world, surrounds the best future use of undeveloped areas.

Regional Cultural History

The cultural history of Latin America provides a vital foundation for building an understanding of the contemporary regional geography. The colonial occupations that dramatically reduced the numbers of Native American peoples and consigned the survivors to an underclass left a legacy of ethnic strife and hindered economic development. Habits and attitudes born under colonialism affected the region's countries for decades after independence.

Pre-European Peoples

The indigenous peoples present in Latin America when the Europeans arrived migrated to the region several thousand years earlier through western North America (see Figure 1.9). While many indigenous groups had a village-based subsistence economy that supported only modest numbers of people, several urban-based civilizations or empires emerged (Figure 10.3). Urban-based civilizations in Latin America prospered through tightly structured societies and large-scale agricultural production. Estimates of the numbers of people living in Latin America before European entry range widely, with 50 million serving as a conservative middle figure.

The Lasting Influence of the Maya, Aztec, and the Inca

The **Maya** civilization rose to regional prominence through agricultural surplus and a rigid social structure. The Classic Period, in which the Maya flourished, occurred from the A.D. 200s through the 900s. The Classic Maya based their regional structure on **city–states,** individually ruled and independently functioning urban centers, primarily located in the Yucatán Peninsula lowlands of southeastern Mexico, the Petén of Guatemala, parts

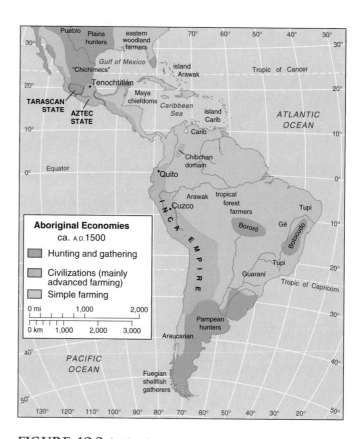

FIGURE 10.3 Latin America: pre-European economies and empires. A variety of Native American groups occupied the region before A.D. 1500. Some lived at subsistence level; others established far-reaching empires.

of Belize and El Salvador, and eventually in parts of western Honduras. During the pinnacle of the Classic Maya civilization, cities flourished with abundant food supplies; extensive construction of urban structures; religious, royal, and athletic ceremonies; and detailed attention to the arts and sciences. City–states of the Maya region conducted trade, competed with each other in athletics, and arranged marital unions among their royal children. The city–states also competed fiercely at times for regional resources and often waged bloody warfare with one another. The growing populations of the Maya cities, coupled with environmental challenges such as drought, placed increasing burdens on the surrounding environment for food and other natural resources. The decline of the civilization occurred over several decades as agricultural harvests diminished, nutritional deficiencies began to take a toll on the crowded urban centers, and resource competition and warfare between some city–states increased. The Maya imprint remains etched in the cultural landscape today in the form of pyramids and other ceremonial structures (see Figure 1.11). There are people throughout the Yucatán and northern Central America today who tie their ancestral heritage to the Maya of the Classic Period.

In the 1100s a group of tribes from central and northern Mexico settled in the area of modern Mexico City. By 1325 the **Aztec** dominated the region with a hierarchical society centered on the massive urban capital of **Tenochtitlán.** Unlike the preceding Maya civilization, where numerous individually ruled city–states existed simultaneously, the Aztec Empire was governed by one king from one capital. The capital of the Aztec empire was built on marshland and lakeshore in the Central Valley of Mexico. This city had many civic, commercial, ceremonial, and residential structures built on islands connected throughout the marshes of the valley by causeways. By the 1400s the Aztec were the dominant empire in Middle America and gathered great tribute riches in their capital. The elite of the empire converted acquired lands and peoples to their systems and used human and natural resources to further their goals. The Aztec forced conquered peoples into slavery, which fueled widespread animosity toward the empire's leadership. In the 1500s, when the Spanish conquistadors reached what is now Mexico, their goal was to topple the Aztec leaders and transform the Aztec cultural landscape to one suited for the glory of Spain. The Spanish conquistadors used the animosity held by many peoples of the region to assist in overthrowing the Aztec Empire.

Far removed from the Aztec Empire, the warlike **Incas** formed a strong society based in the southern part of modern Peru around the mountain city of Cuzco during the early centuries of the second millennium A.D. During the 1400s and early 1500s, the Inca Empire expanded dramatically and covered a region some 4,300 km (2,800 mi.) in length from southern Colombia to central Chile. Inca rulers established a well-connected system of roads running the length of their territory, which facilitated the empire's resource use, military control, and social order. The agriculturally based society also organized gold and silver mining and a system of messengers and transportation that maintained administrative cohesion. The tightly ruled empire functioned from a hierarchical system of noble leadership believed to be the offspring of the Sun God. All resources of the controlled territory were considered to be the property of the Inca leaders. The official language of the Inca Empire, **Quechua,** remains widely spoken today among the indigenous peoples of Bolivia, Ecuador, and Peru.

Subsistence Cultures

The Arawak and Carib groups of the islands and South American mainland areas of the Caribbean gained subsistence by forest hunting and fishing. The Araucarians of modern central Chile and the Pampean tribes of modern Argentina had relatively strong economies. Sharing a similar fate with the larger indigenous empires of Latin America, most of these subsistence groups were decimated by or largely absorbed into the colonial cultures.

Spanish Colonization

Christopher Columbus embarked from Spain in 1492 on the first of four trans-Atlantic voyages. His ships called upon several Caribbean islands as well as a large stretch of the east coast of Central America and the coastal waters of northeastern South America. Spanish military and missionary groups soon followed. Within 50 years of the initial voyage of Columbus, much of the region was conquered and occupied. The only exceptions were some of the smaller Caribbean islands, the far southern extremity of South America, and the territory that became Brazil. Spain and Portugal signed the **Treaty of Tordesillas** in 1494, which established a demarcation line (approximately 46°W longitude) between their global spheres of interest, giving the eastern quarter of South America to Portugal.

The initial Spanish occupation around 1510 centered on the large Caribbean islands of Cuba and Hispaniola. Inspired by tales of gold and other vast riches, Hernán Cortés led a group of conquistadors in 1519 on an expedition to the heart of the Aztec Empire (Figure 10.4). The Aztec thought the conquistadors were gods, which permitted the Spaniards to infiltrate the hierarchy with relative ease. The conquistadors combined the multiple advantages of being revered as deities, having relatively more advanced weaponry, and the accumulated animosity of the numerous enemies of the Aztec leaders to overthrow the empire by 1521. The Spaniards captured Aztec lands and established the former Aztec capital as the core of Spanish colonial America. A Spanish viceroy began to govern the colony of New Spain in 1535. Spain claimed the remainder of Mexico and Central America, northern Colombia, northern Venezuela, and Caribbean islands such as Jamaica and Trinidad. Expeditions from New Spain pushed northward into the southern parts of the modern United States from the 1530s but found no gold.

In 1532 a small Spanish force invaded and captured Cuzco, the capital of the Inca Empire. The Spaniards founded new cities to administer the gold and silver mining industries and facilitate the transportation of resources across the Panamanian **isthmus** (land bridge) and onward to Spain. Spanish expeditions in Southern South America in the 1530s did not lead to material occupation or integration into Spain's colonial system until much later (Figure 10.5). Lima and its nearby port of Callao, which linked the resource wealth of Peru to Spain, became the capital of the Viceroyalty of Peru that included most of northern and western South America. The Viceroyalty of New Granada, centered on Bogotá, was established in 1717 and that of La Plata, centered on Buenos Aires, in 1776.

The Spaniards conquered most of the region in the 1500s and imposed a high degree of control on their colony of New Spain by establishing an oppressive system of agricultural production and tightly connected urban settlements and ports. Several forces hindered the Native American inhabitants in forming strong resistance to Spanish intentions in the Americas. After the conquest, the lack of immunity to European diseases drastically reduced indigenous populations. Surviving communities were converted by the Roman Catholic Church. They were forced to give up their subsistence agricultural practices to provide crops and livestock for the food needs of the foreign city dwellers. Mining settlements in the Andes Mountains used various forms of slavery to coerce the local people.

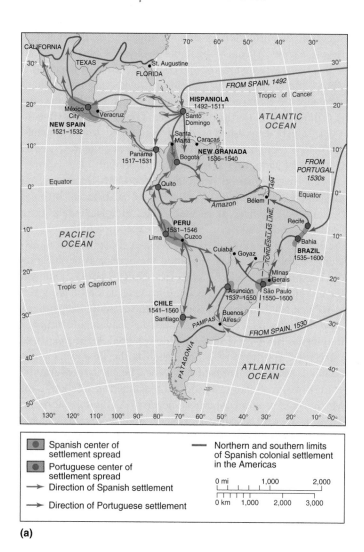

(a)

(b)

FIGURE 10.4 Latin America: colonial conquest and settlement. (a) During the 1500s, Spanish and Portuguese adventurers established bases and conquered huge areas rapidly. The areas of initial conquest became centers of administration and development. Scenes of the Aztec capital of Tenochtitlán (b) depicted in the artwork of celebrated Mexican painter Diego Rivera. **Photo:** © 2002 Banco de México Diego Rivera & Frida Kahlo Museums Trust. Av. Cinco de Mayo No. 2, Col. Centro, Del. Cuauhtémoc 06059, México, D. F. Schalwijk/Art Resource, NY.

The colonial period in Latin America left a legacy of underdevelopment because it primarily focused on mining for export and the imposition of large landed estates with a strict feudal system. The native peoples and many of mixed ancestry were relegated to being landless laborers. By the end of the Spanish colonial period, antagonisms between privileged and underprivileged groups were ingrained. Subsequent history is largely a record of such antagonisms at work.

Spanish Control

Spain granted large land areas to nobles, soldiers, and church dignitaries, in conjunction with the responsibility for political control, establishment of settlements, and exploitation of natural resource wealth. They were given jurisdiction over the native peoples to use them as laborers. This feudal **encomienda** (tribute) **system** was applied throughout the Spanish colonies for more than 200 years. Although some efforts were made in Spain to pass legislation calling for the fair treatment of native peoples, such laws were not applied in a meaningful manner in the colonies. In addition to the subjugation of native peoples, the Spanish eventually imported more than 1.5 million people from the African continent to labor as slaves in mining and agriculture.

Spanish colonial agriculture developed large production estates, or **haciendas,** to cultivate crops and livestock products for local or domestic markets. Landholders often advanced credit to laborers, causing them to be indebted to the hacienda. Coastal areas around the Caribbean were home to plantations, which produced agricultural products for export to Europe and elsewhere outside the region.

Inefficient, corrupt colonial administration of the region established a foundation for pervasive and lasting tension between various socioeconomic groups. The native peoples were dispossessed of land and rarely afforded educational opportunities. At the other end of the socioeconomic scale, the **peninsulares**—Spaniards born in Spain who were living or working in the New World—took the highest offices and largest land grants. **Criollos**—Spaniards born in the colonies—

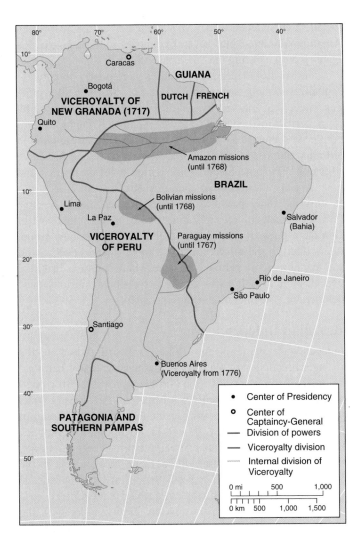

FIGURE 10.5 South America: colonial divisions. The jurisdictions established by Spain and Portugal formed the basis of country boundaries at independence but were poorly defined. The red line divides areas of Spanish (to the west) and Portuguese (to the east) influence. The pink shading highlights early attempts to extend Brazilian territory westward. Buenos Aires became the center of the Viceroyalty of Rio de la Plata.

and the increasing numbers of **mestizos** (mixtures of European and Native American ancestry) had fewer privileges and became resentful as their links to Spain weakened.

Portuguese Colonization

Portuguese colonial efforts in Brazil moved at a slower pace than Spanish efforts in the region until the French, in the 1530s, attempted to colonize in areas previously claimed by Portugal. Then the Portuguese stepped up their settlement and political organization of the region. A Portuguese governor general was installed in 1549. The city of Salvador became the first capital and northeastern hub of Portuguese Brazil, while São Paulo was established much farther south to solidify Portuguese territorial claims and settle southern lands. The Portuguese established a

settlement along Rio de Janeiro Bay after expelling French settlers from that portion of the region. The bayside settlement of Rio de Janeiro became the Portuguese colonial capital in the late 1700s.

In the 1600s the discovery of gold inland of Rio de Janeiro led to increased Portuguese immigration. This in turn set off expeditions to explore and claim the interior. From 1680 various attempts were made to push the Portuguese boundaries southward to the Plata estuary, where Spanish settlement was slow. In 1750 Spain agreed to permit Brazilian interior expansion westward of the Tordesillas line—which was really an expression of Spain's inability to prevent what was already occurring.

The Portuguese process of occupying Brazil resembled that of Spain in its colonies, although it had some distinctive features. Colonial Brazil was economically and socially controlled by an elite class who purchased or captured millions of African slaves and transported them across the Atlantic to work on large sugar plantations along the northeastern coast. The Portuguese were responsible for the forced migration of more people from the African continent than any other single colonial power (see Figure 9.6). Estimates suggest that more than 4 million Africans were shipped to Brazil to fill the labor needs of economic development. Large numbers of present-day Brazilians trace their ancestral heritage to African slaves. Farther south in the São Paulo and Rio de Janeiro areas, Europeans who were not part of elite colonial circles had more opportunity to establish their rights to new lands, leading to a greater degree of personal autonomy.

Other European Colonies

French, Dutch, and British attempts to colonize Latin America came later and were largely resisted by the Spanish and Portuguese. There were limited incursions on lands controlled by the dominant colonizers primarily in and on the periphery of the Caribbean Basin.

The French took the western third of Hispaniola in 1697. A successful slave revolt and independence movement, known as the **Haitian Revolution,** in 1804 expelled the French and led to the creation of the new country of Haiti. The French also took control of the remainder of Hispaniola (Santo Domingo) from Spain in 1797 and maintained a presence until 1809. When Santo Domingo became independent in 1821, Haiti invaded and occupied it until 1844, after which the independent country of the Dominican Republic was declared. Cuba and Puerto Rico remained Spanish colonies until the United States backed a Cuban uprising in 1898 and won the subsequent Spanish–American War. Cuba became independent a few years later. Puerto Rico became a "commonwealth territory" of the United States in 1898, and its people became U.S. citizens in 1917.

Some of the smaller Caribbean islands that were ignored by Spain had their first European settlement by the French, Dutch, or British. Dramatic increases in land area devoted to sugar cultivation and the amount of African slave labor imported to the region contributed to gains in material wealth for many Europeans. Many of the islands changed hands, as when Britain took Jamaica (1670) and Trinidad (1797) from Spain.

The Guianas were within the Spanish realm on mainland South America, but little interest was shown in the area's swampy coasts and forests until the Dutch, British, and French settled them. The Dutch settled first in 1581, eventually establishing what is today the country of Suriname. The British took the western territory in 1814 and established a colony (today the independent country of Guyana). The French settled the eastern part in the 1600s, where the infamous penal colony of Devil's Island deterred further immigration into French Guiana until the prison was closed in 1938.

The British settled along parts of the east coast of the Central American mainland and on islands just offshore. British influence proved to be lasting in the western Caribbean in Belize and parts of Honduras. British control in Belize lasted until well into the 1900s, and English-speaking residents of the Bay Islands of Honduras are today a lasting geographic remnant of British influence in the region.

Independence

The initial wealth that attracted Europeans came from exports of high-value minerals such as gold, silver, and gemstones from Latin American colonies to Europe. When the flows of such wealth slowed in the 1700s, combined with the devotion of resources to fight the Napoléonic conquest of Spain in the 1790s, the colonial power of both Spain and Portugal weakened considerably. The eroding strength of those European colonial empires enabled many Latin American countries to gain independence in the early 1800s. The new countries largely emerged from administrative divisions within the Spanish viceroyalties. The colonial boundaries between these divisions were poorly surveyed, if at all, forming a basis for subsequent disputes.

Independence from Spain

Independence from Spanish rule was often followed by a period of successive attempts at different types of political arrangement. The establishment of the modern countries in the region unfolded slowly. When Mexico separated from Spanish rule in 1821, it covered an area from the present-day southwestern United States through the modern Costa Rica–Panama border. The Central American region separated from Mexico in 1823 as the United Provinces of Central America, which then further divided in 1838 into the Central American republics existing today. Northwestern South America functioned apart from Spanish rule as Gran Colombia from 1822 to 1830. In 1830 the independent countries of Colombia (which included present-day Panama), Ecuador, and Venezuela emerged. Panama did not gain independence from Colombia until 1903, when U.S. interest in building a canal across the isthmus facilitated

Panamanian independence. The remainder of Spanish South America splintered into several countries independent of the Spanish crown during the 1810s and 1820s.

The newly independent countries were poorly prepared to capitalize on their sovereignty. The next 150 years after independence were marked by political instability punctuated by short periods of economic growth. Spain left behind a geographic system based on mineral exploitation, transportation to a limited number of ports, and large estates engaged in raising livestock. The postindependence period created rivalries and boundary disputes between countries. Rigid social class stratification fueled deep and lasting internal resentments. In most former Spanish colonies in Latin America, government control alternated between the elite conservative landowning group, often allied to the military, and the more liberal criollos who wanted more democratic governance. Many countries had long periods of dictatorship after conservative–liberal tensions made democratic government almost impossible.

Brazilian Independence

Independence came in the 1820s to Brazil, which had a stronger foundation in place for modernization than other postcolonial territories. In 1807 the Portuguese royal family evacuated from Europe to Brazil when Napoléon threatened to take their country. Rio de Janeiro became the temporary capital of Portuguese government. Although the royal family went back to Portugal after Napoléon's defeat, the prince regent returned to Brazil in 1816 at a time of flourishing revolutions in the Spanish colonies. In 1822 he proclaimed Brazil's independence from Portugal and himself king as Pedro I. Under King Pedro II (1840–1889), the Brazilian economy grew rapidly with the construction of railroads and ports and the expansion of mining in the east central parts, commercial coffee farming in the south, and rubber collecting in the Amazon River basin. In 1889 a military coup made Brazil a republic, but falling prices of coffee and rubber led to widespread unrest and a period of dictatorship. Brazil encouraged immigration and received many people from Germany, Italy, and Japan in the early 1900s. They moved mainly into the southern parts of the country, where environmental conditions were somewhat similar to those of their homelands.

Economic Colonialism

Although Spanish and Portuguese rule was broken, Europe remained the primary market for Latin America's exports. Britain in particular established a relationship of economic colonialism with several Latin American countries that lasted until the early 1900s. Areas targeted for economic development and control by the British were set up for the export of raw materials and products to supply Britain's industries. The British built railroad networks and port installations aimed at developing the production of minerals, cotton, beef, grain, and coffee. Elite families in Latin American countries who worked within the system often sent their children to schools in Europe and the United States and banked their wealth in those countries.

Continuing External Influences

Import Substitution

The economic depression of the 1930s in the United States and Europe significantly disrupted trade and economic activity for Latin America. Following the global economic depression of the 1930s was World War II in the 1940s. The combination of these two dramatic events caused many decision makers in Latin American countries to strive to be more internally self-sufficient. The countries of the region established the goal of becoming less dependent on selling unprocessed or unrefined raw materials in exchange for high-priced manufactured goods from industrial countries. Under import substitution, Latin American countries attempted to use their raw materials in their own internal production of various manufactures for domestic markets. Governments established high tariffs, quotas, and bureaucratic barriers on goods arriving from countries outside of the region. Many of the new industries were owned by the various country governments, and others, established earlier by foreign interests, were **nationalized,** or taken over by the governments of the Latin American country in which they operated. The process proved to be costly for many Latin American countries as they incurred large amounts of debt to fund the construction of industry and purchase of manufacturing equipment. By the 1970s this policy contributed to the rapid growth of one or two major urban centers in each country. Because of their larger home markets, the most populous countries (Argentina, Brazil, Colombia, Mexico) produced the most under this system. Economic growth continued to be uneven.

Questions to Think About

10A.1 Compare and contrast the location, time period, social structure, and regional influence of the Maya, Aztec, and Inca. What role (if any) did the Spanish conquistadors play in the evolution of each civilization?

10A.2 What were the primary historic events that imposed a Hispanic culture on Latin America? How did that culture affect postindependence changes and attitudes? Which countries did not experience Hispanic influence during the colonial era, and how did this affect their contemporary human geography?

10A.3 Compare and contrast the roles of peninsulares, criollos, mestizos, and African slaves in the social structure and economic development of colonial and postcolonial Latin America.

Key Terms

Maya	encomienda system
city–state	hacienda
Aztec	*peninsulare*
Tenochtitlán	*criollo*
Incas	*mestizo*
Quechua	Haitian Revolution
Treaty of Tordesillas	nationalize industry
isthmus	

Test Your Understanding
10A

Summary

Latin America includes Mexico, Central America, the Caribbean Basin, and South America. Although many parts of the region are a legacy of influence outside of Europe's Iberian Peninsula, a degree of unity exists in contemporary Latin America centered on the Spanish and Portuguese languages and an adherence to Catholicism. While the region is tied by common colonial legacy, it contains dramatic variations in both its human and physical geography. Latin America is plagued by a widening gap of human wealth and economic development in some of the world's largest cities and resource-rich farming regions. Despite considerable growth in many areas, Latin America contains expansive impoverished rural areas throughout the region.

Many Latin American countries are small, especially the Caribbean islands, but historic animosities and the production of competing goods often prevent them from working together for economic growth. The economic and cultural history of Latin America led to slow modernization, internal and international antagonisms, and reliance on one or a few export products.

Natural Environment

Although Latin America extends through almost 90 degrees of latitude—the greatest north–south distance of any major world region—the majority of Latin America lies within the tropical latitudes. When coupled with high mountain altitudes, the latitudinal expanse of the region results in a wide variety of climates, natural vegetation types, and soils.

Tropical and Southern Hemisphere Climates

Middle America and the Caribbean Basin

Nearly all of **Middle America** (Mexico, Central America, and the Caribbean Basin) lies within the tropics (Figure 10.6). East–northeast trade winds dominate much of the Caribbean Basin for about two-thirds of the year, contributing to the area's consistent warmth and humidity. Temperatures remain around 30°C (86°F) through most of the year. The northern Caribbean islands, Mexico, and even parts of northern Central America are occasionally affected by modified cold air masses, locally referred to as **nortes,** that move down from continental Canada and the United States during the Northern Hemisphere's winters. In Middle America, the rainfall

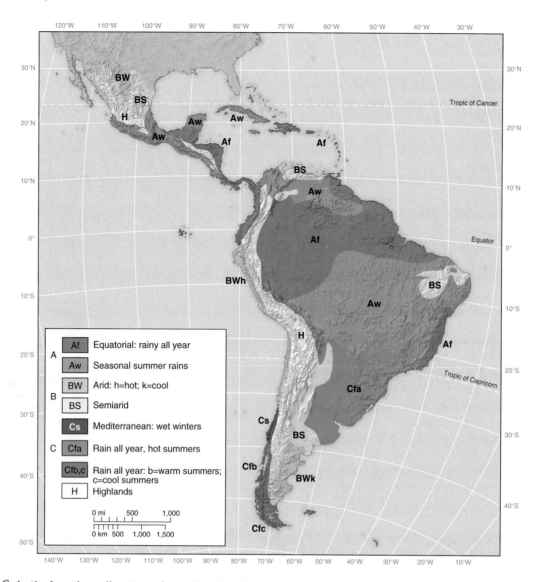

FIGURE 10.6 **Latin America: climate regions.** Note the influence of the Andes Mountains and the tropical latitudes.

generally increases southward, from the arid region that straddles the Mexico–U.S. border toward the coasts of Nicaragua, Costa Rica, and Panama, which receive over 2,600 mm (100 in.) of rain each year.

Most rains fall in the summer and fall, when heating of the lower atmosphere causes humid air to rise, cool, and condense into clouds. When condensed moisture in the clouds becomes heavy, gravity pulls the moisture to the surface as rain or some other form of precipitation. On Caribbean islands with mountains, windward slopes facing the east or northeast trade winds add to the uplift of the humid air and increase precipitation totals. Where the air descends, warms, and becomes less humid, less rain is likely. For example, in Jamaica, the windward northeast coast receives over 3,300 mm (130 in.) a year, while the leeward southern coast requires irrigation for farming in areas where the annual rainfall is less than 750 mm (31 in.).

Middle America is a region of annual hurricane activity. The North Atlantic hurricane season runs from June through November. Early in the season, hurricanes form in the western Caribbean and Gulf of Mexico. During the seasonal peak in August and September, hurricanes form in the eastern Atlantic near the Cape Verde Islands west of the African continent. The storms move westward into the Caribbean Basin and curve toward either the northwest or north (affecting the U.S. mainland or shipping channels) or continue westward, striking Central America or the Mexican Yucatán Peninsula. Late in the annual hurricane season, storms once again form mostly in the western Caribbean and Gulf of Mexico. Only the southernmost countries of Middle America lie outside the hurricane zone. Hurricanes cause both a dramatic loss of life and extensive damage to crops, livestock, personal property, and the vital regional tourism industry. Recovery is often slow in small countries with few resources. For example, in 1998 Hurricane

Mitch caused extensive damage and loss of life that devastated the already poverty-stricken countries of Honduras and Guatemala (Figure 10.7).

Northern South America

All the Northern Andean countries lie within the tropics. Temperatures vary with significant changes in elevation. The amount of rainfall at a location depends on whether the land faces the direction from which moist air arrives. Humid airflow from the Atlantic Ocean into the Amazon River basin carries intense rains far westward across the continent to the east-facing slopes of the Andes. Slopes that face away from the winds, sheltered valleys and basins, and land at high altitude may require irrigation for effective agriculture. On the Pacific coast, rainfall is high on the slopes facing the ocean north of the equator, but southward the offshore winds and cold Peruvian ocean current create an arid climate on the adjacent land.

Northern Brazil experiences an equatorial climate in which the temperatures hover in the low 30s°C (80s°F), humidity is high, and rain falls in all seasons. On the Brazilian and Guiana Plateaus, the rains are more variable and more seasonally concentrated in the high-sun period when evaporation and rising air currents are most intense. The variability is particularly marked in the northeastern corner of Brazil, where severe and prolonged drought occurs periodically. In southernmost Brazil, the climate becomes midlatitude in type, with cool winters and a shorter growing season.

El Niño

The **El Niño phenomenon** is regarded as a significant feature in explaining connections among worldwide climatic environments (Figure 10.8). The major 1997–1998 El Niño event brought drought to parts of Middle America and northern South America, and exceptionally heavy rains to the deserts of Peru and Chile. The same El Niño event was blamed for unusual weather around the world. Droughts in Indonesia and Australia, along with drought-based fires in Florida and unusually hot weather in southern Europe, were indirectly linked to the event. No two El Niño events are the same, and thus the impact on global climatic conditions varies.

In general, the basic features of El Niño are understood. Every two to five years, easterly winds that usually push the cold Peruvian current westward across the tropical eastern Pacific diminish. Warmer water from the western Pacific flows eastward, eventually reaching the coasts of western North, Central, and South America. Warmer and more humid air masses off the west coasts of the Americas alter the regional climate.

FIGURE 10.7 **Hurricane Mitch damage.** Aerial view of a Honduran coastal village damaged by Hurricane Mitch in 1998. **Photo:** © Yann Arthus-Betrand/Corbis.

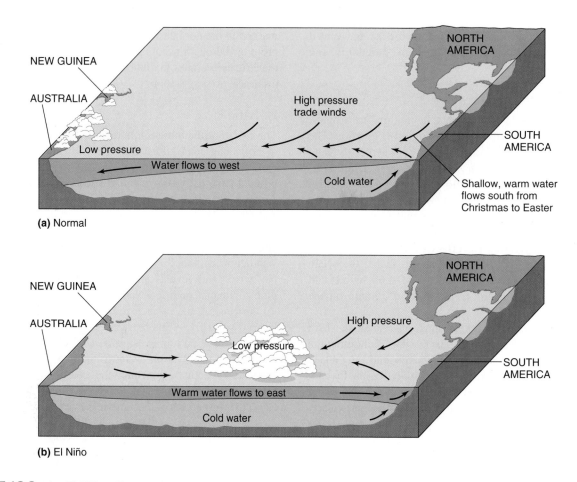

FIGURE 10.8 The El Niño effect. How do the changes in wind direction from a normal circulation pattern (a) to the El Niño conditions (b) affect the surface waters off the coast of western South America? **Source:** From *Foundations of Physical Geography* by Bradshaw & Weaver. Copyright © The McGraw-Hill Companies.

Warm water also dramatically changes the marine ecosystem, and during strong El Niño events, it may wreak havoc on the fishing industries of countries in western South America. With easterly flow diminished across the region, some areas used to abundant precipitation may experience a decrease in humidity, and the benefit of orographic lift is negated during such episodes.

The 1997–1998 El Niño event was the largest since 1982–1983, and its impacts on people's lives in Latin America were diverse. Although some areas enjoyed advantages, such as freshwater reservoir replenishment and a prolonged winter ski season due to increased snow depth on the Chilean Andes, many outcomes were negative. Droughts caused lower farm yields in Central America and northern South America, while floods destroyed crops in Argentina and Uruguay and washed out roads in Ecuador and Peru. Destructive fires burned forest and savanna vegetation on the Brazil–Venezuela border and in Central America. Water shortages disrupted transportation in the Panama Canal as well as the output of some hydroelectric plants, resulting in temporary power cuts.

Southern South America

The climates of Southern South America range from arid regions in northern Chile and parts of Argentina to one of the world's stormiest, wettest regions in southern Chile. The Andes ranges affect the climates of the lands on either side in differing ways.

In northern Chile, the Atacama Desert continues the Peruvian Desert southward between the high Andes and the cold Peruvian current. Winds blow almost parallel to the coast or offshore, pushing the cold current northward along the coast. The cooling effect of the ocean on atmospheric temperatures makes it difficult for air to rise and cause precipitation on the land.

In southern Chile, by contrast, the midlatitude westerly winds bring precipitation and high winds at all seasons of the year. Between the desert and the stormy southern region of Chile is a transition zone of dry summers when the arid climate moves south and wet winters when the storm tracks move north. This zone is similar in climatic regime to the countries around the Mediterranean Sea in Europe (see Chapter 3).

On the eastern side of the Andes is another contrast between northern and southern lands. In the north, the warm Brazilian current in the Atlantic Ocean allows high rates of evaporation offshore, and trade winds blow humid air into the continent. The Patagonia region of southern Argentina is arid. Westerly airflow descends after crossing the Andes, warms up, and becomes drier. Patagonia is often cited as an example of **rain shadow** conditions in the lee of a mountain range, in contrast to the very wet Pacific slopes of the Andes facing the prevailing westerly winds in southern Chile. The strong winds blowing across Patagonia, after crossing the Andes, often produce very dry conditions at the surface.

High Mountains and Island Chains

The major relief features of Latin America (see Figure 10.2) were formed by a combination of clashing tectonic plates along the western and northern margins and of precipitation runoff in major rivers and mountain glaciers. Along the west coast, the South American plate overrides the Nazca plate (Figure 10.9). The convergent plate margin marks the line of the Andes Mountains and causes earthquakes and volcanic eruptions. The tectonic pattern is more complex in Middle America, where the North American, South American, Caribbean, and Cocos plates meet.

Insular and Mainland Middle America

Mainland Middle America is formed of two relief provinces. High-altitude plateau lands between the Eastern and Western Sierra Madres dominate northern Mexico. The collision of the North American and Cocos plates created the plateau, which rises over 2,000 m (6,000 ft.) and contains shallow basins that become more arid with a northward progression. Even in the arid north the plateau's shallow basins are often connected by rivers, which form fertile areas for farming. The slopes of the Western Sierra Madre facing the Pacific Ocean are steep, but those on the east are less steep with wide coastal plains. This province terminates at the Tehuantepec isthmus, where the land narrows in southern Mexico.

South of the Tehuantepec isthmus, the Caribbean plate collides with the Cocos plate, forming a single spine of mountains along and parallel to the Pacific coast. There are very narrow coastal plains and some areas without any flat land between mountain and ocean. Eastward of the mountain spine, west–east ranges are separated by deep basins, often occupied by rivers flowing into the Caribbean Sea. A large limestone platform emerged from the seafloor to form the Yucatán Peninsula. The ridges and basins can be traced eastward across the floor of the Caribbean Sea and beneath the island arcs on its eastern edge.

Middle America is subject to earthquakes and volcanic eruptions where tectonic plates collide. A major earthquake centered off the west coast of Mexico devastated Mexico City in 1985 (Figure 10.10), and earthquakes twice leveled the Nicaraguan capital city of Managua in the 1900s. There are 25 active volcanoes between northern Mexico and Colombia that periodically spew lava and ash on surrounding areas.

Insular Middle America is also affected by the clashes of tectonic plates. The North American plate drove into the Caribbean plate, producing intense volcanic activity along the plate margin and forming the Lesser Antilles arc. The eruption of lava and ash continued into the 1900s at Martinique (1902), Mount Soufrière on St. Vincent (1979), and the Montserrat eruption of 1995 that eventually forced the evacuation of the island. In the Greater Antilles, the plate movements caused a mass of continental rock to founder, leaving only the highest points above sea level. The Bahamas and an outer group of islands including Anguilla, Barbuda, and Barbados are flat limestone islands constructed of coral reefs on top of subsiding former volcanic peaks (Figure 10.11).

FIGURE 10.10 Mexico City: earthquake damage. Rescue workers search for victims in a building completely demolished during an earthquake that devastated parts of Mexico City in 1985. **Photo:** © Owen Franken/Corbis.

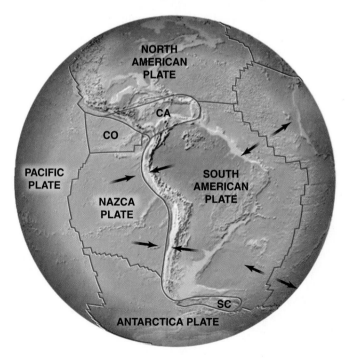

FIGURE 10.9 Latin America: tectonic plates. Tectonic plate interaction and plate boundaries. CO = Cocos plate; CA = Caribbean plate; SC = Scotia plate.

FIGURE 10.11 Caribbean Basin: coral islands. The Exuma Cays, Bahama Islands, are formed of coral reef limestone and rim the Greet Bahama Bank for over 150 km (100 mi.). **Photo:** © Bruce Dale/National Geographic Image Collection/Getty.

Andes Mountains

The Andes Mountains affect all aspects of the physical environment of western South America (see Figure 10.2). The collision of the South America and Nazca plates produced a volcanic and earthquake-prone western mountain range and a folded and faulted eastern range. The Andes rise to over 6,500 m (20,000 ft.) in Argentina, Chile, Peru, and Ecuador. In Bolivia, Peru, and Ecuador, the central Andes have two main ranges, the Cordillera Occidental (west) and Cordillera Oriental (east)

(Figure 10.12). Between the two ranges, a high plateau, the **Altiplano,** is widest in Bolivia and narrows northward into Peru. In Peru, rivers cut deep gorges as they flow northward and eastward to join the Amazon River tributaries.

In Colombia, the Andean ranges splay out northward into three cordilleras—the Occidental, Central, and Oriental. The Atrato River separates these from a coast range, while the Cauca and Magdalena rivers separate the three main ranges. The Pacific coast has only narrow coastal plains. In northern Venezuela, the Cordillera Oriental branches into a further series of lower ranges. Along the Caribbean coast, the mouths of the Colombian rivers, Lago de Maracaibo, and the Orinoco River delta, provide limited areas of lower coastal land between the ranges. Islands such as Trinidad and the Dutch Antilles are extensions of mainland geologic structures.

The southern Andes Mountains dominate the landscapes of Chile and the western parts of Argentina, with their highest points constituting the border between the two countries for most of its length. The highest point in the whole range, Cerro Aconcagua (6,959 m, 22,831 ft.), occurs where the central Andean ranges narrow in width southward from about 500 km (300 mi.) to about 150 km (100 mi.). The lowest point for crossing the Andes between the main centers of population in Chile and Argentina is the Uspallata Pass at 3,841 m (12,600 ft.). To the south of this pass, the Andean peaks get lower, continuing for a further 2,700 km (1,800 mi.) to Tierra del Fuego at the southern tip of the continent. The snow line on the Andes gets lower toward the southern tip of the continent, where glaciers descend to sea level (Figure 10.13). The former more widespread glaciations left behind moraine-dammed lakes and fjords.

On the Chilean side, the Andes come close to the Pacific Ocean in the north. Southward, a coastal range is separated from the main Andes ranges by a series of basins and then a

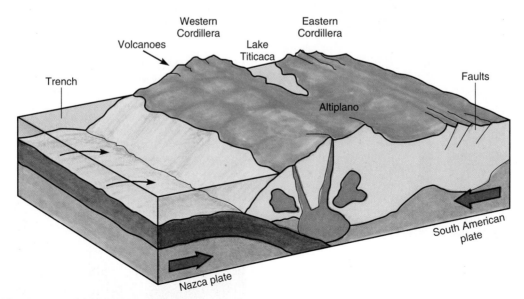

FIGURE 10.12 South America: Andes Mountains. The major features of the central Andes Mountains related to their formation along a destructive plate margin. The Nazca plate plunges beneath the South American plate, causing volcanic activity in the western cordillera along the coast and uplift of the eastern cordillera and the high plateau (altiplano) between the two ranges.

FIGURE 10.13 Southern South America: Andes Mountains. The Moreno glacier, Glacier National Park, Santa Cruz province, Argentina. The glacier is 5 km (3 mi.) wide, and its front melts on entering Lago Argentino. Such glacial features increase in significance in the southernmost part of the Andes Mountains. **Photo:** © James P. Blair/National Geographic Image Collection/Getty.

wide continuous valley south of the Chilean capital, Santiago. The coastal range and the valley get lower alongside the main Andes range and are drowned by the ocean south of Puerto Montt. On the Argentinean side of the Andes, the front ranges in the north are broken by deep, river-carved valleys. Their eastern margins have large alluvial fans formed by the deposition of rock material eroded from the mountains. These fans mark both sides of the Andes throughout Chile and Argentina.

Broad Plateaus

Broad plateaus and wide river valleys dominate Brazil's physical environment. Locally, relief is sharp near physical transition zones, as when one travels from the coast to the first plateau level or from one plateau level to the next. More generally, the traveler in Brazil is struck by small differences in the elevation over great distances.

The main relief features of Brazil consist of the ancient rocks of the Brazilian Highlands, which are topped in the southeast by layers of lava flows, and the similar ancient rocks of the Guiana Highlands on the Venezuelan border to the north. The Guianas have low coastal plains that were formed by the deposition of sediment brought to the Atlantic Ocean by the Amazon River and then moved westward along the coast by offshore currents. Inland, these countries rise to the Guiana Highlands plateau. In southern Argentina, the Patagonia Plateau is cut deeply by rivers draining eastward from the Andes.

Major River Basins

Three major river basins between the high mountains and lower plateaus dominate South America. Tributaries of the Orinoco River primarily drain the largest areas of lower land in

Colombia and Venezuela. The Amazon River tributaries flowing from the Andes are muddy "white water" rivers in contrast to the black rivers, which contain little sediment as they flow from the plateaus. The contrast is visible where the "black" Rio Negro joins the muddy Solimões just below Manaus in the center of the Amazon basin (Figure 10.14). The Amazon River is navigable well into Peru. In Manaus, Brazil, the Amazon River is 2,500 km (1,500 mi.) from the ocean, 15 km (10 mi.) wide, and over 50 m (160 ft.) deep.

In southern Brazil, rivers drain south to the Paraná–Paraguay River system. The southward-flowing rivers present great hydroelectricity potential in the waterfalls at lava plateau breaks. In Argentina and Uruguay, the pampas plains were formed by deposits of the Paraná–Paraguay River system and overlaid by fine windblown loess. These river and wind deposits blanket ancient rocks that occasionally pierce them, forming hilly areas and low ridges in the northern part of Uruguay and on the southern margins of the pampas.

Natural Vegetation

Rain forest is the natural vegetation where tropical rainfall is plentiful and distributed through the year, as in the Amazon River basin, along the northern Pacific and Central American coasts, and on some Caribbean islands (see the Point–Counterpoint: Tropical Forests and Deforestation box on page 450). Such vegetation spreads several thousand meters up the east-facing

FIGURE 10.14 Brazil: Amazon rain forest. A LANDSAT satellite view of the area around Manaus, Brazil, in July 1987 (approximately 150 km, or 100 mi., across). Unbroken tropical rain forest is shown as red. The wide, black river is the Rio Negro that contains little silt. The blue river is the Solimões branch of the upper Amazon, which brings large quantities of silt from its upper reaches in the Andes Mountains. Manaus is the light-colored area just west of the Negro–Solimões confluence, with radiating roads leading to and from it north and south of the river. Small white areas with shadows to the west and east are clouds. **Photo:** Image courtesy of Space Imaging, Thornton, CO, USA.

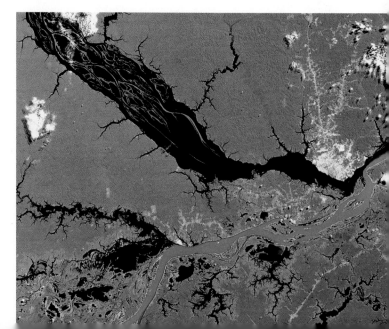

TROPICAL FORESTS AND DEFORESTATION

Deforestation, or the permanent clearing of forest vegetation, is a centuries-old land modification practice. Societies in all world regions and in both temperate and tropical latitudes used forest resources for fuel, shelter, and transportation. Dramatic increases in the permanent clearing of tropical forests, and especially tropical rain forests, in recent decades are causing significant alarm among a growing global body of scientists, medical researchers, government officials, and environmentalists. Global communities concerned with the potential ecological and human health detriment from tropical forest clearing call on the governments practicing or permitting tropical deforestation to cease. Locally, where deforestation is taking place, governments, business communities, and farmers angrily retreat from the global community and assert their sovereign rights to resource use.

The controversy surrounding the permanent clearing of tropical forests is a prime example of the antagonistic relations that may form when globalization forces and local traditions or preferences work against one another. Complicating the struggle between the global and local in tropical deforestation is that globalization forces are specifically one of the major causes of deforestation. Why is tropical deforestation such an emotionally charged issue that seems to pit environmental groups against international development organizations, developing country governments against developed country governments, and global corporations against grass-roots human rights advocates? The answer lies in attempting to understand the scientific significance tropical forest biomes play in terms of global ecology, global climate, and human biology in a politically charged and economically challenged environment.

Tropical rain forests exist in tropical latitude regions where high temperatures and high levels of humidity year-round produce a fairly consistent precipitation and vegetation pattern each month. The three most significant locations of tropical rain forest in the world are in Southeast Asia (see Chapter 6), Central Africa (see Chapter 9), and the largest, the Amazon River basin in South America (Box Figure 1).

Tropical rain forest is the dominant vegetation and climate regime in the Amazon River basin of Brazil and adjacent parts of its neighboring countries, including eastern Colombia, Ecuador, Peru, Bolivia, southern Venezuela, and the Guianas. The tropical rain forest of the Amazon basin is the largest in the world. Although other

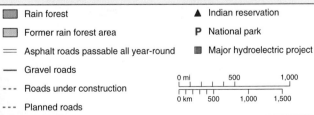

Box Figure 1 (a) Map of the Amazon River basin in Brazil and the extent of the tropical rain forest, (b) Aerial view of the tropical rain forest in Rondonia state, where the forest is cut into as farms are established along the roads. **Photo:** © Michael Bradshaw.

locations in Latin America contain similar vegetation and climatic conditions, including the Pacific coast of Colombia, along the eastern coasts of the Central American countries, and down the northeast coast of Brazil, the Amazon dominates in size and global impact.

Tropical forests contain the highest plant and animal species diversity per unit of land of any ecosystem or biome in the world. Seventy percent of the world's known plant and animal species reside in tropical forests, with hundreds of distinct species existing in relatively concentrated areas. Contemporary global health care depends on the species diversity of tropical rain forests. Existing treatments and promising cures for various forms of cancer come from tropical forest species. Numerous human health issues and aging conditions may be treated, cured, or slowed through medicines derived from the tropical rain forest plants. Medical research communities assert the need to explore the unknown potential hidden in the diversity of tropical rain forest vegetation. International pharmaceutical research, development, and sales generate hundreds of millions in revenue from products related to tropical rain forests. The potential benefit of hundreds to thousands of species present in the tropics has yet to be identified. Permanent clearing could eliminate countless medicinal cures and treatments yet undiscovered.

Trees absorb carbon dioxide from the atmosphere and return oxygen to the atmosphere. Tree respiration takes place in the troposphere, the lowest level of the Earth's atmosphere, where humans live and breathe. Although the scientific community is far from a complete understanding of the relationship between carbon dioxide and the extent of global warming, there is a general consensus that warming is taking place and carbon dioxide is a major culprit. Removing huge tracts of tropical forest eliminates a primary source of oxygen and a mechanism for removing the carbon dioxide at the same time. Burning forests, or drowning them under dammed waters, both common practices in Latin America, contributes to the release of carbon dioxide and methane, both considered culprits in global warming.

Tropical forests provide a number of local environmental stabilizers. Tropical forests hold moisture and release it into the local and downwind environment. They regulate stream flows and prevent downstream flooding. Tropical rain forests provide a source of water when global cycles bring drier periods to a region. Through the process of evapotranspiration, tropical rain forests can provide humidity and rainfall to the region they inhabit as well as far downwind.

Tropical soils are not nutrient rich like many temperate soil regions. The vegetation in the tropical forests stores the life-sustaining nutrients within its green cover, so that clearing the forest removes the local nutrient base and diminishes the potential for future species diversity. The vegetation keeps the local soil intact. In its absence, soil will wash through the watershed and out through the drainage system. The trees on poorer soils maintain their existence by circulating the nutrients without letting them enter the soil: when leaves fall, insects and fungi soon break them down so that roots near and above the surface can capture the chemicals again.

Government debt, crowded population densities in the east, millions of materially impoverished and jobless people, farmers with no land to farm, and a wealth of unexploited natural resources prompted the Brazilian government to promote the development of the Amazon Basin. The Brazilian government is hoping huge mining projects located in forested areas will bring in foreign capital and ease Brazil's massive foreign debt. The government claims sovereign rights to exploit its resources however it deems appropriate. Brazilian political leaders assert the need to develop resource potential fully in order to bring down an indebtedness structure that is so severe they are barely able to make the interest payments. Brazilian officials point to countries such as the United States, where unencumbered resource exploitation led to material wealth and prosperity. Many local and international environmental groups are calling for international reduction of Brazilian debt to help the country shift its development efforts away from the Amazon.

The Brazilian government claims it learned lessons from prior projects such as the Fordlândia experiment to produce rubber in the 1930s and the Jari development for wood pulp in the 1970s, which failed because of plant disease or inadequate soil surveys. However, critics believe that the government will repeat past mistakes and jeopardize the future of the Amazon Basin. Development in the vicinity of mining operations leads to an influx of people to provide services. As supporting industries grow, more people move to the region. As roads are cut into the forest, farmers and ranchers settle along both sides. Poor farming practices coupled with the nature of the soils and vegetation often lead to rapid nutrient depletion, and the farms are either sold to large-scale enterprises or abandoned for new land. As new roads are built, new settlers move in, existing farmers move to more productive land, and the rain forest vegetation slowly disappears.

The government claims such projects produce jobs that employ people who would otherwise have no work or income. People working the mines and providing services to the developing regions are grateful for the opportunities presented. Some farmers claiming land in the region assert their pride in land ownership and their ability to be self-employed. In places where better soils coincide with a good local knowledge and where farmers are not too indebted, there has been some success in growing commercial crops of beans and vegetables. In most parts, the cut areas gave way to cattle ranching, but few cattle can be supported per hectare, and the carrying capacity soon declines to uneconomic levels.

Settlement in the Amazon River basin has often been uncontrolled, without proper surveys or understanding of the rain forest ecosystem, and it often results in armed combat over land rights. The Brazilian government believes development of the region will lead to national stability and will diminish any border discrepancies neighboring countries may have over Amazonian lands. The Brazilian army remains a major influence on Amazonian development, but one that is seldom considered. Arising from the border wars that drained Brazilian manpower earlier in the 1900s, the military encouraged cutting of the rain forest for strategic reasons. When in power, military dictators instigated projects such as the Trans-Amazonica Highway and encouraged families to settle the area as part of the "great march westward." A network of military posts is still being built along the western border of Brazil, linked by ground and air transportation. Neighboring countries fear a loss of their remote lands in the Amazon if Brazilian development pushes too far to the west.

The future of the tropical rain forest in Latin America will be determined by the ability of local and global governments to understand the ecosystem better and cooperatively look for egalitarian uses that may serve the needs of countries like Brazil while sustaining the global ecological and human health value of the rain forest system. The following matrix provides some of the positions asserted by those urging tropical rain forest conservation and even preservation, and those who believe resource exploitation in the rain forest is in the best interest of the local and national communities. Read the conflicting views and consider how you might help to resolve this controversial, emotionally charged, and complex issue. Are there other factors for or against tropical rain forest resource use and deforestation you might include?

CONSERVE TROPICAL RAIN FOREST

Tropical rain forest (TRF) resources provide a sink for carbon dioxide. Burning TRF vegetation adds carbon dioxide to the Earth's atmosphere. TRF areas are a source of oxygen in the lowest levels of the Earth's atmosphere, where humans live and breathe.

There is tremendous biodiversity in the plant life present in TRF ecosystems.

Many medical treatments are derived from TRF products, and many disease cures come from TRF products, including current treatments and potential cures for cancer patients. Destruction may eliminate many undiscovered cures and treatments. TRF-derived pharmaceuticals earn billions internationally each year.

Governments permit the rapid clearing of TRF resources and sell them internationally, claiming rights to destroy domestic resources that impact the entire Earth. Yet the same governments may be corrupt and waste other resources and spend their cash foolishly.

Indigenous tribes and local people are displaced by TRF clearing. In some cases, bloody conflicts ensue while government officials turn a blind eye.

TRF resources provide an increasing tourism revenue potential. During the 1990s, travel to natural areas in the tropics was one of the fastest-growing components of the global travel industry (which is the world's largest industry). Although some governments assert their ability to balance resource clearing for export sales and development goals with conservation for tourism growth, few have shown a true commitment to achieving such a balance.

TRFs provide a natural habitat for species found only in this biome. Removing the TRF would eliminate habitat and cause permanent loss to global species diversity. Loss of species could alter the ecological balance of the Earth.

USE TROPICAL RAIN FOREST RESOURCES

There is incomplete carbon dioxide data for the Earth's atmosphere. Large portions of the Earth's surface are unreported. Ocean exchanges with the lowest levels of the Earth's atmosphere are more significant than TRF exchanges.

There is no conclusive evidence that partial TRF clearing will permanently change the biodiversity of the Earth as a whole.

Medical treatments come from many sources. Many treatments and cures may be synthetically generated in laboratories and do not require the use of naturally growing species from TRF.

Debt-ridden and impoverished countries need to and have the right to use their natural resources for their own best interest. The wealthier countries of the world obtained high material living standards by depleting much of their own and others' resources as they grew. Now those countries want to hold countries with TRF resource wealth back.

Growing countries need to push their frontiers and develop their resources. "Productive" members of society have a right to use land in a manner that will benefit them and their country.

Governments have the right to determine how they will earn revenue from their resources. Governments of TRF resource-wealthy countries assert their ability to balance resource depletion and extraction with conservation and replenishment.

A good source of income in a debt-challenged country with a large, materially impoverished segment in its population is far more important than the conservation of a bird or a tree.

---

slopes of the Andes. The world's largest expanse of tropical rain forest covers most of the Amazon River basin and extends along the eastern coastal lowlands of Brazil to the Tropic of Capricorn. Soils beneath the forest vary, but the areas of good soils are small, apart from the flooded areas close to sediment-carrying rivers.

Tropical grasslands or shrub vegetation communities are present where tropical rainfall is seasonal or significantly lower on average during the year than in rain forest areas. On the Brazilian Highlands, dense deciduous woodland gives way to more open woodland with increasing proportions of shrubs and grasses. Soils are generally poor beneath the natural vegetation and need treatment for agriculture, but some of the lava flows capping the plateau in southeastern Brazil produce easily worked soils.

Cold ocean currents, arid air, and the rain shadow from easterly winds produce deserts along the central west coast of South America. The Atacama Desert here is one of the driest places on Earth. Vegetation ranges from tropical plant species in the north to midlatitude types in the south.

In the southern part of South America, cooler midlatitude conditions coupled with varying levels of annual precipitation create a range from bare desert, through semiarid bunch grasses and drought-resistant plants, to tall grasses and forest in humid areas. Before European settlement, central Chile had natural vegetation of trees and shrubs that could grow in a regime of wet winters and dry summers. The plentiful precipitation of southern Chile supports natural vegetation of beech and pine forests. The pampas region of central Argentina and Uruguay is named after the tall, lush grasses that grew there before the region was plowed.

The majority of the high mountain ranges of Latin America are located within tropical latitudes. The latitudinal position of the Andes Mountains coupled with their very high altitude results in a series of vertical zones with distinctive climate and vegetation regimes. The greatest number of distinctive vegetative zones is at the equator. The **altitudinal zoning** in the region is directly linked to the variety of crops that may be cultivated (Figure 10.15). The lowest 1,000 m (3,000 ft.) have warm to hot conditions and tropical forest in the *tierra caliente*.

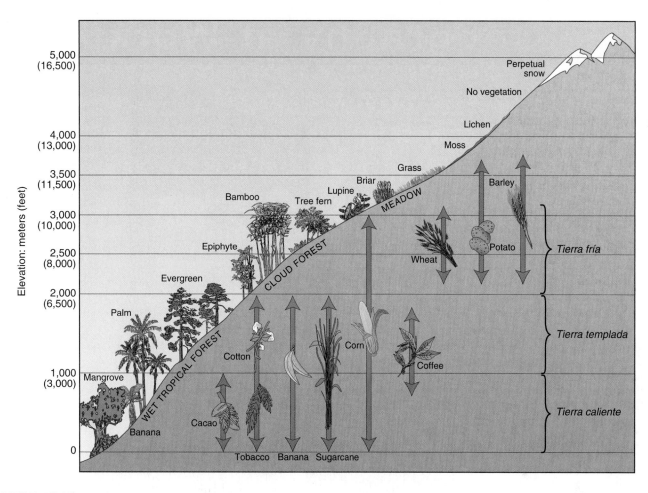

FIGURE 10.15 Northern Andes: altitudinal zoning of vegetation and crops. As they cross the equator, the Andes have maximum height and the greatest number of climatic environments from mountain foot to peak, from coastal mangroves and tropical rain forest to perpetual ice and snow cover. How do the crop zones relate to the vegetation zones?

The next 1,000 m (3,000 to 6,500 ft.) have mild to warm temperate conditions and deciduous forest in the *tierra templada.* Between 2,000 m and 3,000 m (6,500 to 10,000 ft.), the *tierra fría* has cold to mild temperature conditions and pine forests. Above this zone there is grassland, and at the greatest heights, in the *tierra helada,* snow cover is present all year, even on the equator. Climbers of the Andes near the equator pass through vegetation zones in a few kilometers that would require a sea-level trek from the equator to the poles. The similarities to sea-level climatic environments and vegetation zones are not complete, however, because temperatures in mountains near the equator vary little from month to month at all altitudes, while winds increase and the air gets thinner at higher altitudes.

Natural Resources

The natural resources of Latin America include minerals, soils, forests, water, and marine life. The Andes Mountains and the ancient plateau rocks contain considerable resources of metal

ores that were among the early attractions for European settlers. They are still not fully exploited where surface transportation is poor. Rivers flowing from the uplands deposited alluvial concentrations of heavy metal-ore minerals in the adjacent lowlands. Sedimentary rock basins between the mountains and in offshore areas became petroleum and natural gas reservoirs, especially in eastern Mexico, northern Venezuela, Colombia, the offshore areas of northeastern Brazil, and along the eastern slopes of the Andes in Ecuador, Peru, and Argentina.

The alluvial soils in parts of the Amazon River basin, the weathered lava surfaces in southern Brazil, and the pampas of Uruguay and central Argentina form large areas with predominantly good soils for farming. The soils on the eastern plateaus are generally low in plant nutrients, while the steep slopes of the Andes confine potential farming areas to small valleys, high basins, and plateaus.

The water resources of Latin America are huge, including the world's largest river (by volume), the Amazon, which carries more than twice the amount of water of the next largest

GEOGRAPHICAL RESEARCH ON RIVERS IN EASTERN MEXICO

Dr. Paul Hudson (Box Figure 1) is a physical geographer who specializes in fluvial geomorphology, the study of how flowing water interacts with the Earth's surface, particularly as it relates to erosion, flooding, and floodplains. Dr. Hudson's expertise focuses on large lowland coastal plain river systems, and most of his research occurs along the Gulf Coastal Plain. Hudson's current research focus is primarily in eastern Mexico on the Rio Panuco system, which drains portions of the Mexico's Altiplano, Sierra Madre Oriental, and Gulf Coastal Plain. Hudson's ongoing research projects include examining controls on floodplain evolution, river erosion, and flood processes. Mexico's Gulf Coastal Plain is a hot, humid region with extensive swamps and large river systems. However, because the basin is located just south of the Tropic of Cancer, hydrological processes differ from rivers located along the U.S. Gulf Coastal Plain. The Panuco Valley, like most of the world's large river systems, is an agricultural landscape, primarily for sugarcane, citrus, and cattle ranching. The location of land use types within the valley, however, is influenced by the floodplain geomorphology. Older floodplain deposits represent topographic and sedimentological controls on flooding and soils, which spatially constrain specific agricultural activities. The Panuco Valley also represents the heartland for the Huastec culture region, the northern extent of prehistoric complex culture in Mesoamerica. Unlike the Maya and Aztec, very little is known about the Huastec. However, because prehistoric Huastec resided primarily within the floodplain, archaeological materials are found within flood deposits. Thus fluvial geomorphology becomes very important for understanding prehistoric human–environment interaction, as well as the management and preservation of this cultural resource.

To understand hydrologic and sedimentological controls in the lower portions of the basin, it is important to understand land use in the basin headwaters, the eastern Sierra Madre Oriental. A project Dr. Hudson is conducting with one of his graduate students examines the impact of slash and burn (swidden) agriculture on soil erosion. This research is also important because the economy of small communities, many of which are *ejidos,* depends on farming. Soil erosion changes the land suitable for farming and thus directly influences the sustainability of their economic system, in addition to ecological recovery.

Box Figure 1 Dr. Paul Hudson with the University of Texas at Austin–Department of Geography and the Environment. Dr. Hudson (right) on location in the Panuco Valley, Mexico conducting field research.

Dr. Hudson's research projects involve state-of-the-art digital geographic technologies, such as GIS, remote sensing, and GPS. However, while these technologies are useful, the research activities are driven by field evidence. Field research activities frequently involve topographic surveying and sampling floodplain deposits, which are then analyzed in Dr. Hudson's geomorphology laboratory. Other samples may be shipped to laboratories for radiocarbon dating, which provide an age control on flood deposits. Dr. Hudson finds that the opportunity to share field research with students in the classroom is one of the most rewarding aspects of his profession.

river, the Congo in Africa. Many Latin American countries currently generate a large proportion of their power as hydroelectricity, and they continue to explore the expansion opportunities of this power source. Although some areas, such as northern Mexico, northeastern Brazil, the Peru–north Chile coastlands, and southern Argentina, are arid, much of the region has plentiful precipitation to support agriculture. For example, rivers flowing from melting snows on the Andes provide water to the arid coast of Peru and the oasis settlements of northern Argentina.

The combination of soil and water conditions in much of the region fostered the early development of domesticated crops. Pre-Columbian cultures in Latin America cultivated corn, potatoes, sweet potatoes, cassava manioc (cassava), and beans, among numerous others. Many of the crops cultivated in Latin America were not grown in other parts of the world at the time. The early globalization connections created during the colonial era facilitated the diffusion of crops from Latin America to other parts of the world, where they remain major components of agriculture today.

Marine life provides some of the world's richest fisheries along the western coast of Ecuador, Peru, and northern Chile. Other fishing grounds in the Caribbean and off southern Argentina are being developed. Fish production for North American markets is a growing source of income to many countries in Central America and the Caribbean. Such resources are vulnerable, however, to human overuse and natural phenomena such as El Niño.

Environmental Problems

Natural hazards such as earthquakes, volcanic eruptions, and hurricanes bring destruction and death to the Andean and Middle American mountain ranges and to the Caribbean islands. Many environmental problems, however, are related to European colonial patterns and subsequent decades of political instability and corruption. Two of the more serious environmental issues in Latin America are soils and air quality. Perhaps the most serious and globally controversial environmental issue centers on deforestation in the Amazon tropical rain forest.

Soil Erosion

Soil erosion is a major problem in many countries of Latin America, especially where increasing populations place additional pressure on subsistence lands (see Geography at Work on page 454). The primary subsistence crop is corn, which is commonly grown in rows up and down a slope, making it easy for intense rainstorms to wash soil away. As economic and population pressures force more subsistence farmers onto the hillsides, the elimination of the natural vegetation removes one of nature's mechanisms for holding soil in place on steep slopes.

Small Caribbean islands colonized for intensive commercial agriculture became notoriously prone to soil erosion and other forms of environmental degradation. The introduction of sugarcane and banana plantations dramatically altered the local environments. After more than 300 years of commercial agriculture, soil fertility has decreased to the point where large applications of fertilizer are necessary. The development of resort hotels and the construction or expansion of supporting infrastructure on many islands caused vegetation loss and accelerated soil erosion. Clearing natural vegetation on the more hilly islands of the Caribbean causes soil to wash rapidly down the hills and into the surrounding sea, resulting in the accumulation of sediment in the coral reef areas surrounding many of the islands. The **sedimentation** of a reef system may ultimately kill the coral. Fertilizers used to compensate for lost soil nutrients from rapid runoff also enter the sensitive coral reef system, further degrading the health of the reefs.

Air and Water Pollution

Air and water pollution results from mineral extraction and refining, and from the concentration of human activities in urban areas. Mexico City suffers more than any other metropolitan area in Latin America from air pollution. Mexico City is a very densely populated urban center situated in a bowl-shaped depression known as the Central Valley of Mexico. The city sits on the valley floor at approximately 2,200 m (7,000 ft.) above sea level. Mexico City and the valley floor are surrounded by much higher mountains, which virtually form walls around the urban expanse. The air in and just above the city is often trapped in the valley and becomes extremely stagnant. Vehicular exhaust, coupled with industrial and domestic pollutants, collects in the valley's stagnant air mass and settles near the valley floor where millions of people breathe. Naturally occurring **temperature inversions,** or periods during the winter months when cold, dense air remains "trapped" at the surface under warmer air for several days or even weeks, further complicate the air pollution phenomenon of Mexico City. Although migration and natural increase rates of population are now slower than they were a few decades ago in the Central Valley of Mexico, population numbers continue to challenge pollution reduction efforts.

Test Your Understanding
10B

Summary

Latin America's climatic environments are mainly tropical, ranging from the equatorial climates of the Amazon basin to seasonal rainy areas and the deserts of northern Mexico and the Peru–Chile coasts. Midlatitude climatic environments affect the southernmost countries.

Tectonic plate movements gave rise to the Andes Mountains and the complex pattern of mountain ranges and islands in Middle America and the Caribbean Basin. Brazil is dominated by the Guiana and Brazilian Highlands plateaus, which are separated from each other and the Andes by the Orinoco, Amazon, and Paraná–Paraguay River systems.

Latin America has resources of metal ores, petroleum, water, good soils, and major forests. It is subject to earthquakes, volcanic eruptions, hurricanes in the north, and storms in southern Chile. Human activities have degraded water, air, soil, and forest resources.

Questions to Think About

10B.1 What are the primary natural hazards that occur in Latin America? Which areas are prone to these hazards and why?

10B.2 What are the major mineral products of Latin America, and how are they distributed geographically?

10B.3 How do significant mountain barriers, such as the Andes, affect climate patterns and agricultural practices throughout the region?

Key Terms

Middle America	Altiplano
nortes	altitudinal zoning
El Niño phenomenon	sedimentation
rain shadow	temperature inversion

Global and Local Changes

The United States and Latin America

The United States staked its claim on Latin America by formulating the **Monroe Doctrine** in 1823, in which it asserted its role as the geopolitical leader of the Western Hemisphere. From the late 1800s, the United States intervened in the affairs of Cuba, Haiti, and the Dominican Republic on several occasions. Beginning in the early 1900s, the decline in the availability of British capital coupled with growing production in the United States led to a regional shift of economic dominance toward the United States. During the later 1900s, the impact of two world wars, the depression of the 1930s, and political and corporate intervention from the United States caused most Latin American countries to look northward for trade, political support, and elements of popular culture, including clothing fashion, motion pictures, music, and fast food. Miami, Florida, with its large Hispanic population, regional air and sea transportation networks, and growing banking and financial services sector, became a strong cultural link between the United States and Latin America in the second half of the 1900s.

Latin American connections with the United States strengthened in the 1990s. The acceptance by Mexico, the United States, and Canada of the North American Free Trade Agreement (see Chapter 11 Point–Counterpoint: North American Free Trade Agreement box on page 516) enhanced trade and business connections among them. By the end of the 1990s, a unique countertrend changed the dominant unidirectional flow of influence between the United States and Latin America. In the 1980s and 1990s, Latin American popular culture began to rival U.S. homegrown popular culture as the Latin-based segment of the U.S. immigrant population grew rapidly. Latin American influences in the restaurant industry thrived, and Latin American–influenced popular music topped the U.S. charts.

Financial Dependence

In the 1970s geopolitical events in the Middle East contributed to rapidly rising oil prices and diminished supply to world markets. Oil producers initially incurred increased revenues from the higher oil prices. They invested their revenues in European and U.S. banks. These **petrodollars** were urged on Latin American countries in the form of loans. Such loans were used to pay for oil imports in the nonproducing countries and major infrastructure projects in producing countries such as Brazil, Mexico, and Venezuela—and to improve some government officials' overseas bank accounts.

The recession in the more materially wealthy countries caused by the high oil prices led to weakened markets for products from Latin America and much higher interest rates on the loans. The combination of debts and falling export income caused many Latin American countries to default on debt payments by the mid-1980s, resulting in the 1980s distinction as the **Lost Decade.** The Brazilian government had to devote its large overseas trade balance, generated by import-substitution

industries, plus sales of mineral and farm products, to servicing its extensive international debts. Brazil's debt servicing reduced its ability to invest in domestic production, thereby dramatically diminishing economic progress. The high interest rates caused foreign investment and aid to dry up in the late 1980s, and debt liabilities forced countries to emerge from reliance on their internal markets.

Political Change and the Global Economy

The GNI PPP (Figure 10.16) and consumer goods ownership data (Figure 10.17) for Latin American countries reflect a wide range from economic giants such as Brazil to materially impoverished countries such as Honduras. The current global connections of many economies in the region are partially a legacy of past economic institutions and international relations and partially a response to 1990s changes in governmental approaches to participation in the world economy. Elected governments came to power everywhere (except Cuba), and emphasis shifted from inward-looking policies to the need for cooperation. The older policies that focused on government-run industry and the protection of domestic products through tariffs, quotas, and red tape, or **protectionism,** involved high levels of government intervention and highlighted the lack of capital available for internal investments. These policies gave way to those of structural adjustment, urged by the World Bank and the International Monetary Fund, encouraging less government spending, more foreign investment, and more exporting industries. The new policies also opened markets to foreign products and

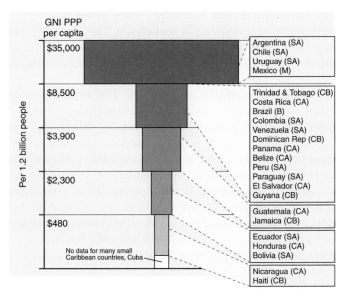

FIGURE 10.16 Latin America: country average incomes compared. The countries are listed in the order of their GNI PPP per capita. Subregions are indicated parenthetically. M=Mexico; CA=Central America; CB=Caribbean; NA=Northern Andes; B=Brazil; SA=South America. No data for many small Caribbean Basin countries or Cuba. **Source:** Data (for 2000) *World Development Indicators,* World Bank, 2002.

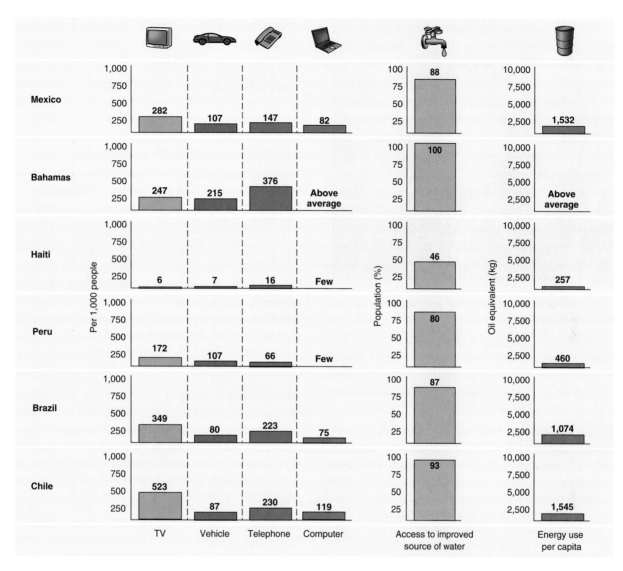

FIGURE 10.17 Latin America: ownership of consumer goods, access to clean water, and energy usage. Contrast conditions in materially wealthy, moderately wealthy, and very poor countries. **Source:** Data (for 2002) from *World Development Indicators,* World Bank, 2004.

privatized government corporations. Structural adjustment policies were partly forced on Latin American countries as a means of reducing debt burdens incurred during the 1970s and 1980s. As with other simplistic approaches to development, this economic restructuring that was superimposed on the Latin American regional culture was subject to potential disasters—as Mexico found in 1994–1995, when its economy opened too rapidly, sucking in imports and capital investments and creating a huge trade imbalance.

Global Cities

Although many cities in Latin America interact and trade on a global level, the region's two giants, Mexico City and São Paulo, are the dominant forces responsible for inserting the Latin American region into globally connected world systems

(Figure 10.18). Both Mexico City and São Paulo have populations of 18–19 million people. Extensive rural-to-urban migration, migration within the metropolitan areas, and high rates of natural increase created very dynamic urban populations for both cities, which challenge demographers in determining exactly how many people reside in each. Rapid economic development in both Brazil and Mexico led to dramatic growth in the largest cities of both countries. The decrease in nationally controlled industries with a corresponding increase in private investment opened a floodgate for the larger cities to establish connections with the industries, governments, and investors of cities, countries, and regions throughout the world.

Both Mexico City and São Paulo have diverse populations, with each housing people from all regions of their respective countries. The cities are representative of the different dialects,

FIGURE 10.18 Global connections and local voices in São Paulo, Brazil. Brazilian business personnel wait outside Brazil's World Trade Center in São Paulo after a bomb threat forced the evacuation of an adjacent shopping and office complex on September 25, 2001. The bomb threat was one of several that followed the devastating attacks on New York City and Washington, D.C., on September 11, 2001. **Photo:** © Reuters New Media/Corbis.

social customs, religious practices, and regional pride existing throughout their respective countries. The expanse of trade and interaction with multinational corporations from all world regions, as well as relationships with government officials from various countries of the world, furthers the diverse composition of people and customs present in each urban center. The size of each, the diversity of activity within, the international trade between each and other major world centers, and the attraction of the cities to business and leisure visitors from across Latin America and around the globe create environments unparalleled in the other metropolitan centers of Latin America.

Mexico City is the political, industrial, financial, and cultural capital of the country. São Paulo is the financial, industrial, and cultural capital of Brazil. Although Brasília is the political capital, São Paulo wields strong political power within the Brazilian system. Both Mexico City and São Paulo are media centers for their countries, and both serve as their countries' media connections to the world. Although each global city–region houses representatives of all subsets of each country, the residents of each take pride in being part of the most influential metropolitan center of their country.

Regional Links

The 1990s saw the rise and regeneration of regional trade groupings. In 1997 trade between Argentina, Brazil, Paraguay, and Uruguay (MERCOSUR) grew by 25 percent, while that in Central America and Andean South America grew by 10 percent. The Caribbean area awaited its request for parity with NAFTA, a proposal that did not progress through the U.S. Congress. The countries of Southern South America opened

talks with the European Union, their biggest trading partner and primary source of investment since 1996. An early 1990s proposal by U.S. President George H. W. Bush to establish a Free Trade of the Americas trade bloc by 2005 remained unrealized in the summer of 2005.

Population Pressures

The Spatial Distribution of People

The population of Mexico is concentrated in the central region from Guadalajara in the west through Mexico City to Veracruz in the east (Figure 10.19). This central plateau and the valleys cutting into it formed both an indigenous and a colonial hearth, and more recently became the center of industrial development and government functions focused on Mexico City. Northern cities, such as Monterrey, and cities along the U.S. border stand out in areas with sparse rural populations. Moderate densities of largely rural population occur on either side of the central urban belt. The areas with fewest people include the arid northwest and much of the Yucatán Peninsula.

In Central America, the main concentrations of people are in and around the largest cities—often in the highlands and closer to the west coast, where temperatures and soils are better for cultivation. Eastern coastal areas and parts of the narrow isthmus of Panama have fewer people in areas of dense forest cover and humid wetlands or mangrove marsh.

In the Caribbean Basin, Haiti and Puerto Rico have very high population densities, whereas other islands are more sparsely populated. Most islands have higher densities near capital cities or major port areas, as in Jamaica.

In countries of the northern Andes Mountains, population distribution reflects the Spanish pattern of colonial settlement. Some concentrations of people are in mountain valleys or on plateaus where midlatitude crops could be grown close to the equator. Other population clusters are on the coasts, having developed as port cities to link the home country of Spain with the interior sources of minerals. The interior lowlands have few people, reflecting the formidable barrier presented by the Andes chain.

The distribution of population in Brazil is a combination of both historical and contemporary regional development goals. The highest densities are in the southeast, around and inland of São Paulo and Rio de Janeiro. Moderate densities occur in a band parallel to the coast from the southeast around Pôrto Alegre to west of Fortaleza in the north. Farther inland, the very low densities of the Amazon rain forest area create a major geographic contrast within the country.

The main population centers in Southern South America are around the Río de la Plata estuary (Argentina, Uruguay) and in central Chile. Smaller centers occur in the irrigated farming oases of northern Argentina and around Asunción, capital of Paraguay.

Natural Increase and Migration Patterns

Mexico's population grew rapidly to around 70 million in 1980 and to 107 million in 2005 (Table 10.2). Although its fertility rates and birth rates are declining (Figure 10.20), Mexico's

FIGURE 10.19 **Latin America: distribution of population.** Explain the distribution of heavily and lightly populated parts of this region. The red lines identify the subregions. **Source:** Data from *New Oxford School Atlas,* Oxford University Press, UK.

population is projected to rise to nearly 130 million by 2025. The age–sex diagram (Figure 10.21a) shows a decline of births that was achieved through a well-developed family planning program, halving total fertility rates from about 6 in 1976 to under 3. The very low death rate and increasing life expectancy, however, maintain a gap between births and deaths that keeps the population totals increasing.

Although rural to urban migration passed its peak in Mexico by the late 1980s, migrants from rural areas continue to contribute to urban growth. Migration from rural areas to Mexico City and to cities and industrial regions in northern Mexico is accompanied by significant numbers of Mexicans emigrating out of the country and into the United States seeking employment opportunities and

a perceived better quality of life. Some migrants to the United States are legal, and others are not. Many Mexicans living in the United States support family members who remain in Mexico. The **remittance** of cash from Mexicans working in the United States to family members back home is a significant source of foreign exchange for Mexico.

The population of Central America is projected to grow by 60 percent in the next 30 years. The 2005 population of nearly 40 million people may grow to 60 million by 2025. Such growth will place increasing stress on the already pressured natural environment and the political, economic, and social systems of these countries. Very high fertility rates in the 1960s and 1970s—over 6 children per woman, apart from Costa Rica

Table 10.2 Latin American Subregions. Data by country, area, population, income (GNI PPP—Gross National Income Purchasing Power Parity), urbanization, Human Development Index (HDI), Human Poverty Index (HPI).

| Country | Capital City | Land Area (km²) | Population (millions) | | GNI PPP 2003 | | % Urban | Index 2002 | |
			Mid 2005 Total	Est. 2025 Total	Total (US$ billions)	Per Capita (US$)	2005	HDI (rank of 177)	HPI (% population)
Mexico and Central America									
United Mexican States	Mexico City	1,956,200	107.0	129.4	915	8,950	75	53	9.1
Belize	Belmopan	22,960	0.3	0.4	1	5,840	49	99	16.7
Costa Rica, Republic of	San José	51,000	4.3	5.6	36	9,040	59	45	4.4
El Salvador, Republic of	San Salvador	21,040	6.9	9.1	32	4,890	59	103	17.0
Guatemala, Republic of	Guatemala City	108,890	12.7	20.0	50	4,060	39	121	22.5
Honduras, Republic of	Tegucigalpa	112,090	7.2	10.7	18	2,580	47	115	16.6
Nicaragua, Republic of	Managua	130,000	5.8	8.3	13	2,400	59	118	18.3
Panama, Republic of	Panama City	75,520	3.2	4.2	19	6,310	62	61	7.7
The Caribbean Basin									
Antigua and Barbuda	St John's	440	0.1	0.1	1	9,590	37	55	No data
Bahamas, Commonwealth of	Nassau	13,880	0.3	0.4	5	16,140	89	51	No data
Barbados	Bridgetown	430	0.3	0.3	4	15,060	50	29	2.5
Cuba, Republic of	Havana	110,860	11.3	11.8	No data	No data	76	52	5.0
Dominica, Commonwealth of	Roseau	750	0.1	0.1	0.4	5,090	71	95	No data
Dominican Republic	Santo Domingo	48,730	8.9	11.0	54	6,210	64	98	13.7
French Guiana	Cayenne	91,000	0.2	0.3	No data	No data	75	No data	No data
Grenada	St. George's	340	0.1	0.1	0.7	6,710	39	93	No data
Guyana, Cooperative Republic of	Georgetown	214,970	0.8	0.7	3	3,950	36	104	12.9
Haiti, Republic of	Port-au-Prince	27,750	8.3	12.9	14	1,630	36	153	41.1
Jamaica	Kingston	10,990	2.7	3.0	10	3,790	52	79	9.2
St. Kitts and Nevis, Federation of	Basseterre	360	0.05	0.1	0.5	11,040	33	39	No data
Saint Lucia	Castries	620	0.2	0.2	1	5,220	30	71	No data
St. Vincent and the Grenadines	Kinstown	390	0.1	0.1	0.7	6,590	55	87	No data
Suriname, Republic of	Paramaribo	163,270	0.4	0.5	No data	No data	74	67	No data
Trinidad and Tobago, Republic of	Port-of-Spain	5,130	1.3	1.3	12	9,450	74	54	7.7

Country	Ethnic Group %	Languages O=Official N=National	Religion Percentages
United Mexican States	Mestizo 60%, Native American 30%, Euro 9%	Spanish (O), local, English	Roman Catholic (nominal) 89%, Protestant, other 11%
Belize	Mestizo 44%, Creole 30%, Native American 11%	English (O), Spanish, Maya	Roman Catholic 62%, Protestant 30%
Costa Rica, Republic of	Mestizo and Euro descent 95%	Spanish (O), English	Roman Catholic 90%
El Salvador, Republic of	Mestizo 90%, Native American 9%	Spanish (O), native tongues, English	Roman Catholic 75%, Protestant 20%
Guatemala, Republic of	Mestizo 56%, Native American 44%	Spanish (O), 20 Native American dialects	Roman Catholic 86%, Protestant, local 14%
Honduras, Republic of	Mestizo 90%, Native American 7%	Spanish (O), local dialects, Creole, English	Roman Catholic 94%, Protestant, other 6%
Nicaragua, Republic of	Mestizo 69%, white 17%, black 9%, Native American 5%	Spanish (O), local, English	Roman Catholic 85%, Protestant 10%
Panama, Republic of	Mestizo 70%, Native American, mixed black 14%, white 10%	Spanish (O), English, Creole, local	Roman Catholic 80%, Protestant 15%
Antigua and Barbuda	Black African 96%, white 3%	English (O), local dialects	Anglican Protestant (main), Roman Catholic, Muslim
Bahamas, Commonwealth of	Black African 85%, white 12%	English (O), Creole	Protestant 75% (Baptist 32%, Anglican 20%), Roman Catholic 19%
Barbados	Black African 90%, mixed 4%, white, other 6%	English (O)	Protestant 67%, Roman Catholic 4%
Cuba, Republic of	Mixed white/black 51%, white 37%, black 11%	Spanish (O)	Nonreligious 55%, Roman Catholic 40%
Dominica, Commonwealth of	Black, Carib natives	English (O), French patois	Roman Catholic 80%, Protestant 15%
Dominican Republic	Mixed 73%, white 16%, black 11%	Spanish (O), French creole, English	Roman Catholic 90%, local, Protestant, other 10%
French Guiana	Mixed white, Native American, black, Arawak natives	French (O), Creole	Roman Catholic, Hindu
Grenada	Black African, South Asian, white	English (O), French patois	Roman Catholic 60%, Anglican 14%, other Protestant 21%
Guyana, Cooperative Republic of	Asian Indian 51%, black and mixed 43%, Native American 4%	English (O), Hindi, Urdu, local dialects	Christian 57%, Hindu 33%, Muslim 9%
Haiti, Republic of	Black 95%, mixed and white 5%	French (O), Creole(O)	Roman Catholic 80%, Protestant 16%
Jamaica	Black African 76%, mixed black/white 15%, Asian Indian 3%	English (O), Creole	Protestant 55%, Hindu, Muslim, Jewish, other 39%
St. Kitts and Nevis, Federation of	Black African	English (O)	Anglican, other Protestant, Roman Catholic
Saint Lucia	Black African 90%, mixed 6%, South Asian 3%	English (O), French patois	Roman Catholic 80%, Anglicans and other Protestants
St. Vincent and the Grenadines	Black African 82%, mixed 14%, whites, South Asian, Carib native	English, French patois	Protestant (mostly Anglican) 76%, Roman Catholic 10%
Suriname, Republic of	South Asian 37%, Creole white/black 31%, Indonesian 15%, Maroon (escaped slaves) 10%	Dutch (O), English, others	Hindu 27%, Protestant 25%, Roman Catholic 23%, Muslim 20%
Trinidad and Tobago, Republic of	Black African 41%, South Asian 40%, mixed 17%	English (O), Hindi, French, Spanish	Roman Catholic 30%, Hindu 25%, Protestant 25%, Muslim 6%

Table 10.2 Latin American Subregions. Data by country, area, population, income (GNI PPP—Gross National Income Purchasing Power Parity), urbanization, Human Development Index (HDI), Human Poverty Index (HPI). *(continued)*

| Country | Capital City | Land Area (km²) | Population (millions) | | GNI PPP 2003 | | % Urban | Index 2002 | |
			Mid 2005 Total	Est. 2025 Total	Total (US$ billions)	Per Capita (US$)	2005	HDI (rank of 177)	HPI (% population)
Northern Andes and Brazil									
Bolivia, Republic of	La Paz	1,098,580	8.9	12.0	22	2,450	63	114	14.4
Brazil, Federative Republic of	Brasília	8,511,970	184.2	228.9	1,322	7,480	81	72	11.8
Colombia, Republic of	Bogotá	1,138,910	46.0	58.3	290	6,520	75	73	8.1
Ecuador, Republic of	Quito	283,560	13.0	17.5	45	3,440	61	100	12.0
Peru, Republic of	Lima	1,285,000	27.9	35.7	138	5,090	73	85	13.2
Venezuela, Republic of	Caracas	912,050	26.7	35.4	121	4,740	87	68	8.5
Southern South America									
Argentine Republic	Buenos Aires	2,766,890	38.6	46.4	419	10,920	89	34	No data
Chile, Republic of	Santiago	756,950	16.1	19.1	155	9,810	87	43	4.1
Paraguay, Republic of	Asunción	406,750	6.2	8.6	27	4,740	54	89	10.6
Uruguay, Eastern Republic of	Montevideo	177,410	3.4	3.8	27	7,980	93	46	3.6

Source: World Population Data Sheet, Population Reference Bureau (2005), World Development Indicators, World Bank (2005), Human Development Report, United Nations (2005), and Encarta Microsoft (2005).

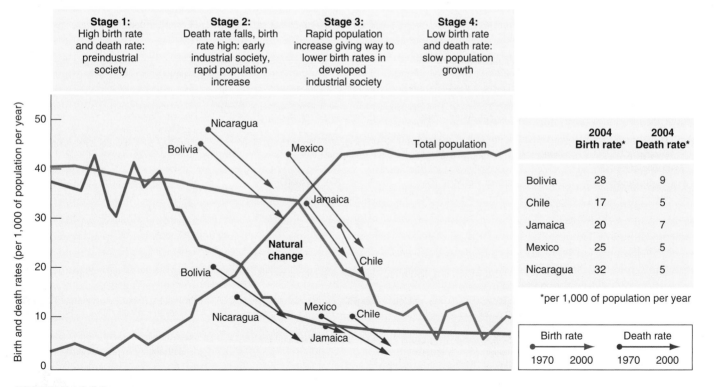

FIGURE 10.20 Latin America: demographic transition. How does the process vary among Latin American countries? **Source:** *2004 Population Data Sheet,* Population Reference Bureau.

Country	Ethnic Group %	Languages O=Official N=National	Religion Percentages
Bolivia, Republic of	Quechua 30%, mestizo 25–30%, Aymara 25%, white 5–15%	Spanish (O), Quechua (O), Aymara (O)	Roman Catholic 88%, Protestant 12%
Brazil, Federative Republic of	White 55%, black African 11%, mixed white/black 32%	Portuguese (O), Spanish, English	Roman Catholic 90%, other Christian 6%
Colombia, Republic of	Mestizo 58%, white 20%, mulatto (white/black) 14%	Spanish (O)	Roman Catholic 95%
Ecuador, Republic of	Mestizo 55%, Native American 25%, white 10%, black 10%	Spanish (O), Native American	Roman Catholic 95%
Peru, Republic of	Native American 45%, mestizo 37%, white 15%	Spanish (O), Quechua (O), Aymara (O), English	Roman Catholic 90%, Protestant and other 10%
Venezuela, Republic of	Mestizo 67%, white 21%, black 10%, Native American 2%	Spanish (O), English, local dialects	Roman Catholic 90%, Protestant and other 10%
Argentine Republic	Descendants European 85%, mestizo, Native American, other 15%	Spanish (O), English, Italian	Roman Catholic 90%, Protestant 2%, Jewish 2%
Chile, Republic of	European and mestizo 95%, Native American 3%	Spanish (O)	Roman Catholic 77%, Protestant 11%
Paraguay, Republic of	Mestizo 95%	Spanish (O), Guarani (O), Portuguese	Roman Catholic 90%, Protestant 10%
Uruguay, Eastern Republic of	European descent 88%, mestizo 8%, black 4%	Spanish (O)	Roman Catholic 77%, Protestant 2%

(fewer than 5)—continue to impact the population growth within the subregion. By 2005, fertility rates significantly decreased for the majority of the countries of Central America, but they remain high relative to those in the United States, Canada, and most European countries. In 2005 rates of natural increase remained between 2 and 3 percent, except in Costa Rica and Belize, where it was 1.3 percent. The age–sex diagrams for Central American countries are similar to those of other materially poor countries throughout the world, with large numbers of young people and small percentages of late middle-aged and elderly groups.

The total population of the Caribbean Basin more than doubled from 17 million to 35 million people from 1965 to 2005. Population pressure is already severe in many islands, with densities of over 600 people per km^2 (1,500 per mi.^{2}) in Barbados and over 200 in many of the other islands. Only in the Guianas does population density fall below 5 per km^2. Jobs available are fewer than the labor force, and there is high unemployment and continuing out-migration.

Population growth rates vary, although the regional averages fell in the late 1900s. In 2005 most Caribbean countries had natural increase rates of around 1 percent. Although total fertility ranged between 1.5 and 3.0 (Haiti's rate was 4.7), out-migration lowered the overall rates of increase. Jamaica's population trends are compared with other Latin American

countries in Figure 10.20. Other factors influencing population changes include high growth rates in islands such as St. Lucia and St. Vincent, which give rise to large proportions of young people (44 percent under 15 years in these islands), declining fertility rates partly balanced by reduced infant mortality, and increasing life expectancy to over 70 years. Only Haitians had a 2005 life expectancy as low as 52 years; the other countries were over 70 years. Jamaica's age–sex diagram (Figure 10.21b) demonstrates the effects of losing population as migrants moved to jobs in the United Kingdom and United States.

The population of the Caribbean Basin is very mobile. In-migration dominated until 1900, when intraregional movement and out-migration became significant. In the early 1900s, many inhabitants of the older British West Indian island colonies moved to new plantations opening in Trinidad and British Guiana, followed by Jamaicans and Haitians moving to Cuba and the Dominican Republic. The cutting of the Panama Canal attracted workers from Caribbean islands such as Barbados. Further movements were toward banana plantations established in eastern Central America, U.S. military bases, and new oil-producing centers in Trinidad and Venezuela.

After World War II, Caribbean residents from British, French, and Dutch colonies were attracted to job openings in Europe. The United States and Canada also became recipients of people on the move. Cubans and Puerto Ricans comprise a

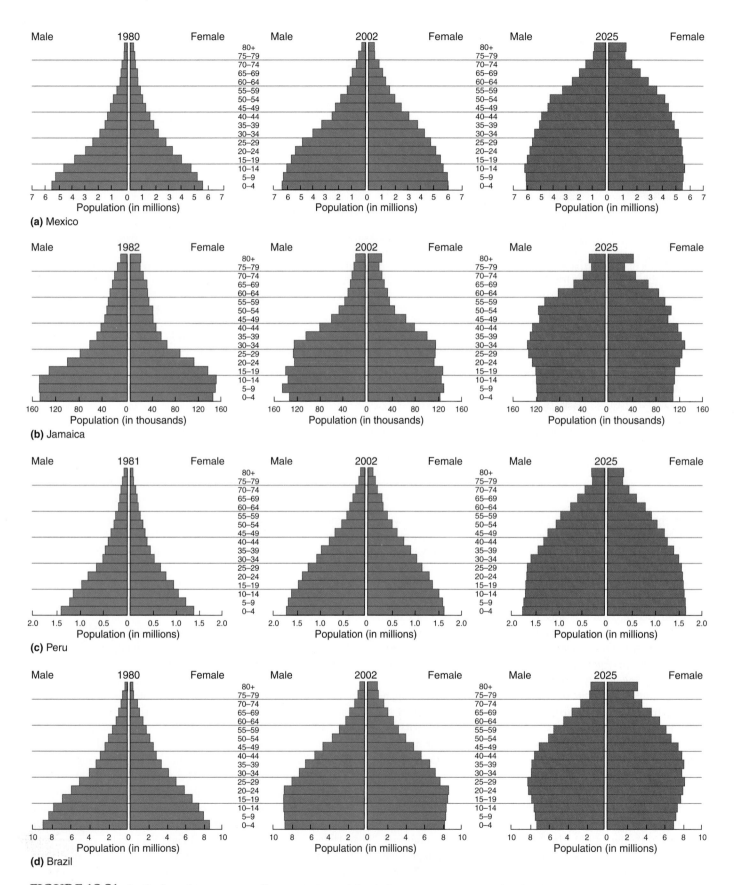

(a) Mexico

(b) Jamaica

(c) Peru

(d) Brazil

FIGURE 10.21   **Latin America: age-sex diagrams.** Source (all): U.S. Census Bureau, International Data Bank.

significant majority of the more than 5 million people of Caribbean origin living in the United States today. The pattern of out-migration eases some of the extensive population pressure on the Caribbean islands and provides a considerable source of foreign cash through remittances, but the exodus often takes away the most active and able young people, leaving behind dependent populations of the very young and the elderly. Although migration to Europe slowed because of immigration restrictions imposed in the 1980s, people from the Caribbean continue to migrate to the United States.

The decline of agriculture as an employer on many Caribbean islands contributed to migration to the largest cities of each country. The push factor of poor rural job prospects was complemented by the pull factors of perceived better employment prospects, social service delivery, and amenities in the towns. Yet most cities were unable to offer the burgeoning population adequate employment opportunities and housing. Congestion, shantytowns, pollution, and crime resulted in many urban centers of the Caribbean.

In the 1960s and 1970s, the Northern Andean countries had annual rates of population increase that were among the highest in the world at 3 to 4 percent. In the 1980s and 1990s, total fertility rates fell from over 5 to around 3 by 2005, except in Bolivia, where they were 3.8. This subregion lags behind others in South America in bringing birth rates and death rates closer together and continues to have annual rates of natural population increase around 2 percent. The total population of 121 million people in 2005 may grow to 160 million by 2025. The large number of young people shown in the age–sex diagram for Peru points to the likelihood of continuing population growth (Figure 10.21c).

Brazil's population is moving out of a period of rapid growth, which produced the relatively young population of today (Figure 10.21d) and a population total that rose rapidly to the 2005 figure of 184 million people (see Table 10.2). Total fertility dropped from nearly 6 to 2.4 between 1970 and 2001, while infant mortality went from 116 to 35 per 1,000 live births between 1960 and 2001. The rate of natural increase fell from 2.4 to 1.4 percent between 1970 and 2005. The fall in birth and fertility rates is linked to rising education levels, increased urbanization, and the impact of the media. Although Brazil is the world's largest Roman Catholic country and that church in Brazil has a conservative hierarchy, its local priests are often liberal on birth control matters. Moreover, a high proportion of women choose to give birth by a cesarean operation followed by sterilization, a service available free from the government. Recent studies suggest a significant impact from television on family size choices for many Brazilians. Television programs depicting happy and stable situations for small families seem to influence Brazilian family planning choices. Life expectancy in Brazil remains similar to most other South American countries at 68 years.

The population of Southern South America is growing more slowly than in the other subregions of Latin America. In 1930 this subregion had nearly 20 percent of Latin America's total population, but it now has less than 12 percent. The 63 million people who lived there in 2005 could rise to around 80 million by 2025 (see Table 10.1). Rates of population growth and fertility are low (see Figure 10.20), apart from Paraguay, where the natural increase rate was 1.7 percent and the total fertility rate was 2.9. Argentina, Chile, and Uruguay had lower natural increase rates around 1 percent, not very different from those in the 1970s. Total fertility rates in 2005 were under 2.6.

In all these countries, immigration from Europe since the late 1800s was of great significance to population growth. Chile grew in population from colonial times, but the growth of commercial farming and mining in the late 1880s led to more rapid increases, including some 200,000 German immigrants between 1881 and 1930. Growth was slower at first in Argentina and Uruguay, but immigration became more important after independence. Between 1821 and 1932, 6.4 million people emigrated to Argentina from southern and eastern European countries, and another million to Uruguay. Paraguay received few immigrants but lost around 60 percent of its population in its wars with Brazil, Argentina, and Uruguay in the 1860s and sustained further losses in the Chaco War with Bolivia in the 1930s. After such losses, Paraguay also encouraged migration—with only moderate success.

Regional Urban Geography

Mexico's population is highly urban. From 1970 to 2005, estimates of the proportion of the Mexican population in urban areas increased from 59 to 75 percent (see Table 10.2). Mexico City grew from a population of 500,000 in 1900 (2.5 percent of Mexico's population) to more than 18.7 million in 2003 (slightly more than one-fifth of Mexico's total population). Mexico City is a classic example of a primate city (Table 10.3). In Mexico, the stimulus for movement of people to cities is the increasing growth of urban-based manufacturing and service jobs. The official rates of unemployment, however, remain high, and many jobs are in the informal sector.

Total urban percentages are lower in Central America than in Mexico, remaining under 40 percent in Guatemala, under 50 percent in Belize and Honduras, and under 60 percent in Costa Rica, El Salvador, and Nicaragua. Most countries have a single major, or primate, city with populations that are several times that of the second largest city. Central America's largest cities are the political capitals of each country.

Although Caribbean countries are smaller in area and total population than most of Latin America's mainland countries, the size of some of the Caribbean's major cities is very impressive, with primacy existing on several islands.

The Northern Andean countries vary in their levels of urbanization. Venezuela was 87 percent urban in 2005, while Colombia and Peru were over 70 percent (see Table 10.2). Ecuador and Bolivia, however, had lower urban percentages of 61 and 63 percent. All these figures show increases of 10 to 20 percent since 1970. Peru and Bolivia each have a single primate metropolitan center. Peru's primate urban expanse is the Lima–Callao metropolitan complex, where more than one-third of the

Table 10.3   Large Urban Centers in Latin America

City, Country	Population 2003	Projected Population 2015	City Role
Mexico City, Mexico	18.7 million	20.6 million	Capital
Guadalajara, Mexico	3.8 million	4.3 million	Central Belt city
Monterrey, Mexico	3.4 million	3.9 million	Northern interior city
Port-au-Prince, Haiti	2 million	2.8 million	Capital, port
Havana, Cuba	2.2 million	2.2 million	Capital
San Juan, Puerto Rico	2.3 million	2.4 million	Capital
Santo Domingo, Dominican Republic	1.9 million	2.2 million	Capital
Lima–Callao, Peru	7.9 million	9.4 million	Capital and port
Guayaquil, Ecuador	2.3 million	3 million	Port
Caracas, Venezuela	3.2 million	3.6 million	Capital
Maracaibo, Venezuela	2.1 million	2.6 million	Port
Valencia, Venezuela	2.2 million	3 million	Port
Bogotá, Colombia	7.3 million	8.9 million	Capital
Medellin, Colombia	3.1 million	3.8 million	Interior center
Cali, Colombia	2.5 million	3.1 million	Interior center
Baranquilla, Colombia	1.8 million	2.3 million	Port
São Paulo, Brazil	17.9 million	20 million	Interior center
Rio de Janeiro, Brazil	11.2 million	12.4 million	Port
Bello Horizonte, Brazil	5 million	6.3 million	Interior center
Recife, Brazil	3.4 million	4 million	Port
Campinas, Brazil	2.5 million	3.2 million	Inland center
Salvador, Brazil	3.2 million	3.9 million	Port
Curitiba, Brazil	2.7 million	3.5 million	Interior center in south
Fortaleza, Brazil	3.1 million	4.3 million	Northern port
Pôrto Allegre	3.7 million	4.2 million	Southern port
Brasilia, Brazil	3.1 million	4.3 million	Capital
Buenos Aires, Argentina	13 million	14.6 million	Capital, port
Santiago, Chile	5.5 million	6.3 million	Capital
Asunción, Paraguay	1.6 million	2.3 million	Capital

Source: United Nations Urban Agglomerations 2003 (2003).

country's population resides. This urban complex extends along the Rimac River valley from Lima to the port of Callao. In Bolivia, metropolitan La Paz has 20 percent of the country's total population.

Leading urban centers in Venezuela include Caracas, Maracaibo, and Valencia. Recent economic growth in the Pacific coast city of Guayaquil, Ecuador, positions this metropolitan region ahead of the established highland capital of Quito. Colombia has a number of major cities that grew up in once-isolated areas and developed in different ways. Bogotá, the capital and largest city, has extensive manufacturing industries as well as the national government bureaucracies. Medellín is another major manufacturing center for Colombia, and Cali is linked to one of Colombia's busiest port areas. Both Cali and Barranquilla are rapidly growing urban centers.

All the major urban centers of the Northern Andes subregion grew dramatically when rural-to-urban migration, beginning in earnest in the 1950s, sparked the rapid establishment of shantytowns. In Lima–Callao, rapid population growth outpaced any attempts to develop urban structural amenities, and the prevalence of uncontrolled street vendors led to the closing of many established shops, loss of local taxes, and deterioration of much of the city's infrastructure. In the 1970s it was hoped that the shantytowns might provide a transition to more organized urban living, but such hopes were largely unfulfilled. The impacts of the Marxist Shining Path guerrilla war in the 1980s and early 1990s, which killed approximately 90,000 Peruvians and caused large-scale abandonment of control in the Andean provinces south of Lima and increased migration into the capital's shantytowns, all complicated attempts to organize urbanization better.

Brazil has two of the three largest metropolitan areas in Latin America and a highly urban population overall. The proportion of the population living in towns or urban settings increased from 56 percent in 1970 to 81 percent in 2005 (see Table 10.2). São Paulo is one of the largest metropolitan areas in the world with 18 million people in 2003 (see Table 10.3). It started in 1522 as an inland mission station and was a base for exploration of the interior by paramilitary bands in the 1600s. Later it became the focus of Brazil's world-dominating coffee industry, and railroads were built westward from it to open up new lands. In the 1900s it developed into the major manufacturing center of Brazil. The state of São Paulo produces nearly half of Brazil's GDP and two-thirds of its manufacturing output. In the late 1990s São Paulo's population growth slowed (from 5 percent to 0.5 percent per year). New industries tended to be sited elsewhere, often farther out in São Paulo state to avoid the high costs of traffic congestion, pollution, and strong union control (high wages and restrictive working conditions). The provision of housing, schools, and health facilities could not keep up with the metropolitan area's growth.

Brazil's second largest metropolitan center is the Rio de Janeiro complex. Rio de Janeiro first grew as a port city for early interior gold mine development, becoming Brazil's largest port and capital city in 1763. Today Rio de Janeiro is a contemporary culture hearth for Brazilian nationals and a major draw for international visitors to the country.

Other major cities include Belo Horizonte, the early center of mining and now the focus of a series of growing mining and metal-manufacturing towns, Recife and Salvador on the northeast coast, Fortaleza on the north coast, and Curitiba and Pôrto Alegre in the south. Manaus, in the center of the Amazon River basin, has over half the population of its state, Amazonas.

Brasília, the city built in the 1950s and established as the Brazilian capital in 1960, now has 800,000 people living in its planned central city area and 2 million included in the surrounding satellite cities (see Figure 2.27).

Southern South America is the most urban subregion in Latin America. Except for Paraguay (54 percent urban in 2005), all the countries of Southern South America had over 85 percent of their populations living in urban environments. As in many countries of Latin America, each country has a primate metropolitan center. Buenos Aires grew from a population of 170,000 in 1870 to around 13 million people, one-third of the Argentinean total, in 2003. Montevideo, Uruguay, Santiago, Chile, and Asunción, Paraguay all contain a significant percentage of their respective country's total population. These cities are the centers of government, manufacturing, and service industries. Few other cities in any of the four countries approach the size or important role of the national primate cities.

Ethnocultural Diversity: Mexico and Central America

Mexico's population is dominated by three groups: Native Americans (around 30 percent), Europeans (9 percent), and people of mixed Native American and European ancestry known as *mestizos* (60 percent; see Table 10.2). The Native Americans are particularly numerous in the southern province of Chiapas, where language dialects of Maya origin are commonly spoken in preference to Spanish, the official language. Almost 90 percent of Mexicans are nominally Roman Catholic, although some personal links to the Catholic church are eroding as established lifestyles are disrupted by moves to urban centers.

The populations of Central American countries comprise racial and ethnic groups that became social classes. Native American, or indigenous peoples, remain in large numbers in western Guatemala (Figure 10.22) and comprise significant minorities in the other countries of the subregion. Many indigenous groups in Guatemala and other Central American locations do not speak Spanish as their first language. Differences

FIGURE 10.22 The peoples of Central America. Indigenous people comprise a significant portion of the population of Guatemala. Patterns and colors present in clothing signify the home village or region for many native Guatemalans. **Photo:** © Bill Gentile/Corbis.

in language, social customs, and livelihood marginalize such communities from those with political power and economic wealth. Costa Rica has the smallest indigenous population of the subregion.

Along the east coast of Nicaragua and the northern and eastern coast of Honduras are many communities of English-speaking Afro-Caribbean peoples, mixtures of Miskito Indians and blacks from Jamaica and other Caribbean islands, and even pockets of English-speaking whites. The Spanish largely ignored coastal activity in this region during the colonial era, which opened the coasts to decades of influence and control by the British. The British forced the migration of blacks from the Caribbean to work primarily in agricultural production. The historical combination of African and Caribbean cultural influence coupled with the imprint of British rule created a very distinctive cultural landscape that remains strong today. Belize also has English-speaking Afro-Caribbean peoples.

European descendants and mestizos hold the political power and economic wealth of Central America. Land tenure was the basis of power and wealth for many years, but during the later 1900s, power passed to the owners of manufacturing industries and financial and legal services in the modern economy—and sometimes to the military. Smaller groups of people within Central American countries include some of African origin who came after escaping slavery in the Caribbean, central and southern Europeans who migrated to Central America between or after the two world wars, and Asians, Jamaicans, and Barbadians who came to work on construction projects such as the Panama Canal or, more recently, to establish factories in Costa Rica or Panama.

Ethno-Cultural Diversity: Caribbean

Hardly any of the indigenous population of the Caribbean Basin survived the early days of European settlement, succumbing especially to the diseases brought across the Atlantic. Although a few Caribs live on Dominica and St. Vincent and some Arawaks on Aruba and in French Guiana, the vast majority of the regional population traces its origin to the colonial occupation. Residents of the Caribbean speak a variety of languages, from French and English to regional hybrids of Patois and Papiamento. Religious practices are equally diverse. Some religious ties to former colonial rule are strong and reflect in a somewhat pure form the religious practices of various European peoples. Many religious beliefs in the region today, however, evolved from mixtures of practices brought from the African continent with religious ideology imposed by the Europeans.

The former Spanish colonies have relatively high proportions of people with European origins. Eighty percent of the population in Puerto Rico counts itself white, as does over 30 percent in Cuba. In the former British, French, and Dutch colonies, this figure is less than 5 percent, and the dominant majority is of African origin. Jamaica, Trinidad and Tobago, Guyana, and Surinam have substantial Asian populations resulting from British and Dutch shipments of workers from

South Asia to replace former slaves after slavery was abolished. Many racial mixtures enhance the cosmopolitan nature of West Indian populations.

Ethno-Cultural Diversity: South America

The Andean countries contain racial and ethnic contrasts that have important economic, social, and political effects. In Colombia and Venezuela, the Native American groups live primarily in highland and interior areas that were not taken over by Hispanic or mestizo groups. In Ecuador, Peru, and Bolivia, indigenous peoples tend to be concentrated in the highlands. Smaller numbers of different Native American tribes occur in the Amazon River lowlands, totally isolated from European intrusions until the 1950s. Current settlement for European descendants and mestizo groups reflects colonial occupations of upland basins and coastal locations that provided overseas links to Spain. Workers from Japan, China, and Africa immigrated to work in the coastal irrigated farmlands of Peru and established neighborhoods near their worksites.

The economic and social differences between European-origin and Native American peoples continue to raise tensions within the countries of the Northern Andes. Although formal attempts to modernize some Native Americans were made, many native groups prefer to retain their traditional cultural values instead of adopting those of the European descendants.

Although the large Brazilian population is extremely varied ethnically, the citizens of the country maintain a strong sense of national pride. The original Native American population is much reduced: in the Amazon River basin, it was probably around 3.5 million in A.D. 1500 but is now closer to 200,000. The traditional culture and people's health are increasingly threatened as outsiders invade their territories.

People of Portuguese and other European heritage make up a major proportion of the present Brazilian population, with highest concentrations in the southern states and elite areas of the cities. The descendants of African slaves form a significant minority proportion of the population along the northeastern coast and have spread into the southern cities. Brazilians of African descent have had a major impact on Brazilian art, food, music, and dance. Immigrants over the last 30 years include large numbers of Japanese, now accepted as a part of the Brazilian people and taking many of the top jobs in business and politics. Japanese Brazilians form the largest Japanese community living outside of Japan. Mixtures of people of African, European, and Native American descent comprise a sizable and growing portion of the Brazilian population (Figure 10.23).

The proportions of immigrant population give different emphases to the racial and ethnic mixes of the Southern South American countries. In Chile around 40 percent of the population claims to be of European heritage, with the remainder considered mestizo. In Paraguay virtually all the people are mestizos apart from a few people of European or

FIGURE 10.23 Brazil: the people. Brazilians watch a World Cup soccer match between their country and the United States on TV. The crowd includes a variety of peoples with origins in Africa, Europe, the Americas, together with those of mixed ancestral heritage. **Photo:** © John Maier, Jr./The Image Works.

African origin. Argentina and Uruguay have the largest proportions of Europeans and the smallest proportions of mixed populations.

The Subregions

Latin America may be divided into several subregions. The most geographically obvious division first is between Middle America and South America. The Middle American subregions consist of Mexico, the seven Central American countries from Guatemala and Belize south through the isthmus to Panama, and most of the islands in and surrounding the Caribbean Sea. The South America component consists of continental South America and some of the islands just offshore in the southern Caribbean.

- *Mexico* dominates Middle America in population, land area, and aggregate economic activity and will be examined as an individual subregion.

- The second subregion consists of *the seven Central American countries* that share a common history, related cultural development patterns, and similar physical geographies.

- Although Central American countries and some of the Caribbean island countries share some similarities, the differences are significant enough to merit examining *the Caribbean area* as a distinct subregion.

- The countries of the *Northern Andes* subregion include Colombia, Bolivia, Ecuador, Peru, and Venezuela.

- *Brazil* encompasses nearly half the area and population of South America, supporting the need to examine Brazil as an individual subregion.

- Paraguay, Uruguay, Chile, and Argentina comprise the countries of the final subregion, *Southern South America*.

Mexico

Regional Icon

The combination of Mayan and Aztec ancestry, Spanish colonial influences, and, more recently, interaction with the United States has created a distinctive Mexican culture that diffused into surrounding areas of the southwestern United States and Central America. The human landscape of Mexico's cities exhibits a blend of modern office and residential towers, mixed with historic streets, plazas, and churches reflecting the Spanish colonial influence on the subregion (Figure 10.24). Many public and government buildings are adorned with beautiful murals painted on the exterior walls facing garden courtyards. Other nationalistic trends are expressed in music, motion pictures, and literature. Many Spanish-speaking countries in the Caribbean and Central America look to Mexico for elements of popular culture. Mexican soap operas (*telenovelas*) are commonly viewed in places like Honduras and Puerto Rico. Mexican food, based on corn tortillas, local vegetables, and chilies, is one of the most distinctive national cuisines.

The burgeoning population of Mexico reached 107 million in the year 2005, making Mexico the most populous country in Middle America and the second largest after Brazil in all of Latin America (see Table 10.2). Mexico's economy is the largest in Middle America due in part to its physical size, industrial development history, and vast oil reserves. Although Mexico serves as a regional leader in Middle America, the government continues to look northward for economic growth prospects. The population growth and economic size of Mexico result in an increasing amount of global attention focused on its role in the larger Latin American sphere.

Political Changes

Mexican independence from Spain came in 1821 through a series of rebellions. Political instability resulted in numerous power struggles and dozens of government turnovers in the decades after the end of Spanish rule. Stability came in 1871 with the harshly repressive dictatorship of Porfirio Díaz at the cost of many social freedoms. Oppression during this period, referred as the **Porfirioto,** planted the seeds for revolution that would shape Mexico's political culture for most of the 1900s. Revolutionaries who overthrew the Porfirian government in 1911 created a new land tenure system with the purpose of easing rural poverty. The plots of land created in 1917 became known as *ejidos,* which were state-owned rural cooperatives developed to provide landless peasants with a degree of control over the land they farmed. The government granted the right to use land to individual *ejido* farmers, or *ejiditarios,* as well as the theoretical right to pass land use to their offspring.

FIGURE 10.24 Colonial architecture. The Spanish colonial imprint is visible throughout the cultural landscape of Mexico. Colonial plazas and churches, as here in Guanajuato, dominate contemporary urban settings. **Photo:** © R-F/Corbis.

However, ownership of the land remained with the government. The *ejido* system resulted in the fragmentation of landholdings as the government took control away from some landowners and gave it to *ejiditarios*. In many cases the government placed already impoverished farmers on marginal land, exacerbating their situation. Although rights to individual property ownership were defined more clearly in the 1990s, many rural or materially poor farmers continued to struggle while large-scale agribusiness took control of more land.

Another long-lived institution arising from the Mexican revolutionary period from 1910 to 1920 is the Institutional Revolutionary Party (*Partido Revolucionario Institucional*, **PRI**). The PRI grew out of the revolutionary movement into the most powerful and lasting political party in the history of Mexico, dominating Mexican politics for more than 70 years. The party controlled the government, nationalized industries, and allegedly fixed many political votes to remain in power. In July 2000 a historic election overturned the PRI's dominance of Mexican politics with the election of Vicente Fox, who represented the National Action Party (*Partido de Acción Nacional*, **PAN**). Fox and the PAN inherited a vastly corrupt system supported in part by graft and money from illegal drug smuggling to the United States.

Economic Shifts

Serious debt in the 1980s contributed to a reversal of the government's protection of industry, leading to new policies permitting foreign competitors into Mexican markets. The Mexican government significantly reduced tariffs and trade restrictions; cut inflation; attempted to reduce deficit spending; privatized telecommunications, banking, and agriculture; and reduced a wide range of central government controls. Mexico signed the North American Free Trade Agreement (NAFTA) with Canada and the United States in 1992, and after passage in the respective legislatures, the pact was formally implemented in 1994.

NAFTA increased the domestic demand for imports, making them much less expensive due to significant tariff reductions, and caused a rapid rise in Mexico's overseas trade deficit. By insisting on keeping an exchange rate with U.S. dollar parity, Mexico could not adjust to the demands placed on it. Foreign investors lost confidence and began to withdraw their capital. In 1993 Mexico attracted US $75 billion of foreign investment, mainly from the United States and Japan. In 1994 it attracted $60 billion, but in 1995 the net flow was just over $1 billion. At the end of 1994, Mexico had to devalue its currency. Investor confidence appeared to be recovering in the early 2000s through further U.S.-supported reform measures and increased demand for inexpensive Mexican goods in overseas markets.

Social Issues

Other factors of Mexican social, political, and economic conditions, however, may also discourage foreign investments. While a few people increased their wealth, the materially poor gained little from the changes. Larger companies were privatized to the advantage of their new owners but shed labor to become more productive. It was also difficult for many Mexicans to adjust to the new norms of world market competition after decades of state control, protection, and provision. The three areas in which Mexico needed to make rapid changes to compete in world markets—increase productivity throughout the economy, decrease the budget deficit, and maintain civil peace—all confronted contrary forces that slowed progress and pointed to the long-term nature of Mexico's problems.

The 1994 challenge mounted by the Zapatista National Liberation Army in the southern province of Chiapas drew attention to the social costs of economic reform. The **Zapatistas** made the point that the Native Americans, who comprise one-third of the Chiapas population and were already very materially poor, were further disadvantaged by the economic restructuring and NAFTA. They called NAFTA a "death sentence." The Native Americans in Chiapas are only a part of the 13.5 million Mexicans who live in extreme poverty and a further 23.6 million who are poor. In the early 2000s, the Chiapas situation remained unresolved. Reforms promised in 1996 bought time for the government but were not implemented. In response, Zapatistas set up their own autonomous municipalities in defiance of government orders. The Mexican government maintains a large military presence in the region but does little to stimulate dialogue. **Revolutionary tourists**—supporters from other countries who travel to the region and offer their fighting services to the cause—add to the perpetuation of the Zapatista revolution.

Regional Diversity in Mexico

Mexico extends from the subtropical latitudes along its border with the United States to the tropics in the southern reaches of the country (Figure 10.25). Mexico is a country of mountain ranges, plateaus, basin-shaped lowlands, and coastal plains. Most of southern Mexico receives adequate rainfall; much of northern Mexico along and near the U.S. border, however, is quite arid.

Mexico's dynamic human geography continues to evolve new spatial patterns. From the 1500s to the early 1800s, ranchers and miners settled northern areas as part of the colonial expansion of empire that took Spanish soldiers and Roman Catholic priests from the Mexico City area into California, New Mexico, and Texas. Today foreign-owned factories are widespread. The western coastlands have fishing, commercial agriculture, and a burgeoning tourist industry on the Baja Peninsula. Central Mexico is the home of the bulk of Mexico's population and is the seat of Mexican political, cultural, and economic activity, centered on Mexico City. Urbanization and industrialization extend westward to Guadalajara and eastward to Veracruz. The east coast plains around the Gulf of Mexico and the Yucatán Peninsula became a region of oil and natural gas production and tourist opportunity in the second half of the 1900s. The indigenous people of the Yucatán region and the Mexican state of Chiapas do not feel as though they are included in the

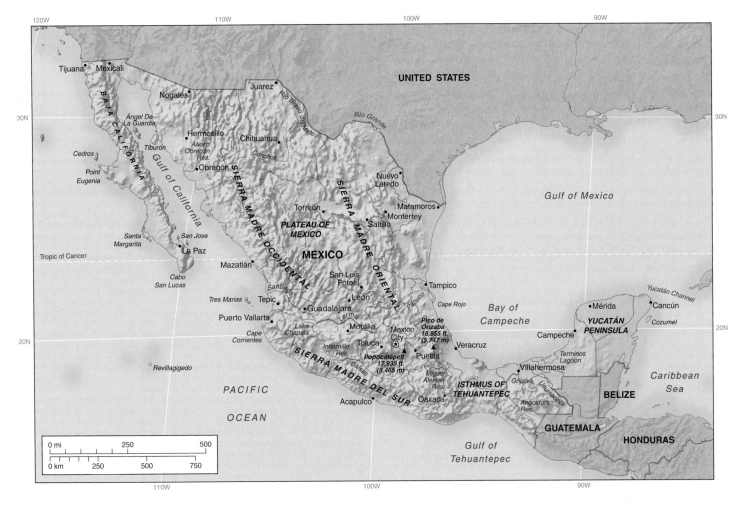

FIGURE 10.25 Mexico. Northward, the climatic environments are drier, but economic opportunities increase close to the U.S. border. Southward, the land narrows and localities become increasingly remote from the influence of the Central Valley of Mexico.

Mexican mainstream. The southern mountains and the border region of Chiapas are the remotest parts of the country. Slow economic development and feelings of political and cultural isolation contributed to the Zapatista uprising in the region.

Mexico City

Mexico City is by far the largest urban agglomeration in Middle America. Its site was a major population center before the arrival of the Spaniards and even before it became the Aztec capital. It continued to be the focus of road and rail networks within New Spain and independent Mexico. Mexico City is the political capital and media center of the country, and has three-fourths of Mexico's manufacturing industry and nearly all of its commercial and financial establishments.

During the 1950s, 1960s, and 1970s, up to 1 million people per year migrated from rural areas in Mexico to metropolitan Mexico City. Housing supplied by government efforts combined with that from private construction, yet development projects could not nearly provide adequate accommodation for almost one-third of the rapidly growing city's population. The

city attracted rural migrants with the expectation of greater economic opportunities, better education, more diverse recreation and cultural choices, and more substantial health-care services. Squatter settlements exploded in many parts of the Central Valley of Mexico in and around the city. The massive shantytown city of **Nezahualcóyotl,** now with more than 1.5 million people living on flood-prone lands, is the valley's most notorious squatter settlement resulting from Mexico's migration decades. Squatter neighborhoods often lack amenities such as electricity, water, sewage, and even paved streets. In some cases, once such shantytowns were established, the authorities began to pave the streets, put in utilities, and provide access to schooling and health care; but amenities are often slow to arrive for many, and millions live well below poverty standards.

The overall result of such rapid growth is a combination of overcrowding, congestion, and air and water pollution. The physical geography of the Central Valley of Mexico, coupled with the intensely crowded living conditions, makes the urban center one of the most polluted, in terms of air quality, in the world. Although residents of metropolitan Mexico City have greater access to health care than rural

Mexicans, infant mortality rates and other social health indicators are among some of the worst in the country due in large part to poor air and water quality. Depletion of underground water resources during the past several decades presents another problem for the city. Many areas within the urban system are subsiding, some more than 5 meters. Engineers struggle to stabilize historic structures that have been slowly sinking for decades. As job markets also failed to cope with the population explosion of the city, the informal sector of the economy grew from around 4 percent to 26 percent of people of working age from 1980 to the 1990s. Government efforts from the 1980s through the early 2000s to attract jobs and migrants to other regions in Mexico helped to slightly alleviate population pressure in Mexico City, but overcrowding and employment remain formidable challenges in the Federal District.

Economic Development and the Human Landscape

Mexico is by far the most economically developed of the countries of Middle America. In 2003 Mexico had a total GNI PPP that was 77 percent of Middle America's total GNI. Mexican GNI per capita is one of the highest in Latin America and near the top of the World Bank's upper-middle-income group. Ownership of consumer goods was well above that in the countries of Central America apart from Costa Rica (see Figure 10.17). Mexico's HDI and GDI are much higher in rank than Brazil's but not as high as Argentina's.

Mexico has a diversified economy. Peasant farmers cultivated corn before the Mexican Revolution of the early 1900s, while most of the country was covered by large wheat or livestock-producing haciendas (Figure 10.26). In the middle

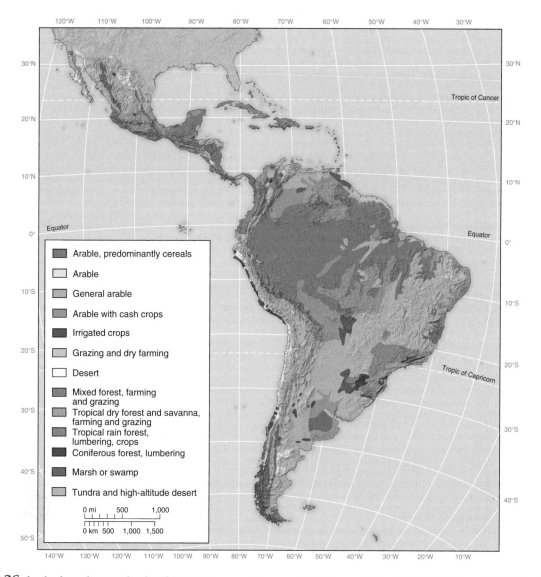

FIGURE 10.26 Latin America: major land uses. Compare this map with that of population distribution in Figure 10.19. Are there any links? What factors are missing to explain the distribution of population? Compare it also with the climate map (Figure 10.6). **Source:** Data from *New Oxford School Atlas,* p. 98. Oxford University Press, UK, 1990.

decades of the 1900s, rapid urbanization occurred, as well as some land redistribution and agricultural irrigation. Farming became commercial with the growing urban market at hand, while foreign companies built packing plants for canned and frozen vegetables. Farming remains important today, employing one-fourth of Mexico's labor force.

Manufacturing became a major source of employment together with increasing numbers of jobs in services and government. Tourism is a major and growing source of income for Mexico, with 20 million visitors in 2002 (up from 17 million in 1990), which is millions more than anywhere else in Latin America. Mexico is a regional leader in developing a tourism industry and is one of the first Latin American countries to have a separate ministry of tourism with dedicated funds within its federal governmental structure. Mexico City, Pacific coastal resorts such as Cabo San Lucas, Mazatlán, and Acapulco, well-preserved ruins of Mayan city–states, and the Caribbean coastal resorts of the Yucatán Peninsula are major tourist attractions for Mexicans, Latin Americans, and visitors from North America, Europe, and Asia.

Variations in the development of manufacturing define economic regions within Mexico. Nearly two-thirds of the Mexican population lives in the urban region surrounding Mexico City that includes Guadalajara, Veracruz, and León. The majority of Mexico's economic power and activity is centered in this megalopolis-type urban expanse, which is similar to the northeastern United States and to parts of Japan and Western Europe. Northern Mexico became the main growth area in Mexico's economy for new industrial development in the 1970s and 1980s, building on already established industries such as steel and iron production in Monterrey.

Maquila

More recently, parts of northern Mexico and towns along the U.S. border, from Tijuana in the west to Matamoros in the east, experienced very rapid growth (Figure 10.27). After 1965 the *maquiladora* program made it possible for foreign-owned factories (*maquila*) sited in Mexico to import components for assembly without customs duties. Such factories used cheap local labor and took advantage of Mexico's fewer labor and environmental regulations to assemble goods that could be exported, duty free, across the Mexican border and back into the producing country (most often the United States). U.S. corporations involved in textiles, apparel, electronics, and wood products built their factories in these towns, where there was rapid growth through the 1990s. Asian and European corporations joined U.S. facilities in producing under *maquila* laws in the northern zone.

The concentration of so much economic activity along the border, however, brought increased air and water pollution. Medical research continues to indicate a strong correlation between the industrial expansion of the maquiladora region and increased rates of cancer and other diseases among the people on both sides of the Mexico–U.S. border. Both Mexico and the United States are implementing programs to reduce industrial pollution from both sides of the border.

The *maquila* industrial zone enhanced global connections, especially those established with the United States. A human landscape now exists in which U.S. popular culture is incrementally prevailing upon the Mexican people of the zone. Local Mexican voices are talking about Halloween instead of *Día de Los Muertos* (Day of the Dead) and Beyonce instead of Mexican pop star Thalia, and they are increasingly using Spanglish, a Spanish–English hybrid, or English to do so.

The combined forces of the maquiladora program and NAFTA opened a floodgate of economic globalization in Mexico. Ford, GM, Nissan, IBM, Whirlpool, Kodak, and Caterpillar are just a few of the hundreds of major foreign companies operating within Mexico's borders. Furthermore, the irrigated northern coastal areas became a primary center of modern agriculture, producing winter vegetables for U.S. markets in addition to cotton, sugarcane, and rice.

Oil and Natural Gas

While diversified urban–industrial economies develop in Mexico's center and north, the east coasts facing the Gulf of Mexico supply much of Mexico's wealth from large oil and natural gas fields. Although discovered early in the 1900s, this wealth was not effectively tapped until the 1970s, when oil went from 2 to 80 percent of Mexico's foreign exchange earnings. During the high oil-price period of the late 1970s, Mexico borrowed heavily, and falling prices and demand, coupled with very high interest rates in the 1980s, left the country deeply in debt. Oil production had a major impact on the economy of the coastal area between Tampico and Campeche, and recent finds extended this affected area inland. During the 1980s new refineries and petrochemical industries were built. From the late 1990s oil formed less than 40 percent of exports as more oil was used by growing industries within Mexico, and the products of those industries made up a higher proportion of exports.

FIGURE 10.27 Contemporary urban landscape in northern Mexico. Maquila development in northern Mexico fueled growth in the city of Monterrey, where colonial and contemporary human landscapes merge. **Photo:** © Sergio Dorantes/Corbis.

The Mexican South

Southern Mexico is less developed than northern areas of the country and more reminiscent of the countries of Central America. Most of the region is remote and populated by Native American subsistence farmers who grow corn on the hillsides and some wheat on the valley floors. Overgrazing by sheep is common. Poor transport facilities slow development in this region. Contrasting with this poverty, the government-planned tourist resorts such as Acapulco bring American tourists to the coast, and new highway links connect the resorts to airports. Unfortunately, sewage disposal from the hastily built hotels and surrounding shantytowns has polluted the bay at Acapulco, and a 1997 hurricane devastated the city's materially impoverished neighborhoods.

Traditional Native American culture is least disturbed in Chiapas, Yucatán, Oaxaca, and the southernmost Mexican states. Over a million Native Americans in Chiapas do not speak Spanish. Rapid tourism development in parts of the Yucatán Peninsula is dramatically altering the human geography for some while furthering a feeling of isolation for others (Figure 10.28). The far south remains very remote, with few people or commercial activities. In recent years, government investment in Chiapas resulted in the irrigation of farmland from the Angostura Dam waters that occupy 90 km (60 mi.) of a once-farmed valley. Chiapas also houses thousands of refugees who fled earlier civil disturbances in Guatemala. The combination of dispossessed refugees and local disillusionment with the Mexican treatment of these states contributed to the Zapatista rebellion and regional unrest.

Central America

Central America is a land bridge, or **isthmus,** that narrows in width from the Mexican border in the northwest to the Colombian boundary in the southeast. The Central American isthmus provides a narrow physical boundary separating marine environments of the Caribbean Sea/Atlantic basin to the east and the Pacific Ocean to the west. Seven independent countries occupy the land of Central America (Figure 10.29). These states have a common Hispanic culture and geographic proximity. Belize was a British colony from the late 1800s until 1981 and has a relatively high percentage of people who trace their ancestry to the African continent. Its mainland location, increasing use of the Spanish language, and Roman Catholic populace link Belize to Central America rather than to the Caribbean Basin.

The seven Central American countries had a total population of nearly 40 million people in 2005 and are among the most materially poor countries in the Americas. Many of the countries of this subregion depend on the foreign capital generated from a limited production of export crops. They are sometimes termed **banana republics** because of their economic dependence on one or two export crops, among which bananas figure strongly. Dependence on one primary export crop and slow economic growth were linked to decades of regional political instability. Although Central America forms a land bridge between North and South America, the physical and human geography of the subregion presents a barrier to the development of strong travel and communication connections between the core areas of the two continents. Periods of instability, dire rural poverty, drug trafficking, and the nearly impenetrable

FIGURE 10.28 Cancún resort development. Tourism growth supports the diffusion of resort hotel development along the Caribbean coast of Mexico's Yucatán Peninsula, while economic conditions worsen in some interior villages. Mexico is one of the world's major tourist destinations with 20 million visitors in 2002. **Photo:** © R-F/Corbis.

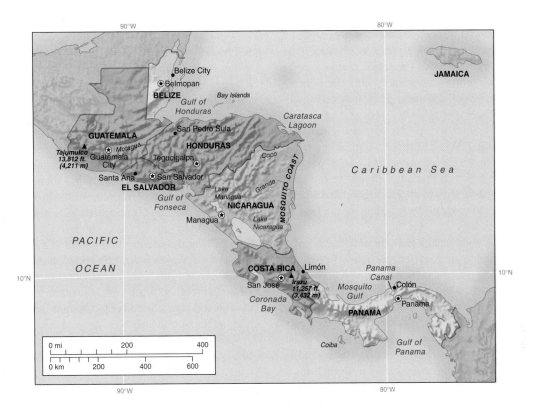

FIGURE 10.29 Central America: major geographic features, countries, and cities.

environmental conditions in the Darién of eastern Panama complicate efforts to form substantial links to South America. In the late 1990s further setbacks to these countries included widespread forest fires linked to an extended period of drought that destroyed crops and the 1998 devastation following Hurricane Mitch.

Countries

Regional independence from Spain came with a brief union between contemporary Mexico, El Salvador, Guatemala, Honduras, Nicaragua, and Costa Rica as the Mexican Empire from 1821 to 1823. In 1823 the incipient Central American countries separated from Mexico as the United Provinces of Central America, which lasted until 1838, when they each became independent. Panama existed as a province of Colombia until a U.S.-brokered canal deal created a separate country in 1903. Belize remained within the British sphere until 1981. Since 1838, unity has not figured prominently in the interactions among the countries of the subregion.

The four largest countries to the south of Mexico—Guatemala, El Salvador, Honduras, and Nicaragua—had 2005 populations ranging from Nicaragua's nearly 6 million to Guatemala's nearly 13 million (see Table 10.2). Coffee plays an important role in the cash crop export-oriented economies of each. Civil unrest and natural disasters have wreaked havoc on the social fabric, physical infrastructure, and economies of all

four countries. Deadly conflict, stemming in part from dramatic inequalities in land ownership in Guatemala, El Salvador, and Nicaragua, focused attention on disparity issues but did little to bring about an egalitarian solution.

In Guatemala, Native Americans comprise approximately half the population yet have little economic or political clout. The subregion's largest indigenous population percentage exists in Guatemala, where attempts by native peoples to improve their living conditions sparked fierce government repression. In 1980 and 1981, 150,000 of Guatemala's indigenous peoples fled into Mexico as government forces burned down hundreds of their villages.

Although El Salvador experiences frequent earthquakes, land reform conflict in the 1980s proved far more costly to the country than any natural disaster. Civil war erupted when landowners of vast coffee estates challenged changes imposed by a reformist government. Thousands of people migrated in an attempt to escape the bloodshed. Such large numbers of war refugees moved to the capital, San Salvador, that the economy and infrastructure of the city virtually collapsed. Recovery continues today.

Decades of oppression under the **Somoza** family dictatorship in Nicaragua instigated revolutionaries to overthrow Somoza control. The Marxist-oriented **Sandinista** revolutionaries took control in 1979 and inflicted another form of oppression and corruption on the people of Nicaragua for the next 11 years. The United States imposed a trade embargo on the

Sandinista government, which added to the country's economic disruption and already impoverished social conditions. The United States also backed armies **(Contras),** partially composed of former Somoza military personnel, who were fighting against the Sandinistas. The Sandinista party was eventually voted out of office in Nicaragua in 1990.

Civil unrest in El Salvador, Guatemala, and Nicaragua eased considerably in the 1990s. The three countries began the slow process of repairing the physical and social damage inflicted by years of conflict. It remains difficult for the governments of these three countries to attract external investment because of their recent violent histories, and tourists are slow to return and spend their money.

Belize, Costa Rica, and Panama have the smallest populations of the Central American countries. Belize is by far the smallest country in the region, having only 300,000 citizens. The country became independent in 1981 after a long period of British colonial rule. Although years of British tenure established an English language base and some cultural elements different from those of neighboring countries, the Hispanic influence of the region exerts strong forces on Belize and is causing the development of a more regionally uniform human geography. Many Guatemalans assert a historic right to the lands and waters of Belize. Guatemala's formal claims eased in 1991, and the issue seems to be fading with time, although oil deposits off the Belize coast could renew claims for some Guatemalans.

Panama became a separate country when its territory was carved out of Colombia in 1903 in preparation for the construction of the Panama Canal. The newly independent country had an immediate economic and military dependence on the United States, which resulted in a relationship in which Panama functioned more like a U.S. colony than as an independent country. The construction of the canal brought in people from the Caribbean islands, Asia, and southern Europe. The Canal Zone created a dramatic dichotomy in the human landscape for Panama. The corridor along the canal is relatively wealthy and developed in contrast to areas of Panama to the east and west. Panama City, the country's capital, retains a cosmopolitan population and an international role in trade and finance. The United States virtually ran the country until 1979 and then invaded in 1989 to protect strategic and economic interests.

Human Activity and Economic Development

In contrast to Mexico's diversifying economy, the Central American countries have a much narrower economic base, resulting in slower economic and social infrastructure development. In Central America, the major difference in economic geography is between the more populous centers of Native American and Hispanic culture in the uplands and the less populated Caribbean coastal areas with their American-owned plantations growing bananas, other fruits, and sugarcane.

Guatemala is a country of three parts. First, in the uplands northwest of Guatemala City, the largest Central American concentration of Native American peoples lives in subsistence conditions, often landless and in extreme poverty. An uprising in the early 1980s resulted in government repression and destruction of what little infrastructure existed. Second, people in the southern part of the country adopted a more European lifestyle. This area's commercial agriculture is based on coffee on the Pacific slopes and bananas and sugarcane near the Caribbean (Figure 10.30). Guatemala City has manufacturing industries, but recent influxes of people swamped the job markets, and half of the working-age population is unemployed. Third, the lowland **Petén** area of the north is forested and less populated than parts of the country to the south. Dramatic archaeological relics of the Maya civilization attract numerous tourists (nearly 1 million in 2002) and scholars from all global regions. Although tourism brings some revenue to the Petén, much of the forest resources are being cleared, which could ultimately hurt the tourism industry.

El Salvador has the densest population in the subregion. Most of the people are concentrated in the western highlands, where the dominant crop, coffee, is grown. Light manufacturing industries were established from the 1960s. Civil war over land redistribution since the 1980s disrupted both rural and urban economies. Thousands of Salvadorans fled the country during the civil war and took up residence in the United States.

FIGURE 10.30 Central America: Pacific coast. An eastward-looking space shuttle view of Guatemala and southern Mexico. The main cities are in the highlands, where temperatures are modified by elevation. **Photo:** NASA.

These foreign-based communities send regular payments to relatives remaining in El Salvador. Such remittances are currently El Salvador's largest source of foreign revenue.

Honduras is one of the most materially poor countries in the Western Hemisphere. Barely 25 percent of its land can be used for farming. The ranching economy established in colonial times persists in the parts of the west, and large banana and pineapple plantations dominate the Ulua River valley and the northern coastal plain. Coffee became a primary export in the 1960s, and after Hurricane Mitch devastated the banana industry, coffee is the leading source of foreign income today. Shrimp farming on the Pacific Ocean coast is a new export specialty that is finding markets in the United States. Tourism is a growing industry, attracting more than 500,000 visitors in 2002, centered on Maya ruins in Copán, nature tourism mountain hikes and river rafting, and a relatively thriving scuba diving and beach attraction along the north coast and in the Bay Islands. Much of the former forest cover was logged, and even the secondary growth of pines is being removed rapidly. Manufacturing was encouraged after 1950, and a growing maquiladora industrial belt is forming around the city of San Pedro Sula.

Nicaragua developed a mixture of agriculture and modern manufacturing industries. The U.S.-imposed trade embargo that followed the Sandinista revolution in 1979 devastated the economy, giving it the lowest GNI in the subregion in the late 1990s and undercutting Honduras as the poorest country in Central America. Coffee, cotton, bananas, and sugarcane are important export crops. The population is concentrated around the lakes of Nicaragua and Managua in the structural depression that is subject to earthquakes. Other parts of the country have poor transportation links to this region. Tourists, once frightened away by conflicts and natural disasters, are slowly returning to the country. The potential for tourism growth is significant, and the government is encouraging investment and exploring development opportunities. It may be some years until the stigma of warfare is removed enough for large numbers of visitors to enjoy the diversity of the country.

The United States relinquished control of the Panama Canal Zone in 1999 to the government of Panama. The country has economic contrasts between the urban belt, including Panama City and the corridor along the **Canal Zone,** and the rural areas on either side of this developed belt. Panama City and the corresponding Canal Zone are a cosmopolitan center of global trade, where banking and financial services create twice the GNI of agriculture. In 1991 Taiwanese investors established a free-trade zone in Panama City in which some 8,000 people are employed in clothing, furniture, toy, and bike factories. The products are sold in the U.S. market. In the rural districts to the west, coffee is grown on uplands, and rice and palm oil are produced along the coasts. To the east of the Canal Zone, the Central American Highway tapers off in the impenetrable **Darién,** an area of extremely dense vegetation between the canal and the Colombian border that prevents completion of effective north–south highway links.

Many believe completion of the highway would dramatically increase the flow of illegal drugs northward from the Northern Andes region of South America.

Panama is dealing with the problem of an aging canal that is now too small for many ships (Figure 10.31). Work began in 1991 to widen and lengthen the main locks, and a pipeline now conveys oil across the isthmus to the west of the Canal Zone, virtually eliminating the need for oil tankers to pass through the canal. Although the United States formally closed its military bases in Panama in 1999, it maintains close ties with Panama in an effort to continue the U.S. war on drugs in the region.

Costa Rica, with 3.7 million people in 2001, is the only country in the region that has had a long-term democratic government. This may be related to the legacy of its initial colonial settlement, when there were few Native Americans to be dominated and the people of European origin took up medium-sized farms. As a stable democracy, Costa Rica attracted manufacturing and free-trade zones financed by U.S. and Taiwanese corporations. Costa Rica has a more prosperous and diversified farming industry than the other countries, based mainly on coffee production along the Pacific slopes, bananas and cacao on the eastern coast, and a range of new crops such as tropical fruits, ornamental plants, cut flowers, and tropical nuts. Areas of former rain forest are used for livestock production. Costa Rica has little civil strife. It does not have an army, but the militarized police and relatively good quality of life help to maintain internal peace. Costa Rica has many successful industries, including Microsoft facilities, textiles and pharmaceuticals for export, and cement, tires, and car assembly for the domestic market. Costa Rica is one of the top tourist destinations in Latin America. Costa Rica's beaches, volcanoes, and culture help the country to attract more than a million tourists per year. An extensive system of national parks incorporates a variety of ecological habitats and microclimate zones (through altitudinal zonation) and protects a vast array of flora and fauna. In the 1990s Costa

FIGURE 10.31 Central America: Panama Canal. The Miraflores locks with a cruise ship entering to be raised to the level for traveling through the canal. Note the land uses on either side of the canal. These locks are now too narrow for the world's largest ships. **Photo:** © Morton Beebe-S.F./Corbis.

Rica increasingly became home for U.S. retirees seeking to take advantage of its climate, stability, and lower cost of living than in the United States. In the early 2000s more than half a million U.S. retirees lived in the country.

Test Your Understanding
10C

Summary

Mexico is one of the wealthiest countries in Latin America, having a diverse economy, but it has problems adjusting to increased involvement in the world markets and NAFTA. Mexico City is one of the world's largest metropolitan areas. The contrasts of internal development among regions and groups of people within Mexico cause social tensions and fluctuating patterns of economic growth that threaten to slow the process of human development.

Central America comprises several small countries: Belize, Costa Rica, El Salvador, Guatemala, Honduras, Nicaragua, and Panama. Their peoples, with the exceptions of Costa Rica and Panama, are among the poorest in Latin America. Central America's environment consists of mountain ranges and swampy tropical coasts. Its countries produce coffee, bananas, and a few other crops for fluctuating world markets, but many economies have been affected badly by civil wars.

Questions to Think About

10C.1 How do the main features of the population and economy of Mexico compare with those of the Central American countries?

10C.2 What impact did civil war have on the societies, population distributions, and economies of the countries of Central America? What is the situation today with respect to the regional conflicts?

10C.3 Do Mexico and the countries of Central America count as the "near abroad" of the United States? (See Chapter 4 for the usage of this term with reference to Russia.)

Key Terms

Monroe Doctrine	revolutionary tourist
petrodollars	Nezahualcóyotl
Lost Decade	*maquiladora*
protectionism	isthmus
remittance	banana republics
Porfirioto	Somoza
ejido	Sandinista
ejiditario	Contra
PRI	Petén
PAN	Canal Zone
Zapatista	Darién

The Caribbean Basin and Environs

The contemporary human and physical geography of the Caribbean Basin reflects dramatic transformations imposed on the region by European countries during the colonial era. The loss of indigenous people to disease, the forced migration of Africans to the basin, and the establishment of extensive sugar cultivation on the islands laid the foundation for the social, environmental, and economic conditions present in the subregion today. The Caribbean Basin consists of a few large islands, several small islands, and three political units on the South American mainland (Figure 10.32). The political geography of the Caribbean is one of the most diverse of any subregion. Caribbean countries vary in size and population from Cuba, with more than 11 million people in 2005, to hundreds of tiny island countries with few inhabitants (see Table 10.2). Regional patterns in the ownership of consumer goods, access to potable water, and energy consumption (see Figure 10.17) present a picture of the basin's economic diversity, from poverty-stricken Haiti to the more affluent Bahamas. In 2005, 35 million people lived in the Caribbean Basin. The majority of the Caribbean's residents live on one of the four largest islands, known as the **Greater Antilles:** Cuba, Hispaniola (consisting of two countries: Haiti and the Dominican Republic), Puerto Rico, and Jamaica.

The smaller Caribbean islands are commonly referred to as the Lesser Antilles. The **Lesser Antilles** chain is an arc of small islands around the eastern edge of the Caribbean Sea, with the Leeward group in the north (Virgin Islands to Guadeloupe) and the Windward group in the south (Dominica to Grenada). Many of the small island countries of the Caribbean have only a few thousand residents.

The Bahamas, situated north of Cuba and southeast of Florida, include more than 700 islands. Another group of islands, including Trinidad and Tobago, lies along the northern shore of South America. Also included in this subregion are the three Guianas—Guyana, Suriname, and French Guiana. Although situated on the South American mainland, their histories, people, and present economies are more connected to those of the Caribbean Basin than to those of other mainland South American countries.

The islands of the Caribbean Basin are situated between continental North and South America. Increasing global connections have made the islands of the Caribbean strategically significant from the colonial era to the present. In the 1500s they were on the route for Spanish treasure ships between Europe and the mainland of the Americas, and from the later 1600s they became a major source of European colonial wealth through sugarcane cultivation. The opening of the Panama Canal in 1914 placed these countries on a major world route, further elevating the strategic importance of regional waterways. The involvement of the United States in the Cuban Missile Crisis of 1962, plus U.S. interventions in the Dominican Republic in 1965, Grenada in 1983, and Haiti in 1994,

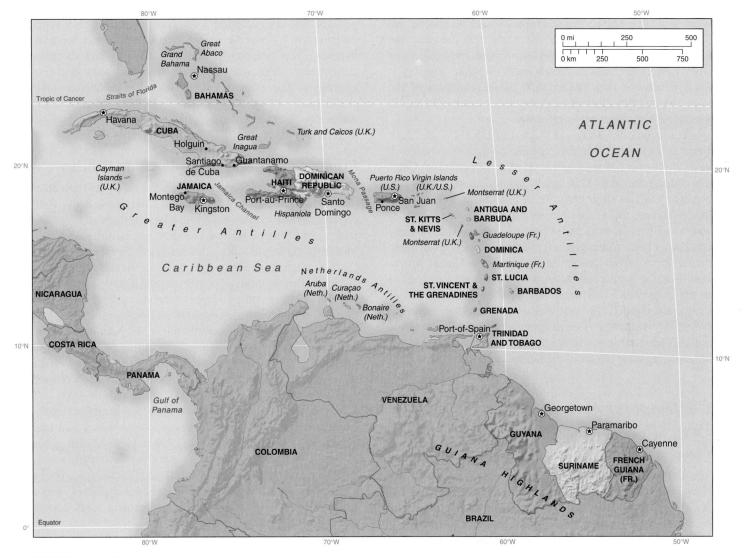

FIGURE 10.32 **The Caribbean Basin.** Numerous island countries and some overseas colonies.

highlighted the sensitivity of Americans to neighboring events. The continued presence of U.S. military bases, including Guantánamo Bay in Cuba, and concerns over drug traffic through the region, emphasize the geopolitical significance of the Caribbean Basin.

Countries

The people of the Caribbean have a very strong sense of belonging to the individual island or country of their birth, as opposed to strong feelings of regional unity or identity. Although the majority of the Caribbean residents are of African ancestral heritage, most people identify with their fellow countrymen or women and perceive readily apparent differences between their nationality and those of neighboring countries. A Jamaican would identify with other Jamaicans and see their national people as different from those of Barbados, rather than perceive any unity from ancestral heritage. National pride is often expressed through rivalries with neighboring island countries. For example, sporting events, such as cricket and soccer, among the people of Jamaica, Barbados, Trinidad, and Guyana are very competitive, prestigious, and emotional affairs.

A major source of variety within the human geography of the region arises out of the different colonial histories of the countries. While Middle America has an almost exclusively Hispanic history, the Spanish, French, British, Dutch, and Danes all influenced the contemporary geography of the Caribbean. Although the Spanish were the first to discover and claim the largest islands, their main areas of interest shifted to mainland Middle America and to South America after their invasion of Mexico in 1519.

Spain retained colonies in a few of the Caribbean islands to supply meat to the mainland empire of New Spain and sugar and tobacco to the home country in Europe. Ports were fortified to protect trade routes. In the 1600s the Dutch, British, and French established colonies on Bermuda, the Bahamas, the Lesser Antilles, the Guianas, and the offshore South American islands. At first some were merely bases for pirates preying on Spanish treasure ships. After 1650 the development of the sugarcane industry dramatically elevated the value placed on the region by the European colonizers.

Throughout the 1800s, as the rest of the Latin American countries gained independence and the slave trade was banned, most of the sugar-producing islands remained colonies. Colonial connections declined in significance as European countries began to cultivate sugar beets. The United States took a growing interest and played an increasing role in the area. The U.S. war with Spain in 1898 ended Spanish colonial rule of Cuba and Puerto Rico. Cuba took full political independence, while Puerto Rico moved politically closer to the United States. Nearly all the British and Dutch colonies became independent in the mid-1900s, but the remaining French islands—Martinique and Guadeloupe—together with French Guiana remain to this day within the French political and economic arena, sending representatives to the French parliament and being officially regarded as part of the European Union.

Contemporary Legacy of the Colonial Past

Colonial Farming Heritage

The Caribbean colonies of Spain, France, Britain, and the Netherlands assumed a huge significance in the global economy when the Caribbean Basin took over the leading sugarcane production role from Mediterranean countries in the early 1600s. From 1500 to the 1800s, some 10 million African slaves were shipped to the Americas, of whom half went to the Caribbean Basin, 39 percent to Brazil, and under 5 percent to British American colonies (including the United States). The Africans brought their own social customs, religious beliefs, handicraft skills, and various forms of artistic expression. At the height of sugar cultivation prosperity in the 1700s, the average plantation unit was 200 acres, used 200 slaves, and produced 200 tons of sugar each year.

The British navy stopped the slave trade across the Atlantic after 1807, and slaves were emancipated in British colonies by the 1830s, in French colonies by 1848, and in Spanish colonies by the 1880s. The plantations declined in numbers and output and faced increasing competition from sugar production in other parts of the world. When new plantations opened in Trinidad and the Guianas, labor was imported from Asia. Cuba increased its sugar production in the 1800s, becoming the largest producer of all, and attracted labor from other Caribbean islands. In the 1900s the family-owned plantations on the sugar islands sold out to major corporations, such as Tate & Lyle (U.K.) in Jamaica and Gulf & Western (United States) in the Dominican Republic. Since the mid-1900s many island governments have taken ownership of their sugar industries.

The colonial occupations and economies often took major decisions away from the region and placed them in countries overseas. The interests of external countries and company shareholders often assumed greater importance than local considerations. The colonies were restricted to producing single commercial crops for the needs of overseas markets, and a white elite controlled the imported African- and Asian-origin working populations.

Economic Strategies Following Independence

The majority of the independent countries emerging from colonial rule were small and had limited natural resources. Their economies depended on exporting a limited range of crops and minerals and importing manufactured goods, food, and fuel. Multinational corporations often marketed the export products. Demand, as measured by world prices, for the main products—bananas, other fruit, sugar, petroleum, and bauxite—fluctuated.

Caribbean countries adopted economic policies similar to those of other developing countries. At first manufacturing was promoted through import substitution. Caribbean-based factories were protected from external competition by tariffs, which often led to high prices and low local profits. Together with the small size of local markets, this strategy had little impact on the economies and employment levels of individual countries.

In the 1980s policies switched to making local sites attractive to investment by overseas manufacturers that would take advantage of cheap locations and labor to produce such goods as clothing, sporting goods, and electronics for export, principally to markets in the United States and Canada. This policy proved more effective in economic growth but did not reduce external dependence. Manufacturing remains a small part of most island economies.

The Challenge of Cooperation

The need for economic integration among the small Caribbean Basin countries is clear, but progress has not been easy. Countries of the Caribbean cultivate the same crops, produce the same goods, and largely compete for the same markets. The **Caribbean Community and Common Market (CARICOM)** formed in 1973, taking over from previous attempts at free trade and comprising 13 former British colonies with a total of 5.5 million people. The Caribbean Basin Initiative of 1984 was set up partly to promote economic growth in the Caribbean Basin and partly to protect U.S. interests. The implementation of NAFTA in 1994 fueled Caribbean fears that the region would be shut off from investment and trade opportunities from the United States and Canada. Mexico, the Central American states, Colombia, Venezuela, and countries of the Caribbean formed the **Association of Caribbean States** in 1995. It was proposed by the CARICOM group, and although it comprises the fourth largest trading bloc in the world, progress is stifled by numerous challenges. Mexico's NAFTA relations with the United States and Canada, a lack of unity in Central America, strained relations between Hispanic countries and English-speaking countries within the Caribbean Basin, dramatic variation in country size, population, and resource base, and the United States' refusal to lift its trade embargo on Cuba are some of the many challenges facing the cohesion of the association.

Tourism

Tourism is increasingly important in the Caribbean Basin, attracting around 15 million visitors in 2002. Localities within the Caribbean Basin cater to a range of tourist interests from the elite to the mass markets. The gleaming beaches and turquoise water of the region are its number one attraction (Figure 10.33). The islands of the Caribbean also provide unique historical and cultural experiences, and a wide range in topography from the flat Cayman Islands to the rugged hills of Jamaica and Dominica. The United States now provides two-thirds of the tourists, while most of the rest come from Canada, Europe, and Japan. Puerto Rico, the Bahamas, the Dominican Republic, Cuba, and Jamaica attract the greatest number of visitors, while the Virgin Islands, Antigua, Guadeloupe, and Martinique continue to grow in significance.

Initial tourism development consisted largely of foreign-owned and -managed resorts, many of which were all-inclusive. Foreign control of tourist infrastructure resulted in severe rates of economic leakage for many areas. **Economic leakage** results when the revenue generated flows to the foreign owners and foreign employees of tourism infrastructure (hotels, restaurants, gift shops, bars, and dive shops) and thus "leaks" from the local economy. Although many foreign-owned all-inclusive resorts remain in the Caribbean, some governments are increasingly taking control of tourism infrastructure and marketing and are attempting to involve local communities in tourist infrastructure investment and management.

The number of cruise ship passengers visiting Caribbean ports continues to rise annually. Although a cruise ship may unload thousands of money-spending tourists into a port town, many cruise visitors participate in prepackaged events where revenues generated go to the cruise lines and only a couple of local businesses fortunate enough to be working with the ship's activity planners. Island residents view resort and cruise tourism with mixed emotions. Some see the industry as economically beneficial, providing jobs, housing, and schools, while others do not reap any direct economic benefits and resent the crowds, resource consumption, and pollution that accompany tourism growth in many areas.

Tourism is an economic activity that is extremely susceptible to external forces. Economic recessions in the wealthier countries from which visitors originate, destination-based crime or terrorism, or media reports of natural disasters such as volcanic eruptions or hurricane damage all may produce immediate reactions in the form of dramatic declines in tourist visits. Although tourism brings in foreign exchange, much of it is used to buy the foods and equipment expected by the tourists.

Island and Country Specialties

Puerto Rico is not an independent country, nor is it a state within the U.S. political system. An official relationship between Puerto Rico and the United States beginning in the early 1900s became more formally structured in 1952 with the establishment of **commonwealth status.** Puerto Ricans are considered U.S. citizens, yet they do not have true representation in Washington, D.C. Puerto Rico's representatives in the U.S. Congress do not have the right to vote on the passage of bills. Puerto Ricans have open access to migration into the United States, and many have taken advantage of this since the 1950s. U.S. statehood remains an issue with distinct political views regarding Puerto Rico's status. The smallest group would like Puerto Rico to become an independent country, free of any U.S. authority; another group supports statehood; and the largest group currently is content to maintain the commonwealth status. Although the Puerto Rican government declared Spanish the official language in 1991, English as a second language is mandatory in the Puerto Rican public school system.

FIGURE 10.33 St. Thomas, Virgin Islands. While tourists enjoy the sea, sand, and sun of the Caribbean, much of the revenue generated from their vacations may not reach the local economy or community. **Photo:** © Neil Rabinowitz/Corbis.

Puerto Rico has a much higher per capita GNI than other Caribbean islands. Although its sugar industry expanded, its economy remained narrow until after World War II, when farming shifted to dairying and manufacturing increased rapidly. Machinery, metal products, chemicals, pharmaceuticals, oil refining, rubber, plastics, and garments provide a diverse product mix and have grown under special U.S. tax laws for the island. Tourism became a leading economic sector for Puerto Rico with 3.1 million visitors in 2002.

The *Dominican Republic,* comprising the eastern two-thirds of the island of Hispaniola, became independent in 1821 and subsequently suffered from economic stagnation and political instability. As in Cuba, U.S. corporations formed links with the plantation owners, and the U.S. government intervened with military occupations in 1916–1924 and 1965. Spurts of economic development occurred in the mid-1900s, when construction began on irrigation and road infrastructure, and U.S. corporations made industrial investments. Agriculture remains the dominant economic activity, based on sugar, while the Standard Fruit Company operates large banana plantations and the Dole Corporation administers pineapple plantations. Meat is exported from cattle ranches, and coffee is grown on the uplands. There is some mining, and oil reserves are present. Most manufacturing is the assembly of imported components for export. The Dominican Republic dramatically increased its tourism capacity from 1970 through the early 1990s in preparation for the 500th anniversary of Columbian voyages to the Americas. In the 1990s and early 2000s, tourism continued to grow, with 2.8 million visitors in 2002. Through the gradual diversification of its economy, the Dominican Republic now has one of the highest incomes of the larger Caribbean countries.

Haiti, occupying the smaller western part of the island of Hispaniola, is the poorest country in the Americas. Life expectancy is low, infant is mortality high, and 50 percent of its adults are illiterate. Over three-fourths of the people are crowded onto poor-quality lands resulting from an imbalanced division of land resource holdings. Independence in 1804 was followed by the political instability that continues today. The sugar plantations deteriorated, and economic stagnation and decline set in. The United States occupied the country from 1915 to 1934 and again in 1994 to end military takeovers of the government. In between these dates, corrupt and repressive governments produced little economic development, leaving most people dependent on their subsistence plots of land. Some commercial farming for coffee and cacao produced export income; but a 1980 hurricane destroyed many trees, and recovery was slow. Tax incentives in the 1970s attracted U.S. corporations to set up factories in Haiti. Exports of clothes, electronics, and sports equipment became more valuable than farm products. Haiti's mineral resources, however, remain undeveloped. Tourism is not developing in Haiti as it is in other Caribbean countries due to the extreme material poverty, political instability, and a high incidence of HIV/AIDS.

Following the military coup against a democratically elected government in 1992, an international embargo on Haitian products caused the economy to collapse, including the loss of vital aid, default on its public debt, and closure of the new assembly industries. Exports and imports halved. A democratic government was reinstated in 1994 after U.S. intervention. Haitians expect rapid improvements from the new government but face a bleak future with an economic gap that is being filled by drug traffic–related corruption.

Jamaica gained historic significance as the largest British colony in the Caribbean Basin. Sugar plantations expanded until the abolition of slavery; bananas then became important on the plantations and groups of cooperative farms. Agriculture now provides less than 10 percent of Jamaican GNI, although it still employs one-fourth of the labor force. After World War II, the Jamaican economy diversified. Some 60 percent of its export income now derives from bauxite mines and aluminum-refining plants. Although overseas competition from Australia, Brazil, and Guyana led to a fall in this source of income in the 1980s, the 1990s saw increased output. Jamaican manufacturing is becoming more important, based on the duty-free imports of parts to be assembled for export and textile products. An extensively developed tourism industry along Jamaica's northern coast supported 1.3 million visitors in 2002 and contributed income equivalent to half the total export of goods.

Of the Lesser Antilles, *Trinidad and Tobago* has the largest area and population. Trinidad and Tobago has an oil-based economy. A unique ethnic mixture in the local population includes many who tie their heritage to South Asia. The British took Trinidad from Spain in 1797, and British settlers developed it as a sugar island. Only one-seventh of the land was settled by the time of slave emancipation in 1834–1838, and only one-fifth of that was cultivated. Most plantation development occurred later in the 1800s, when British economic interests shipped indentured labor from South Asia. After serving a set number of years on the plantations, these laborers were paid off with land of their own, and today their small farms produce most of the sugar grown. The increasing quality of its product enables Trinidad and Tobago to expand sugar output as other countries diminish their production.

Natural gas and oil production, together with the refining of oil from Southwest Asia, is a major sector of Trinidad and Tobago's industrial base. In the mid-1990s oil and natural gas output rose by 10 percent. Manufacturing industries include petrochemicals and steel, employing 15 percent of the labor force, while government and other service jobs provide 50 percent of employment. Tourism is not yet a rapidly developing industry.

The other islands of the Lesser Antilles are small and have restricted economies. Only 5 percent of the workforce on Barbados is currently employed by the sugar industry due to a continual decline in cultivation and production, while 80 percent of the workers are in service jobs (dominated by the tourism industry). Bridgetown has some light industries, and its high level of literacy attracts U.S. data processing facilities. Commercial agriculture is on the decline in many other former British colonies, such as Antigua and Barbuda, St. Kitts and Nevis, Montserrat, Anguilla, Dominica, St. Lucia, St. Vincent, and Grenada. Most of these islands are attempting to build their tourism industries.

The former French colonies of *Martinique, Guadeloupe,* and *French Guiana* are political subdivisions of the country of France, known as "overseas departments." Their residents are French citizens and members of the European Union. They once were wealthy sugar producers, but output declined with a shift to bananas, melons, and pineapples for the French market in line with European Union policies. The French government is the largest employer, and tourism is increasing. External contacts tend to be with France rather than with other Caribbean islands.

The former Netherlands colonies of *Aruba, Bonaire,* and *Curaçao* off the northern coast of Venezuela are low-lying and arid, obtaining their water supplies from desalination plants. They remain administratively linked to the Netherlands today. Although Aruba has had a separate status since 1986, it is moving back toward closer political ties with the Dutch government. The people are exceptionally cosmopolitan, including descendants of Africans, Indians, Dutch, Portuguese, Danes, and Jews. Residents are able to speak an old trading language, Papiamento, as well as Dutch (the language of their government), English (the language of tourist visitors), and Spanish (the language of neighboring Venezuela). Oil refineries linked to the Dutch development of Venezuelan oil, along with tourism and offshore banking, are among the main sources of income.

Guyana (former British Guiana), *Suriname* (former Dutch Guiana), and *French Guiana* all have small populations on relatively large areas of land. Dutch drainage engineers in the early 1800s made the coastal plains of the Guianas habitable. Indentured laborers were shipped in from South Asia to work on the vast sugar plantations near the coast. In both Guyana and Suriname, living standards declined after independence as the result of civil strife. Guyana and Suriname export bauxite from inland mines, while French Guiana exports timber to the EU. The reopening of gold mines by Canadian companies in Guyana caused exports to quadruple in the 1990s.

U.S. corporate ownership of the sugar industry dominated *Cuban* economic geography in the early 1900s. When Fidel Castro led a Communist takeover in 1959, ties with the United States and U.S. companies were severed, and dependence shifted to the former Soviet bloc countries. Large state farms and cooperative farms replaced confiscated private plantations. Although some farms diversified to produce citrus fruit for Eastern Europe, overall productivity remained relatively low. Manufacturing included import-substitution units for locally needed goods such as textiles, wood products, and chemicals. Cuba exported sugar, tobacco (cigars), and some strategic minerals to the Soviet bloc in exchange for oil, wheat, fertilizer, and equipment. During the 1980s, 85 percent of Cuba's trade was with the Soviet bloc, but this close link was broken with the 1991 dissolution of the Soviet Union. Cuba suffered as much as other Communist countries from the breakup of the Soviet bloc. From 1989 to 1994, Cuban GNI fell by 34 percent as it moved cautiously toward market economy policies. The sugar crop fell from 8.4 million tons in 1990 to 3.4 million in 1995 because of poor organization, bad weather, and shortages of fuel, fertilizer, and willing labor. Most farmworkers found growing fruit and vegetables for the black market to be more profitable. The United States maintained its trade embargo. Tourism, however, began to grow as a source of foreign currency, with 1.7 million visitors in 2002. Canadian and other non-U.S. mining companies took over Soviet-instigated mining projects, primarily for nickel, cobalt, and gold. In the short term, Cuba faces major economic and social problems.

Northern Andes

The Andes Mountains are a dominant feature in all five countries of the Northern Andes subregion consisting of Bolivia, Colombia, Ecuador, Peru, and Venezuela (Figure 10.34). The world's second highest mountain range creates a multitude of local environments at different heights throughout the subregion. The geologically active Andes are in the process of uplift, resulting in frequent earthquakes, landslides, and occasional volcanic eruptions. The dramatic topography of the Andes Mountains isolates large interior sections of the Orinoco and Amazon River basins from easy global connections.

Many people in the Northern Andes subregion live with a pervasive interaction of traditional culture and globalization. Elements of mountain culture groups exhibit centuries-old social customs played out in quiet agricultural towns tucked into the valleys of the steep terrain. Cash from the illegal drug trade creates a unique elite with dramatically different lifestyles from those in rural mountain villages. Political stability is occasionally challenged by powerful criminal elements and intermittent border disputes, such as the disagreement between Ecuador and Peru over Amazonian territories.

Economically, the countries of this subregion range from South America's poorest (Bolivia) to some of South America's more materially wealthy countries (Colombia and Venezuela) (see Table 10.2). In the later 1990s, only Colombia increased its income per capita; consumer goods ownership remained relatively low (see Figure 10.17), while all aspects of human development have modest scores.

Countries

Although rugged terrain and high relief are common features in each of the Northern Andes countries, topographic and environmental variety is present throughout. The high altitude of the Andes in the western part of the subregion contrasts with and isolates the lower lands to the east. Eastern areas are largely covered by tropical rain forest vegetation and experience high rainfall throughout most of the year. The interior reaches of Colombia, Venezuela, and Bolivia (see the Personal View: Bolivia box on page 485) experience more seasonal rains and a tropical grassland vegetative cover. The most dramatic contrast of the subregion exists on the Pacific coast of Peru, where air circulation patterns, cold currents, and the rain shadow of the rugged Andes combine to produce one of the most arid climates in the world.

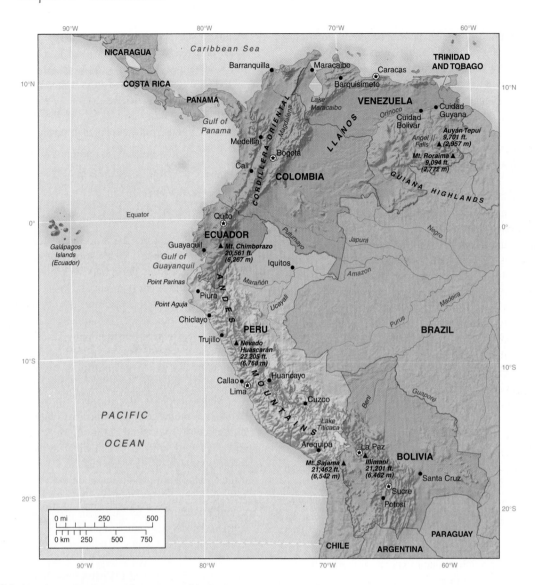

FIGURE 10.34 Northern Andes: main geographic features. Note the position of the Andes Mountain ranges in each country and how they isolate interior lowlands from the main centers of population and trade.

Colombia, Peru, and Venezuela have the highest incomes in the subregion (see Table 10.2). Venezuela's income is directly related to its dominant oil and mineral exports that continue to be in demand. Debts accrued during the expansion of oil production in the late 1970s restricted development in the 1980s, when oil prices declined. Colombia's income strength in the subregion relates to its diverse mix of economic products, including mineral extraction, agriculture, manufacturing, and the illegal drug trade. Dramatic economic growth in Colombia is restricted by political instability and civil strife. Ecuador suffers from a small home market. Peru's economy experienced a series of booms and busts as it developed different export products. Bolivia's mining output of tin and silver is less in demand on world markets than it used to be. It lacks an ocean outlet since a war over phosphate deposits in the 1800s resulted in Chile taking over its route to the sea at Arica. Bolivia has natural gas reserves that are second only to Venezuela's in South America. Bolivia and Chile are beginning efforts to construct a natural gas pipeline leading from Bolivian reserves to Chilean port facilities. The joint venture will export Bolivian natural gas to California.

Although the formal economies of the Northern Andes countries are not thriving, the informal and illegal economies based on a globally significant role in the production of coca and cocaine grew rapidly in the 1970s. Increasing demand from the wealthy U.S. market fostered dramatic growth and continues to support coca cultivation and cocaine production. Government efforts thus far have not been successful in eradicating coca production. The number of routes from Colombian processing centers through the Caribbean Basin to the United States is rapidly increasing. Coca-growing areas are diffusing from the eastern slopes of the Andes into Brazil.

Personal View

BOLIVIA

Bolivia is one of the poorest countries in Latin America. Although it is five times larger than the United Kingdom, it has only one-tenth of the population. It is a mountainous, landlocked country that long depended on exports of tin and silver for its economy (Box Figure 1). People living in Bolivia experienced important changes after the 1950s that opened new opportunities to them. The following is the personal view of living in Bolivia from Dr. Salinas, who practiced medicine in a mining community for many years.

The 1951 revolution led to huge political changes. Until that time, British mining companies virtually ruled Bolivia, but they were dispossessed when the mines were nationalized in 1952. Although the old mining companies were criticized for the slavelike conditions in which they held their laborers, they left an infrastructure heritage of housing, schools, railroads, and medical facilities that were scarcely improved in the next 50 years.

The new government also instituted land reforms, dividing huge landholdings among those who worked them and setting up cooperative arrangements for purchasing and marketing. This policy was an attempt to diversify the economy away from the reliance on mine products by expanding commercial farming. People of Native American origin, who make up around 50 percent of the Bolivian population, were given the vote for the first time and now wield a major influence in government. For them in particular, the revolution brought access to education for a group that was 90 percent illiterate and had no hope of advancement. Today many Native Americans are qualifying in the professions, and one has been vice president.

The revolution and its aftermath were not a total success. In the 1980s the government shut down mines when the costs of producing tin rose to $17 per pound while the world price was only around $4. In the mining town of Catavi (40 miles west of La Paz),

where Dr. Salinas served the mine laborers and the town, the number of mine workers declined dramatically, from 5,000 in 1982 to 500 in 1986. Dr. Salinas's salary fell in real value at a time of rapid inflation, but the mining unions forced him to attend to his patients until his contract expired. In the late 1980s and 1990s, many of the former government-owned mines were privatized, often bought by Japanese companies, but the new owners employed few miners.

Bolivia as a whole remains a country of contrasts. There is an affluent class of people that includes businesspeople, professionals such as doctors and lawyers, and military officers, many living in the capital, La Paz (Box Figure 2). Some teachers are in this group if they teach in private schools as well as in the poorly paid public school jobs. Most of the elite claim Spanish origins, but many are of mixed ancestry. They live in good houses, and many own luxury cars bought mainly from Brazilian suppliers. If not traveling by car, they travel in special-class trains or well-equipped buses.

There is also a working class of factory, mine, and shop workers. They have regular jobs, but wages are low. They travel in the back of trucks or on buses. There are also many landless and jobless people in Bolivia. Many are rural Native Americans who still do not have access to public education. The lack of facilities in rural areas set off a major migration to the urban areas, especially for access to schooling. When such people reach urban areas, they do not have jobs or housing. They move into the shantytowns, or *barrios pariféricos,* often close to people from their home district, and do whatever they can to earn some money. In the barrios, electricity is available for those who can pay, while water standpipes are shared among some 50 houses. There are stories of Native American families arriving from the country and the parents working in domestic service while the children go to school and become successful professionals. But most Native Americans do not rise so quickly, if at all. Many unemployed peasants obtain a living by begging.

Box Figure 1 Map of Bolivia.

Box Figure 2 La Paz. One of the capitals of Bolivia, built in a valley in the Andean ranges below Mount Illimani. **Photo:** © Wolfgang Kaehler/Corbis.

Drug-related instability in Colombia worsened during the 1990s and early 2000s. Allegations of a political power base that is increasingly controlled by drug money are rampant. Clashes between some government officials who openly oppose the drug cartels and guerrilla groups supporting narcotic cultivation regions often result in kidnappings and executions on all sides. The United States injects large amounts of money into fighting the Colombian drug war, yet poor government relations between the two countries and corruption within the Colombian political system dilute the value of any external assistance efforts. Cease-fire efforts between the Revolutionary Armed Forces of Colombia, a formidable guerrilla group whose base is in the heart of a prime drug cultivation region, and the Colombian government crumbled in 2002, leaving many in the region and around the world wondering if full-scale combat would follow.

Economic Development

Land use and trade patterns implemented during the colonial era established lasting trade relations in which most countries of the region relied on cash crop and mineral resource exports to pay for imports that only the most materially wealthy segment of the population could afford to buy. The governments of several Andean countries convened to explore the formation of an economic union to reverse lasting trade patterns and boost internal economic growth. In 1969 five countries (Bolivia, Chile, Colombia, Ecuador, and Peru) formed the Andean Group of trading partners. In the 1970s Venezuela joined and Chile left. At first this group saw its purpose as creating an internal trading area and keeping out foreign goods. It faltered as members disagreed over which country should produce what goods. In 1988 it was relaunched as the **Andean Common Market** with a greater emphasis on integration with world trade patterns.

Export-Led Underdevelopment

All five of the countries of the Northern Andes subregion suffer from **export-led underdevelopment.** In the late 1800s and early 1900s, the countries of this subregion relied on a series of export-based booms as specific mining and agricultural products came into demand in Europe and North America. Local people and foreigners invested in these products, while the government taxed the exports as the basis of its own income. Any hard currency earned went to pay for imports, on which tariffs were not charged. In the 1930s, attempts at changing this process by encouraging import-substitution manufacturing at home were not very successful because illegal imports undercut local products and only a restricted range of goods was made in local factories. As a result, the Northern Andean countries found that fuller involvement in the world economic system made them vulnerable to external control. Today's dependence on smuggling coca and cocaine arose from the need to find a high-value product with continuing demand in world markets.

Patterns of Economic Diversification

Early colonists in *Peru* established mines in the cordilleras and cultivated crops along the coast to support the people and pack animals that delivered mined silver and gold to the coast. Later Peruvian export booms from the mid-1800s involved less valuable minerals, such as lead and zinc from the cordilleras, guano fertilizer from offshore islands, and sugarcane and cotton from irrigated coastal lands. Oil from the north, iron from the south, and fishmeal from what became the world's second largest fishery (Figure 10.35) added new aspects to Peru's income from the 1950s. Although this output appears diverse, each product comes from a particular area and had a limited period of importance.

Peru dramatically opened its economy to external investment in the 1990s, creating a boom in mining exploration. In 1992 Peru opened Latin America's largest gold mine at Yanacocha near Cajamarca. It began production at other mines by the late 1990s and continued to explore opportunities. Peru's rapid economic growth in the mid-1990s was later slowed by the combination of expanding demand for imports, increasing indebtedness, and continuing poverty among a large sector of the population. Peru experienced a recession pattern similar to that which occurred in Mexico. Peru is among the world's largest producers of cocaine. Huge cocaine-generated revenues may cushion some of Peru's economic challenges, but wealth distribution related to the drug trade is extremely uneven.

Exports of silver and tin dominated the *Bolivian* economy for decades. High elevation limited extensive agricultural development and impeded economic expansion or diversification. A 1980s fall in world tin prices hurt Bolivia's limited economy. The Bolivian economy today depends on other

FIGURE 10.35 Northern Andes: fishing industry. Peruvian fishing boats working the rich grounds of the cold Peruvian current, where nutrients support the food chain from plankton to fish and birds. El Niño events bring warm waters that often kill fish and reduce fishing catches. **Photo:** © Bates Littlehales/Corbis Images.

metallic ores, such as gold, which comprise up to 50 percent of Bolivia's exports. New development is occurring in the eastern interior plains, where nutrient-rich soils and improved transportation make it possible to move more soybeans from field to consumer each year. The crops are trucked across the Andes to Peruvian ports for export. Oil from the eastern Andean slopes makes Bolivia self-sufficient in energy.

Bolivia is one of the world's largest producers of cocaine. U.S. pressure on the Bolivian government to crack down on production in the Chaparé Valley east of Cochabamba has met little success due to internal political disagreement and an economically challenged population. Bolivian economic reform, based on the encouragement of foreign capital investments in state corporations and Bolivian public institutions, has not created a financial environment worthy of challenging the cocaine industry.

Ecuador is a small country with limited upland and interior economic potential. Economic growth occurs primarily on the coast around Guayaquil, where the expansion of fishing for tuna and whitefish, together with shrimp farming, is adding to an increasingly diversified base of light manufacturing, commercial farming for sugarcane, bananas, coffee, cacao, rice, and tourism. The construction of an oil pipeline in the 1970s from the Andes to the port of Esmeraldas increased overall oil production. Petroleum products make up nearly half of Ecuador's exports. New laws opened mining prospects for foreign corporations from Canada, South Africa, France, and Belgium that are testing the viability of gold, silver, lead, zinc, and copper deposits.

Although shrimp farming diversifies the economic base of Ecuador by providing a non-oil export, further expansion is slowed by disputes with environmentalists who claim it is destroying coastal mangroves. The farmers counterclaim that they have to preserve the mangroves in which the shrimp begin life and that they are providing jobs and services in areas that have great poverty, malaria, no electricity, and no health services.

Global attention remains focused on Ecuador's **Galápagos Islands,** where unique animal and plant species attract scientists and tourists from all parts of the world (Figure 10.36). The government of Ecuador is caught between the desire to fully exploit the tourism potential of the islands and the need to preserve a world-class natural heritage site as well as to maintain the conditions that would attract tourists in the first place. Visitor quotas, established by the government to protect the ecological balance of the islands, have been increased several times and are often exceeded. The economic potential of the islands could easily be destroyed if too many visitors degrade the health of the animals and natural resources that make the Galápagos such an attractive destination. An even greater challenge to the Galápagos environment may be taking shape as Ecuadoran migrants move from the mainland coast to the islands in search of tourism jobs or to illegally fish in the marine reserve's waters. Uncontrolled population growth and the fishing of protected resources may pose a more destructive and rapid threat to the fragile balance of the Galápagos ecosystem.

FIGURE 10.36 Galápagos. The volcanic landscape of the Galápagos Islands. **Photo:** © David Zurick.

In the early 1900s *Venezuela* became a major oil producer, raising its income to the highest in the Northern Andes subregion. Oil production in the Lake Maracaibo region of Venezuela increased during World War II and again when world oil prices rose in the 1970s. The government of Venezuela nationalized the oil industry in 1976 and, under the expectation of increasing oil revenues, simultaneously acquired extensive foreign debt for the purpose of domestic investment. The 1980s brought a period of higher interest rates globally, decreased oil demand, and lower oil prices. Countries such as Venezuela faced the seemingly impossible task of making payments on loans out of significantly reduced oil revenues. Venezuela could barely pay the interest on the loans, thus worsening its debt situation. National debt caused investment in oil exploration to lag in the 1980s. In the 1990s, exploration opened to private companies that are linking to joint production and marketing arrangements with the government as discoveries are proved.

The Caracas area is the focus of manufacturing and service industries. Southern Venezuela is a source of mineral wealth from iron and bauxite mines. The mines provide a basis for a manufacturing zone in Ciudad Guyana on the Orinoco River. The manufacturing zone is powered by hydroelectricity that is generated in the Guiana Highland valleys. This region has potential for future economic development but remains isolated by poor transportation facilities and is affected by Venezuela's shortage of investment capital.

Colombia was Latin America's first democracy, and today it has a more diversified economy than the other Northern Andean countries. Colombian cities such as Bogotá and Medellín grew from colonial settlements established in elevated basins in the Andes mountains. Local manufacturing developed in the 1800s to supply a growing market. Medellín and the capital, Bogotá, became the main industrial centers for textiles, steel, agricultural equipment, and domestic goods. In the lower valleys and near the northern coast in contemporary Colombia, large plantations raise tropical cash crops, including cotton, sugarcane, bananas, cacao, and rice for export. Colombia is the world's second largest producer and supplier of coffee, after Brazil. Coffee, which is Colombia's largest legal export, thrives in the optimal conditions of well-drained mountain slopes, fertile soil, and sufficient rainfall. Colombia began to grow new varieties of corn and rice in the mid-1990s, which were bred for higher yields in the poor soils of the eastern savannas. The new corn and rice, coupled with livestock and locally grown cattle feed, are meeting with success in this sparsely settled area. Marijuana, coca, and poppies are major crops in the more rugged and secluded reaches of the mountains. Colombia is also a mining country with considerable resource stocks of iron, coal, oil, natural gas, gold, and emeralds. Colombia's important overseas trade is regularly hampered by continuous internal political strife. In the 1980s and 1990s guerrilla warfare, based partly on political repression and partly on strife among the drug syndicates, defeated government attempts to capitalize on the country's strengths.

The Northern Andes and the International Drug Trade

The countries of the Northern Andes comprise one of the world's primary drug-producing regions (Figure 10.37). The initial system in the subregion centered on cultivation areas in Peru and Bolivia, with processing or production centers based in Colombia. The expansion of Bolivian cultivation coverage slowed in the early 1980s, and the Peruvian coverage declined in output following a fungus infestation of the crop. Colombian cultivation increased and expanded to coca, opium poppies, and marijuana, and is now greater than that in Bolivia.

The global demand for cocaine from consumers in the United States, Europe, and Asia has a devastating impact on the local politics, agriculture, and social fabric in the Northern Andes. An excess of production over market demand in the early 1980s caused the price of cocaine to fall by 75 percent. The huge drop in price dramatically opened the international market to those with less disposable income, resulting in a subsequent rapid increase in demand. The United States led efforts to prevent drug production in these countries, but its efforts were diluted by limited cooperation from the governments and farmers of Bolivia and Peru. Although Bolivian farmers planted other commercially valuable crops on former coca-growing land as a result of U.S. aid, the acreage under coca remained the same. Peruvian government attempts to halt cocaine production are limited in part by fears of the resurgence of guerrilla groups among disaffected former coca growers. Subsistence and small-scale farmers are attracted to working on large coca farms where average wages are far in excess of what they would otherwise earn.

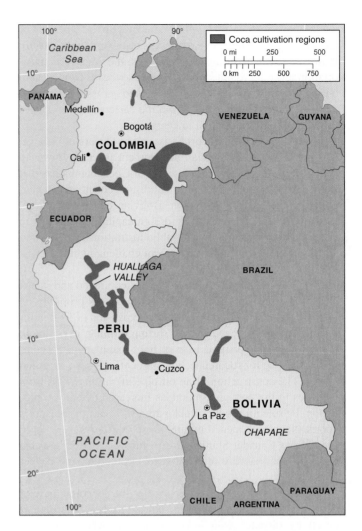

FIGURE 10.37 Northern Andes: coca-growing areas. The dominance of Peru and Bolivia in the 1980s gave way to rapid increases in cultivation lands in Colombia in the 1990s and early 2000s.

Other farmers are forced by threats of violence to become part of the drug labor force, and in many cases, their farms are forcibly taken over by the drug cartels.

In 1995 Colombia attempted to adopt a tougher stance and announced its intention to eradicate production within two years. Programs of eradication, however, have to compete with the high prices farmers get for their drug crops and with the guerrilla-backed disorder that such farmers may support, including the strategic occupation of isolated oil-pumping stations or airports. Chemical defoliation of coca-growing areas paralleled a doubling of the output of coca. By 1998 the failure of such programs led the United Nations Drugs Control Program to try to establish a new global drugs control convention. It has to contend with the income from coca growing that lifts farmers out of poverty and the long-established custom of coca-leaf chewing among Native Americans living at high altitudes in the Andes.

Test Your Understanding
10D

Summary

The Caribbean Basin includes four large islands (Cuba; Hispaniola, with Haiti and the Dominican Republic; Jamaica; and Puerto Rico), many small islands, and the Guianas on mainland South America. They were mainly settled by Europeans to develop sugar plantations and by imported African slaves.

Most of the former colonies are now independent, and their economies are based on more varied agricultural products, tourism, local mineral products, and some light manufacturing. Some have developed offshore financial services mainly linked to the United States.

The Northern Andean countries of Bolivia, Colombia, Ecuador, Peru, and Venezuela are dominated by the high mountains that isolate internal plains covered by savanna and tropical rain forest from the main centers of population. Their economies are based on mineral production, including increasing oil discoveries, and export crops. These countries include major drug producers, especially of cocaine but increasingly of marijuana and opium.

Questions to Think About

10D.1 How are the physical and human geographies of the Caribbean Basin different from those of Mexico and the Central American countries?

10D.2 What is the impact of the Andes Mountains on the human geographies of the subregion's countries? What role does physical geography play in drug cultivation?

10D.3 What local and global relationships are involved in Andean drug cultivation?

10D.4 Why are the Galápagos Islands important? What issues does the Ecuadoran government face in trying to preserve the Galápagos environment?

Key Terms

Greater Antilles	economic leakage
Lesser Antilles	commonwealth status
Caribbean Community and Common Market (CARICOM)	Andean Common Market
	export-led underdevelopment
Association of Caribbean States	Galápagos Islands

Brazil

Brazil is the largest country in both area and population in all of Latin America (see Table 10.2). Brazil has three times the area of Argentina, the second largest South American country. In 2005 Brazil's population exceeded 184 million, which is much larger than the total population of any other Latin American subregion. The people of Brazil have a strong sense of national pride and a passion for cultural expression. Although the Portuguese language and Roman Catholic church provide some national homogeneity, the diverse contemporary human geography and cultural expression of Latin America's largest country are tangible extensions of the historical mixture of European, African, and indigenous influences. The physical expanse of Brazilian territory incorporates a vast array of topography, climate, and vegetative cover, and holds a diverse and abundant supply of natural resources. Brazil presents an immense economic market, has the greatest economic diversity of any Latin American country, and was among the world's 10 largest economies in 2004. Despite decades of rapid economic growth between 1940 and 1980, and the size of its economy, Brazil faced increasing problems from the 1980s that resulted in high inflation, declining economic growth, and growing numbers of materially poor people.

The establishment of sugar plantations along the northeast coast of Brazil during the colonial era began the course of European-induced human and physical geographic change experienced in so many locations in Latin America. What most distinguishes the colonial history of Brazil from that of other Latin American countries is the lasting influence of the Portuguese rather than the Spanish. The Portuguese language and unique blend of Portuguese, African, Caribbean, and indigenous traditions give Brazil a national flavor that is distinctively different from Spanish Latin America. Portuguese economic institutions forced the migration of millions of Africans to Brazil. The contemporary human mosaic includes the largest African and Afro-Caribbean heritage of any Latin American country.

Regions of Brazil

The expansive size and geographic diversity of Brazil dictate further subdividing the country for geographic examination (Figure 10.38). Politically, Brazil is divided into states that are part of a federal government system. Further human geographic patterns relate to natural environment contrasts, the history of European settlement, and political emphases among the different states.

- Nearly 60 percent of Brazil's territory is drained by the Amazon River network. Much of this area is covered by tropical rain forest. Although the Amazon River basin has relatively low population densities, development efforts led by the Brazilian government are increasingly encroaching on tropical forests of the region, and deforestation is rampant in some parts of the Amazon Basin.

- The northeastern coastlands and plateau were the first settled areas of Brazil. Sugarcane plantations prospered until competition with the Caribbean Basin began in the later 1600s. Periodic droughts devastate the interior of this region (as in the late 1990s), causing many to emigrate to other parts of Brazil.

FIGURE 10.38 Brazil: major geographic features. The states, rivers, and major cities.

- The main economic development in Brazil occurred in the southeast around Rio de Janeiro, the colonial and national capital for 200 years, and São Paulo. The early industrial focus of the region centered on mining and commercial agriculture (based on coffee cultivation). Today this region is a major center of manufacturing and financial service activities for Brazil.

- The three southern states of Paraná, Santa Catarina, and Rio Grande do Sul became centers of growing population and agriculture from the 1930s. After 1950, the coffee crop exhausted the soils of the area west of São Paulo, causing cultivation to spread to the west–southwest along new railroads into northern Paraná until bursts of cold wintry air from the south demarcated its limits. Cattle and a variety of temperate and subtropical agricultural products, including oranges, are produced in these states, settled largely by immigrants encouraged to move from Germany, Italy, and Japan. Hydroelectricity generated on the Paraná River and its tributaries powers manufacturing in the area.

- Inland of São Paulo, straddling the drainage divide between the Amazon and Paraná–Paraguay river systems, a new area of farming development in Campo Cerrado grew in significance from the 1970s. With improved road and communication links, this area became one of the world's main soybean producers, with some 6 million hectares (about 15 million acres) growing 12 million tons per year in the 1990s. Farm management, use of machinery, fertilizer inputs, and marketing procedures are highly organized, and yields rival the best in the U.S. Midwest. New towns grew to over a quarter of a million people within 10 years, and Brazil now exports a greater value of soybeans than of coffee.

Economic Development

Brazil is a leader in Latin America in targeted economic modernization and development. Brazil's rates of consumer goods ownership (see Figure 10.17) are similar to those of the countries in southern South America, but human development indicators for Brazil are not as high as those for Argentina or Chile. Economic and related social problems stem in part from a long-term dominance of public involvement in the economy and the high rates of population increase in the 1960s and 1970s. Other factors contributing to lower human development indicators include the shorter-term impacts of the 1985 constitutional changes that enshrined provisions for social programs and job security. The constitutional provisions maintained high levels of people in employment but placed a huge burden on the country's finances. In addition to being responsible for social provision administration, the federal government had to distribute a high proportion of its income to Brazilian states. The combination of international debt servicing, the need to cover constitutional demands, financing domestic development projects, and numerous cases of government corruption led to inflation that rose to 50 percent per month in early 1994. In the mid- and late 1990s the Brazilian government made changes that reduced the inflation rate and altered the constitution to remove many of the 1985 provisions.

Varied Economic History

Brazil's colonial economic development began with coastal sugarcane plantations along the northeast coast, interior cattle ranches, mining for gold, and trading for Amazon River basin forest products. In the 1800s the area west of São Paulo became the world's chief coffee producer. This required the recruitment of workers from Europe and the building of railroads through British financing. The Amazon River basin became a rubber producer of world importance in the late 1800s, but competition with Asian producers ended that boom at the outset of the 1900s.

By the early 1900s, the southeastern coffee-producing states contributed over 70 percent of the Brazilian GNI. São Paulo and its port city, Santos, grew as the hub of this development (Figure 10.39), but by the mid-1900s, the soils of the older coffee lands became exhausted and new coffee plantations

FIGURE 10.39 Brazil: coastal lifestyles. Guajara beach, near the port of Santos, southeastern Brazil. The many apartment blocks around the beach house families mainly from prosperous São Paulo, 80 km (50 mi.) inland. **Photo:** © Michael Bradshaw.

sprang up westward along the railroads into Paraná state, with growth especially around the city of Londrina. Brazil's development of primary products for export paralleled that in many other Latin American countries. The sugarcane, gold, and rubber export surges lasted for a while and then declined; coffee production was maintained only by shifts to new lands.

Economic stagnation in the 1920s and the world economic depression of the 1930s caused a crisis during which the military took over the government of Brazil. From this point, the Brazilian economy diverged from that in many other Latin American countries by basing internal economic growth on a form of state capitalism. The federal government assumed the ownership of major economic activities and established tariffs to protect internal manufacturing. Import substitution reached its greatest level of activity in Brazil aided by the huge internal market. Durable goods such as automobiles and domestic appliances that were made abroad were subject to an 85 percent import duty, while some, such as computers, were excluded altogether.

In 1937 the new government nationalized the iron mines in the state of Minas Gerais. The huge mines around Itabira became 80 percent government-owned, and production was increased with 70 percent now being sold overseas, delivered by mining company ships. A steel industry grew up along the railroad from Itabira to the port of Tubarão, with the largest integrated mills at Volta Redonda being established in 1940 and extended in 1970 to double their output.

Modern Mining

The national iron ore company is developing the iron ore mining Carajás Project in the eastern Amazon River basin. A railroad built in 1985 takes the ore to a port near São Luis on the northern coast for export to Japan, the United States, and Europe, while the Tucurui Dam generates hydroelectricity for mining and industrial needs. The controversial project required the clearance of 3 million hectares (7.4 million acres) of rain forest, and much of that area is now used for ranching and small farms.

Other Brazilian mining developments include the production of manganese (used in hardening steel), tin, and bauxite. One-third of the world's bauxite resources occur east of Manaus in the Amazon River basin. In the upper reaches of the basin, small deposits of gold attract thousands of independent miners, but the use of mercury to separate the gold pollutes the rivers, and this activity forms a major intrusion in the lives of Amazon tribes such as the **Yanomami.**

Brazil is increasing its production of petroleum. The oil consumers' crisis of the 1970s hit Brazil hard and was responsible for much of its high debt. An immediate response was the development of **ethanol fuel** from sugarcane. The 1975 policy for ethanol development guaranteed a price at 65 percent of gasoline prices. The aim was for one-third of all cars to use this fuel. Early technical problems were overcome, although ethanol-using cars still have low resale prices. The ethanol policy was hampered by the lack of rural pumps selling the fuel, the cutback in sugarcane production as a result of lower subsidies, and the development of Brazilian oil products. It is now necessary for Brazil to import sugarcane to make ethanol, and the pump prices for ethanol fuels are higher than those for gasoline. Petrobras, the Brazilian state oil company, was established to import, refine, and distribute oil products but now finds itself an oil producer with offshore wells along the eastern and northern coasts and major reserves in the western Amazon. By the mid-1990s, over 60 percent of Brazil's oil needs were produced from domestic sources.

Farming Today

The government continues to encourage agricultural production, but agriculture's proportion of GNI fell as that of manufactured goods rose. Coffee and sugarcane remain important exports, but coffee, once the mainstay of the economy, now makes up less than 10 percent of the agricultural exports. New developments, such as the growing of oranges and other citrus in the south and tropical fruits and nuts farther north, added to the diversity of commercial farming. The major new development in agriculture is the efficient commercial production of

soybeans that shot Brazil to the top of international traders in this crop (Figure 10.40). Government-based research, advice to farmers, and marketing facilities contributed to this success. In many localities, however, one commercial crop tends to be dominant, and so the problems of reliance on monoculture in the context of shifting world markets are still felt.

Manufacturing Expansion

The largest manufacturing sector in Brazil is the production of automobiles and trucks, concentrated in the southern São Paulo suburbs. Autolatina, in which Ford and Volkswagen shared ownership with the government and produced cars specifically for the Brazilian market, made about half the annual output of 1 million cars until 1995, when Ford and Volkswagen split to take advantage of new sales opportunities in Brazil and Argentina. General Motors and Fiat (near Belo Horizonte) have similar but separate manufacturing arrangements. In the 1990s the lowering of import tariffs led to further competition, falling prices, and expansion of Brazilian car sales and production, especially of small cars. Output of 900,000 cars in 1992 doubled by the mid-1990s. By 1995 Renault (France) and Hyundai

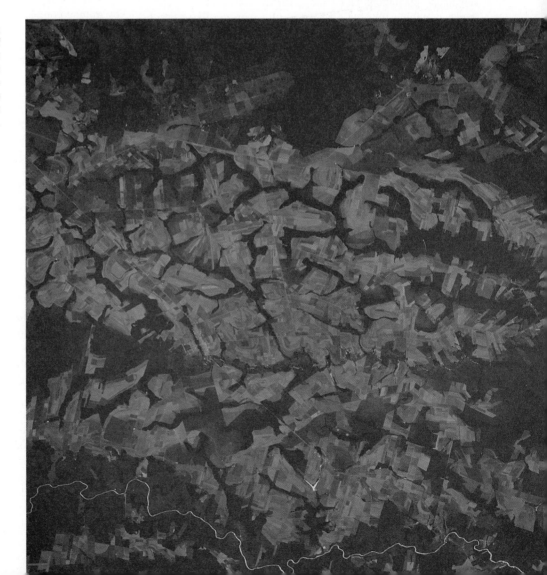

FIGURE 10.40 Brazil: western Mato Grosso. A space shuttle photo of new cattle ranches and soybean farms in a 60 km² (40 mi.²) area. The plateau surface has been cleared and the steep, floodable intervening valleys left in forest. Lands along the Río Sangue (bottom) are flooded up to 10 m (40 ft.) deep for three months in the summer. **Photo:** NASA.

FIGURE 10.41 Brazil: hydroelectricity. The spillway at the Itaipu hydroelectricity station on the Paraná River between Brazil and Paraguay. The 18 massive turbines make this Brazil's largest hydroelectricity investment. Although the electricity is shared between the two countries, Paraguay sells nearly all of its quota to Brazil, where there is greater demand. **Photo:** © Michael Bradshaw.

FIGURE 10.42 Brazil: Amazon River. The city of Manaus has an opera house that was built early in the 1900s to cater to 2,000 or so rubber barons who lived there and controlled trade in the valuable commodity. Now refurbished, the opera house is in the center of a city of over a million people, with high-tech industries and improved communication by river, road, and air. **Photo:** © Michael Bradshaw.

(South Korea) decided to build new factories in Brazil to take advantage of the growing market and the Mercosul free-trade agreements.

The Brazilian government owns many large corporations that employ thousands of citizens. Electrobras is a nationalized corporation that produces and distributes electricity. Small hydroelectricity projects on the plateaus in the east gave way to huge projects in the interior (Figure 10.41). The world's largest hydroelectricity project is at the **Itaipu Dam** on the Paraná River at the border with Paraguay. It was opened in 1983 and produces one-fifth of Brazil's power (and much of Paraguay's).

Economic globalization connections flourish in the **free-trade zone** established in the Amazon River city of Manaus in 1966 (Figure 10.42). In this zone, foreign companies may import materials for assembly without tariffs and export the assembled products to other countries. Manaus is now Brazil's fastest-growing city and the second in manufactured goods value after São Paulo. The local human and physical landscapes are dramatically different from outlying forest areas. There are more than 6,000 factories in the huge industrial parks of Manaus, many of which are the production facilities of foreign-owned companies, including Honda, Sharp, Kodak, Olivetti, Toshiba, Sony, and 3M. Manaus was transformed by the number of visitors from elsewhere in Brazil coming to purchase goods that they were not allowed to import into Brazil. The "electronics bazaar" occupies a maze of streets in the city

center. People come to buy the foreign-made computers and domestic electronic goods that Brazilian companies, protected by high tariffs, make poorly and sell at high prices.

Development Issues in the Amazon

The main failure of government efforts is in the Amazon River basin. Efforts to settle the lands along the Trans-Amazonica Highway from the 1950s under the regional development ministry, SUDAM (*Supertendencia para o Desenvolvimento da Amazonia*), saw little progress. Land was not surveyed properly, families with little farming experience were recruited, and bureaucratic convenience caused most of the grants to go to large-scale ranches. When a new road, BR 364, was built into the states of Rondônia and Acre in the western Amazon, thousands of settlers entered the area on their own initiative and cut plots in the forest without government encouragement. Dramatic increases in forest clearing and burning aroused international outrage. Many of these settlers failed to grow anything on their plots. The alternative attraction of the rapid development of soybean farming in the Campo Cerrado in the early 1990s led to a slowing of such settlement along with the linked forest burning and destruction in the western Amazon. In early 1998 devastating fires reversed this downward trend by destroying rain forest in the north of the Amazon River basin along the Venezuelan border.

Another aspect of government activity in the Amazon is the Indian rights agency, **FUNAI** (*Fundação Nacional do Indio*), which seeks to provide education, health care, and support for Native Americans. Unfortunately, it is underfunded and, although it has taken some steps to protect its Indian reserves, ranchers and miners frequently penetrate tribal lands and begin to develop them before FUNAI can act. Numbers of indigenous communities in Brazil are diminishing along with the area of the land on which they are sequestered.

Indebtedness and New Policies

Following its huge infrastructure and state-owned enterprise investments of the 1970s and 1980s, Brazil found itself deeply indebted to world banks in the late 1980s, reaching over $120 billion in the 1990s. The cost of oil imports and the huge public works projects to develop iron mining and hydroelectricity, to build Brasília, and to construct new roads and airports, together with reported political corruption, contributed to this national debt. Despite Brazil's being by far the most prosperous country in Latin America and having a trade balance in which exports were double that of imports until the early 1990s, most of the surplus income went to the payment of debt interest. Defaults on loan repayments caused wider foreign investments in Brazil to collapse. **Hyperinflation,** or the process in which rising prices (and often wages) feed upon themselves and potentially spiral out of control in a vicious cycle, is another financial problem that is an obstacle to further development because it stifles savings. Hyperinflation may significantly devalue a currency and wreak havoc on a country's economy. Brazilians became used to living in an environment of high inflation. In early 1994 inflation reached 50 percent per month (over 1,000 percent per year), but renewed attempts to bring inflation under control marked government financial policy at the end of the 1990s.

The 1990s also saw the start of new government policies moving away from the import substitution and state ownership of manufacturing that had been major factors in economic growth within Brazil. Tariffs on goods not produced in Brazil were lowered. For example, the 85 percent levels on foreign autos and appliances fell to 20 percent. Some products, including computers, remained illegal imports unless purchased in Manaus. Foreign investment was encouraged, and a sign of these changes is that Japanese car manufacturers are establishing outlets in the major cities.

Brazilian economic diversity places it on a par with some of the newly industrialized countries of East Asia. It also hides the gap between the materially rich and poor within Brazil that is reflected in the uneven distribution of land, the high levels of unemployment, and the strength of the informal economy. Brazil has the capability to develop its economy further, but current economic and social problems and environmental issues make the future difficult to assess.

Favelas: Urban Strife in Brazil's Cities

Within the Brazilian cities, the shantytowns, known as **favelas** in the southern parts of the country, house millions of materially poor people who cannot be accommodated by the formal patterns of housing construction, infrastructure, and service provision (Figure 10.43). It is estimated that over 7 million people in São Paulo and up to 6 million people in Rio de Janeiro live in favelas. Favelas vary in their character. In and around the city centers, they occupy gaps in the built environment, in which families erect their own minimal accommodations. It is rare for these favelas to have water, electricity, waste disposal, or anything resembling a road. They provide no security for the inhabitants and are often centers of disease and crime. Many such favelas have been in existence for 10 years or more. Old apartment blocks in the city centers, which were abandoned and taken over by impoverished people, resemble favelas in many ways, with several families occupying each room. On the outskirts of the cities, different types of favelas are pushing

FIGURE 10.43 Rio de Janeiro. The dramatic physical geography of Rio de Janeiro contributes to the strong pride Brazilians feel for this city (a), which also attracts millions of visitors from around the world. (b) Contrasting with expensive hotels and condos along Copacabana beach, the Rocinha favela crowds thousands of materially poor Brazilians into marginal structures on a steep hillside. **Photos:** (a) © R-F/Corbis, (b) © Stephanie Maze/Corbis.

(b)

(a)

the built-up area outward at a rapid pace. These favelas spring up almost overnight and grow by thousands of people within months. Once established, favelas may be provided with some basic roads and may soon acquire utilities, shops, and schools.

The favelas are also the primary site where thousands of **Brazilian street children** base their nomadic and often shelterless existence. This serious issue focuses attention on global connections and local voices in a negative manner. Economic, political, and social globalization forces have neither eliminated nor meaningfully diminished the social conditions that combine to place hundreds of thousands of Brazilian youth in jeopardy. Some argue that multinational corporations, international banking and development institutions, and intergovernmental trade relations have increased Brazilian poverty rates rather than reduced them. Brazilian government debt and resource-export economic goals also figure prominently in the social environment that produces so many children of the streets. The socioenvironmental conditions produced from such domestic and foreign relationships include extreme disparities in wealth, high poverty rates, relatively high

death rates for women in childbirth, high rates of teenage pregnancy, high illiteracy rates, and high jobless rates. Street children either literally live on the streets with no shelter or work the streets at a young age, trying to earn money for their survival or that of their families. The street children often end up in prostitution or an illegal narcotics trade, and many contract life-threatening diseases. The most challenging situation is the reported periodic murder of groups of street children (often believed to be the work of vigilantes), whose bodies are abandoned in ravines or concealed areas. Although local and global human rights groups are persistent in their attempts to attract significant attention to the plight of the street children, the problem remains very serious.

Southern South America

Argentina, Chile, Paraguay, and Uruguay form the southernmost part of South America and are sometimes called the "Southern Cone" because of their combined shape on a map (Figure 10.44).

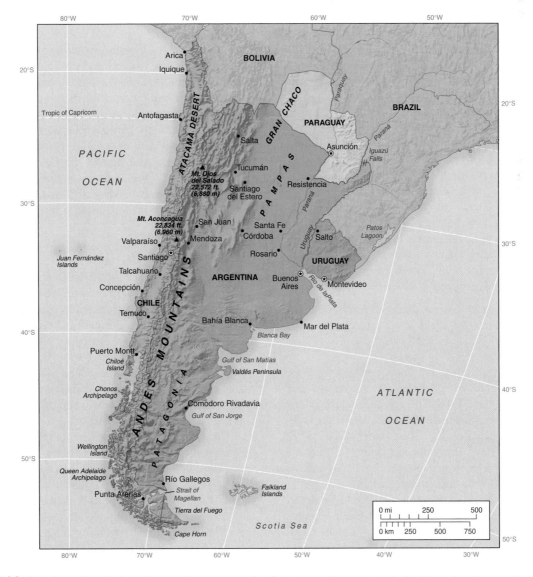

FIGURE 10.44 Southern South America: major geographic features. The countries, Andes Mountains, rivers, and major cities.

Physically, this long and narrow subregion extends far south into the midlatitudes. Average temperatures cool and rainfall amounts increase progressing from the north to the south in the subregion. The Hispanic cultural imprint is a strong common element of the human geography of Southern South America. The majority of the population of each country speaks Spanish as their first language and adheres to Roman Catholicism.

Southern South America has the highest percentage of European descendants of the subregions on the South American continent. Sparse indigenous populations and climatic environments more reminiscent of Europe led to the establishment of pervasive European-based populations in Argentina, Chile, and Uruguay. The levels of consumer goods ownership, human development, access to clean water, and energy usage are among the highest in Latin America (see Figure 10.17), although Paraguay does not reach the same levels as the other three countries. Argentina's strong economy experienced a major economic crisis in 2001 and 2002. Although the Argentine economy showed signs of slow rebounding, recovery will take several years. Argentina's population dominates the four countries of the subregion, with 60 percent of Southern South America's people living here in 2005 (see Table 10.2).

Countries

The diverse physical and human geographic spatial patterns of Southern South America relate in part to the subregion's extensive latitudinal coverage from north to south, a relatively narrow width, the rugged Andes Mountains and geologic instability of the west, and the influences of the cold Pacific and warm Atlantic Ocean currents. All the physical properties of the region support a wide range of temperature and precipitation regimes, resulting in varied settlement and land use patterns. The physical geography of Southern South America, coupled with the historic goals of Spain, created a vibrant contemporary human mosaic in the subregion's countries: Argentina, Chile, Paraguay, and Uruguay. The Andes Mountains proved an important dividing factor between types of settlement.

Economic Development and Human Activity

In 1991 Argentina, Brazil, Paraguay, and Uruguay formed the trading group of **Mercosur** following discussions between Argentina and Brazil since 1988. The countries agreed to cut internal tariffs as intraregional trade increased, despite the problems of high inflation rates in Brazil. The group produces half of Latin America's GNI, and the European Union—Mercosur's main trading partner—is showing interest in negotiating a closer agreement. Chile and Bolivia are associate members.

Argentina, Chile, and Uruguay followed economic development patterns that are somewhat typical of Latin America. Paraguay always lagged behind the others. Until the mid-1900s, Argentina and Uruguay were the most prosperous. Since the 1950s the Argentine people have been through a series of economic boom-and-bust periods. By the 1990s the Argentine

economy was very robust, and the citizens of the country were enjoying a relatively prosperous standard of living. It appeared to the global economic community that Argentina's troubles were far behind it. A series of both local and global actions combined to devastate the economy in Argentina by the end of 2001 and bring the middle classes into the streets of Buenos Aires in protest. The combination of government policies, global trade relations, and government corruption necessitated devaluation of the Argentine currency. The Argentine economy now may be in an incipient stage of recovery, but any positive change is likely to be slow.

Growing Engagement with the World Economy

In the late 1800s Britain and other European countries recognized the potential of the pampas grasslands as a source of grain and livestock for the expanding demand in the industrial Northern Hemisphere countries. By then, refrigerated ships could carry meat products to these countries. British capital built railroads around Buenos Aires and Montevideo and encouraged the owners of large estates to move from extensive livestock management to intensive grain farming.

To cope with such economic expansion, the governments of Argentina and Uruguay in particular encouraged immigrants from Europe to come and work on the estates. Many of the immigrants soon moved into the towns. The diffusion of commercial farming methods led to the final control of the warlike Pampas Native Americans, who had kept settlement at bay until the 1880s. In similar fashion, German immigrants to the southern part of central Chile in the mid-1800s removed the Araucarian Native Americans, who had prevented occupation of the land south of the Río Bío Bío. By the early 1900s, the lands around Buenos Aires and Montevideo vied with some European countries in both the quantity and quality of life.

Economic Isolation

The economic depression of the 1930s and World War II in the 1940s shifted these countries away from dependence on the farm and mine products demanded by Europe. The countries of Southern South America introduced a broader base of manufacturing protected by high tariffs to reduce dependence on Europe and the United States. Import substitution was at first successful, but its impact was reduced after 1950 as the United States and Europe reestablished their manufacturing markets with better and cheaper goods, while prices for primary products fell because of overproduction.

The economic problems were compounded by a series of governments, both civilian and military, that tried to maintain their protected industries in order to reduce dependence on primary product exports. Their reliance on import-substitution products tied to the needs of their own country's limited market, however, made it difficult for them to sell these goods abroad the way Brazil, with its huge internal market and diverse range of products, managed to do in selected sectors. Primary products still constitute a large proportion of the exports of all these countries, although they have also been diversified. Manufacturing constitutes over 25 percent of employment and GNI in Chile, Argentina, and Uruguay but under 20 percent in Paraguay.

Chile

In Chile, the dictator Augusto Pinochet's regime that lasted from 1973 to 1989 established an economy based on open trading, increasing exports, good financial management that withstood the fallout of the Mexican collapse in 1995, and low unemployment. After the 1989 democratization of government, the new government followed similar policies with the addition of a greater emphasis on social programs to reduce the numbers of poor people.

At the same time, Chile's economy diversified. Copper, which made up 80 percent of Chile's exports in the 1970s, was down to 44 percent in the late 1980s, although the amounts increased. Agricultural products such as apples, soft fruits, grapes, and wine, together with fish, lumber, and new minerals, grew in proportion. The agricultural products and lumber come from central Chile, where a large proportion of the population lives. Copper mining is important in the Atacama Desert of northern Chile, with the world's two largest mines at Chuquicamata and Escondido along with a series of new large mines being prospected and developed in the late 1990s. After the highest-grade copper deposits were mined out by the early 1900s, Chile used U.S. technology to mine lower-grade ores. Its companies now lead the world in refining technology.

Paraguay

Paraguay remains a materially poor country. The larger portion of the country to the west of the Rio Paraguay, the mainly semi-arid Gran Chaco, remains lightly settled apart from military camps and lumber operations. The quebracho ("ax-breaker") tree is its commercial timber product. The tree is very hard and in demand for railroad ties and its tannin extract.

The eastern part of Paraguay is more developed, with forest products being diversified by commercial agriculture. Livestock products and some industrial crops are exported. Paraguay gets its power from hydroelectricity and earns foreign exchange from its share of the Itaipu project on the border with Brazil (see Figure 10.41) by selling most of its power to Brazil. Transportation connections with Brazil have improved, and that is now the main direction of trade. Paraguayan border towns such as Ciudad del Este sell cheap consumer goods to Brazilians and are involved in smuggling between Brazil and Argentina.

Uruguay

Uruguay remains a largely agricultural country with meat and wool exports; but the industrial developments in and around Montevideo, powered by hydroelectricity generated along the Rio Negro, draw workers from the rural areas. Over 80 percent of Uruguayan trade passes through Montevideo, which has over half the country's economic product and population. It developed an offshore banking industry. It attracts tourists to its beaches in an industry that contributes more than either farming or manufacturing to GNI, with 2.2 million visitors in 2002 (up from 1.2 million in 1990).

Argentina

In 1995 Argentina negotiated a cooperative venture with the United Kingdom to develop the oil potential of the continental shelf area southwest of the Falkland Islands. Although efforts have not produced oil, the area is believed to hold significant deposits. Drilling began in 1998 north of the islands (outside the Argentinean area), but commercial production is many years away.

Argentina has a larger manufacturing base than Chile. The upturn in world economic activity and economic restructuring in the 1990s resulted in more investment, mainly from foreign sources, in Argentinean manufacturing. Argentina became the fastest-growing economy in Latin America as it established what appeared to be a stable financial system.

Most Argentinian factories are based in and around the Buenos Aires metropolitan area. The farm products of the pampas are still important to Argentine trade, but oil and gas resources fuel new industries, mainly on the coast. Argentinean service industries developed as part of this growth. In 2002 nearly 3 million foreign tourists visited centers on the coast and in the mountains. Some of the older settlements, such as the northern cities of Tucumán, Córdoba, and Mendoza, where the economy is still based on irrigation agriculture, do not grow as rapidly because of the dominance of Buenos Aires in attracting manufacturing investment. Some of the inland centers, such as Jujuy near the Bolivian border, remain poor and overwhelmed by Bolivian migrants. In the late 1990s, the Mexican crisis and fears of open competition with Brazil as Mercosur develops led to a slowing of growth in Argentina, which contributed to the economic crisis of 2001.

Buenos Aires, Argentina

The rapid growth of Buenos Aires followed patterns that are common to other cities in Southern South America. After slow growth in its early history, Buenos Aires expanded in the late 1800s as commercial farming took hold in its pampas hinterland. The built environment developed a trading and industrial waterfront and a central thoroughfare at right angles along the route inland. Rapid growth occurred in both the economy and the immigrant population around 1900. It resulted in the middle class of skilled workers and office workers moving out to new suburbs, while the poor and most affluent remained in the inner city. Amenities came slowly to the suburbs. During the later 1900s, increasing rates of population growth produced squatter settlements and rising inner-city population densities as apartment blocks replaced mansions. Population growth stagnated in the coastal industrial areas. From the 1960s, much of the commercial and industrial activity moved out of central Buenos Aires to the city edges. Attempts were made to divert the overcrowding in the Buenos Aires metropolitan area by placing new projects and development in other centers farther inland from Buenos Aires, but they have not been successful. By 2003, the total metropolitan population of 13 million made Buenos Aires the third largest city in Latin America.

Test Your Understanding
10E

Summary

Brazil is the largest country in Latin America and contains one-third of the region's population. Its economy is in the world's top 10, but extremes of riches and material poverty exist side by side. Brazil consists of contrasting regions: the sparsely settled tropical rain forest of the Amazon River basin, the industrial cities around São Paulo and Rio de Janeiro, the poverty-stricken northeast, the commercially farmed plateau interior around Brasília and Campo Cerrado, and the southern farming and industrial states. Much of Brazil's economy is controlled by the government through import duties and ownership of corporations that are responsible for utilities, encouraging farming developments, and the high-tech sector.

Southern South America (Argentina, Chile, Paraguay, and Uruguay) has midlatitude climatic environments. The Andes form a distinctive boundary between Chile and Argentina. These countries, apart from Paraguay, have greater proportions of European-origin peoples and more diversified economies than other parts of Latin America outside of Brazil, but they suffered from unstable political conditions and protectionism until the 1990s.

Questions to Think About

10E.1   What are the main economic uses of the Amazon River basin? How have attempts to settle the area solved the problems faced?

10E.2   How would you summarize the main features of the distribution of population, ethnicity/ancestral heritage, agriculture, and manufacturing in Brazil?

10E.3   Why did the European settlement of Southern South America occur later than in other parts of Latin America? What effect did this have on those countries' human geography?

Key Terms

Yanomami	hyperinflation
ethanol fuel	favelas
Itaipu Dam	Brazilian street children
free-trade zone	Mercosur
FUNAI	

Chapter 11

North America

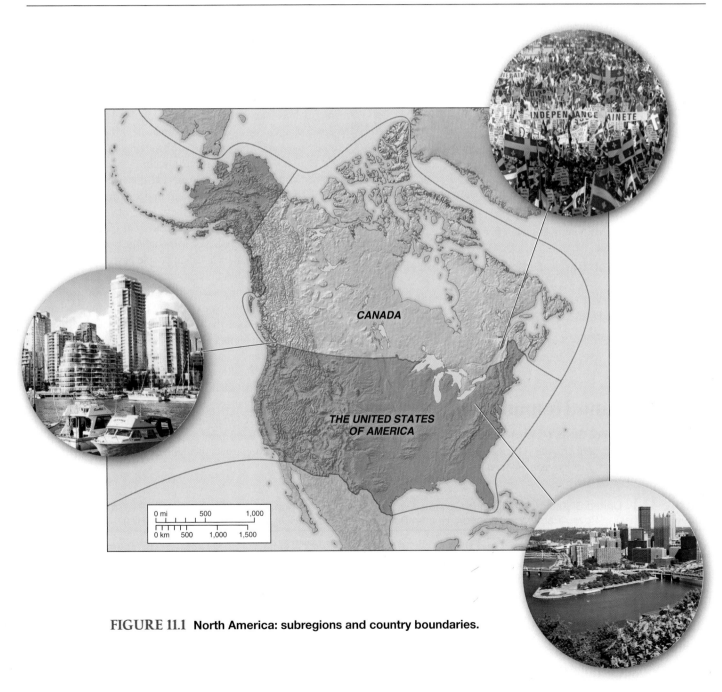

CANADA

THE UNITED STATES
OF AMERICA

| 0 mi | | 500 | | 1,000 |
| 0 km | 500 | 1,000 | 1,500 |

FIGURE 11.1 North America: subregions and country boundaries.

Themes in This Chapter

The high living standards enjoyed by the residents of Canada and the United States are very strong pull factors for migrants from all world regions who come to North America seeking prosperity and stability. The majority of immigrants congregate in the vast urban areas existing in both countries, where opportunities are greatest. The dynamics of North America's migration patterns are not just centered on those immigrating from other countries, but are also strong for the region's domestic population, especially in the United States. The young adults of the United States are an especially mobile group. Young Americans frequently move to the cities or locations where they perceive their job opportunities to be the greatest. The major themes are:

A region of past and present cultural intersection.

Diverse natural environments and abundant natural resources.

Global connections to the region.

People on the move: internal mobility and immigration.

The material wealth and high living standards of the region.

A region of large cities.

Two subregions (Figure 11.1):

- United States of America: material wealth, global influence, urbanization, and diversity.
- Canada: resource wealth, southern population concentrations, high latitudes, devolutionary pressure and national identity, and living in the shadow of a giant.

Point–Counterpoint: North America Free Trade Agreement

Point–Counterpoint: Hurricane Katrina and Metropolitan New Orleans

Geography at Work: Using Geography to Aid Public Policy: The White House

Personal View: Canada

Territory and Human Settlement

Canada and the United States of America are two of the world's largest countries in area (Figure 11.2). Canada is the second largest country after the Russian Federation, and the United States is fourth after China (Table 11.1). Although the vast size of each country plays a key role in their respective contemporary geographies, it is important to consider their territorial expanse with other geographic characteristics such as population numbers and settlement patterns. Canada's huge land area is sparsely inhabited with a 2005 population of 32.2 million. The United States is the world's third largest country in population, yet its 2005 figure of 296.5 million is substantially smaller than the Earth's two population giants, China and India, each with more than 1 billion residents. Both Canadians and U.S. residents are mostly urban peoples. Almost 80% of the people of both countries live in or near urban metropolitan areas, which leaves extensive rural lands with little population (Figure 11.3). Thus the huge output of Canadian and U.S. industries comes from just 5 percent of the world's population, making the majority of the people who live in the region extremely affluent relative to global averages.

North America as a world region is distinguishable from Latin America through a number of cultural historical developments such as language and political ideology. Although Mexico has formed strong economic alliances with Canada and the United States and continues to influence the human geography of the region significantly, it has greater cultural affinities with the other countries of Latin America that were colonized by Spain in the 1500s and is discussed in detail in Chapter 10.

Canada is linked closely to the United States in many ways but asserts its individuality through different global connections and distinctive internal policies. Although the two countries share many commonalities, the Canadian people have a strong sense of national pride, which has thus far prevented their Canadian identity from being overwhelmed by their giant neighbor.

This chapter begins by examining the ways in which European cultures took over the land from indigenous peoples and set the stage for the human and economic development that created the diverse and connected countries of Canada and the United States. Next, the natural environmental conditions of both countries are detailed, leading into a discussion of issues stemming from human–environment interaction in the region. Finally, the global connections and economic prominence of the region are outlined prior to a detailed examination of the chapter's two subregions: the United States of America and Canada.

Table 11.1 United States of America and Canada. Data by country, area, population, income (GNI PPP—Gross National Income Purchasing Power Parity), urbanization, Human Development Index (HDI), Human Poverty Index (HPI).

| Country | Capital City | Land Area (km²) | Population (millions) | | GNI PPP 2003 | | % Urban | Index 2002 | |
			Mid 2005 Total	2025 Est. Total	Total (US$ billions)	Per Capita (US$)	2005	HDI (rank of 177)	HPI (% population)
Canada	Ottawa	9,976,140	32.2	36.0	941	29,740	79	4	12.2
United States of America	Washington, DC	9,809,460	296.5	349.4	10,914	37,500	79	8	15.8

Source: World Population Data Sheet, Population Reference Bureau (2005), World Development Indicators, World Bank (2005), Human Development Report, United Nations (2004), and Encarta Microsoft (2005).

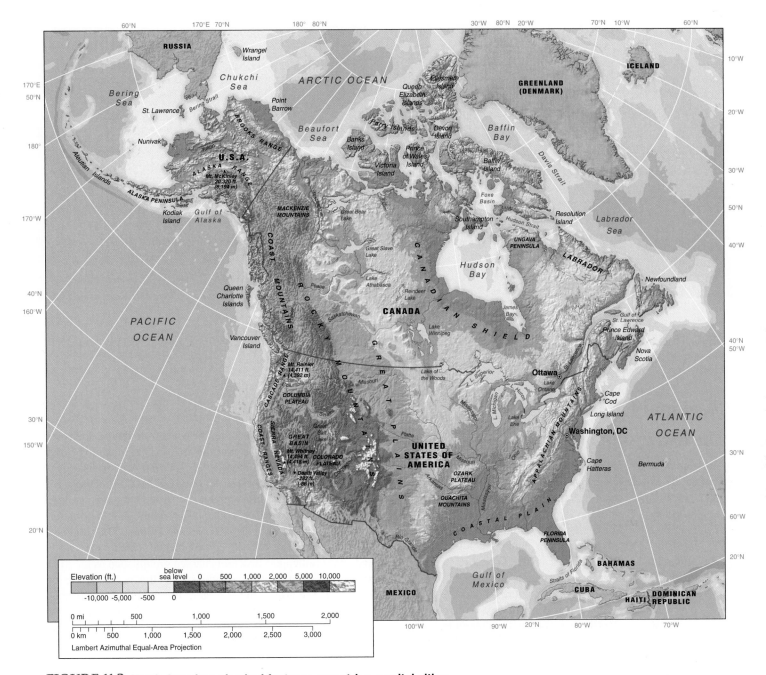

FIGURE 11.2 **North America: physical features, countries, capital cities.**

Country	Ethnic Groups Percentages	Languages O=Official N=National	Religion Percentages
Canada	British origin 35%, French origin 25%, other European 20%, First Nations 3%	English (O), French (O)	Roman Catholic 45%, Protestant 28%
United States of America	White 83%, African American 12%, Asian 3%, Native American 1%	English, Spanish	Protestant 58%, Roman Catholic 26%, Jewish 2%

(a)

(b)

FIGURE 11.3 North America: urban and rural landscapes. The urban setting of Toronto, Canada (a), contrasts with the open wilderness of Zion National Park in the U.S. State of Utah. **Photos:** © Joseph P. Dymond.

Regional Culture History

The Emergence of a Region

Although archeologists believe Viking explorer Leif Erickson reached the eastern shores of present-day Canada around A.D. 1000, North America and its indigenous population were largely unknown to the rest of the world before Columbus's sailings beginning in 1492. The indigenous peoples of North America lived in societies based on agriculture, hunting, trading, and the local communal sharing of resources. Global connections began with the exploration of the region by various European interests. The indigenous Americans throughout the region were overwhelmed by the **cultural hegemony,** or forced conversion of their social customs to the vastly different technologies and social tenets of the Europeans.

In the three centuries after the European "discovery" of the region, North America became a series of colonies and occupied territories governed by the French, Spanish, British, Dutch, Russians, and Swedish. The British became the dominant power by the mid-1700s. Almost immediately, settlers south of the St. Lawrence River valley fought to become the independent United States of America by signing the Declaration of Independence in 1776, forcing Britain to recognize U.S. sovereignty.

Canada arose mostly out of the colonial territories first established along the St. Lawrence and the Great Lakes and remained a British colony far longer than the United States. It achieved a degree of independence through the British North America Act in 1867, while maintaining legal ties to Britain until 1982, when it gained control of its constitution. Today Canada is a fully independent country that enjoys membership in the Commonwealth of Nations.

Native Americans

Indigenous groups of people, who are referred to today as **Native Americans** (**First Nations** in Canada), inhabited North America for centuries before 1500. The first Americans probably migrated from Siberia to Alaska over 20,000 years ago. By 1500 A.D. they lived in hierarchically structured ethnic groups, commonly called "tribes," adapted in culture and distribution to the physical conditions of terrain and climate. The combination of natural resources, environmental conditions, and a sedentary settlement structure enabled groups such as the Mississippians in present-day southwestern Illinois to flourish for centuries. The groups in the eastern forests raised corn, beans, and squash, hunted and fished, and lived in village settlements. Those in the warmer lower Mississippi Valley used a surplus from farming to support urban development such as Natchez. Hunting groups, such as the Dakota, inhabited the prairie environment on the plains between the Mississippi River and the Rocky Mountains and killed bison to supply their food, clothing, and shelter needs. Their tepee tents could be moved to follow the bison herds. The introduction of horses by Spanish colonists facilitated hunting, travel, and communications for many western hunting-based societies, thus enabling them to grow in both significance and numbers of people.

Tribal numbers were even sparser farther west in the more arid parts of the mountains and high plateaus and farther north in the colder parts of northern forests and Arctic lands. Some locations in the arid southwest favored the development of irrigation farming and the building of distinctive villages by the pueblo cultures (Figure 11.4). Along the Pacific coast from present-day northern California to southern Alaska, small tribes fished plentiful seafood, including salmon and tuna. They

carved the region's tall trees into dugout canoes, plank houses, and totem poles. The Arctic and north Pacific coasts of the continent supported small numbers of Aleuts and Inuits, who hunted and fished the nearby ocean waters.

After the arrival of the Europeans, many Native Americans were killed by the introduction of diseases to which they had no immunity. The survivors were increasingly pushed to marginal lands of the region. This process began in the east and southwest in the 1600s and lasted into the 1800s in the central and northwestern parts of North America. Although many Native American trails evolved into major travel routes for the regional settlers, the indigenous groups had much less of an impact on the evolution of the regional landscapes once the Europeans began their dramatic land use transformations.

European Settlers

Several European countries began to settle the region as early as the 1500s, including Britain, France, the Netherlands, Russia, Spain, and Sweden.

The French settled the mouth of the St. Lawrence River in present-day Canada from the early 1500s. They established river-related farm settlements along the banks of the St. Lawrence valley. Others explored the interior, helped by Native Americans, to hunt for beaver and other furs. The French traded in and claimed lands through the Great Lakes and along the Mississippi River.

The Spanish settled parts of present-day Colorado, California, Arizona, New Mexico, Texas, Florida, and the Carolinas from the 1500s. Spanish territory, known as the colony of New Spain, centered on the area of modern Mexico. When areas, such as southern North America, did not deliver the

gold that they hoped for, the Spanish left much of the land to be cared for by Roman Catholic missions or to be used for cattle ranches.

The British first targeted the tidewater lands surrounding the Chesapeake Bay for colonial development. In 1607 they established their first permanent settlement of Jamestown in present-day Virginia, after an earlier failed attempt on Roanoke Island (present-day North Carolina). The tidewater settlers found the local climate and soils well suited to tobacco cultivation—the lucrative commercial crop needed to fund the settlement and satisfy investors in England. The settlers also adopted Native American crops for subsistence. Sales of tobacco back in England attracted more settlers and led the British government to take over the lands as a crown colony. This early success led to the diffusion of plantations and colonies southward along the Atlantic coastal plain. An increasing range of commercial crops were grown, including indigo, rice, sugarcane, and cotton. In 1619 the first Africans were imported to perform the labor-intensive fieldwork. Slavery became a significant institution by the 1700s, fueling the development and expansion of agriculture south and west from the southeast Atlantic coastal plain.

Religious freedom–seeking migrants from England on their way to Virginia brought a second wave of British settlers in 1620. They landed at Cape Cod in present-day Massachusetts and moved across the bay to Plymouth, beginning the settlement of the region that became known as New England. Other groups followed and established a community-based township settlement pattern founded on self-governing villages and a subsistence economy. New England's community orientation contrasted with the individual family-owned plantations of the southern settlers, which produced fewer towns.

In the 1630s the Dutch were the first to settle in the Middle Atlantic between New England and Virginia around their port town of New Amsterdam (modern New York City). Swedish settlers took small areas along the lower Delaware River but were quickly ousted by the Dutch. Next the British drove the Dutch from their settlements in the 1660s. The British Duke of York received charge of the Dutch lands and gave his name to the largest city, changing it from New Amsterdam to New York. He paid off his war and land acquisition debts by selling most of his remaining lands to William Penn, the founder of Pennsylvania. This Middle Atlantic area attracted settlers who wished to set up family homes with the least amount of external control in a new country. Many Scots–Irish and Germans came to the region in the early 1700s, establishing a farming system based on growing corn and raising livestock, and spreading it southward along the Appalachian valleys.

After independence in 1783, the three areas of settlements along the Atlantic coast with their different economic and social systems formed springboards for thousands who moved westward within the United States as new lands were acquired. The New England community subculture diffused into areas around the Great Lakes. The Pennsylvania system of individual family farms became the basis of farming in the southern Midwest. The plantation system first established in the Virginia tidewater, along with the accompanying use of slaves, was extended south and then southwestward along the Gulf lowlands to Texas. The North–South tensions that came to a head in the Civil War of the 1860s arose from the spatial diffusion of these early differences.

Wealth of Natural Resources

The early settlers of North America discovered a seemingly limitless wealth of natural resources, including vast open lands, good soils, a variety of minerals, fur-bearing animals, fish, and timber. The resource wealth of the region provided a solid foundation for the settlers to establish commercial farming and other resource-related industries.

Primary Products and New Lands

At first, in the 1600s and 1700s, the wealth was unearthed by an agricultural society. The long, warm summers of the widening coastal plain southward from New York encouraged the commercial farming of subtropical crops such as tobacco, sugarcane, rice, and indigo, and later cotton. The growing slave trade provided an abundant source of relatively low-cost labor. Slave labor coupled with good climate and soil conditions enabled planters to sell crops to Europe, make payments to investors, and pocket or reinvest the proceeds. New England's timber and pitch provided new ships for the British navy. The coastal harvest of fish in Canada and New England was followed by an inland harvest of animal furs, which further diversified developing patterns of global trade.

After establishing complete independence in 1783, the United States tripled its area by 1850, buying land from France and Spain and acquiring the western third by negotiation and military conquest (Figure 11.5). To ensure the rapid occupation of these new areas, land was surveyed and sold by the U.S. government at lower and lower prices. The **Homestead Act** of 1862 provided families with very inexpensive or even free

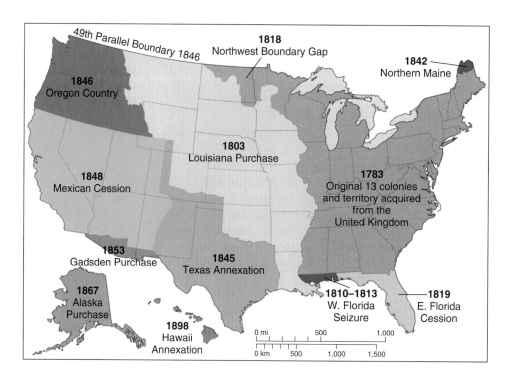

FIGURE 11.5 United States: rapid expansion of territory, mainly between 1800 and 1850. Lands first occupied by Native Americans, and later assumed by European colonial empires, were gradually incorporated into the expanding United States. The eastern lands were acquired from the United Kingdom at Independence in 1783. The Louisiana Purchase bought lands that France had just won back from Spain. Texas claimed independence after ousting the Mexicans in the early 1830s but then joined the United States in 1845. The western lands of the conterminous United States were added by 1850, including the southwestern lands acquired through the 1848 Mexican Cession after a rapid U.S. military victory. Alaska and Hawaii came later.

farmland, although much land near the railroads was sold by the railroad companies to pay for the track, so homesteading land was often less accessible and poorer in quality. Settlers spread to the vast interior plains. Many found that although the land was free, there were numerous costs associated with the move and the establishment of farms (including clearing vegetation, marking boundaries, raising buildings, creating roads, and buying livestock and seeds). Families able to overcome the expense of starting farms in the interior took advantage of the fertile soils and the warm, moist summer climate, a combination that proved ideal for growing corn and wheat and raising cattle and pigs. From the early 1800s, the United States exported a range of farm produce to Europe, competing directly with European farmers in bulk grain markets as well as continuing to export crops such as cotton and tobacco that could not be grown in northern Europe.

Slower Canadian Changes

The settled region along both sides of the St. Lawrence River and estuary (part of modern Canada) was a materially poor and often environmentally hostile remnant left to British authority after the United States declared its independence. British loyalists who voluntarily emigrated or were forced out of the incipient United States joined small numbers of Native Americans and French-speaking European settlers. Because the French settlers along the lower St. Lawrence resented British rule, those loyal to the British Crown mostly settled on the east coast or farther inland around the northern and western shores of Lake Ontario. Westward of the Great Lakes there was little settlement in an area still largely administered by the **Hudson's Bay Company,** which was created by a Royal British Charter in 1670 to administer the development of nearly one-third of present-day Canada. The interior remained the realm of Native Americans and isolated groups of French settlers attempting to escape the watchful eye of British rule. When the Hudson's Bay Company fur trappers began to compete with Americans for land along the lower Columbia River near the coast of the Pacific, negotiations in 1846 set the boundary between the two countries west of Lake Superior at the 49th parallel.

Until the middle of the 1900s, the natural resources of Canada primarily comprised Atlantic and Pacific marine life, the good soils and temperate climate of southern Ontario, the vast prairies opened to farming settlement in the early 1900s, and the timber resources of the St. Lawrence River valley and British Columbia. In the post–World War II era, rich stocks of metal ores in Canada's north; coal, oil, and natural gas in Alberta; and hydroelectric power generation in northern Québec ignited new geographic directions of development and provided the basis for growing Canadian affluence.

Resources for U.S. Manufacturing

The first Industrial Revolution began in Western Europe (see Chapter 3), coinciding with the United States' independence from Great Britain. Independence and British hostility motivated Americans to establish their own manufacturing industries, providing goods that were at first not obtainable from Britain. Textiles, metal goods, and leather goods were among the first industries to develop, often powered by water mills in New England, New Jersey, and eastern Pennsylvania. By 1850 the coal in eastern Pennsylvania and around Pittsburgh replaced charcoal in iron making and, later, steel industries that set off further industrial revolutions. The northeastern United States soon emerged as a dominant region of the country, housing the bulk of the population with a growing material wealth and related political power based on manufacturing industries.

The primary symbol of the early steel-based Industrial Revolution—the railroad—spread across the country to the Pacific Coast. The Manufacturing Belt of the northeastern United States, which initially developed based on the regional resources of coal, iron, and the agricultural produce of the Midwest, later drew in mineral resources from other parts of the country. Railroads linked the mines of the western mountains to the eastern economic and political core. The mining of gold, silver, and then copper and zinc in the new western lands provided capital for further development and widened the scope of the late-1800s metal industries.

In the 1900s the availability of such natural resources, the utilization of rivers for irrigation, navigation, and hydroelectricity, and the discoveries of oil and natural gas drew more people and manufacturing corporations southward and westward from the country's core. Although the emphasis of most Americans with regard to natural resources was on the potential for economic gain, influential groups from the late 1800s persuaded the federal government to set aside large areas of attractive wilderness or places of historic significance as national parks, often before much settlement had occurred. Most of these are in the West, but smaller areas of eastern lands have also been purchased and designated as national parks (Figure 11.6).

Human Resources

The variety and richness of the natural resources—such as minerals, climate, and soils—available within a territory as expansive as North America combined with a diverse stream of immigrants to build a framework of exploration, expansion, and development.

New People, New Skills

Immigrants came initially from northwest Europe. They brought European cultural traditions and individual skills and initiative. Tensions among Protestants, Catholics, and Jews were worked out in the context of economic competition and residential congregations of like-minded people. There was ample space within the developing United States for a variety of ethnically, nationally, and religiously diverse groups to establish themselves. Africans forced to migrate to the region to serve primarily as slaves in the agriculture

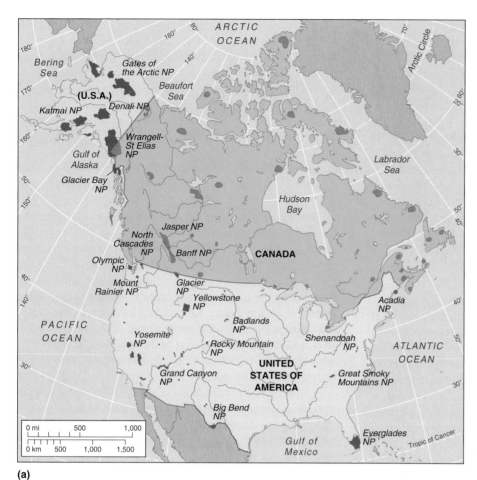

(b)

FIGURE 11.6 North America: national parks. (a) Some major parks in the United States and the intended expansion of national parks in Canada. In the United States, there are also extensive wildlife refuges and national forests, in which varying levels of commercial use are allowed. (b) Grand Canyon National Park. **Photo:** (b) © Joseph P. Dymond.

(a)

industry remained the target of debilitating discrimination even after the Civil War (1861–1865) guaranteed freedom to most in the population.

Education and Technology

The new United States sought to develop conditions in which its whole population could flourish by making the most of their freedoms under the democratic Constitution. The United States established compulsory education much earlier than European countries. Its emphasis on technology transfer began with the late 1800s land grants for establishing engineering and farming universities in many states.

Management of Manufacturing

The manufacturing structures and processes in which the developing United States excelled quickly brought it prominence in world economic activity by the late 1800s. The expanding internal market of the United States provided the demands that stimulated many of the developments.

1. Americans achieved **economies of scale** by building larger factories for increased output, together with access to larger markets for their growing range of products. Such economies cut the cost of individual items by manufacturing large numbers of each item in a single factory.

2. **Horizontal integration** occurred when financiers bought up several producers of the same product, giving the new owners a large share of the total production of goods and enabling them to set market prices.

3. **Vertical integration** took this process further by bringing together, in a single corporation, the producers of inputs to a product and the users of the same completed product. As an example of these processes, Andrew Carnegie built larger steel mills to gain economies of scale; then he bought out other steelmakers in a time of economic recession in the 1870s to integrate the steel industry horizontally. In the next phase of his vertical integration scheme, he purchased both coal and iron mines that provided the raw materials and the heavy engineering corporations that used the steel. By 1900 Carnegie's United States Steel Corporation dominated the industry.

4. Henry Ford took the concept of the factory **production line** to a new level. He applied it to the assembly of a wide range of components in the production of automobiles. His company produced thousands of a limited number of models for a growing market. His methods, called **Fordism,** were widely adopted by other industries and in other countries.

5. The making of machine tools to produce machines that manufacture and assemble parts was another feature that made the United States the world leader in output of manufactured goods by the mid-1900s.

Canada Emerges

Not until the late 1800s did Canada begin the process of becoming a major industrial country. British rule that suppressed internal development, as well as a lack of capital and expertise, hindered the creation and expansion of Canada's global connections. In 1867, Canada became a dominion within the British Empire with increased responsibility for its own affairs. At that stage, Canadian leaders began to integrate their vast country by building transcontinental railroads to encourage the settlement of the prairie grasslands and the Pacific coast. Manufacturing industries, including wood processing and steelmaking, were established behind tariff walls to protect them from American competition in particular.

Test Your Understanding
11A

Summary

The United States and Canada are two of the largest countries on Earth in terms of area, with approximately 15 percent of total land, yet together they house only 5 percent of the global population. North America is the world's most affluent region, producing nearly 30 percent of the world's goods and services.

European settlers migrated in waves from various parts of Europe to North America from the 1500s to the 1900s. As European settlement expanded across the North American landscape, North American indigenous groups were increasingly displaced and pushed onto marginal lands. Native groups in both Canada and the United States today continue to work toward greater recognition and equality.

Questions to Think About

11A.1 What role did Europeans play in shaping the human geography of North America?

11A.2 What impact did European decisions play in shaping current human geography patterns for Native and African Americans?

11A.3 What role did the natural resource wealth of North America play in shaping the contemporary human geographies of both Canada and the United States of America?

Key Terms

cultural hegemony	economies of scale
Native American	horizontal integration
First Nations	vertical integration
Homestead Act	production line
Hudson's Bay Company	Fordism

Physical Geography and Human–Environment Interaction

While the natural environments of most of the United States are conducive to human settlement, Canada's natural environments present significant challenges because so much of the country lies near or beyond the margins of productive or habitable land. Most human settlement in Canada is restricted to a narrow zone just north of the U.S.–Canada border.

Tropical to Polar Climates

Climatic environments within North America are mainly midlatitude in type (Figure 11.7). The western coast is dominated by the midlatitude west coast and Mediterranean climates that give way eastward to midlatitude interior and midlatitude east coast climates. This is a sequence similar to that found across Europe and northern Asia (see Chapters 3 and 4). The west coast climates of Canada and the United States are restricted in their inland impact by the north–south mountain ranges.

Although the U.S. Southwest is situated north of the Tropic of Cancer, the tropical arid climatic environment present in northern Mexico extends into the region. Southernmost Florida has tropical, humid conditions, and the Gulf Coast from Texas to Florida and the Atlantic Coast as far north as the Virginia tidewater have subtropical conditions that bring hot, humid summers with many thunderstorms and the annual threat of hurricanes. In northern Canada and Alaska, the climate is deeply Arctic in winter and has only short summers.

The central plains and lowlands enable bursts of Arctic air to move southward in winter, bringing extreme cold especially to the Midwest and northeastern United States but also extending in shorter bursts to the Gulf Coast. Warm, dry southwestern air or humid air from the Caribbean and Gulf of Mexico moves northward in spring and summer. The frequent confrontation of cold Arctic air and humid subtropical air in the central lowlands leads to the formation of active cold fronts with thunderstorms that may produce local violent tornadoes mainly in spring and summer. Tornado activity moves northward from Texas in February to the Great Lakes in June. The United States experiences more tornadoes each year than any other country.

Past climatic changes left their impacts on the present environment. During the Pleistocene Ice Age, the last 2 million years of geologic time, much of Canada and the northern parts of the United States were covered by advancing ice sheets centered on the Hudson Bay area and the western mountains. The greatest effects occurred in the last half million years, but most of the ice melted beginning about 15,000 years ago, leaving glaciers only in the highest and coldest areas. Distinctive ice-eroded landscapes are relic features in high mountains and in northern Canada, while the melting ice sheets left depositional features mainly in the Midwest.

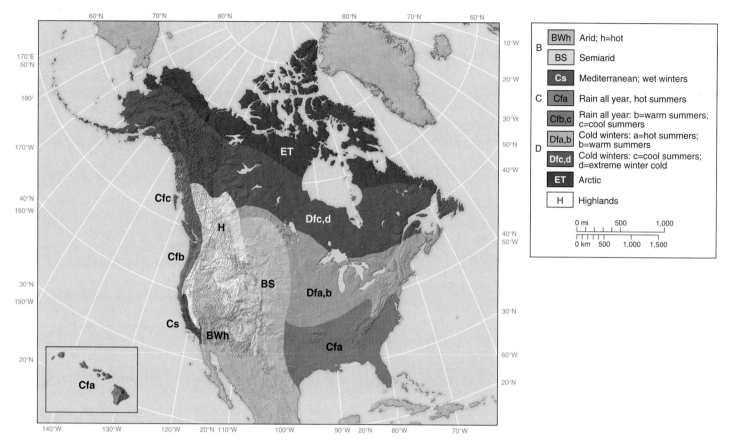

FIGURE 11.7 **North America: climates.**

Mountains and Plains

North America is part of the North American plate (Figure 11.8). New plate material forms along the divergent margin at the Mid-Atlantic ridge, forcing the plate to move westward and clash with the Pacific plates at the convergent and transform plate margins along the western coast of North America. The younger and more geologically active west coast of North America has higher mountains and is affected by frequent earthquakes and periodic volcanic activity, while the less disturbed eastern coasts are low-lying or hilly with older mountains inland and infrequent earthquake activity.

Western Mountains

The western third of Canada and the United States is primarily composed of rugged mountains consisting of several different elements. The highest peak in the region is Mount McKinley in the Alaskan Range (6,194 m, 20,230 ft.). The Rocky Mountain ranges extend from Alaska and northwestern Canada southward to New Mexico. To their west in the United States are extensive high plateaus, from the Colorado Plateau in the south to the lava layers of the Columbia Plateau just south of the Canadian border. West of the plateaus are the Sierra Nevada mountains of California and the Cascade ranges (Northern California through Oregon and Washington to the U.S.–Canada

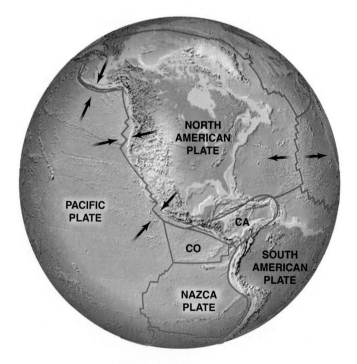

FIGURE 11.8 **North America: plate interaction.** The North America plate dominates the region's plate structure and interacts, sometimes violently, with several plates around its margins.

FIGURE 11.9 United States: Mount Saint Helens eruption, 1980. Image of Mt. St. Helens after the 1980 eruption, with the mountain top blown away and a new small cone forming in the crater. New seismic and volcanic activity occurred in 2004 and 2005. **Photo:** U.S. Geological Survey.

border). Some of the many volcanoes in the Cascades are active, such as Mount Saint Helens (Figure 11.9), and there are others in southwestern Alaska. Frequent transform plate interaction (parallel plate movement) between the Pacific Plate and the North American Plate makes California susceptible to earthquakes along the San Andreas Fault system.

Canadian Shield

About half of Canada has the ancient rocks of the **Canadian Shield** at the surface. These are the oldest rocks in North America and contain major deposits of mineral ores. Covered by ice sheets during the last Ice Age, most of the soil was scraped from their surface. Deposits left by the melting ice sheets over the last 15,000 years are scattered and varied in nature and often separated by large lakes that fill hollows gouged out by the ice. A line of very large lakes follows the approximate boundary between the shield rocks and the overlying rocks, from Great Bear Lake in northwest Canada to the Great Lakes straddling the U.S.–Canada border.

Interior Lowlands

Lowlands with little relief dominate southern Ontario, the Prairie Provinces, and the central United States. Layers of sedimentary rock cover the shield rocks, forming plateaus and escarpments such as that over which the Niagara Falls plunge. In the U.S. Midwest and in Ontario, Canada, most of these rocks are covered by deposits from the melting ice sheets. Similar deposits occur over part of the Canadian prairies, but in areas not covered by the ice sheets, distinctive features such as groups of small mounds and depressions were formed by exposure of the land to an atmosphere of extreme cold.

East of the Rockies, the Mississippi drainage basin comprises more than a thousand kilometers (620 miles) of lowland river valley, which rises gently eastward and westward from the river to the Rockies and Appalachians. The northern lowlands were covered by rock fragments, particles dropped by the ice sheets blanketing the area during the Pleistocene Ice Age, and by the finer particles blown farther afield by winds and deposited as loess as far south as the states of Louisiana and Mississippi. West of the Mississippi River, the lowlands rise to the Great Plains in the United States and the prairies in Canada.

Appalachian Mountain System

East of the Mississippi lowlands, the Appalachian Mountains form a continuous chain of rolling hills and mountains extending from northern Georgia into the Adirondacks of New York, the Green and White Mountains of New England, and the Atlantic Provinces of Canada. Few Appalachian mountain ridges exceed 2,000 m (6,500 ft.). The rocks were deformed and uplifted by ancient plate tectonic movements and subsequently eroded. In the northeastern United States and eastern Canada, the glaciers scraped away much of the surface rock and soil, carried it southward, and deposited it along the east coast in the low ridges (moraines) that form much of Long Island and Cape Cod. West of the Mississippi, the Ozarks and Ouachitas form similar upland areas.

Major Rivers and the Great Lakes

The physical and human geographies of North America are greatly affected by the region's surface **hydrology** (surface water drainage). In particular, the major rivers and the Great Lakes provide sources of fresh water and transportation routes. The Mississippi River was the basis of early interior transportation and continues that role for bulk materials, though diminished in significance. Its tributary, the Ohio River, was a transportation route at the heart of manufacturing developments in the United States during the late 1800s. The Colorado River in the arid Southwest is harnessed for power and irrigation water (Figure 11.10). The Columbia River is used to irrigate farmland and to generate hydroelectricity in the northwestern United States. In the 1950s the Great Lakes–St. Lawrence Seaway brought easier trade and associated industries deep into the interior of Canada and the United States, although this function is much reduced owing to the increased size of ocean ships (Figure 11.11). In northwest Canada, the Mackenzie River is a major summer transportation route.

The largest area of lowland and fertile farmland occurs in the combined drainage basin of the Mississippi–Missouri–Ohio rivers and the Great Lakes, covering nearly one-third of Canada and the United States. Although this is a continental interior region, the rivers provide relatively easy outlets to world ocean trade routes to the Gulf of Mexico in the south and along the St. Lawrence and Hudson Rivers to the Atlantic Ocean in the east.

FIGURE 11.10 The Hoover Dam along the Arizona—Nevada border is a regional source of power generation. The dam controls the flow of the Colorado River and creates Lake Mead, which is a main source of fresh water and recreation in the region. **Photo:** © Joseph P. Dymond.

FIGURE 11.11 The St. Lawrence River remains an important waterway for recreation and commerce in Canada. A large container ship is docked along the St. Lawrence in Montréal Canada. **Photo:** © Joseph P. Dymond.

Much of the lower relief along the Mississippi Valley formed by sand and clay deposition from meltwater, as the ice sheets dissipated (Figure 11.12) and the ocean level rose.

The Great Lakes formed as Pleistocene ice sheets advanced over Canada and the United States. Lobes of ice gouged depressions between festoons of marginal moraines built from rock fragments dropped at the margins of melting ice masses. As the ice sheet retreated northward at the end of the last glacial phase, meltwater accumulated in the depressions, which became the Great Lakes. At first the meltwater overflowed southward into the Mississippi system; later the lake water drainage was limited to flowing eastward into the

(1) 20,000 years ago
Sea level—120 meters below present
Mississippi Trench

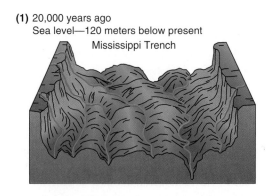

(2) 10,000 years ago
Sea level—30 meters below present
Fans   Braided channel
Sand and gravels

(3) 5,000 years ago
Sea level—6 meters below present
Backswamps   Braided
Braided stream and backswamp deposits

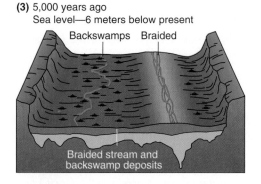

(4) Modern
Sea level as now   Backswamps
Meander belt
Zone of meander migration   Overbank deposits

FIGURE 11.12 **North America: impact of ice meltwater on the landscape.** Stages in the evolution of the lower Mississippi River valley. (1) Large flows and low sea level resulted in the Mississippi River carving a deep channel at first. (2) and (3) As the sea level rose and the flows lessened, the channel filled with gravel, sand, and mud. Finally (4), flows from the Great Lakes were diverted eastward, and the Mississippi flow was reduced to its present levels.

Hudson and St. Lawrence Rivers. Today the Great Lakes function as inland seas with coastlines, ports, and recreation areas. In 1959 the St. Lawrence Seaway was opened as a cooperative project between the Canadian and United States governments. Through it, oceangoing ships can reach Chicago and Duluth.

From the early 1900s, the Mississippi–Missouri River system was controlled for flood protection, improved water transportation, and hydroelectricity generation by building a series of dams. The flood protection system prevents most floods from spilling onto surrounding land; however, extreme rain events are too much for the levees and dams to handle (Figure 11.13). The dams prevent silt and clay reaching the mouth and so reduce deposition on the delta south of New Orleans. In recent years the delta surface subsided along its southern margins because it did not receive sediment to build it to sea level.

Natural Vegetation and Soils

The combinations of terrain and climate produced a range of natural vegetation and soils distributed in a north–south pattern similar to that of Russia and its neighbors (Figure 11.14), although the east–west extent is less in North America. The hot deserts of the Southwest support mainly drought-resistant varieties such as cactus and low shrubs and little else. The subhumid Great Plains and Prairies of the western Mississippi River basin and south central Canada had a natural vegetation of prairie grasslands that are now largely plowed to make use of the underlying black earth soils. Eastward, the more humid conditions, from tropical southern Florida to the cool, temperate Northeast, supported **deciduous** (broadleaf) forest, which gave rise to brown earth soils of moderate to good fertility. North of the deciduous forests and the prairies, a wide band of **coniferous** (needleleaf) forest has poor podzol soils and gives way to the tundra along the Arctic Ocean shores.

Today the eastern mountains have thin soils on steep slopes. Before extensive clear-cutting by loggers in the late 1800s and early 1900s, the eastern mountains supported a profusion of trees containing a much greater variety of species than those in Western Europe. The best farming soils formed under forest and grassland in the interior plains, where the old glacial and wind-blown materials combining with the moderate amounts of precipitation produced rich brown (forest) or black (grassland) soils. In the southeastern United States, the frequent presence of sandy soils with low nutrient content caused some areas to be dominated by pine trees. Along the western coast of the United States, north of San Francisco, huge firs and cedars, growing over 100 m (300 ft.) tall, formed a massive timber resource.

Natural Hazards

A greater range of natural hazards than in any other country in the world marks the natural environments of the United States. The conterminous United States is affected by hurricanes, severe thunderstorms with accompanying tornadoes, lightning, hail, earthquakes, volcanoes, floods, blizzards, ice storms, and wildfires. Alaska and Hawaii are affected by various natural hazards including extreme winter weather and earthquakes in Alaska and nearly continuous volcanic activity in Hawaii. All the natural hazards experienced within the borders of the 50 United States have a major impact on the human activities of U.S. residents. Each hazard has a geographic area of greatest impact:

- Hurricanes threaten the mid- and southern Atlantic coast and Gulf coast states in late summer and fall, when water temperatures are warmest.

- Tornadoes are most common in the Plains states, the Mississippi Valley, and Florida, where the necessary climate ingredients meet to support their formation. Tornadoes affect Texas in February and move northward to the Great Lakes area by June.

- River floods occur most commonly in the Mississippi River valley following the spring melting of snow on the surrounding hills or heavy summer rains. Flash flooding occurs in all regions of the United States and is particularly dramatic in arid regions in the West after sudden rains fill dry streambeds.

- Lightning is a year-round event across the southern part of the United States and is especially prevalent in Florida.

- Severe winter storms affect the upper Plains states, the Midwest, the western mountains, and the northeast coast of the United States. Ice storms pelt many areas east of the Rocky Mountains.

- Earthquakes and volcanoes occur mainly along the west coast from Alaska southward through California. The meeting of the North American Plate and the East Pacific rise causes horizontal displacement along the San Andreas Fault, which produces frequent earthquake activity (see Figure 2.25). Volcanic activity is nearly continuous in Hawaii, and it also occurs in the Cascades where the Juan de Fuca minor plate plunges beneath part of the North American Plate.

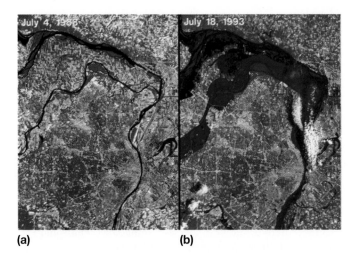

(a)   **(b)**

FIGURE 11.13 United States: flood hazard. The 1993 floods on the Mississippi River near St. Louis. (a) Before the floods. (b) During the floods. The reddish areas are highways and buildings. The darkened areas on (b) show the extent of floodwater cover. **Photo:** Courtesy of Space Imaging, Thornton CO, USA.

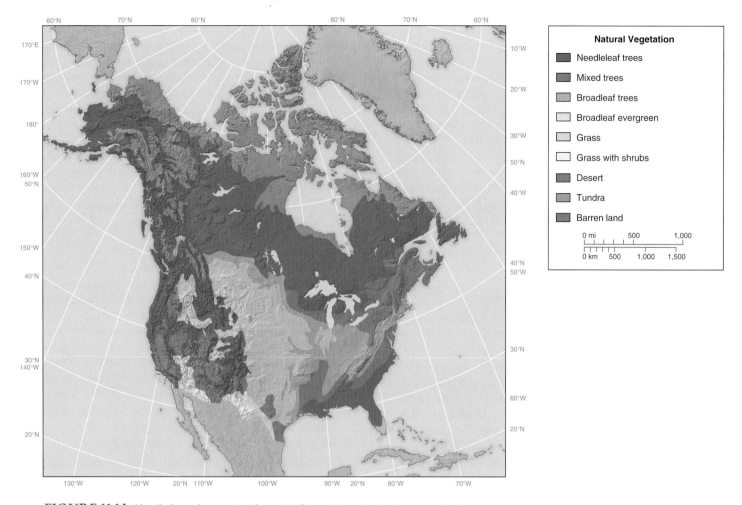

FIGURE 11.14 **North America: natural vegetation.** Relate this distribution to the climatic environments (see Figure 11.7) and land use patterns (see Figure 11.26).

Canada is less troubled by most of the natural hazards plaguing the United States, primarily due to its position in the high latitudes of the Northern Hemisphere. The waters off the Canadian coasts are much too cold to support hurricanes. The extreme temperature and moisture contrasts needed for severe thunderstorms and accompanying tornadoes, hail, and lightning are also less common in Canada due to its high latitudinal position. Canada does experience earthquake activity and the threat of tsunamis (tidal waves) along the west coast, and very cold temperatures and high snowfall totals are common across the country during the winter.

The Canadian provinces of Québec, Ontario, and New Brunswick were devastated by an ice storm in January 1998, when 7–11 centimeters of ice (3–4 inches) accumulated over six days. The heavy ice brought down trees, power lines, utility poles, and transmission towers. Millions of Canadians went without power and electric heat for days (some had no power for a month). The 1998 ice storm was Canada's costliest natural disaster.

Regional Environmental Issues

The emphasis on expansion and economic growth led to many environmental challenges in Canada and the United States before the era of concern that developed through the 1900s. The mining of fuels and metallic ores, the outputs of manufacturing industries, and the plowing of land across the Canada–United States region resulted in spectacular examples of environmental degradation.

- The plowing of subhumid grasslands in western Oklahoma, Kansas, and Texas resulted in the **Dust Bowl** disaster of the 1930s, in which winds blew away the dried finer soil particles and left piles of sand behind. Even in the more humid Midwest, the soils thinned as winds and surface runoff removed the topmost layers.

- Water from aquifers and the Colorado River facilitated the expansion of farming into arid areas of the southwestern United States. As southwestern populations grew, competition and controversy erupted between local states over the use of fresh water and the depletion of groundwater resources.

- Population settlement in arid areas where natural burning cycles frequently occur in the vegetation presents a growing hazard. Wildfires are increasingly affecting populated areas, such as the Los Angeles Basin in southern California, where businesses and homes may be consumed by fire in a matter of minutes.
- The construction of homes on steep hillsides along the California coast contributes to the devastating impact of mudslides that often occur during periods of brief heavy winter rains.
- Mining on a large scale dug huge pits for extracting copper and other metal ores in the West. The strip-mining of Appalachian hills for coal after 1950 devastated large areas of eastern Kentucky and West Virginia.
- Cities like Washington, D.C., Baton Rouge, and Los Angeles became plagued by smog and high concentrations of ground-level ozone, generated from the exhaust gases of increasing numbers of vehicles and thermal power plants.
- Acid rain, derived from power plant emissions, particularly along the Ohio River valley, affected trees and caused rivers and lakes downwind in the northeastern United States and eastern Canada to become more acidic.

Greater public awareness and increased conservation education led to the enforcement of environmental legislation and a corresponding reduction in the impacts of soil erosion, poor water management, and urban air quality, despite increases in population and intensity of human activities.

Canada, with a much smaller population in a slightly larger area, has fewer environmental problems than the United States. Industrial pollution largely created in the United States travels through air currents and reaches the ground in parts of Canada in the form of acid rain. Industrial effluents from both Canada and the United States foul the waters of the Great Lakes. Although Canada has reduced polluting outputs from smelters on its territory near the British Columbia border and along the northern shore of Lake Superior, more significant reduction is needed from both countries to ensure lasting health of the region's fresh water systems. Canadian attempts to diminish industrial effects on the natural environment through federal legislation resulted in the imposition of heavy financial fines on pollution-creating industries such as the nickel smelters at Sudbury in Ontario.

The increasing size of hydroelectricity projects in Québec and elsewhere led to environmental and Native American protests over the resulting hydrologic and land use changes. Canadian provinces respond individually to their internal environmental problems. For example, Québec moves ahead with its huge hydroelectric projects against the wishes of local indigenous communities; Alberta encourages the extraction of coal, oil, and natural gas; Ontario is more concerned about environmental impacts.

Test Your Understanding
11B

Summary

North America contains a nearly complete range of the world's natural environments. Its climates range from tropical to Arctic and its terrains from young folded mountains to wide river plains and ancient shield areas. Natural vegetation and soils reflect climatic differences, with deserts in the southwest, grasslands and black soils west of the Mississippi River, forests and brown soils in the east, and forests with podzols in the north, giving way to tundra in northern Canada and Alaska.

The natural hazards affecting North America rival the richness and variety of the region's natural resources. They include hurricanes, tornadoes, earthquakes, volcanoes, and floods. Other environmental problems result from intensive human use of the soils, forests, and mineral resources and from pollution of air and water.

Questions to Think About

11B.1 What role did glaciers play in shaping the current physical geography of North America?

11B.2 What is the spatial distribution of natural hazards in North America? What role do latitude and proximity to water play in the distribution of climate-related natural hazards in the region?

11B.3 How has human activity intensified or alleviated human susceptibility to natural hazards?

Key Terms

| Canadian Shield | deciduous | Dust Bowl |
| hydrology | coniferous | |

World Roles

The United States is uniquely positioned in the world today due to its political, economic, and military prowess. The global influence and economic connections of the United States are highly sought by many governments, national groups, nonprofit organizations, and corporate leaders. The very same connections and influence are also often met with protest and even violence. Special interest groups throughout the world disagree with U.S. government policy or intervention in events occurring outside U.S. borders. Other groups maintain an antagonistic position with respect to U.S. multinational corporations whose external connections exist through foreign operations and investment. Groups who feel threatened by the global prominence of the United States attempt to raise their local voices to the international agenda through protest or demonstration. Canada often finds itself involved in controversy through association with the United States more than through its own policies or actions, which may differ significantly.

The United States and Canada participate in a number of intraregional and interregional diplomatic, economic, and military alliances. Some of the most significant global connections today are directly linked to this region through formal structures in which the United States and Canada engage with European states and other countries worldwide. Both the United States and Canada, along with Japan, Germany, the United Kingdom, France, Italy, and Russia, are part of the **Group of Eight (G8),** which is an economic discussion forum of the world's eight most materially wealthy countries. The United States wields significant influence in the United Nations (UN), the North Atlantic Treaty Organization (NATO), the World Bank, the International Monetary Fund (IMF), the Organization of American States (OAS), and the World Trade Organization (WTO), among many others. The UN, World Bank, IMF, and OAS are all headquartered in the United States.

The United States and the United Nations

North America has a worldwide role in most areas of the global political economy and international security. The United States enjoys an especially distinctive role as the host country of the UN. The international headquarters of the UN is in New York City. The **UN Security Council** is the most powerful branch of the organization. The purpose of the Security Council is the "maintenance of international peace and security." The United States (along with China, France, the United Kingdom, and Russia) is one of five countries granted a permanent seat on the Security Council, while the remaining member countries of the UN hold rotating seats. The permanent members of the Security Council have the power to veto council decisions. The decisions of the Security Council are binding for the entire UN General Assembly.

The decisions and actions of the United Nations often simultaneously result in praise and sharp criticism. Domestically, actions of the Security Council produce healthy debate among U.S. citizens concerning the role the UN should play globally and the degree to which the United States should be involved in that role. Internationally, countries and national groups benefiting from actions of the UN in general and Security Council specifically applaud the execution of decisions made by the organization. Other groups, not directly benefiting from specific measures, criticize the organization and often the United States for interfering. Critics question the validity of the UN due to the unequal distribution of power on the Security Council. Controversy is furthered by the U.S. reluctance to promptly pay its financial dues and to fully participate in the United Nations World Court.

Globalization: The Good, the Bad, and the Ugly

The diffusion of U.S. fast-food chains around the world is often used as an example of the widespread impact U.S. cultural and economic practices have on global homogenization. Globalization is much more than the residents of Moscow eating a hamburger at a fast-food restaurant owned by a U.S. company. There are countless global connections in communications, the media, transportation, politics and diplomacy, athletic competition, entertainment, recreation, and leisure and business travel. The globalization of culture diffuses in all directions. Few aspects of life outside the region are not touched in some way by North American connections, and few aspects of life within the region are not touched by international relationships.

Although disagreement and controversy easily bond with many globalization trends, both the United States and Canada enjoy numerous positive relationships within the globalization frame. The space shuttle program under the wing of the U.S. National Aeronautics and Space Administration (NASA) is an example of positive global cooperation. The production of the various components of the shuttle and its operation are not solely generated by the United States. Research from Canada, countries in Europe, and Russia plays prominently into the success of the shuttle missions. The flight paths of shuttle missions present one of the more unique and positive aspects of a globalization relationship. During critical takeoff and landing windows as well as while the shuttle orbits Earth, the United States has the cooperation of several countries forming a complete ring around the globe to assist in monitoring shuttle flights and, perhaps more important, to ensure safe landing sites should something go dramatically wrong.

Although the positive aspects of globalization are numerous for the United States and Canada, it has seemed in recent years that the negative side of globalization relationships garners the greatest attention of the media. Some of the most dramatic manifestations of antiglobalization sentiment are the increasingly violent protests staged at the meeting sites of the G8 or the WTO, which is the primary international organization governing global trade. Such protests are often aimed directly at the United States because the U.S. government and U.S.-based multinational corporations enjoy considerable influence with the WTO. Supporters of the WTO assert the organization opens barriers, creates jobs, provides development opportunities, and infuses capital where there otherwise would be none. Protesters argue the WTO favors materially wealthy countries, such as the United States and Canada, over more impoverished countries. Protest groups also argue that the decisions of the WTO are detrimental to the rights of laborers around the world. The annual meeting of the WTO in 1999 in Seattle, Washington, was engulfed in dramatic protest that resulted in rioting, property destruction, and human injuries. When the G8 met in Genoa, Italy, in 2001, demonstrators clashed with Italian police forces, resulting in the death of a protester.

Global Role of the Regional Economy

The United States had the world's largest economy in 2003, over twice the size of China's, the second largest. Canada's aggregate GNI, when adjusted for purchasing power parity, is eleventh. Economic growth and victory in World War II gave the United States political superpower status alongside the Soviet Union. After the 1991 breakup of the Soviet Bloc, the United States remained the sole superpower.

U.S. corporations expanded abroad, taking opportunities to do so after World War II during the economic recoveries in Europe and East Asia, and in the context of Cold War competition in developing countries. After 1950 the United States established a huge lead in the initial development and use of computers. Although Western Europe and especially Japan now challenge this economic and technological superiority, no other single country is close to rivaling the total GNI PPP of the United States or the size of its internal market for products (Figure 11.15).

Canada lagged behind its neighbor through the early 1900s, partly because of restrictions resulting from its colonial ties to Britain and partly because of its smaller home market for goods and protection of its own industries. It gradually became more closely enmeshed with the U.S. economy and dependent on it for financing and markets. In the post–World War II era, rich stocks of metal ores in Canada's north; coal, oil, and natural gas in Alberta; and hydroelectric power generation in northern Québec ignited new geographic directions of development and provided the basis for growing Canadian affluence. Canada's centers of manufacturing production around Toronto and Vancouver, and to a lesser extent around Montréal and in Alberta, now rival individual centers in the United States in size and diversity of output. Canadians enjoy material living standards almost equal to those in the United States, and many who live in Canada strongly support the notion that they enjoy an even better quality of life. Some Americans prefer the Canadian way of life and live across the border from Detroit or Buffalo. The United Nations human development and gender-related development (HDI and GDI) indexes both placed Canada in the top five countries with Northern Europe and Australia.

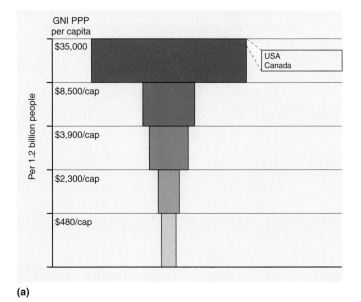

(a)

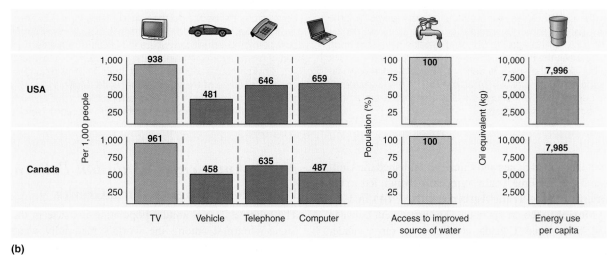

(b)

FIGURE 11.15 North America: country incomes compared. (a) The countries are listed in the order of their GNI PPP per capita. (b) Ownership of consumer goods, access to clean water, and energy usage are among the highest world levels. **Source:** (a) Data (for 2000) from *World Development Indicators,* World Bank, 2002. (b) Data (for 2002) from *World Development Indicators,* World Bank, 2004.

NORTH AMERICAN FREE TRADE AGREEMENT

The formal implementation of the North American Free Trade Agreement (NAFTA) in 1994 created one of the world's largest trading blocs. The United States, Canada, and Mexico moved toward an extensive elimination of trade barriers to increase the economic activity between them and strengthen their economic and political positions at the global scale. The three countries tied by NAFTA are now each other's largest trading partners.

NAFTA has many proponents and critics in each of the three countries. The similar living standards in the United States and Canada contrast sharply with poverty conditions common to millions in Mexico. Proponents of NAFTA claim it creates jobs, strengthens and expands business and industry, diversifies the economies of each of the three country partners, enhances foreign revenues, and fortifies each country's ability to compete in a global trade arena. Those opposing NAFTA assert the agreement results in the exportation of jobs and revenues from Canada and the United States to Mexico, the perpetuation of harsh labor conditions for Mexican workers, the degradation of the natural environment in all countries, and the granting of nearly supreme powers to big business and industry at the expense of workers, consumers, and those living in the shadow of production facilities.

The North American Free Trade Agreement supersedes the U.S.–Canada Free Trade Agreement implemented in 1989. Political leaders from the three NAFTA partners (Mexican President Carlos Salinas de Gortari, 1988–1994; U.S. President George H. W. Bush, 1989–1993, and, later, U.S. President Bill Clinton, 1993–2001; and Canadian President Brian Mulroney, 1984–1993) signed the 1992 trade pact. The respective legislative bodies each then subsequently passed the agreement, which was formally implemented in 1994. The agreement effectively removed most tariffs immediately and established a timeline by which all tariffs would be eliminated within 15 years of implementation. NAFTA eroded barriers, restrictions, and red tape, making it nearly effortless for companies from the three countries to conduct business and sales anywhere in the region.

The United States led efforts to push NAFTA forward in hopes of creating a competitive trade bloc to the European Union. The majority of NAFTA proponents, who pushed through an accelerated legislative process in the United States, were mostly representatives of large corporations. Labor unions, human rights groups, and environmentalists were strongly united against the rapid development of a trade agreement between the three countries and hoped to stop fast-track legislation, but argue they were largely not included in the negotiations. Environmental critics of NAFTA claim many of the key negotiators rapidly pushing NAFTA through the U.S. Congress were among the United States' biggest polluters.

Proponents of NAFTA assert that the trade agreement forces the equal treatment of corporations and industry throughout the region. Business leaders claim the package promotes fair and equal business opportunities for all companies in the three countries. Corporate leaders also assert that equal treatment reduces the risks of governmental interference in free market flows. Mexico has a long history of protectionism, or imposing restrictions, taxes, and quotas on goods and services produced outside Mexico, and nationalization of industry (government takeover and control of business and industry, thus removing private ownership). Business proponents claim protectionist policies would be heavily curtailed under NAFTA. Business leaders firmly believe that capitalism thrives best in NAFTA-created conditions in the region and this will promote and fortify economic and political stability for all three countries.

Strong opposition to NAFTA emanates from labor unions in the United States. Union leaders argue that the agreement exports thousands of jobs from the United States to Mexico. Working conditions in production facilities in Mexico are dramatically different from those in the United States. Mexican wages are, on average, one-eighth to one-tenth of U.S. wages. Benefits and other labor costs are nonexistent or significantly lower in Mexico also. Laws restricting or regulating environmental pollution from production facilities and those concerning labor safety are fewer and less enforced in Mexico. United States–based companies are able to manufacture products in Mexico with significantly lower labor costs and with far fewer environmental or labor restrictions. U.S. labor leaders say these factors are far too attractive for U.S. companies to forgo, and they claim numerous companies are shifting production to Mexico and therefore replacing U.S. workers with Mexican ones. Unions claim thousands of jobs have been lost already in the United States and project thousands more as NAFTA progresses and removes any remaining trade barriers.

The Mexican Border Industrial Program, first implemented in 1965, slowly opened the door for foreign firms to establish production facilities in Mexico. The idea of the program was to stimulate growth in northern Mexico, create jobs, and infuse capital into a region that had operated under protectionist and nationalist economic policies for decades. The program created a manufacturing zone across northern Mexico known as the *maquila* zone (the factories are referred to as *maquiladoras*). The idea was that foreign firms, such as U.S. companies, could import raw materials or parts, pay Mexican workers low wages to refine or assemble them into finished goods, take the goods back across the border (to the United States for a U.S. firm) relatively tax free, then sell them to consumers. Although maquila enterprise is now permitted virtually anywhere in Mexico, the greatest concentration of maquiladoras remains in the border region in the north.

Environmental conditions in the maquila zone in northern Mexico are among some of the worst in the world. Multidisciplinary research suggests an increasing array of health problems and ecological damage due to the concentration of factories in a country with loose envi-

In 1988 the United States and Canada established the United States and Canada Free Trade Agreement, which led to the **North American Free Trade Agreement (NAFTA)** in 1994 by adding Mexico to the arrangement. The agreement between the United States and Canada continues to thrive under NAFTA, despite a sequence of disputes over individual items (Point–Counterpoint: North American Free Trade Agreement box above).

North American Population Patterns

Population Change: Natural Growth

The relatively rapid annual population increase in the United States—unusual among the world's materially wealthy and technologically advanced countries—is due in part to the young immigrant communities within the United States who have larger families and higher birth rates than other groups.

ronmental and worker safety laws. The health and environmental issues present in northern Mexico also exist across the border in the southwestern part of the United States. Cancer rates along the Texas–Mexico border are some of the world's highest. When NAFTA discussions began, proponents argued the agreement would ease the problems critics claim were created by maquila enterprise for both Mexico and the United States. Since NAFTA's implementation, critics of the agreement argue that production in the border region of northern Mexico is growing and thus increasing the workers' and residents' health problems, as well as the related environmental damage.

The controversy surrounding NAFTA is far from over. Further trade barriers will be reduced or eliminated in the next several years in accordance with the progression plan of the agreement. Activist groups both for and against NAFTA will continue to lobby the respective governments, especially the U.S. government, to promote their views on the future of the agreement. There are proponents and critics in each of the three countries involved. Depending on the source, convincing statistics touting its value and success, or its detriment to society, the economy, and the health of the region's citizens all make compelling arguments.

SUPPORTERS OF NAFTA	CRITICS OF NAFTA
NAFTA opened new markets for the three countries.	NAFTA primarily opened new markets for Mexico and Canada.
Competition of lower-priced goods produced in Mexico forces down the price of goods produced in the United States and Canada; thus U.S. and Canadian consumers "win" with lower-priced goods.	Low labor costs in Mexico and lower-priced goods in the United States and Canada cause U.S. and Canadian companies to move their production operations to Mexico; thus thousands of jobs are lost ("exported") to Mexico, hurting employment in the United States and Canada.
NAFTA strengthens the global economic weight of the three countries and makes them better able to compete with the EU and other trade blocs and countries of the world.	Significant economic disparity exists between the affluent United States and Canada and the relatively materially impoverished Mexico, placing Mexico at a disadvantage and creating more of a service role for Mexico to Canada and especially the United States, rather than truly making it an equal trade partner and stronger international economic player.
NAFTA promotes democracy and political stability in Mexico and strengthens the Mexican economy, thus ensuring greater stability for North America.	Perceptions by some Mexicans of heightened economic disparity in their country due to NAFTA result in political instability such as the Zapatista uprising (see Chapter 10).
NAFTA creates thousands of jobs in Mexico. Cities and towns in northern Mexico, where the majority of NAFTA-related production (maquila) takes place, enjoy much higher living standards and higher rates of employment than most other parts of the country.	Cultural distinctions are blurred in all three countries. U.S. culture may overpower parts of Canada and especially northern Mexico. Fast food is replacing traditional food; U.S. holiday celebrations are replacing traditional celebrations. Areas in the U.S. Southwest are developing a watered-down culture that is a mixture of U.S. and Mexican elements. The increased use of the Spanish language in parts of the United States, especially areas in the Southwest, increases tension with some English-speaking residents.
NAFTA strengthens Mexican environmental conditions through environmental side agreements negotiated along with the primary trade agreement, resulting in a healthier environment for Mexico and especially the border region; it reverses environmental damage on the U.S. side of the border.	The side agreements negotiated with NAFTA fall far short of strengthening environmental laws and enforcement in Mexico. U.S. companies are further attracted to relocating in Mexico due to lax enforcement. The wording of NAFTA facilitates environmental abuse by companies in all three countries as NAFTA protects the companies' rights to free trade over the rights of people living in areas polluted by factories and other production facilities.
NAFTA forces the equal treatment of corporations in the three countries.	Corporations are too powerful under NAFTA.

The U.S. population of 296.5 million could rise to 350 million by 2025. The demographic transition for the United States suggests the above-average population increase will maintain population growth for the country in the short term (Figure 11.16). The age–sex diagram for the United States details both a maturing population and new growth in younger age groups due to immigration and higher birth rates among some ethnic populations (Figure 11.17).

In the 114 years from 1891 to 2005, the Canadian population increased by approximately 27 million, from 4.8 million to 32.2 million. Over the same period, the U.S. population increased by over 200 million (63 million in 1890 to 296.5 million in 2005). Both countries received immigrants in the early 1900s, but two world wars separated by a major economic depression in North America deterred further large-scale increases until after World War II. In the post–World War II era, population growth in

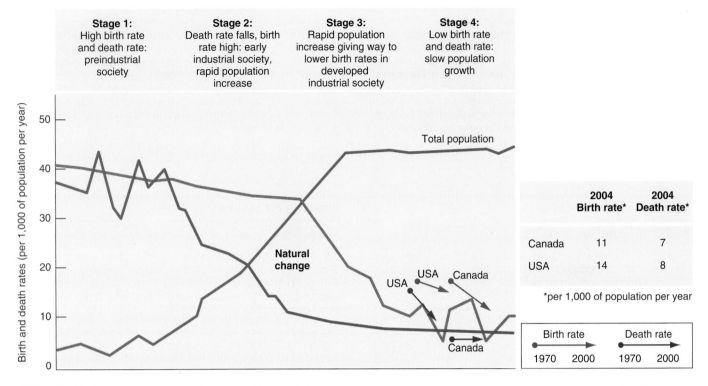

FIGURE 11.16   North America: demographic transition.   Both Canada and the United States have relatively high birth rates for affluent countries, a sign of the continuing immigration of young families. **Source:** for 2004 upgrade: *Population Data Sheet,* Population Reference Bureau.

Canada was rapid, doubling from 1941 to 1971 before slowing to a one-fourth increase from 1971 to 1991. Canada's total fertility rate of just under 2 is less than that of the United States, but its population growth rate is similar. Canada has lower death rates than the United States, and its continued immigration rates almost match those of the United States. Canada's age–sex diagram (Figure 11.18) resembles that of the United States.

Population Distribution: Increasing Urban Density

The United States of America is a highly urbanized country, with over three-quarters of its population living in urban areas and just over half in metropolitan centers of over 1 million people (Table 11.2). More than 90 percent of the U.S. population lives within a two-hour drive of a large city of over 300,000 people. More than 40 million people live along the eastern seaboard in an urban corridor from Washington, D.C., to Boston, Massachusetts, known as the **Megalopolis,** and another 30 million people live in the metropolitan areas of the western Manufacturing Belt and Midwest. In addition to the Megalopolis, urban clusters exist around the Great Lakes from Chicago to Detroit, in Florida and westward along the Gulf coast to southeastern Texas, and along the Pacific coast in the west (Figure 11.19). Large cities are the essential feature of the country's contemporary geography. Urban population clusters are more numerous in the eastern contiguous United

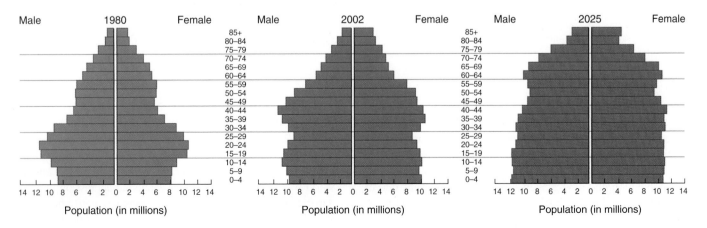

FIGURE 11.17   United States: age–sex diagram.   The 1980 and 2002 graphs show how the baby boom of the 1950s and 1960s produced a bulge in the age groups. **Source:** U.S. Census Bureau, International Data Bank.

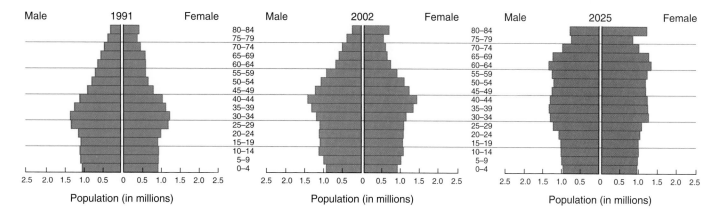

FIGURE 11.18 Canada: age–sex diagram. The 2002 graph details a maturing population with the largest numbers of people in the 35- to 44-years age groups. **Source:** U.S. Census Bureau, International Data Bank.

States. Vast areas of the western United States, away from the Pacific coast, contain very low population densities. Hawaiian population is clustered near Honolulu and resort towns, and Alaska, which is the largest U.S. state in area, has the lowest population density in the country.

Canada has approximately one-tenth the population of the United States on a significantly larger land area, which gives Canada a much lower average population density than that of the United States. It is necessary to consider the distribution of people in Canada in order to better understand Canada's population geography. The vast majority of Canadians are concentrated in a belt across the southern part of the country nearly parallel to the border with the United States. Thus portions of this populated belt across southern Canada have high population densities, while most of the remainder of the country is virtually empty. Canada is experiencing a trend toward metropolitan expansion. The two largest metropolitan areas, Toronto and Montréal, rival many U.S. metropolitan areas in size. Together with Vancouver on the west coast and Ottawa, the capital, these cities contain one-third of Canada's population. Winnipeg, Edmonton, and Calgary are other major Canadian cities. Canadians are less divided by segregation than Americans, but congregation is important among newer immigrant groups in big cities such as Toronto and Vancouver.

Table 11.2 North America: large cities

City, Country	2003 (millions)	2015 estimate (millions)	City, Country	2003 (millions)	2015 estimate (millions)
Atlanta, USA	4.0	5.3	New York, USA	18.3	19.7
Austin, USA	1.0	1.4	Orlando, USA	1.2	1.5
Baltimore, USA	2.1	2.4	Ottawa, Canada	1.1	1.2
Boston, USA	4.2	4.8	Philadelphia, USA	5.3	5.7
Calgary, Canada	1.0	1.3	Phoenix, USA	3.2	4.0
Chicago, USA	8.6	9.4	Pittsburgh, USA	1.8	1.9
Cincinnati, USA	1.6	1.7	Portland, USA	1.7	2.1
Cleveland, USA	1.8	2.0	Providence, USA	1.2	1.4
Columbus, OH, USA	1.2	1.4	Riverside–San Bernardino, USA	1.6	1.9
Denver, USA	2.1	2.6	Sacramento, USA	1.5	1.8
Detroit, USA	4.0	4.2	San Antonio, USA	1.4	1.6
Houston, USA	4.1	4.9	San Diego, USA	2.8	3.1
Kansas City, USA	1.4	1.5	San Francisco–Oakland, USA	3.3	3.6
Las Vegas, USA	1.6	2.2	San Jose, USA	1.6	1.8
Los Angeles, USA	12.0	12.9	St. Louis, USA	2.1	2.3
Memphis, USA	1.0	1.2	Tampa–St. Petersburg, USA	2.2	2.5
Miami, USA	5.2	6.0	Toronto, Canada	4.9	5.8
Milwaukee, USA	1.3	1.4	Vancouver, Canada	2.1	2.4
Minneapolis–St. Paul, USA	2.5	2.8	Virginia Beach, USA	1.4	1.6
Montréal, Canada	3.5	3.7	Washington, DC., USA	4.1	4.6
New Orleans, USA	1.0	1.1			

Source: United Nations Urban Agglomerations 2003 (2003).

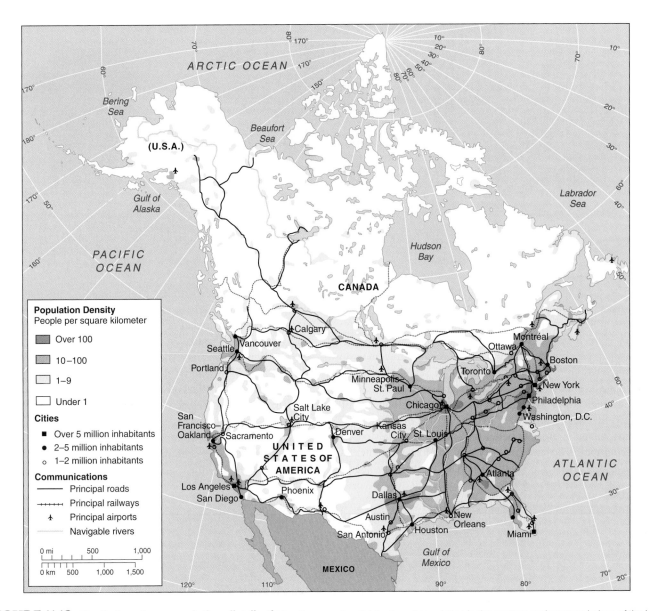

FIGURE 11.19 North America: population distribution. Compare the distribution of population in the eastern and western halves of the United States and note the closeness of most Canadians to the U.S. border. **Source:** Data from *New Oxford School Atlas,* Oxford University Press, UK.

Patterns of Migration in the United States

The continuation of relatively high rates of population increase for the United States is primarily a result of immigration, which has been vital to the country's economic growth since the days of European exploration. Immigration currently accounts for a third of U.S. population growth. Following the early domination by people of British origin, Irish and German people immigrated in greater numbers during the middle of the 1800s (Figure 11.20). In the later 1800s the highest numbers of immigrants came from southern Europe, particularly Italy; and at the turn of the century, large numbers came from the Slavic countries of eastern Europe, the Balkans, and Russia.

Peoples from the African continent were forced to migrate to North America in the 1700s. During the 1800s the African American population continued to grow through natural increase after the end of the Atlantic slave trade in the 1860s.

From the mid-1900s Hispanic peoples, from Middle and South America, and Asians became major sources of immigrants. The Chinese, who came to the United States as laborers in the late 1800s, and the Japanese, who moved in as farmers in the early 1900s, were later joined by other Asian groups from 1960, particularly from Vietnam, Korea, and India. Most recently, influxes of Africans and the 1990s immigration of Russians following the demise of the Soviet Union further diversified the cosmopolitan American population.

The United States faced a dilemma during the 1900s: whether it should encourage immigration as a continuing feature of its dynamic society or discourage it with the perceived objective of preserving the standard of living of its people. Many European countries adopted the latter approach, but they now experience shortages of people in the working age groups (see Chapter 3). Immigrant groups often maintain a younger age for

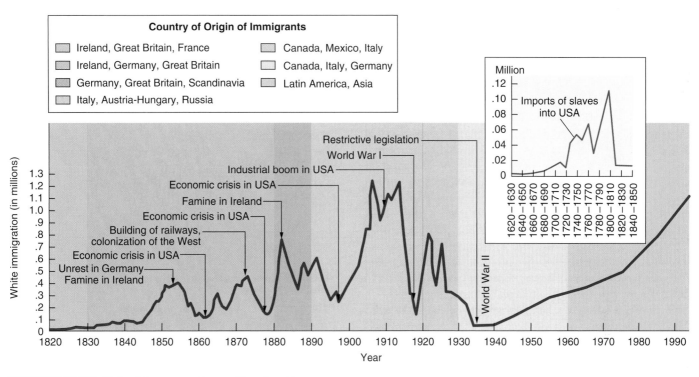

Country of Origin of Immigrants

- Ireland, Great Britain, France
- Ireland, Germany, Great Britain
- Germany, Great Britain, Scandinavia
- Italy, Austria-Hungary, Russia
- Canada, Mexico, Italy
- Canada, Italy, Germany
- Latin America, Asia

FIGURE 11.20 United States: pattern of immigration. The effect of push and pull factors on immigration fluctuations and the changing geographic sources of immigrants. Compare the size of forced slave migration to the levels of white immigration up to the mid-1800s. **Source:** Data from National Academy of Sciences.

the working population and have higher birth rates than segments of the population established generations earlier. Beginning in the 1920s, U.S. immigration laws gave first preference to Europeans, but from the 1960s, they opened access more widely on the basis of family ties or specific skills. Today, however, several million illegal immigrants, or "undocumented aliens," mainly of Hispanic origin, live in the United States. An examination of immigration trends throughout the development of the United States indicates that immigrants respond more to boom or recession in the United States economy than to legislation.

Increasing numbers of people living in countries outside North America apply for immigrant status each year. Democratic freedoms and the attractive quality of life in the world's most affluent region are strong pull factors creating migration streams from all regions of Earth. Total numbers of legal immigrants and refugees living in the United States—the latter being given special status for a limited period because of adverse political conditions in their own country—topped 800,000 by 1997. Illegal immigrant totals were estimated to be well over 1 million. Overall, immigration helped to maintain a U.S. population growth of around 1 percent per year, compared to increases of under 0.5 percent in many countries of Europe and Japan.

Contemporary Immigration Patterns in the United States

The contemporary immigration picture for the United States is extremely diverse. Immigrants today come from all corners of Earth for a variety of reasons. Immigrant residence and ethnic diversity in the United States have established distinctive spatial patterns throughout the country (Figure 11.21). Many highly

educated people migrate to the United States to fill employment demands in high-tech and computer-related industries, medical research, and numerous fields within the hard sciences. Other immigrants come to the United States with no material wealth or education in search of a life different from the environment in their country of origin. Impoverished immigrants often take low-wage positions in agriculture or as housekeeping or maintenance staff in hotels, restaurants, and retail business.

The 2000 federal census counted 281 million people in the United States. Of these, almost 200 million were of European origin, 34.6 million (12.3 percent) were African Americans, 2.5 million (0.9 percent) were Native Americans, and just over 10 million (3.6 percent) were Asians. Some 35 million (12.5 percent) were Hispanics, an ethnic designation based mainly on speaking Spanish.

Over half of the African Americans in the United States live in the South in both rural and urban areas. The majority of the remaining African American population is clustered in northern and western metropolitan centers. The African American percentage of the total U.S. population is diminishing as the percentages for Hispanic, Asian American, and other groups continue to increase.

In the United States, Spanish is used as a home language by 5 percent of the total population of the country, which amounts to more than 14 million people. Spanish-speaking peoples are located mainly in the states bordering the Latin American region, including Florida, Texas, New Mexico, Arizona, and California; most large cities of the northeastern part of the United States; and in the expansive urban reaches of metropolitan Chicago. The Hispanic population is the fastest-growing ethnic group within the United States, and Hispanic

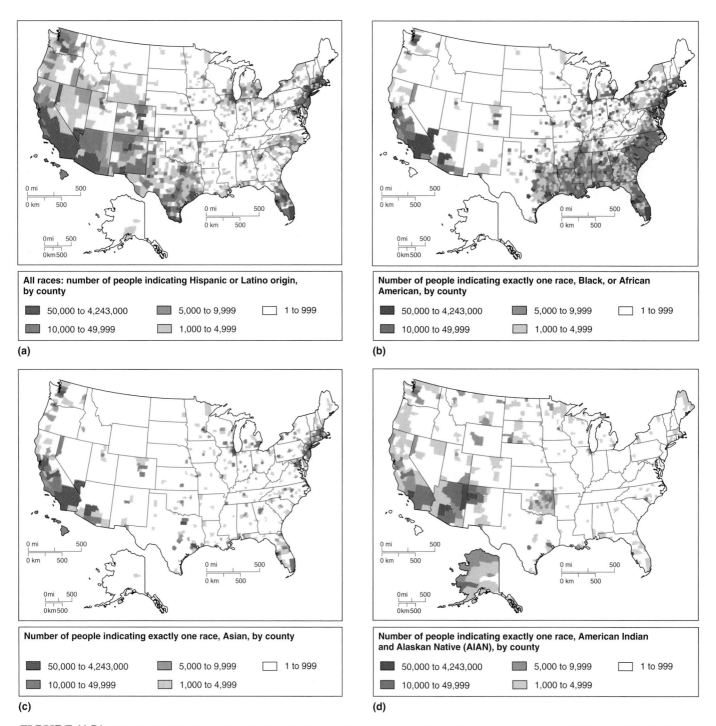

FIGURE 11.21 Ethnic distribution in the United States. (a) Primary concentrations of Latin Americans include California and the Southwest, Florida, Chicago, and the cities of the Megalopolis. (b) Primary African American concentrations include California, the Southeast from east Texas through North Carolina, Florida, major cities of the Great Lakes, and cities of the Megalopolis. (c) Primary concentrations of Asian Americans include the Pacific Coast, Chicago, and cities of the Megalopolis. (d) Primary concentrations of Native American and Alaskan Native peoples include Alaska, California, Arizona, New Mexico, and Oklahoma. **Source:** Data from U.S. Census Bureau, Census 2000 Redistricting Data (PL 94-171) Summary File. Cartography: Population Division U.S. Census Bureau, American FactFinder at factfinder.census.gov provides census data and mapping tools.

Americans are emerging as the largest ethnic minority within its borders. The Hispanic population in the United States grew 58 percent during the 1990s, to 35 million.

Increasing immigration of people from many Asian countries further intensifies the ethnic mosaic of the United States. Asian immigrant settlement patterns are in concentrations along the West Coast of the United States, especially in the dynamic urban centers. There are also large concentrations of Asian Americans in parts of the upper Midwest and in large cities in the eastern United States. The climate and ecosystems of the Gulf Coast region attracted and now house a significant population of people who have emigrated from

Vietnam. Vietnamese Americans living along the Gulf Coast engage in agricultural and fishing practices similar to those of Vietnam.

Native Americans and the United States Government

In the 1970s the Native American indigenous groups began to increase their proportion of the U.S. population for the first time since a census was taken in 1790. Many Native Americans concentrated on federally sequestered reservation lands that were often arid and hostile for agricultural productivity. Native American peoples within the United States are among the country's most materially impoverished communities. Complicating the reservation lifestyle was an administration structure imposed by and monitored directly from the federal government rather than state or local levels. By the 1970s some 50 percent of Native Americans were unemployed, and 90 percent were on welfare.

Attempts were made in the later 1900s to reduce the poverty imposed on Native Americans, who had often been forced to live on marginal and largely unwanted lands. Helped by the clarification of their rights in Alaska arising from statehood and the selection of state lands from the federal holdings, Native Americans elsewhere in the United States raised questions about the ownership and management of natural resources on their reservations, including minerals and water. Although some indigenous communities lived on lands with timber and coal resources, others filed lawsuits seeking compensation for being placed on resource-deficient land. Lucrative deals with governments and water projects, however, did little to change the economic structure of the Native American reservations.

New enterprises brought limited satisfaction. Some Native Americans developed tourist facilities on their reservations, but others chose to forgo such opportunities—often to avoid copying the commercial practices of white Americans. Casino gambling is now a significant source of income for some Native American groups. Among the casino-related challenges are managing the influx of capital, discouragement of low-wage jobs, combating gambling addiction, avoiding increasing dependence on a single employment and revenue stream, and numerous social ills that often accompany a gaming-based culture. While casino-related development on reservations remains controversial, such enterprises may encourage some Native American groups to engage with the wider United States economy. To capitalize on their water, minerals, and tourist opportunities, Native Americans need to agree on how to administer and develop the lands that are held in trust for them.

Internal Migration in the United States

The dynamics of population geography in the United States involve internal migrations as well as international immigration (Figure 11.22a). During the 1800s, the main internal migrations were from the Atlantic seaboard westward to the interior and then on to the West Coast. Within the South, the plantation economy was transferred westward from the Atlantic plain to the Mississippi Delta area, taking with it planters and slaves. The northern (mainly white) and southern (mainly black) streams remained distinct from each other during the 1800s. After the Civil War in the 1860s, many African Americans were able to migrate independently, creating a small stream of northward movement that remained relatively insignificant until the twentieth century. The majority of African Americans remained working on southern farmland for several decades following the Civil War.

From the early to mid-1900s, large numbers of African Americans moved from the rural and urban South to northern cities. Those from the Atlantic coastal plain mostly moved to Washington, D.C., Baltimore, Philadelphia, New York, and Boston; those from the Mississippi Delta moved primarily to Cleveland, Cincinnati, Indianapolis, and Chicago. Smaller numbers of African Americans moved to the West Coast. In 1900 nearly 90 percent of African Americans lived in the South, but this proportion declined to just over half by the early 2000s. The movements also stimulated the dissemination of cultural features of African American society, such as the varied types of blues music (Figure 11.22b).

Most of the movements of African Americans out of the South were completed by 1970. After that, smaller **counter-migration** or return movements, often on retirement, balanced or exceeded the northward flow. Movements of European Americans from the Northeast to both the West Coast and the South continued into the 1980s, but recessions in that decade slowed southern economic growth and reduced such migrations.

Canadian Patterns of Ethnic Integration

Government policy protecting multiculturalism and well-known high living standards combine to make Canada an extremely attractive destination for peoples emigrating from their homelands. The Canadian federal government instituted policies outlawing ethnic discrimination and protecting the rights of immigrants. For several decades, Canada has been one of the leading immigrant-receiving countries of the world. Contemporary immigration trends for Canada exhibit significant increases in the numbers of people migrating from Asia and the Americas (including the Caribbean) and an overall percentage decrease in the European immigrant population.

The large urban centers of Toronto, Vancouver, and Montréal are experiencing dynamic cultural processes that increase the diversity of each metropolitan area. Canada's largest city, Toronto, attracts more immigrants from all reaches of planet Earth than any other Canadian city. Asian migration is most notable along the west coast of the country. The city of Vancouver and its hinterlands in the province of British Columbia are the hosts to two types of immigrants from Asia: those who settle year-round and others who use the city and environs as a second home or home for their families while they "commute" to work in Hong Kong.

Centrifugal Force: The Challenge of Québec and the Rights of Indigenous Canadians

The greatest ethnicity-related challenge facing the Canadian federal government and the unity of the country is the devolutionary pressure from an increasing movement of French-speaking

FIGURE 11.22 **United States: internal migration.** (a) The major movements of African and European Americans in the 1800s and 1900s. (b) Compare the developments in blues music and their diffusion to the movement of African Americans.

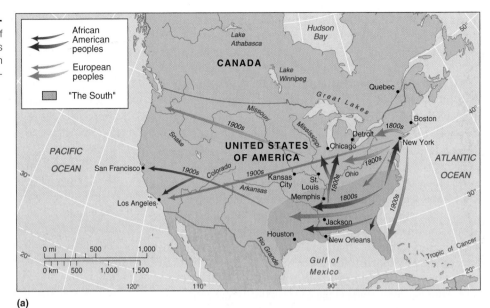

(a)

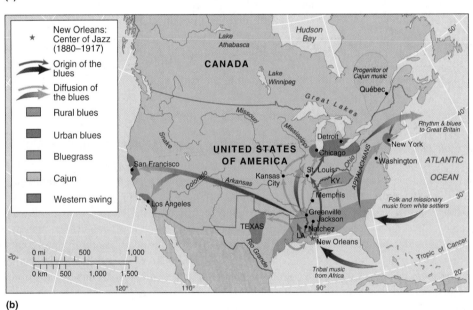

(b)

Canadians toward greater autonomy. Approximately 25 percent of Canadians are French-speaking descendants of the French settlers who came to the area during the earliest stages of European development in the region. Most French Canadians are distributed along the St. Lawrence River, primarily in the province of Québec. Most **Francophones,** or French-speaking Canadians, have an extremely strong sense of being a separate national group within the larger Canadian political frame. They see themselves as having a different ethnic history, different religion, and different culture than the English-speaking Canadian majority (Figure 11.23).

Although the majority of the French descendants adhere to Catholicism, which is in contrast to the primarily Protestant English Canadians, Québec's Francophones see their French language as the strongest symbol they have for discerning their uniqueness from the English-speaking Canadian majority. The perception of national identity loss is enhanced by the choices

new immigrants make in metropolitan Montréal, Québec's largest city and the largest French-speaking city outside of France. Many immigrants settling in Montréal, who spoke neither English nor French as their first language, are choosing to learn English rather than French and are choosing to conduct business in English and to send their children to private English-language schools.

The Québec separatist movement gained momentum during the 1980s and became especially strong during the 1990s. Decades of perceived oppression from the federal government as well as the perception of a threat to their cultural identity and language have unified members of the **Parti Québécois,** a political party formed with the sole purpose of achieving Québec's independence from the federal government. The Parti Québécois would like to see an independent country of Québec. A 1995 referendum pushed for by the Parti Québécois resulted in an almost 50 percent vote by the residents of Québec province in favor of separation.

FIGURE 11.23 **National Pride in Québec, Canada.** French Canadians proudly wave the Québec flag and independence signs during a parade in Montréal. **Photo:** © Reuters New Media, Inc./Corbis.

Several groups inside the provincial borders of Québec do not share the enthusiasm for separation. English-speaking and other business owners fearing revenue losses are adamantly opposed to independence. The more recent diverse immigrant population residing in the metropolitan Montréal area also wishes to remain part of Canada. One of the most spatially organized internal challenges to the Francophones' movement for independence comes from an indigenous Canadian people known as the **Cree,** who claim the northern third to northern half of Québec as their ancestral land. The Cree do not wish to be part of an independent Québec and threaten to secede from any newly formed Québec state and remain with Canada. The northern half of Québec is the site of significant natural resource potential. Most notably, the generation of hydroelectric power is so successful in northern Québec that the province is able to export energy to the United States and generate profitable revenue streams.

The Québec issue raises questions of whether the varied regions, separated by great overland distances, can continue to provide a complementary unity or will embark on a course that will tear the country apart. Other provinces in Canada have grown weary of the federal government's attempts to appease the provincial government of Québec. The provincial governments of the Prairie Provinces have successfully lobbied the federal government for greater control over their resources and decisions using the perceived "special treatment" given to Québec as a bargaining tool.

Indigenous Canadian issues throughout all Canadian provinces and territories are important for federal government consideration in addition to the situation in Québec. In 1973 the Canadian government opened the possibility of negotiating land claims with organizations representing native peoples. Until that date, little had been done in much of Canada to implement treaties negotiated with native peoples in the 1800s. Now several areas are identified for a degree of local government (Figure 11.24).

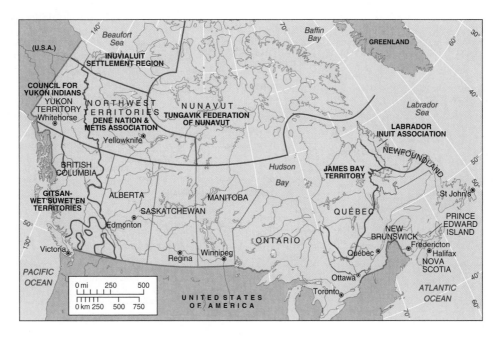

FIGURE 11.24 **Canada: contemporary First Nation claims.** Major Native American land claims in the Arctic north, Québec, and British Columbia, late 1990s. Some of these areas may obtain a greater degree of self-government than others.

The largest area, Nunavut ("Land of the People" in Inuit language), became a new territory with its own elected government in 1999, although it will remain subject to federal control. Other agreements were reached in northern Québec and with the Inuvialuit people in the northwest Arctic. Further discussions are under way, although many of the smaller claims may take several years to resolve. Some are complex because of overlapping land claims and because bargaining involves the often-opposing interests of native groups, the federal government, and provincial governments.

Subregions of North America: The United States

The United States at a Glance

The United States of America dominates the midlatitudes of continental North America with 48 of the country's 50 states situated between the Atlantic and Pacific Oceans to the east and west, respectively, Canada to the north, and Mexico to the south (Figure 11.25). Alaska and Hawaii are separated from the conterminous 48 states by Canada and the Pacific Ocean, respectively.

Metropolitan areas of the United States house the majority of the country's transportation and communications connections, the larger manufacturing facilities, corporate headquarters, and financial and business services. Some 85 percent of U.S. real estate value occurs in the 2.5 percent of the country's land that is occupied by urban areas.

The United States market economy, which is the world's largest, significantly influences economic activity across the globe. The United States remains a global leader in high technology and is the world's entrepreneurial leader. In 1996 the United States was first among the countries of the world in terms of raising venture capital. Of the more than $10 billion raised that year in the United States, California's share was $3.7 billion, and Massachusetts raised over $1 billion. Both states were ahead of the United Kingdom, France, and Germany—the next three—and 10 U.S. states raised enough to be ranked in the top 20 world "countries."

Problems of Affluence

The country's material wealth does not reach all within its borders. Those with advanced educations or double-income families continue to garner greater material wealth, while those who

FIGURE 11.25 **The United States of America: the 50 states and major cities.**

are located in impoverished regions or places with fewer educational opportunities struggle to acquire the basic necessities for nutrition, shelter, and health care.

Increasing Gap between the Materially Wealthy and Materially Poor

The gap between the materially wealthy and materially poor in the United States widened from the 1970s to the early 2000s. A large proportion of the increasing inequality is explained by two factors. The first is the lightly regulated labor markets that react to world conditions by depressing wages for unskilled jobs in competition with poorer countries and increasing salaries for skilled workers as demand from service industries and professions rises. The second factor is the changes in U.S. households that polarized the increasing incomes in a greater number of two-income homes against growing numbers of single parents, who make up more than a third of the poorest 20 percent of the population.

The materially poor areas of many inner cities have lower-quality buildings and services such as education and health because of the inability of small jurisdictions with predominantly poor populations to support such services. Such effects create a downward spiral of living conditions. The contrasts produced by uneven development are illustrated in the example of Boston, where some of the best public secondary education in the country is found in the suburbs while some of the worst is in Boston's inner city—both ends of the spectrum in the same metropolitan area.

Although the poorer groups in American society gain some sympathy, there is less pressure for taxing the materially wealthy further to support the poor than is found in many other wealthy countries. Although a percentage of tax revenues is used for welfare, public health, and infrastructure programs, redistribution of income on a scale of European countries does not occur. This is partly because an unusual level of mobility still exists between economic groups. It is increasingly difficult, however, to climb out of poverty into the next stage of more secure income where most upward mobility begins.

Congregation and Segregation

The income and material wealth gaps are often linked to perceived ethnic and racial differences. High income–earning people have many choices with respect to where they live. Many choose to congregate with those of similar socioeconomic standing and cultural tastes in trendy inner-city areas or suburban neighborhoods. They may live close to shopping, ethnic restaurants, cinemas, places of worship, or recreation facilities. Less affluent groups in U.S. society have fewer opportunities for choosing where they live and become segregated into communities that more affluent groups avoid. Such segregated groups often occupy inner-city areas that contain high proportions of African American or Hispanic people and have poor access to high-quality education and job opportunities.

Environmental Impact of Affluence: Disproportionate Consumption

Other problems of affluence result from the environmental impacts of intensive use and extraction of resources. In the early 2000s the United States, with just under 4.6 percent of Earth's total population, consumed 40 percent of the world's oil production.

In the beginning of the 1970s, the United States passed stringent environmental legislation to improve its air and water quality. Following the rise of the environmental movement in the United States and related legislative changes, air and water quality standards increased from the 1980s. Such regulations, however, sometimes led to the relocation of the most polluting industries to more materially impoverished countries.

The limits of resource usage occasionally became an issue. In 1972 a scare arose over whether world grain supplies could supply the needs of growing populations. In 1973 the oil-exporting countries raised their prices. To many it seemed that the "American Dream" of access to increasing affluence was coming to an end. The worst, however, did not happen, and new sources of materials, often at cheaper prices, enabled Americans to enjoy increasing affluence.

Economic Development

The United States of America is the world's most developed country in terms of economic prosperity and influence. Although its economy is subject to growth fluctuations, the United States maintains the largest total GNI PPP of any country in the world.

Commercial Farming Basis

Commercial farming regions developed as the combination of natural resources (climate, land, soils, water) and economic factors led to areas of specialization (Figure 11.26). The early farming areas of the East Coast suffered soil erosion, and, unable to compete with more fertile inland areas, much of the land reverted to woodland. The Midwest grain–livestock area, the Mississippi Delta cotton area, the Great Plains feedlots, and the irrigated lands of California became among the world's most productive areas. Drier and higher parts of the West became low-productivity grazing land.

In the 1990s many farmers in the U.S. heartland found that falling or static prices for their output of grain, meat, or milk could not sustain their debts on sophisticated equipment. Some went bankrupt. A few, especially on family-owned farms, returned to lower-cost farming using less equipment and fertilizer and relying on crop rotation to maintain soil fertility. The extra labor previously used on farms before the machinery revolution was not available, so those managing family farms worked harder.

In 1996 the federal government ended its controls on farming. Planted acreages surged in response to rising prices and increased exports to Mexico and, especially, China. This process is likely to exaggerate the move to increased

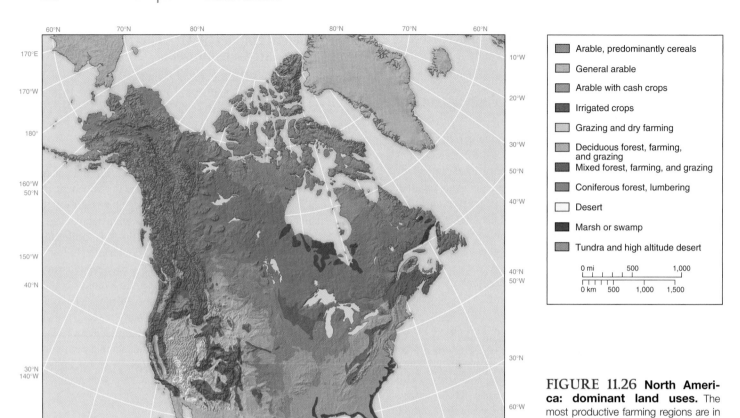

FIGURE 11.26 North America: dominant land uses. The most productive farming regions are in the Midwest of the United States, the Prairie Provinces of Canada, and parts of California, Texas, and Florida. What factors make farming difficult in other areas? **Source:** Data from *New Oxford School Atlas,* Oxford University Press, UK.

Legend:
- Arable, predominantly cereals
- General arable
- Arable with cash crops
- Irrigated crops
- Grazing and dry farming
- Deciduous forest, farming, and grazing
- Mixed forest, farming, and grazing
- Coniferous forest, lumbering
- Desert
- Marsh or swamp
- Tundra and high altitude desert

specialization. The government interventions had helped to moderate the swings of farm income. The farmers' problem remains the better management of their output.

Manufacturing Becomes Central

Until the 1990s, many economists saw the development of the U.S. economy in the 1800s and 1900s as a model for others to follow. Although the conditions necessary for economic development vary significantly depending on time and place, it is worth tracking a history of U.S. economic development to help an understanding of the regional differences within the country.

Manufacturing developed from local crafts (pottery, smithing, weaving) in the early 1800s in southern New England, New Jersey, and eastern Pennsylvania. It was based mainly on water mill power and produced metal goods, textiles, and leather goods. Numerous small mill factories required new transportation facilities to take their products to widespread markets. At first, water transportation was also vital. Local capital, accumulated by merchants, was invested in the early factories, often under family ownership.

By 1860 railroads reached west to Chicago (Figure 11.27a) from several East Coast cities, and their construction formed the basis of iron industry expansion. In eastern Pennsylvania, the use of charcoal gave way to coal in the iron-smelting process, then diffused westward to metropolitan Pittsburgh.

A major growth in manufacturing occurred after the 1860s with the adoption of steel production and the new possibilities it opened in heavy engineering. Many of the manufacturing industries of this phase were tied closely to their raw materials, such as coal and iron ore. New industrial areas emerged inland from the primary markets on the East Coast. New markets also developed as the interior of the United States was settled.

Major federal investments in transport infrastructure stimulated much of the spread of economic growth. Beginning in the 1860s, the transcontinental railroads were financed by federal and state governments granting lands along the routes to the railroads, which sold them and encouraged homesteading.

In 1860 the value of U.S. manufactured goods exceeded the value of commercial farm products for the first time. Agriculture became industrialized with the increasing use of mechanization and chemicals; markets for crops and livestock products were linked to the growing railroad network (Figure 11.27b).

Until 1950, manufacturing was the primary engine fueling the expansion of the U.S. economy. New products developed in consumer goods and transportation vehicles, including cars, trucks, and airplanes. These industries were less tied to sources of raw materials than the 1800s industries, and many were market-based or assembly industries where the best locations were central to a range of component producers or large markets.

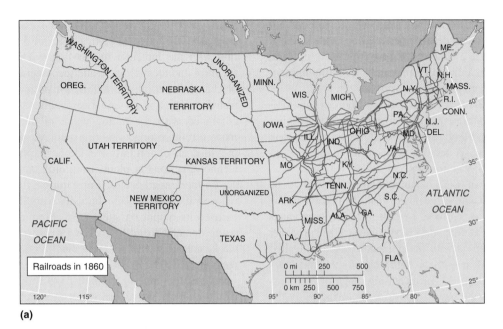

(a)

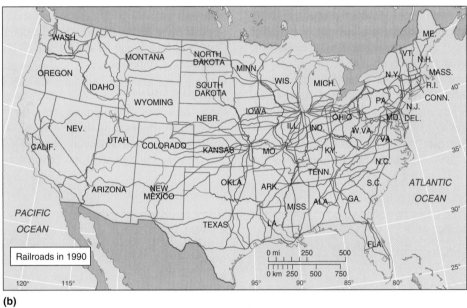

(b)

FIGURE 11.27 United States: railroad networks. (a) 1860. (b) 1990. The network was virtually complete by 1900; the extension across the continent between the late 1860s and 1900 was linked to the settlement of the U.S. West, although the building of transcontinental routes was held back until after the Civil War of 1861–1865. In the second half of the 1900s, many lines in the East were closed as truck transportation increased on the interstate highway system. **Source:** From P. Guinness and M. Bradshaw, *North America.* Copyright © 1985 Hodder & Stoughton Education, London, Reprinted by permission.

The Manufacturing Belt, however, was both the main market area and the location of basic metal industries and continued to be where most of these developments occurred. Factories were built in established market locations, such as the New York City area and around Chicago with its central location on the national railroad network, or in new locations, such as in Detroit, on which the auto industry was centered.

After World War II, manufacturing industries expanded along the West Coast and in the southern United States, where they were located during the war. The previous concentration in the Northeast became more widespread, satisfying new

markets in what had been the periphery of the country and relating more clearly to growing markets overseas in Latin America and Asia.

Constructing interstate highways, distributing electricity to rural areas, and developing a network of airline routes facilitated the wider geographic diffusion of manufacturing industries. The interstate highway system, begun in 1956, created nearly 80,000 km (50,000 mi.) of limited-access highways across the United States, opening many previously isolated parts of the country to economic development. The construction of airport facilities with federal grants increased the amount of traffic possible. Combined with the wide availability of high-quality telecommunications, these developments reduced the costs of distance between rival locations for economic development. They provided a vital infrastructure base for continuing economic growth, but their maintenance is costly.

While products diversified and became more technologically sophisticated, production and management techniques developed. American-based multinational corporations took their products to the world and opened factories in many countries in Europe and other continents.

In the later 1900s, U.S. manufacturing continued to be important, although it employed fewer than 20 percent of the workforce compared to nearly 40 percent in 1950. Its products were even more diverse, with some heavy industry and producer units moving to countries with lower labor costs. The United States is a world leader in applying high technology to manufacturing and service industries. An increasing range of high-tech goods is produced in the newer industrial areas, including the globally known Silicon Valley of California, metropolitan Boston (Figure 11.28), metropolitan Washington, D.C. (primarily Fairfax County, Virginia), the Research Triangle region in North Carolina, metropolitan Austin, Texas, the Denver–Boulder region in eastern Colorado, and areas of the Pacific Northwest. These all have national as well as regional markets.

As new manufacturing industries developed, older established industries either perished or reinvented themselves. Many older industries slimmed their workforce, installed new machinery, and became more specialized. Some changed their focus to accommodate new trends. For example, the steel industry, which until the 1950s had been dominated by the

huge, inflexible integrated complexes that resulted from economies of scale and vertical integration, lost out in competition with the Japanese in the 1960s. The Japanese had new facilities and technologies, and the American steel mills closed or changed. During the 1980s, the policy of voluntary export restraints for foreign countries selling to the United States, combined with the falling value of the U.S. dollar—which made foreign products more expensive in the United States—provided an opportunity for the large plants to upgrade their technology. Smaller plants (minimills) established themselves across the United States. For example, the USX Corporation reduced its jobs at Gary, Indiana, from 30,000 to 10,000 and focused its output on sheet steel. Productivity increased, and rising demand from the early 1990s helped much of the U.S. steel industry to meet international competition as the export restraints came to an end.

While comprising a small proportion of U.S. GNI today, the primary sector contributes important raw materials and fuels. Mining, forestry, fishing, and agricultural outputs were often linked to the manufacture of metal goods, energy supplies, paper, fertilizer, textiles, and food products. They were commonly produced by multinational corporations and involved high inputs of capital and technology with declining numbers of employees.

Service Industries

U.S. economic development from the late 1900s, however, focused increasingly on the growth of information-using occupations, including financial services, publishing, computer services, and design services. Tourism and leisure-based serv-

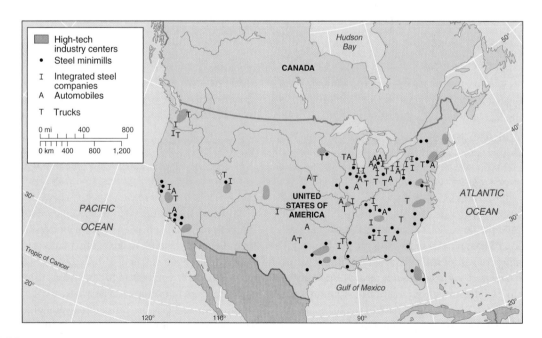

FIGURE 11.28 United States: distribution of some major manufacturing industries. The continuing concentration of steelmaking and vehicle manufacturing in the Manufacturing Belt contrasts with the widespread nature of high-tech industry centers. Steel production in minimills has become more widely distributed since 1970.

ices also became a leading part of many local economies. In 2002, 42 million tourists from other countries visited the United States, while 56 million went from the United States to other countries, and many millions of Americans were tourists in their own country. Although the number of visitors to the United States decreased in 2002 from those prior to September 11, 2001, the decrease appears to be temporary as distance from the September 11th tragedy increases. Such industries locate economic activity away from the workshop and factory. The United States, with its economic and technological lead, has so far been able to take greater advantage of this new economic base than any other country and is investing to maintain the lead.

Regional Policies

During the 1900s, rapid growth in some parts of the United States coincided with slower growth and even poverty in others. **Uneven development** reflected the concentration of capital investments in some regions and cities rather than in others. The favored and less fortunate regions changed over time. Until the Civil War, some of the South was prosperous for a privileged group of planters whose wealth was possible only through the enslavement of people originally forced to migrate from Africa. After the Civil War, the South became a large and long-lasting region of poverty for the United States. Southern Appalachia was also extremely poor, with hill farmers living on small plots of eroded land. The plight of southern Appalachia was first attacked after 1933 by the establishment of the **Tennessee Valley Authority (TVA),** one of the few federal programs designed to stimulate economic growth in lagging regions. TVA was initially viewed as a possible model for action in other regions, but the dominant political views in the United States were against government-funded regional aid. The huge investments in hydroelectricity and other forms of inexpensive electrical power generation in the Tennessee Valley provided a basis for economic growth and population growth—and delivered some serious environmentally detrimental side effects. For example, the rising demand for inexpensive electricity caused TVA to invest in first thermal and then nuclear power plants that became increasingly costly and dangerous to the environment. In the early 1990s the TVA nuclear program ended.

By the 1960s, however, it was recognized that a more extensive Appalachian region was the largest of several rural areas that were lagging behind the rest of the country. The main achievement of the **Appalachian Regional Commission (ARC),** established in 1965, was to make available to this region the many federal grant aid financial packages, particularly in the late 1970s. In the 1980s the administration of President Ronald Reagan reduced ARC's funding and effectiveness.

The rural problem grew, however, as the metropolitan centers of the United States assumed increasing dominance in the economic and social life of the country. The wealthiest farming regions in the United States continue their efforts to overcome the impact of overinvestment in the 1970s, and subsequent periods of drought.

Urban Landscapes

Urbanization within the United States produces distinctive landscapes through the types of building, the differentiation of land uses, and the distribution of groups of people. Human variation in the landscape includes the materially wealthy and the materially poor, African Americans, Hispanic Americans, European or white Americans, Asian Americans, and many others. American cities grew from colonial ports and inland market centers to 1800s industrial and 1900s commercial metropolitan centers. Large cities increasingly dominated the life of the country. Although Americans are often quick to knock down the old and build new, most cities retain landscape relics of each historic period that combine with the natural environment to give each city a specific character.

Preindustrial U.S. Towns

The colonial ports of New England and the Middle Atlantic were small, with scarcely 10,000 people each and a limited number of functions. Most people lived in the surrounding rural areas. By the mid-1800s, a pattern of market towns and small villages, linked to each other by road, canal, and railroad, covered the farmed plains of Ohio, Indiana, and Illinois. In the Northeast, small factory towns huddled around water mills. South of Chesapeake Bay, plantation agriculture in coastal areas produced individual homes for the planters, surrounded by slave dwellings and the homes of poor white workers. There were few towns, although places such as Charleston, South Carolina, developed as a regional port and important political center. Atlanta, Georgia, became the first inland rail junction in the South and began to grow in size before it was destroyed during the Civil War in the 1860s.

Some of the oldest towns grew under the Spanish occupation of the southwestern parts of the modern United States, with relics still in Santa Fe and Albuquerque, New Mexico, and San Antonio, Texas. All the early towns comprised relatively small buildings, some in stone or brick, but most in wood. The historic areas composed of such buildings take up smaller areas of modern cities than those in Europe (see Chapter 3).

Industrial and Commercial Cities

In the later 1800s and early 1900s, many cities in the east and center of the United States expanded into large conurbations of housing, factories, shops, and offices with the rapid growth of manufacturing industry and the railroad network. The larger factories and mills required many workers, who were accommodated in surrounding housing units. Areas of worker housing became differentiated from the better housing areas of the managers and owners. New **central business districts (CBDs),** where shops, banks, and other financial services became concentrated, were easily accessible by streetcar to and from the industrial areas and leafy suburbs.

By 1920 American cities often evolved into a **concentric pattern of urban zones** around the CBD, with poorer housing in inner suburbs and more affluent zones beyond

(Figure 11.29a). Some of the inner-city areas housed distinctive segregated groups such as Slavic or Jewish people in areas that became known as **ghettos** (after Jewish quarters in medieval Italian cities). In cities built across a local relief of hills and valleys, the development of railroads, roads, and commercial activities was concentrated along the valleys, giving the city geography a pattern of wedge-shaped urban sectors of different land uses (Figure 11.29b). In some cities the distinction between land uses was clear, but the distribution was not in such regular patterns, establishing multiple nuclei (Figure 11.29c) in which industrial and commercial activities occurred in several specialized and separated zones, instead of around a single CBD.

During the 1930s and 1940s, American cities grew more slowly than they had in the previous 50 years. The economic depression of the 1930s and World War II made capital either in short supply for house building or directed to military purposes. By the mid-1940s, many families shared cramped inner-city housing units.

Post-1945 Cities

An explosion of house and road building opened up huge suburban areas of single-family homes for those with rising incomes after World War II as the United States dominated the world economy. New federal roads linking major cities, followed by the interstate highway system established in 1956, made greater mobility possible and encouraged suburban housing and commercial development.

These trends gave rise to new forms of cities that developed from the monocentric (single center) before 1940 to the multicentric pattern of today (Figure 11.30). The CBD became more specialized around financial businesses as many of its shopping and industrial functions moved to the suburbs. High land prices led to the construction of tall office buildings with 50 and more floors to accommodate banks and other commercial corporations. Tall buildings formed the skyline of the central area, which sometimes expanded into surrounding warehouse and poor housing districts. New highways, often elevated or in tunnels, circled the CBD.

Beyond this sector of intense commercial activity, the pre-1940 suburbs, crossed by railroad yards and older highways with commercial strip developments, were often taken over by the expanding African American, Hispanic, or other ethnic groups. Such groups either did not have the resources to move to newer suburban homes, faced discrimination by financial institutions or home sellers when they tried to do so, or preferred to stay in an area that was familiar to them. The African American ghettos of northern cities did not appear until influxes of African Americans from the U.S. South combined with the suburb-bound movements of the European American population in the 1950s.

The post-1945 suburbs were quite different from the inner suburbs built before 1930. They were laid out with more space to allow single-family homes and yards. They are interrupted by shopping centers, light manufacturing factories, and office buildings that associate together and may include more economic activity than in the CBD. Such suburbs are linked by

FIGURE 11.29 United States: urban landscapes, 1900 to 1950. (a) The concentric zone model, based on a 1922 study of Chicago. At that time, the "ghetto" was not related to African Americans. (b) The sector model, based on 1930s census information, illustrating the impact of railroads and highways on the location of different land uses. (c) The multiple nuclei model of 1945 still has a single center, but commercial and manufacturing districts are not in close proximity to it. **Source:** From C. D. Harris and E. L. Ulman, "The Nature of Cities" In *The Annals of the American Academy of Political and Social Sciences*, Vol. 242, 1945. Reprinted by permission of the American Academy of Political & Social Studies.

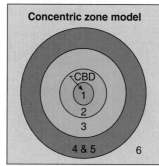

(a)

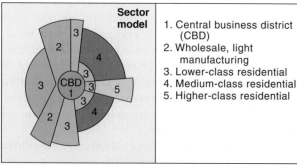

(b)

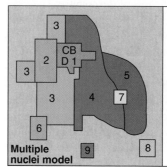

(c)

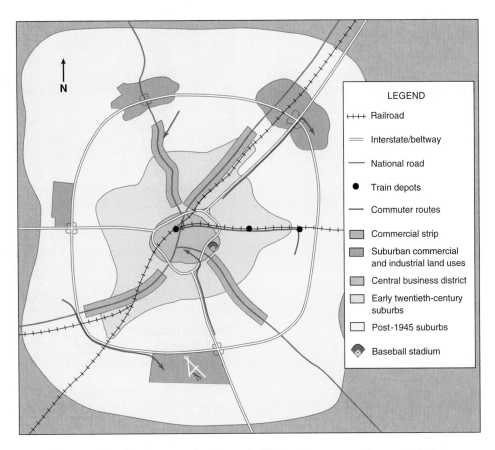

FIGURE 11.30 United States: urban landscapes in the early 2000s. The monocentric pre-1940 city is now small compared with the post-1945 suburbs in which interstate and other highway links and junctions provide bases for new centers of retail, office, warehouse, and manufacturing employment—the multicentric American city. Some of these centers produce as much as the CBD. The international airport is the center of a distinctive suburban commercial zone. Commuter routes can still be from suburbs to CBD, but many go in other directions. The baseball stadium shown may also be a location for other major sports arenas or the symphony hall.

interstate highways and beltways. These types of suburban centers in the larger metropolitan complexes often contain the city's international airport, which itself may be the size of a small city with multiple economic activities surrounding it, from hotels to warehouses and factories.

The changing face of U.S. cities since 1950 resulted in many movements of people within the cities. The growth of the suburbs involved people moving out of the older areas to new single-family homes. This trend was labeled "white flight" because it resulted in low-income African Americans and other minority groups being left in the older areas that lost many services and jobs to the growing suburbs. Property values in the older areas declined, sometimes leading to abandonment and dereliction. The social result in Detroit has been described in graphic terms as a city with an empty center like a donut, or as the "city of death" with high levels of abandonment (Figure 11.31).

Postindustrial Cities

From the late 1970s, expanding American urbanism spread new land uses into former rural areas. The term "edge city" was coined for new exurban developments relying primarily on car and truck transport and secondarily on air travel. Edge cities include large developments of shopping malls, offices, warehouses, and factories, located on the edge of major metropolitan areas (Figure 11.32). Since the 1970s, there were some movements of people from the suburbs farther out from the city into surrounding rural areas and other movements back to the city center.

Gentrification

Gentrification is the movement of higher-income families and individuals into materially poor areas of inner cities, leading to the physical and economic improvement of property and potentially dramatic changes in the quality of life in the districts affected. It was first noticed in British cities in the early 1970s and is now common in urban centers of both the United States and Canada.

Gentrification most commonly involves higher-income men and women, typically in the age groups of mid-twenties to mid-thirties and mid-fifties to mid-sixties. They are mostly singles or couples with few children. Demographic trends toward later marriage and child-bearing in the younger groups,

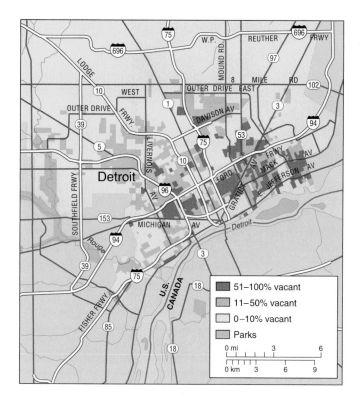

FIGURE 11.31 United States: urban landscape in Detroit.
The "city of death" shown in the high levels of housing vacancies in the inner city and older industrial areas, compared to the "fat city" of the suburbs.

FIGURE 11.32 An edge city: Bethesda, Maryland. The characteristics of new blocks of offices and apartments intermingling with parks, highways, and shopping centers are typical of these new urban developments located adjacent to, or just outside, large cities. **Photo:** © Jon Feingersh/Stock Boston.

together with large "empty-nest" homes in the suburbs, rising divorce rates, and widowhood, provide a large, affluent group of small households.

The availability of inexpensive property, coupled with the proximity to in-town jobs, attracted many suburbanites to move back into inner-city areas in the 1970s. The proximity of inner-city neighborhoods to a wealth of urban amenities (access to leisure-time facilities and theaters, symphony halls, opera houses, museums, restaurants, improved parks and waterfront areas, and major sports facilities) fueled dramatic urban renewal and inner-city growth for many metropolitan centers through the early 2000s. The primary drawbacks consisted of higher crime rates and less funding for public schools. Crime rates have decreased in many inner-city areas, resulting from changes in police and city government attitudes as well as changes brought about by the gentrification process itself. Schools have improved in some areas, while in others, it remains common for younger affluent families who begin living in the inner city to move out to suburban locations when their children are of school age, or to send their children to private schools.

The movements of people with high incomes into city centers were at first heralded as an "urban renaissance," and much was made of the process in the context of the years of declining and abandoned inner-city areas in the 1950s and 1960s. These cries were muted, however, when it was realized that many

materially poorer people were displaced by the process and had nowhere to go. Quality buildings erected at the end of the 1800s, for instance, often became rooming houses before gentrification turned them into more costly accommodations and displaced the previous occupants. Poorer-quality housing was swept aside and replaced by new apartment blocks. In neither case was provision made for the previous inhabitants.

The extent of gentrification is great in terms of the number of cities affected by it but often modest in relation to the total built environment of each inner city—several blocks rather than larger tracts. In Manhattan, New York, for instance, an area of the Upper West Side between Central Park and the Hudson River and near the Lincoln Center and museums experienced gentrification. Similar areas have been identified in Baltimore, Boston, Philadelphia, Pittsburgh, Washington, D.C., Atlanta, and San Francisco.

During the movement to the suburbs in the 1950s and 1960s, large sectors of the older parts of American cities were left behind by middle-class whites and taken over by black and other materially poor minority groups. Housing values dropped, and many financial institutions drew lines around areas in which they would not support house purchases with their loans—a process known as "redlining." Redlining led to further deterioration of inner-city areas. Once gentrification began to bring higher-income people back to the city center, the process was reversed ("whitelining"). The financial institutions that had contributed to inner-city decline by refusing loans gained much from the revival of inner-city living, backed by government support in renovation grants. They invested heavily in financing the building of new apartments in the 1970s and 1980s.

Opinion is divided as to whether gentrification is a small-scale process of individual choices over short timescales that affects a few people in each city or whether it is symptomatic of large-scale forces at work in remodeling the cities as part of uneven development. It has not involved the numbers of people that took part in the suburbanization of the 1950s and 1960s, but it has been significant in forming a partial reversal of that trend. It may be the outcome of manipulation by financial institutions, but most people who change their housing locations would not see that as a factor they considered. Certainly gentrification is linked to the continuing changes within urban environments in response to economic shifts, such as the changes in the central business district, demographic changes in family sizes and age structure, social changes in people's living habits and groupings, and perceived improvements in the environmental conditions of city center areas.

Regions of the United States

The large areal extent of the United States makes it a country of many internal geographic regions that developed from the interaction of people with the natural environments and resources. The historic pattern began along the east coast and spread westward. The geographic distribution of its varied natural resources and human conditions resulted in internal regional units being recognized by groups with widely differing interests, such as geographers, novelists, and government and business administrators.

Many different regional divisions of the United States have been proposed, but the one adopted here is based on the human geographic regions that emerged in the 1800s and early 1900s. Current geographic processes are changing their characters and external linkages, but the regional entities established at that time still form an essential basis for understanding the human geography of the country. The overlapping set of regions in the northeast (Figure 11.33) reflects the outcomes of the initial geographic concentration of European settlements and subsequent industrialization and urbanization.

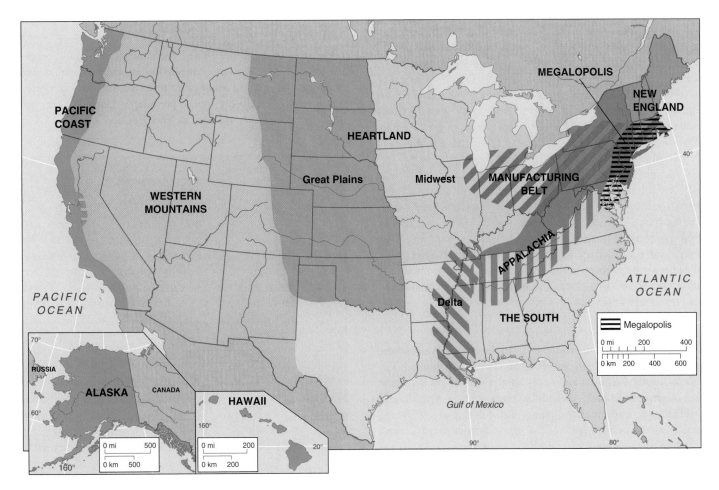

FIGURE 11.33 United States: the regions. They are based on differences in natural environments and on patterns of human occupation. The overlap of regions in the northeast reflects the early concentration of people and changing economic and social emphases.

New England

New England comprises six states in the northeastern corner of the United States. The first European immigrants sailed to the region in 1620 on the *Mayflower* seeking religious freedom. Settlers from Europe first established ports and townships in the southern parts of New England. Secular politics supplanted religious motivations as New Englanders played an important role in the revolution that brought the United States independence in 1783. The southern part of the region then became the first part of the United States to be industrialized in the early 1800s, based on water mill power, local labor, merchant capital, and access to markets by water transport. The early metal, textile, and leather industries grew through the 1800s. Possessing no coal or oil, the region was at a disadvantage when new industries based on steam and electrical power developed later in the 1800s and early 1900s.

By the mid-1900s, New England's established manufacturing industries were declining. Cities such as Fall River and Lowell, Massachusetts, had many derelict textile mills, high unemployment, and out-migration. During and after World War II, many of the traditional industries were replaced by high-tech computer design and manufacturing, airplane engine–related industry, and others favored again by high levels of defense spending throughout the Cold War period. At this time, the large number of higher education institutions in the region combined research innovations with the local availability of entrepreneurial capital as a basis for a fresh boom of economic development. This is sometimes referred to as the "Massachusetts Miracle" because of the numbers of new jobs created in a declining area. Although periods of struggle and recovery occurred in the mid- and late 1990s for some companies along Boston's growth corridor, the area remained an important region of high-tech industry concentration.

Another industrial boom may be passing, but many parts of New England retain evidence of wealth as its financial and professional services continue to grow. New England's environment of rocky and sandy coasts, wooded hills, interior hills and mountains, winter sports facilities, and well-preserved early American history, together with the sophisticated provision for entertainment in the cities such as Boston, appeals to many Americans as a place to vacation or as a high-quality living environment (Figure 11.34).

Megalopolis

The Megalopolis region stretches from the greater Washington, D.C., area (by some definitions, as far south as Richmond, Virginia) at the southwest end of its long urban arm to Boston at the northeastern end. The urban chain of the Megalopolis includes the huge metropolitan areas of Baltimore, Philadelphia, and New York, as well as many smaller urban centers positioned between metropolitan Washington, D.C., and Boston (Figure 11.35). Approximately 50 million people—nearly one-sixth of the total population of the United States—live in this region. The Megalopolis overlaps the Middle Atlantic and New England regions. In many places, urban land uses follow the roads and railroads from one city to another

FIGURE 11.34 New England life. Bostonians enjoying neighborhood festivities on St. Patrick's Day. **Photo:** © Ted Splegal/Corbis.

without a break. Even the rural areas in this belt are subsidiary to the urban activities in the cities. Farmers produce milk and vegetables for the local urban markets and specialty crops, such as tobacco, for industrial processing. Many rural homes set in wooded lots house commuters, while the coasts and hills provide recreation centers for the urban population.

The Megalopolis phenomenon was first identified in the United States in the late 1950s. The concept is the same as the European conurbation or connected urban chain phenomenon. Such a closely spaced set of cities reflected the historic competition among the closely spaced preindependence colonies and the newly independent states in New England and the Middle

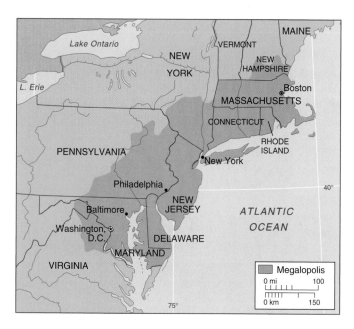

FIGURE 11.35 Megalopolis: spatial distribution. The Megalopolis region is defined by largely built-over land that has closely spaced metropolitan cities and a predominance of urban and urban-linked activities. Some areas within this definition are rural but linked closely to the urban functions of the region (commuters, truck farming, dairying).

Atlantic. Although New York always retained its early advantage as the largest port city, Baltimore grew rapidly in the early 1800s as an exporter of wheat. From 1830, the city fathers of Philadelphia built railroads westward to compete with the New York State Barge (Erie) Canal and New York railroads for access to the newly settled lands in the Midwest. Washington, D.C., emerged as one of the largest U.S. cities when the functions of federal government multiplied in the 1900s. The suburban counties around Washington, D.C., contain some of the most affluent communities in the country.

Despite the migration of some corporation headquarters out of the Megalopolis cities to other parts of the United States, this region still dominates the American economy, federal politics, and the modern media industry. It is the internal core of the United States. Other groups of cities within the United States have been identified as new versions of the megalopolis phenomenon, but none matches this region in its numbers of people or its economic, political, or social leadership.

Manufacturing Belt

Between the 1800s and mid-1900s, an area stretching from Boston on the east coast through New York and Philadelphia westward to Pittsburgh, Cleveland, Detroit, and Chicago became so dominated by manufacturing industries that it was termed the "Manufacturing Belt." The textile and metal goods industries continued to develop in the east as the Pittsburgh–Cleveland area became important in steel and heavy engineering, Detroit became the center of the automobile and truck industries, and the Chicago area produced a range of engineering and farm machinery. The largest cities became the centers of complexes of linked mining and mill towns with high densities of population, high proportions of blue-collar workers, and close networks of railroad and road links. From Ohio westward, intensive farming produced raw materials for food-processing industries, while other factories made the chemicals and machines needed on the farms. This region produced over two-thirds of U.S. manufactured products until challenged by the Pacific Coast and southern states during and after World War II.

The region remains dominant in manufacturing output within the United States, but older industries are less concentrated here, and many have been replaced by modern consumer product manufacturing industries taking up sites close to the country's largest markets for such goods. As manufacturing growth slowed from the 1970s, service industries grew in significance in this region. Chicago became a major center for financial services arising out of its futures trading in agricultural products. Some of the smaller towns became higher education centers, where facilities expanded to accommodate 30,000 and more students on each college campus.

After two decades of slower industrial growth, the western part of the Manufacturing Belt experienced a recovery of its manufacturing industries in the 1990s. Steel output along the southern shores of Lake Michigan rose rapidly to supply demands for sheet steel from expanding auto and consumer goods industries, which had a positive impact on unemployment. Since the late 1970s, the economy of the region diversified within each sector (farming, manufacturing, and services) with rising productivity. In the 1990s manufacturing employment grew faster there than in the rest of the United States. Autos remained central to the region's factory output, and some Japanese manufacturers sited their factories in the region. The problems of the region changed from high unemployment to a potential shortage of skilled labor.

Pittsburgh, once considered the industrial furnace of the United States, underwent a dramatic transformation during the 1970s and 1980s. Pittsburgh played a prominent role in the growth of the United States through the development of an intense concentration of iron and steel production and related heavy industry. In addition to being known for its significant industrial output, the city also gained a reputation for its pollution. During the city's renaissance of the 1970s and 1980s, many manufacturing jobs were replaced with service positions. Selected factories closed to make way for *Fortune* 500 companies. The city emerged as an international center for medical research. Pollution-filled air and the glow from the blast furnaces have been replaced with a dramatically modern and vertical skyline (Figure 11.36).

Appalachia

The hilly region of the Appalachian Mountains spans the middle part of the Manufacturing Belt and extends into the U.S. South. Each part of Appalachia has distinctive economic problems.

Eastern Kentucky, West Virginia, western Virginia, and eastern Tennessee form central Appalachia and remain one of the poorest parts of the United States. Coal mining continues to be important but only in areas where the coal has a low sulfur content, making some people materially wealthy from this industry. After 1950, the lack of other employment opportunities in central Appalachia caused many younger families to leave it for the prospect of jobs elsewhere. They often found getting jobs difficult because of their poor education and limited skills. The early 1980s boom in coal output when energy prices were high, together with improved infrastructure, health, and education provision, brought people back to central Appalachia, but conditions deteriorated again as coal prices slumped in the mid-1980s.

FIGURE 11.36 The modern skyline of Pittsburgh, Pennsylvania. Photo: © Joseph P. Dymond.

Although southern Appalachia was one of the poorest parts of the United States in the early 1900s, its economy became diversified with the availability of cheap electrical power through the Tennessee Valley Authority from the late 1930s. Major aluminum companies and federal facilities, such as the Oak Ridge atomic laboratories and the Huntsville rocket center, were joined by many other manufacturing concerns. Cities such as Asheville, Knoxville, Chattanooga, and Huntsville expanded, developing a wide range of service industries. Contrasts of material wealth, infrastructure, health care, and educational opportunities remain between the urban centers and the surrounding isolated rural mountains.

U.S. Heartland: Midwest and Great Plains

The lowland area between the Appalachians and Rockies that is north of the Ohio River and the Ozark Mountains is an essentially agricultural region, including 8 of the top 10 U.S. farming states, and is sometimes called the "breadbasket" of the country because of its production of grain crops. From the 1970s, this region produced sufficient food to make up for periodic shortages elsewhere in the world. When foreign markets needed less grain, land in the Midwest and Great Plains was idled to prevent overproduction.

The term "Midwest" is mostly applied to the eight states of the Corn Belt and Great Lakes area. The eastern part overlaps with the Manufacturing Belt, so that cities such as Chicago, Detroit, Cleveland, and Cincinnati stand amidst the world's most productive farming region. The Great Plains include North and South Dakota, Nebraska, Kansas, Oklahoma, and parts of Montana, Wyoming, Colorado, and northern Texas. They have a sub-humid climate and rise in elevation to over 1,500 m (5,000 ft.) above sea level at the foot of the Rockies (Figure 11.37).

The settlement of the heartland area in the early and mid-1800s rapidly established the suitability of its environments for different combinations of crop and livestock products. The growing season declines in length toward the north, and average annual precipitation decreases toward the west. The brown soils in the eastern states of Ohio and Indiana developed mostly beneath deciduous forest, while the black soils farther west developed beneath grassland; both are fertile and workable.

The patterns of farming established in the 1800s lasted until after 1950, when productivity increased rapidly. Agribusiness—the close commercial linking of inputs to farming, farm activities, and the processing and marketing of farm products—made American agriculture an integral part of a much larger industry. On the farm, agribusiness brought greater mechanization and the use of fertilizers. New varieties of corn became available that would ripen in a shorter growing season, new crops such as soybeans became prominent, new consumer tastes for more vegetables and less dairy foods and meat were expressed, and more rapid road transport made distance costs less significant. World markets for grain opened, especially in the former Soviet Union. By the 1980s, a major shift occurred in the Corn Belt output from grain-fed meat to grain for cash. Soybeans rivaled corn in the southern half of the region, while corn took over from grass in the dairying regions of the north. Many dairy farmers added to their income by growing vegetables and other valuable cash crops.

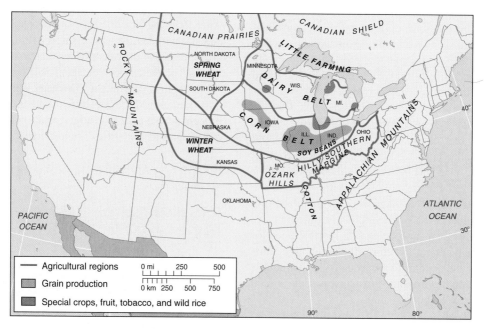

FIGURE 11.37 United States Heartland: agricultural regions. The Corn Belt is the most productive and versatile sector, where corn and soybeans ("grain production" areas on map) are grown for direct sale or for feeding livestock (beef in west, pigs in east). In the cooler north, dairying is important, and in the drier west, wheat and other small grains are dominant, while cattle feedlots dominate toward the south. Farther west, the Great Plains have cattle feedlots and produce wheat in drier conditions. High-value crops are grown on costly land close to the main urban areas. Special conditions favor fruit growing on the eastern shores of the lakes where spring comes late, delaying the opening of buds until after killing frosts. Poor soils, short growing seasons, and mountains mark the boundaries of this productive farming region.

In the drier areas west of the Missouri River, the development of center-pivot "circle" irrigation led to more corn and other crops, such as sugar beets, being grown in formerly wheat-growing areas. Huge feedlots for cattle were established, with the largest at Sioux Falls, South Dakota. Both irrigation and feedlots required increasing capital inputs at a time of fluctuating prices.

The dependence on groundwater for the expanded use of irrigation raised questions about the limits of water availability. Some streams dried up, and the High Plains aquifer—the main source of groundwater—was drawn down. Water levels in the wells of western Kansas fell by 62 percent between 1940 and 1991, raising pumping costs and the specter of using up all the groundwater. Better methods of water use later slowed rates of depletion, and the High Plains aquifer still contains a substantial proportion of the water it contained 50 years ago. While some parts of this area are short of water, others have plenty for the foreseeable future.

Despite the high and increasing farm productivity of this region, major problems arose. Many farmers had to obtain off-farm jobs, rent some of their land to other farmers, or sell their lands to banks and other financial businesses in order to make loan payments and cover their costs. Much land that had resided in family farms became corporately owned. Problems of declining population and few alternative occupations that had plagued other rural regions for decades extended to America's richest farming region during the 1980s.

The Great Plains area, particularly the zone where it meets the Rocky Mountains in Colorado, became a major center of high-tech industries, initially based around the federal facilities in Denver and the University of Colorado at Boulder. The area's population grew rapidly in the early 1990s but slowed into the early 2000s as air pollution, water shortages, and crowded schools deterred some from moving into the region.

The South

The American South extends from the southern Atlantic coastal plain westward to the southern Mississippi River and beyond into eastern Texas. Western Texas has more in common with other states along the Mexican border. The northern panhandle of Texas has some cultural ties to the South but others to the Great Plains and western United States. Florida is physically part of the South, but economically and culturally, it has ties to the southeast and northeast United States, as well as to Latin America.

The region's common cultural identity stems from colonial migration patterns and agricultural practices, as well as membership of the Confederacy states in the Civil War. Most of the South was dominated by the plantation economy, established in colonial times. It received an economic boost with the development of cotton growing from the 1790s. All of the southern states suffered after the Civil War, experiencing decades of poverty while the northern states made huge strides in industrial expansion and garnered significant material wealth. The black slaves were freed but soon became debt slaves, bound to the sharecropped land they cultivated. They did not gain political and social rights for 100 years.

Economic modernization, including industrialization and the adoption of better farming technology and management, began after World War I. The real changes in the South did not occur until the civil rights of African Americans were acknowledged from the 1960s. This development coincided with the building of the interstate highway system, which opened up much of the South to new economic opportunities, and the expansion of cities such as Miami, the urbanizing area from Charlotte to Atlanta, New Orleans, Houston, Dallas–Fort Worth, Memphis, and Louisville. Service industries, as well as manufacturing, moved into the area, and its changing fortunes were highlighted by local boosters who took up the term "Sun Belt" to refer to the U.S. South with its mild winters and sunnier climate—in contrast to the "Frost Belt" of the U.S. Northeast. Although this term was also applied farther west and much of the economic growth with which the Sun Belt became associated was geographically patchy—more a set of "sun spots" than an evenly prosperous belt—the changes in outlook in the region were real and lasting.

The economic advantages enjoyed by the South since the 1960s include improved and more focused agriculture, using the better lands for cotton, corn, and peanuts, and leaving the poorer lands to woodland. The demand for cotton increased after experiencing a period of market decline. Cotton products fought back against artificial fibers while bad weather and insect infestations affected other world producers. The United States' 1994 record cotton harvest was one-fourth of the world total, reflecting increasing yields. While South Carolina's previous highest crop in 1877 was grown on 6.9 million hectares (2.8 million acres), the 1994 crop came from 568,000 hectares (230,000 acres). Exports took half of the U.S. crop. The raising of poultry in factory units is another major capital-intensive farm activity in the region.

Rural areas remain the poorest parts of the South, with the Mississippi Delta lowlands forming one of the most economically challenged areas in the United States. The Delta lowlands extend up the Mississippi River valley from Louisiana and Mississippi through Arkansas and into southeastern Missouri (see Figure 11.33). The Delta was a major area of cotton production until the mid-1900s. Today the region remains materially poor, with relatively less infrastructure and fewer educational opportunities.

Some new manufacturing jobs have eased the economic deficiencies in parts of the region. Mississippi County, Arkansas, once the leading cotton county in the country, is now one of the largest steel producers. It has two mills making sheet and I-beam steel using scrap metal brought in by river barge. Other industries also moved into this county, but its unemployment is still twice the Arkansas average.

Most Southerners now work in manufacturing and service industries. In the 1920s textile mills sprang up in a belt northeast of Atlanta on the basis of cheap local land and labor as the cotton farms were mechanized or abandoned. By the 1990s many of the textile mills had closed as textile industries moved to Latin America and Asia. In the environs of Greenville, South Carolina, such jobs were replaced by those in a variety of new industries, including Michelin tires (from France), BMW cars (from Germany), Lucas automotive electronics (from Britain), and Hitachi electronics (from Japan). Such investments were based on the perception of this region as having skilled labor, often without labor unions, good local support for industry, and pleasant environmental conditions for management. In Texas,

HURRICANE KATRINA AND METROPOLITAN NEW ORLEANS

Early morning, Monday, August 29, 2005, Category 4 Hurricane Katrina made landfall in southeastern Louisiana and then pushed up into the border region where Mississippi and Louisiana meet the Gulf of Mexico. Katrina's harsh assault began late Sunday as the storm's approach created above normal tides and its squalls produced gusty winds and heavy rains along the north-central Gulf Coast. Katrina's size, intensity, and direction of approach, coupled with the configuration of the Mississippi and Louisiana coasts, pushed record storm surge tides into coastal Mississippi communities destroying highway bridges, hotels, shopping centers, gambling casinos, and entire subdivisions of houses. The storm surge significantly elevated the water levels of Lakes Borgne and Pontchartrain, tidal estuaries east and north of the city of New Orleans. The levee system in metropolitan New Orleans could not withstand the pressure created by the storm's surge and several levee failures, beginning as early as Monday morning, during the height of the storm, spewed water into the city of New Orleans, and neighboring St. Bernard Parish to the east. By late Tuesday approximately 80% of New Orleans, and 100% of St. Bernard Parish was flooded (Box Figure 1). Hurricane Katrina left more than 1,000 people dead, and untold billions of dollars in damages.

Widespread communications failure, confused government coordination between local, state, and federal authorities, and the overwhelming scale of the hurricane's destruction produced a chaotic and even deadly aftermath in the flooded metropolitan area of New Orleans (a metropolitan area with a pre-storm population between 1.3 and 1.5 million people). Thousands of people stranded in New Orleans made their way to the Louisiana Superdome and the New Orleans Convention Center, designated as shelters of last resort. For the next several days millions throughout the United States and around the world watched on their televisions in horror as

Box Figure 1 Widespread flooding, produced by Hurricane Katrina, inundates homes, businesses, and major transportation infrastructure in metropolitan New Orleans. **Photo:** © Kyle Niemi/US Coast Guard via Getty Images.

the accumulation of oil wealth was paralleled by manufacturing and services development around Houston, Dallas–Fort Worth, and Austin partly related to the National Aeronautics and Space Administration center in Houston and other high-tech research industries.

The service industries in the South include tourism, with Florida leading the subregion in visitor numbers and tourism revenue, health care, education, and financial services (Figure 11.38). Some service industries pay higher wages than semi-skilled jobs in manufacturing.

Several strong hurricanes made landfall in southern states along the U.S. Gulf Coast in 2004 and 2005. The costliest storm in United States history, Hurricane Katrina, struck southeast Louisiana and southern Mississippi on August 29, 2005. Katrina

left more than 1,000 people dead and produced tens of billions of dollars in damages (see Point–Counterpoint above). It will take Louisiana and Mississippi several years to recover from the storms's destruction.

Western Mountains

The western one-third of the United States is a mountainous region. The highest ranges are the Rockies on the east and the Sierra Nevada of California and the Cascades on the west. They are separated by broad plateau areas (see Figure 11.2) and areas of basin and range. The mountains block much of the humid westerly winds flowing into the region from the Pacific Ocean, resulting in vast arid rain shadow areas in the east.

a stranded population, at these shelters and elsewhere in the city, went without food, water, adequate security, and medical attention. Several elderly storm victims who had survived the immediate effects of the storm passed away in the following days from dehydration, heat exhaustion, or lack of medical care. The New Orleans police force, themselves victims of the storm, tried to maintain order while awaiting state and federal assistance. Those glued to their televisions, watching the post-storm human disaster unfold, witnessed people "looting" items ranging from television sets to survival necessities such as shoes, food, water, and diapers. Local police and medical professionals also were forced to "loot" area stores for food, water, and medical supplies.

The political firestorm that erupted in the days immediately following the landfall of the hurricane raised questions covering a range of issues from what appeared to most as a lack of preparedness and a deadly slow response from the United States' federal government (a response typically coordinated through the United States Federal Emergency Management Agency, or FEMA), to development strategies and urban planning, environmental conservation, and social issues such as the material poverty of the majority of the victims. There even were allegations that racial indifference was a factor in the level of assistance provided to the victims of the storm.

Well documented, published research and computer model simulations, some existing for decades, detailed the probability of levee failure and catastrophic flooding in New Orleans should a storm such as Katrina move through the metropolitan area. Did the U.S. federal government fail New Orleans by not funding fortification of the levee system such that it could withstand a Category 4 or 5 hurricane? If the government had funded needed levee improvements, would U.S. taxpayers living far removed from this metropolitan area have willingly supported such federal expenditures?

Would a better understanding of the human and physical geography of metropolitan New Orleans, on the part of civic and government decision makers at all levels, and the voting public of the United States have lessened the impact of Hurricane Katrina? Could it help to create an informed reconstruction strategy? Although the levee system along the Mississippi River was created to mitigate potential disaster from river flooding in cities like New Orleans it may have exacerbated New Orleans susceptibility to flooding from the Gulf of Mexico. The construction of walls along the river prevents the natural seasonal flood process, which normally places tons of sediment onto adjacent lands in the floodplain. A lack of natural sediment recharge since levee construction in metropolitan New Orleans contributed to the sinking, or subsidence, of parts of the city such that much of New Orleans is lower now in elevation than it was when the French first settled here in the early 1700s. A lack of sediment placement along the coast of Louisiana and corresponding saltwater intrusion, is fostering the rapid erosion of the state's protective coastal wetlands (Louisiana has the highest rate of wetland losses in the United States). The Louisiana coastline and the Gulf of Mexico are migrating westward and getting closer to the city of New Orleans each year. The combination of subsidence and coastal erosion with potential sea level rise from global warming complicate New Orleans' vulnerable situation and must be considered in the reconstruction plans of the metropolitan area.

Economic development in metropolitan New Orleans fostered a spatial pattern of settlement such that the most materially impoverished people in the city and its environs live on the most vulnerable land. Areas such as New Orleans East and the 9th Ward, suburban Chalmette, and most of St. Bernard parish are on the most flood prone and lowest elevation land in the metro area, while wealthier business and residential neighborhoods such as the historic French Quarter and "Uptown" New Orleans are situated on the city's highest lands. The port of New Orleans and coastal Louisiana industries are material components of the U.S. economy. The annual tonnage processed through the port of New Orleans is among the largest in the United States. Regional oil, natural gas, and seafood production and processing are extremely significant to the U.S. economy. The losses in industrial productivity from the storm, combined with the profoundly more significant economic cost of insured and uninsured losses and the expense of reconstruction (costs affecting everyone in the United States) far exceed the amount that would have been incurred for wetlands restoration and levee fortification prior to Katrina's landfall.

An integrative geographic awareness facilitates a more informed strategy. What lessons do you think we will learn from Hurricane Katrina? How might professional and government leaders and the voting public of the United States alter their future behavior in response to lessons from this natural and human disaster?

Settlement focuses on irrigation farming and towns that grew up as markets, mining centers, or nodes on the transcontinental railroads. A large proportion of the land between such centers is still sparsely settled. Much is owned by the federal government and designated as national parks, national forests, or grazing lands. Even these uses in a sparsely populated area generate conflicts when the interests of ranches and farms interact with environmental conservation or tourism.

The western mountain region is characterized by high-altitude lands and a predominantly arid climate. Most of the water available falls on the mountains as snow and runs off in swollen streams during the spring season. Many federally funded irrigation projects manage this water supply, with the largest groups of projects along the Colorado River in the south and the Columbia River in the north. Irrigated farming and cheap electricity created a series of productive oases that formed the basis for the growth of cities such as Las Vegas, Nevada, Phoenix, Arizona, and Salt Lake City, Utah. Las Vegas continues to capitalize on its gaming and entertainment industries, sunny and dry climate, and proximity to national parks and government military bases, which serve as large employers (Figure 11.39). The rapidly growing metropolitan area is enjoying a healthy business climate and a relatively low cost of living.

In the southern part of the region, Hispanic people make up a large proportion of the population. Defined in part by the Rio Grande, the border with Mexico is largely a mountainous desert where few people live outside the border towns. The border cities are twinned (Figure 11.40) and hold three-fourths

FIGURE 11.38 Central Florida capitalizes on its tourism potential. The Disney World experience is viewed as a cultural rite of passage for many children in the United States. Visitor numbers to the Disney parks in central Florida continue to increase, and the parks' areal extent continues to expand south of metropolitan Orlando. Although many U.S. families find the Disney World experience to be an extremely pleasant one, the Disney icon is often used negatively to argue against the globalization of U.S.-style capitalism or to criticize the creation of cultural geographies contrived for tourism purposes. **Photo:** © Joseph P. Dymond.

FIGURE 11.39 New York, New York hotel and casino: Las Vegas, Nevada. New York, New York is one of several themed resort and casino complexes built as Las Vegas gaming and tourism developers attempted to "reinvent" the city to attract more visitors. In the early 2000s, developers began to move away from the themed properties and back to more traditional resort-style development. **Photo:** © Joseph P. Dymond.

of the Americans and Mexicans living within 80 km (50 mi.) of the border—a population that rose from 9 million to over 15 million people since 1980.

From 1965 the Mexican maquiladora ("mills") program enabled United States and other foreign companies to assemble imported products tax-free in northern Mexico for immediate reexport without duties (see Chapter 10). In the early 2000s around 2,000 such factories employed half a million Mexicans and brought in one-third of Mexico's foreign exchange. On the U.S. side of the border, many Hispanic people find new homes, making up a high proportion of the local population, which is rising so fast that squatter shanty-towns form in southern New Mexico and southern Texas. Mexicans like to buy U.S. goods, spending around half of the wages earned at the maquiladoras.

Although the U.S.–Mexican border areas remain poorer than the U.S. average, people living there have more than twice the incomes of people in Mexico. Environmental conditions in this region are also a serious concern. Numerous studies suggest a strong correlation between pollution from border factories and high incidences of cancer. The impact of NAFTA on this region is still uncertain, and it will take some years for maquiladora jobs to move to other parts of Mexico (see the Point–Counterpoint: North American Free Trade Agreement box, on page 516).

Pacific Coast

From the Seattle area around Puget Sound in the north, to San Diego near the Mexican border in the south, the American Pacific Coast has become a second national core, close in economic importance to that of the region between Boston and Washington, D.C. Settlement increased after the arrival of transcontinental railroads in the 1870s and 1880s. Before World War II, the main products were timber in the north and farm products and the growing motion picture industry in the south. National defense needs for World War II in the Pacific then placed manufacturing and military centers along the coast. Puget Sound in the north and San Diego in the south became major naval centers. Seattle, once the corporate headquarters of the Boeing Aircraft Company, grew as the company produced bombers and jet airliners. The Columbia River became a staircase of huge hydroelectricity stations that support 11 aluminum smelters and the Hanford nuclear weapons facility, among other industries—but reduced the salmon catch. Los Angeles in the south also had major aircraft manufacturers, Lockheed and McDonnell-Douglas (both now linked to Boeing), together with a variety of manufacturing and service industries.

FIGURE 11.40 Western mountains: the United States–Mexico borderlands. The contacts between the two countries led to a series of twinned towns with linked economies. U.S. manufacturers site factories in Mexico to take advantage of cheap labor costs, as well as less regulation and enforcement of worker safety standards and environmentally polluting effluent.

After World War II, manufacturing prospered and service-men returned to live in new communities along the coastal Pacific. San Francisco (Figure 11.41) was the only city along the west coast until the early 1900s, when Los Angeles rapidly out-stripped its growth. Both cities attracted growing financial service businesses after World War II. Defense cuts in the 1990s led to lost jobs and plant closures, but only a mild recession resulted in California, which was growing again by the mid-1990s as a result of the shift to high-tech jobs, a more diversified range of industries, and increased overseas exports. Over 400,000 new jobs were created in California from mid-1996 to mid-1997, and personal incomes rose. The Los Angeles economy alone is greater than that of South Korea, and the city now contains a reinvigorated motion picture industry, a growing major port, and more computer software jobs than Silicon Valley, the best-known high-tech center situated just south of San Francisco. California has the world's eighth largest economy and is America's most productive farming state. The Seattle area became home to many software firms, including Microsoft.

Important to economic growth in the southern half of California was the supply of water to its largely arid environs (Figure 11.42). Federal and state government funds supported huge water storage and distribution projects following the early appropriation of water by the cities of Los Angeles (Owens Valley) and San Francisco (Hetch-Hetchy).

Years of drought in the 1980s and 1990s stretched limited water supplies. In early 1991, after four consecutive drought years, the major state-level water project reservoirs held less than one-fourth of their capacity, and farmers lost their supplies from that source; the federal project rationed farmers to one-third of normal amounts. Cities still receive a water supply, but most pay more for it. There are now no further sources of water to be tapped by California. It is likely that farmers, who use 85 percent of the water to produce less than one-tenth of the state's economic output, will have to pay more in the expectation that they will use it more carefully. At present, farmers pay only half the cost of delivering water to them, while city dwellers pay 20 times as much as the farmers.

Arizona, California, and Nevada all compete for fresh water from the Colorado River basin. The states have signed an agreement that limits each to an allocated percentage of fresh water from the river. Arizona is in the process of creating its own distribution system, the Central Arizona Project. The city of Las Vegas in southern Nevada is one of the fastest-growing metropolitan areas in the United States. The water needs for metro Las Vegas increase significantly each year as both the population, and the number of tourists visiting the region, continue to grow (Figure 11.43). Southern California is home to two very large, growing metropolitan areas: Los Angeles and San Diego. As Southern California's water needs increase, officials are looking for alternative sources in the water-scarce region.

The West Coast developed a new importance through an increasing volume of business and personal movements with Asia and other Pacific countries. Seattle and especially Los Angeles/Long Beach are major world ports. From its development as a frontline base for war operations against Japan during World War II and later in Korea and Vietnam, this region is now America's door to the world's most rapidly growing countries. The West Coast attracts most of the growing Asian investment in the United States. Five of California's top 11 banks are Japanese owned; 30 of the smaller ones are backed by Chinese money. Asian capital, particularly Japanese, backs car design centers, media industry corporations, and leisure industries. Half of the 1,400 Taiwanese companies in the United States are based in California, and research–marketing–manufacturing links across the Pacific Ocean are increasing.

The region also has trading links to Europe and Latin America, and close links with British Columbia in Canada to the north and Mexico to the south. In 1990 just under half of

FIGURE 11.41 Pacific Coast: San Francisco Bay area. San Francisco occupies the partly cloudy peninsula in the foreground of this space snuttle photo. The Golden Gate inlet is obscured by low clouds forming over the cold water offshore. The line of the San Andreas Fault can be traced from left (north) to right (south) across the foreground. Inland, the urbanized bay shore from Richmond to Berkeley and Oakland, the reclaimed areas of tidal lands, and the complex waterways of the Sacramento–San Joaquin River delta stand out clearly. **Photo:** NASA.

California's exports went to Asian countries, around 30 percent went to Europe, and a further 18 percent was shared between Canada and Mexico. Such links attracted Hispanic and Asian people to live and work in the region. Southern California has the highest proportion of Hispanic Americans in the country, with Los Angeles having doubled its middle-class Hispanic population since 1980 to more than half a million; its Hispanic businesses have doubled since 1993.

Alaska and Hawaii

In 1959 Alaska and Hawaii (Figure 11.44) were added to the 48 conterminous states of the mainland United States. They contribute to the extreme variety of environments and resources within the United States but have their own issues.

Alaska is a huge area of northern land that was bought from Russia in 1867. High mountains, icefields, and glaciers mark its southern coast. Inland, the Yukon lowlands have long cold winters, and to the north, the Brooks Range and North Slope leading down to the Arctic Ocean are even colder. Any potential for commercial farming is severely limited by climatic conditions. The Yukon lowlands widen westward to the Bering Sea, but there is little mining or manufactured output to trade in that direction. The dramatic mountain scenery and abundant wildlife in the state support a growing tourist industry.

After purchasing Alaska in 1867, the federal government tried to resist attempts to settle and develop it, but gradually, salmon fishers and gold miners whittled away this policy. Further exploration showed that Alaska contained resources of

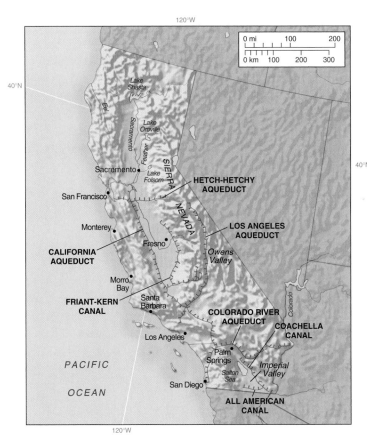

FIGURE 11.42 Pacific Coast: water flows in California.
Water flows from the high-precipitation areas of the north, where reservoirs store the water, toward the arid south. Water supplies for San Francisco (Hetch-Hetchy) and Los Angeles (Owens Valley) in the early 1900s were later enhanced by federal and state water projects in the Central Valley. The federal Imperial Valley project and Colorado River Aqueduct in the far south took water from the Colorado River to irrigate dry lands and provide water for Los Angeles and San Diego.

FIGURE 11.43 Growth and development in the desert.
Rapid growth and development in metropolitan Las Vegas expands across the desert valley floor. A new golf course community (a) and several new subdivisions (b) required large amounts of water from very limited resources. Nevada, Arizona, and California all compete for fresh water from the Colorado River. **Photos:** © Joseph P. Dymond.

(a)

(b)

fuels and metal ores on a large scale, but only a few thousand people lived in the territory. During World War II, when the Japanese invaded the western islands of the Aleutian chain, the strategic importance of Alaska was recognized, and the military presence in Alaska grew. Military activity increased further through the Cold War because the territory was close to the former Soviet Union. Statehood in 1959 gave Alaskans the opportunity to select lands for development from the 99 percent owned by the federal government.

The discovery of oil on the North Slope forced the state and federal governments to develop policies for Alaska's native peoples, who live mainly in coastal areas, and for the other settlers, most of whom live in and around Anchorage in the south. The oil from the North Slope made many of Alaska's 500,000 people wealthy in the early 1980s. Residents of the state receive payments from the state government from taxes on oil production. World oil prices fell from the mid-1980s to the late 1990s, with oil production peaking in 1988. The oil discoveries, and especially the oil spill from the Exxon *Valdez* tanker, made Alaska a focus of

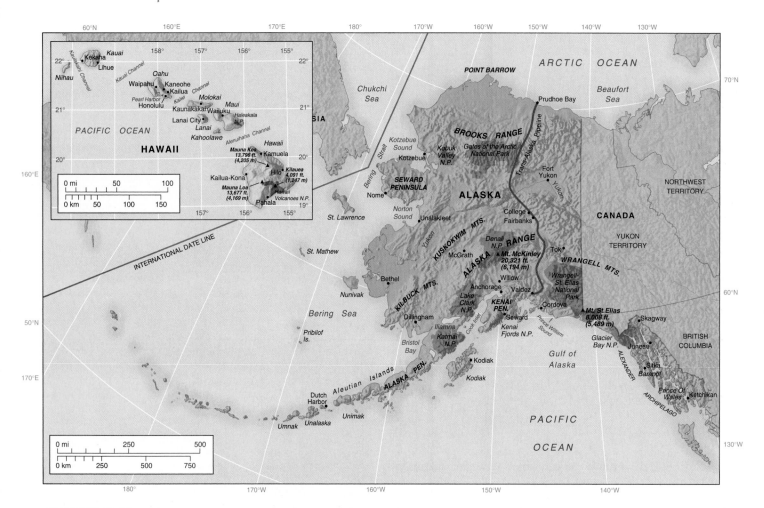

FIGURE 11.44 Alaska and Hawaii: the last two states to join the USA. They contrast in size, climate, and resources. Explain why the smaller Hawaiian Islands have twice the population of Alaska.

environmental concern. Energy budget decisions made by the U.S. legislature in the spring of 2005 brought the United States closer to controversial drilling in the Arctic National Wildlife Refuge on the north coast east of Prudhoe Bay. Parts of the National Petroleum Reserve to the west of Prudhoe Bay were opened for bids in 1998, with pumping expected to begin in 2010 (Geography at Work: page 547).

Hawaii was a long-time hinge-point of U.S. trade in the Pacific, a naval port, and a source of agricultural products such as pineapples and sugar. Involvement in World War II brought its existence home to many more Americans, and it was admitted as the 50th state just after Alaska in 1959. Tiny compared to Alaska, but with more people, it has a major international tourist industry. Hawaii experienced economic challenges in the late 1990s and early 2000s with high unemployment. Its tourist industry suffered from cutbacks on travel by the Japanese as the yen lost value. Sugarcane plantations closed as Asian competition and changing U.S. consumption habits reduced the market. Hawaii also suffers from a poor business climate due to many regulations, poor schools, a high union presence, and its geographic isolation from globally connected business systems.

Test Your Understanding
11C

Summary

The United States of America is a city-based country thriving on high-tech manufacturing and growing financial and business services industries. Its federal government structure led to power moving from the states to Washington, D.C., during the 1900s. The variety of regions within the United States reflects the uneven distribution of natural resources and the historic processes of occupation, industrialization, and developing transportation and telecommunications facilities.

The population of the United States is becoming increasingly multicultural as the European groups and African Americans who peopled it until the early 1990s are added to by peoples from Latin America and Asia. Immigration continues to be important in population growth.

The U.S. economy has the largest GNI in the world and is market-oriented. The country's farming is highly productive and commercially organized, its diversified manufacturing output reacts to changes in technology and world demand, and its service industries are now the dominant employers.

USING GEOGRAPHY TO AID PUBLIC POLICY: THE WHITE HOUSE

A growing opportunity for geography students in the coming decades is to find employment within the federal government using geography skills to inform public policy decisions. As policymakers face more complex decisions and are called on by stakeholders to use the best available science and information, an increasing number of federal agencies are recognizing their needs for someone with a geography background. Geographers have desirable skills in the creation and updating of geographic databases, selecting and evaluating data for potential use from a variety of specified sources, producing maps and charts, and documenting sources and procedures. Geographers can also produce graphical and tabular data that demonstrate relationships between policy and outcomes and can prepare visual representations to show policymakers how best to carry out activities in support of their agency's mission.

For example, pursuant to President Bush's Executive Order 13212, Dr. Bryan Hannegan (Box Figure 1) heads an interagency group working to improve the permitting process associated with a new oil and gas pipeline testing and repair program enacted by Congress in 2002. By using Geographic Information Systems (GIS) to display pipeline maps overlaid on maps of ecologically sensitive areas, ranges of threatened and endangered species, and other useful information, pipeline operators may be better informed regarding what permits will be required, what information they will need to gather, and which federal and state agencies will be involved. The database will also support visual display of the best management practices intended for that area, and allow pipeline operators to communicate seamlessly with the permitting agencies to indicate when a repair need has been indicated by testing, as well as when a repair has been completed.

This database, when completed, could also provide information for several other policy areas that rely on complete, verifiable environmental data. For example, by highlighting where ecologically sensitive areas overlap with the range of a threatened or endangered species, the database would allow a Fish and Wildlife Service biologist to quickly prioritize land conservation areas or management practices in

Box Figure 1 Geography and Public Policy. Dr. Bryan Hannegan utilizes Geographic Information Systems technology and geographic skills to better inform the public policymaking process.

the highlighted areas, thus providing the most protection with the least economic impact. Because these data would be available to the public, stakeholders could easily identify priority areas for cooperative conservation efforts and work with state and federal officials to create a proper policy framework to protect those areas.

However, it is not simply the environmental policy area that will benefit from expanded use of geographic tools. GIS offers tools to display any kind of data in a format useful to policymakers, whether it be drug use incidence for the Office of National Drug Control Policy or fundamental census data, which the Department of Commerce and a number of other federal agencies use to allocate funding and set priorities. As the federal government looks more and more to integrate scientific data into its decision-making process, the need for trained geographers to help communicate these data to policymakers will only increase.

Questions to Think About

11C.1 What are some of the ways in which the United States influences global activity?

11C.2 How do the regions within the United States compare and contrast in terms of population distribution, economic activity, physical environment, natural resource availability, and global connections?

11C.3 Which are the two primary regions of dense population and economic, political, and international activity in the United States? How have they changed over time, and what are the key factors in shaping these regions?

Key Terms

Group of Eight (G8)

UN Security Council

North American Free Trade Agreement (NAFTA)

Megalopolis

counter-migration

Francophone

Parti Québécois

Cree

uneven development

Tennessee Valley Authority (TVA)

Appalachian Regional Commission (ARC)

central business district (CBD)

concentric pattern of urban zones

ghetto

gentrification

Subregions of North America: Canada

The 3,000 miles of distance from Victoria, British Columbia, in Canada's southwest to Halifax, Nova Scotia, on Canada's southeast coast, coupled with the narrow width of the concentrated region of population along the country's southern border, results in a series of regionally individualistic groups of people who have to try to interact domestically to make Canada function as a country. Geographic logic might suggest the likelihood of Canadians interacting more easily with cities and systems across the border in the United States. Although Canadians do interact with urban systems across their border, the people of the country function internally in domestic subregions and on a country level through their federal government. The distinctive subregional groups within Canada's borders include the people of the Atlantic Provinces, the French-speaking peoples of Québec, the multicultural population of Ontario and especially of Toronto, the scattered farming populations and industrial cities of the Prairie Provinces, the people of Vancouver and the British Columbia coast who have an increasing outlook toward Asia, and the indigenous Canadians, or First Nations, of the northlands (Figure 11.45). The Canadian provinces were created a century after the U.S. states, are generally much larger, and have greater internal political control than U.S. states with respect to the federal government.

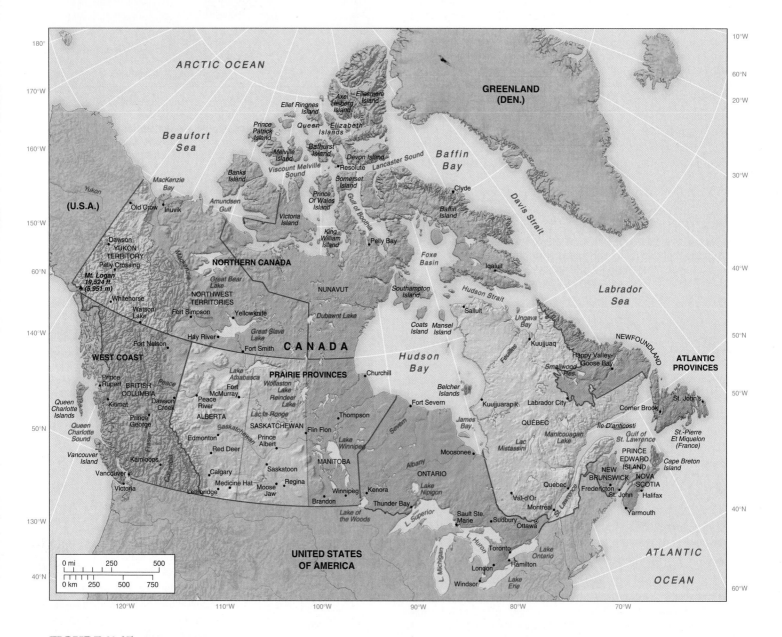

FIGURE 11.45 Canada: regions, provinces, territories, and cities. The Canadian provinces have greater powers than the individual states of the United States. Their geographic distribution in a line north of the U.S. border often makes communications difficult among different parts of the country. The red lines indicate the division of regions within Canada.

In the 1980s a substantial minority of independence-seeking people in Québec attempted to secede from Canada, potentially fracturing the Canadian state into at least two individual countries. In addition, the closeness of the major centers of population and commercial development to the United States and the beaming of media across the border from the United States have powerful influences on modern Canadian life that could blur the clarity of a cohesive Canadian national identity. The unique population distribution of Canadians, combined with living in the shadow of the globally mighty United States, actually appears to provide strong incentive for, rather than detriment to, the formation of a national identity (the Québec issue aside). Perceptions of people living outside the region, as well as many residents of the United States, that Canada is nothing more than a "satellite" or "subsidiary" of the United States may provide momentum for Canadians to strive to show the world how they are separate and different from the United States.

The United States continues to be seen as both a friend and an opponent. Canadians enjoy many of the benefits of affluence and national security because of their close integration with the American economy and defenses. Canada, along with the United States, is a member of NATO. Canada has a history of military alliance and cooperation with the United States. Both countries worked together to maintain a line of radar warning stations facing the Soviet Union across the Arctic Ocean during the Cold War. Canada and the United States participate in open free trade and (with Mexico) are each other's most important trade partners. The two countries first entered into formal open trade relations with the United States–Canada Free Trade Agreement of 1987–1988, which eliminated most tariffs and trade barriers. Both countries then joined with Mexico to sign and formally participate in the North American Free Trade Agreement in 1993–1994.

Political and economic ties are not the only areas in which Canadians cooperate with the United States. The two countries have integrated forms of entertainment. Much of Canadian television programming is U.S. produced. The two countries are integrated into many of the same professional sports leagues, including the National Hockey League, the National Basketball Association, and the professional baseball leagues. Hockey, basketball, and baseball teams representing the larger cities in Canada participate equally with the U.S. professional teams representing U.S. metropolitan areas.

Although Canada benefits from its geographic proximity to and societal commonalities with the United States, situations result from the same proximity that elicit criticism from the Canadian population toward the United States. Many Canadians object strongly to being the recipients of acid rain from pollution generated by heavy industry in the Ohio River valley of the United States. From the 1990s, problems arose on the west coast over salmon fishing yields in Alaska, British Columbia, and Washington. Canadians enjoy very low crime rates and have willingly accepted more stringent gun-related legislation from their government. There is a perception in Canada that the United States is exceptionally crime ridden, and many Canadians joke that everyone in the United States is walking around with a gun.

Although living next to an economic giant generates some resentful and even fearful attitudes among Canadians, a few Canadian activities present challenges for the U.S. government. One area of Canadian activity that is of great concern to the United States is the smuggling of high-quality marijuana into the United States from its production area around Vancouver. In addition, Canada's external affairs policies during the Cold War were sometimes at variance with those of the United States. Canada developed relations with some countries in the Soviet camp and some nonaligned, materially poor countries. In the 1990s Canadian companies took over the development of Cuban mining from the former Soviet groups, despite a U.S. embargo. Nonetheless, the positive aspects of the juxtaposition of the two countries, the shared media and trade, the political cooperation, and the generally friendly attitudes the residents of the two countries have toward one another far outweigh any negative attitudes or friction emanating from either country.

Canadian City Landscapes

Of Canadian cities, only Québec has a historic heart with buildings older than the 1800s. Most cities from Toronto westward were built almost entirely in the 1900s. Other major differences between Canadian and American cities include the level of planning involved and the relationships of government units within metropolitan centers.

Toronto is the largest Canadian city and the center of Canadian financial services and manufacturing industries, as well as the provincial capital of Ontario, Canada's wealthiest province (Figure 11.46). When Toronto began to extend suburbs into

FIGURE 11.46 Canada: urban landscape in central Toronto. The skydome, CN Tower, and downtown commercial buildings as seen from Lake Ontario. Toronto's lake frontage made it a major port with the opening of the St. Lawrence Seaway in 1959. In the late 1900s many dock areas became derelict, and the waterfront was redeveloped to provide public access to recreation opportunities. **Photo:** © Wolfgang Kaehler/Corbis.

surrounding jurisdictions in the 1950s, the province of Ontario required that these jurisdictions and the city plan cooperatively to make possible the amalgamation of services and regional road construction (Figure 11.47). Further growth in metropolitan Toronto was encouraged around hubs outside downtown, including North York, the area adjacent to the international airport, and a new center with coordinated and concentrated development. A major project was also undertaken to redevelop the waterfront.

Toronto was able to cope with increasing immigrant groups moving into older inner suburbs: Italian, Greek, Portuguese, and Chinese districts form distinctive ethnic enclaves of lively congregation but not segregated ghettos. The city capitalizes on the increasing diversity of its residents. Restaurants, stores, art galleries, and festivals reflecting the vast international heritage

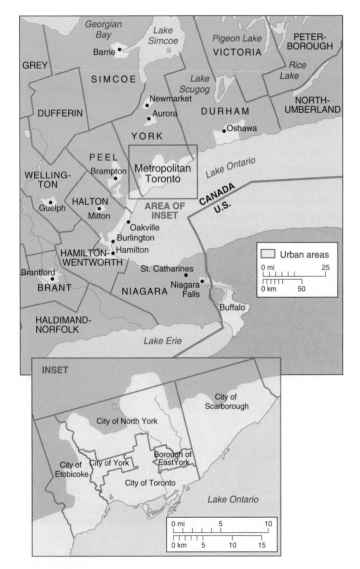

FIGURE 11.47 Canada: extent of metropolitan Toronto.
The constituent jurisdictions and surrounding counties in southern Ontario. Toronto's development is controlled by the province of Ontario and has been planned carefully.

of the citizenry continue to multiply throughout the city. Locals and tourists alike crowd ethnic business establishments, helping the diversifying economy of the metropolitan region. Toronto retains a busy downtown and surrounding older suburbs, and it has the lowest homicide rate of large North American cities, uniformly good schools, good public transportation systems, and controlled urban sprawl. Many cities in the United States envy Toronto's development situation because they have problems of multiple jurisdictions and limited overriding planning controls that affect only a few functions. Some U.S. citizens, however, find Toronto too ordered and even monotonous.

Economic Development

Canadian economic development generally followed that of the United States but often with a lag of several years. Today, Canada combines a continuing emphasis on its natural resource base with being an affluent, high-tech society. It has one of the highest proportions of trade per capita in the world. Canada's GNI per capita rivals that of many European countries but is still not quite equal to that of the United States.

Until the mid-1900s, Canada's economy depended mainly on primary products such as grain, timber, and minerals. Canada remains a major world producer of newsprint, wood pulp, and timber, and is one of the world's leading exporters of minerals, wheat, and barley. Canada produces 30 percent of the world's newsprint, which, with wood pulp, is manufactured mainly along the lower St. Lawrence River valley. Most timber output comes from the west coast forests. The minerals that place Canada in the forefront of world mining countries include coal, oil, and natural gas from Alberta, iron ore from Labrador and Québec, uranium from Ontario and Saskatchewan, nickel from Ontario and Manitoba, and zinc from several places. Agriculture remains significant in the Prairie Provinces, southern Ontario, and the specialized fruit-growing districts of British Columbia, where wine production is gaining international attention (see Figure 11.26).

Industrialization in Canada was a much later phenomenon than in the United States. Beginning before World War II and developing rapidly during the 1940s, the production of aluminum, vehicles, and consumer goods became important as Canada developed import-substitution industries to supply its own markets and avoid total economic domination by the United States. Hamilton at the western end of Lake Ontario became a steelmaking center. Montréal, Toronto, and Vancouver became centers of financial services and a wide range of commercial enterprises. The signing of NAFTA augmented trade between the two countries, resulting in increased U.S. investment in Canadian industries.

Within Canada exists a greater political consciousness of regional disparities and a greater will to do something about it than in the United States. Canadian regional policy enables the federal government to stimulate economic growth in the more materially poor provinces, such as the Atlantic Provinces and the largely agricultural Prairie Provinces, by supporting some services and financial investments in infrastructure. Such policies are popular in the recipient provinces, but their impacts on both source and recipient regions are controversial.

Regions of Canada

Canada has a federal government that links the 10 provinces, the Northwest Territories, the Yukon, and Nunavut—the country's main political regions. This arrangement was decided by Britain in the **1867 Act of Confederation.** That act united the different parts of what became modern Canada in the face of a perceived military threat from the United States after its Civil War ended in 1865. In 1982 Canada ended its legal ties to Britain, although it remained within the Commonwealth of Nations. Canada was then free to determine the constitutional roles of federal and provincial governments. That was not easy. Several attempts to reconcile the wishes of the people of Québec with those of other provinces highlighted the problems of a federal constitution in which the provinces have, in many ways, greater powers than the central government. This situation contrasts with that in the United States, where the Constitution was generated internally nearly a century earlier, in 1787. The U.S. states are generally smaller than Canadian provinces and lost powers to the federal government during the 1900s.

Atlantic Provinces

Canada's political tensions are enhanced by its regional economic and cultural differences. The east coast was settled first and forms the hilly Atlantic Provinces of Newfoundland (which has jurisdiction over the almost uninhabited Labrador), Nova Scotia, Prince Edward Island, and New Brunswick. The small-scale economy of these provinces, based at first on fishing and farming, was augmented locally by mining and manufacturing and by the naval base at Halifax. These provinces remain the primary recipients of federal regional aid.

In the 1980s and early 1990s, the region was hit badly by declining fish stocks on the Grand Banks. Some 30,000 fishers and fish plant workers lost their jobs as cod stocks virtually disappeared. Unemployment in Newfoundland rose to over 20 percent. Few new industries were attracted by government efforts. One source of hope could be to switch some fishing capacity to commercial sealing, although environmental concerns have to be assessed against the economic plight. Other prospects, such as pumping oil from the Hibernia field 315 km (200 mi.) offshore in the main iceberg lanes, would be very costly and environmentally risky. The development of nickel and cobalt mining at Voisey Bay, Labrador, scheduled to start production in the early 2000s, will bring wealth to the owners and possibly a few local workers but will have little other local impact. It is significant on a world scale, however, that the cobalt from Voisey Bay replaces the falling output from the previously dominant mines in the Democratic Republic of Congo.

Although the Atlantic Provinces contain some groups of French speakers, particularly in New Brunswick, sympathy for them is eclipsed by the potential outcomes of Québec leaving Canada. An independent Québec would sever direct ground contacts between the Atlantic Provinces and the rest of Canada. The possibility forces the Atlantic Provinces to consider independence for themselves, a seemingly impossible prospect given the level of federal economic support. If the Québec rift occurred, the Atlantic Provinces might look south: New England's shopping opportunities and events in Boston are closer than those in Toronto and already draw many visits across the border.

Nova Scotia, Prince Edward Island, and, to a lesser degree, New Brunswick are developing tourist industries in an attempt to replace lost components of the economy. Nova Scotia has a well-structured and geographically comprehensive tourist industry covering all parts of the province. New jobs in the region may be found in retail, hotel, and restaurant services.

Québec

The province of Québec was settled shortly after the first Atlantic coast settlements by French people who developed a distinctive type of **long-lot** land settlement along either side of the St. Lawrence River. The French long-lot pattern, which first appeared around Québec City and later diffused upstream to Montréal, was designed to maximize land ownership along the river. Québec became part of British Canada following General James Wolfe's defeat of the French army under Montcalm at Québec City in 1759. The French settlers and their descendant Québecois resented living under their conquerors. The economics of their culture were barely above subsistence, but traders bringing furs from far inland supported a merchant class during the 1800s. The combination of French language and Roman Catholic religion forged a strong loyalty that shifted to political activism in the 1900s as a reaction against Anglo-Canadian control. The erosion of French speech in the rest of Canada as new immigrants settled the prairies fostered a resolution that this would not happen in Québec.

As a result of such political activities, French was accepted as an equal national language, and Québec gained other concessions from the remainder of Canada despite not acknowledging English as an alternative language within its province. Québec looks beyond Canada to other French-speaking countries and takes a leading role in developing a new global French technical language rather than simply accepting English words.

The province of Québec covers a large area, extending northward to include most of the peninsula east of Hudson Bay (Figure 11.48). The majority of the population lives along the St. Lawrence River estuary. The two largest cities, Québec City and Montréal, have a range of manufacturing and service industries that place them among the world's great cities (Figure 11.49). A series of industrial towns use local hydroelectricity to power timber industries, pulp and paper mills, and aluminum refineries along one of the world's main shipping lanes. With global market production of wood pulp extending to tropical areas, the prices—and hence the well-being of

FIGURE 11.48 Canada: Québec's geography. The major cities, mining areas, industrial centers, and hydroelectricity sites. Nearly all economic activity crowds into the St. Lawrence River lowlands.

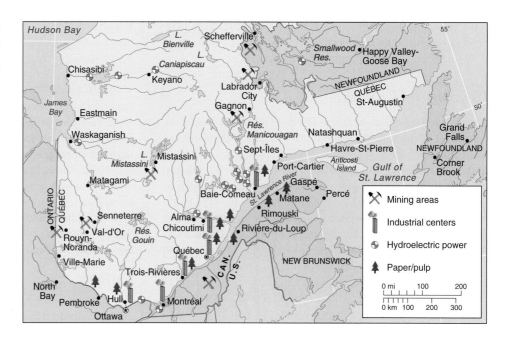

Québec's major industry—fluctuated since the late 1980s. Expansion of global demand in Asia, Europe, and North America does not always balance the production output.

North of the St. Lawrence River estuary, the land is bleak, eroded by former ice sheets and covered by coniferous forest, lakes, and tundra. The rocks contain large mineral resources that are gradually being exploited as world markets and transport facilities make this possible. One of the major potential resources is hydroelectricity, and Québec is investing heavily in new facilities to export power to the United States. In doing this, it came up against problems of Native American land rights that are stalling parts of the plan. Overall, Québec produces 25 percent of Canada's manufacturing output and has a strong economy in its own right.

FIGURE 11.49 Québec City. The provincial capital of Québec, Québec City, is situated at the confluence of the St. Lawrence and St. Charles rivers. In the foreground, the waterfront area is referred to as Lower Town. Upper Town looms on a bluff in the background with the imposing Château Frontenac hotel dominating the cityscape. Québec City was designated as a World Heritage site by the United Nations. **Photo:** © Joseph P. Dymond.

Ontario

The province of Ontario contrasts with Québec. Although it was settled later than the provinces to the east, Ontario became the center of British rule after 1776 and possessed the best farming land. Southern Ontario between Lakes Huron, Ontario, and Erie has a relatively mild climate, considering its interior location on the North American continent. The regional climate is modified by a maritime influence from the very large lakes and from being situated farther south than most other parts of Canada. Soils and climate suitable for pasture, grain crops, and tobacco made this the most attractive part of Canada for immigrant farmers in the 1800s. The city of Toronto arose at that stage; but its main development occurred from the middle of the 1900s, when the opening of the St. Lawrence Seaway turned it into an ocean port, the French-language policies of Québec drove English-speaking businesspeople westward from its former rival, Montréal, and its role as capital of Ontario expanded. Toronto is the largest city of Canada and continues to grow rapidly.

Northern Ontario extends to the shores of Hudson Bay, but its ancient rocks were scraped bare of soil by ice sheets and are covered partially by coniferous forest. Local deposition of clays in meltwater lakes provided usable soils, but growing seasons are short. Most of the settlements in northern Ontario are mining settlements, including the huge complex around the nickel mines of Sudbury. Settlements in the areas west of Sudbury and north of Lakes Huron and Superior are few and far between and constitute a major gap in the linear east–west Canadian settlement pattern.

Prairie Provinces

The Prairie Provinces of Manitoba, Saskatchewan, and Alberta were settled following the building of the Canadian Pacific and Canadian National Railroads in the late 1800s (Figure 11.50). In

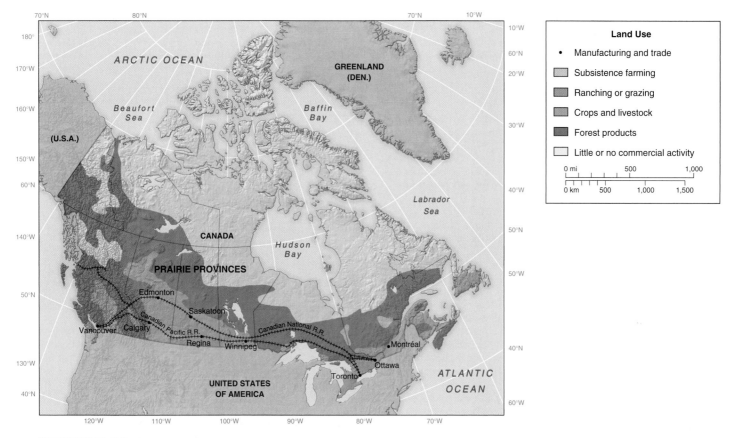

FIGURE 11.50 Canada: Prairie Provinces in the context of land use patterns. The building of the transcontinental railways was linked to the farming development of the prairies and the growth of the major cities west of Toronto.

the early 1900s they became major wheat-growing areas. The crop was planted in the spring and ripened in the short growing season, which was as little as 90 days on the northern margin of the plowed area. Toward the west, there was insufficient moisture for this, and cattle ranching took over. The area is essentially a northward extension of the Great Plains in the United States. The ground rises westward in a series of steps, further reducing the growing season and precipitation. Many of the agricultural districts had out-migrations of people since 1950 as a result of the mechanization of farming and the closing of marginal farms.

Winnipeg, Regina, Edmonton, Saskatoon, and Calgary were railroad towns that became grain and meat markets, and three of them became provincial capitals. After 1950, the discovery of coal, oil, and natural gas in Alberta brought extractive and manufacturing industry wealth and more people to that province. The large distances east and west to Canadian ports raised the price of exports and led to some of the main energy markets being southward in the United States.

West Coast

British Columbia (see the Personal View: Canada box on page 554) faces the Pacific Ocean and includes all the mountainous west of Canada apart from a range shared with southwestern Alberta. It is a province of geographically concentrated economic activity, including pockets of mining, fruit growing, and tourism in small lowland areas that are separated by high mountains. Most

of the people of the province live in the Vancouver area, extending to Vancouver Island. Attempts to develop the port of Prince Rupert farther north were not so successful.

The people of British Columbia may distance themselves at times from the rest of Canada. Some act more British than the British themselves with elaborate rituals of afternoon tea common in Victoria (Figure 11.51), while

FIGURE 11.51 Victoria, British Columbia. A distinctive British imprint is visible in the cultural landscape of Victoria in the architecture, red double-decker buses, and numerous places to take afternoon tea. **Photo:** © Joseph P. Dymond.

CANADA

Gwen and Jim own and operate a bed and breakfast in Vancouver, British Columbia, Canada. The two are proud to be Canadian, and although they assert the existence of strong ties between the peoples of Canada and the United States, they see a distinction between the cultures of the two exhibited in the respective social structure and systems of each. Gwen and Jim marvel at the vast expanse of their country; the abundant wildlife; the cleanliness, diversity, and safety of Canada's rapidly growing cities; the unique native history and rural livelihood of people thriving in high latitudes on some of the world's least populated lands; and the seamless fusion of tradition with modernization. When asked about their personal view of Canada, they highlight the diversity of societal roles and cultural traditions as being distinguishing elements of Canadian life and nationality. They discuss the coexistence of traditional fishermen in the Atlantic Provinces of eastern Canada with new service providers for the region's growing tourism industry. They point out the contrasts between the semisubsistence lives of Canada's many native groups, some living very far to the north where summers and winters bring 24 hours of light and 24 hours of darkness, respectively, with Canadians living in large metropolitan areas. They describe Québec and Ontario, the two most populated provinces, as regions where the cities are as cultured and as sophisticated as any in the world. Although they are not French Canadians, they seem to take pride in the remarkable human landscape where Anglo-Canadian and French cultures merge in Québec. They say that in Québec, you could mistake parts of Montréal for Paris and Québec City for a fortified town like St. Malo in northern France. And they will tell you the three Prairie Provinces—Alberta, Saskatchewan and Manitoba—in addition to being a good source region of oil and natural gas for the country, have some of the wealthiest agricultural lands in the world and produce excellent grains.

National pride fully surfaces in Gwen and Jim when they discuss their home province of British Columbia and especially the region surrounding the city of Vancouver (Box Figure 1). The following is their personal view of life in British Columbia.

British Columbia (BC) is a mixture of people from all over the world. The first residents of BC were the First Nations people who arrived from Asia thousands of years ago. The First Nations culture is still an important part of BC life and is exhibited in many forms of art, literature, song, dance, and food. The Spanish and English were the first Europeans to explore the region. The British estab-

Box Figure 1 **Vancouver waterfront.** The cultural landscape of Vancouver includes many vertically oriented commercial and residential buildings that capitalize on views of the surrounding mountains and waterfront. **Photo:** © Joseph P. Dymond.

lished fur trading posts that eventually evolved into permanent settlements, together with towns founded near mining, fishing, and logging sites. Compared to the rest of Canada, European settlement in BC is very new.

Most adults living in BC moved here from other places. In the 1950s, 1960s, and 1970s, Europeans were the primary immigrants to the region, while in the 1980s and 1990s, Asians came in large numbers. In the 2000s Americans are moving to BC in record numbers. These Americans seem to love the region's natural beauty and very low crime rates. The steady growth of the BC economy over the last 50 years created employment opportunities that attracted people from all over the world, resulting in an increasingly plural culture in the region. The depth of multicultural expansion in BC is especially evident in the province's largest city, Vancouver. Metropolitan Vancouver is home to people who trace their ancestral heritage or national origin to more than a hundred different countries. The Chinatown section of the city is the second largest ethnic Chinese neighborhood in North America. Multilingual signage, varied architecture (especially distinc-

others emphasize the Pacific connections of Vancouver, which has an increasing Asian population and growing economic ties to East Asia. Asian migrants brought money to metropolitan Vancouver, which facilitated the growth of many metropolitan industries. The growing Asian community further diversifies the culture of the city. In some Vancouver neighborhoods, however, tension developed between some British Canadians and Asian migrants. Many Asians living in Vancouver garnered wealth in business endeavors in Hong Kong, and they tore down traditional dwellings in

parts of Vancouver and erected large Asian-style homes in their place, which frustrated long-time residents who preferred the traditional architectural styles of the residential areas of the city.

Northern Canada

The northern parts of all the provinces from Saskatchewan to British Columbia end in the barren rock, trees, and tundra that characterize northern Canada. The Northwest and Yukon Territories long formed the largest part of Canada—a federally

tive cathedrals, mosques, and temples), and dozens of multiethnic festivals celebrating the best aspects of cultures from around the world are some of the many visible elements of Vancouver's diverse cultural landscape. Vancouver's world-class restaurants blend North American, European, Asian, and East Indian flavors. In Canada these cultures are not melted down into one but are individually preserved, treasured, and celebrated.

One main difference between BC and the rest of Canada is the climate. Most people in BC live in the southwest corner of the province, where the climate is mild and snow is relatively rare (except in the surrounding mountains, where abundant snow supports a thriving winter sports industry). BC people wash their cars on New Year's Day just so they can phone their snowbound relatives in Edmonton, Winnipeg, Toronto, and Montréal and boast about the mild temperatures. Also on New Year's Day, thousands of Vancouverites go for a swim in English Bay.

Like most Canadians, people in British Columbia enjoy the wilderness. One is never far from the outdoors in BC, and area residents constantly take full advantage of the encompassing wilderness. Downhill and cross-country skiing, hiking, rock climbing, bird watching, and fishing are only minutes away from downtown Vancouver. The mild climate allows the enjoyment of outdoor sports all year. The climate of the city, coupled with the high-elevation mountains nearby, creates conditions in which residents could enjoy boating and snow skiing in the same day. The sheltered waters of the Strait of Georgia make water activities very popular. Fishing, sailing, power boating, and kayaking are enjoyed year-round.

Vancouver is the largest city in British Columbia and one of the best places to live in the world, according to UN surveys. The city is situated on Burrard Inlet with the snow-capped Canadian Rocky Mountains in the background. Visitors describe the natural beauty as "stunning." The city is clean, relatively crime-free, prosperous, and cosmopolitan. Vancouver's location makes it a convenient port for goods coming and going to Asia. Vancouver International Airport is the busiest hub in North America for flights to and from Japan, China, and Korea. Recently Vancouver became a major center for making motion pictures, TV shows, and commercials. Among the numerous recent productions filmed in the area was the popular and long-running *X-Files* science fiction TV series. Production crews are attracted to the varied scenery and unique combination of environments and climate, the availability of a highly skilled and appropriately trained labor force of technicians, stable politics, low crime rates, and a low Canadian dollar relative to the U.S. dollar.

Each provincial government in Canada funds its own public education system, creating conditions in which small towns and inner cities have the same quality of schools and teachers as the most materially wealthy suburbs. British Columbia has universities and colleges in all areas of the province, enabling people to study close to home. The Vancouver area has two universities, the University of British Columbia (UBC) and Simon Fraser University. Both offer undergraduate and graduate degrees. UBC is acknowledged as one of the finest research universities in Canada, with many researchers and scholars who are recognized worldwide for their outstanding contributions. Canada's national magazine, *Maclean's,* has consistently ranked Simon Fraser University among the country's top comprehensive universities. Vancouver has also become a world-famous center for learning English as a second language (ESL). Besides having excellent ESL teachers and schools, Vancouver offers a low-crime environment and a beautiful setting for students to enjoy when studying here. There are over 65 ESL schools and colleges in the city. These schools are especially popular with students from Asia and South America. Many of these students love the city so much that they end up living there permanently.

Canadians are proud of the national and provincial health care systems. British Columbia, like the rest of the country, is striving to maintain the standards of its system. Increased costs are challenging these efforts, but Canadians place a high value on the coverage and availability of health care. People from all walks of life in Canada are treated the same way in hospitals and doctors' offices—regardless of their financial status. Canadians envision their system as a national treasure, are willing to work hard to keep it open to everyone, and are constantly striving to improve it so future generations will continue to enjoy and value the system.

Canada's proximity to the United States creates a sharp awareness of U.S. political and economic actions on the part of Canadians. Americans are well liked in Canada and are often thought of as extended family. Canadians are often challenged when the U.S. media and political leaders seem to ignore them. George W. Bush's September 20, 2001, speech to Congress is a good example. When he cited a long list of countries rallying to the U.S. side after the attacks of September 11, he did not mention Canada. Canadians had opened their homes and hearts to Americans at this disastrous time and felt quite rebuffed. Many Canadians are also disappointed when U.S. trade barriers are placed to prevent a free flow of trade between Canada and the United States. The stipulations of NAFTA are sometimes ignored by the United States. For example, the tariffs the U.S. government placed on BC's lumber critically damaged the industry and added an unnecessary expense to homeowners in the United States. Canadians love Americans but are sometimes disappointed by U.S. politicians' actions regarding Canada.

Vancouver is a prosperous, clean, friendly city surrounded with spectacular natural scenery. British Columbia is an exciting and breathtakingly beautiful province. Canada is a kind and caring country. These are just a few of the countless reasons why Gwen and Jim have a very fond personal view of living in Canada.

---

controlled zone where few people live permanently apart from American Indian and Inuit (Eskimo) groups of Native Americans. In 1999 Nunavut became a new territory (see Figure 11.50).

Mining settlements produce a range of metallic ores, and there are few communications in this wilderness area apart from those linking mines to their markets. In the late 1990s a new diamond mine near Yellowknife, Northwest Territories, was expected to produce up to 3 percent of the world's diamonds. The diamond mine needs to address local community concerns about caribou herd migration paths and threats to hunting and fishing if it is to establish a lasting positive impact. On a global scale, the mine, together with others outside South Africa, could reduce De Beers's control of diamond markets. Native Americans attracted to the mining settlements are often unable to relate to the contrasting lifestyles and in the past have created a social problem that the Canadian government did not tackle well until the 1980s. The future of this region is not bright as mining activities and federal subsidy incomes decline. A possible alternative income is from tourism

Test Your Understanding
11D

Summary

Canada is closely tied to the U.S. economy but politically is distinctive—partly because of internal tensions over the future of French-speaking Québec and partly because of its more liberal trading and aid policies. Internal differences are increased by the structure of large provinces that have more powers than do states in the United States.

Canada's regions each have an economic heart close to the U.S. border and huge expanses of sparsely settled northern lands. Canada's economy was largely based on agriculture and forestry until around 1950; since then, Canada has become high-tech and industrialized, and one of the world's major mining countries.

Questions to Think About

11D.1   What is the relationship between Canada and the United States? Do Canadians have a strong sense of national identity? Explain.

11D.2   What issues might cause Canada to fracture into two or more separate countries?

11D.3   What areas of Canada are growing and why? How might growth affect the future cultural landscape in Canada and the Canadian sense of nationalism?

Key Terms

1867 Act of Confederation   long lot

Chapter 12

A World of Geography

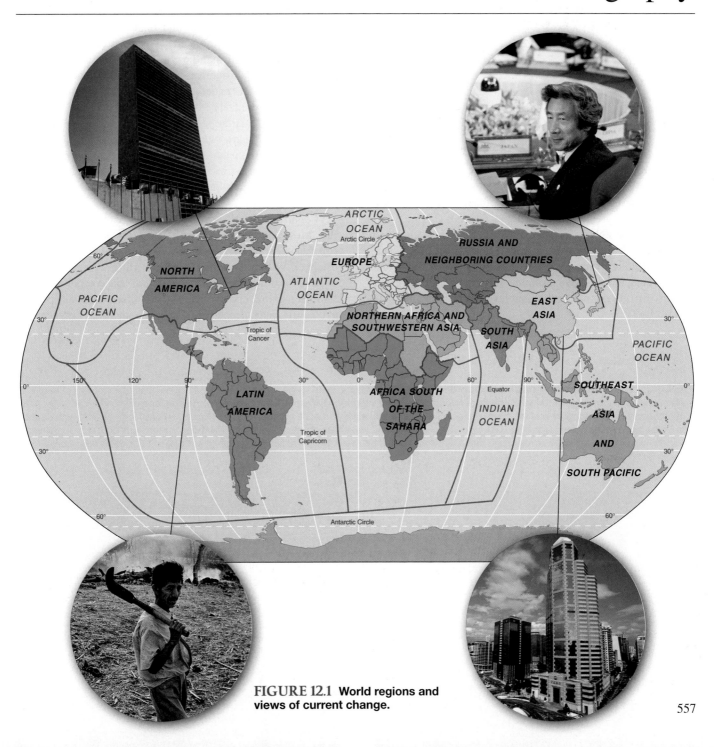

FIGURE 12.1 World regions and views of current change.

This chapter provides a conclusion to world regional geography studies that combines ideas about globalization with what we have learned and understood about local responses in Chapters 3 through 11. The major themes are:

Understanding global and local geographies in our world today in the context of varied political, economic, cultural, and environmental forces.

Evaluating the historical development of global connections and the continuity of local responses.

A case study of global connections and local voices: southern China.

The future of world regional geography.

Geography at Work: AAG President in Iran: Reconciling Differences

Eyes Opened

You have examined world regions (Figure 12.1) and should have a better understanding of people and countries around the globe, as well as a means of assessing the forces of change impacting our world. We can now evaluate whether we are moving toward greater global order or disorder (Figure 12.2) and what that might mean for people in different regions. In this chapter we focus on developments in aspects of globalization that need to be understood as the daily news of world and local events unfolds. We examine the emergence of local, country, world regional, and global levels of geographic scale (Figure 12.3).

Diverse Worlds and Geographies of Conflict

Human activities, often mixtures of political, economic, and cultural policies and attitudes, interact with natural environments and generate different and changing regional geographies. Today these interactions are increasing in extent and intensity.

Political Activity: Countries Act

Politics is about power and the ways in which countries are governed, allocate resources, and relate to each other. From 1950, the Cold War set the United States and its allies against the Soviet Union and its allies, a conflict that dominated the global political environment. Although many feared nuclear attacks, the Cold War ended in 1991 after never coming to a full "hot war." It has since been discovered that the Cold War rhetoric between the main opponents was more effective in maintaining political fears rather than a greater reality based on economic power. In the 1990s political activity shifted to a fresh emphasis on the roles of individual countries. A new context of intensifying global connections heightened concern for local interests.

Countries are self-contained entities with established borders. Most countries continue to be seen as nation–states, a term that suggests every people (or nation) has its own country (or state). Within its borders, the government of a country is sovereign: no other being or institution can compel it to act, although external bodies such as the United Nations or powerful countries such as the United States sometimes intervene, as

FIGURE 12.2 Global order or global disorder? (a) United Nations building, New York. The United Nations has membership of nearly all world countries. It carries out a range of worldwide programs that are intended to improve the well-being of children and environmental sustainability. It organizes military support to some countries facing civil strife. However, it is not a global government. (b) Demonstrators against globalization face police in downtown Genoa, Italy, in 2001, when the leaders of the wealthiest countries had a summit meeting. Tens of thousands of protestors streamed toward the prohibited zone around the conference venue, but were met by police firing tear gas. **Photos:** © (a) Joseph Sohm/Corbis. (b) © AP/WideWorld Photos.

(a)

(b)

Geographic Scale	Political	Economic	Cultural	Environmental
Global/ Worldwide	United Nations Nongovernmental organizations (NGOs) (no united will for a global government)	Borderless Capitalist Market System: World Bank, IMF, OECD, G8, multinational corporations	Spreading Western and Islamic cultures; Westernization resisted. Olympic Games	Global warming, ozone hole, ocean resources
World Region/ Subregion	European Union, Commonwealth of Independent States, original ASEAN	Regional trading groups: EU, NAFTA, Mercosur, ASEAN, APEC, etc.; Regional emphases within the capitalist system	Cultural Groupings: N Africa/SW Asia; Africa S of Sahara; S Asia; China; Japan; Russia; Latin America; North America; etc.	Acid rain, tropical rain forest destruction
Country (Nation–State)	Basic Political Unit: accepted borders; taxing, defense, international relations powers	Fiscal and monetary policies; rich/poor countries	"Nation–state" concept	Country-Based Issues: public health; air and water quality; conservation of soils, forests; national parks
Local Region/ Major City	Devolution of administration; world cities; federal states within countries	Regional product specialization; economic and physical planning; rich/poor areas	Local elements combining with external long-term traditions; distinctive way of life; basis for political pressures	Physical planning impacts

FIGURE 12.3 **Geographic facets of globalization.**

in Kosovo, Afghanistan, and Iraq. Country governments have powers to tax, provide for defense, negotiate with other countries, and develop their own legal system. Within countries, local issues and pressures, as much as international relations, occupy politicians. Few country governments are willing to give up any of their nation's sovereignty. What are seen as attacks on sovereignty come both from the nationalisms of minority groups and also from attempts by larger organizations such as the European Union or the United Nations to overrule their own decisions. Countries remain the dominant political units of our world.

Although the polarized Cold War ended in 1991, a *global level* of government did not follow. The Cold War conflict was replaced by a vacuum of political ideas, in which a simplistic conservatism in the West faced challenges from increasingly polarized causes actively resisting Western influences in other parts of the world. The United Nations achieved only modest results in reducing country rivalries. It could not control international problems such as terrorism and the drugs, arms, and slave trades. It found it difficult to deal with internal country strife, often where fervent nationalist governments put down internal opposition. Although the military dominance of the United States and the wealthy European countries enabled them to intervene in the Gulf War of 1991, Bosnia, Kosovo, and Afghanistan, they did little to help in the Rwandan conflict or that between Armenia and Azerbaijan. In Somalia, U.S. intervention ended as its forces had to leave.

At the *world regional* level after 1991, countries that formed groups for defensive purposes during the Cold War period in the second half of the 1900s sought new roles. The

United States and Canada had joined with European countries in the North Atlantic Treaty Organization (NATO) to defend western Europe against potential attacks from the Soviet Union. Once the Cold War ended, NATO had to find new roles or cease to exist. By the early 2000s it was clear that new members from the former Soviet satellite countries in East Central Europe wished to be part of continuing military links and protection—but with NATO rather than Russia. Indonesia, Malaysia, Singapore, and the Philippines joined the Association of Southeast Asian Nations (ASEAN) to resist the advance of Communist governments. After 1991 the ASEAN countries shifted to economic objectives and included the former Communist countries of Cambodia, Laos, and Vietnam and even the Myanmar dictatorship. In the early 2000s they traded with countries in East Asia (Figure 12.4). Up to the early 1990s, the countries neighboring South Africa formed the Southern Africa Development Coordination Conference (SADCC) to oppose apartheid (separation of whites and blacks) and resist South African attacks. After the end of apartheid, this group of countries accepted South Africa as a member and opted for a greater economic emphasis.

At the smaller *local scale,* political rivalries and conflicts among countries and nationalist causes within them continue. The political causes shifted from the ideological political–economic conflict between capitalism and communism to the emergence of smaller countries that were deemed likely to threaten world peace (such as Afghanistan, Iraq, and North Korea). Cultural and environmental issues attained greater prominence. Some commentators talked about the "New Global Disorder" rather than a "New Global Order." Governments

FIGURE 12.4 Political links: ASEAN+3. Japanese Prime Minister Junichiro Koizuni at the 2002 conference round table in Phnom Penh, Cambodia. The "+3" refers to Japan, China, and South Korea. **Photo:** © Reuters/Jason Reed/Corbis.

and international bodies at all levels struggled to meet challenges from terrorism and trade in guns, drugs, and slaves. Extremist groups with strong nationalist (Chechnya, Palestinian Territories, Basque areas of Spain) or ethnic (Indonesia and Philippines, Northern Ireland, Kashmir, Solomon Islands) agendas grew in significance at local levels and fed into country and world regional issues. To date, there is disagreement about how to tackle such events in the political realm, especially because the United States, as the sole remaining superpower, is often at odds with the United Nations.

Economic Activities: Global Trends

Economics is the study of how people and groups use resources to meet their material needs through producing, distributing, and consuming goods and services. Economic geographers are particularly concerned with the differences of economic well-being at local, country, world region, and global scales. Materially wealthier people in wealthier countries enjoy a good life, with security of income, housing, food, health care, and education, together with environmental safety, confidence in the future, and involvement in local and national decision making. They use a disproportionate share of Earth's resources to maintain such lifestyles, including big houses, gas-guzzling vehicles, expensive vacations, and high-cost medical services.

Poverty is basically a lack of material wealth that has wider outcomes. Poorer people do not have security like wealthier people and suffer from malnutrition, poor housing, exhaustion, disease, rejection, and vulnerability. Poor people are deprived of a living wage, education, health care, housing, clothing, food, and access to jobs. They will not likely have a long life and cannot fully participate in community affairs. Furthermore, some people may not be income-poor but are kept out of the mainstream of society by physical disability, ethnic or racial discrimination, or destructive behaviors associated with alcohol or drug abuse.

Economic activities are increasingly *global.* Global connections enhance living conditions for millions of people. International movements of people are driven by the possibility of better living conditions, trade, capital investment, and information exchanges. While parts of the world enjoy greater global connections, others remain isolated or marginalized by poor links to main economic centers. Increasing interconnections or continuing isolation underlie differences among places. Some peoples in isolated places have a better quality of life than the connected ones. For example, small groups of people living in the Amazon rain forest, large areas of Africa, and in places such as Papua New Guinea engage little in global economic activities, but may have better family and community-based lives than those, including children, working in sweatshops.

As world income increased in the later part of the 1900s, more people enjoyed material well-being as middle classes grew wealthier in poorer countries. One indicator of poverty—the numbers of people living on less than $1 per day—increased from 900 million people worldwide in 1820 to 1.4 billion people in 1980 but then fell to 1.2 billion in 2000. Although this is a large number of very poor people, the proportion in the whole world population living on less than $1 per day fell (from 85 percent in 1820 to 30 percent in 1980 and 20 percent in 2000). In the 1990s and early 2000s, some *world regions*, including former Soviet Union countries and Africa South of the Sahara, experienced increases in poverty, while East Asia saw major reductions in poverty. In the early 2000s, as poverty continued to affect so many people, the World Bank and United Nations focused economic and related policies on attempts to reduce poverty further.

In the 1990s the uneven spread of expanding global economic activities caused groups of countries to enter or revive regional economic agreements at the world regional level, mainly through trade. The best known and farthest advanced is the European Union, but others include the North American Free Trade Agreement (NAFTA), Mercosur (southern South America), the Association of Southeast Asian Nations, and the Southern Africa Development Conference. In East and Southeast Asia, the "Asian Way" of free-market economic activity works through family linkages and government–business liaisons more commonly than through the independently verified banking and legal systems regarded as most important in the United States and Europe. The "European Way" is another distinctive regional approach, in which social welfare provides support to people unable to benefit from advanced economic activity.

Within world regions, *countries* differ in their engagement levels with world trade. In the 1990s, the United States, the countries of Western Europe, and Japan controlled nearly all the investment, production, and consumption of goods. However, by the early 2000s, China, India, and Brazil increased their involvements.

Variations in resource provision, government efficiency, education and health levels, and the availability of investments to develop local resources also affect *local regions* within countries. Urban areas tend to attract most investment, while rural areas are often ignored or subject to the exploitation of people and the environment.

Cultural Activities: Identities in Major Regions, Local Voices

While political and economic factors remain highly significant in geographic differences, cultural influences came to the fore in the 1990s. The culture of a group of people results from the learned behavior in the ideas, beliefs, and practices they hold in common and pass on from one generation to the next. Religion, language, social organization, the design of the items they make, responses to the natural environment, and the level of technology involved are expressions of a shared culture (Figure 12.5). Different cultures often express values through communal or family life, and attitudes toward human rights. For example, cultural values are reflected in the role of women, the number of working hours, food and meal times, the nature of recreational activities, and the design and decoration of clothing, houses, and buildings. The culture of a people gives them an identity that they resist losing.

Although there are some indications that we may be heading toward "one world" culturally, many of the trends seem to imply that Western cultural norms can (or should) be extended to the rest of the world. Differences among places on a *global*

scale now need to take account of the links between cultural factors and political–economic processes. For example, people in many countries still adopt material aspects of the Western culture they find attractive, such as clothing, food, music, and magazines. At the same time, Muslims and many people in other Asian countries also wish to improve their material wealth but resist Westernization, which they see as having many negative characteristics. Many people associate Western cultures with emphasizing materialism, consumerism, and other superficial values that lack ethics or spirituality. Western cultural trends are seen as threats to the social fabric, especially of families and communities. Some argue that such trends must be opposed, even violently if necessary.

A closer examination shows that the "Coca-Cola-ization" of eating and drinking habits, the spread of Western TV, motion pictures, and popular music, and the global markets for some consumer goods have not wiped out local cultural differences. On the contrary, local cultural practices and preferences exert themselves and even blend with global culture, adding to the diversity of places. In India, for example, only the wealthier elites can afford to buy imported goods and the new multinational fast food. The more numerous middle-class groups buy cheaper Indian-made goods and eat at local restaurants. Most Indians prefer their homemade "Bollywood" films. Moreover, the major Western multinational corporations make their products acceptable to local people, with Coca-Cola making its drinks sweeter or more carbonated; McDonald's includes distinctive vegetarian food in India, and its restaurants have prayer rooms in Indonesia. Once again, global connections are met by local voices.

In the 1990s cultural activities gained in prominence at the *world regional level*. Religious beliefs and languages remain central to motivations for retaining cultural traditions. Western cultures, Christian or secular, which come from Europe and North America, have different priorities from the mainly Muslim cultures of different varieties in the Arab world and parts of South, Central, and Southeast Asia. In other parts of the world, Hindu, Jewish, Buddhist, animist, communist, and secular cultures have prominence. However, the missionary spread of Christianity, Islam, and Buddhism over the centuries brings such cultures face-to-face today in many countries. Local languages are increasingly challenged by English, other European languages, Arabic, Hindi, and official (Mandarin) Chinese. Ingrained cultural outcomes lead to conflict in some places, but people in other places seek to work together. Such cultural (or ethnic) diversity is commonly discounted by political leaders and theorists, who thus isolate minorities, including indigenous groups, within society. They may subject them to oppression, including religious persecution, ethnic cleansings, and everyday exclusion from education, jobs, and political participation. Those actions and attitudes are replaced in some places by identity politics, which recognizes the contribution of a rich ethnic variety to a country's future.

Few, if any, *countries* can claim to contain only a single political, religious, linguistic, or pure "national" culture, although some have tried to impose one on their inhabitants.

FIGURE 12.5 Cultural expression. St. Mary's of Zion Coptic Church, Lalibela, Ethiopia. An Ethiopian bible from the A.D. 900s shows detail of biblical events. **Photo:** © Gallo Images/Corbis.

Localities within countries often have cultural groups that feel underprivileged or discriminated against economically and politically. Examples include the Tamils in northern Sri Lanka, the Basques in northern Spain and southwestern France, the Kurds in parts of Iran, Iraq, and Turkey, and the Native Americans in Canada (where they are called "First Nations") and the United States.

Environmental Issues at Varied Scales

Natural environments create differences among regions. Hurricanelike storms, acid rain, and damage from river floods and volcanic eruptions present issues at world regional, country, or local scales. At the global scale, effects such as global warming, El Niño (Figure 12.6), the ozone hole over Antarctica, and the destruction of tropical rain forests are subjects of international debates. In the 1990s world environment conferences at Rio de Janeiro, Brazil (1992), and Kyoto, Japan (1997), adopted policies to avert environmental crises, but it is left to individual countries to implement the policies. The United States recently decided against ratifying the Kyoto recommendations. Hurricane Katrina in 2005 brought such huge destruction and loss of life to New Orleans and the Mississippi–Alabama coasts that the world was alerted to the increasing number of high-intensity hurricanes, probably linked to global warming.

One impact of the globalization of telecommunications is that people around the world are more aware of human interactions with the natural environment. Disasters, such as the Indian Ocean tsunami at the end of 2004, and Hurricane Katrina in August 2005, alert people around the world to the environmental and natural resource issues that are likely to increase at global and regional scales in the 2000s. They have political, economic, and cultural impacts.

People modify all natural landscapes in which they live, leaving their human (anthropogenic) impacts on the natural world. They produce cultural landscapes containing buildings, fields, managed woodlands, and transportation links. In doing so, they modify local climates, increase surface rain runoff from towns that also become "heat islands," change soils by cultivation, and add gases to the atmosphere. Some human modifications improve the productivity, landscape quality, and livability of a place, whereas others degrade a place's attractions, resources, and future prospects. Over thousands of years, people gradually modified landscapes by removing forests to expand food production and support growing populations. The rapid industrialization in the 1800s and 1900s involved digging huge mine pits, building larger and larger factories, and extracting oil and gas. These human activities polluted air and water. In some places the buildup of wastes and toxic gases from factories and cities overwhelmed the natural environments' abilities to deal with them, requiring countries to pass laws that penalize polluters.

The world's growing populations and improved living standards have rising impacts on the resources available. World population rose from 1.6 billion people in 1900 to over 6 billion in 2000, and it could rise to around 9 billion by 2050. The population increase will not be uniform across the globe. Almost all of it will be concentrated in the world's poorer countries. World population conferences in the 1990s suggested that population growth should be curbed by birth control, but some countries with conservative religious beliefs resist restrictions on family size.

Natural resources provide benefits for people whether used at source or sold abroad to other countries. In the past, regional geographies were linked closely to local resources. Increasing globalization through the late 1800s and 1990s led to exports of metallic minerals, oil, and tropical plantation crops: producer countries sold to consuming manufacturing countries. Such trade is often accompanied by frictions that could escalate into major conflicts in the 2000s. The region or Northern Africa and Southwestern Asia could be a center of such conflict because it involves countries worldwide in competition for its oil, but also has internal disputes over its water resources. The world's oil demands focus political interests in this world region, and some political analysts blame U.S. oil interests for the Gulf War of 1991 and the invasion of Iraq in 2004. Other oil-centered conflicts arise in the South China Sea, where China, Vietnam, and the Philippines dispute ownership of the Spratley Islands; in Central Asia; and between Australia, Indonesia, and East Timor.

FIGURE 12.6 Environmental disaster: El Niño. A farmer, Heriberto Castro, examines his dry field near Las Ovejas, Guatemala, in 2002 after two years of drought linked to the El Niño phenomenon. **Photo:** © Reuters/Corbis.

Globalization and World History

Throughout our studies of world regions, it has been clear that today's geographies emerged from often complex histories that gave distinctive cultural characters to the world regions, led to specific types of political divisions and economic development, and had varied interactions with natural environments. Although some historians trace globalizing elements of "capitalist" free-trade markets back 5,000 years, major developments began around A.D. 1500 and increased in intensity after around 1750.

Phase I: Partly Formed Globalization before 1850

Until around 1500, most parts of the world had economies based on feudal relationships. The beginnings of a "capitalist" trading system were found in a few places. In particular, the Arab trading system (of approximately A.D. 1100 to 1300) dominated the Indian and the Chinese interests in South, East, and Southeast Asia and was important in linking widespread places. However, there was limited penetration of each other's areas of influence, and the scale of exchanges was often limited to goods of high value. After around 1500, the Chinese withdrew from ocean trading, but Chinese land expansionism and powerful groups in Japan, Mughal India, and the Ottoman Turks continued to have strong regional political, economic, and cultural impacts.

By 1500 European explorers and merchants extended their economic influences to the newly discovered Americas and also to Africa and Asia. Cosmopolitanism progressed alongside growing national loyalties and the emergence of nation–states, or countries with defined (but still rather porous) borders in Europe. The coastal boundaries of island Britain brought an early consciousness of "Britishness" alongside London's expanding overseas trade and the wish to bring Protestant Christianity to other parts of the world. The European countries, moreover, had greater financial resources to support military adventures in the new worlds, and developed service industries to administrate foreign lands.

After around 1750, European global influence increased. Early industrialization in Europe both provided technological superiority in transportation and armaments, and also demanded new sources of raw materials. Private trading companies developed worldwide networks despite a confusion of protectionist and often predatory countries fighting each other inside Europe. Although some historians suggest that Asian decline enabled European expansionists to triumph, others record continuing or reviving economic prosperity in isolationist China and Japan, in increasing local trade in Southeast Asia, and in freshly established states within India. Furthermore, both Buddhism and Islam spread energetically into Asia and Africa through the 1700s, enabling the spread of cultural and trading diasporas. Such trends—not local weakness—provided a base for later European and United States–dominated trading patterns and linked cultural and political impacts. A diversity of forces contributed to later global connections.

Phase II: Beginnings of Modern Globalization, 1850–1970

By the mid-1800s there was a shift from mercantile capitalism to industrial capitalism. It was enhanced by free-trading policies and particularly by technical developments in communications (telegraph and telephone), transportation (larger oceangoing ships, railroads, road vehicles, and then airplanes), an international capital market (London), expanding worldwide trade in goods, and large-scale migrations (as from Europe to new colonial territories and sources of raw materials). European and American imperialism brought Africa, India, and Southeast and East Asia into the capitalist free-market system, sometimes forcibly under political control. There was increasing economic convergence and migration, especially of Europeans to new countries.

These trends were interrupted in 1914 by World War I, when European imperial competition led to massive conflict and withdrawal from globalizing trends. The war drew to Europe armed contingents from many parts of the world, but had adverse effects on world trading economies, including both colonies and noncolonial territories. Attempts to return to the pre-1914 system in the 1920s were not successful, and a worldwide economic depression led to further isolationism in the 1930s. The tension between nationalism and cosmopolitanism came to a head again in World War II (1939–1945) as German and Japanese expansionism was resisted by the rest of the world.

Global forces began to reestablish themselves after World War II, although this took some decades. Large numbers of new countries (with their own nationalisms) emerged from decolonization. The Cold War divided countries into three worlds. On the one hand, global links increased through trading agreements, the United Nations agencies, and multinational corporations. On the other hand, nationalisms persisted, and regional identities emerged from Western and non-Western diasporas, including the Chinese in Southeast Asia and Indian-subcontinent people in the Persian Gulf countries.

Phase III: Globalization Today, 1970 Forward

From the 1970s, the end of the Cold War, worldwide dependence on increasingly open trade, and the shifts of occupation and services to the postindustrial society led to a greater intensity of globalization. The rise of foreign direct investment, the expansion of multinational corporations (especially from the United States, Europe, and Japan), and the falling costs of communications and transportation brought challenges to national sovereignty, but did not end it. In fact, globalizing forces often met resistance at country or local levels.

There is debate over the significance of these trends. Some see globalization as a crisis confronting the world, developing from the mid-1970s (see the Point–Counterpoint: Facets of Globalization box on page 12). Germany and Japan, and later China, reemerged as economic powers to challenge U.S. dominance in the "free" world. Oil price rises increased global financial exchanges but increased the poverty of many countries. Countries lost powers to global capitalist practices.

At the other end of the spectrum of opinions, some see globalization as progressive, bringing mainly positive outcomes. From the 1990s, when there were no alternative options, prosperity began to spread along with democracy and quality of life in Asia and Latin America. However, it is not easy to accept either view as a black-and-white alternative. For example, the growing Asian economies are linked to authoritarian governments in China, South Korea, and Singapore; and African countries have not prospered, however they approach the future.

A Case Study: Southern China

Southern China is a region of changing fortunes, prominence within China, and wider geographic linkages. Following the 1978 Chinese Communist Party Congress in Beijing that initiated huge reforms affecting central government policies and local opportunities, it experienced massive economic, political, cultural, and environmental changes. A move to an "open policy" of export orientation in the economy began with experiments that became the accepted basis for Chinese future economic development. Southern China had the first four special economic zones, based in Guangdong and Fujian provinces. Although largely ignored by previous Chinese development, these provinces were close to the major ports of Hong Kong and Macau and their established links to the outside world. These provinces were the source of most Chinese who moved abroad over the last few centuries. A study of this region illustrates both global and local forces that influence geographic regions.

Marginal Part of Greater China

For over 2,000 years, China was governed from northern centers ranging from X'ian to Beijing, where the essential Han culture based on rigid Confucian controls evolved. The southern parts were regarded as different. Northerners traveled by donkey or horse, grew wheat and millet, looked to river-based linkages, and avoided coastal locations. Southerners grew rice in a warmer, wetter climate and developed a coastal outlook. For long the southern parts were seen by other Chinese as populated by strange peoples with more leisurely lifestyles and softer-spoken dialects and languages. The south was more open to outside influences, from the variety of religions to economic linkages.

A Maritime Frontier

A series of small ports grew along the lower parts of southern river valleys as they reached the ocean. They were markets for local farmers, fishers, and coastal traders. Apart from Xiamen, built on an offshore island, other ports (Fuzhou, Quanzhou, Ningbo, Guangzhou) at first preferred the more sheltered sites a few miles inland. Seasonal winds brought ships from the north in winter and blew them back in summer. Buddhist monasteries occupied much of the best land and carried out land reclamation, intensified rice farming, and some commerce. The Han people migrations into southern China and increased northern political control from the A.D. 700s to the 1200s, combined with local resources, brought prosperity. Overseas commerce established tribute-paying centers throughout Southeast Asia in the following years. The maritime industries of the south—making boats, establishing port facilities, and commercial relationships—expanded to other Chinese ports farther north.

Western Influences: Open Ports and Treaty System

Up to 1851, a combination of Chinese and Japanese wholesale merchants visited Fuzhou as a ritual, bringing long-established tribute payments. Western and, to an extent, Islamic traders changed this, looking for free-trade rights. Under the treaty system, coastal centers opened to foreign residence and trade through the 1800s. Hong Kong and Shanghai became the main international ports, while other parts remained local in orientation. The opium trade arose out of British purchases of silks, porcelain, and tea, making a trade deficit that the opium sales covered.

Chinese notions of imposing tribute payments (along with *kowtow* respect to China) clashed with developing free trade. After 1759 China tried to restrict foreign trade to Canton (modern Guangdong), but in 1839 the British fleet blockaded the port, leading to the Opium War of 1839–1842.

In 1842 the Treaty of Nanking allowed Western imperialism to dominate Chinese external trade. This mainly affected coastal cities and their hinterlands. British subjects, their families, and firms were based in Canton, Amoy (Xiamen), Foochowfoo, Ningpo, and Shanghai, while Britain took Hong Kong island along with money to cover lost trade and war expenses. Through the later 1800s European countries also gained rights in another 10 ports. Traders and Christian missionaries penetrated inland. Local rebellions against these incursions were put down.

As Hong Kong and Shanghai expanded, ports such as Fuzhou (a center of antiforeign resistance), Amoy (the focus for overseas Chinese family links), and Ningbo (a center of coastal trade at the end of the Grand Canal) grew slowly. Most open ports had separate areas for foreigners and provided stages for opposing Western dominance and culture. In Fuzhou, however, Christian missionaries established a presence despite opposition and later worked with local leaders against the opium and tobacco trades. They established Western-style schools that educated later nationalist leaders.

The overall outcome of these trends was to expand urbanization and commercial facilities, including banking, based on trade in the southern ports. Despite the political instability after the fall of the imperial regime in 1911 and the faltering progress of the new republic, investment in the cities continued with construction of commercial districts and transportation facilities that built on links with other countries.

Revolution and the Peoples' Republic

Southern China acted as a base for the revolutionary movements that determined China's history through the 1900s, with centers in Guangzhou and Shanghai. Sun Yat-sen grew up in Guangzhou and gained education abroad, returning to China in 1912 to form the Nationalist Party (Guomindang); in 1927 he moved the capital southward from Beijing to Nanjing. Chang Kai-shek, who succeeded Sun, came from a Ningbo merchant family and spent time in Japan. The Nationalists dominated the coastal cities, linking abroad to the Western democracies. The Communist Party had an interior base, depending on rural peasant support as it took their part against landlords. After World War II the Communists took over the whole of China. Many Nationalists fled to Taiwan, with Xiamen acting as the final port of retreat. Under the Communist People's Republic of China, the southern port cities were labeled as "parasitic Western" places. Their links to foreign exchange continued to be useful, but there was little or no new economic development.

The combination of overseas trade over 200 years and flight from Communist rule since 1949 led to the migration of the overseas Chinese—30 million people worldwide with 75 percent in Asia. Most came from southern China, especially Guangzhou and Fujian provinces. Taiwan became the largest center of overseas Chinese. Although persecuted at times, the Chinese populations in Southeast Asian countries often dominated commerce internally and externally. Many became materially wealthy. For example, political jealousies over the large Chinese population caused Singapore to leave the Malaysian Federation. The Chinese in Indonesia and Malaysia are important as business leaders.

Reform, Industrialization, and Wider Linkages

After Mao Zedong's death in 1976, the reform policies under Deng Xiaoping brought new economic prominence to southern China. The encouragement of foreign investment and export industries through special economic zones and open ports led to huge developments in Guangdong province, inland of Hong Kong. This development spread through the other coastal ports until Shanghai took over as the main center of new economic development in the 1990s.

At the same time, China changed in other ways. The loosening of the *hukou* household registration system, once required for urban residence, enabled people, especially single women, to move to urban areas where they were employed in export industries at low wages. The rural household responsibility system gave women new opportunities, while the one-child birth planning policies enabled more women to enter the labor force before or after having a baby. The restructuring of state-owned enterprises brought local layoffs of workers, many of whom moved to the new privately owned industries. In many of these trends, the previous (pre-Communist and Maoist) male bias in Chinese life remained, but the movements of women to new urban jobs enabled them to develop their own futures.

Investment from abroad was not only in manufacturing. In 1993 a Chinese Indonesian corporation built a theme park on Meizhou Island (Fujian province) to attract Taiwanese tourists brought by increased ferry crossings. In 2004 a new hospital being built in Dongguan (Guangdong Province) was 25 percent funded from Taiwan to produce a center of excellence for the 100,000 or more Taiwanese living and working in the city in thousands of Taiwanese-funded companies. These continue the inter-Chinese links established across Asia over the centuries.

The huge growth of urban centers along the southern Chinese coastlands from Guangzhou to Shanghai provides the latest links to globalization as centers of trade, financing, information, production, foreign links, and population mobility (Figure 12.7). Shanghai and Hong Kong have become increasingly prominent in these relationships.

Conclusions

Globalization trends affecting southern China do so in the context of spatial transformations in its economy and society. Globalization processes built on previous linkages and internal developments. Some of the outcomes have strong historic influences. Others are a result of Chinese internal policies. The regional past of mercantile and maritime economies, from tribute bearers to imperialists, made this China's main

FIGURE 12.7 South China cities. (a) Guangzhou factories and apartments expand into farmland. The farming family works hard to produce two crops a year, but earns only around US $1,000. (b) Macau: the Bank of China's new downtown building. **Photos:** (a) © David Butow/Corbis. (b) © Viviane Moos/Corbis.

(a)

(b)

AAG PRESIDENT IN IRAN: RECONCILING DIFFERENCES

Geography provides fundamental insights into the nuances of globalization, whether economic, political, or cultural. Globalization is creating tensions between peoples as some cultural practices spread and challenge local cultures and voices. Here too, geography not only concerns itself with this phenomenon but also can build bridges and help ease cultural tensions and misunderstandings. For example, Alexander B. Murphy, from the Department of Geography at the University of Oregon, traveled to Iran in 2004 when he was president of the Association of American Geographers (Box Figure 1). While in Iran, he addressed the "Second International Congress of Geographers of the Islamic World"; he talked to students and faculty at two of Tehran's major universities; and he visited different parts of the country with Iranian geographers. His visit was covered in the *Tehran Times,* and he was a guest on an Iranian television interview show.

The governments of Iran and the United States may be at odds with one another, but Professor Murphy had the chance to see why the Iranians have a reputation for being so hospitable. Everywhere he went, people were extraordinarily nice, helpful, and friendly. The trip also reminded him of how distinctive Iran is compared to its neighbors. Murphy had traveled in other parts of the Muslim Middle East—especially Egypt, Jordan, and Palestine. There are some continuities with these places, but Iran is clearly different—in language, in culture, in social norms, and much more. To visit Iran is to understand the fallacy of treating the "Islamic world" as if it were a monolith.

The presentations Murphy gave at the two universities led to interesting academic discussions with faculty and students. Very few Americans go to Iran these days, and this opportunity allowed Murphy to give Americans a human face in Iran. At the same time, he gained insight into the diversity of opinion in Iran on political and social issues. Iranians express views that span the political spectrum. This is not surprising given the rapid transformations we are all confronting in the contemporary world. What particularly impressed Murphy, however, was how well informed Iranians seem

Box Figure 1 Professor Murphy, the tallest person, with colleagues at the Iranian geographers' conference.

to be—gaining information not just from government news sources but also from the Internet and from friends and relatives in other parts of the world. Many Iranians are strongly critical of stances taken by the U.S. government on particular issues, but Murphy was never held personally responsible for those stances.

The Western press sometimes paints a picture of Iran as an isolated place with little personal freedom and much cultural and political radicalism. Iranians are often confronted with an image of the United States as a country where Donald Rumsfeld represents the political norm and Eminem represents the cultural norm. Murphy's trip revealed how shallow these stereotypes are. Murphy saw how much Iranian society has opened up in recent years, and his discussions with people and his media appearances highlighted a side of America that Iranians rarely see.

region of mobility and the basis for the Chinese diaspora, distinctive local attitudes, and global linkages. A region that was long seen as different than the center of Chinese culture and control became a place where "experiments" led to new trends across the country. It illustrates the role of regions within a globalizing world. Each territory experiences the interaction of internal and external political, economic, cultural, and environmental processes that bring new geographic orientations and features, new regional consciousness and identity, and new attitudes to environmental sustainability. Such regions as southern China occur within countries, subject to internal policies, but contain many transboundary

elements that force the country's government to recognize them and adjust policies accordingly. That is happening in all of China, but it started in the south.

World Regional Geography: Final Point

World regional geography is always relevant and always changing. The ideas and principles studied in this course should help you to assess world and local events in a context that makes them easier to understand. Of course, our world is not just a changing one but also a very complex set of human and natural environmental processes interacting with each other. Fifty years ago, it was still possible to study a local region or country with little reference to its surroundings; but in the 2000s the wider world regional and global interconnections are increasingly relevant for us all (Geography at Work: AAG President in Iran: Reconciling Differences, on page 566).

Before World War II, American politicians counseled President Franklin D. Roosevelt to keep out of the war that was erupting in Europe. No one could attack the United States, they asserted, and it would be best for it to remain aloof from the conflict. This ignored the potential threat from the opposite direction, which was realized when Japan attacked Pearl Harbor, Hawaii. Since that time, triumph in World War II and the Cold War encouraged U.S. feelings of supremacy and confidence in its ability to fend off potential problems. The threats of

nuclear war from the Soviet Union were diminished by the equal threat of retaliation and subsequently by the breakup of the Soviet Union.

In the late 1990s some Americans hoped the United States would retreat from some forms of global interaction and decrease its controversial role as a global police force. However, it is geographically impossible for a country with the political, economic, and cultural influence of the United States to shut itself off from the rest of the world, especially when it actively exports its goods, services, and culture and generally argues for increased global trade. In the increasingly connected global system present today, the actions of all governments, large nongovernmental organizations, and multinational corporations have far-reaching impacts. None may operate in isolation. A better understanding of world regional geography facilitates the ability to connect cause and effect, action and impact, and places and peoples throughout the world.

Glossary of Key Terms

Each term is defined and linked to the chapter and Test Your Understanding box in which it was a key term.

Aborigine (6A). Race of people whose ancestors inhabited Australia before the arrival of Europeans.

absolute location (1A). Location of a place on Earth's surface as defined by latitude and longitude or by distance in km (mi.) from another place.

acid deposition (3B). Dry or wet deposition of acidic material from the atmosphere, often resulting from sulfur and nitrate gases and particles emitted into the air from coal combustion in power plants.

Act of Confederation (Canada) (11D). The act of the United Kingdom in 1867 that established Canada as a united and militarily protected entity.

afforestation (6D). The replanting of previously cut forest.

agglomeration economies (3C). The total economies achieved by a production unit because of a large number of related economic activities located in the same area.

agribusiness (3C). The large-scale commercialization of agriculture that places farming within the broader context of inputs of seeds, fertilizer, machinery, and so on, and of outputs of processing, marketing, and distribution.

alluvial fan (7A). A fan- or cone-shaped river deposit, often formed where a stream issues from mountains into a plain.

alluvial layers (6B). Deposits of sand, clay, and boulders dropped by rivers in the lower parts of their valleys.

altiplano (10B). High plateau in Peru and Bolivia.

altitudinal zonation (10B). Significant changes in altitude produce corresponding climate and vegetative changes. Higher altitudes produce cooler temperatures, variation in soils and precipitation, and distinctive vegetation regimes. Different crops may be cultivated in the changing environmental zones.

For example, in the Andes Mountains of Latin America near the equator, the *tierra caliente* in the lowest 1,000 m (3,000 ft.) has warm-to-hot conditions and tropical forest. The *tierra templado* in the next 1,000 m has mild-to-warm temperate conditions and deciduous forest. The *tierra fria* between 2,000 m and 3,000 m has cold-to-mild conditions, pine forests, and grasslands. The highest zone is the *tierra helado,* with some grassland but mostly permanent snow and ice cover.

Andean Common Market (10D). A free-trade area among the countries of the Northern Andes—Bolivia, Colombia, Ecuador, Peru, and Venezuela.

Angkor Wat (6A). Place in Cambodia with many Hindu temples.

animism (6A). Traditional religious beliefs based on the worship of natural phenomena and the belief in spirits separable from bodies.

Antarctic Treaty (6E). Treaty, signed in 1961 by 39 countries, agreeing on rules for nonmilitary scientific cooperation, environmental safeguards, and international control.

Appalachian Regional Commission (11C). U.S. federal–state agency established in 1963 to enable the poverty-stricken region to access federal grant aid.

Arabic (8A). Language spoken and written by Arabs, diffused from Arabia with trade, conquest, and religious use. The only language allowed in the Islamic Qu'ran and prayers.

Arab League (8B). Organization created in 1945 to encourage the united action of Arab countries for their mutual benefit.

archipelago (6C). An island group, such as the components of Indonesia.

artesian conditions (6B). Groundwater that flows naturally out of springs and wells under pressure that is imposed by the geologic arrangement of water-bearing rocks.

Asia–Pacific Economic Cooperation (APEC) (6C). A group of Asian and Pacific countries dedicated to trade liberalization.

Association of Caribbean States (10D). Formed in 1995 by West Indies countries, Mexico, Central American countries, Colombia, and Venezuela as a response to the formation of NAFTA.

Association of Southeast Asian Nations (ASEAN) (6A). Established in 1967 as a defensive alliance among Indonesia, Malaysia, Philippines, Singapore, and Thailand in response to the advance of Communism in Southeast Asia. Now increasingly a trading group that admitted Vietnam, a Communist country, as a member in 1995, and Myanmar in 1997.

Aztec (10A). A group of Native Americans who moved from northern Mexico and established control by A.D. 1325 over much of the surrounding region from its base at Tenochtitlán, the site of modern Mexico City.

badlands topography (7A). Closely spaced networks of deep gullies, often carved by occasional streams in soft sediments unprotected by a vegetation cover. These destroy the usefulness of the land.

banana republic (10C). A country that relies on a single export crop (such as bananas), is dependent on world core country markets, and has a poor government that is often a dictatorship.

barrier reef (6B). A coral reef structure surrounding an island and separated from it by a lagoon. The Great Barrier Reef lies off Queensland, Australia.

Benelux (3D). An acronym for **Be**lgium, the **Ne**therlands, and **Lux**embourg. The term was coined in recognition of the close working relationship that these countries have with one another.

Bengali (7C). Language of 98 percent of Bangladesh people; also spoken in eastern India.

Berber (8A). Language of African origin spoken, but little written, in North Africa.

Biharis (7C). People of non-Bengali origin and Urdu-speaking who migrated to East Pakistan (Bangladesh) at the partition of India in 1947.

biome (2B). A world-scale ecosystem type, such as tropical rain forest or savanna grassland.

birth rate (2A). The number of live births per 1,000 of the population in a year.

black earth (4B). Highly fertile soil type in which organic matter accumulates near the surface, commonly beneath midlatitude grassland communities.

Black Triangle (3B). A heavily polluted industrial area straddling the Polish, Czech, and German borders.

brain and skills drain (1B). The migration of professional or highly educated people from one country to another for better economic and career prospects.

Brazilian street children (10E). Abandoned children with a nomadic and shelterless existence in Brazilian cities.

British East India Company (7A). A British company that traded with and conquered much of the Indian subcontinent. After an 1857 mutiny in India, the British government took over political control.

British Indian Empire (7A). Established after 1857 on the Indian subcontinent, including Ceylon and later extended to Burma. Lasted until independence and partition in 1947.

brown earth (4B). Fertile soil type in which plant matter replenishes nutrients in the upper layers, commonly forming beneath midlatitude deciduous forest.

Buddhism (7A). A religion that began in South Asia but became the major religion of East Asia. Followers of Buddhism have a greater social openness than do followers of Hinduism and accommodate other philosophies and religions such as Confucianism and Shinto.

buffer state (5A). A country that stands between major world powers and helps to reduce direct conflicts between them.

Burmese (6A). The dominant people in Myanmar (Burma).

Canadian Shield (11B). The ancient rocks of northern Canada, the oldest in North America, which contain many mineral deposits.

Canal Zone (10C). Area close to the Panama Canal, formerly held by the United States, but relinquished to Panama in 1999. A cosmopolitan area of global trade.

capital city (2A). The city in which central government functions are concentrated; sometimes the largest city but often one that is specifically designed for the purpose.

capitalism (3A). An economic system in which goods and services are produced and sold by private individuals, corporations, or governments in competitive markets. The means of production are owned by those investing capital, to whom workers sell their labor. Linked to democratic government and increasing trade among places and countries.

Caribbean Community and Common Market (CARICOM) (10D). Formed in 1973 by 13 former British colonies to provide special entry to U.S. markets.

cartel (8B). An organization that coordinates the interests of producers (such as OPEC).

caste order (7A). A social class system associated with Hinduism that is based on the supremacy of Aryan peoples. The Aryan castes include priests (Brahmans), warriors, and other Aryan people; non-Aryan castes include cultivators, craftspeople, and untouchables.

Celts (3A). European tribes that diffused skills in metal-making (bronze and iron) as they moved from the Alpine area around 1000 B.C. into Spain, France, and Britain.

central business district (11C). The area in the center of a city where shops, offices, and financial services are concentrated, normally to the virtual exclusion of residential facilities.

central planning (4A). The Soviet practice in which the government decided how many goods and services were needed by society, almost without cost considerations.

Christianity (8A). A religion that developed out of Judaism, based on the belief that God came to Earth in the form of Jesus of Nazareth. Main religion of the Western world.

city–state (10A). An individually ruled and independently functioning urban center.

class (2A). A stratification of society that is based on economic, religious, or social criteria.

climate (2B). The long-term atmospheric conditions of a place.

collectivization (5D). The transformation of rural life in Communist countries such as China and the Soviet Union, in which individual farmers were grouped in cooperatives that took ownership of their land and labor.

colonialism (3A). The system by which one country extends its political control to another territory to improve local conditions and/or economically exploit the human beings and natural resources of the subordinate territory.

command economy (4A). The Soviet practice whereby the government ran the economy, owned all industries, and set quotas, favoring heavy industry over the production of consumer goods.

Commonwealth of Independent States (CIS) (4A). A political and economic organization created in 1991 by 11 republics of the former Soviet Union. Members included Russia, Belarus, Ukraine, Moldova, Armenia, Azerbaijan, Kazakhstan, Kyrgyzstan, Tajikistan, Turkmenistan, and Uzbekistan. Georgia became the 12th member in 1993. The CIS coordinates relations between member countries, including issues involving economics, foreign policy, and defense matters.

commonwealth status (10D). The political status of Puerto Rico within the United States of America. Its people have citizenship and open migration access, but elected representatives cannot vote in the U.S. Congress.

Communauté Financière Africaine (CFA) (9B). Economic links between France and its former colonies in Africa. From independence until 1994, it involved links to the French franc, but this arrangement ended with the devaluation of the CFA franc.

commune (5D). The organization that controls rural life in Communist countries, including agriculture, industry, trade, education, local militia, and family life.

communism (3C). A system in which the workers govern and collectively own the means of economic production. When spelled with a capital "C," Communism refers not to the system as it was originally envisioned but rather to the totalitarian systems adopted in countries such as the Soviet Union and China, where small elite groups rule or ruled under the guise of communism.

concentration (of agriculture) (3C). Agricultural production carried out on fewer and larger farms and limited to smaller areas of higher productivity.

concentric pattern of urban zones (11C). A pattern of urban geography with a central business district in the middle, surrounded by a hierarchy of residential zones.

Confucius (Kong Fuzi) (5A). A Chinese administrator who established a system of efficient and humane political and social institutions that became the basis of procedures and ways of life in much of East Asia.

coniferous (11B). Plants with needle-shaped leaves that are mostly evergreen.

containment (6A). The U.S. policy designed to stop the spread of communism during the Cold War.

continentality (4B). Especially cold winters and hot summers resulting from locations on landmasses that are far from the moderating effects of large water bodies such as oceans.

Contra (10C). United States–backed former Somoza military personnel who fought the Sandanistas in Nicaragua.

convective rain showers (6B). Rain formed when warmed humid air rises, condenses into clouds, and is precipitated back to Earth in almost the same place.

convergent plate boundary (6B). In plate tectonics, where two plates move toward each other, causing earthquakes, volcanic activity, and mountain building as they meet.

copra (6E). The dried white "meat" inside a coconut.

coral atoll (6B). Coral islands with a surface just above sea level and a central lagoon.

cottage industry (7C). Manufacturing based on processes that are carried out in dispersed homes instead of in a central factory.

counter-migration (11C). Movement of people returning to regions they once left, as in the return of African Americans from northern U.S. cities to the South.

counterurbanization (5B). The movement of people from metropolitan urban areas to take up residence and employment in small towns or rural areas.

country (2A). A self-governing political unit having sovereignty within its borders and recognized by other countries.

Cree (11C). A "First Nation" of people, who claim the northern part of Québec province in Canada.

creole language (9A). A mainly trading form of communication that combines several languages.

criollo (10A). Person living in the Americas during the colonial era whose sole ancestral heritage was tied to Spain or Portugal. People of African heritage living in the Caribbean during the colonial era were referred to as Black Creoles.

crony capitalism (6C). The economic system in which entrepreneurs gain advantages through close links to government officials and ministers.

cultural fault line (2A). A line or zone of tension between contrasting cultural regions across which conflict may occur.

cultural geography (2A). The study of spatial variations in cultural features such as material traits, social structures, languages, or belief systems.

cultural hearth (1B). A small region of the world that acted as a catalyst for developments in technology, religion, or language, and as a base for their diffusion.

cultural hegemony (11A). When a dominant culture group forces another group of people to reject their own culture and accept that of the dominant group.

Cultural Revolution (5D). The attempt by Mao Zedong between 1966 and 1976 to change the basis of Chinese society. The disruption held back the country's economic development.

cultural rights (2C). The rights to protect one's cultural traditions.

Daoism (5A). The teachings of Chinese philosopher Laozi, who disliked the organized and hierarchical social system of Confucius and advocated a return to local, village-based communities with little outside interference.

Darién (10C). The area of Panama east of the Canal Zone, where dense vegetation and difficult terrain make land transportation difficult.

death rate (2A). The number of deaths per 1,000 of the population in a year.

deciduous (11B). Plants with broad leaves that mostly lose their foliage in cold or dry seasons.

deindustrialization (3C). A rapid fall in manufacturing employment and the abandonment of factories in a once-important industrial region.

democratic centralism (3C). The practice of sole governance by the Communist Party, the political party of the working class, because it is believed that only the Communist Party is the true representative of the people.

demographic transition (2A). The process in which the population of a materially poor country moves from high birth and death rates toward the low birth and death rates of materially wealthier countries, through a period of high birth rates and low death rates—when the population increases rapidly.

demography (2A). The study of human populations in terms of numbers, density, growth or decline, and migrations from place to place.

density (2A). The frequency of a phenomenon within a unit of land area: for example, people per hectare.

deposition (2B). The dropping of particles of rocks carried by rivers, wind, or glaciers when they stop flowing or blowing, or melt, respectively.

desalination plant (8A). A mechanism that extracts fresh water from seawater by evaporation and condensation.

desert biome (2B). Major biome-type characterized by a discontinuous plant cover or none.

desertification (2B). Processes that destroy the productive capacity of an area of land.

deurbanization (6C). The process adopted by the Communist government of Vietnam after 1975 to reduce the size of Ho Chi Minh City and return people to rural areas.

development (2C). The process by which human societies improve their quality of life, including economic, cultural, political, and environmental aspects.

devolution (3C). The process by which local peoples desire less rule from their national governments and seek greater authority in governing themselves.

diaspora (2A). The scattering of a people, such as the Jews and Chinese, to other countries.

direction (1A). The position of one place relative to another, measured in degrees from due north or by the points of the compass.

distance (1A). The space between places, measured in kilometers (miles), travel time, or travel cost.

distribution (2A). The spatial spread of a phenomenon, as in lines, clusters, or at random.

diversified economy (8D). An economy in which manufactures are more important than primary products and where there is a variety of manufactured products and a growing service sector.

domino theory (6A). The political theory suggesting that the countries of Southeast Asia would fall one by one to Communist pressures. Thailand and Malaysia were "dominos" that did not fall.

Dust Bowl (11B). The area of Oklahoma, Kansas, and Texas, USA, where plowing of subhumid grassland resulted in massive wind erosion in the dry years of the 1930s and since.

Dutch East India Company (6A). The company established in the Netherlands in the A.D. 1500s to control trade in the East Indies (approximately equal to Indonesia today).

---

economic geographer (2A). A geographer studying the spatial aspects of material wealth and poverty, the use of resources, and the production of goods.

economic leakage (10D). Foreign ownership and control of industry, such as tourism in many parts of Latin America, where the revenue generated by the industry flows to the foreign owners and foreign employees and largely bypasses local communities and thus "leaks" from the local economy.

economies of scale (11A). Increased productivity gained by building larger factories and gaining access to larger markets.

ecosystem (2B). The total environment of a community of plants and animals, including heat, light, and nutrient supplies.

ecotourism (6C). Tourism based on viewing natural wonders such as wildlife and rock formations.

ejiditarios (10C). The individual *ejido* farmers who were granted the right to use the land by the Mexican revolutionary government created following the overthrow of dictator Porfirio Díaz. Although the farmers were also granted the theoretical right to pass land use to their offspring, ownership of the land remained with the government.

ejido (10C). Land tenure system created by Mexican revolutionaries who overthrew the Porfirian government in 1911 with the purpose of easing rural poverty. The system created state-owned rural cooperatives, which provide landless peasants with a degree of control over the land they farmed. The *ejido* system resulted in the fragmentation of landholdings as the government took control away from some landowners and gave it to *ejiditarios*. In many cases, the government placed already impoverished farmers on marginal land, exacerbating their situation.

El Niño phenomenon (10B). The process by which the warm equatorial waters push back the cold Peruvian current off the western coast of tropical South America, producing local fish kills and having wider world climatic effects.

encomienda system (10A). The Spanish colonial system employed in Latin America in which lands were allotted to

Spaniards who were responsible for exploiting their wealth and had jurisdiction over native peoples.

entrepôt (5D). A port that collects goods from several countries to trade with the wider world and distributes imports to its immediate area or hinterland. Examples include Hong Kong and Singapore.

erosion (2B). The wearing away of rocks at Earth's surface by running water, moving ice, the wind, and the sea to form valleys, cliffs, and other landforms.

ersatz capitalism (6C). The economic system that implies an inferior substitute for locally based economic development when the capital, skills, and management are all imported to produce goods for export.

estuary (3B). A wide river mouth experiencing tidal changes in water level and quality.

ethanol fuel (10E). Motor fuel derived from sugarcane; used in Brazil.

ethnic cleansing (3E). The process by which a dominant group of people in a country causes another ethnic group to leave a region, often using threats or military force (*see* **genocide**).

ethnic group (2A). A group of people with common racial, national, religious, linguistic, or cultural origins.

ethnic religion (2A). A religion that is linked to a particular ethnic group, such as Judaism and Hinduism.

European Union (EU) (3C). Name adopted by the European Community in 1993, suggesting both an expansion to other European countries following the end of the Cold War and the possibility of a future closer political federation.

Euroregions (3C). Border areas of differing countries within European Union countries where the people within them work together to make trans-boundary movement easier.

export-led underdevelopment (10D). The process by which countries export their raw materials while their governments charge taxes on exports to pay for their imports. There is no stimulus to develop manufacturing and so diversify the economy to bring more wealth into it.

extensification (of agriculture) (3C). The production of fewer livestock or crops from the same area.

Farsi (Persian) (3A). The main language of Iran: a modern form, written in Arabic script. The ancient form came from Eurasia around 3000 B.C.

favela (10E). A shantytown in a Brazilian city.

federal government (2A). A division of central government functions among states or provinces.

feudalism (1B). A political system based on lord–vassal relationships in which the vassals gave homage and service in return for protection.

First Nation (11A). Indigenous or Native Canadians.

First World (2C). The Western countries, led by the United States.

five-year plans (4A). A series of comprehensive economic plans in the Soviet Union. These plans were followed from 1928 until 1991.

fjord (3D). A formerly glaciated valley that was flooded with ocean water after the sea level rose in the postglacial age.

Fordism (11A). The application of production lines in the assembly of a wide range of components in various manufacturing industries based on the production system established by Henry Ford for the automobile industry.

forest biome (2B). A major biome-type characterized by closely spaced trees.

formal economy (2C). The economic sector in which workers have recognized or licensed jobs, receive agreed-upon wages, and pay taxes.

Francophone (11C). French-speaking Canadian.

free-market capitalist system (2A). The economic system that is based on free competition and pricing of goods determined by the market. It is the basis of capitalism, but market "freedom" is often reduced by government and major producer actions.

free-trade zone (10E). An area within a country where components can be imported without tariffs for assembly with a view to exporting the finished goods (*see also* **special economic zones**).

friction of distance (1A). The relative difficulty of moving from one place to another, which increases with kilometers (miles), cost, or travel time.

FUNAI (10E). Amazon Indian Rights Agency (*Fundação Nacional do Indio*), which seeks to provide education, health care, and support for indigenous peoples in the Amazon basin of Brazil.

Galápagos Islands (10D). A group of islands in the Pacific Ocean near Ecuador that attracts scientists and tourists to view its unique plant and animal species.

gender (2A). The cultural implications of being male or female, with particular reference to the inequalities suffered by females in human societies.

genocidal rape (3E). The rape of women of an ethnic group or race in the belief that the women's ethnic group or race will be exterminated after the women are impregnated by the perpetrators' own genetic "seed."

genocide (3A). The systematic extermination of an ethnic group, nation, racial, or religious group.

gentrification (3C). The movement of higher-income groups to occupy and improve residences in older and poorer central parts of Western cities.

geographic inertia (3C). Once capital investments are made in factories and infrastructure that give a region agglomeration economies, production will continue there for a period of years after other areas emerge with lower production costs.

geographic information system (GIS) (1A). The computer-based combination of maps, data, and often satellite images that is a foundation for geographic analysis.

geography (1A). The study of spatial patterns in the human and physical world: where and how the human and natural features of Earth's surface are distributed, are related to each other, and change over time.

Germanic peoples (3A). A broadly defined group of peoples from northern Europe who began to move south into Germany and other areas of Europe around 500 B.C. Modern Germans, Austrians, Dutch, and the Scandinavians (Danes, Norwegians, Swedes) are the most numerous of today's Germanic peoples.

ghetto (11C). An urban area in which a particular group of people such as Jews or African Americans is segregated.

glasnost (4C). The Soviet Union policy of the late 1980s designed to create greater openness and exchange of information.

global choke point (1A). A point at which surface transportation routes converge that is vulnerable to the control by the country in which the choke point occurs; for example, the Suez Canal (Egypt) or the Strait of Hormuz (Iran).

global city–region (2A). A region that is dominated by one or more cities with major involvements in the global economy.

globalization (1B). The growing interconnections of the world's peoples and the integration of economies, technologies, and some aspects of cultures.

global warming (2B). The process by which average temperatures in Earth's atmosphere rise over a period of several decades or centuries, leading to the melting of ice masses and a rising sea level.

Gold Coast (6D). The Queensland, Australia, coast, a center of tourism.

Gondwanaland (6B). The former huge continent, from which the southern continents and southern Indian peninsula broke away.

governance (2A). The coordination and regulation of human activities at different levels of geographic scale, often outside the powers of sovereign countries.

grassland biome (2B). A major biome type characterized by grasses, with few, scattered trees.

Great Australian Desert (6D). The arid area covering almost all the Australian continent.

Great Dividing Range (6D). Mountain range parallel to the east coast of Australia.

Greater Antilles (10D). The four largest Caribbean islands: Cuba, Hispaniola (Haiti and Dominican Republic), Puerto Rico, and Jamaica.

Great Leap Forward (5D). The attempt by the Chinese Communist government in the late 1950s to increase the pace of industrialization. It failed by ignoring food production at a time when bad weather brought poor harvests and famine.

Greeks (3A). Historically, a group of peoples who formed a foundation of European ideas and established a major empire extending to India around 300 B.C.

greenhouse effect (2B). The natural process of heating Earth's atmosphere. Solar rays of short wavelength reach Earth's surface and are absorbed and reradiated as long wavelength (heat). This radiation is partially absorbed by and heats the lower atmosphere containing water vapor and carbon gases. When humans add to the carbon gases, they enhance this process, raising temperatures above "natural" levels.

Green Revolution (6C). The result of introducing high-yielding strains of wheat and rice. The outputs of commercial farms in South and East Asia increased by several times, but the costs of seeds, fertilizers, and pesticides were too high for smaller farmers.

gross domestic product (2A). The total value of goods and services produced within a country in a year. Often expressed as GDP per capita, when the total GDP is divided by the country's population.

gross national income (2A). The total value of goods and services produced within a country in a year, together with income from labor and capital working abroad less deductions for payments to those living abroad.

Group of Eight (11C). An economic discussion forum consisting of the world's eight most materially wealthy countries; it includes the United States of America, Canada, Japan, Germany, the United Kingdom, France, Italy, and Russia.

guest worker (3D). A foreigner who has permission to reside in a country to work but is not a citizen of that country. From the German word *Gastarbeiter.*

gulag (4C). Short for the Russian name ***Glavnoe upravlenie ispravitel no-trudovykh lagerei*** ("Main Directorate for Corrective Labor Camps"), a collection of prison camps in the Soviet Union where criminals and those who opposed the Communist government were sent to perform hard labor as punishment.

Gulf Cooperation Council (8D). Formed in 1981 under the leadership of Saudi Arabia to focus on political problems raised by the Iran–Iraq War.

---

hacienda (10A). Large estates developed during the Spanish colonial reign in the Americas; usually owned by a single family with substantial material wealth or by the church.

Haitian Revolution (10A). A slave revolt in 1804 that expelled French colonists and established the independent country of Haiti.

Han Chinese (5D). The largest group (94 percent) of people in China. An ethnic grouping based on the administrative culture spread by the Chinese empire in the A.D. 200s and 300s.

heartland (4D). The area of a country that contains a large percentage of the country's population, economic activity, and political influence.

Hebrew (8A). A language related to Arabic that is the official language of Israel.

Hindi (7B). One of the official languages of India but spoken by only 30 percent of the population, mainly in the north.

Hinduism (7A). A religion of South Asia, observed mainly in India, that includes the worship of many gods related to varied historic experiences and is associated with the caste system.

hinterland (4D). The areas of a country that lie outside the heartland. The hinterland usually has a relatively small percentage of a country's population, economic activity, and political influence compared to the heartland, though it may be well-endowed in natural resources.

Homestead Act (11A). The U.S. act of 1862 that provided land cheaply or freely to families settling the U.S. West.

horizontal integration (11A). The combining of producers of the same product in a single corporation to create economies of scale, including the control of prices.

household responsibility system (5D). The replacement for communes in rural China after 1976, returning ownership and decision making to individuals and groups that could sell surpluses in open markets.

Hudson's Bay Company (11A). A company set up by Royal British Charter in 1670 to administer the development of one-third of modern Canada.

human development (2C). A broader view of development that focuses on people rather than economic change, which is regarded as a means to the end of enabling people to enlarge their capabilities so that they can enjoy the richness of being human.

human development index (HDI) (2A). A measure of human development based on income, life expectancy, adult literacy, and infant mortality.

human geography (1A). The study of geographic aspects of human activities, often with a population, political, economic, cultural, or social focus.

human poverty index (HPI) (2A). A measure of human poverty, linked to HDI, that indicates levels of personal deprivation.

human rights (2C). The rights that should be a normal part of human experience, including justice, a decent standard of living, personal security, and freedom of thought and speech.

Hutu (9B). The majority ethnic group of Burundi and Rwanda with an agricultural culture.

hydrology (11B). The surface water drainage of a region, including lakes and rivers.

hyperinflation (10E). The process in which rising prices (and often wages) feed upon themselves and begin to increase rapidly.

---

imperialism (3A). The practice of extending the rule of an empire over foreign lands.

import substitution (2C). A policy in which countries develop manufacturing industries to fulfill internal market demands, often protected by high tariffs to exclude foreign competition.

Incas (10A). A group of people based in the area of modern southern Peru, South America, that extended their empire northward to modern Ecuador and southward to Chile in the A.D. 1400s.

Indian National Congress Party (7A). Formed by a mainly Hindu elite in 1885, it widened its base to become a major force in the independence movement and then for many years India's main political party.

indigenous people (2A). The first inhabitants of an area or those present when the area is taken over by another group.

Industrial Revolution (3A). The period of the late 1700s and early 1800s when increasingly complicated machines and chemical processes fueled by inanimate power sources such as water and coal replaced traditional ways of making goods by hand with simple tools. The mass production of goods resulted, as did the need for raw materials. The Industrial Revolution began in England and then spread to other areas of Europe and the world.

infant mortality (2A). The number of deaths per 1,000 live births in the first year of life.

informal economy (2C). The economic sector in which workers act outside the formal sector.

inselberg (9A). A bare, rocky hill characteristic of much of Africa.

intensification (of agriculture) (3C). The increased output of crops or livestock per area unit of land.

irredentism (3E). The desire to gain control over lost territories or territories perceived to belong rightfully to a group; associated with nationalism.

Islam (8A). A religion of Northern Africa and Southwestern Asia, and parts of South and East Asia, based on the teachings of Muhammad as recorded in the Qu'ran. *Islam* means "submission to the will of God."

isthmus (10A). Naturally occurring land bridge.

Itaipu Dam (10E). World's largest hydroelectric dam (until the Three Gorges Dam in China is completed in 2009) on the Panama River on the Brazil–Paraguay border.

Jainism (7A). A religion, mainly in India, that involves a nonviolent code and has laws against harming animals (thus forbidding farming).

Judaism (8A). The religion of the Jewish people who worship Yahweh as the creator and lawgiver.

Karen (6A). Minority group in Myanmar.

Khmer (6A). A people who moved into the Cambodian area from the north.

Khmer Rouge (6C). Communist group that ruled Cambodia from 1975 to 1979, noted for many atrocities.

kibbutzim (8D). A communal farming village in Israel, the social and spiritual basis of the new Israeli nation after 1948 but now the home of fewer than 5 percent of the Israeli population.

Kurdish (8A). A language akin to Persian spoken by a group of tribes in the hilly lands of eastern Turkey, northern Iraq and Syria, and western Iran. The tribes see themselves as a nation and wish to establish their own country.

Kyoto Protocol (2B). Statement adopted by the United Nations Convention on Climate Change in 1997, setting targets for reducing primary greenhouse gases.

---

land breeze (6B). As air over land cools at night, it sinks and flows out to sea.

language (2A). The means of communication among people by speaking, writing, and signing.

Lao (6A). The dominant people in Lao People's Republic.

latitude (1A). The distance of a place north or south of the equator, measured in degrees.

Lesser Antilles (10D). Smaller Caribbean islands: a chain around the eastern edge of the Caribbean Sea (Leewards in the south; Windwards in the north).

localization (1B). The geographic differentiation of places among and within countries.

location (1A). A place's location is defined by its position on Earth's surface in terms of latitude and longitude (absolute location) or by its level of interaction with other places (relative location).

loess (3B). Fine-grained and fertile soils developed from windblown deposits.

longitude (1A). The distance of a place east or west of 0° longitude, measured in degrees.

long lot (11D). The unit of land settlement by French colonists in Anglo America. Each plot of land had a narrow river or road frontage and often extended through different types of land suitable for crops, pasture, or woodland.

Lost Decade (10C). In the 1980s, country economies in Latin America stalled because of high interest rates and debts.

---

mallee (6B). A vegetation community occurring in Australia that is composed of drought-resistant eucalyptus shrubs forming impenetrable thickets.

Maori (6A). A people who settled New Zealand from the wider South Pacific around the A.D. 800s and continue to form an important component of the New Zealand population.

map (1A). The representation of the features of Earth's surface on paper at varying scales.

maquiladora (10C). Mexican government program that encourages foreign-owned factories to be sited in Mexico by not charging import duties on raw materials for assembly.

market gardening (3E). The commercial production of high-cash-value, specialty fruit and vegetable crops such as table grapes, raisins, oranges, grapefruits, apples, and lettuce.

marsupial (6B). A mammal that raises its young in a pouch instead of a womb. Mainly found in Australia, these include the kangaroo and koala.

Maya (10A). A group of Native Americans whose civilization centered on the eastern lowlands of modern southern Mexico, Guatemala, and central Honduras. Main period from A.D. 200s to 900s, with wealth based on a surplus of corn.

medina (8B). The crowded streets of the older sections of Arab towns in Northern Africa and Southwestern Asia.

mediterranean climate (3B). A climatic environment that occurs between midlatitude west coast and tropical arid climatic environments, having wet winters and dry summers.

megalopolis (5B). An expanded urbanized area that includes several metropolitan areas with over a million people and dominates the economy of surrounding areas. First identified in the northeastern United States covering the area between Boston and Washington, D.C.

Melanesian people (6A). "Dark-skinned" peoples who inhabit the South Pacific islands.

Mercosur (10E). Trading group established in 1991 among Argentina, Brazil, Paraguay, and Uruguay.

meridian of longitude (1A). A line joining places of the same longitude on Earth's surface.

mestizo (10A). A person of mixed Native American and European stock in Latin America.

Micronesian people (6A). Peoples of the South Pacific who inhabit the "small islands" of the northwestern area.

Middle America (10B). Mexico, Central America, and the Caribbean.

midlatitude climates (2B). Climates typical of the zone between the tropics and polar regions, having summer–winter temperature contrasts.

midlatitude continental interior climate (3B). A climate environment that is marked by very cold and often dry winters and hot summers with thunderstorms.

midlatitude cyclone (6B). A weather system of low pressure, characterized by fronts that lift air, causing cloud formation and precipitation. Winds circulate around the central low pressure in a clockwise flow in the Southern Hemisphere and anticlockwise flow in the Northern Hemisphere.

midlatitude east coast climate environment (5A). A climatic environment with warm-to-hot wet summers and cool-to-cold drier winters.

midlatitude west coast climate (3B). A climatic environment dominated by cyclonic weather systems linked to oceanic heat and moisture sources, interspersed with anticyclones. Typically has cooler summers and milder winters than continental interiors on the same latitude.

migration (2A). The long-term movement of people into (in-migration, immigration) or out of (out-migration, emigration) a place.

Ministry of International Trade and Industry (MITI) (5C). The Japanese government ministry charged with promoting Japanese trade and products abroad.

modernization (2C). The theory of development in which poorer countries attempt to follow the stages through primary and secondary to tertiary sectors that the countries of Western Europe and North America followed in the A.D. 1800s and early 1900s.

Mon (6A). A people who moved into the Cambodian area from the north.

Monroe Doctrine (10C). An 1823 declaration asserting U.S. rights to control activity in the Americas over those of countries outside the region. In the Monroe Doctrine, the United States vowed to resist any intervention from outside countries in Latin American affairs.

monsoon climatic environment (5A). A tropical climatic environment in which there are wind shifts between summer and winter, bringing heavy rains from oceanic air in the summer and dry winds of interior continental air in winter.

Mughal (Mogul) dynasty (7A). Turkish invaders of India from Persia in the A.D. 1500s who conquered most of the region and left a heritage of magnificent buildings such as the Taj Mahal.

Muhajirs (7C). Muslim people who migrated to Pakistan at the time of partition in 1947. Many still speak Hindi.

multinational corporation (2A). A corporation that makes goods and provides services in several countries but directs operations from headquarters in one country.

Muslim League (7A). Formed in India in 1906 to uphold the position of Muslims in the British Indian Empire.

Muslims (8A). "Those who submit to Allah." Followers of Islam.

---

nation (2A). An "imagined community" in which a group of people believe that they share common cultural features, often linked to a specific area of land.

nationalism (2A). The desire of a nation to become a self-governing country.

nationalize (10A). The assumption of ownership and control of private companies by governments or government institutions.

nation–state (3A). The linking of a separate and distinct people ("nation") and a politically organized territory with its own sovereign government ("state").

nation–state ideal (3A). The belief that each people ("nation") must have its own country ("state") in order to be free and govern itself as it desires.

Native American (11A). People who inhabited Anglo America before the European arrival, including Amerinds and Inuits (Eskimos).

natural environment (2B). The world as it might be without humans, including climate, landforms, plants, animals, soils, and oceans.

natural hazard (2B). A natural event, such as a volcanic eruption, earthquake, tornado, hurricane, or flood, that interrupts human activities by causing extensive damage and deaths.

natural resource (2B). Materials present in the natural environment and recognized by humans as of practical worth (minerals, soils, water, building stones, timber).

new rice technology (6C). Improved strains of rice and methods of growing that developed in the Philippines as part of the Green Revolution.

New Zealand film industry (6D). A growing industry making major movies and TV series.

Nezahualcóyotl (10C). Shantytown in Mexico City where over 1.5 million people live on flood-prone land.

Nile Waters Agreement (8B). Agreement between Egypt and Sudan in 1959 to share Nile River waters, with 70 percent allocated to Egypt.

node (1A). A place where flows of people, goods, information, money, innovations, or ideologies begin, intersect, or end.

nongovernmental organization (2A). Groups of people who act outside government and major commercial agencies, mainly in advocacy roles such as delivering aid and lobbying for particular causes.

nonrenewable resource (2B). A natural resource that is used up once it is extracted; for example, coal or metallic minerals.

nortes (10B). Cold air masses from the north affecting weather in Middle America.

North American Free Trade Agreement (11C). An economic agreement among Canada, Mexico, and the United States, signed in 1994.

North Atlantic Treaty Organization (NATO) (3C). A military alliance of non-Communist European countries and the United States, founded in 1949 to counter the military threat of the Soviet Union. In recent years, former Communist countries have joined the alliance, and Russia has formed a partnership with NATO.

northern coniferous forest (taiga) (4B). A forest composed of coniferous trees (firs, pines, cedars) common in the northern parts of midlatitude continental interior climatic environments.

---

ocean biome (2B). The oceans as ecosystems with energy pathways, nutrient cycles, and food chains.

Organization of Petroleum Exporting Countries (OPEC) (8B). Established in 1960 to further the interests of oil and gas producers throughout the world, often in materially poorer countries and often to resist the overriding power of multinational oil corporations.

Organization of the Islamic Conference (OIC), (8B). Established in 1970 by foreign ministers of Muslim countries throughout the world. It has 45 members but advances individual countries' interests rather than pursuing a common agenda.

orographic rain (6B). Hilly areas, usually facing oceanic moisture sources, cause uplift of air and enhanced precipitation totals.

ozone hole (6E). The thinning of the ozone in Earth's stratosphere at the and of the Antarctic winter in October. It allows harmful ultraviolet rays through.

Pacific Rim (6C). The countries that border the Pacific Ocean, including those in the South Pacific, East Asia, North America, and Latin America.

padi (6C). Wet-rice cultivation in which fields are inundated with water during the growing season.

Palestine Liberation Organization (PLO) (8B). An organization that promotes the reestablishment of a country of Palestine. It has a secular, left-wing political basis.

PAN (10C). The National Action Party (*Partido de Accion Nacional*) of Mexico. In July 2000, the PRI's dominance of Mexican politics was overturned with the election of PANcandidate Vicente Fox.

Pan-Arab country (8B). The idea of Arab countries joining to form a single country. Egypt and Syria were joined for a short while in the 1960s, but few attempts have been made to achieve this end since.

pantheistic religion (5A). A religion that equates god with universe forces and/or includes many gods or religions.

parallel of latitude (1A). A circle joining places of the same latitude on Earth's surface.

Parti Québécois (11C). A political party formed with the purpose of achieving Québec's separation from the Canadian people and political independence from the Canadian federal government.

peninsulare (10A). Used in reference to persons living or working in the New World who were born in Spain or Portugal on the Iberian Peninsula (used more for those born in Spain). *Peninsulares* held the highest offices and received the largest land grants in the New World.

perestroika (4C). The Soviet Union policy of the late 1980s designed to reconstruct the political and economic structure of the country so that it could compete with or in the capitalist world economic system.

permafrost (4B). Permanently frozen ground extending several hundred meters below the surface in Siberia and northern Canada. In summer, water in the surface active layer, 50 to 100 cm deep, melts.

Petén (10C). Forested area of northern Guatemala.

petrodollars (10C). Dollars invested by Middle East oil producers during high oil prices in the 1970s.

physical geography (1A). The study of geographic aspects of natural environments.

physiological density (2A). Numbers of people per unit of cultivable land.

place (1A). A point or area on Earth's surface having a geographic character defined by what it looks like, what people do there, and how they feel about it.

planned economy (3C). The Communist practice of the government, rather than the free market, deciding what goods and services need to be produced within a country.

plantation (6C). The large-scale and concentrated commercial cultivation of crops often associated with the tropics and colonial attempts to produce such crops for the home countries.

plateau (9A). An upland area with flat-topped hills or having a tablelike form.

podzol soils (4B). Soils of low fertility in which plant nutrients are removed by water passing through. Commonly develop beneath midlatitude coniferous forest and on sandy soils.

polar biome (2B). Major biome type where plant growth is inhibited by extreme cold.

polar climates (2B). Climates typical of the polar regions: extremely cold all year.

political geography (2A). The study of how governments and political movements influence the human and physical geography of the world and its resources.

political rights (2C). The rights to vote and participate in one's own government.

Polynesian people (6A). Lighter-skinned peoples of the South Pacific islands who inhabit mainly the eastern islands.

poorer people (1A). People who have below-average possession of material goods.

population density (2A). The numbers of people per given area.

population doubling time (2A). The time in years taken to double the numbers of people at a place by natural and migration changes.

population–resource ratio (2A). The relationship between the quantity and quality of an area's material resources and the size and technical competence of its population.

PRI (10C). The Institutional Revolutionary Party (*Partido Revolucionario Institucional*) arose from the Mexican revolutionary period of 1910–1920. The PRI grew out of the revolutionary movement into the most powerful and lasting political party in the history of Mexico, dominating Mexican politics for more than 70 years. The party controlled the government, nationalized industries, and allegedly fixed many political votes to remain in power.

primary production (2A). The sector of an economy that produces output from natural sources, including mining, forestry, fishing, and farming.

primate city (6C). A city that contains a large proportion of the urban population of a country, often several times the population of the second city.

producer goods (3C). Industrial goods used by other industries to make consumer goods.

producer services (3D). Service industries that are involved in the output of goods and services, including market research, advertising, accountancy, legal, banking, and insurance industries.

production line (11A). The system of manufacturing in which components are made and assembled into the final product in a sequence of factory-based processes.

productive capacity (3C). The total amount of goods a country's industries can produce during a given period.

productivity (3D). The measure of the amount of product generated or work completed per hour of labor.

protectionism (10C). Governmental policies that focused on government-controlled industry and the protection of domestic products through tariffs, quotas, and red tape.

Punjabi (7C). A language spoken by two-thirds of Pakistan's population.

purchasing power parity (2A). The measure of GNI or GDP that is based on internal country costs of living rather than external exchange rates related to the U.S. dollar.

---

quaternary production (2A). The sector of an economy that specializes in producer services, including financial services and information services.

Quechua (10A). The official language of the Inca Empire. Quechua is spoken today by many indigenous groups in Bolivia, Ecuador, and Peru.

Qur'an (8A). The holy book of Islam.

---

race (2A). A biologic stock of people with similar physical characteristics or a group of people united by a community of interests.

rain shadow (10B). Low rainfall in an area to the lee of a mountain range, where winds warm and get drier as they descend after flowing across the mountains.

region (1A). An area of Earth's surface distinguished from others by its physical and human characteristics and interacting with other regions in trade and the exchange of people and ideas.

regional geography (1A). The study of different regions at Earth's surface in their country and global contexts.

relative location (1A). The direction and distance of a place relative to others, often affected by factors that slow or increase contacts among people.

relief (2B). The physical height and slope of the land, as in hills, mountains, and valleys.

religion (2A). An organized system of practices that seeks to explain our purpose on Earth and may include a set of values and/or worship of a divine being.

renewable resource (2B). A resource that is replaced by natural processes at a rate that is faster than its usage.

revolutionary tourist (10C). Supporters of *Zapatista* (in Mexico) and other causes who travel from foreign countries to offer their fighting services to the local cause.

rift valley (9A). A deep valley caused by the rocks of Earth's crust arching and cracking to let down a section of crust to form the valley floor.

Romans (3A). Historically, a group of people who established an empire around the Mediterranean Sea that extended to northwest Europe from the first century B.C.

rural area (2A). Land outside urbanized areas, often having an economic emphasis on farming, mining, and/or forestry. Such areas may dominate population distribution in materially poorer countries.

Russification (4A). Policies directed at making non-Russians into Russians by encouraging or forcing non-Russians to adopt Russian cultural characteristics such as the Russian language.

Sahel (9A). The zone immediately to the south of the Sahara Desert in Africa that suffers droughts as the arid area expands by natural or human-induced actions.

salinization (8A). The process by which soils become unproductive because of an accumulation of alkaline salts near the surface. Often associated with poorly managed irrigation systems in arid areas.

Sandinista (10C). Marxist-oriented revolutionaries who controlled Nicaragua from 1979 to 1990.

scale (1A). The relationship of horizontal ground distance to map distance, quoted as a fraction (1/10,000) or as a ratio (1:10,000) in which one unit on the map represents 10,000 units on the ground.

sea breeze (6B). As air rises over heated land, cooler air from the ocean replaces it at ground level.

Second World (2C). The Communist countries, led by the Soviet Union until 1991. The term is now redundant.

secondary production (2A). The sector of an economy that changes the raw materials from the primary sector into useful products, thus increasing their value, as in chewing gum or parts for airplanes.

sedimentation (10B). The deposition of rock debris, including in offshore areas, where too much fine debris may kill coral reefs.

Shan (6A). Minority group in Myanmar.

shantytown (5B). An unplanned residential sector of urban areas in poorer countries. Housing is often built of any materials that come to hand and does not have electricity, water, or waste disposal.

Shia Muslims (8A). Also known as Shiites. Muslims who are partisans of the imam Ali (not acceptable to Sunni Muslims) and look to his return. They make up 90 percent of the Iranian population and 60 percent of Iraqis.

Shinto (5A). The traditional religion of Japan, built on ancient myths and customs that promote the national interests and identity.

Sikhism (7A). A Hindu-related religion with a strict code of conduct. Its temple kitchens provide food for all.

Sindhi (7C). A language spoken by 12 percent of Pakistan's population.

Sinhalese (7C). A language spoken by the Buddhist majority of people in Sri Lanka.

Slavs (3A). A broadly defined group of people who migrated from the east, settling primarily in East Central Europe between approximately A.D. 400 and 800. The Slavs developed into three distinct subgroups: western Slavs (Poles, Czechs, Slovaks, Sorbs), southern Slavs (Slovenes, Croats, Bosnians, Serbs, Montenegrins, Macedonians, Bulgarians), and eastern Slavs (Russians, Belarussians, Ukrainians).

social rights (2C). The rights to have a job and earn a living with basic material standards.

soil (2B). Weathered rock material that develops by the actions of water, animals, and plants into a basis for plant growth.

Somoza (10C). A family that held long-term dictatorship power in Nicaragua.

South African Development Conference (SADC) (9B). Established by countries in Southern Africa opposed to South Africa's apartheid policy to organize alternative trade outlets. Now encouraging trade among the constituent countries, including South Africa.

South Pacific Forum (6D). Established in 1971 among 13 countries in the South Pacific, including Australia, New Zealand, and several island countries. Aims to promote regional cooperation and confronts regional problems of external exploitation.

spatial analysis (1A). The study of linkages between places at Earth's surface, often in terms of points, lines, and areas, together with statistical associations.

spatial view (1A). A geographic view that focuses on differences among places.

special economic zones (5D). Zones established by China in 1979 to encourage foreign investment and export-oriented manufacturing in the southern coastal provinces. Similar zones are found in other countries (*see* **free-trade zone**).

specialization (of agriculture) (3C). The concentration on fewer commercial products within a farming region.

state (3A). A country, or division of a country within a federal government.

state socialism (3C). The Communist Party actively running the political, social, and economic activities of the people.

steppe grasslands (4B). Midlatitude grasslands typical of the transition between forest and arid areas of midlatitude continental interior climatic environments.

structural adjustment (2C). The theory of development in which a country becomes more involved in the world economic system by producing export goods, reducing tariffs, encouraging privatization, and developing good government and a balanced budget.

subduction (6B). In plate tectonics, where two converging plates clash and one dives beneath the other, causing earthquakes and volcanoes.

suburbanization (5B). The movement of people and corporations from older central parts of urban areas to new residential and commercial areas on the city outskirts.

Sunni Muslims (8A). Also known as Sunnites. Traditional, conservative followers of Islam, forming the majority in most Muslim countries. Includes strict sects such as the Wahhabi.

supranationalism (3C). The idea that differing nations can cooperate so closely for their shared mutual benefit that they can share the same government, economy (including currency), social policies, and even military.

sustainable forestry (6D). Practices that result in a continuing production of timber.

sustainable human development (2C). A level of development in which resources are exploited at a rate that is sustainable for future generations.

---

tectonic plate (2B). A large block of Earth's crust and underlying rocks approximately 100 km thick and up to several thousand kilometers across. Earth's interior heat causes plates to move apart and crash together, forming major relief features including ocean basins, mountain systems, and continental areas.

temperature inversion (10B). Naturally occurring periods during the winter months when cold dense air remains "trapped" at the surface under warmer air for several days or even weeks. The cold air is usually trapped by mountains or some related physical barrier. Temperature inversions may produce stagnant air, which may complicate pollution problems in urban areas.

Tennessee Valley Authority (TVA) (11C). Established by the U.S. government in 1933 to stimulate economic growth in southern Appalachia through the damming of rivers to improve transportation, flood control, and electricity costs.

Tenochtitlán (10A). The urban capital of the Aztec Empire located in the Central Valley of Mexico. Modern Mexico City originated on the site of the Aztec capital during the Spanish colonial reign.

Terra Australis (6A). The "Southland" mythical continent of the South Pacific, now Australia, that was discovered by Dutch and British explorers in the 1700s.

tertiary production (2A). The sector of an economy concerned with the distribution of goods and services, including trade, professions, and government employment.

Thai (6A). The majority people in Thailand.

Thar Desert (7A). Arid area across India–Pakistan border, often known as the "Great Indian Desert."

Third World (2C). The poorer countries, mostly in the Southern Hemisphere, not part of the First or Second Worlds.

toponym (11A). Place name.

total fertility rate (2A). The number of births per woman in her childbearing years.

transform plate margin (6B). In plate tectonics, where two plates move horizontally alongside each other, often causing earthquakes.

transmigration (6C). The name given to the process through which Indonesia attempted to ease overpopulation on Java by moving large numbers of people to the less inhabited islands.

transnational Chinese economy (5B). Economic and trading links established by Chinese people living outside China, especially in East Asia and North America.

Treaty of Tordesillas (10A). The 1494 demarcation line (approximately 46° W longitude) dividing land rights in the Americas between the Spain and Portugal. The line was created under the authority of the pope. Spain gained control of lands to the west of the line, and Portugal gained control of lands to the east.

tribe (9A). An ethnic group with strong kinship and territorial links. The name is often applied to groups of people in formerly colonial territories.

tropical climates (2B). Climatic environments typical of the tropical zone, having high temperatures all year.

tsunami (6B). A huge wave generated by an earthquake that changes the ocean floor shape suddenly. The wave travels fast across the ocean and causes destruction on entering shallow coastal waters and confined valleys.

tundra (4B). Ecosystem type occurring in cold polar environments, consisting of mosses, grasses, and low shrubs.

Turkish (8A). A language family spoken by peoples in Central Asia and extended to Turkey in Turkey, the language is written

in Roman script, but Azeris use Arabic script.

Tutsi (9B). Originally a cattle-herding group that settled in the area of Burundi and Rwanda, assuming a leadership position over the more numerous agricultural Hutu tribe.

typhoon (6B). A tropical storm of hurricane type experienced in Southeast and East Asia.

---

uneven development (11C). The increase in the gap between poor and wealthy regions in a country and the shifting locations of economic growth and decline over time, seen by Marxists as an outcome of capitalism.

unitary government (2A). Government of a country administered from a single center.

universalizing religion (2A). A religion that seeks to be global in its application, such as Islam and Christianity.

UN Security Council (11C). One of the principal divisions of the United Nations. The purpose of the Security Council is the maintenance of international peace and security. The United States, China, France, the United Kingdom, and Russia are the five countries granted permanent seats on the Security Council. The permanent members of the Security Council have the power to veto council decisions. The decisions of the Security Council are binding for the entire UN General Assembly.

urban area (2A). An area with high densities of people, buildings, transportation linkages, and human activities of a high economic, political, and cultural order. Urban areas dominate population distribution in materially wealthy countries.

Urdu (7C). The official language of Pakistan, although spoken by only 7 percent of the population.

---

vertical integration (11A). The combining of producers of raw materials, manufacturers that process the materials, and those that assemble the products in a single corporation to achieve economies of scale.

Vietnamese (6A). The dominant people in Vietnam.

Vikings (3A). A group of peoples who spread out from Scandinavia from the A.D. 700s, conquering much of northwest Europe, Iceland, and Greenland and influencing events in Russia.

Virgin Lands Campaign (4C). A Soviet agricultural campaign begun in the 1950s. It promoted farming in lands where it had never taken place before, primarily in lands that were very marginal because the soil was poor or not enough water or heat was present to grow crops. Much of the land was in the semidesert and desert areas of southern Siberia and Central Asia, especially in the Kazak Republic.

Wallace Line (6B). The line between the Australian and Asian plant species drawn by the botanist Alfred Russell Wallace in the mid-1880s.

wealthier people (1A). People who have above-average possession of material goods.

weathering (2B). The action of atmospheric forces (through water circulation and temperature changes) on rocks at Earth's surface that breaks the rocks into fragments, particles, and dissolved chemicals.

white Australia policy (6C). The policy designed to exclude mainly Asian people from Australia. Ended in 1972.

world region (1B). This text recognizes nine world regions that each include a number of countries linked by cultural, political, economic, and environmental conditions.

Yanomami (10E). Group of people in the northern Amazon rain forest whose territory is threatened by gold miners.

Zapatista (10C). Indigenous revolutionaries existing primarily in the Mexican state of Chiapas. The *Zapatistas* assert a lack of representation for indigenous communities in Mexico. Many seek independence from the Mexican government.

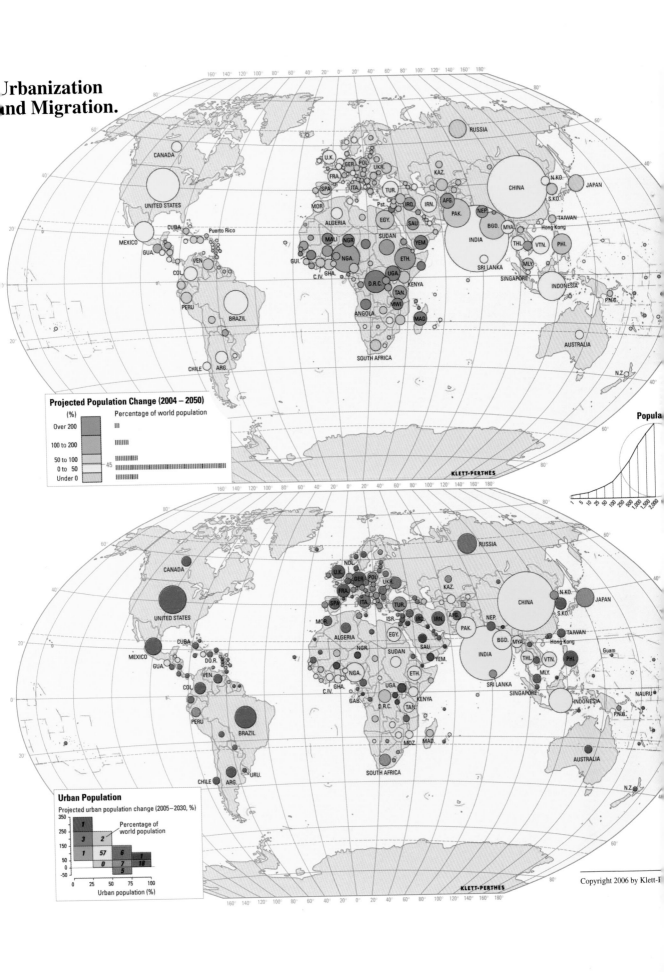

Urbanization and Migration.

Projected Population Change (2004 – 2050)

(%)	Percentage of world population
Over 200	|||
100 to 200	|||||||
50 to 100	|||||||||||
0 to 50	45
Under 0	|||||||

Popula

Urban Population

Projected urban population change (2005 – 2030, %)

Percentage of world population

	1		
3		2	
1	57	6	
1	0	7	1
		5	18

Urban population (%)

KLETT-PERTHES

Copyright 2006 by Klett-P

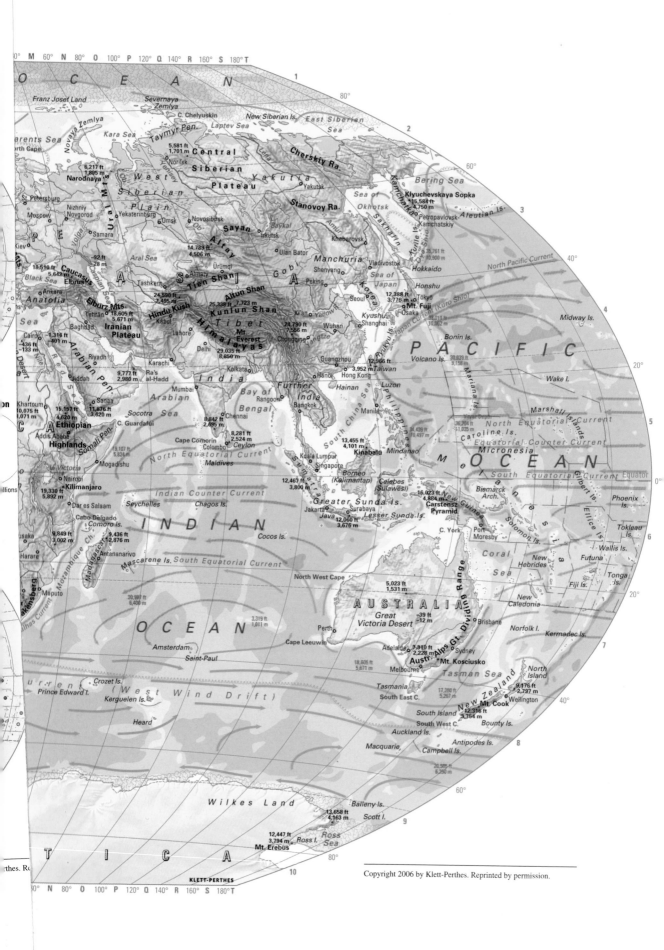

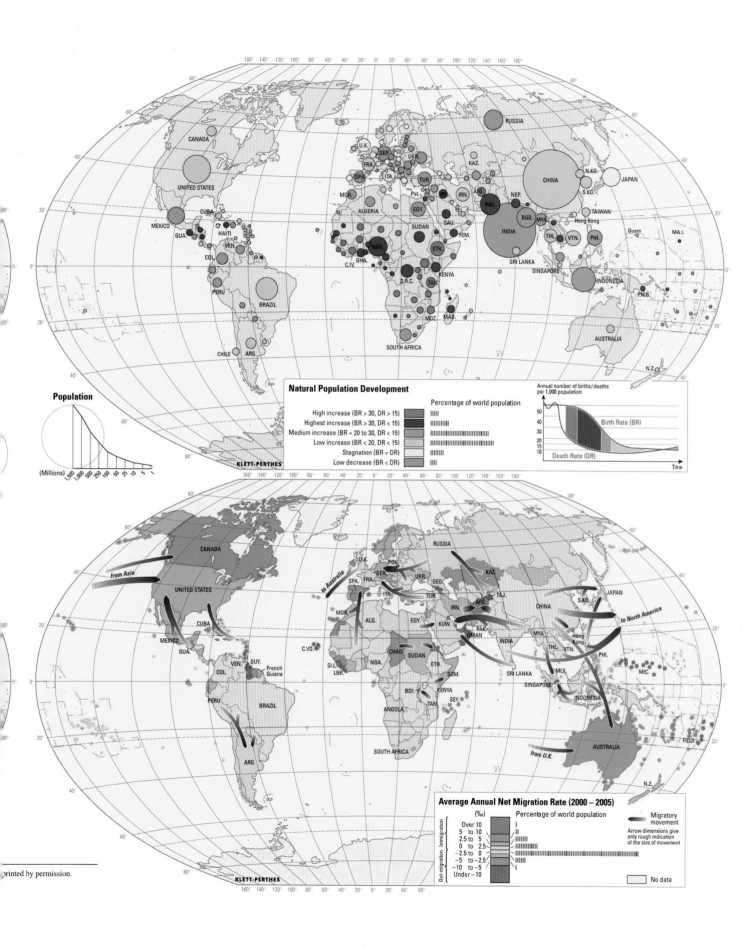

Population

(Millions) 1,500 1,000 500 250 100 50 25 10 5 1

Natural Population Development

Percentage of world population

High increase (BR > 30, DR > 15)
Highest increase (BR > 30, DR < 15)
Medium increase (BR = 20 to 30, DR < 15)
Low increase (BR < 20, DR < 15)
Stagnation (BR = DR)
Low decrease (BR < DR)

KLETT-PERTHES

Annual number of births/deaths per 1,000 population

Birth Rate (BR)

Death Rate (DR)

Time

Average Annual Net Migration Rate (2000 – 2005)

(‰) Percentage of world population

Over 10
5 to 10
2.5 to 5
0 to 2.5
−2.5 to 0
−5 to −2.5
−10 to −5
Under −10

Immigration / Out-migration

Migratory movement

Arrow dimensions give only rough indication of the size of movement

No data

KLETT-PERTHES

printed by permission.

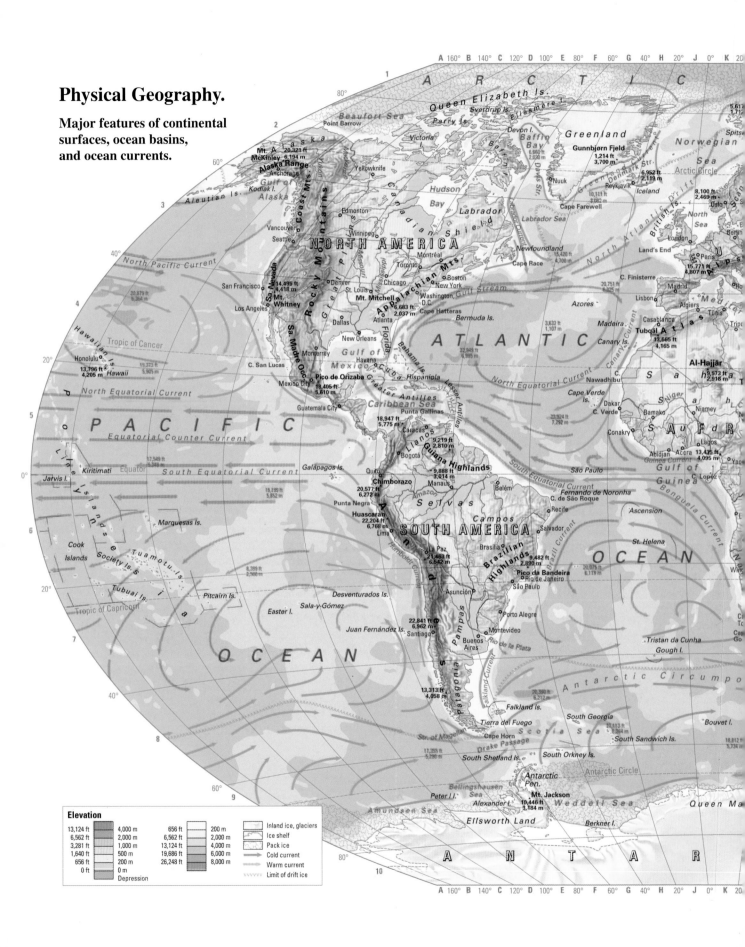

Physical Geography.

Major features of continental surfaces, ocean basins, and ocean currents.

Elevation

13,124 ft	4,000 m	656 ft	200 m		Inland ice, glaciers
6,562 ft	2,000 m	6,562 ft	2,000 m		Ice shelf
3,281 ft	1,000 m	13,124 ft	4,000 m		Pack ice
1,640 ft	500 m	19,686 ft	6,000 m	→	Cold current
656 ft	200 m	26,248 ft	8,000 m	→	Warm current
0 ft	0 m			vvvvvv	Limit of drift ice
	Depression				

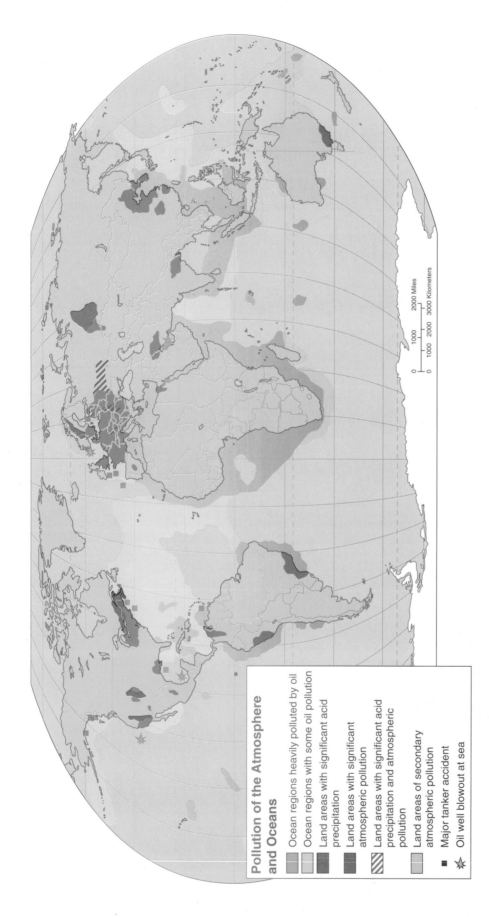

Pollution of the Atmosphere and Oceans

Ocean regions heavily polluted by oil

Ocean regions with some oil pollution

Land areas with significant acid precipitation

Land areas with significant atmospheric pollution

Land areas with significant acid precipitation and atmospheric pollution

Land areas of secondary atmospheric pollution

■ Major tanker accident

★ Oil well blowout at sea

0   1000   2000 Miles
0 1000 2000 3000 Kilometers

Air and Water Quality

Pollution of the atmosphere and ocean are both vital aspects for future and expanding populations on Earth. The circulations of the atmosphere and the oceans control many features of Earth's natural environments. Land areas downwind of industrial concentrations emitting sulfur and nitrogen gases are subject to acid deposition that damages forest and lake ecosystems; atmospheric pollutants may also be damaging to human health; forest destruction and burning cause "secondary" atmospheric pollution. Oil spills in particular affect the surface plankton basis of life in the oceans. Much of our planet is affected by pollution—a lowering of air or water quality.

Source: Reprinted with permission from *Student Atlas of World Politics,* 4th ed. by John L. Allen. © 2000 The McGraw-Hill Companies, Inc. All rights reserved.